Building Contract Claims

# Building Contract Claims

*Guy Higginbottom*
*David Chappell*

Sixth Edition

WILEY Blackwell

This edition first published 2026


*Edition History*
*2011, 5e; 2005, 4e; 1998, 3e*



*Registered Offices*
John Wiley & Sons, Inc., 111 River Street, Hoboken, NJ 07030, USA
John Wiley & Sons Ltd, New Era House, 8 Oldlands Way, Bognor Regis, West Sussex, PO22 9NQ, UK

For details of our global editorial offices, customer services, and more information about Wiley products visit us at www.wiley.com.

The manufacturer's authorized representative according to the EU General Product Safety Regulation is Wiley-VCH GmbH, Boschstr. 12, 69469 Weinheim, Germany, e-mail: Product_Safety@wiley.com.



***Library of Congress Cataloging-in-Publication Data applied for:***

Hardback ISBN: 9781394210008

Cover Design: Wiley
Cover Image: © kanpisut/Shutterstock

Set in 9.5/12.5pt STIXTwoText by Straive, Chennai, India

# Contents

# List of Cases

## List of Figures

## List of Tables

# Preface to the Fifth Edition

It was with pleasure, flecked with trepidation, that, in 1997, I accepted the invitation to enlarge the scope and update the third edition of this popular book.

Much of the original text was retained, but updated in the light of case law; the number of case references was almost doubled and, at the wish of the late Professor Powell-Smith, footnotes were included; a change which has proved very popular. Some restructuring took place within chapters to establish a comprehensible system of sub-headings and a few topics, such as liquidated damages (the subject of many disputes), global claims, causation and concurrency, were given greater importance. The scope of the book was extended to include more contracts.

In the fourth edition, the opportunity was taken to carry out further fundamental changes to the structure of the book, bringing general principles to the beginning of the book and dealing with their application to specific contracts later. The text also was substantially revised and almost a hundred additional cases added. New contracts were added including the JCT Construction Management and Major Projects contracts, the JCT Standard Form of Domestic Sub-Contract and the Engineering and Construction Contract.

In this fifth edition, the structure of the book has been slightly amended to give greater prominence to important topics such as notices, mitigation and the measure of damages and more has been said about the way in which a contractor should put together a claim. Account has been taken of the JCT 2005 suite of building contracts and sub-contracts. The Constructing Excellence, Measured Term and the ACA Project Partnering contracts have been included for the first time, and the latest NEC contract has been considered. As before, when dealing with JCT contracts, the style has been to use the JCT Standard Building Contract 2005 (SBC) as the basis and highlight important differences in the other forms. In some instances there are few similarities. Reference has been made to more than a hundred additional legal cases. In previous editions, the text of relevant contract clauses was reproduced. In this edition, the decision has been taken to remove them. The reason is twofold: it removes from the book many pages of clauses which many readers will not require, and it is assumed that readers have a copy of the relevant contract beside them.

At the time of writing, the latest official amendments have been taken into account as follows:

Main contract forms:

| | |
|---|---|
| SBC | Revision 2 |
| IC and ICD | Revision 2 |
| MW and MWD | Revision 2 |
| DB | Revision 2 |
| PCC | Revision 2 |
| MC | Revision 2 |
| MP | Revision 2 |
| CM/TC | Revision 2 |
| CE | Revision 1 |
| MTC | Revision 2 |
| GC/Works/1 (1998) | |
| ACA 3 | 2003 Revision |
| PPC2000 | 2008 Amendment |
| NEC 3 | June 2005 (with 2006 amendments) |

Sub-contract forms:

| | |
|---|---|
| SBCSub/C | Revision 2 |
| SBCSub/D/C | Revision 2 |
| ICSub/NAM/C | Revision 2 |
| ICSub/C | Revision 2 |
| ICSub/D/C | Revision 2 |
| DBSub/C | Revision 2 |
| MCWC/C | |
| ACA/SC | 2003 Revision |

Reference has also been made to the *Housing Grants, Construction and Regeneration Act 1996* as recently amended by Part 8 of the *Local Democracy, Economic Development and Construction Act 2009* and the *Arbitration Act 1996*, where appropriate. It should be noted that, at the time of writing, the 2009 Act is not yet in force.

It should also be noted that in the reproduction and commentary on the JCT standard forms, for brevity 'Architect' is used to stand for 'Architect/ Contract Administrator'. Throughout the book, contractors and sub-contractors have been assumed to be corporate bodies, and they have been referred to as 'it'.

Building contract claims combine a good understanding of the law and of building practice. Certain principles can be discerned, and this book is an attempt to explain the principles and to show how those principles should be applied to the popular standard contracts. Standard contracts not only set out the powers and duties of the parties and of various consultants, they also often give procedures which must be followed to enable the participants to carry out the duties and exercise the powers. However, it must not be thought that standard building contracts are excused from the operation of the general law of contract. They are contracts, like any others and subject to the same rules. Thereby hangs the solution to many perceived problems. In my experience, many claims fail because the basic principles are misunderstood, and contractors and sub-contractors do not appreciate the amount of effort required to properly substantiate a claim. A loss of money or lack of profit alone cannot sub-stantiate a claim, although it is usually the trigger. The book is addressed to all parties involved in construction. It is not always possible to give a definitive answer to every question, either because the courts have not considered the matter or because there have been apparently conflicting judgements. Where there is doubt, the doubt is expressed and, if practicable, I have taken a view of the situation.

This book was the idea of the late Professor Vincent Powell-Smith LLB(Hons) LLM DLitt FCIArb DSLP MCL FSIArb and John Sims FRSA FRICS FCIArb MAE. These two eminent practitioners in this field were responsible for the first and second editions, and their names were kept on the cover in recognition of this in the third and fourth editions, although Vincent had died when I came to write the third edition and John took no part in the writing thereafter. It is a pleasure to formally acknowledge their inspiration and work, without which, of course, there would be nothing of which to write a fifth edition. However, nothing of the original text now remains and it has been thought appropriate for their names to be taken off the cover for this edition.

One of the great perks of writing a preface is that it provides the opportunity to thank the people who have assisted me. I am extremely grateful for the help given to me by Michael Cowlin LLB(Hons) DipOSH Dip Arb FCIArb Barrister (not practising) who has assisted me by locating cases and quotations, commenting on various portions of the text and making many helpful suggestions. Michael Dunn BSc(Hons) LLB LLM FRICS FCIArb provided relevant citations and contracts and many useful comments. He went far and beyond what I could expect by giving me the benefit of a very detailed criticism of the material included in the Appendix and suggestions for its improvement. I am grateful for their expertise, but the responsibility for using or not using their suggestions is mine. I am grateful also to Caroline Dalziel LLB(Hons) Solicitor who was meticulous in preparing the table of cases. My wife, Margaret, has shown her usual patience throughout the long writing process.

I have endeavoured to state the law from available sources at the end of February 2011.

February 2011
Wakefield

*David Chappell*

## Preface to the Sixth Edition

The title 'Building Contract Claims' distinguishes this book from one providing claims guidance for civil engineering or process engineering projects. The approach for this sixth edition, therefore, remains as for compiling 'building contract' claims – now under the Joint Contracts Tribunal (JCT) 2016 and NEC4 main contracts, their corresponding sub-contracts and for professional consultants' additional fee claims. Together with Wiley's 'Evaluating Contract Claims', authored by R Peter Davison and John Mullen (now in its third edition), guidance for claims – from both the 'making' and 'assessing' standpoints – is comprehensively provided.

I first encountered Wiley's 'Building Contract Claims' in the late 90s when working as a contractor's quantity surveyor. I purchased its third edition, which heralded the first involvement of my co-author David Chappell. Through the third and fourth editions, David expanded upon the first two, retaining much of the original text but updating it to reflect new case law and updated versions of the standard forms of building contract published during the intervening years. When David completed the fifth edition (published in 2011), he included many of the newer contracts that were published by the JCT in 2005 and the increasingly popular NEC3 contracts that were first published in 2007.

Since the fifth edition of Building Contract Claims, many of these newer JCT contracts remain in publication, but their usage remains relatively rare when compared to other JCT contracts such as the various versions of the Standard Building Contracts (SBC/Q, SBC/XQ and SBC/AQ), the Intermediate Contracts (IC, ICD), the Minor Works Contracts (MW and MWD) and the Design and Build Contract (DB). Given that the vast majority of the principles in the JCT 2005 contracts are retained by the 2016 editions mentioned earlier, I felt that by focussing on the more widely used JCT contracts and by consolidating these principles, they could be covered more economically. It is perhaps a pity that the JCT chose to update their 2016 contracts and only recently release them in two collected volumes (in December 2024 and July 2025), when most of the drafting for this book had been finalised. However, in terms of claims, the principles in the JCT 2016 editions remain relatively unchanged.

The popularity of NEC contracts has also increased. NEC3 was amended in 2013 and then replaced by the equally popular NEC4 in 2017 (itself, amended in 2023).

The sixth edition's main areas of expansion include (the contracting party's obligations as to) 'Time' (see Chapter 3) – providing guidance for the various recognised methods of delay analysis, legal developments concerning damages and liquidated damages and the new section for professional fee claims (see Chapter 15).

The consolidation exercise took the common principles found in the JCT 2016 (Relevant Events and Relevant Matters) and NEC4 contracts (compensation events) and relies on the widespread use of tables and figures to show how (and where) those principles apply to the main contracts, sub-contracts and consultant's appointments considered by this sixth edition.

I am grateful to David for his kindness in providing me with a plentiful source of material from which to compile this sixth edition and for his generosity in essentially giving me free rein in doing so. I must also thank the Wiley teams in the United Kingdom, the United States and India once again for their invaluable assistance.

October 2025 *Guy Higginbottom*

## List of Abbreviations

| | |
|---|---|
| ACA | Association of Consultant Architects |
| ACE | Association for Consulting and Engineering |
| ACE PSA | Association for Consulting and Engineering's Professional Services Agreement 2017 |
| AC, App Cas | Law reports, Appeal Cases |
| All ER | All England Law Reports |
| ALJR | Australian Law Journal Reports |
| ALR | Australian Law Reports |
| APCLR | Asia-Pacific Construction Law Reports |
| BCL | Building and Construction Law |
| BCLC | Butterworth's Company Law Cases |
| BLM | Building Law Monthly |
| BLR | Building Law Reports |
| BURR | Burrow |
| Camp | Campbell's Reports |
| CDM | Construction (Design & Management) Regulations 2015 |
| CDP | Contractor's designed portion (JCT 2016 contracts) |
| C&F | Clark & Finnelly's House of Lords Cases |
| Ch App | Chancery Appeal Cases |
| Ch, Ch D | Law Reports, Chancery Cases/Division |
| CIOB | Chartered Institute of Building |
| CILL | Construction Industry Law Letter |
| CLC | Commercial Law Cases |
| CLD | Construction Law Digest |
| CLR | Commercial Law Reports (Canada) |
| Com Cas | Commercial Cases |
| Con LR | Construction Law Reports |
| Const LJ | Construction Law Journal |
| CP | Common Pleas |
| CPA | Critical Path Analysis |
| CSIH | Court of Session Inner House |

| | |
|---|---|
| CSOH | Court of Session Outer House |
| Ct Cl | Court of Claims Reports |
| DAB | Dispute Avoidance Board (NEC4 option W3) |
| DLR | Dominion Law Reports |
| EG | Estates Gazette Cases |
| EoT | Extension of Time |
| ER | English Reports |
| ER's | Employer's Requirements (JCT 2016 contracts with contractor's design) |
| EWCA (Civ) | England and Wales Court of Appeal Civil Division |
| EWHC (TCC) | England and Wales High Court (Technology and Construction Court) |
| EWN | Early warning notice (e.g. Under NEC4 clause 15) |
| Ex, Exch, Ex D | Law Reports, Exchequer Cases/Division |
| F&F | Foster & Finlayson's Nisi Prius Reports |
| FIDIC | Federation Internationale Des Ingenieurs-Conseils |
| FSR | Fleet Street Reports |
| GWD | Green's Weekly Digest |
| H & C | Hurlstone and Coltman's Exchequer Reports |
| H & N | Hurlstone & Norman's Exchequer Reports |
| HLC | Clark & Finnelly's House of Lords Reports New Series |
| HGCRA 1996 | The Housing Grants Construction and Regeneration Act 1996 (as amended) |
| JCT | Joint Contracts Tribunal |
| JCT DB 2016 | JCT Design & Build Contract (2016 Edition) |
| JCT DBSub/C 2016 | JCT Design & Build Sub-Contract (2016 Edition) |
| JCT IC 2016 | JCT Intermediate Building Contract (2016 Edition) |
| JCT ICD 2016 | JCT Intermediate Building Contract with Contractor's Design (2016 Edition) |
| JCT ICSub/C 2016 | JCT Intermediate Sub-Contract (2016 Edition) |
| JCT ICSub/D/C 2016 | JCT Intermediate Sub-Contract with sub-contractor's design (2016 Edition) |
| JCT ICSub/NAM/C 2016 | JCT Intermediate Named Sub-Contract (2016 Edition) |
| JCT MW 2016 | JCT Minor Works Contract (2016 Edition) |
| JCT MWD 2016 | JCT Minor Works Contract with Contractor's Design (2016 Edition) |
| JCT MWSubD 2016 | JCT Minor Works Sub-Contract with sub-contractor's design (2016 Edition) |
| JCT SBC/AQ 2016 | JCT Standard Form of Building Contract – with Approximate Quantities (2016 Edition) |
| JCT SBC/Q 2016 | JCT Standard Form of Building Contract – with Quantities (2016 Edition) |
| JCT SBCSub/C 2016 | JCT Standard Building Sub-Contract (2016 Edition) |
| JCT SBCSub/D/C 2016 | JCT Standard Building Sub-Contract with sub-contractor's design (2016 Edition) |

| | |
|---|---|
| JCT SBC/XQ 2016 | JCT Standard Form of Building Contract – Without Quantities (2016 Edition) |
| JCT ShortSub 2016 | JCT Short Form of Sub-Contract (2016 Edition) |
| JP | Justice of the Peace & Local Government Review |
| JPL | Journal of Planning Law |
| KB | Law Reports, King's Bench |
| LAD's | Liquidated and ascertained damages |
| L&E | Loss and Expense |
| LGR | Local Government Reports |
| LJCP | Law Journal Reports, Common Pleas |
| LJQB | Law Journal Queen's Bench |
| LJ Ex | Law Journal Reports, Exchequer New Series |
| Lloyd's Rep | Lloyds' Reports |
| LPCDA | Late Payment of Commercial Debts (Interest) Act 1998 (as amended) |
| LR | Law Reports |
| LT | Law Times Reports |
| M & W | Meeson and Welsby's Exchequer Reports |
| NBS | National Building Specification |
| NE | North Eastern Reports |
| NEC | New Engineering Contract |
| NEC3 | NEC4 Engineering and Construction Contract (3rd Edition) |
| NEC4 / NEC4 ECC | NEC4 Engineering and Construction Contract (4th Edition) |
| NEC4 PSC | NEC4 Professional Services Contract |
| NRM | New Rules of Measurement |
| NRM2 | New Rules of Measurement – Detailed measurement for building works |
| NSWLR | New South Wales Law Reports |
| NY | New York Court of Appeal |
| NZLR | New Zealand Law Reports |
| P & CR | Planning and Compensation Reports |
| PD | Law Reports, Probate Division |
| QB | Law Reports, Queen's Bench |
| R | Rettie, Crawford & Melville, Session Cases |
| RIBA | Royal Institute of British Architects |
| RIBA PSC | RIBA Professional Services Contract 2020 |
| RICS | Royal Institution of Chartered Surveyors |
| RICS SFCA | RICS Standard Form of Consultant's Appointment 2022 |
| SCL | Society of Construction Law |
| SCR | Supreme Court Reports |
| Sc LR | Scots Law Reports |
| ScotCS | Scottish Court of Session |
| ScotHC | Scottish High Court Decisions |

| | |
|---|---|
| SMM7 | Standard Method of Measurement – 7th Edition (superseded by NRM) |
| (TCC) | Technology and Construction Court |
| TLR | Times Law Reports |
| UCTA | Unfair Contract Terms Act 1977 |
| UKHL | United Kingdom House of Lords |
| UKPC | United Kingdom Privy Council |
| UKSC | United Kingdom Supreme Court |
| WAR | Western Australian Reports |
| WASCA | Western Australia Supreme Court: Court of Appeal |
| WCD 98 | JCT Standard Form of Building Contract with Contractors Design (1998 Edition) |
| WLR | Weekly Law Reports |
| WWR | Western Weekly Reports (Canada) |

# Part I

# 1

# Introduction

## 1.1 General Introduction to 'Claims'

In construction contracts, the word 'claim' is often synonymous with the contractor seeking additional time and money for what's likely to be perceived as a contentious reason. 'Contentious' in this context means arising from circumstances which fall outside those that are usually encountered when carrying out the works, which may not be immediately understood, and almost certainly require additional explanation. 'Claims' in this context are also misunderstood. Many experienced consultants and contractors have been in situations where a party may threaten to raise a 'claim', either to dissuade the other from a course of action or to elevate his own bargaining position when negotiating. These kinds of 'claim' references conjure up images of a contractor spending months working up a voluminous set of documents in many lever arch files, to triumphantly submit to the employer at the eleventh hour. However, in the case of modern building contracts, such as the Joint Contracts Tribunal (JCT) 2016 editions and New Engineering Contract (NEC4) contracts considered in this book, that scenario is not only far from reality but also unlikely to benefit the contractor. Claims under building contracts should be promptly notified and submitted as soon as possible, so they can be properly assessed and resolved – and the claiming party may benefit from receiving payment as early as is practicable. Whilst it may not be possible to submit a finalised claim within a short period of time, some building contracts permit the claiming party to update their claims, so the principles can be agreed in the first instance, and the quantum (or time) is finalised and agreed at a later date.

A 'claim' can be defined[1] as

- to assert that something is the case; or
- formally request or demand.

In this book, the word 'claim' is taken to mean the assertion of an alleged right (usually by the contractor) to an extension of the contract period and/or to additional payment arising under the express or implied terms of a building contract. However, the forms of building contract considered in this book do not use the word 'claim'. Instead, they include specific provisions for seeking additional payment and/or an extension of time (in the case of the contractor), or payment of damages (in the case of the employer), which operate

1 Oxford Dictionary of English (3rd Ed) 2010.

*Building Contract Claims*, Sixth Edition. Guy Higginbottom and David Chappell.

in a similar way to the provisions for obtaining payment for additional work. If it was not for 'claims clauses' in building contracts, the contractor would be obliged to fall back on a common law claim for damages, usually, for breach of contract. In that sense, claims clauses may also be considered (albeit, not entirely correctly) as a contractual procedure for dealing with damages.

A party may also make other types of 'claims', such as those where a 'declaration' is sought. These can include obtaining decisions (or declarations) that its works comply with the contract or are being carried out safely. However, this book only considers claims that entitle a party to payment of money or additional time to complete their obligations.

### 1.1.1 General

For employers, 'claims' are often seen as a problematic and, in certain employer sections of the industry, as an irritation. It is easy to understand why. Claims are often unexpected, seek something that seems initially intangible and often results in the accepted budgets and construction periods being exceeded. The employer's irritation is likely to be compounded by feelings that neither he nor his advisers have control over a project when a claim arises. This perceived 'loss of control' is perfectly normal but is not always easy to accept unless the employer (or the contractor in some cases) and his consultants have experience of dealing with claims. If they have, they should understand that claims develop because the required substantiation may only be obtained over time. Common examples include a contractor being reliant upon his sub-contractors or his own consultants to provide supporting information, or an employer collating the costs that he's incurred. This, often gradual nature of claim development, is frustrating for the recipient, as they want to know at the earliest opportunity if they are liable, and if so, to what extent. The only way to 'shortcut' the claims process is to come to an early agreement, but this can leave the parties feeling that they have traded immediacy for certainty, and perhaps could have obtained a better deal for themselves if they'd waited for more information to be provided.

There is firm evidence[2] that since 2012, around 30% of parties to building contracts in the United Kingdom prefer a 'quicker' resolution to a claim, instead of perhaps a more 'thorough' resolution – judging by the number of projects let under the NEC3 and NEC4 contracts.[3] However, this leaves approximately 70% of building UK contracts being let under other contracts (such as those published by JCT) – under which a far longer period[4] is available for the parties to assess and agree contractor's claims.

### 1.1.2 Contractual Claims

Claims for both time and money under the terms of the contract feature in many building projects. These 'contractual claims' are relatively simple to generate, as the building contract will set out the specific grounds that entitle the contractor or employer to bring a claim

2 National Construction Contracts & Law Reports 2012 to 2022. Approximately, 30% of contracts in 2012 were let under NEC contracts, peaking at over 50% in 2015 and returning to around 30% in 2022.

3 Refer to Section 4.1.2 of this book.

4 Typically, JCT contracts do not conclude 'claims' (for extensions of time and direct loss and/or expense) until between 8 and 14 months after the works have become practically complete, allowing for the rectification period (typically either 6 or 12 months) to expire and a further period (typically 2 months) for the 'Final Certificate' (JCT SBC/Q 2016) or 'Final Statement' (JCT DB 2016).

against the other. However, the parties respective obligation to prove a claim often differ. If an employer is seeking to claim liquidated damages (see Section 8.2), he need only comply with the contract's procedures, and does not have to prove the rate of liquidated damages as this is already agreed. However, contractor's claims are far more difficult to substantiate, and therein lies the employer's protection. Where the claim is brought within the contract procedure, the contractor must also show that it has followed the administrative procedures provided by the contract's terms. Failure to comply with the procedure precisely may create difficulties for the contractor and in some cases may negate the claim. However, that may not mean that the contractor is entirely without a remedy. Above all, claims must be founded on facts and these facts must be substantiated by the contractor. Merely because a contractor is losing money on a particular contract does not mean that it is entitled to look to the employer to make good its loss. It must be able to establish that the loss results directly from some act or default of the employer (or those for whom the employer is responsible in law) or else is referable to some express term of the contract entitling the contractor to reimbursement. Common types of contractual claims are summarised in Section 1.2.1 (submitted by the contractor) and in Section 1.3 (submitted by the employer).

### 1.1.3 Extension of Time and Loss and/or Expense

Contractual claims for time or money must stem from a particular clause in the contract, and some notable examples are:

- The JCT standard forms all contain provisions for extensions of time, additional costs and variations as follows
  - Extensions of time[5] in the JCT SBC/Q 2016, JCT DB 2016, JCT IC 2016 and JCT MW 2016 contracts are dealt with under separate clauses to
  - Direct loss and/or expense[6] (additional costs); and
  - Variations are also dealt with separately – providing a definition of a 'variation' and which method of valuation applies in the circumstances, with the exception that
  - JCT MW 2016 provides for 'direct loss and/or expense' to be included within the valuation of variations.
- NEC4 includes all the provisions for claiming additional time and money, along with 'variations'[7] together as 'Compensation Events[8]' in clause 60.1 (not separately like in the JCT contracts).

It should be noted that an extension of time does not always entitle the contractor to loss and/or expense. This is often a point that is overlooked in construction projects. The giving of an extension of time is not always linked to a right to loss and/or expense either contractually or otherwise in law[9] – an obvious example being an extension for an event that is the fault of neither the employer nor the contractor – such as adverse weather affecting progress.

5 Refer to section 3 (*Time*) and section 11 (*Extensions of Time under JCT standard form contracts*).

6 Refer to section 4 (*Financial Compensation*) and section 12 (*Loss and expense under JCT standard form contracts*).

7 Refer to Chapters 3 (*Time*), 4 (*Financial Compensation*) and 5 (*Claims for: Variations* … etc.).

8 Refer to section 13 (*Contractor Claims under NEC4*).

9 *H. Fairweather & Co Ltd v London Borough of Wandsworth* (1987) 39 BLR 106, and *Methodist Homes Housing Association Ltd v Messrs Scott & McIntosh*, 2 May 1997, unreported.

### 1.1.4 Variations

Variations or changes to the work can also be considered to be 'claims', but are usually far less contentious. This is often because the 'variation' is tangible and something that the employer can immediately recognise, such as the contractor carrying out an additional quantity of work or an item of work being substituted for another of a different specification. However, a 'variation' may also occur if the way in which the original work is changed, but the scope (in terms of its specification and / or quantity) of the work remains unaltered. For example, changes in site logistics may require the contractor to change its method of working. The effect of these less tangible changes may prove to be more contentious.

### 1.1.5 Unexpected Problems

If the contractor is simply finding the work more difficult, then unless the building contract contains some express (or implied) entitlement (such as a target cost or cost reimbursable contract), the contractor usually has no basis for a claim. If the contractor encounters physical or logistical problems with the work or incurs additional expense, then if the contract is for a fixed price, it's likely that he will be responsible for any additional time needed for completing the works and will also have to bear the additional costs himself.

Some claims are produced because the contractor underestimated the cost of doing a job. It may become apparent from the contractor's correspondence during a project that he intends to make a claim at a later date. Further, it is not uncommon to find that there are some contractors who are intent on making claims right from the outset. Indeed, a claim may form part of a contractor's commercial strategy before any work is carried out on site. An employer should therefore take care when selecting contactors for a project. Modern pre-qualification processes and obtaining references from the contractor's previous clients (and their consultants and advisers) should go some way to identifying more unscrupulous contractors. In addition, thorough analysis of the contractor's tender using appropriate cost data can identify any obvious shortcomings or anomalies in the contractor's price. On the other hand, there are many examples of contractors delayed and disrupted by the actions or inactions of the employer or the professional team, but who find it quite difficult to recover the appropriate loss and/or expense.

The contractor should therefore be fully aware of the physical, logistical and commercial risks that he accepts when agreeing the terms of a building contract. Problems can arise if a contractor mistakenly believes that he's entitled to additional time or money, but no such entitlements are found in the contract. This may be for a number of reasons:

- The contractor may simply be mistaken over what entitles him to make a claim. An example of this could be if a contractor is more familiar with contracts where the employer is responsible for design but then carried out work under a design and build contract; or
- If the contractor mistakenly believes that he's included a qualification or caveat in the contract that entitles him to claim, but it's not been included in (or incorporated into) the finalised contract. In one case,[10] the contractor believed that the contract

10 *Davis Contractors Ltd v Fareham Urban District Council* [1956] 2 All ER 145 HL.

price was subject to adequate labour being available, as it had set this out in a letter during negotiations. However, the letter had not been incorporated into the contract, so the contractor was bound by the contract price and was not entitled to payment on a *quantum meruit* basis. The Court's reasoning included:

- Just because a contract became more difficult to perform than initially envisaged was no reason to excuse the contractor from further performance; and
- '*We could seriously damage the sanctity of contracts if we allowed a builder to charge more, simply because, without anyone's fault, the work took him much longer than he thought*'.[11]

## 1.2 Standard Forms of Contract Considered in This Book

Previous editions of this book have referred to a variety of building contracts. By the 'building contract' nature of this book, there are no dedicated chapters or sections that focus on civil engineering contracts.[12] For many years since the first edition of this book in 1983, the most common forms of building contract used in the United Kingdom have been those published by the 'Joint Contracts Tribunal' (JCT). Previous editions of this book considered the more common JCT contracts, which have been derived from the 'Standard Form of Building Contract' (currently the JCT SBC/Q 2016[13] for its 'with quantities version'). Other JCT contracts have been derived from what's now become the JCT SBC/Q 2016[14] contract, such as

- The Design and Build Contract (JCT DB 2016)
- The Intermediate Contract ('JCT IC 2016')[15]
- The Minor Works Contract (JCT MW 2016).[15]

The previous editions of this book also included sections covering other JCT standard forms of contract, such as the Management Building Contract, Prime Cost Building Contract, Construction Management Trade Contract, Major Project Construction Contract, Measured Term Contract and Constructing Excellence Contract. However, this edition only focusses on the most popular versions: JCT SBC/Q 2016, JCT DB 2016, JCT IC 2016 and JCT MW 2016.

Since the mid-1990s, the New Engineering Contract ('NEC') has become increasingly popular. The NEC contracts are now published in their fourth edition (NEC4) and include the following versions:

- NEC4 Option A: Priced Contract with Activity Schedule
- NEC4 Option B: Priced Contract with Bill of Quantities

11 Ibid as per Denning LJ at 278.

12 Such as the Institute of Civil Engineers Contract ICE 7th (1999) now replaced by the Infrastructure Conditions of Contract 2014, or one of the contracts published by the Federation Internationale Des Ingeniurs-Conseils ('FIDIC'), such as the Conditions of Contract for Building and Engineering Works designed by the Employer ('Red Book'), Conditions of Contract for Plant Design-Build ('Yellow Book') and 'Silver Book'.

13 Previous versions were 'JCT 63', 'JCT80', JCT98, JCT 2005 and JCT 2011.

14 The 'Standard Building Contract' also includes variants for 'Without Quantities' (JCT SBC/XQ 2016) and with 'Approximate Quantities' (JCT SBC/AQ 2016).

15 There are also versions of these contracts where elements (or all) of the design are the responsibility of the contractor – JCT ICD 2016 and JCT MWD 2016.

- NEC4 Option C: Target Contract with Activity Schedule
- NEC4 Option D: Target Contract with Bill of Quantities
- NEC4 Option E: Cost Reimbursable Contract

Certain key aspects of NEC4 contracts operate in fundamentally different ways to their JCT counterparts and place great importance on matters that may not have been addressed in any detail by JCT contracts. Early into their more widespread use, it was not unusual to see parties attempting to administer an NEC contract in a similar way to a JCT contract – a clear mistake which was often incompatible with the NEC ethos. Since the third editions of the NEC contracts (NEC3) were introduced in 2005, the building industry had become more accustomed to the operation of NEC contracts, so these earlier mistakes are less likely to be repeated. NEC contracts have now largely replaced the General Conditions of Government Contracts for Building and Civil Engineering Works (GC/Works/1) for most public projects.

Other contracts that were covered in previous editions included the Form of Building Agreement (ACA 3) and the Standard Form of Contract for Project Partnering (PPC 2000) both published by the Association of Consultant Architects. Like GC/Works/1, these forms of contract have either fallen out of common use (ACA3) or are only used for a small number of projects, relative to those procured under a JCT 2016 contract or under NEC4. Whilst this edition focusses on the JCT 2016 contracts and the NEC4 contracts, many of the principles relating to claims under these current building contracts will apply to claims under the older forms. Previous editions of this book is still available from the publisher.

This book also considers sub-contractor claims under their JCT 2016 and NEC4 counterparts. Many of the principles applicable to the JCT 2016 and NEC4 main contracts will apply to claims under their respective sub-contracts.

A new chapter covering claims for professional fees has been introduced. This includes claims under the NEC4 Professional Services Contract (many aspects of which are similar to claims under the NEC4 main contracts and sub-contracts), along with claims under other forms of consultant's appointments published by the RIBA, ACE and the RICS.

### 1.2.1 JCT 2016 Contracts

The JCT contracts considered in this book have been developed over time. JCT published its standard form of contract in 1963 (JCT 63), where the employer remained responsible for the project's design (this is often called 'traditional procurement'). JCT 1963 was revised in 1980 (becoming 'JCT 80') and its minor works contract (MW80) was also published. Design and Build contracts, as we now know them, did not appear until 1981 when the JCT published the 'with Contractor's Design' contract (CD 81), which followed the format of JCT 80. In 1984, JCT published its 'Intermediate' form of contract (IFC84) – for projects whose values were in between those usually procured under the JCT80 and MW80 contracts. In 1987, JCT then published its Management Contract (MAN87). Between 1984 and 1998, developments in case law and statutory law caused JCT 80 (then consisting of 42 clauses) to be extensively amended, with around 18 amendments being published. The JCT contracts were further revised in 1998 and again (adopting the current 9 section format of the standard building contract) in 2005, 2011 and 2016.

It is important to summarise the operation of JCT 2016 contracts, and differentiate between this, and the way that NEC4 contracts operate, as this will aid understanding of both suites of building contract. The aims of JCT 2016 contract can be summarised as the parties wanting to get on with the works and sort out the precise details at the end. This approach is evident, partly from (1) the manner in which claims for the value of work done, extensions of time and loss and expense are dealt with as the work proceeds (i.e. on a progressive basis), but mainly from (2) the 'concluding steps' that each of the JCT 2016 contains, be that in the form of a Final Certificate (found in JCT SBC/Q 2016, JCT IC 2016 and JCT MW 2016) or the relevant Final Statement (in JCT DB 2016). These 'concluding steps' take place in the months following practical completion of the works and depend upon the length of the agreed 'Rectification Period' (they may take place between a total of 8 and 14 months after[16]). This length of time can be problematic for both parties. First, from the contractor's perspective, he will have completed the works (save for making good any defects), and his cash flow may be suffering because he has not concluded the values of his work (most likely the values of variations) and/or loss and expense with the employer. Conversely, from the employer's perspective, he may have obtained his completed project, but may feel that he's overpaid for the project, and may have to wait until the 'concluding steps' to recover any overpayment. The JCT 2016 contracts all provide the means for the precise answer to be obtained; however, it may take more time than the parties may expect.

### 1.2.2 NEC4 Contracts

The first edition of the NEC was published in 1993. Although from its title, one might be excused for believing that it was not suitable for use for building works, but as we have seen in the following years, it is used for both civil engineering and building projects. Its second edition was published in 1995, changing its title to the Engineering and Construction Contract although it was still known by the initials 'NEC'. Amendments were published that dealt with the Housing Grants, Construction and Regeneration Act 1996 and a third edition ('NEC3') was published in June 2005 and subsequently re-published with further minor amendments in June 2006 and again in April 2013. NEC4 was published in June 2017, with some further amendments published in January 2023. The NEC philosophy remains very different from the JCT 2016 contracts and is sometimes thought of as a 'project manager's' building contract.

Although the form was praised by Sir Michael Latham, as containing the kind of provisions advocated in his report 'Constructing the Team'[17], it has been the subject of some criticism by legal commentators.[18] Some perceived shortcomings are that the syntax and grammar is not good. This is partly because the present tense[19] is used almost exclusively even where one would normally expect to the see past or future tenses. NEC contracts are supposed to be a model of simple English, but sticking almost exclusively to one tense may not always assist understanding of what can be an essential part for a legally binding

16 On the basis that neither party commences adjudication (or other) proceedings against the other to resolve any valuation or payment disputes before the 'concluding steps' take place.

17 HMSO July 1994.

18 *D G Valentine, The New Engineering Contract* (1996) 12 Const LJ 305.

19 *'I have to confess that the task of construing the provisions in this form of contract is not made any easier by the widespread use of the present tense in its operative provisions. No doubt this approach to drafting has its adherents within the industry but, speaking for myself and from the point of view of a lawyer, it seems to me to represent a triumph of form over substance'. Anglian Water Services Ltd v Laing O'Rourke Utilities Ltd* [2010] EWHC 1529 (TCC) at 28 per Edwards-Stuart J.

contract. It can be difficult to know whether something is being expressed as a power or a duty or just as a fact.

Another perceived shortcoming of NEC contracts is that many commentators feel that their provisions have not been robustly tested in the same way that provisions of JCT contracts have been. The example of 18 amendments to JCT80 is mentioned above, with a number of these being implemented following new legislation and decisions of the Court. There are only a few reported cases dealing with the NEC forms contract. Consequently, and whilst there are some excellent texts devoted to NEC contracts, the relative absence of cases that specifically address issues exclusive to NEC can leave the reader feeling that some textbook guidance may be rather 'general' in its nature.

The operation of NEC contracts (now NEC4) significantly differ from that of a JCT 2016 contract (or its predecessors) in the following fundamental ways:

- Any variation to the work, or other matter entitling the contractor to additional payment or additional time (collectively referred to as 'compensation events'), cannot be the fault of the contractor, so he cannot lose out; and
- The vast majority of the contract's administrative functions are concluded during the course of the works, i.e. before completion. Very little is left to be determined once the work is completed so there is no detailed 'final account' stage (or 'concluding steps') that are found in JCT 2016 contracts.

A relevant example for the first difference could involve the contractor being instructed to carry out an additional quantity of work. Under a JCT 2016 contract, if the applicable unit rate or price for valuing this work was a loss-leader for the contractor, his losses would increase by carrying out the variation. However, this would not be the case under NEC4, as the basis of valuing the additional quantity of work would be based upon the relevant 'defined cost' of the additional quantity of work, so the contractor should not lose any money.

The effect of the second difference is that NEC4 requires many of its administrative functions to be carried out contemporaneously. This means that the values for variations, extensions of time and any loss and expense are agreed as the work proceeds, and once agreement has been reached, any changes to the price for the works or to the time taken for the works cannot (under most circumstances) be adjusted or revised. Comparing how NEC4 contracts operate against JCT 2016 contracts shows that instead of concluding all matters under the contract at a date sometime after completion, NEC4 requires matters to be progressively concluded to an acceptable detail (i.e. during the works and before completion). In order to do this, and to emphasise the 'project management' nature of the contract, NEC3 and now NEC4 place time constraints (also called 'time bars') on a number of important administrative functions. Some examples include using the contractor's programme as a key project management and monitoring tool (requiring it to be regularly updated and 'accepted'), default acceptances of the contractor's programme and quotations for compensation events if not accepted within the agreed time constraints, and perhaps the most controversial – a condition precedent that removes the contractor's entitlement to additional payment and/or time if he fails to notify compensation events within the specified period.

It is not the purpose of this book to take sides over whether the JCT 2016 contracts or NEC4 contracts are better or worse. It is not difficult to envisage projects where NEC4 contract would be more suitable than a JCT contract, and vice versa.

## 1.3 Types of Claims by the Contractor

It is useful to classify claims by the parties to a building contract into the following categories:

- Contractor claims:
  - contractual claims,
  - common law claims,
  - quantum meruit claims and
  - ex gratia claims.

### 1.3.1 Contractual Claims

Contractual claims are those based on clauses in the contract which expressly provide for the contractor to make a claim in certain situations. Some examples of contractual claims made by the contractor are as follows:

- JCT Standard Building Contract with Quantities 2016 (JCT SBC/Q 2016):
  - Extension of time[20] clauses 2.26–2.29.
  - Loss and expense clauses 4.20–4.24. With the exception of the Minor Works contract, other JCT contracts follow a similar format to SBC/Q in setting out the grounds for extensions of time and loss and/or expense in separate sections.
  - Variations (and their valuation) clauses 5.1–5.10.
- NEC4:
  - Compensation Events clauses (typically clauses 60–66, depending upon which Main Option Clauses are selected).

Such claims make use of the clauses in the contract to first establish an entitlement, then set out how the parties process the claim and finally produce a result. The principal reason for having such provisions in the contract is to provide the parties with the certainty of knowing the precise circumstances and events that give rise to a claim. This avoids the necessity for either party to have to seek redress at common law and the inevitable expense involved for both parties in doing so. Many standard form contracts preserve the contractor's right to seek damages at common law if it is not satisfied with its reimbursement under the contract.

### 1.3.2 Common Law Claims

Common law claims are claims for damages, usually but not exclusively, for breach of contract under common law. The breach giving rise to the claim will normally be a breach of one of the 'implied terms' of the building contract. Implied terms are essentially those setting out obvious or 'common-sense' terms that every contract ought to contain. There are a number of these implied terms and these are explored in more detail in Chapter 6 (*Basis for common law claims*). Common law claims may also embrace claims for breach of some other aspect of the law such as tortious claims or claims for breach of statutory duty. Most standard forms expressly reserve the contractor's right to make such claims, two examples

20 Referred to as 'Adjustment of the Completion Date'.

being clause 4.22.5[21] of JCT SBC/Q 2016 and clause 60.1(18)[22] of NEC4. A common law claim may also be made when it is impossible or difficult to make the claim by relying upon the contract clauses, or it may be for another breach which is not expressly covered by the contract's express conditions. A contractor may raise a 'common law claim' for damages for the following:

- A breach of the building contract by the employer that is not covered by the extension of time, loss and expense,[23] compensation event[24] or risk clauses in the contract; or
- A breach by the employer of one of the implied terms of the building contract. An 'implied term' is one which applies to most contracts for reasons of common sense – for example, the employer shall not hinder or prevent the contractor from carrying out the works (for more details, see Sections 3.2.2 and 6.2.2).

Common law claims are sometimes referred to as 'ex-contractual' or 'extra-contractual' claims. These terms are sometimes confused with the term *ex contractu* – a term rarely found in legal textbooks when referring to claims which arise from the contract.

### 1.3.3 *Quantum meruit* Claims

'Quantum meruit' (meaning 'as much as he deserves[25]') enables a party providing a service (such as building work) to recover a sum in payment (see Section 4.5).

### 1.3.4 Ex gratia Claims

An ex gratia claim (meaning 'done as a matter of favour[25]') is a claim which has no contractual or legal basis. Consequently, if a contractor raises one, the employer is not obliged to consider it, let alone pay for it. Hence it is often referred to as a 'hardship claim'. Despite there being no strict obligations on the employer to consider or pay such a claim, making an 'ex gratia' payment to a contractor may have an external commercial benefit to the employer. For example, if a contractor is suffering cash-flow difficulties on another project, it may be in the employer's interest to make an 'ex gratia' payment to the contractor, so he completes his project (compared to the expense of appointing another contractor). This situation may arise where the works are almost complete and will swiftly provide the employer with income (e.g. from rental or purchases). However, such payments should be made with caution, particularly if the contractor is at risk of becoming insolvent.

The employer's representative normally has no powers under standard form contracts to consider or certify[26] 'ex gratia' claims, unless given express authority by the employer. An employer's representative would likely to become liable to the employer if an 'ex gratia' payment was certified, then made but was then irrecoverable if the contractor became insolvent.

---

21 JCT SBC2016, clause 4.22.5 – loss and expense for *'any impediment, prevention or default ... by the Employer'*.
22 NEC4 – clause 60.1(18) – a compensation event arising from a *'breach of contract by the Client'*.
23 JCT 2016 contracts.
24 NEC4 contracts.
25 Oxford Dictionary of Law (2018).
26 Via a 'payment notice' complying with s110A(1)(a) of the Housing Grants Construction and Regeneration Act 1996 (as amended) or its equivalent.

## 1.4 Types of Claims by the Employer

Clearly, a contractor may be in breach of the building contract and this will entitle the employer to make a claim against him. Typical claims by employers often include:

- Delay damages - for late completion of the work (or parts of the works).
- Damages for specific matters (e.g. failure to construct the works to a specified size or achieve a particular performance output).
- Reduction of the contract price for non-compliant work.
- Balance owing on a final certificate (or equivalent).

### 1.4.1 Delay Damages

The most common claim an employer raises against the contractor is damages arising from the contractor's failure to complete the works (or parts of the works if appropriately divided into sections) by the agreed completion date. This kind of claim is known as 'delay damages' (see Section 8.2). The employer's entitlement to claim and often the method of claiming damages from the contractor for such a breach are often relatively straightforward, although there is usually a very formal administrative process for the employer to follow. It is usually self-evident that the contractor has failed to complete by the completion date, so in most cases, it will not be difficult for the employer to demonstrate the contractor's breach:

- Upon demonstrating the contractor's breach, the employer then becomes entitled to claim damages from the contractor immediately, or after some administrative process has been concluded – such as serving a formal notice (or notices) confirming the contractor's failure and/or notifying the contractor of the employer's intent to claim damages;
- In many cases, the employer can recover damages by deducting them from sums that are otherwise due and payable to the contractor, instead of subsequently claiming payment from the contractor; and
- There is often no disputing the amount (or quantum) of the employer's damages as it is calculated by applying a rate (£ per week or £ per day) that has already been agreed by the contractor (referred to as 'liquidated damages'). If no such rates have been agreed, the employer will have to demonstrate its actual costs arising from contractor's breach (referred to as 'un-liquidated damages').

The relative simplicity for an employer to claim delay damages (particularly liquidated damages) should not be taken for granted. Even if the works are continuing after the completion date and a rate of liquidated damages has been agreed, the contractor may challenge the employer's claim on several grounds (see Section 8.4). The employer may also be able to deduct damages if the contractor fails to meet the conditions for a particular event to occur (such as achieving a 'Key Date' under NEC4 – see Section 8.6.2).

### 1.4.2 Low Performance Damages

A number of civil engineering or process engineering contracts allow the employer to claim damages from the contractor if the completed works do not meet the levels of performance required by the contract (see Section 8.6.2). Typically, the contractor is constructing something with a clear function or purpose – such as a power station, a chemical plant or a highway or railway, where the function (or purpose) can easily be measured in terms of its output. NEC4 contracts contain an option[27] for 'low performance damages', where the contractor has to pay a sum of money to the employer if the works' performance falls below a particular level.

### 1.4.3 Reduction of the Contract Price for Non-Compliant Work

Many forms of contract allow reduction (also called 'abatement') of the contract price if the standard of work performed does not satisfy that specified in the contract. Building contracts are no different (see Section 8.5). The employer has a common law right of reduction (or abatement), and this is expressly confirmed in standard forms of contract as follows:

- JCT contracts[28] permit the employer to accept a defect in return for reduction of the contract price; and
- NEC4[29] not only permits the acceptance of a defect in return for a reduction in the 'Prices', but it provides that the works' specification is adjusted (reduced) accordingly so the contractor's work is then compliant.

### 1.4.4 Other Claims

The types of employer claims described in Section 1.4 are expressly set out in the JCT and NEC4 contracts. However, the employer may have other 'claims' to raise against the contractor, such as

- Recovering an overpayment. Paying the contractor a sum that exceeds that which ought to be due to it, may arise for a number of reasons, including:
  - Subsequent discovery of defects which reduce the value of the contractor's work; and
  - Negligent over-valuing or over-certifying the value of the contractor's work.

  If the contractor has not completed the works and remains entitled to receive interim payments, then the employer may simply be able to correct the amount properly due to the contractor for the next interim payment:
- Damages for the contractor breaching any other terms of the contract. If the contractor breaches its principal obligations to complete on time and/or achieve the specified standard of work, the employer is likely to suffer a loss, which could be significant. However, if the contractor commits lesser breaches of the building contract, then the employer may not suffer significant loss at all. NEC4 provides for the contractor to complete discrete parts of the works by 'Key Dates[30]' (see Section 8.6.2). In the event that the contractor fails to complete the work for a Key Date, then clause 25.3 of NEC4 provides that the employer may recover his 'additional cost'.

27 NEC4 – Option X17: Low Performance Damages.
28 JCT SBC/Q 2016 – clause 2.38, JCT DB 2016 – clause 2.35, JCT IC 2016 – clause 2.30, JCT MWD clause 2.11.
29 NEC4 clause 45 (Accepting Defects).
30 NEC4 clause 11.2(11) defines a 'Key Date'. The work required to achieve a Key Date is not treated in the same way as if the works (as a whole) are divided into sections. The Employer is not permitted to deduct liquidated damages from the contractor for failure to achieve a 'Key Date'; instead, the Employer may recover his actual costs.

## 1.5 Structure of This Book

The book has been arranged into the following parts:

| | |
|---|---|
| • Part One | Introduction, general principles of claims, claims for time and money (including common law and global claims), claims by the employer, employer's Representative's duties and preparation and substantiation of claims |
| • Part Two | Extensions of time and loss and expense under JCT main contracts and compensation events under NEC4 |
| • Part Three | Sub-contractor claims and professional fee claims |
| • Appendices | Examples of claims under JCT 2016 and NEC4 contracts |

# 2

# General Principles

## 2.1 Entitlement

A party's entitlement to claim anything, be that an interim payment, an extension of time, loss and expense, variations or damages, will either arise from the other party's breach of the contract or from a particular provision in the building contract. When referring to a 'breach of contract', it is important to note that the breach may not only relate to the 'express' (or written) terms found in the building contract itself but also may be:

- A breach of one of the 'implied terms' (briefly described in Section 1.2.2 and explored in detail in Section 6.2); or
- An entitlement to payment on a 'quantum meruit' basis (see Section 4.5).

Entitlement arising from a breach of the express (or written) terms of the contract is considered in this chapter.

### 2.1.1 Contractual Entitlement

The contractor's contractual entitlement is set out in specific clauses in the building contract itself. Sections 1.1.3 and 1.2.1 briefly summarised the typical locations of these specific clauses as follows:

- JCT 2016 contracts. The more detailed forms of contract[1] include the provisions for extensions of time, loss and expense and variations[2] in separate sections;
- NEC4 collates the grounds for claiming an extension of time, loss and expense and variations under a single heading – 'Compensation Events'.

Table 2.1 (below) summarises and compares the contractual entitlements for what would be extensions of time, loss and expense and variations under JCT SBC/Q 2016, against the 'compensation events' in NEC4.

In terms of the JCT 2016 contracts considered in this book, Table 2.1 only provides a cross-reference to the clauses in JCT SBC/Q 2016. Sections 4.1, 5.2 and 5.3 and Chapters 11, 12 and 13 contain reconciliations of the relevant 'claims' clauses found in the JCT SBC/Q 2016, JCT DB 2016, JCT IC 2016 and JCT MW 2016 contracts, along with 'compensation events' in NEC4.

1 JCT SBC/Q 2016, JCT IC 2016 (Including their variants) and JCT DB 2016.
2 Variations are referred to as "Changes" in JCT DB 2016.

*Building Contract Claims*, Sixth Edition. Guy Higginbottom and David Chappell.

**Table 2.1** Comparing 'claim' events under JCT SBC/Q 2016 with NEC4.

| **NEC4 Compensation Event – clause 60.1**<br>**(for all Main Option Clauses A, B, C, D and E)** | **Equivalent matter under JCT SBC/Q 2016[a)]** |
|---|---|
| (1) Project manager's instruction ('PMI') to the contractor to change the scope | Variation 5.1<br>EoT 2.29.1/2.29.2<br>L&E 4.22.1/4.22.2 |
| (2) The Client prevents access to parts of the site by the 'access date' or relevant dates on the Accepted Programme | EoT 2.26.7<br>L&E 4.22.5 |
| (3) The Client fails to provide something by the Accepted Programme's date | EoT 2.29.3, 2.26.7<br>L&E 4.22.5 |
| (4) PMI not to start or stop work | Variation 5.1<br>EoT 2.29.1/2.29.3<br>L&E 4.21.1/4.21.2 |
| (4) PMI to change a Key Date | N/A |
| (5) The Client (or other contractors) do not do their work in accordance with the Accepted Programme | EoT 2.29.7, 2.29.8[b)]<br>L&E 4.21.5 |
| (5) The Client (or other contractors) do not work in accordance with the Scope | EoT 2.29.7<br>L&E 4.21.5 |
| (5) The Client (or other contractors) undertake work that is not set out in the Scope | EoT 2.29.7 / 2.29.8<br>2.29.14[c)]<br>L&E 4.22.2.5, 4.22.5 |
| (6) The PM[d)]fails to respond to a communication within the period for reply | EoT 2.29.7<br>L&E 4.21.5 |
| (7) PMI dealing with an object of archaeological value or that is of interest found on the Site | EoT 2.29.4[e)]<br>L&E 4.22.3 |
| (8) The PM changes a previous decision | Variation 5.1<br>EoT 2.29.1/2.29.2<br>L&E 4.22.1/4.22.2 |
| (9) The PM's reason(s) for not accepting something is/are not set out in the contract | EoT 2.29.7<br>L&E 4.21.5 |
| (10) The contractor is instructed to search for a defect but there is not one | EoT 2.29.2.2<br>L&E 4.22.2.2 |
| (11) Testing causes an otherwise avoidable delay | EoT 2.29.2.2/2.29.7<br>L&E 4.22.5 |
| (12) The Contractor encounters unforeseen physical conditions within the site, which are not weather conditions | N/A |
| (13) The contractor encounters adverse weather conditions within the meaning of the contract | EoT 2.29.9 |
| (14) The Client causes an event for which he is liable | EoT 2.29.7<br>L&E 4.22.5 |
| (15) PM certifies 'taking over' before 'Completion' and the 'Completion Date' | N/A |
| (16) Client fails to provide necessary facilities, equipment and materials for tests | EoT 2.29.7<br>L&E 4.22.5 |

**Table 2.1** (Continued)

| **NEC4 Compensation Event – clause 60.1 (for all Main Option Clauses A, B, C, D and E)** | **Equivalent matter under JCT SBC/Q 2016[a)]** |
|---|---|
| (17) The PM corrects an assumption he made about a previous Compensation Event | N/A |
| (18) Client commits a breach of the contract which is not a Compensation Event | EoT 2.29.7<br>L&E 4.22.5 |
| (19) An event occurs (which is not another Compensation Event) which prevents the contractor from completing the works[f)] that is outside of each party's control or that the contractor could not have reasonably foreseen | EoT 2.29.10,[g)] 2.29.11, 2.29.12, 2.29,13, 2.26.15 |
| (20) The PM does not accept a contractor's quotation for a proposed instruction | N/A |
| (21) Any additional Compensation Events set out in the Contract Data Part One occur | N/A |
| Options B and D – differences between the final quantities of work and the quantities in the Bill of Quantities:<br>● If it does not arise from a change in the Scope;<br>(all clause 60.4) | Variation 5.1[h)]<br>2.29.1, 2.29.5<br>4.22.2.3 |
| ● If it causes the Defined Cost to change (see Table 31 in Section 8.9.3.1); or | N/A |
| ● If the Bill of Quantities for that rate exceeds 0.5% of the total of the Prices. | 2.29.5[i)]<br>4.22.2.3 |
| ● If the difference in quantity delays Completion or a Key Date (clause 60.5) | 2.29.1, 2.29.5<br>4.22.2.3 |
| ● If the Project Manager corrects a mistake in the Bill of Quantities arising from a departure from the method of measurement, ambiguity or inconsistency (clause 60.6) including between the Bills of Quantities and another document (clause 60.7) | Variation 5.1,[h)]<br>EoT 2.29.1<br>L&E 4.22.2.1, 4.22.2.3 |
| Y2.4 Contractor suspension | EoT 2.29.6<br>L&E 4.13.2 |

a) JCT SBC 2016 – references "Variation," "EoT" means extension of time, and "L&E" means direct loss and / or expense.
b) JCT SBC/Q 2016, clause 2.29.8 ascribes an EoT (Relevant Event under clause 2.29) to Statutory Undertakers failing to carry out their work, but no entitlement to L&E (see the 'Relevant Matters' under clause 4.21).
c) JCT SBC/Q clause 2.29.14 permits an EoT if a "Named Specialist" becomes insolvent.
d) Including those working on the Project Manager's behalf – such as the Supervisor.
e) JCT SBC/Q – clause 2.29.4 refers to instructions given under clause 3.22.1 or 3.22.1 (Antiquities).
f) NEC4 clause 60.1(19) – the event can also prevent the contractor from completing by the Accepted Programme's planned completion date.
g) JCT DB 2016 – EoT under clause 2.26.9 is for one of the "Specified Perils" which is an event covered by the employer's Joint Names Policy for insuring the Works and Existing Structures (and their contents) owned by him under Insurance Option C (see also clause 6.9).
h) JCT SBC/Q – correction of the Contract Bills under clause 2.14.1 (departure from measurement method) is a Variation.
i) JCT SBC/Q – clause 2.29.5 execution of work for which an Approximate Quantity is not reasonably accurate.

**Table 2.2** Duty, breach, causation and loss.

| Step | Claim for 'breach' | Claim for non-breach |
|---|---|---|
| • Duty | • Claimant must establish that the respondent owed him a duty under the contract.[a)] | • Claimant must establish that the contract terms[a)] entitle it to raise a claim. |
| • Breach | • Claimant must show that the respondent is in breach of that duty. | • Claimant must show that the circumstances or event giving rise to that entitlement have either occurred or (subject to the contract terms) are likely to occur; |
| • Causation | • Claimant must show that the respondent's breach of duty caused it to suffer a loss; and | • Claimant must show that the occurring event or circumstance giving rise to the entitlement has caused (or is likely to cause) it to either suffer a loss or be entitled to additional payment and/or time; and |
| • Loss | • Claimant must demonstrate the amount (or quantum) of its loss caused by the respondent's breach of the duty owed to the Claimant. | • Claimant must demonstrate the amount (or quantum) of its loss or entitlement to additional payment and/or time arising from the occurring event or circumstance, ordinarily in accordance with the relevant building contract's terms. |

a) Under the express or implied terms.

### 2.1.2 Duty, Breach, Causation and Loss

In order to demonstrate an entitlement to a claim, there are generally four key steps for the claimant to establish. The nature of those steps may slightly vary, depending upon whether the claim arises from a breach of the contract or from one of the contract's express terms, but their principles generally remain consistent. Table 2.2 summarises these four key steps in relation to claims for 'breach of contract' and from non-breaches – e.g. a 'variation' – arising under the contract's provisions.

The following are two simple examples to illustrate how the four key steps could apply to claims for a breach and for a non-breach:

- Claim for a 'breach' example (employer fails to give access to the site for two weeks):
  - The employer has an express '**duty**' to provide access on particular date;
  - The employer's failure to do so is a clear 'breach' of its duty;
  - The employer's breach 'causes' the contractor to be on site for two weeks longer than contracted for and incurs an extra two weeks of time-related (preliminaries) costs;
  - The contractor demonstrates the amount (or quantum) of 'loss' by providing invoices for the additional two weeks of preliminaries costs.
- Claim for a 'non-breach' example (e.g. the works' progress is delayed by one week due to an instruction for additional work, or for the occurrence of bad weather):
  - In the 'non-breach' example, the equivalent of a 'duty' is the contractor's express right to additional entitlement – e.g. for a variation increasing the quantity of work (entitling

the contractor to both additional time and money), or additional time only, in the event of bad weather;

- The equivalent of the 'breach' is the instruction for the aforementioned variation or the occurrence of bad weather;
- The 'causation' equivalent is the contractor demonstrating that it is entitled to an addition payment for the variation, or that it cannot carry out any work on site for the bad weather's duration; and
- The 'loss' will be the amount (or quantum) of the additional payment for carrying out the variation, or one week of lost progress, or recovering any delay damages for which the contractor would otherwise be liable, and to make good the loss, the contactor is entitled to a two week extension of time.

## 2.2 The Burden of Proof

### 2.2.1 The Burden of Proof

It is well established that when making a claim, 'he who asserts must prove;'[3] therefore, the burden of proof rests with the party making a claim. There are some instances where the facts supporting a claim appear so obvious that the burden shifts to the other party to, in effect, disprove the claim (see Section 2.2.2 below). Such instances are rare in building contracts, so it is usually for the contractor to prove its claim. Many employer's representatives (such as an architect, contract administrator or project manager) will state that the contractor's task is to convince them that the claim is both viable and worth as much as the contractor states. That is clearly correct; however, the courts have clearly set down the standard of proof in different circumstances.

There is some misconception surrounding the standard of proof which the claiming party must satisfy. In criminal cases, the Crown must prove its case 'beyond a reasonable doubt', and it is not uncommon for stubborn employers and their representatives to misapply this higher standard in civil cases, when the contracting party has made a claim. The correct standard of proof in civil cases (such as building contract claims) is that the claimant must prove 'on the balance of probabilities'. This is a much lower standard than 'beyond a reasonable doubt. In simple terms, the employer or his representative must be satisfied that it is more likely than not that the contractor is entitled to claim for the reasons it has given.

### 2.2.2 Res Ipsa Loquitur – It Speaks for Itself

The literal translation is 'the thing speaks for itself.'[4] Although earlier it has been said that the onus of proving is on the party making a claim, there are some situations where the facts are so clear, and any reasonable interpretation can only point one way, then the burden of proof can effectively be reversed, so the other party must then show that it is not at fault.[5] A slightly macabre example might be where scaffolding is erected next to a public highway and a passer-by is found on the floor with head injuries. A blood-stained brick of the type being

3 *Joseph Constantine Steamship Line v Imperial Smelting Corporation Ltd* [1942] AC 154 at 174 per Viscount Maugham.

4 Oxford Dictionary of Law (2018).

5 *Scott v London and St Katherine's Docks Co.* (1865) 3 H & C 596.

stacked on the scaffolding is found beside the injured person. The facts clearly indicate that a brick has fallen from the scaffolding onto the head of the passer-by, causing the injuries. In any legal action by the injured person, the burden of proof is likely to lie with the contractor to show that it was not responsible. Rarely, some elements of a contractor's claim may fall into this category. For example, if the employer or his representative postpones all work on the critical path for a period of a week, then it is highly likely that the completion date will similarly be delayed by one week.

## 2.3 Causation

Causation has been summarised in Section 2.1.2 (above). It is the relationship between cause and effect and is an extremely important concept in the context of liability. A wrongful act may trigger a series of events which eventually result in damage (or loss) being suffered. This is called the 'chain of causation' (see Figure 2.1 below):

The loss or damage must be direct[6] in the sense of not being remote, and also in the sense that the chain of causation remains unbroken. The matter on which the contractor seeks to rely must be linked, without interruption, to the eventual loss suffered. Therefore, if the cause is not the initiating event, but an intervening event,[7] the chain of causation will be broken, and there may be no liability or claim. To put the situation another way, the loss and/or expense must have been caused by the breach or act relied upon, and not merely be the occasion for it.[8]

The following two simple examples may be contrasted:

- A variation is instructed which necessitates plant lying idle for some days. The plant is needed for the original work, but at a very late stage, the work is varied and so the plant is not required for a period of time. Suppose the contractor has hired the plant. The contractor's plant hire charges, subject to any re-letting or the plant owner accepting an early return, would be a direct loss and, therefore, reimbursable.
- Another variation substitutes clay roof tiles for slates. The contractor promptly orders the substituted slates to be provided by an agreed date, in order to maintain his programme. However, problems are encountered at the slate quarry, causing the supply of slates is interrupted and putting the supplier is in breach of the supply contract. The delay and

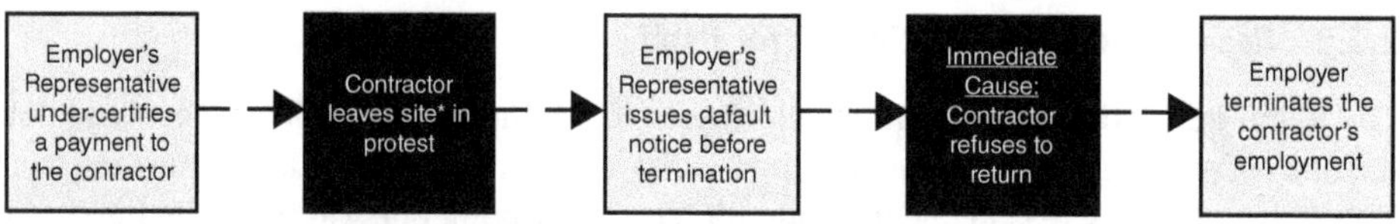


* The Contractor is not entitled to leave site for under-certification, and on receiving the 'default notice' should have returned to site. Failure to do so puts the contractor in breach and the employer is entitled to terminate the contractor's employment. Ending (or termination of) the contractor's employment was caused by the contractor unlawfully leaving site, and not the 'under-certification' of payment.'

**Figure 2.1** Causation example.

6 JCT contracts refer to the "loss and/or expense" being "direct."

7 The legal term used to be expressed as *novus actus interveniens* - a new act coming in between.

8 *Weld-Blundell v Stevens* [1920] AC 956.

disruption to the contract works consequent upon the delayed slate supply is clearly a direct consequence of the breach of the supply contract and only an indirect consequence of the variation. It is in fact the direct consequence of an intervening event – the supplier's breach. In such a case, the appropriate remedy is for the contractor to seek recompense from the slate supplier.

The principles of causation are further explained in the following statements:

- *'It seems to me that there is no abstract proposition, the application of which will provide the answer in every case, except this: one has to ask oneself what was the effective and predominant cause of the accident that happened, whatever the nature of that accident may be.'*[9]
- *'This choice of the real or efficient cause from out of the complex of facts must be made by applying common-sense standards. Causation is to be understood as the man in the street, and not as either the scientist or the metaphysician, would understand it.'*[10]
- *'Causation is a mental concept, generally based on inference or induction from uniformity of sequence as between two events that there is a causal connection between them .... The common law, however, is not concerned with philosophic speculation, but is only concerned with ordinary everyday life and thoughts and expressions.'*[11]
- Causation has been held to be purely a question of fact to be decided on the basis of common sense.[12] In *P & O Developments Ltd v Guy's & St Thomas' National Health Service Trust*, Judge Bowsher aptly summarised the position so far as the building industry was concerned:

  > *'The test is what an informed person in the building industry (not the man in the street) would take to be the cause without too microscopic analysis but on a broad view.'*[13]

Everything depends on the facts and circumstances. Some situations are very complex, and it will be important to identify the damage from which it was intended to protect a party.[14] An example of the concept of causation is to be found in a case[15] where negligent architects issued defective interim certificates and the contractors withdrew from site. The contractor lost its claim against the negligent architects because it broke the chain of causation by suspending the works despite the employer serving a preliminary notice of default (prior to determination). The contractor's default, not the architects', was responsible for the termination of the contract. Although the architect's negligence was the source of the events, it was overtaken and overwhelmed by the contractor's breach by unlawfully suspending work.

---

9 *Yorkshire Dale Steamship Co. Ltd v Minister of War Transport* [1942] AC 691 at 698 per Viscount Simon LC.

10 *Yorkshire Dale Steamship Co. Ltd v Minister of War Transport* [1942] AC 691 at 706 per Lord Wright.

11 *Monarch Steamship Co. Ltd v Karlshamns Oljefabriker (AB)* [1949] AC 196 at 228 per Lord Wright.

12 *Leyland Shipping Co. Ltd v Norwich Union Fire Insurance Society Ltd* [1918] AC 350.

13 [1999] BLR 3 at 9.

14 *Skandia Property (UK) Ltd v Thames Water* [1999] BLR 338.

15 *Lubenham Fidelities & Investment Co. v South Pembrokeshire District Council and Wigley Fox Partnership* (1986) 6 Con LR 85. This case pre-dated the payee's statutory right of suspension for non-payment under s112 of the Housing Grants Construction and Regeneration Act 1996 (as amended).

Another case concerned a claim against insurers,[16] following the Hatfield train disaster in 2000. The disaster was caused by a broken rail which itself was caused by a particular form of cracking. Speed restrictions were imposed by Railtrack as an emergency measure at sites where that particular form of cracking was known to exist, but various train companies alleged that the restrictions had caused them to lose money. The insurers relied upon an exclusion clause in the insurance policy which excluded damage caused by wear and tear. The court at first instance had held that the wear and tear was simply the occasion for the loss, but that actual cause was the speed restrictions. However, the Court of Appeal disagreed and held that there was neither a break in the chain of causation nor any intervening event; the wear and tear (i.e. the cracking) caused the speed restrictions to be imposed which, in turn, caused the losses.

In relation to extension of time claims, the court commented on causation as applied to extensions of time:

> *'There may well be circumstances where a relevant event has an impact on the progress of the works during a period of culpable delay but where that event would have been wholly avoided had the contractor completed the works by the previously-fixed completion date. For example, a storm which floods the site during a period of culpable delay and interrupts the progress of the works would have been avoided altogether if the contractor had not overrun the completion date. In such a case it is hard to see that it would be fair and reasonable to postpone the completion date to extend the contractors' time.'*[17]

Where there appear to be concurrent causes (see Section 3.7), one being the responsibility of one party and the other being the responsibility of the other party, the correct test to apply is not whether one of the causes is the sole cause or the dominant cause, but whether a cause is an effective cause.[18] Where there are several possible causes, the burden of proof on the contractor is to show that one cause is more likely that the others.[19]

## 2.4 Categories of Claim

Provided that the contractor can fulfil the four key steps of showing 'duty, breach, causation and loss' (see Section 2.1.2), there are few limits to the categories of claim that an ingenious contractor may bring under the terms of the contract. Therefore, in referring to prolongation and disruption claims, it is not suggested that these are the most common categories of claims that are made (other categories of claim, such as for variations are commonplace – see Chapter 5). However, in practice, claims for prolongation and disruption are the two most common categories of more contentious financial claims. Unlike when claiming a variation, a contractor making a claim under these categories usually has to convince the employer or his representative that he is entitled to make such a claim under the terms

16 *Midlands Mainline Ltd and Others v Eagle Star Insurance Co. Ltd* [2004] EWCA Civ 1042.
17 *Balfour Beatty Ltd v Chestermount Properties Ltd* (1993) 62 BLR 1 at 34 per Coleman J.
18 *Loftus-Brigham v Ealing LBC* (2004) 20 Const LJ 82.
19 *Plater v Sonatrach* (2004) CILL 2073.

of the particular contract before starting the process of drafting and substantiating the claim itself. Both categories are summarised below:

- A prolongation claim may arise when completion of the works (or a section[20]) is delayed beyond the contract date for completion – as the overall duration for the works (or a section) has been prolonged. Simple examples of the contractor's losses, claimed under 'prolongation' could be:
  - Site operations or time-related 'preliminaries' costs for the period of prolongation;
  - A contribution towards the contractor's head office overheads that but for the prolongation, could have been earned by the same resources on another project; and
  - Interest on retention that but for the period of prolongation, would have been paid earlier.
- A disruption claim differs from a prolongation claim because irrespective of completing the works on time, the contractor may have suffered losses. Examples of a 'disruption' claim include:
  - The contractor is instructed to vary high-level works that already require a tower crane. The critical path of his programme shows a 12-week period for the original high-level work, which includes 4 'spare' weeks (float[21]), as the contractor believes he can complete the work in eight weeks. His price allows for a tower crane for eight weeks. He completes the varied works in 11 weeks, and requires use of the tower crane throughout. There is no prolongation because the varied work is completed in 11 weeks within the available 12 weeks for that activity before a critical delay is caused, but the contractor has incurred an extra three weeks of tower crane costs. The 'disruption' claim would include the extra three weeks' costs for the tower crane; and
  - A contractor has priced some masonry repair work to the external wall of a three storey building, which does not fall on the critical path. The labour element of the £/m$^2$ rate is based upon working in a constant sequence, from 2nd floor, down to the ground floor – and this sequence is shown on the contractor's programme. However, the employer's representative subsequently instructs the contractor to repair only the areas he identifies during piecemeal inspections, carried out on an ad-hoc basis. The contractor is entitled to a variation if there is a change in the conditions under which he carries out the work. Whilst the masonry repair works are using the same quantities of materials as if they were carried out to the contractor's programmed sequence, the labour cost is now three times what was originally priced. The 'disruption' claim would be the 200% increase in masonry repairs' the labour element.

It is often contended that there is no such thing as a claim for disruption as part of loss and expense. Rather, it is said that instead, such a claim should be considered as part of the valuation of variations, because most standard form contracts provide for adjustment of rates and even the creation of new rates to deal with changes in the conditions under which the work is carried out of circumstances (see the second example above). There is some merit in such arguments, but they do not successfully deal with all types of disruption and especially cannot cover disruption which is not the result of a variation (see Section 4.4.3).

---

20 Completion in sections in JCT SBC/Q 2016, JCT DB 2016, and JCT IC 2016 (but not in JCT MW 2016), and NEC4 Option X5.

21 See Section 3.6 (Float).

### 2.4.1 Prolongation

A prolongation claim is probably the common category (or 'head') of claim. It is also often the simplest and most straightforward for the contractor to prepare and for the employer (or his representative) to understand. In essence, it simply states that either the employer (or a party acting on his behalf) has acted (or failed to act) in such a way that the contractor has been unable to complete the Works by the contractual date for completion, or some neutral event (such as weather) has had the same effect. Therefore, he is forced to complete later and the contract period has been prolonged. Moreover, the actions or inactions for which the employer is responsible fall under grounds set out in the loss and/or expense clause in the relevant contract.[22]

The contractor then proceeds to quantify the loss and/or expense he has suffered. Despite, or perhaps because of, its perceived simplicity, many contractors are slovenly when preparing prolongation claims. The typical steps when preparing a prolongation claim are as follows:

- First, establish a claim for an extension of time. This is the usual precursor to a prolongation claim, to the extent that many contractors and architects believe that unless the contractor is given an extension of time first, it is not eligible for a prolongation payment, but this is not always the case.[23] Once the employer or his representative has received the extension of time claim, the contractor has to establish the loss and expense suffered as a result. Ultimately, it should be the actual loss and actual expense which must be ascertained, although for the purposes of an interim payment, using the time-related prices in the contractor's preliminaries may temporarily suffice. However, eventually, the actual cost to the contractor of being on site for the prolonged period must be ascertained.

In considering the amounts due as a result of delays, the court set out the important difference between a contractor's claim for damages for delay and a claim for extension of time on account of the delay, as follows:

> *'When an extension of time of the project completion date is claimed, the contractor needs to establish that a delay to an activity on the critical path*[24] *has occurred of a certain number of days or weeks and that that delay has in fact pushed out the completion date at the end of the project by a given number of days or weeks, after taking account of any mitigation or acceleration measures. If the contractor establishes those facts, he is entitled to an extension of time for completion of the whole project including, of course all those activities which were not in fact delayed by the delaying events at all, i.e. they were not on the critical path.*
>
> *But a claim for damages on account of delays to construction work is rather different. There, in order to recover substantial damages, the contractor needs to show what losses he has incurred as a result of the prolongation of the activity in question. Those losses will include the increased and additional costs of carrying out the delayed activity itself as well as the additional costs caused to other site activities as a result of the delaying event. But the contractor will not recover the general site overheads of carrying out all the*

22 See Sections 1.2.1 and 2.1.1 and Table 1.

23 *H Fairweather & Co Ltd v London Borough of Wandsworth* (1987) 39 BLR 106.

24 The "critical path" and "critical delay" are explained in Chapter 3 (Time) at Sections 3.3,3.4 and 3.5.

*activities on site as a matter of course unless he can establish that the delaying event to one activity in fact impacted on all the other site activities. Simply because the delaying event itself is on the critical path does not mean that in point of fact it impacted on any other site activity save for those immediately following and dependent upon the activities in question.*

*It seems to me that [the contractor's] claim in respect of its prolongation costs has fallen between the two stools described above. The claim is put on the basis that the delays to the foundation works caused critical delay*[24] *to the whole project of over 12 weeks and the whole project's general site costs are claimed on that basis. Those costs are evaluated as at October 2002 to January 2003 and not at the end of the project which occurred well over two years later in May 2005. But no evidence has been called to establish that the delaying events in question in fact caused delay to any activities on site apart from the [Rapid Gravity Filter] and [Inlet Works]. That being so, it follows, in my judgment, that the prolongation claim advanced by [the contractor] based on recovery of the whole of the site costs of the … site, fails for want of proof.'*[25]

Here, the court highlights a very important point that non-critical activities may not be affected by a delay to the critical path and, therefore, no damages are claimable for them. This sets out a much stricter test for ascertaining loss and/or expense than has been common practice. It is therefore important for the contractor to demonstrate that the prolongation costs were incurred during the period of critical delay and not some other period (e.g. simply 'tacked on' to the period of prolongation after the original completion date).

Although the period of prolongation tends, for obvious reasons to be measured as starting from the contractual completion date and extending on as appropriate, this is not necessarily the time frame within which the costs should be ascertained because the true costs resulting from a delay will usually follow the date the delay occurred. In the case of several critical delays, the relevant periods may be scattered throughout the contract period. The few weeks at the beginning and end of a contract will be characterised by a build up then a reduction respectively of site-related costs. Therefore, if the costs of prolongation were to be ascertained in respect of the final weeks of a contract, the contractor may get much less than its true entitlement as shown in Figure 2.2 (below).

Figure 2.2 shows that for a critical delay occurring mid-way through the project (e.g. between months 5 and 6), the monthly costs may be much higher than the monthly costs towards the end (during months 8 to 9).

Therefore, if a contract is prolonged for a period of two months which the contractor can establish is a direct result of a matter that entitles it to financial compensation,[26] two months may be made up of several delaying occurrences taking place at differing times during the contract period (in the above example, during month 5 and month 8). The task of identifying each delay and its monetary consequences is not always easy, but it must at least be attempted. Many employers or their representatives may agree to take a sample of the appropriate number of weeks prolongation from somewhere in the middle of the

25 *Costain Ltd v Charles Haswell & Partners Ltd* [2009] EWHC B25 (TCC) at 183 to 185 Richard Ferneyhough QC sitting as a Deputy Judge of the High Court.

26 Such as under clause 4.22 of JCT SBC/Q 2016 or clause 60.1(1) of NEC4.

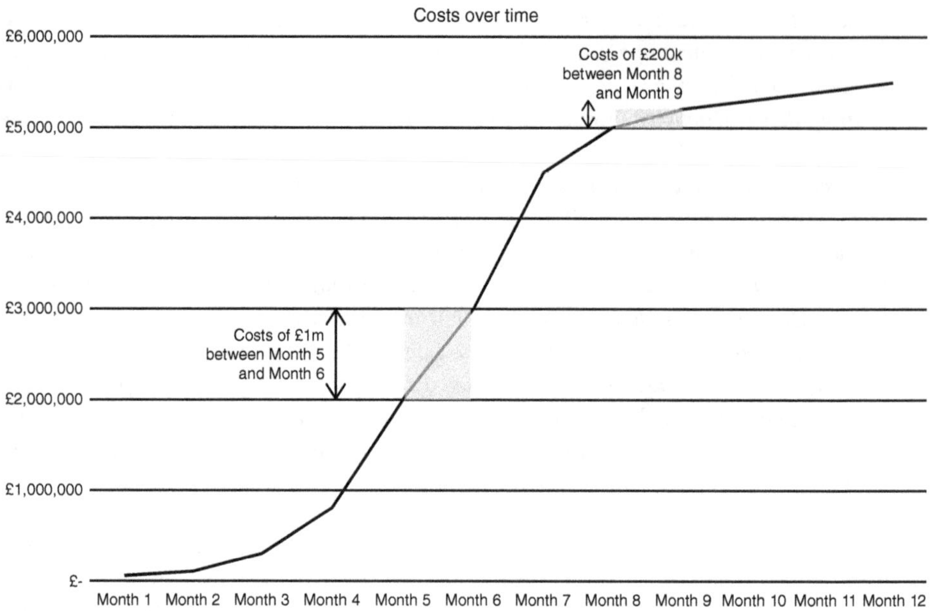


**Figure 2.2** 'S curve' – example of a contractor's costs over time.

contract period for convenience. This seems, at first sight, to be a reasonably good method of establishing appropriate costs, but it is not 'ascertainment' its proper sense and may be difficult to substantiate if challenged. However, in practical terms, it is not uncommon that both the employer and the contractor may agree that an approximate system is acceptable, and there is nothing more to be said. Indeed, such an approximate method may be employed by the Project Manager and contractor when quoting for and assessing compensation events under NEC4.[27]

### 2.4.2 Disruption

Disruption is usually claimed separately from prolongation. It may be present with or without prolongation. Disruption claims rely upon keeping good records and are difficult to establish with any precision without them. Often, a contractor's claim for disruption has simply been calculated by subtracting the anticipated costs of labour (e.g. from the contractor's tender) from the actual costs of labour that were incurred, with little attempt made to precisely relate this to the facts. This simple approach does not bear consideration and it has been roundly condemned[28] (refer to Chapter 7 – Global Claims). There may be many reasons for the actual costs of labour being greater than those anticipated for reasons other than for which the employer or his representative are responsible.

27 NEC4 refer to clauses 62.1 (Quotations for compensation events), 63.2 and 63.8 (Assessing compensation events).

28 *London Borough of Merton v Stanley Hugh Leach Ltd* (1985) 32 BLR 51.

Commonly, disruption amounts to delays in non-critical parts of a project, but not to the extent that those parts cause critical in programming terms (refer to the disruption example in Section 4.4.3).

One accepted method of evaluating disruption is to compare the value to the contractor of the work done per man during a period of no disruption with the value per man doing the disrupted period and then to apply the ratio to the total cost of labour.[29] However, for this method to work, it must be possible to identify a period free from disruption, and the compared outputs must relate to similar work. This may not always be the case, particularly where a project's progress has been adversely affected from the start.

## 2.5 The 'Knock-on' Effect

The term 'knock-on' effect is often used to describe a secondary effect on the contractor's progress of work, which is directly caused by an initial or preceding effect on progress. The chain of causation between the initial, or preceding event, and the secondary event must be maintained (see Section 2.3).

A common example of the 'knock-on' effect is a claim for 'winter working' claim, although the principle could apply to any other situation where an initial delay inevitably causes the works to be carried out in a situation which is less felicitous than originally envisaged. In the case of winter working, the problem for the contractor is that a critical delay which entitles it to financial compensation may also postpone the affected programme of work so that work originally programmed to be carried out in good weather may have to be carried out during a period of bad weather. An obvious example being activities programmed for the summer months now to be carried out in the winter. Examples of work affected in this way could include:

- Fewer daylight hours are available unless the contractor introduces additional lighting measures;
- It may be unsafe to carry out high-level work during periods of high winds (particularly if a tower crane is utilised);
- Frozen ground may prolong groundworks;
- Concreting work may be prevented from being carried out because sub-zero temperatures may affect the concrete's curing time and its structural integrity; and
- Internal finishes may be prolonged because of higher moisture content and lower temperatures affecting setting times – unless the contractor either introduces temporary heating or dehumidifying measures.

Under JCT 2016 contracts, a contractor is entitled to an extension of time only for *exceptionally adverse* weather conditions,[30] but he does not benefit from a corresponding entitlement to loss and expense. However, in this example, as a result of one relevant event, the 'knock-on' effect is that some work may be pushed back into a period of unfavourable weather. Account should be taken of the weather conditions when assessing the extension

29 *Whittal Builders Co. Ltd v Chester-Le-Street District Council* (the 1985 case) [1996] 12 Const LJ 356.
30 JCT SBC/Q 2016 – clause 2.29.9, JCT DB 2016 – clause 2.26.8 and JCT IC 2016 clause 2.20.8.

of time, as this is the 'knock-on' effect of an initial or preceding delay and not caused by a delay arising from exceptionally adverse weather.

There will be occasions, of course, in which the contract sum (or contract price) allows for work to be carried out at a difficult time of year and a delay caused by the employer may actually improve the working conditions. However, the question of whether the employer would be entitled to a saving in such circumstances would depend upon the building contract:

- Under a JCT contract, only 'additional or substituted work' can be revalued[31] if not executed under 'similar conditions' to work forming part of the contract sum, and the contract sum would not be adjusted because the postponement of the relevant work arose because of a critical delay and not because the employer instructed a variation (e.g. by directing the contractor to carry out the work in a specific order)[32] but
- Under NEC4's target cost contracts (Options C[33] and D[34]), the employer would benefit from the Defined Cost[35] being less than it would otherwise have been, owing to the improved working conditions.

Whether the contractor can make a claim for financial compensation will depend on whether the chain of causation between the initial or preceding delay and the 'knock-on' delay can be maintained. In this respect, it is important to remember that it is the originating delay which is crucial. It does not matter that the period of bad weather in which the contractor finds itself working is not exceptional. It may be quite normal for the time of year. The point is whether the period is worse than that for which the contractor allowed in its price.

The case which is much relied upon in this situation is the Canadian case of *Ellis-Don v The Parking Authority of Toronto*.[36] The basic facts of the case were simply that a contract period of 52 weeks was delayed by a further 32 weeks, of which, 7 weeks were caused by the employer. This led to a 'knock-on' delay in starting up and a total delay of 17. Five weeks was blamed on the employer. Part of the work had to be carried out in winter, and the contractor was successful in claiming additional payment.

In essence, however, the principle is simply damages for breach of contract.[37] A court has accepted a winter working claim where a contractor was claiming in respect of the architect's repudiation of its contract with the contractor. The principle is just the same as if the contractor had been claiming under the building contract:

> *'The effect of working through the winter months was that the work was undertaken during extended periods of wet weather rather than in the significantly drier summer months. This was particularly so in the winter 2000–2001 which records showed was the wettest winter since 1766. The effect of this was to require [the contractor] to work*

31 JCT SBC/Q 2016 clause 5.6.1.2, JCT DB 2016 clause 5.4.2, JCT IC 2016 clause 5.3.1.2.
32 JCT SBC/Q 2016 and JCT DB 2016 and JCT IC 2016 - clause 5.1.2.4,
33 NEC4 Option C (Target Cost with Activity Schedule).
34 NEC Option D (Target Cost with Bill of Quantities).
35 NEC4 – "Defined Cost" is defined in clause 11.2(24) in Options C and D.
36 (1978) 28 BLR 98.
37 See Section 5.2 above and the view of the House of Lords in *Koufos v Czarnikow Ltd (The Heron II)* [1969] 1 AC 350.

*in chalk ground conditions which had turned into slurry and in excavations where the sides had a tendency to collapse. There were, in consequence, four additional heads of loss.'*[38]

In another case, the judge postulated a knock-on scenario thus:

> '*Assume the following facts: A contract is entered into in this form of contract in May for one year for completion on 31 July of the next. The work is of a tunnelling nature. No tunnelling can be carried out from 1 November to 31 March for seasonal reasons but during that period the contractor will have expensive equipment lying idle. In early April when the works were on course for completion on 31 July the architect issues an instruction under 11(1) requiring a variation the execution of which will add three months to the contract period. At the same time on the contractor's application he grants an extension of time for completion to 31 October. A fortnight before 31 October when the works as varied are on course for completion in due time a strike occurs which continues until 31 March. The contractor recommences work on 1 April but because he had no opportunity to protect his machinery during the six months period it then takes the contractor two months not two weeks to complete. There has been no fault on either party. If the architect grants an extension of time of eight months only under 23(d) I can see no reason why the contractor under the contract cannot still recover all his direct loss and expense under 11(6).'*[39]

The case concerned a JCT 63 contract, so the principle would be the same under JCT SBC/Q 2016. In the accompanying commentary, the editors of *Building Law Reports* note that it is useful to demonstrate why there is no necessary link between the grant or the refusal of an extension of time and the success of an application for loss and/or expense. They proceed, however, to disagree with the judge. In their view, the contractor's costs flowed directly from the strike, not from the variation. They point out that a contractor takes the risk that a strike may occur not only during the original contract period, but also during any period of extension. They agree that the contractor would be entitled to an extension of the contract period for the strike. This example very clearly highlights the difficulties in considering knock-on claims. The question to be asked in each case is 'What is the direct cause of the winter working?' If the contractor cannot show that there is a direct connection between the cause of the delay and the necessity to work in conditions which are less advantageous than anticipated at the time of entering into the contract and that there is no compensating saving, its claim is likely to fail.

Some of the difficulties in formulating a claim for winter working were highlighted in *Costain Ltd v Charles Haswell & Partners Ltd*[25] where the contractor (Costain) was claiming against a specialist consulting civil engineer (Haswell). Part of the claim related to delays

---

38 *CFW Architects (a firm) v Cowlin Construction Ltd* [2006] EWHC 6 (TCC) at 156 per Judge Thornton, who identified the four additional heads of loss as additional lean mix concrete (at 158), additional "Type 1 filling under slabs" (at 164), "6F2 materials" (at 167) and crushed concrete (at 170).

39 *H. Fairweather & Co Ltd v London Borough of Wandsworth* (1987) 39 BLR 106 at 118 per Judge Fox-Andrews. Despite the calculations of the 31 July completion date and the "six month period" is actually five months, the principle remains unaffected.

caused by winter working. The court quickly went to the heart of the problem when dealing with the conclusions of Costain's expert who, in cross-examination, frankly accepted that this claim and his calculation of it was purely theoretical since he had done no research into the actual effect of winter working on the productivity of works

> *'I have no hesitation in rejecting this part of [the expert's] analysis. It is wholly theoretical and based on nothing but the meteorological records for the relevant period and [the expert's] experience and hunch. It seems to me to be unlikely that, as a matter of course, productivity of outside building works in October and November is always measurably lower than for, say, the months of August and September. In this country the productivity of outside work depends to a great extent upon the weather which can be changeable at any time of year and there can be no presumption that it will be generally worse in October and November than in any other month. In the absence of hard facts and figures to support such a claim related to the facts of this case, which do not exist, in my judgment, this claim has not been established on the balance of probabilities.'*[40]

## 2.6 Damages

The measure of damages is the *amount* of damages or the loss suffered. Therefore, the loss and/or expense should not be some notional or estimated figure, but the actual amount lost or spent by the contractor.

The principal function of damages, assuming that the loss is not too remote (see Section 4.2), is to put the injured party into the same position, so far as money can, as if the contract had been performed without breach.[41] This is the basic principle – the right of the injured party to receive precisely what it bargained for. If a party contracts to, and pays for having six apple trees planted but only five are planted, it is clear that the party is entitled to have the sixth tree delivered and planted. If, for some reason, the supplier cannot or will not supply and plant the remaining tree, it is open to the purchaser to take action to recover the reasonable costs of purchasing a tree and having it planted. That is a principle which is easy to understand, but it is not always rigorously applied. An employer who tries to insist on a house being demolished and rebuilt because there is a small error in the overall dimensions is likely to be disappointed. In certain cases, the courts have resisted the principle's strict application if some more proportionate remedy would suffice, and if the cost of strict entitlement is out of proportion to the benefit to be gained thereby:

> *'What constitutes the aggrieved party's loss is in every case a question of fact and degree. Where the contract breaker has entirely failed to achieve the contractual objective it may not be difficult to conclude that the loss is the necessary cost of achieving that objective. Thus if a building is constructed so defectively that it is of no use for its designed purpose the owner may have little difficulty in establishing that his loss is the necessary cost of reconstructing. Furthermore in taking reasonableness into account in*

40 *Costain Ltd v Charles Haswell & Partners Ltd* [2009] EWHC B25 (TCC) at 199.
41 *Wertheim v Chicoutimi Pulp Co.* [1911] AC 301 PC.

*determining the extent of loss it is reasonableness in relation to the particular contract and not at large. However, where the contractual objective has been achieved to a substantial extent the position may be very different.*'[42]

The amount of damages recoverable where the contract has been substantially performed depends upon the particular circumstances of each case. It is sometimes said that the proper measure of damages in a building case is the diminution in value of the property. However, that will not always result in a just outcome. The position has been put like this:

'*It is a common feature of small building works performed on residential property that the cost of the work is not fully reflected by an increase in the market value of the house, and that comparatively minor deviations from specification or sound workmanship may have no direct financial effect at all. Yet the householder must surely be entitled to say that he chose to obtain from the builder a promise to produce a particular result because he wanted to make his house more comfortable, more convenient and more conformable to his own particular tastes: not because he had in mind that the work might increase the amount which he would receive if, contrary to expectation, he thought it expedient in the future to exchange his home for cash. To say that in order to escape unscathed the builder has only to show that to the mind of the average onlooker, or the average potential buyer, the results which he has produced seem just as good as those which he had promised would make a part of the promise illusory, and unbalance the bargain.*'[43]

In general, there are no rules which can be applied in all circumstances in order to arrive at the amount of damages recoverable. Each case requires careful analysis.

## 2.7 Mitigation

This principle is also known as mitigation of damage, and the idea is to reduce the extent of the loss. In construction projects, the following types of mitigation are most common:

- Mitigation of delays (i.e. aiming to reduce the length or effect of a delay to progress) and
- Mitigation of financial loss.

### 2.7.1 Mitigation of Delay

In the event that the progress of work is delayed, either by the employer or the contractor, or by some neutral event, some building contracts impose express obligations on the contractor to mitigate the delay. This is covered in Section 3.1.3 (Progress obligations) of this book.

42 *Forsyth v Ruxley Electronics and Construction Ltd & Others* (1995) 73 BLR 1 at 12 per Lord Jauncey.
43 *Forsyth v Ruxley Electronics and Construction Ltd & Others* (1995) 73 BLR 1 at 14 per Lord Mustill.

### 2.7.2 Mitigation of Costs

Considering the simple example of water is dripping through a defective roof covering, it would be inadvisable for the building owner simply to leave it to drip in the knowledge that the cost of repairing or reinstating all the damage to the furnishings can be recovered from the contractor responsible. The building owner would, at the very least, be expected to put a bucket to collect the drips and take steps to reduce or temporarily repair the leak.

There is much confusion about the principles of mitigation of loss. They are quite simply stated:

- A party cannot recover damages resulting from the other party's breach of contract if it would have been possible to avoid any damage by taking reasonable measures;
- A party cannot recover damages which it has avoided by taking measures even if such measures were greater than what might be considered reasonable; and
- A party can recover the cost of taking reasonable measures to avoid or mitigate (reduce) its potential damages.

This is said to give rise to a duty to mitigate,[44] although it is perhaps more accurate to say there is an incentive to mitigate. A failure to mitigate will not give rise to a legal liability; it will simply reduce the recoverable damages to what they would have been if reasonable mitigating measures been taken. A common example of mitigation in standard forms of contract[45] is that, the contractor is to be permitted to remedy, at its own cost, defects which arise in the works during a period after completion.[46] An employer who, for no good reason, refuses to allow the contractor to remedy such defects may only recover what it would have cost the contractor to do the remedial work.

That is not to say that the claimant must do absolutely everything that is possible. It need not do anything other than an ordinary prudent person in the course of his or her business would do.[47] Moreover, the innocent party, if faced with different ways of mitigating, does not have to act reasonably in exercising a choice.[48]

The application of this principle is illustrated by the following example:

- Plant is standing idle as a result of a variation order. The contractor would not be entitled simply to accept the situation, but would be obliged to make reasonable endeavours to try and off-hire the plant or use it productively elsewhere. In the first instance, the costs of, say, moving the plant to be used on another site would be recoverable as a part of a direct loss claim, provided of course that this sum did not exceed the costs which would have been otherwise incurred. However, although the injured party must only take reasonable measures and not unreasonable measures, the courts usually will not look too critically in hindsight at his actions in attempting to mitigate. The crucial question is whether, in attempting to mitigate, the injured party acted reasonably.[49] Even if the actions of the injured party resulted in an increase in loss, the cost may be recoverable if the party

44 *British Westinghouse v Underground Railways Company* [1912] AC 673.

45 JCT SBC/Q 2016 clause 2.38, NEC4 clause 44.2.

46 JCT contracts refer to the "Rectification Period" and NEC4 contracts refer to the *"defect correction period"* – typically between 6 and 12 months after the works are completed.

47 *London and South England Building Society v Stone* [1983] 1 WLR 1242.

48 *Strutt v Witnell* [1975] 1 WLR 870.

49 *Banco de Portugal v Waterlow & Sons Ltd* [1932] AC 452.

acted reasonably.[50] Further, the injured party must be given a reasonable time in which to decide how to mitigate the loss.[51] The reasonable time may of course begin immediately, as in a case where a contractor's piece of machinery develops a fault which results in it causing damage. The contractor must start to mitigate by immediately ceasing to use it.

It has been argued that the existence of a duty or incentive to mitigate favours the contract breaker because the party in breach is not obliged to pay damages equivalent to the whole of the loss, but only such sum as represents the loss after mitigating steps have been, or should have been taken. Therefore, it may sometimes be more cost efficient for a contractor to withdraw its resources from one contract in order to place them in another, more lucrative, contract. Although in breach of the original contract, the contract breaker may overall be better off when the effect of the duty to mitigate is taken into account.

A similar in principle, but opposite, situation can arise where an employer wrongfully terminates the contractor's employment and the contractor claims damages for the repudiatory breach. Although the employer wrongfully terminated the contractor's employment, the contractor can now put more resources into another project or even to commence a new project which he would otherwise have found difficult to resource. The amount of damages which the contractor is able to claim for the repudiation would have to take these circumstances into account and may amount to nothing.

The position is, in fact, spelled out specifically in the extension of time clauses of many standard form contracts, requiring the contractor to use its best endeavours to prevent delay occurring and to mitigate the effects of a delay once encountered.[52] The position has been clarified where the court addressed the question of mitigation in relation to delay:

> '*It is difficult to see how there can be any room for the doctrine of mitigation in relation to damage suffered by reason of the employer's culpable delay in the face of express contractual machinery for dealing with the situation by extension of time and reimbursement of loss and expense. However that may be as a matter of principle, what is plain is that there cannot be both an extension to the full extent of the employer's culpable delay, with damages on that basis, and also damages in the form of expense incurred by mitigation, unless it is alleged and established that the attempt at mitigation, although reasonable, was wholly ineffective.*'[53]

It appears that 'best endeavours' clauses like clause 2.28.6.1 of JCT SBC/Q 2016 not only apply to delays which entitle the contractor to an extension of time but also to delays caused by neutral events and the contractor's culpability. It is for the party receiving the claim to show that the claimant has failed to mitigate.[54] Therefore, when the contractor submits an application for direct loss and/or expense, it is for the employer or his representative to show that the contractor has not mitigated its losses. However, although that burden falls on the employer (or his representative), he is entitled to seek relevant information from the contractor in order to form an opinion about the matter.

50 *Melachrino v Nicholl & Knight* [1920] 1 KB 693.

51 *C Sharpe & Co Ltd v Nosawa* [1917] 2 KB 814.

52 JCT SBC/Q 2016 – clause 2.28.6.1, JCT DB 2016 – clause 2.25.6.1, JCT IC 2016 – clause 2.19.4.1.

53 *Ascon Contracting Ltd v Alfred McAlpine Construction Isle of Man Ltd* (2000) Const LJ 316 at 332 per Judge Hicks.

54 *Garnac Grain Co. Inc v Faure & Fairclough* [1968] AC 1130.

## 2.8 Betterment

'Betterment' means the improving of something.[55]A common example is when repair or restoration work results in a something which is better in quality or standard than which existed beforehand or was originally specified.

The ordinary position is for damages to reinstate the injured party into the same position, so far as money can, as if the contract had been performed without breach[41] (see Section 2.6). Therefore it is unlikely that any additional sum will be awarded for work that results in betterment, unless the party making the claim had no reasonable choice, but to follow that course.[56] For example, if a party decides to re-build a damaged property to a better standard than before the damage took place, it will be unable to claim cost of the improvement. However, where a party is claiming the cost of rebuilding premises which are destroyed as a result of another party's breach, the injured party is entitled to the full cost of rebuilding even though the effect will be to get new in place of old. In one case, the position was put like this:

> '*I think that the law can be shortly summarised. If the only practicable method of overcoming the consequences of a defendant's breach of contract is to build to a higher standard than the contract had required, the plaintiff may recover the cost of building to that higher standard. If, however, a plaintiff, needing to carry out works because of a defendant's breach of contract, chooses to build to a higher standard than is strictly necessary, the courts will, unless the new works are so different as to break the chain of causation, award him the cost of the works less a credit to the defendant in respect of betterment.*[57]'

The question has been dealt with many times in the courts, and it was considered again in the case of *Witts and Others v Montgomery Watson Ltd.*[58] The claimants were the owners of properties along the seafront at Putsborough Sands in Devon and commissioned the defendants to produce recommendations for the construction of sea defences. The recommendations were accepted, a design was prepared and the defences were built, but they were subsequently badly damages by storms. The claimants took action and liability was settled. The question remaining was the amount of damages payable to the claimant. Clearly, the sea defences ought to have been extensive, and it was agreed that the amount required to rectify them was more than £500 000, which was far more than the original cost. The defendants argued that (1) they should only have to pay the original (and much lower) price and (2) if the new defences were better than the old, the claimant should pay the additional cost of reinstating to the original standard. The court considered the effect of the defendant's negligence and found that it was reasonably foreseeable that the negligent design would cause the defences to fail. The damages would be the cost of rectifying the failure, but the only practical way of doing so was much more expensive than the original scheme. The

---

55 Oxford Dictionary of English (2010).

56 *Voaden v Champion* [2002] 1 Lloyd's Rep 623.

57 *Richard Roberts Holdings Ltd v Douglas Smith Stimson Partnership and Others* (1988) 5 Const LJ 223 at 227 per Judge Newey.

58 *Witts and Others v Montgomery Watson Ltd* [1999] 11 BLISS 5.

owners had no real alternative and, therefore, they were entitled to the whole cost. In this case, the court held that there was no element of betterment.

If a party is in doubt how to act (considering a course of action that may lead to arguments of betterment), but relies on proper legal advice, it will usually be assumed to have acted reasonably.[59]

## 2.9 Notices

Many contracts require one party to notify the other when making a claim or informing the other that a particular state of affairs exists. Some examples of the parties' notifications are as follows:

- Interim payments – the employer (or his representative) must issue the following notifications[60] to the contractor:
  - Issue a 'payment notice' and/or
  - A 'pay-less notice'
- Compensation events under NEC4:
  - By the Project Manager or the Supervisor[61] when issuing an instruction, notification, changing a previous decision or issuing a certificate; or
  - By the contractor, if an event which he believes to be a compensation event (and has not already been notified by the Project Manager or Supervisor) has happened or is expected to happen[62];
- JCT contracts require the contractor to notify:
  - If progress of the works is being delayed, particularly if completion is likely to be delayed beyond the Completion Date[63] (extension of time); and of
  - Loss and/or expense.[64]

The contractor should not perceive that the requirement for giving notices, particularly in respect of compensation events (NEC4) or for extensions of time and/or loss and expense (JCT) is meant to be a hurdle, or a barrier imposed by the employer to prevent the contractor becoming entitled to such claims. Instead, the purpose of notices is to provide an advance warnings[65] to the employer, allowing him an opportunity to avoid or mitigate (reduce) any negative effects on the works. A notice often has the following effects:

- It triggers a procedure of evaluation, leading to a decision being made by the employer (or by his representative) over the notice's subject matter (e.g. an extension of time,[66] or loss and expense); and/or

59 *Lodge Holes Colliery v The Borough of Wednesbury* [1908] AC 323.

60 Where UK law applies, the notifications must comply with s110A (payment notice) and/or s111(3) (pay-less notice) of the Housing Grants Construction and Regeneration Act 1996 (as amended).

61 NEC4 clause 61.1.

62 NEC4 clause 61.3.

63 JCT SBC/Q 2016 – clause 2.27, JCT DB 2016 – clause 2.24, JCT IC 2016 – clause 2.19, JCT MW 2016 clause 2.7.

64 JCT SBC/Q 2016 – clause 4.21.1, JCT DB 2016 – clause 4.20.1, JCT IC 2016 – clause 4.16.1.

65 Not to be confused with the "Early warning" procedure under clause 15 of NEC4.

66 Upon receipt of the contractor's notice and particulars, the employer (or his representative) must commence the procedure to fix a later completion date – JCT SBC/Q 2016 – clause 2.28, JCT DB 2016 – clause 2.25.

- It may preserve the a party's entitlement to claim time and/or money
  - Under JCT contracts, the employer must issue a notice or certificate of non-completion[67] in order to deduct liquidated damages and
  - Under NEC4 contracts, the contractor must notify the Project Manager of a compensation event[62] within eight weeks of becoming aware of the event that may be a compensation event.

### 2.9.1 Requirement for Notice

Where the contract provides that a notice must be given, a question which often arises is whether the employer (or his representative) is entitled or obliged to attend to matters concerning compensation events, extensions of time or loss and/or expense in the absence of a notice, or where notice has not been given precisely in the form or manner or at the time prescribed in the contract? The answer depends upon the following two points:

- If the contract's procedure for evaluating the particular entitlement provides that the employer's (or his representative's) obligation commences upon receipt of the relevant notice (and possibly other documents)[66]; and/or
- If the requirement for the particular notice are a condition precedent to the entitlement to additional payment (in the case of compensation events, or loss and expense) or time.

For the first point, the employer (or his representative) could argue that in the absence of a notice from the contractor, that their obligations to assess and decide upon an extension of time or loss and expense (under a JCT contract) simply do not commence. This is particularly relevant for an Architect, Contract Administrator or Employer's Agent under a JCT 2016 contract, as they may become liable to the employer if they grant an extension of time without receiving the contractor's triggering notice beforehand – as they will have no obligation to do so.[66]

The second point considers a 'condition precedent' to a party's entitlements to additional payment and/or additional time. These entitlements may expire within a period of time, if the party seeking that entitlement fails to provide a notice at all or fails to provide a non-compliant notice within the required period of time. Perhaps, the most common 'condition precedent' in building contracts is clause 61.3 of NEC4 that requires the contractor to provide its notice of a compensation event within eight weeks of becoming aware of the compensation event, otherwise he will not be entitled to additional time and/or money. Therefore, if the contractor fails to provide his notice within eight weeks, the condition precedent is not satisfied, and the contractor is not entitled to any additional payment or time, so there is no need for the Project Manager to assess the extent of any extension of time or payment of additional costs (including those arising from a variation).

A condition precedent is a condition which makes the rights or duties of the parties depend upon the happening of an event – in the previous example, the giving of a valid notice within a stated period. A term is often a condition precedent if it expressly states that it is so. However, even if so expressly stated, the courts may decline to hold that a term is

67 JCT SBC/Q 2016 – clauses 2.31 (Non-Completion Certificate) and 2.32.1.1, JCT DB 2016 – clauses 2.28 (Non-Completion Notice) and 2.29.1.1, JCT IC 2016 – clauses 2.22 (Certificate of non-completion) and 2.23.1.1.

a condition precedent if to do so would be contrary to commercial sense in a special situation.[68] If a notice provision is to be a condition precedent, it has been held that it must state a time for service and make clear that a failure to serve will mean loss of rights.[69] In practice, this strict requirement may not always be enforced. Clearly, where the requirement for notice is a condition precedent, the absence of a notice will be fatal to any claim to which that notice relates.

### 2.9.2 Time Bar Clauses

Time bar clauses are limitation clauses, as they are intended to limit a party's rights if they fail to comply with the clause – most commonly, a requirement for the contractor to notify the employer of particular circumstances or events. The consequences of failing to comply with a time-bar clause may be extremely serious, so a limitation clause of this nature must be expressed in very clear words[70] and is often interpreted *contra-proferentum*[71] (i.e. against the party putting it forward). There are three essential aspects to a time bar clause which must be clearly, and unambiguously[72] expressed:

- What must be done and by whom;
- The period in which the thing must be done; and
- The sanction imposed upon the party who fails to comply.

There are several examples of time-bar clauses found in standard forms of building contract:

- JCT 2016 contracts:
  - Unless the contractor commences proceedings within 28 days after the Final Certificate (or its equivalent) is issued, he will be unable to challenge the conclusiveness of certain aspects of the Final Certificate[73]; and
  - Under Part 1 of the Supplemental Provisions of JCT DB 2016,[74] the contractor's failure to provide an estimate of the value of any Changes[75] (variations), extension of time or loss and/or expense[76] within 14 days[77] of the date of the employer's instruction, the contractor will not be entitled to receive payment of the same in an Interim Payment and must wait until the final payment.

68 *Koch Hightex GmbH v New Millenium Experience Company Ltd* [1999] EWCA Civ 983.
69 *Bremer Handelsgesellscaft MBH v Vanden Avenne-Izegem PVBA* [1978] 2 Lloyds Rep 109.
70 *Dairy Containers Ltd v Tasman Orient Line CV* [2005] 1 WLR 215.
71 *Hollins v Davey (J) Ltd* [1963] QB 844.
72 *Steria Ltd v Signma Wireless Communications Ltd* [2007] EWHC 3454 (TCC) "*...the principle which applies here is that if there is genuine ambiguity as to whether or not notification is a condition precedent, then the notification should not be construed as being a condition precedent, since such a provision operates for the benefit of only one party, i.e, the employer, and operates to deprive the other party (the contractor) of rights which he would otherwise enjoy under the contract.*" at para 89.
73 JCT SBC/Q 2016 – clause 1.9.2, JCT DB 2016 (relating to the contractor's Final Statement, or the employer's Final Statement) – clause 1.8.2.1, and JCT IC 2016 – clause 1.9.2.1.
74 JCT DB 2016, Schedule 2 contains the Supplemental Provisions includes contractor's estimates for "Valuation of Changes" and Loss and expense". The Contract Particulars must state that these Supplemental Provisions apply (the default position is that they do not apply).
75 JCT DB 2016 – Schedule 2, Supplemental Provisions Part 1, para 2.6.
76 JCT DB 2016 – Schedule 2, Supplemental Provisions Part 1, para 3.6.
77 JCT DB 2017 – Schedule 2, Supplemental Provisions Part 1, paras.

- NEC4:
  - Clause 61.3 provides that if the contractor's fails to notify[78] the Project Manager of a compensation event within eight weeks of when he believes the event is a compensation event, he will not be entitled to any change in the Prices (additional payment), the Completion Date, or to any Key Date; and
  - Clause 64.4 provides that following the contractor's quotation for a compensation event, if the Project Manager fails to assess a compensation event within two weeks of the contractor's notification of that failure, then the contractor's notification is treated as acceptance of the earlier quotation.

All of these time-bar clauses include the three essential aspects of a time-bar clause that are mentioned above.

In a Scottish case,[79] the court considered a heavily amended JCT 80 contract. The relevant amendments included a time bar clauses:

- Clause 13.8.1 required the contractor to notify the architect within 10 working days of receiving an instruction, and such notice had to include estimates of the adjustment to the contract sum and completion date, any additional resources required, a method statement, estimates of any extensions of time and loss and expense, along with any other information the architect may require; and
- Clause 13.8.1 stated: '*If the Contractor fails to comply with one or more of the provisions of Clause 13.8.1, where the Architect has not dispensed with such compliance under Clause 13.8.4, the Contractor shall not be entitled to any extension of time under Clause 25.3.*'

The contractor argued that clause 13.8.5 constituted a penalty, but this was rejected by the court:

> '*In our opinion, the contractor would not commit a breach of contract in either respect. Clause 13.8 does not impose any obligation on the contractor when he receives an architect's instruction. If the contractor receives such an instruction, he has to consider its likely effects, and in particular its likely effect on the duration of the building period. He may, for reasons of his own, decide to accept the instruction without resistance and hope that he will be able to complete the work, as varied by the instruction, within the building period. But, if he wishes an extension of time, he must comply with the conditions precedent that clause 13.8 provides for in these specific circumstances (cf. cl. 13.8.5). In particular, he must serve notice on the architect of his estimate of inter alia the probable cost. In the light of that the employer may take the opportunity to withdraw the instruction. But if the contractor fails to take the steps specified in clause 13.8.1, then unless the architect waives the requirements of the clause under clause 13.8.4, the contractor will not be entitled to an extension of time on account of that particular instruction.*'[80]

---

78 NEC4 – clause 61.3, unless the Project Manager or the Supervisor has notified the contractor under clause 61.1.

79 *City Inn Ltd v Shepherd Construction Ltd* [2003] ScotCS 146 (20 May 2003) and [2003] BLR 468.

80 *City Inn Ltd v Shepherd Construction Ltd* [2003] ScotCS 146 (20 May 2003) and [2003] BLR 468 at para 23.

### 2.9.3 Whether a Condition Precedent

Traditionally, the requirement for the contractor to give notice of delay under JCT contracts has not been treated as a condition precedent. It has been held that the employer's representative (such as an architect or contract administrator) ought to give an extension of time if it seemed warranted, even where the contractor had not submitted a notice of delay, in order to avoid the possibility that time would become at large and the employer would be deprived of the opportunity to recover liquidated damages. The right to an extension of time was not to be unaffected by the contractor's failure to give notice. The architect was entitled to take the contractor's breach of obligation into account when deciding upon the extension of time by ensuring that the contractor did not get a greater extension of time than would have been the case if notice had been timeously given.[81]

The idea that a contractor was entitled to an extension of time if the employer committed any act of prevention whether or not notice had been given was considered in an Australian case[82]; but subsequent decisions[83] by English courts have declined to follow that decision. It is now clear that there is no obstacle to making such a notice a condition precedent if that is what the parties agree in clear words.

In an English case from 2007,[84] the judge concluded:

> *'I consider that a notification requirement may, and in this case does, operate as a condition precedent even though it does not contain an express warning as to the consequence of non-compliance. It is true that in many cases (see for example the contract in the Multiplex Constructions [UK] case itself) careful drafters will include such an express statement, in order to put the matter beyond doubt. It does not however follow, in my opinion, that a clause – such as the one used here – which makes it clear in ordinary language that the right to an extension of time is conditional on notification being given should not be treated as a condition precedent. This is an individually negotiated sub-contract between two substantial and experienced companies, and I would be loathe to hold that a clearly worded requirement fails due to the absence of legal 'boilerplate.*[85]*'*

In later English case,[86] the court considered whether the phrase, *'provided always that'* was indicative of a condition precedent. This case was not only of interest because the phrase occurred in a standard trade contract which the parties had amended but also

81 *London Borough of Merton v Stanley Hugh Leach Ltd* (1985) 32 BLR 51.

82 *Gaymark Investments Pty Ltd v Walter Construction Group Ltd* [1999] (2005) 21 Const LJ 71.

83 *Multiplex Constructions (UK) Ltd v Honeywell Control Systems Ltd* [2007] EWHC 447 (TCC), *Steria Ltd v Sigma Wireless Communications Ltd* [2007] EWHC 3454 (TCC) and *CIS General Insurance Ltd v IBM United Kingdom Ltd* [2021] EWHC 347 (TCC).

84 *Steria Ltd v Sigma Wireless Communications Ltd* (2007) 118 Con LR 177 at 203 per Judge Stephen Davies.

85 Boilerplate clauses are standard contract terms that are agreed to with little or no negotiation. These are usually found toward the end of the agreement and help to define the rights and obligations of the parties and affect how disputes will be solved. Oxford Dictionary of Law (2018).

86 *WW Gear Construction Ltd v McGee Group Ltd* (2010) 131 Con LR 63 at 75 per Akenhead J.

that it occurs in JCT contracts[87] for the employer (or his representative) deciding upon an extension of time:

> '*This type of wording is often the strongest sign that the parties intend there to be a condition precedent. What follows such a proviso is usually a qualification and explanation of what is required to enable the preceding requirements or entitlements to materialise.*'

As the court said, this is probably one of the clearest indications of the presence of a condition precedent. However, in JCT contract, the words '*provided always that*' only trigger the employer's (or his representative's) obligation to consider the fixing of a later completion date upon receipt of the contractor's notice and particulars. Those words do not operate to limit the employer's liability for extensions of time and/or loss and expense in the same way as failure to comply with a time bar clause (described in Section 2.9.2) may limit a party's rights.

The key message from these cases seems to be that where the giving of notice is a condition precedent to an extension of time, then ordinarily, the failure to give notice will prevent the giving of an extension of time because the following procedure to assess an extension of time will not be triggered (in the case of extensions of time before practical completion in JCT 2016 contracts – see Sections 11.3.2 and 11.3.3). Unfortunately, there is no golden rule which can be automatically applied to these clauses in order to decide whether there is a condition precedent or not. However, it is suggested that where extension of time or loss and/or expense clauses are concerned, the contractor providing its notice (and possibly other actions such as provision of specific information) will be a condition precedent if:

- the requirement for notice is preceded by the words 'it is a condition precedent that', or clear wording is used to show that the parties intend the requirement to be a condition precedent;
- the requirement for notice is preceded by the words 'provided always that' or probably 'provided that'; and
- There is a procedure for the employer (or his representative) to determine if, and if so to what extent the contractor may be entitled to an extension of time or loss and expense, and that procedure can only begin once the contractor has provided the relevant notice.

### 2.9.4 Waiving the Requirement for Notice

Although the giving of notice may be a condition precedent, the party entitled to rely on the condition may waive this requirement at its discretion. For such a waiver (also referred to as forbearance) to be effective, the following elements[88] must be present:

- There must be a legal relationship (such as a contract or sub-contract) between the parties setting out rights and duties between them;
- A clear or unequivocal promise or representation by one party that it will not enforce its strict legal right (i.e. the condition precedent) against the other; and

87 JCT SBC/Q 2016 – clause 2.28.6, JCT DB 2016 – clause 2.25.6, JCT IC 2016 – clause 2.19.4.
88 Chitty on Contracts (30th Ed) para 3–086.

- An intention that the other party will rely upon that unequivocal promise or representation made by the party making the waiver.

In practice, a party way waive it's rights in a number of ways:

- Expressly – for example, confirming in writing that it does not intend to, or will not rely upon the condition precedent provision; or
- By its conduct[89] – for example, it may ignore the condition precedent and proceed to grant an extension of time or assess and pay an amount of money. A party may have repeatedly acted in such a way, creating a 'course of dealings'.

The assertion of a waiver (or forbearance) can cause particular problems between main contractors and sub-contractors. A sub-contractor may deal with a number of the main contractor's personnel (such as a director project manager, site manager or engineer) on a day-to-day basis – but far less frequently than with the main contractor's commercial staff. Members of the main contractor's staff outside its commercial team may have effectively waived the main contractor's right to insist on a particular provision of the sub-contract. For the opposite reason, waiver is less of a problem for employers as the main contractor will only deal with the employer's representative, so there are less points of contact (on the employer's side) from whom a waiver may be given.

It is therefore common for main contracts and sub-contracts to be amended to include a 'no waiver' clause. These types of clauses usually act in two ways:

- They provide that only a waiver expressly given in writing will be valid and no implied forbearance, or other failure to insist upon, or enforce the contract's legal rights will amount to a waiver and
- Only designated personnel (either identified by name, or by class – e.g. a director) have the authority to issue a written waiver. Such a provision reduces the likelihood of a waiver being made by a junior member of staff.

### 2.9.5 Essential Elements of a Notice

Whether any communication qualifies as a notice (for the purpose of the relevant contract provision) is a question which can only be answered by reference to the document itself, and at times, to the prevailing circumstances. Of course, such questions would not arise if every notification was correctly, and clearly headed as '*This is a notice under clause XX of the contract*', but this is not always the case. Often, the recipient must examine the communication for the following criteria:

- The manner of service (i.e. in which the notice is sent) and
- The communication's 'form' or content.

The manner of sending a notice may be specified in the contract. Ordinarily, notices that deal with the usual administration[90] of the contract must be in writing, and can be sent by

89 *Donnison v The Employers' Accident and Livestock Insurance Co. Ltd* (1897) 24 R. 681 at 686 per Clerk LJ.
90 Such notices may include (i) notices of delay, or loss and expense from the contractor, or (ii) payment, pay-less notices, notices of defective works or a completion notice by the employer (or by his representative).

any means – i.e. by email,[91] first class post, actual delivery[92] or delivered in person. Notices that could lead to more serious consequences – such as termination, or in some cases issuing a concluding document (such as the equivalent of a JCT Final Certificate) often carry special requirements for service[93] to prove when they were sent and received, such as being sent:

- By First class registered post – so the dates of sending and receipt can clearly be evidenced; or
- By Actual delivery – i.e. delivery by hand, in person; and
- To a particular address, such as:
  - Specific addresses for each party for the service of notices.[94] These may be the parties' addresses or the address of an authorised representative such as a lawyer;
  - The registered address if the recipient is a limited company; or
  - The recipient's last known address.

Care should be taken attempting to issue a notice, where the contract calls for the notice to be served in a particular manner, as the following examples illustrate:

- If the contract distinguishes 'ordinary' notices which can be served by any means, from specific notices which have be served by particular means. In one case,[95] the contract distinguished the former from the latter by defining the types of notice as an 'informal notice' or a 'formal notice'. The recipient had to confirm receipt of the 'informal notice' otherwise it would be considered ineffective and the other party would have to send a 'formal notice' instead.
- The wording of the relevant notice clause should be read carefully. For example, if the clause provides that a notice 'may' be sent in a particular way (as opposed to saying 'shall be sent…') then it gives the sender the option of choosing the manner of sending the notice,[96] provided that it is not less advantageous to the recipient.[97]
- Where the contract makes clear that only notices served in the manner and time prescribed will be considered valid, notices sent by any other method will usually be invalid. For example: 'If a claimant is required to serve X and, mistakenly purports to serve Y, the mere fact that Y informs X of the purported service so that X knows of it, cannot convert Y's receipt of the documents into good service upon X.'[98]

What must be included in the notice ordinarily depends on what the contract's notice clause requires, however, as long as a reasonable recipient has no reasonable doubt as to the notice's meaning, then the precise form of words used many not be important.[99] Nevertheless, the starting point for the content of the notice must be the clause itself:

---

91 *Bernuth Lines Ltd v High Seas Shipping Ltd* (2006) 1 Lloyd's Rep 537.

92 *Construction Partnership UK Ltd v Leek Developments* [2006] EWHC B8 (TCC) – considering sending by facsimile.

93 JCT SBC/Q 2016 – clause 1.7 (Notices and other communications).

94 For example, JCT contracts contain an address for service of notices for the employer and the Contract Particulars.

95 *Orchard (Developments) Holdings plc v Reuters Ltd* [2009] EWHC Civ 6.

96 *Ener-G Holdings plc v Phillip Hormell* [2012] EWCA Civ 1059.

97 *Yates Building Co. Ltd v Pulleyn and Sons (York) Ltd* (1976) 237 EG 183.

98 *Lantic Sugar Ltd v Baffin Investments Ltd* [2009] EWHC 3325 (Comm) at paragraph 40 per Gross J.

99 Hudson's Building and Engineering Contracts (14th Ed) 2020 at 8–047.

> *'In my judgment, the requirement for [the sub-contractor] to give notice of the circumstances giving rise to the delay cannot be extended to include a requirement that the notice must make it clear that it is a request for an extension of time under cl 6.1, or to include a requirement that it gives an assessment of the delay. Both would involve reading into the clause words which are not there, and which do not meet the stringent requirements for implication of such terms. I do not however accept [the sub-contractor's] argument that all that is required is a notification that particular relevant circumstances have occurred. In my judgment it is necessary for [the sub-contractor] to notify [the contractor] first that identified relevant circumstances have occurred and second that those circumstances have caused a delay to the execution of the sub-contract works. In my judgment the latter is required, either by a process of purposive construction or by a process of necessary implication, because otherwise it seems to me that the notice would not achieve its objective…'*[100]

It is often asked whether the minutes of a site meeting that mention delays provides sufficient notice for the purposes of the contract. Whilst it is not uncommon for contracts (predominantly sub-contracts) to exclude meeting minutes, progress reports and other specified forms of communication from qualifying as notices in this sense, the ordinary position can be illustrated as follows:

> *'I also consider that the written notice must emanate from [the sub-contractor]. Thus for example an entry in a minute of a meeting prepared by [the consultant] which recorded that there had been a delay by [the employer] … and that as a result the sub-contract works had been delayed, would not in my judgment by itself amount to a valid notice under cl 6.1. The essence of the notification requirement in my judgment is that [the contractor] must know that [the sub-contractor] is contending that relevant circumstances have occurred and that they have led to delay in the sub-contract works.'*[100]

In another case,[101] it seems to be suggested that a note in site meeting minutes recording an oral representation by the contractor during that meeting is not a valid notice, but that will depend upon the forms of communication that are permissible by the contract.

A contract may also state the circumstances in which a notice must not be given. For example, JCT contracts[102] state that a notice terminating the contractor's employment shall not be given 'unreasonably or vexatiously'.

### 2.9.6 Giving, Issue and Receipt of Notices

Most building contracts refer to the giving of notices and the issuing of certificates. The word 'issue' suggests a more formal process than 'giving', but in each case there is the sense that something passes from the giver (or the issuer) to the recipient. However, the precise wording of the notice clause must be examined and understood, as the following examples illustrate:

---

100 *Steria Ltd v Sigma Wireless Communications Ltd* (2007) 118 Con LR 177 at 201 per Judge Stephen Davies.

101 *John L Haley Ltd v Dumfries and Galloway Regional Council* (1988) 39 GWD 1599.

102 JCT SBC/Q 2016, JCT DB 2016, JCT IC 2016 – clause 8.2.1, JCT MW 2016 – clause 6.2.1.

- Termination of the contractor's employment under JCT 2016 contracts[103] takes place upon receipt of the relevant notice, not on the giving of that notice; or
- On the other hand, if a contract refers to the employer's representative issuing a certificate within 10 days of an event, it simply means that the certificate must be sent within 10 days. 'Issue' ordinarily means 'supply or distribute'[104]; therefore, it may be inaccurate to use the words indiscriminately. The issue of a notice is neither the same as the receipt of the notice, nor may it be the date written on the face of the purported 'notice'.[105] If the contract refers to a notice being issued, it is the date when it was sent out which is the relevant date no matter what date is written on the notice itself.[106]

A notice can neither be issued nor given if it does not leave the person issuing or giving it. Therefore, a notice which is properly made out, signed and dated, but put in a drawer, has no effect whatsoever:

> *'In the ordinary meaning of the word 'issue', it seems to me plain that something more is needed for a certificate to be issued … than the mere signature of the architect upon it, whether he be employed under a contract of service or as an independent agent.'*[107]

103 JCT SBC/Q 2016, JCT DB 2016, JCT IC 2016 – clause 8.2.2, JCT MW 2016 – clause 6.2.2.
104 Oxford Dictionary of English (2010).
105 *Glen v The Church Wardens & Overseers of the Parish of Fulham* (1884) 14 QBD 328.
106 *Cantrell and Another v Wright and Fuller Ltd* (2003) 91 Con LR 97.
107 *London Borough of Camden v Thomas McInerney & Sons Ltd* (1986) 9 Con LR 99 at 111 per Judge Esyr Lewis.

# 3

# Time

## 3.1 General Points

### 3.1.1 Time Obligations in Building Contracts

The contractor's obligations as to 'time' are often considered one of the essential terms of a building contract.[1] Completion of the works, or sections of them, or achieving particular 'milestones' may be the central criteria to a project's success. Typically, the contractor's time obligations will be as follows:

- Commencement
  - The contractor either commences all the work on an agreed date,[2] as set out in the contract;
  - He commences sections of the work on agreed dates;
  - He commences work in separate stages: work offsite and then on site, on different dates[3] set out in the contract; or
  - He commences work within an agreed period following notification by the employer (or his representative).
- Carrying out the works
  - The contractor may be obliged to work at a particular rate of progress.
- Completing the works
  - The contractor usually has to complete the works 'on or before' the agreed completion date; or
  - He may have to complete the works in sections, each section having its own completion date.
  - He may have to complete discrete parts of the works or ensure functional parts of the work achieve a state of readiness by a 'milestone' date.

In the absence of either an agreed completion date, or a means of calculating one, the contractor is usually entitled to complete within a reasonable time.[4]

1 The 'essential' terms of a building contract also include the parties' identities, the description (or scope) of the works and the contract price (or a means of calculating it).
2 JCT SBC/Q 2016 – clause 2.4 on the "Date of Possession".
3 NEC4 – the contractor may commence work off-site (e.g. design) on the "starting date", and commence work on site on the first "access date" – clause 30.
4 s14 Supply of Goods and Services Act 1982.

*Building Contract Claims*, Sixth Edition. Guy Higginbottom and David Chappell.

The employer also has obligations to perform at specific times. For example, he is obliged to issue payment notices and make payments and, depending upon the project, may have other obligations to provide certain facilities, access or materials to the contractor at particular times.

### 3.1.2 Commencement Obligations

The date upon which the contractor commences work (and usually commences work on site) is very important as it generally signifies the date from which the contractor's major obligations begin. The most obvious will be the remaining time in which the contractor has to complete the construction (and perhaps the design) of the works. Other obligations, such as the contractor's obligation to insure the works, may also begin on this date. Different building contracts deal with commencement in slightly different ways.

The parties to a JCT 2016 contract are expected to have agreed on the commencement date (called the 'Date of Possession[5]') and included it in the Contract Particulars. Subject to any obligations to complete the works in sections, on the 'Date of Possession', the employer shall give possession of the site to the contractor,[6] who shall then commence the physical works. If the work is to be completed in Sections,[7] the entries for the 'Dates of Possession of Sections' (again, in the Contract Particulars) for each 'Section' must be completed. The description of the Sections is set out in the entry for the relevant 'Recital' in the Contract Particulars.

The date of possession is also relevant to the following:

- Deferment of possession. The employer may delay giving possession of the site to the contractor if the Contract Particulars state that deferment applies.[8]
- Interim payment interim valuation dates.[9] If the first 'interim valuation date' is not set out in the contract particulars, then by default, the first interim valuation date is 1 month after the date of possession;
- Retention Bond. The contractor is encouraged to provide the retention bond[10] on or before the date of possession;
- Insurance. Whilst the date of possession is not specifically mentioned, the period of insurance in clause 6.7.2[11] (applying to either the employer or the contractor in insurance options A, B or C[12]) appears to begin on the date of possession.

5 "Date of Possession" is defined in clause 1.1 (Definitions): JCT SBC/Q 2016 and JCT IC 2016 – clause 2.4, JCT DB 2016 – clause 2.3. JCT MW 2016 – clause 2.2 refers to the "Works commencement date".
6 See also Section 12.4.4.
7 JCT SBC/Q 2016 – Sixth Recital, JCT DB 2016 – Fifth Recital, JCT IC 2016 – Eighth Recital and corresponding entries in the Contract Particulars. There is no sectional completion in JCT MW 2016.
8 JCT SBC/Q 2016 and JCT IC 2016 – clause 2.5, DB 2016 – clause 2.4, the employer may defer for a maximum period of 6 weeks.
9 JCT SBC/Q 2016 and JCT IC 2016 – clause 4.8, JCT DB 2016 – clause 4.7.2, JCT MW 2016 – clause 4.3.
10 JCT SBC/Q 2016 – clause 4.18.2, JCT DB 2016 – clause 4.17.2.
11 Clause 6.7.2 under JCT SBC/Q 2016, JCT DB 2016 and JCT IC 2016.
12 JCT DB 2016: Schedule 3, Insurance Option A (New Buildings – All Risks Insurance of the Works by the Contractor), Insurance Option B (New Buildings – All Risks Insurance of the Works by the employer), and Insurance Option C (Joint Names Insurance by the employer of Existing Structures in or Extensions to them).

- Termination. The start of the contractor's entitlement to terminate his employment if the employer suspends performance for longer than the period stated in the contract particulars.[13]

NEC4 provides two distinct dates from when the contractor is able to commence elements of the work. The first is the 'starting date', and it seems clear that the latter will be the first 'access date'. The distinction between the two is that the contractor may commence any other work after the 'starting date', whereas he is not permitted to start work on the 'Site' until the first of the access dates. Such 'other work' may include off-site tasks such as preparing his design for the Project Manager's approval (under clause 21.2), procuring long lead-in items, with the following obligations commencing after the 'starting date':

- Early warning risk register issued and first early warning meeting[14]
- Activities carried out at the intervals stated in the Contract Data:
  - Revising the programme[15]
  - Providing forecasts of defined costs[16] (under Options C, D, E and F)
- Providing insurance,[17] including professional indemnity under Option X15 (the Contractor's design[18])

Reading the NEC4 contract as a whole indicates that the 'starting date' is intended to occur before the first 'access date'. This allows the contractor to carry out the obligations mentioned above, and by virtue of clause 50.1 (Assessing the amount due), the contractor is entitled to be paid for carrying out certain obligations that occur after the 'starting date', but before the first 'access date' (i.e. when the contractor may commence work on site).

Contract Data Part One (section '3 Time') provides for several 'access dates' for particular parts of the Site. This provides an 'access date' for each part of the works in the following circumstances:

- When the contractor will progressively be given access to parts of the works but must complete the whole of the works on or before the 'completion date'; or
- Where Option X5 (Sectional Completion) is selected, 'access dates' can be allocated to each of the sections, as only the sections' 'completion dates' are provided in contract data part one (against the entries for Option X5).

### 3.1.3 Progress Obligations

Progress, in general, means to develop something with the aim of reaching an advanced or completed state. In the context of construction and engineering projects, it is often

13 JCT SBC/Q 2015, JCT DB 2016 and JCT IC 2016 – the Contract Particulars entry for clause 8.9.2 (if no period is stated, the period of suspension is 2 months). JCT MW 2016 – clause 6.8.2 entitles the contractor to terminate if the employer suspends the works for 1 month.
14 NEC4, clause 15.2.
15 NEC4, clause 32.2.
16 NEC4, clause 20.4 for Option C (Target Contract with Activity Schedule), Option D (Target Contract with Bill of Quantities), Option E (Cost Reimbursable Contract) and Option F (Management Contract).
17 NEC4, clauses 83 and 84 (insurance by the contractor), clause 86 (insurance by the Client).
18 NEC4, Option X15, clauses X15.5 and X15.6.

expressed both as a 'rate' of progress (particularly where carrying out the physical works is linear or repetitive in nature[19]) and as a manner of progress.

- The rate of progress:
  - Fairly rudimentary construction economics refers to an 'S' curve, where the value of work completed is plotted on the Y-axis, against the passage of time on the X-axis (refer to Table 2.2[20] in Section 2.4.1). Following the shape of the resulting flattened 'S', it can often be observed that progress expressed as 'value/time' is slower (flatter) towards the project's start and finish, but usually increases rapidly (i.e. showing a steeper gradient) to a peak, somewhere in the between. Owing to the varied nature of construction projects, it is not always possible to express or measure an overall rate of progress as many different activities will proceed concurrently, but at different rates, which may not be comparable;
  - The rate of progress can be expressed in subjective terms, often in relation to the project's completion date or in relation to intervening periods during which work has been completed. Comparing progress to a programme can be a useful tool, but it is rare that the contractor is bound to complete the work strictly in accordance with the programme. Usually, his principal obligation to complete the works is to complete on or before the completion date, so any progress he makes in order to do that is usually considered satisfactory and will not be a breach of the contract. If the contractor is at fault for the works progressing too slowly, he risks failing to complete by the completion date and may face delay damages, so the incentive for him to maintain sufficient progress is clear;
  - Working strictly in accordance with a programme exposes both parties to additional risk. If the contractor is falling behind on a particular task, but isn't jeopardising the completion date, then without additional controls,[21] it is unlikely that the employer can seek compensation. Conversely, if the employer fails to provide something to the contractor (such as sufficient or unimpeded access or items of plant and materials) and prevents him from maintaining progress in accordance with the programme, then the employer will be in breach. If the contractor's progress is interrupted by the employer but the completion date is still achieved, the employer may be exposed to a disruption[22] claim from the contractor. If the employer seeks to control the contractor's progress, he can impose individual targets for parts of the works. He can do this by dividing the works into sections, so that the contractor has to complete the sections in a particular order or by particular dates. He can also impose milestone (or 'Key Dates[23]'), where the contractor must complete a particular task – such as a part of the works being ready for testing or receiving equipment.

19 For example, progress of a construction project such as a hotel refurbishment could be expressed in terms of "rooms completed per week", or the progress of an engineering project such as a road or railway, could be expressed as "km per week".

20 Table 2.2 plots "costs over time" but is also representative of "value over time".

21 Such as working to Key Dates (NEC4).

22 Disruption claims are usually made by the contractor to recover the additional expenditure (caused by the employer's breach) of carrying out the work when the completion date is achieved (refer to Section 2.4.2).

23 NEC4, "Key Date" is defined in clause 11.2(11).

- The manner of progress is to describe the way in which the contractor proceeds with the work. Ordinarily, the contractor is entitled to perform the work in any way he pleases,[24] but the contract may impose some general restrictions. These are likely to oblige the contractor to carry out the work competently and properly and applying a consistent effort:
  - Competently and properly – carrying out the work in accordance with the contract (e.g. with employer's requirements, drawings and specification), any applicable laws (such as the building regulations) and often, a further obligation of working in accordance with 'good practice[25]' is added or implied.
  - The 'frequency' obligation intends that the contractor's efforts should be consistent and uninterrupted. The employer will not want to see periods of inactivity on site.

Further details of the manner of progress are usually provided in documents such as method statements, the contractor's programme and in technical information. As a general rule, the contractor's method of work is his own, and this usually means he is entitled to adopt the most economical method of working for the particular project, provided that his chosen method complies with the contract, any restrictions imposed by it (e.g. working hours), and health and safety legislation.

When compared to NEC4, JCT contracts contain far less detailed requirements in terms of monitoring the progress of works to completion. The contractor's principal obligation as to time is that he must complete the works, or a section before the relevant completion date, otherwise he will be liable for delay damages (if he is not entitled to an extension of time). The manner of progress is set out in Table 3.1 (applying to all the works, either per section or as a whole):

The obligation to '*...carry out and complete the works...*' echoes the implied obligation[26] that services will be carried out with 'reasonable care and skill'. However, the obligation to proceed *'regularly and diligently'* is much more wide-ranging, as the contractor must

**Table 3.1** Progress obligations under JCT contracts.

| Obligation | JCT SBC/Q 2016 | JCT DB 2016 | JCT IC 2016 | JCT MW 2016 |
|---|---|---|---|---|
| '...carry out and complete the works in a proper and workman-like manner' | 2.1 | 2.1.1 | 2.1 | 2.1.1 |
| '...regularly and diligently proceed with and complete the [Works] on or before the relevant Completion Date.' | 2.4 | 2.3 | 2.4 | — |
| '...constantly use his best endeavours to prevent delay' | 2.28.6.1 | 2.25.6.1 | 2.19.4.1 | — |

24 *John Barker Construction Ltd v London Portman Hotels Ltd* (1996) 60 Con LR 43.
25 "Good practice" is a subjective term which can mean *"The accepted method of building, with the standard and quality being satisfactory to all requirements and regulations."* Oxford Dictionary of Construction, Surveying and Civil Engineering (2020). Importantly, it should be added that "good practice" will vary over time, with some previously accepted methods becoming obsolete or superseded.
26 Supply of Goods and Services Act 1982, s13.

proceed both 'regularly' and 'diligently' and failure in each respect may put him in breach. The contractor's obligations were described as follows:

> *'Taken together, the obligation upon the contractor is essentially to proceed continuously, industriously and efficiently with appropriate physical resources so as to progress the works steadily towards completion substantially in accordance with the contractual requirements as to time, sequence and quality of works…'*[27]

It appears that an obligation to proceed regularly and diligently would prohibit the contractor from working in a disjointed, 'stop-start' or haphazard manner. The obligation as to quality is also expressly set out in the JCT 2016 contracts, as the contractor must also carry out the works in accordance with the Contract Documents.[28]

The contractor's *'best endeavours'* obligation is also important as it imposes a further obligation to prevent delay, irrespective of who's fault that delay may be. Seemingly, the 'best endeavours' obligation applies to preventing 'likely' delays,[29] as well as preventing an existing delay from becoming worse. Employers should not assume that 'best endeavours' means that the contractor should take every imaginable step, irrespective of cost.[30] However, it seems that the contractor should neither remain passive nor make a minimal or token effort. He should take prudent steps such as advising the employer of delays, along with rescheduling the works and the programme if appropriate. However, it seems that the extent of using 'best endeavours' is measured by reference to someone acting in their own interests[31] and not fulfilling a contractual obligation. In other words, the contractor's best endeavours are focussed on helping himself avoid delays. The scope of the contract's variation clause provides a useful guide to whether the contractor has used constant 'best endeavours' as he will be entitled to additional payment if the 'endeavours' fall within the scope of the variation clause.[32] Under a design and build contract, a 'best endeavours' obligation may place more risk with the contractor than under a 'traditional' contract,[33] as variation clauses under the former may be less detailed and more restrictive than under the latter, reflecting the contractor's additional responsibility to carry out the design.

Unlike JCT 2016 contracts, NEC4 does not expressly set out the manner or rate of progress that the contractor must achieve. Instead, clause 30.1 appears to imply that the contractor must achieve sufficient progress in order to complete the works on or before the relevant, contractual 'completion' dates set out in contract data part one.[34] This rate of progress should also be sufficient to meet key dates (clause 30.3) and complete any sections of the works if option X5 (sectional completion) is selected.

---

27 *West Faulkner Associates v London Borough of Newham (1994) 71 BLR 1.*

28 JCT DB 2016, "Contract Documents" is defined in clause 1.1 as the Agreement, the Conditions, the employer's Requirements, the Contactor's Proposals, the Contract Sum Analysis, and any applicable BIM Protocol.

29 When read alongside JCT DB 2016 clause 2.24.1.

30 Keating on Construction Contracts, 10th Edition at 20-152.

31 Noble A, "Contractors' and Sub-Contractors' Constant Best Endeavours to Prevent Delay", (1997) 13 Const LJ 293.

32 JCT DB 2016, clause 5.1.2.

33 A contract where the employer is responsible for the design.

34 NEC4 – Contract Data Part One, section 2 (The Contractor's main responsibilities) *"key date"*, and section 3 (Time") the *"completion date for the whole of the works."* If Option X5 (Sectional Completion) is selected, the *"completion date"* for each *"section."*

The employer can monitor progress through the contractor submitting programmes for the project manager to accept (under clause 31.2), but an insufficient rate of progress is not one of the grounds for rejection. For each revised programme, the contractor is obliged to show the actual progress for each programmed activity,[35] and as he must show any measures to overcome delays, his programme ought to identify steps to increase the rate of progress, provided that the contractor's insufficient progress is the cause of delay.

### 3.1.4 Completion Obligations (Including Partial Possession)

In general terms, 'completion' signifies the end of RIBA Stage 5 (Manufacturing and Construction), so the contractor's obligations to complete the physical works are at an end, and his obligations to rectify defects or non-compliant work or carry out further testing will begin (depending upon the particular form of contract).

JCT contracts express 'completion' as 'practical completion', which rather unhelpfully remains undefined in the 2016 editions of all of the JCT building contracts. The term 'practical completion' has been used since the 1963 (then RIBA) edition of the standard building contract, and its meaning has been considered by the Courts as follows:

- 'Practical completion' means completion of all the work that has to be done[36];
- 'Practical completion' can occur, notwithstanding the presence of latent defects[37]; however, the presence of patent defects ordinarily prevents the certificate (or 'statement' in the case of JCT DB 2016) of practical completion being issued[37]; and
- The supervising officer (in this case, the architect or contract administrator) has discretion to certify practical completion where there are minor items of incomplete or defective work (the de minimis principle).[38]

Ordinarily, all of the physical work save for latent or very minor (*de minimis*) defects or items of incomplete work must be complete. The Courts' interpretation of 'practical completion' places more emphasis upon the extent of completion and possibly an aesthetic element, rather than focussing more on the building's 'practical' or 'functional' aspects. For example, a dwelling house may fulfil the function of being fit for habitation,[39] but if it is partly incomplete, or its aesthetic standard of finish may be such to prevent a sale or lease at the desired price, it may not be deemed to have achieved a state of 'practical completion' as required by the building contract.

JCT 2016 contracts provide that the employer (or his representative) issues the 'practical completion certificate'[40] (or in the case of sectional completion,[41] a 'section completion certificate') when the Works as a whole (or in the relevant section) have reached a state of practical completion. This has the following effects:

35 NEC4 clause 32.1. This is usually shown as a "drop line", recording progress ahead or behind the baseline programme dates, or by comparing the baseline programme with activities' current progress.
36 *Jarvis & Sons v Westminster Corp (1979) 1 WLR 637 at 646D-F.*
37 *Kay v Hosier & Dickinson (1972) 1 WLR 146.*
38 *HW Nevill (Sunblest) v William Press (1981) 20 BLR 78 at 87*, see also *Emoson Eastern (in receivership) v EME Developments Ltd (1991) 55 BLR 114* and *Walter Lilly & Co Ltd v Mackay & Another (No. 2) (2012) EWHC 1773 (TCC) at 372.*
39 As per s1(1) of the Defective Premises Act 1972.
40 JCT SBC/Q 2016 – clause 2.30.1, JCT IC 2016 – clause 2.21.1, JCT MW 2016 – clause 2.9, whereas under JCT DB 2016 – clause 2.27.1, a "Practical Completion Statement" is issued.
41 JCT SBC/Q 2016 – clause 2.30.2 and JCT IC 2016 – clause 2.21.2, whereas under JCT DB 2016 – clause 2.27.2 a "Section Completion Statement" is issued.

- The contractor's obligations to carry out and complete the physical works and its design are at an end, save for rectifying any latent defects (i.e. those which could not reasonably be discoverable), during the rectification period;
- If retention has been deducted from amounts due to the contractor in interim payments, then half of the retention may be deducted from subsequent interim payments until the certificate of making good is issued[42];
- The rectification period commences. This will be as per the entry in the contract particulars,[43] or if no period is entered, a default period of 6 months applies (or different rectification periods may apply if the works are being completed by sections);
- The contractor's obligation to insure the works under insurance option A (all risks insurance of the Works by the contractor) shall cease[44]; and
- The employer is no longer entitled to instruct changes as the Works are complete.

'Practical completion' is also 'deemed' to occur if the employer takes 'partial possession' of part of the works (or of a section).

JCT 2016 contracts permit the employer to take 'partial possession' of parts of the works, or of a section before practical completion.[45] If the employer intends to take partial possession, the contractor shall provide the following[46]:

- His consent to the employer taking partial possession and
- A notification to the employer of the following that identifies:
  - The part or parts of which the employer has taken possession (referred to as the 'Relevant Part') and
  - The date on which the employer took partial possession (referred to as the 'Relevant Date').

When the employer takes partial possession, 'practical completion' is deemed to have occurred[47] on the Relevant Date for the Relevant Part, and this has the following effects:

- The rectification period commences from the Relevant Date for the Relevant Part,[48] and the employer must issue a separate notice that all defects have been rectified for each Relevant Part;
- For the Relevant Part, half of the retention is released[49];

42 JCT SBC/Q 2016 – clause 2.39, JCT IC 2016 – clause 2.31, JCT MW 2016 – clause 2.11, whereas under JCT DB 2016 – clause 2.36, the "Notice of Completion of Making Good" is issued.
43 Contract Particulars' entries against CT SBC/Q 2016 – clause 2.38, JCT DB 2016 – clause 2.35, JCT IC 2016 – clause 2.30 and JCT MW 2016 – clause 2.10 (default period is 3 months).
44 JCT DB 2016, clauses 2.33 and 6.7.2 (Insurance Options and Period) – also see Schedule 3 (Insurance Options), Insurance Option A, paragraph A.1 (Contractor to effect and maintain a Joint Names Policy).
45 JCT SBC/Q 2016 – clauses 2.33 to 2.37, JCT DB 2016 – clauses 2.30 to 2.34, and JCT IC 2016 – clauses 2.25 to 2.29. There is no provision for 'partial possession' in JCT MW 2016.
46 JCT SBC/Q 2016 – clause 2.33, JCT DB 2016 – clause 2.30 and JCT IC 2016 – clause 2.25.
47 JCT SBC/Q 2016 – clause 2.34, JCT DB 2016 – clause 2.31 and JCT IC 2016 – clause 2.26.
48 JCT SBC/Q 2016 – clause 2.35, JCT DB 2016 – clause 2.32 and JCT IC 2016 – clause 2.27. The entry in the Contract Particulars against states the Rectification Period is a period of months (if uncompleted, the default period is 6 months for the Works, or for a Section). In most cases, the parties agree that the Rectification Period is 12 months.
49 JCT SBC/Q 2016 – clause 4.19.2 and JCT DB 2016 - clause 4.18.2 (Retention – amounts and periods). JCT IC 2016 does not provide for 'retention' per se; instead clause 4.9.1 provides that the 'applicable

- Insurance under options A, B or C2 shall cease for the Relevant Part[50]; and
- Liquidated damages are reduced for the Works (or for a section) proportionally[51] to the value of the relevant part (against either the Contract Sum, or relevant Section Sum[52]).

The employer should carefully consider his options before seeking to take partial possession. He may be taking possession of a part of the works which, under normal circumstances, would not have met the criteria for practical completion (e.g. the Relevant Part may be uncompleted or have patent defects present). As soon as the employer takes partial possession, that Relevant Part is deemed to be practically complete, so the contractor has no obligation to complete any outstanding work or rectify any patent defects, and the employer cannot instruct any variations to the Relevant Part of which he has taken possession.

It is entirely possible for the employer to progressively take 'partial possession' of all parts of the Works. Whilst this presents the contractor with the following advantages:

- Having Rectification Periods commencing earlier (and together, possibly lasting for a shorter overall period);
- Progressive and possibly earlier releases of the first half of retention for the Relevant Parts;
- A similar reduction in his overall insurance liability; and
- Rates of liquidated damages are reduced proportionately;

but as the employer may never issue either a practical completion certificate or statement, this may affect the contractor's release from any bonds. The contractor will have to ensure that the wording of any bond for which the surety's liability expires at 'practical completion' also accounts for partial possession, and provides for its liability to expire in relation to the Relevant Part, from the Relevant Date.

Similar to the provisions in JCT 2016 contracts for 'partial possession', clause 35 of NEC4 provides that the Client may 'take over' the Works or 'parts' of them in the following circumstances:

- Before 'Completion' the Client may take over the Works or 'part' of it except[53]:
  - If stated in the Contract Data[54];
  - When using the Works benefits the contractor's method of working; or
  - The reasons for using that 'part' of the Works are set out in the Scope; or
- The Client takes over the works not later than two weeks after 'Completion'.[55]

Within one week of the Client doing so, the Project Manager must certify[56]:

- The date when the Client took over the Works or a 'part' and
- The extent (or 'part') of the Works taken over by the Client.

---

percentage' in 'Contract Particulars' (at clause 4.9.1) shall be applied to the amount due to the contractor. In the event that the parties do not agree the 'applicable percentage', the default percentages are 95% before practical completion, and 97.5% after practical completion.

50 JCT SBC/Q 2016 – clause 2.36, JCT DB 2016 – clause 2.33 and JCT IC 2016 – clause 2.28.

51 JCT SBC/Q 2016 – clause 2.37, JCT DB 2016 – clause 2.34 and JCT IC 2016 – clause 2.29.

52 The following entries in the Contract Particulars sets out the Section Sums: JCT SBC/Q 2016 – clause 2.37, JCT DB 2016 – clause 2.34 and JCT IC 2016 – clause 2.29.

53 NEC4, clause 35.2.

54 NEC4, clause 35.1 – see Section 3 (Time) in Contract Data Part One.

55 NEC4, clause 35.1.

56 NEC4, clause 35.3.

Owing to NEC4's ability to be used as a 'hybrid' contract, for both 'construction' and 'engineering' projects, it is unsurprising that its procedures for completion contain similar steps to those found in JCT contracts.

First, 'Completion' is defined in clause 11.2(2) as follows:

- Where the Work required to achieve Completion is set out in the Scope;
  - It must all be completed by the Completion Date[57]; and
  - No notified defects remain which would prevent the Client (or Others[58]) from using the Works.[59]
- However, if the Scope does not describe the work that the contractor must complete by the Completion Date, then the contractor must carry out all the work 'necessary' for the Client (or Others) to use the Works.

Clause 30.2 provides that the project manager decides when the 'date of completion' occurs and that he must certify it within 1 week.

Following completion, the project manager must provide the following:

- If Option X16 (retention) is adopted, then half of the retention must be included in the subsequent assessment of the amount due to the contractor (under clause 50.1)[60] – see Section 8.7.3 and / or
- If either of the target cost options (C or D[61]) are adopted, the Project Manager must provide the 'preliminary assessment of the contractor's share[62] – see Section 9.6.3.

Table 3.2 summarises the 'completion' procedures under JCT SBC/Q 2016, JCT DB 2016 and NEC4:

### 3.1.5 Time at Large

Where a building contract does not provide any contractual mechanism for fixing a new date for completion, time may become 'at large' if the contractor suffers delay due to some action, inaction or default on the part of the employer or persons for whom the employer is responsible. In such instances, the contractor's duty will be to complete the Works within a reasonable time.[63] Provided the contractor has not acted unreasonably or negligently, he will complete within a reasonable time despite a protracted delay if the delay is due to causes outside its control.[64] In such circumstances, time is said to be 'at large'. However,

57 Similar to the standard required to achieve "practical completion" in JCT DB 2016.

58 NEC4, clause 11.2(12) defines "Others".

59 Similar to the standard required to achieve "Taking Over" in FIDIC Yellow Book 2017 (save for the contractor's administrative obligations). It appears that defects that have not been notified, and which affect the use of the Works would not prevent "Completion" – however, given the Client's (via the Supervisor) and contractor's obligations under clause 43.2 (Searching for and notifying defects), it is unlikely that any significant defects would go unnoticed.

60 NEC4, Option X16 (Retention), clause X16.2.

61 NEC4, Option C (Target Contract with Activity Schedule), or Option D (Target Contract with Bill of Quantities).

62 NEC4, Option C - clause 54.3, or Option D – clause 54.8.

63 *Wells v Army and Navy Co-operative Stores* (1902) 2 HBC 4th Edition (vol 2) 346.

64 *Pantland Hick v Raymond & Reid* (1893) AC 22.

**Table 3.2** Summary of obligations/procedures on completion.

| Procedure/Obligation | JCT SBC/Q 2016 | JCT DB 2016 | NEC4 |
|---|---|---|---|
| Administrative<br>'Completion' is recorded in writing | 2.30.1[a],<br>2.34[b] | 2.27[c],<br>2.31[d] | 30.2[e] |
| Period of retaining contractor's documents is known (Option X15) | — | — | X15.4[f] |
| Employer's rights and obligations<br>Employer takes over the works | — | — | 35.1<br>(2 weeks after) |
| Employer can no longer instruct certain remedial works | 2.38.1 | 2.35.1 | — |
| Contractor's obligations<br>Period for rectifying defects begins | 2.38 | 2.35 | 44.1, 44.2[g] |
| Payment<br>Half the retention is released | 4.19.2 | 4.18.2 | X16.2 |
| Early completion bonus calculated (Option X6) | — | — | X6.1 |
| Preliminary assessment of contractor's 'share'<br>(Options C & D) | — | — | Options<br>C – 54.1<br>Option<br>D – 54.7 |
| Delay damages cease to apply | 2.32.2 | 2.29.2 | X7.1 |
| Responsibility for the Works/ Insurance<br>Contractor's obligation to insure ceases | 2.36,<br>6.7.2 | 2.33,<br>6.7.2 | 80.1 |
| 'End of liability' date is known (Option X18) | — | — | X18.6[h] |
| Reporting/monitoring/forecasting<br>No further programmes are provided | — | — | 32.2 |
| Monthly progress reports cease | — | — | 15.1 |
| Delay notifications/early warnings cease:<br>• EWN delaying completion<br>• EWN meetings | — | — | 15.2 |
| Contractor's forecasts for Defined Costs cease (Options C, D, & E) | — | — | Options C, D & E<br>20.4 |
| Early contractor involvement (X22)<br>• Contractor's forecasts cease<br>• Budget incentive prepared | — | — | X22.2(5)<br>X22.7(2) |

a) JCT SBC/Q 2016, clause 2.30.1: the Practical Completion Certificate is issued for the Works, or (under clause 2.30.2) a Section Completion Certificate for each Section of the Works

b) Where "Partial Possession" applies, JCT SBC/Q 2016, clause 2.34 provides that "practical completion" of the Relevant Part is deemed to have occurred from the Relevant Date

c) JCT DB 2016, clause 2.27: the Statement of Practical Completion is issued for the Works, or a Sectional Completion Statement for each Section of the Works

d) JCT DB 2016, clause 2.31 provides that "practical completion" of the Relevant Part is deemed to have occurred from the Relevant Date

e) NEC4, clause 30.2 the Project Manager certifies completion

f) NEC4, Contract Data Part One, entry for Option X15 (The Contractor's Design)

g) NEC4, clause 44.2 the "defect correction period" begins at Completion

h) NEC4, Contract Data Part One, entry for Option X18 (Limitation of Liability)

both JCT 2016[65] and NEC4[66] contracts contain provisions that enable extensions of time to be given under most circumstances, so time becoming 'at large' is a relatively rare occurrence.

Time may also be at large from the beginning of the contract if the parties have not agreed any date for completion, and whilst the contractor's obligation will be to complete within a reasonable time, the determination of a reasonable time in such circumstances may be a difficult task. In one case, where the parties had corresponded regarding the date for completion, the court held that there was an agreed date, but ventured the opinion that in any event, a reasonable date for completion would be not later than the date which the defendant had consistently put forward.[67]

The question of time being 'at large and the relationship between the extension of time clause and liquidated damages provisions in JCT contracts has been stated[68] in this way:

> *'1. The general rule is that the main contractor is bound to complete the work by the date for completion stated in the contract. If he fails to do so, he will be liable for liquidated damages to the employer.*
> *2. That is subject to the exception that the employer is not entitled to liquidated damages if by his acts or omissions he has prevented the main contractor from completing his work by the completion date - see for example Holme v Guppy (1838) 2 M & W 387, and Wells v Army and Navy Co-operative Society (1902) 86 LT 764.*
> *3. These general rules may be amended by the express terms of the contract.*
> *4. In this case [which involved a contract in terms identical to JCT 63] the express terms of clause 23 of the contract do affect the general rule …'*

In 2002, JCT introduced an amendment to the 1998 versions of its standard forms of contract, which added a relevant event that was essentially 'catch-all' provision to prevent time from becoming 'at large'. This now[65] provides that any impediment, prevention or default of the employer, his representative or others for whom the employer is responsible, entitles the contractor to an extension of time (provided of course, that completion is likely to be delayed beyond the date for completion). The inclusion of this relevant event has enabled the JCT to dispense with some of the other relevant events which could be said to fall into this category, for example, the late provision of information by the architect/contract administrator.

However, in some cases, even where a building contract contains terms providing for extension of the contract period, time may yet become at large either because the terms do not properly provide for the delaying event or because the employer (or his representative) has not correctly operated the terms. Some examples are:

---

65 Refer to the following Relevant Events for *"impediment, prevention or default…by the* Employer": JCT SBC/Q 2016 – clause 2.29.7, JCT DB 2016 – clause 2.26.6 and JCT IC 2016 – clause 2.20.6. JCT MW 2016 does not include any equivalent "Relevant Event" but clause 2.7 entitles the contractor to an extension of time if a delay is beyond his control.

66 NEC4 clause 60.1(18) *"A breach of the contract by the Client which is not one of the other compensation events in the contract."*

67 *J & J Fee Ltd v The Express Lift Co Ltd* (1993) 34 Con LR 147.

68 *Percy Bilton v Greater London Council* (1982) 20 BLR 1 at 13 per Lord Fraser of Tullybelton, delivering the unanimous decision of the House of Lords.

- 'Acts of prevention by an employer do not set time at large, if the contract provides for extension of time in respect of those events.'[69]
- If the employer's representative was late in delivering necessary drawing information to the contractor, but failed to give any extension of time. This is a clear case of the employer's representative not taking advantage of the available mechanism.
- Another example is where the contract provision sets out a timetable within which the employer or his representative must operate to give an extension of time. If the power is not exercised within the relevant period, the employer's (or his representative's) power to give an extension will end and time will become at large.

Contractors sometimes argue that time is at large and, therefore, liquidated damages are not recoverable. Although, in appropriate circumstances, that is a very strong argument, the consequences of time becoming at large are not necessarily to prevent the employer from recovering damages. Unliquidated damages, with a ceiling on recovery equal to the rate of liquidated damages, could be recovered.[70] When these questions go before an adjudicator or the court, the issue ought to be presented as a claim[71] that the employer or his representative should have given an extension of time to the date of practical completion and to decide upon a reasonable date for completion.[73] Therefore, for all intents and practical purposes, the decision will put the parties in the position as if the employer or his representative had issued an extension of time to that date, thus permitting liquidated damages to be claimed thereafter. The occasions when a tribunal is asked for a declaration that time is at large appear to be rare.

### 3.1.6 Time of the Essence

If a contract contains a terms stating that 'time is of the essence', it may be taken to mean that if the contractor (or supplier) fails to comply with its obligation to perform the contract by the specified date (or within the specified time), then this breach is sufficiently serious to 'go to the root of the contract', and the injured party may terminate the contract.[72]

'Time of the essence' clauses are often found in contracts for supply of goods or where performance of the contract can be achieved by a single action, such as a goods delivery or the making of a payment. However, they are rarely appropriate, or indeed effective, when used in building contracts for the following reasons:

- Most construction contracts already contain sophisticated and wide-ranging provisions dealing with time, extending time, and the consequences for the contractor if he fails to complete on time[73] and

69 *Multiplex Constructions (UK) Ltd v Honeywell Control Systems Ltd (No. 2)* (2007) EWHC 477 (TCC) at para 56(ii). Further considered in *Stellite Construction Ltd v Vascroft Contractors Ltd* (2016) EWHC 792 (TCC), a case concerning whether or not an adjudicator had breached the rules of natural justice in deciding that time was 'at large' – however, it was unnecessary to investigate the adjudicator's reasoning for this issue with any depth.

70 See the discussion on this point in Chapter 3, Section 3.12.

71 *Stellite Construction Ltd v Vascroft Contractors Ltd* (2016) EWHC 792 (TCC).

72 *Lombard North Central plc v Butterworth* (1987) 1 QB 527 CA.

73 *Lamprell v Billericay Union* (1849) 18 LJ Ex 282; *Babacomp Ltd v Rightside* [1974] 1 All ER 142.

- When considering the practical aspects of the transaction as a whole, if the contract does not require a 'time of the essence' clause, then the courts are reluctant to endorse such clauses, particularly if termination of the contractor's employment would, in most instances, provide no practical relief to the employer.[74]

Time of the essence clauses may still be found in building contracts, but not in standard forms. They are often found in bespoke building contracts, or in purchase orders (for the delivery of goods) that have been unwittingly or mistakenly used by a contractor's site team to procure minor items of sub-contract work.

## 3.2 Extension of Time Clauses in Building Contracts

### 3.2.1 Basic Principles

It may seem surprising, but the chief benefactor of an extension of time clause is the employer and not the contractor. In the event that the contract did not include an extension of time mechanism and the employer had caused a critical delay, then he would lose the ability to claim liquidated damages. To a lesser extent, the contractor benefits by receiving protection from a claim for general (or unliquidated) damages. If the parties intend that the contractor pays liquidated damages to the employer if he fails to complete the Works, a date for completion[75] must clearly be stated in the contract. That is because there must be a definite date from which to calculate liquidated damages.[76] If a building contract did not set out any express reasons why an extension of time could be given, there are implied terms in every contract that the employer will do all that is reasonably necessary to co-operate with and not to hinder the contractor in the performance of his obligations (refer to Section 6.2.3). Breaches of these implied terms create the grounds upon which a contractor may be entitled to an extension of time, and these, along with other grounds, are expressly captured in the JCT 2016 contracts as 'Relevant Events', and in NEC4 contracts as 'compensation events'. A summary of these 'grounds' for extensions of time is shown in Table 2.1 (in Section 2.1.1).

It is now accepted that in the vast majority of cases, the contractor will be entitled to an extension of time in the following circumstances:

- A delay event, be that a relevant event (JCT), a compensation event (NEC4) or an act of prevention by the employer (or someone acting on the employer's behalf) is likely to occur, or has occurred;
- That delay event is likely to delay completion of the works (or a section of the works) beyond the contractual completion date[77];

---

74 *Astea (UK) Ltd v Time Group Ltd* (2003) EWHC 725 (TCC). However, also see *Peak Construction (Liverpool) Ltd v McKinney Foundations Ltd* (1970)1 BLR 114 at 120 per Salmon LJ, where the employer was permitted to terminate the contractor's employment on the expiry of the period extended by the architect.
75 Or a period for completion commencing upon a certain commencement date.
76 *Miller v London County Council* (1934) 50 TLR 479.
77 The extension of time position reflected under the standard forms of building contract is that the delay (entitling the contractor to an extension) will or is likely to delay completion beyond the 'Completion Date' (JCT 2016 contracts) or *"planned Completion is later than planned Completion as shown on the Accepted*

- The contractor has complied with the relevant provisions in the particular building contract (such as giving notices and other required information); and
- The contractor can show that the chain of causation (i.e. between the delay event and completion of the works taking place later than the contractual completion date) remains unbroken.

### 3.2.2 The Prevention Principle

The 'prevention principle' can be described as having the following aspects:

- That a party will neither prevent the other from performing a contract nor delay him in performing the contract[78] and
- Neither party can take advantage of its own wrongdoing, such as hindering[79] the other party in the performance of a contract.[80]

It has been argued that if the employer committed any act of prevention, the contractor was entitled to an extension of time irrespective of whether it had complied with any notice requirements. That meant that if no extension was given, time would become at large and the contractor was relieved of its obligation to complete by the completion date in the contract and complete the Works within a reasonable time. In an Australian case,[81] the employer claimed liquidated damages from the contractor for delay. The cause of delay was substantially down to the employer, but the contractor had failed to operate the contract's strict requirements for giving notice. An arbitrator found in favour of the contractor and this was upheld by the Supreme Court of the Northern Territory of Australia:

> *'Acceptance of [the employer's] submissions would result in an entirely unmeritorious award of liquidated damages for delays of its own making…'*

However, where there is a clear requirement for the contractor to provide a notice of the delay, it is going too far to argue that the 'prevention principle' overrides any failure by the contractor to comply with a clear duty to serve such a notice. In the United Kingdom, it has since been held that if a contractor ignores a notice provision which is a condition precedent (see Sections 2.9.1 and 2.9.2), there will be no entitlement to extension of time, even though there would otherwise be a clear basis for such on the grounds of the employer's acts of prevention.[82] Although the court was not called upon to decide the point, it provided a succinct analysis of the position:

> *'Whatever may be the law of the Northern Territory of Australia, I have considerable doubt that the Gaymark Investments case represents the law of England. Contractual*

---

*Programme current at the dividing date"* (NEC4, clause 63.5). Ordinarily, the contractor is not entitled to an extension of time if he planned to complete before the 'completion date' but was delayed so he completed later than planned, but still before te 'completion date'. See Section 3.6.2 ('Time' effects of using up float).

78 *Barque Quilpé v Brown* (1904) 2 KB 264. See also *Holme v Guppy* (1838) 3 M & W 387 at 389 per Parke B – "*…and there are clear authorities, that if the party be prevented, by the refusal of the other contracting party, from completing the contract within the time limited, he is not liable in law for the default.*"

79 *London Borough of Merton v Stanley Hugh Leach Ltd* (1985) 32 BLR 51.

80 *Roberts v Bury Commissioners* (1870) LR 5 CP 310.

81 *Gaymark Investments Pty Ltd v Walter Construction Group Ltd* [1999] (2005) 21 Const LJ 71.

82 *Multiplex Constructions (UK) Ltd v Honeywell Control Systems Ltd* (2007) EWHC 477 (TCC).

> *terms requiring a contractor to give prompt notice of delay serve a valuable purpose; such notice enables matters to be investigated while they are still current. Furthermore, such notice sometimes gives the employer the opportunity to withdraw instructions when the financial consequences become apparent. If the Gaymark Investments case is good law, then a contractor could disregard with impunity any provision making proper notice a condition precedent. At his option the contractor could set time at large.'*[83]

If the employer's act of prevention in *Gaymark* has been simply to hinder or prevent the contractor for complying with the delay notice provisions, then the prevention principle (albeit, preventing a different action) would not have been over-stretched.

### 3.2.3 Onerous Clauses

A contract may contain onerous clauses, the intention of which is to hold the contractor to its obligation to complete by the completion date, even when faced with employer variations, strikes or adverse weather conditions.[84] It is now well-established that very clear words will be needed in order to bind a contractor to a completion date if the employer is the cause of the delay.[85]

### 3.2.4 Further Delays During a Period of Culpable Delay

It is suggested that if the contractor, through its own culpable delay, is thrown into a season when weather conditions are less beneficial than during the original contract period, the contractor will have to put up with it. However, if during the period of culpable delay, the conditions are exceptional for that time of year, the contractor will be entitled to an extension of time if exceptionally adverse weather conditions are a ground for extension of time under that contract. In some contracts, there is a term permitting the employer or his representative to extend time if any delays which are the responsibility of the employer (or of his representative) occur after the contractual date for completion, but before practical completion. It is thought that the inclusion of such an express term precludes the employer from extending time if a neutral relevant event occurs after the completion date during a period of culpable delay. Initially, this approach appears to have received judicial approval,[86] although older authority from the Court of Appeal, however, suggests the contrary.[87]

A question which often arises concerns the period around Christmas and New Year, when many contractors traditionally take two weeks holiday. If a proposed extension of time takes

83 *Multiplex Constructions (UK) Ltd v Honeywell Control Systems Ltd* (2007) EWHC 477 (TCC) at 105 per Jackson J.

84 *Wells v Army and Navy Co-operative Society Ltd* (1902) 86 LT 764.

85 *Percy Bilton Ltd v Greater London Council* (1982) 20 BLR 1. See also *Multiplex Constructions (UK) Ltd v Honeywell Control Systems Ltd (No 2)* [2007] EWHC 447 (TCC) at 56, and *North Midland Building Ltd v Cyden Homes Ltd* (2018) EWCA Civ 1744 at 39.

86 *Balfour Beatty v Chestermount Properties Ltd* (1993) 62 BLR 1, in which an amended JCT 80 was under consideration.

87 *Amalgamated Building Contractors Ltd v Waltham Holy Cross UDC* (1952) 2 All ER 452.

in the Christmas period or would end during the period, should the whole of the two weeks be added to the extension? There are two views of this:

- One view is that the whole of the two weeks should be added because the Christmas holiday period is well-established as a time when all the contractors are on holiday or
- The other, and better, view is that the contractor is only entitled to the public holidays (Christmas Day, Boxing Day and New Year's Day) just like any other public holidays, but not to any other days, because the decision to take two weeks as the Christmas holiday is simply a choice made by many contractors. It is not a statutory holiday and many people in the construction industry, including some contractors, do work during this time.

## 3.3 Critical Path Analysis and Planning Software

It is almost universally accepted, both in and outside of the United Kingdom, that if the nature of a project's activities are generally linear, that some form of critical path analysis should be employed in order to ascertain the effects of a delaying event. The methods of assessing delays will vary depending upon the necessary sequence of carrying out a project's scope of works. Two differing sequences are as follows:

- A 'linear' sequence, where each subsequent activity (or a significant number of them) must necessarily follow its preceding activity. Examples of projects with a 'linear' sequence are highways, an airport runway, a railway or straightforward construction projects (e.g. a warehouse unit).
- A 'non-linear' sequence, such as a project with a definite completion date, but where the scope of work comprises many activities (or sub-sets of activities) that are often repetitive in nature. These activities can largely be undertaken independently from one another, but may be constrained by the available resources (usually labour). Examples of a 'non-linear' project could involve repairing potholes on a highway, refurbishing rooms in a hotel, or repairs to local authority housing stock.

Critical path analysis and other forms of delay analysis are by their nature, complex tasks. Delay analysis also comes with its share of technical terms. Whilst many of these terms can seem like 'jargon', it is important to employ these terms in a consistent manner, as they can often be misunderstood.

The appropriate terminology used in critical path analysis is worth stating here:

- Programmes (in Gantt[88] chart format):
  - 'Baseline'. This is usually the contractor's initial programme or a programme against which progress and delays can be measured. Some planning software applications show the 'baseline' programme as narrower bars, underneath the bars for the 'updated' programme activities, so a simple, visual comparison between the two can be made (see Figure 3.3);

88 *"Gantt chart – A schedule of activities shown on a time-line, by a horizontal bar chart. The chart shows the projected start and finish, and can be shaded to show the actual start, finish and progression. Links from one activity to another, and associated resources may also be shown."* Oxford Dictionary of Construction, Surveying and Civil Engineering (2nd Ed) 2020.

  - 'Updated'. Initially the 'baseline' programme is updated to reflect current progress and new events (such as variations and progress) to become the first 'updated' programme, and each subsequent updated programme becomes the current updated programme;
  - 'Rescheduled'. After inputting the relevant information (such as variations, progress or delays) into the previous version of the programme (beginning with the 'baseline' programme), the planning software can then 'reschedule' the programme, so the Gantt chart shows any changes to the activities' sequence, timing and completion, that arise from the inputted information.
- Critical path:
  - The Society of Construction Law ('SCL'), Delay and Disruption Protocol[89]:

    > *'The longest sequence of activities through a project network[90] from start to finish, the sum of whose durations determines the overall project duration. There may be more than one critical path depending upon workflow logic. A delay to progress of any activity on the critical path will, without acceleration or re-sequencing, cause the overall duration to be extended, and is therefore referred to as a 'critical delay'.*

  - The Chartered Institute of Building ('CIOB') Guide to Good Practice in the Management of Time[91]:

    > *'The longest sequence of logically linked activities from the start of a schedule, to its finish that determine the earliest possible achievement of a key date, sectional completion date, or the completion date.'*

- Float[92] (see Table 3.4 and Section 3.6):
  - SCL: *'The time available for an activity in addition to its planned duration.'*
  - CIOB: *'The degree of flexibility available for the scheduling of an activity…The amount by which an activity can be delayed without causing delay to subsequent activities or the completion date.'*
- Mitigation measures
  - Acceleration (see Section 3.8)
  - Re-sequencing the programme
  - Recovery (achieving better progress than planned or anticipated).

The widespread use of sophisticated programming or planning software (now commonplace with construction and engineering projects) has made delay analysis much easier to depict and understand. Provided that the user generates a logical sequence of work (or 'network of activities') and links the relevant activities appropriately, most planning software

89 Second Edition (2017).
90 "Project network" and "Schedule" are other names for a Gantt chart programme.
91 Second Edition (2018).
92 This also includes "Free Float" – which the CIOB define as *"The period by which an activity may be delayed without delaying the start of any of its successor activities"* and "Total Float", the CIOB definition of which is *"The amount of time an activity can be delayed without delaying a key date, sectional completion date or the completion date, without contravening a constraint."* The CIOB define a "constraint" as *"A restriction on the ability of a critical path network schedule to obey the activity and / or resource logic of the schedule."* For more detail see Section 3.6 (Float).

can automatically show the critical path of activities, the float and any critical delay. This is referred to as critical path analysis (CPA) – covered in more detail in Section 3.5 (Methods of delay analysis).

The extent of difficulty in ascertaining the effect of a delaying event can significantly vary, depending upon the complexity of either the project (as a whole), the quality of availability records or the interrelationship of the affected activities. Examples at each end of this 'scale of difficulty' are as follows:

- A simple example would be if the employer fails to provide the contractor access to the site for 4 weeks (so that he's unable to carry out any physical work), then on the basis that (i) the work can be progressed from one activity to the next in a single series of steps, (ii) the work is continuous[93] and (iii) it is realistic to expect the project would otherwise have been completed on time, then it seems highly likely that the entire project will be delayed by 4 weeks.
- A more complex example could be where a variation (change) to the specification of some steelwork-mounted plant has been instructed. The structural loading for the steelwork may need to be changed, so this could require more time to re-design the structural steel members and their connections, along with procuring, fabricating and installing the new steelwork. Whilst it sounds like progress could be delayed, the following may not be immediately clear and require investigating to ascertain the following:
  - If a delay to the steelwork activity will similarly affect the progress of subsequent activities (e.g. diverting design resources to work on the steelwork change);
  - If the delay can be accommodated within the overall duration of the steelwork activities because the contractor has made sufficient time-risk allowances (float) within his programme for such delays;
  - If the delay can be naturally accommodated within the overall steelwork programme because it occurs at the same time as other steelwork activities (i.e. concurrently); or
  - The contractor is able to mitigate any delays.

Critical path analysis functions best when supported by plenty of factual information. If the contractor keeps records of progress (such as photographs, time-lapse video, allocation sheets for labour and plant, delivery tickets, marked-up programmes and drawings, substantiation for payment applications etc), then the programme's depiction of progress should be fairly accurate. If the majority of the relevant logic links between activities can be verified from the project's drawings and specification, then the CPA exercise becomes less subjective, and there should be less scope for disputes.

For non-linear projects, it is often necessary to analyse the resources used for each individual task (or sub-set of tasks if the mini-sequence is repetitive). A 'line of balance diagram'[94] is often used in tandem with an appropriate CPA, where the focus is on the quantity and distribution of resources.

93 The contractor has neither planned for any "breaks" nor any "float" when progressing the work.
94 CIOB – Guide to Good Practice in the Management of Time in Major Projects – 2nd Ed (2018) defines Line of balance as *"An illustration of sequence of repeated tasks across a number of work areas identifying resource, location and time used to optimise the use of resources across all trades working in defined areas"* p203.

The use of programming or planning software is commonplace – particularly with main contractors. A project's programme usually begins with an outline tender programme, which is developed further when suppliers and sub-contractors provide additional information, such as times for manufacture, dates for delivery, and the time required for installation. This may then become the programme that shows the contractor's initially intended sequence and timing of completing the works that reflects the start and finish dates in the contract. This programme may be referred to as the 'master programme', the 'baseline programme' or the simply the 'programme[95]', and then become the 'Accepted Programme'[96] or the 'revised programme'[97] under NEC4.

The use and understanding of programming software is seemingly far greater with contracting organisations that it is with the employer's representatives or other consultants. Unfortunately, consultants often misunderstand planning software and the information it conveys mainly because their use is less common in consulting organisations like architects and quantity surveyors. The use of planning software is more common with project management consultancies or engineers.

Software such as Elecosoft UK Ltd's Powerproject, Microsoft Project are commonly used to programme (or schedule) many construction and engineering projects. Oracle Corporation UK Ltd's Primavera is often the standard software for large engineering and infrastructure projects, although this is more of a database for project controls, which has a Gantt chart function (albeit, a very good one) for programming.

Modern programming (also referred to as scheduling) software greatly assists construction professionals by providing a single-platform tool to generate, monitor and analyse the sequence and timing of a project's activities.

## 3.4 The Society of Construction Law Delay and Disruption Protocol

In October 2002, the SCL published its 'delay and disruption protocol' ('the 2002 Protocol'), the aim of which was to:

> '*...provide useful guidance on some of the common issues that arise on construction contracts, where one party wishes to recover from the other an extension of time and/or compensation for the additional time spent and the resources used to complete the project. The purpose of the Protocol is to provide a means by which the parties can resolve these matters and avoid unnecessary disputes.*'[98]

The SCL neither intended that the Protocol was to be incorporated as a contract document nor was it intended to override the express terms of a building contract. It includes 21 'core

95 NEC4 – there is no definition of the "programme" (in clause 11.2) but detailed information is provided in clauses 31.1 and 31.2.

96 NEC4 "Accepted Programme" is defined in clause 11.2(1) and referred to in clause 31.3.

97 NeC4 – "revised programme" is referred to in clauses 32 (Revising the programme), 36 (Acceleration), 45.2 (Accepting defects), and Option X21 (Whole Life Cost).

98 SCL Delay and Disruption Protocol, October 2002, Introduction, para A.

principles'[99] along with guidance notes for each, covering both their time and financial aspects. The preceding decade had seen a number of delay cases come before the court, the effects of which had prompted some standard forms of building contract to be revised.

The 2002 Protocol gave rise to considerable debate, including further SCL papers that gave their appraisals, including from foreign jurisdictions.[100] One of its guidelines[101] stated that the preferred technique for dealing with retrospective delay analysis was 'time impact analysis' (see Table 3.3). This was interpreted by some commentators as a criticism of the other delay analysis methods.

In 2017, the SCL published a revised version of the Protocol. As stated in Section 3.3, the perceived use of 'jargon' can delay analysis difficult to understand. Where appropriate (and in particular in Section 3.5), the terminology used in the Protocol has been adopted by this book.

## 3.5 Methods of Delay Analysis

There are essentially four CPA methods, which are summarised in Table 3.3[102]:

Each of the four methods in Table 3.3 relies upon factual information being employed in each analysis. However, the degree of success for each method will depend upon the amount and accuracy of factual material available for the particular analysis. In general, the methods shown in Table 3.3 become progressively more dependent upon factual information, with 'impacted as-planned' being the least reliant and 'collapsed as-built' being the method that is entirely reliant on factual material to show the effects of delays. The intermediate methods of 'time impact analysis' and 'as-planned v as-built' are heavily reliant upon factual information, but they also utilise 'planned' or 'forecast' data – particularly if used to assess progress and delays during the construction stage of the project, instead of retrospectively.

### 3.5.1 Initial Steps: The Programme

Before embarking upon any of the 'impacted as-planned', 'time impact analysis' or 'as-built v as-planned' methods of delay analysis, there must be a programme. In most cases, the

99 The 2002 Protocol's "core principles" are: (1) Programme and records, (2) Purpose of extension of time, (3) Entitlement to extension of time, (4) Procedure for granting extension of time, (5) Effect of delay, (6) Incremental review of extension of time, (7) Float as it relates to time, (8) Float as it relates to compensation, (9) Concurrent delay – its effect on entitlement to extension of time, (10) Concurrent delay – its effect on entitlement to compensation for prolongation, (11) Identification of float and concurrency, (12) After the event delay analysis, (13) Mitigation of delay and mitigation of loss, (14) Link between extension of time and compensation, (15) Valuation of variations, (16) Basis of calculation of compensation for prolongation, (17) Relevant tender allowances, (18) Period for evaluation of compensation, (19) Global claims, (20) Acceleration, and (21) Disruption.

100 SCL papers: Pickavance, K (Sept 2004) D52 Putting the SCL Delay and Disruption Protocol into practice – is this what the industry wants, Lowe, Evans, Barba and Lare (May 2007) D78 – A view from across the pond: An American perspective on the SCL Delay and Disruption Protocol, Ganer (Aug 2007) – D82 A South African view on Float and other aspects of the Delay and Disruption Protocol.

101 SCL Delay and Disruption Protocol (2002), Guidance Section 4: para 4.8.

102 Also see the SCL Delay and Disruption Protocol – 2nd Ed (2017) at paragraph 11.5.

**Table 3.3** Critical Path Analysis (CPA) methods.

| CPA Method | Description of method | Required resources |
|---|---|---|
| Impacted As-Planned | • The relevant delay is inserted into the original (or baseline) programme to calculate the revised completion date<br>• Critical path and delay impact may be determined prospectively | • Original contract (or baseline) programme<br>• Record of the delay(s) duration |
| Time Impact Analysis[a)] | • The 'baseline' programme is updated periodically[b)] to include actual progress and the insertion of delay events (either actual or forecast durations). The programme is then updated (or re-scheduled) after each period (or 'window') to calculate the updated completion<br>• The critical path is determined contemporaneously, and the impact of delay can be determined prospectively or retrospectively | • Logic-linked baseline programme<br>• Updated progress information (including delay events and accurate records of progress) |
| As-Planned v As-Built | • A comparison is made between the original or baseline programme (the 'As-Planned programme), and a programme showing the actual sequence and timing of all work activities (the 'As-Built' programme'). The differences between the programmes show the extent of delay and (on further analysis) indicates who is responsible for individual delays<br>• The critical path can be determined contemporaneously, but the delay impact is usually determined retrospectively | • Original or 'baseline' programme<br>• Updated progress information (including delay events and accurate records of progress) |
| Collapsed As-Built[c)] | • An accurate 'as-built' programme with detailed logic links is compiled, from which delaying events are removed, so the programme can be 'collapsed' to show the net effect of the delay events, and any contractor delays relative to the contract's completion date(s)<br>• The critical path and impact of delay are determined retrospectively | • Logic-linked as-built programme<br>• Updated progress information (including an accurate record of progress)<br>• Does not require a baseline programme |

a) May also be referred to as "Time slice analysis", as there are notable similarities between the two methods.
b) For example, in "windows" of 1 month.
c) Also may be referred to as "Retrospective Longest Path Analysis.

programme will be produced by the contractor, but this may not always be the case. If the programme is generated by the employer (or his representative), the contractor must satisfy himself that it is realistic, achievable and shows the works being carried out in a sequence that he is happy with.

The differences between JCT's reliance on the contractor's programme and those for NEC4 are stark. Clause 2.9 of JCT SBC/Q 2016 requires the contractor to provide a 'master programme', which has the following characteristics:

- To identify the critical paths (but only if stated the entry against clause 2.9.1.2 in the contract particulars);
- Clause 2.9.2 provides that the effects of any extension of time (fixing of a later completion date under clause 2.28.1 or agreement of a pre-agreed adjustment[103] under clause 2.26.2) must be shown; but
- Clause 2.9.3 provides that the 'master programme' does not override the obligations imposed by the contract documents.[104]

Unlike NEC4 contracts, the 'master programme' in JCT SBC/Q 2016 only records the effects of an extension of time; it is not used to determine extensions of time. This remains something that the employer's representative (Architect / Contract Administrator) must do in accordance with clause 2.28.1. Other JCT 2016 contracts contain few references to a programme.[105]

NEC4 contains extensive requirements for what the contractor's programme must show. Under NEC4, the programme becomes central to calculating extensions of time (i.e. delays to the completion date or to a key date[106]).

Table 3.4 shows what is and what is not required under NEC4 for the contractor's programme:

Whilst JCT SBC/Q 2016's conditions hardly contain any of the requirements for the contractor's programme that are found in NEC4, it is rare for the contractor to escape having to provide any programme information at all. The contract bills,[107] employer's requirements[108] and specification or work schedules[109] will often oblige the contractor to provide a programme showing much of the equivalent information as is required under NEC4 (as summarised in Table 3.4).

---

103 JCT SBC/Q 2016 – clause 2.26.2: agreement of a "Pre-Agreed Adjustment" means agreeing the fixing of a later date as the Completion Date (for the whole Works or a Section thereof) arising from a Variation Quotation (see clause 5.3 and para 1 of Schedule 2), of an Acceleration Quotation (see para 2 of Schedule 2).

104 JCT SBC/Q 2016 – clause 1.1. defines the "Contract Documents" as consisting of the Contract Drawings, the Contract Bills, the Agreement (comprising the Recitals, Articles and Contract Particulars), and if there is a Contractor's Designed Portion (see the Ninth Recital) the Employer's Requirements, the Contractor's Proposals, and the CDP Analysis, along with any BIM Protocol.

105 JCT contracts only contain reference to the contractor's programme for (Cost savings and value improvements): JCT DB 2016, para 7 of Part 1 of Schedule 2 (Supplemental Provisions), and JCT IC 2016 para 3 of Schedule 5 (Supplemental Provisions).

106 NEC4 – clause 63.5 (Assessing compensation events).

107 JCT SBC/Q 2016 (or Bills of Quantities in JCT IC 2016). For example, see NRM2 at section 1.7.2.1 (Programme/progress) which sets out the programme's format and content, any exclusions, any document control requirements (e.g. provision of design information by the employer) and any programme submission requirements.

108 JCT DB 2016 (or in JCT SBC/Q in the CDP Documents).

109 JCT IC 2016 and JCT MW 2016.

**Table 3.4** Requirements of the Contractor's programme (NEC4).

| The Contractor's programme must show: | NEC4 (clause 31.2) |
|---|---|
| The particular planning software to be used | No[a)] |
| The specific format of the programme | Yes |
| The programme must show the following information: | |
| • Logic links and the Critical Path | No[b)] |
| • Earliest and latest start and finish dates for activities | No[c)] |
| • Commencement Date | Yes |
| • Date for Completion | Yes |
| • Planned Completion[c] | Yes |
| • Key dates (including the contractor's planned dates for meeting the condition for a key date) | Yes |
| • Dates for possession of or access to the Site (or each part of the Site) | Yes |
| • The sequence and timing of the works, including: | Yes[d)] |
| – Each design stage | Yes |
| – Information to be provided by others | Yes |
| – Preparing and submitting the contractor's design documents | Yes |
| – Procurement | Yes |
| – Manufacture and inspections | Yes |
| – Delivery | Yes |
| – Construction | Yes |
| – Testing further inspections and commissioning | Yes |
| – Work carried out by or on behalf of the employer | Yes |
| – Time contingencies[e)] (including float) | Yes |
| – Health and safety obligations | Yes |
| • Periods for reviewing documents or submissions | Yes |
| • Any other requirements set out in the Employer's Requirements or Scope | Yes |

a) By way of comparison, these are requirements which may be incorporated in the Scope (NEC4). Other contracts (such as the FIDIC suite of contracts - Red book and Yellow Book 2017 editions) prescribe the planning software to be used.

b) Contrast with JCT SBC/Q 2016, clause 2.9.1.2 that may require the "master programme" to show the critical paths.

c) NEC4. Whilst here is no definition of "planned Completion", it means the date when the contractor is aiming to complete the work, and this may occur before the Completion Date, it be the same as the Completion Date, or if the progress of work is delayed by events which are not Compensation Events, it may occur after the Completion Date.

d) NEC4 – Clause 31.2 provides that the contractor must show all "operations" that comprise carrying out the Works. This clause is wide enough to cover all the specifically mentioned stages or operations set out in clause 8.3 of FIDIC Yellow Book 2017.

e) NEC4 – clause 31.2 requires the contractor to include time contingencies in his programme for 'float' and 'time risk allowances'. An example of a 'time risk allowance' could be the inclusion of adverse weather conditions (a 'weather measurement') that whilst are seemingly extreme, have occurred more than once in 10 years. – see clause 60.1(13) and section 13.5.13. If a 'weather measurement' is recorded for weather that either occurred only once, or has never occurred within 10 years, then the contractor is entitled to a compensation event under clause 60.1(13).

NEC4 (and to a lesser extent, JCT SBC/Q 2016[110]) tries to ensure that the contractor keeps the programme up-to-date and accurate. It aims to achieve this by the contractor revising the programme and submitting it for acceptance by the Project Manager. NEC4's Managing Reality[111] gives a pragmatic explanation of why a contractor may want to revise his programme. It considers that the contractor's initial programme (or 'first programme' in NEC4[112]) may only be an extension of the contractor's preliminary or tender programme. Once the contractor commences work (and particularly when the design is developed further and becomes more co-ordinated), he may find that the original sequence and timing of work needs to be amended to ensure the works can be constructed logically and efficiently. NEC4 allows the programme to 'evolve' so it is an effective monitoring tool during each stage of the works' construction.

Many less experienced consultants, particularly those unfamiliar with the concepts of the NEC contract, find dealing with a changing (or evolving) programme to be difficult, as they often treat an initial (or even the tender) programme as the contractor's binding commitment to undertake the works in precisely the sequence and timing shown thereon. This is rarely the case, as in general, the contractor is not obliged to work strictly in accordance with a programme, but must complete the works by a particular date (see Section 3.1.4) or dates if the works are sub-divided into sections. Consultants should remember that ordinarily, the contractor's programme evolves, and he is neither obliged to work in accordance with his current programme and certainly not in accordance with the programme he initially provided.

NEC4 provides that in addition to the contractor submitting a revised programme within the 'period for reply'[113] (following the Project Manager's request), he must also provide a revised programme in any event, at the interval specified in the Contract Data. Table 3.5 summarises the procedures for revising the programme under NEC4:

The project manager cannot arbitrarily object to or refuse to accept the contractor's revised programme, although the available grounds for rejection under NEC4 may, at first, seem fairly wide. Essentially, the contractor must accurately show the project's current progress, along with maintaining all of the other requirements that are either provided in the contract, or in the Scope. The reasons for the project manager not accepting the contractor's revised programme are summarised in Table 3.6.

NEC4 contains default provisions if the project manager fails to properly respond to the contractor's revised programme within a set period of time – the contractor's revised programme effectively becomes accepted in the absence of the project manager's timely dissention.

A brief description of how a contractor's programme may evolve over time is given in Section 3.5.6, and a simple example of a programme for constructing a warehouse (that should satisfy the requirements of JCT SBC/Q 2016's 'master programme' and those of NEC4) is shown in Figure 3.1.

The key for activities in Figure 3.1 is as follows.

---

110 JCT SBC/Q 2016, clause 2.9.2.

111 NEC4 Managing Reality – Book Three (Managing the Contract) (3rd Ed) 2017, at 2.6.2.

112 NEC4, clause 31.1.

113 NEC4, clause 13.3 (Communications).

**Table 3.5** Requirements for revising the Contractor's programme (NEC4).

| **Requirement of revising the contractor's programme** | **NEC4(clause 32)** |
|---|---|
| Revised programmes should be submitted for acceptance:<br>• Period after the engineer/project manager notifies or instructs the contractor to do so:<br>• Revised programmes submitted in any event:<br>– Period for the contractor to submit: | Yes – clause 32.2[a)]<br>Yes – clause 32.2[b)]<br>Prescribed interval |
| Revised programmes should show or include:<br>• Remedial works (correcting defects)<br>• Actual progress including delays<br>• Effect of actual progress on remaining activities<br>• Contractor's proposed changes to the programme | Yes – clause 32.1<br>Yes – clause 32.1<br>Yes – clause 32.1, 62.2<br>Yes – clause 32.1 |
| The programme is used to calculate extensions of time (to either the date for completion, including sections, or key dates) | Yes – clause 63.5 |

a) NEC4. The "period for reply" is set out in Contract Data Part One (Data Provided by the Client).
b) NEC4. The prescribed interval are set out in Contract Data Part One (Data Provided by the Client) and are expressed as *"intervals no longer than"*, which permits the contractor to provide revised programmes before the expiry of the relevant interval.

**Table 3.6** Reasons for rejecting the Contractor's programme – and procedure (NEC4).

| **Project Manager's review and acceptance of the contractor's revised programme** | **NEC4 (clause 31.3)** |
|---|---|
| Period for responding to the contractor's revised programme | 2 weeks |
| Reasons for requesting a revised programme:<br>• It fails to comply with the contract<br>• Inaccurately shows actual progress<br>• Is not consistent with the contractor's obligations<br>• Is not a reasonable depiction of the contractor's plans | <br>Yes<br>Yes<br>Yes<br>Yes |
| Default provisions for Non-Objection to or acceptance of the revised programme | Yes[a] |

a) NEC4, clause 31.3. Should the Project Manager fail to respond to the contractor's revised programme within 14 days, the contractor notifies him accordingly. If the Project Manager fails to respond within 1 week after the contractor's notice, then the contractor's revised programme is deemed to be "accepted".

Figure 3.1 assumes that the employer is responsible for the project's design, so no periods for design stages, for others to provide information, submitting the contractor's design documents, reviewing submissions or acceptances[114] are shown. However, the programme includes the following elements required by NEC4:

- The 'starting date' and 'access date' – activity 1
- Planned completion – activity 26
- The Completion Date – activity 28

114 NEC4, clause 31.2 required to be shown on the contractor's programme.

| Line | Name | Start | Finish | Duration |
|---|---|---|---|---|
| 1 | Date of Possession / starting date | 06 Jun 22 | 06 Jun 22 | |
| 2 | Site set up | 06 Jun 22 | 10 Jun 22 | 1w |
| 3 | Strip topsoil / remove trees | 13 Jun 22 | 17 Jun 22 | 1w |
| 4 | Excavate to reduce levels | 17 Jun 22 | 30 Jun 22 | 2w |
| 5 | Excavate foundations | 23 Jun 22 | 01 Jul 22 | 1w 2d |
| 6 | Concrete foundations | 01 Jul 22 | 07 Jul 22 | 1w |
| 7 | Bolt boxes | 08 Jul 22 | 08 Jul 22 | 1d |
| 8 | Steelwork procurement and manufacture | 06 Jun 22 | 24 Jun 22 | 3w |
| 9 | Steel frame erection | 11 Jul 22 | 22 Jul 22 | 2w |
| 10 | Roof safety netting | 25 Jul 22 | 27 Jul 22 | 3d |
| 11 | Profiled roof sheet and skylights | 28 Jul 22 | 10 Aug 22 | 2w |
| 12 | Wall cladding | 04 Aug 22 | 17 Aug 22 | 2w |
| 13 | Pre-cast concrete mezzanine floor | 25 Jul 22 | 29 Jul 22 | 1w |
| 14 | Pre-cast concrete stairs | 01 Aug 22 | 03 Aug 22 | 3d |
| 15 | Ground floor slab | 04 Aug 22 | 12 Aug 22 | 1w 2d |
| 16 | Internal blockwork walls | 15 Aug 22 | 26 Aug 22 | 2w |
| 17 | External windows and doors | 18 Aug 22 | 24 Aug 22 | 1w |
| 18 | 1st fix M&E | 30 Aug 22 | 05 Sep 22 | 1w |
| 19 | Plastering to offices | 06 Sep 22 | 12 Sep 22 | 1w |
| 20 | Ceilings | 13 Sep 22 | 15 Sep 22 | 3d |
| 21 | 2nd fix M&E | 16 Sep 22 | 22 Sep 22 | 1w |
| 22 | Decorations | 23 Sep 22 | 29 Sep 22 | 1w |
| 23 | Form external roads and car park | 18 Aug 22 | 08 Sep 22 | 3w |
| 24 | Surfacing and lining | 09 Sep 22 | 15 Sep 22 | 1w |
| 25 | Approval under s278 Highways Act 1980 | 30 Sep 22 | 30 Sep 22 | |
| 26 | Planned completion (on site) | 30 Sep 22 | 30 Sep 22 | |
| 27 | Testing, commissioning & snagging | 30 Sep 22 | 06 Oct 22 | 1w |
| 28 | Date for completion | 07 Oct 22 | 07 Oct 22 | |

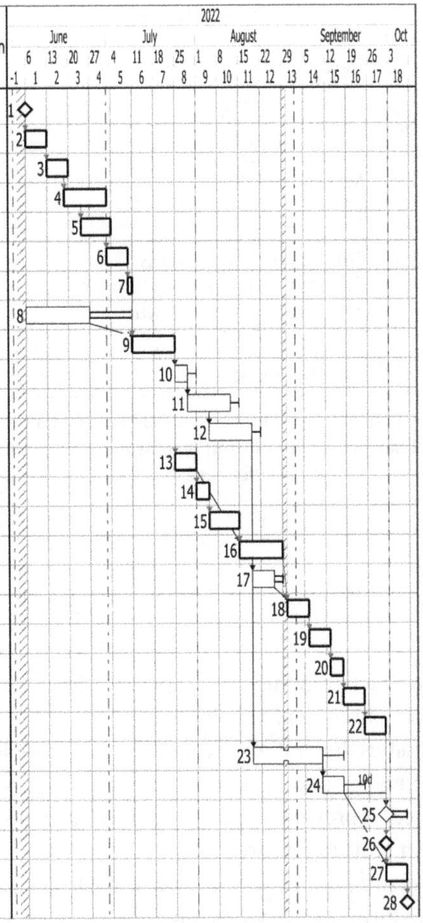


**Figure 3.1** Warehouse programme example (compliant with JCT SBC/Q 2016 and NEC4).

- The critical path[115] (activity no's 1 to 7, 13 to 16, 18 to 22 and 26 to 28)
- Key Date (including the date when the contractor plans to meet it) – activity 25
- Float (or time risk allowances):
  - 'Free float' – activity 8 (10 days) and activity 25 (5 days);
  - 'Total float' – activities 10, 11 and 12 (2 days each), activities 23 and 24 (5 days each)
- Health and safety requirements – activity 10.

For all four methods of delay analysis discussed in sections 3.5.2 to 3.5.6, it is essential to establish the critical path. For simplicity, the critical path shown in this chapter's examples was generated by the planning software used to produce the programmes. However, the accuracy of a critical path's should not be assumed to be correct, just because the software

115 The critical path is not expressly required by NEC4 clause 31.2, although the programme must show the *"order and timing of the operations…in order to provide the Works."* The critical path may have to be shown on the "master programme" under JCT SBC/Q 2016 if stipulated by the Contract Particulars' entry for clause 2.9.1.2.

**Table 3.7** Key for programme activities.

| Symbol | Description |
|---|---|
| ▭ | Planned activity |
| **▭** | Critical planned activity |
| ◇ | Milestone activity |
| ⟶ | Logic link (e.g. finish to start) from one activity to another |
| ——\| | Total float for an activity: The amount of time an activity can be delayed without delaying a key date, sectional completion date or the completion date, without contravening a constraint[a)] |
| ══\| | Free float for an activity: The period by which an activity may be delayed without delaying the start of any of its successor activities[a)] |

a) CIOB – Guide to Good Practice in the Management of Time in Major Projects – 2nd Ed (2018) defines Line of balance as *"An illustration of sequence of repeated tasks across a number of work areas identifying resource, location and time used to optimise the use of resources across all trades working in defined areas"* p203.

says so. The following should be carefully considered when establishing or verifying a critical path:

- The contract drawings and specification – in order to verify a logical and realistic sequence of physical activities;
- Site logistics. Even if a logical sequence of construction can be established from the contract drawings and specification, it may not be possible to carry out the works in that sequence, if prevented by logistical constraints (e.g. limitations for site access, working space, or working hours); and
- The procurement of subcontractor packages and materials, goods or equipment that is likely to form the activities found on (and adjacent to) the critical path.

### 3.5.2 Impacted as-Planned

The 'impacted as-planned' method of delay analysis requires the following information:

- The contractor's programme and
- The duration for the delaying event, and when it occurred (or is likely to occur).

Of the four methods of delay analysis in Table 3.3, the impacted as-planned method is perhaps the least reliant on facts for the following reasons:

- The 'contractor's programme':
  - The programme used for an impacted as-planned analysis may not have been updated and may even remain as per the contractor's 'tender programme' if no changes have been made after entering into the building contract;
  - Little or no adjustments or updates may have been made to the programme (e.g. in the form of applying actual progress, accounting for variations, adding extra activities or removing unnecessary or omitted works).

**Table 3.8** Employer delay activity.

| Symbol | Description |
|---|---|
|  | Employer delay activity |

Therefore, the ‘programme’ used in the ‘impacted as-planned’ analysis may show little more than the contractor’s initially intended sequence and timing of the works, before he’s even commenced work. As a factual document, the programme may be completely outdated and of little value.

- The duration for the delaying event:
  - This may be a forecast of how long a delay may last or
  - It may be an actual and verifiable duration.

The disadvantages of the ‘impacted as-planned’ method should be immediately clear. As a method of analysis, it may be wholly inconsistent. A verifiable, factually correct period of delay may be used, but if this is applied to an outdated programme; one perhaps showing a completely incorrect sequence of work, then the analysis is likely to be flawed. The absence of any updated information for the activities that are not affected by any delays, may render the contractor’s programme useless for any kind of analysis. If the employer can show that the analysis is based upon a programme that may be flawed from the outset, then the impacted as-planned analysis is likely to be worthless.

However, the ‘impacted as-planned’ method should not be ignored, as it may be appropriate to use it in circumstances where little factual information may be available. Consider the situation before the works have started on site, the employer informs the contractor that he won’t be able to grant possession of the site until 3 weeks after the date when the contractor planned to begin work on site. Figure 3.2 shows the effect of such a delay on the warehouse example programme (Figure 3.1) using the ‘impacted as-planned’ method:

The key for the Figure 3.2 programme remains as for Figure 3.1, only with the addition of an activity to show an employer delay.

When the ‘Date for Completion’ (now activity 29) is compared with its counterpart in Figure 3.1 (activity 28), it has been postponed in Figure 3.2 by 3 weeks, from 7 to 28 October 2022. This 3-week postponement is solely caused by the expected employer delay of giving delayed possession (activity 1 in Figure 3.2). The only notable effect is that float period for steelwork procurement (activity 9) has increased from 10 to 25 days (or from 2 to 3 weeks).

In this example, and order to comply with notification requirements of both JCT SBC/Q 2016 or NEC4, the use of an ‘impacted as-planned’ programme would be acceptable in the following circumstances:

- JCT SBC/Q 2016:
  - Notify the architect/contract administrator under clause 2.27.1 that progress of the works is ‘likely’ to be delayed, owing to the employer’s failure to give possession for 3 weeks (a Relevant Event under clause 2.29.3 and/or 2.29.7);
  - Include the ‘impacted as-planned’ programme as the ‘particulars’ and ‘estimate’ of delay under clause 2.27.2; and
  - Issue the ‘impacted as-planned’ programme as a revision of the ‘master programme’ in accordance with clause 2.9.2.

| Line | Name | Start | Finish | Duration |
|---|---|---|---|---|
| 1 | Delay in providing site access | 06 Jun 22 | 24 Jun 22 | 3w |
| 2 | Date of Possession / starting date | 27 Jun 22 | 27 Jun 22 | |
| 3 | Site set up | 27 Jun 22 | 01 Jul 22 | 1w |
| 4 | Strip topsoil / remove trees | 04 Jul 22 | 08 Jul 22 | 1w |
| 5 | Excavate to reduce levels | 08 Jul 22 | 21 Jul 22 | 2w |
| 6 | Excavate foundations | 14 Jul 22 | 22 Jul 22 | 1w 2d |
| 7 | Concrete foundations | 22 Jul 22 | 28 Jul 22 | 1w |
| 8 | Bolt boxes | 29 Jul 22 | 29 Jul 22 | 1d |
| 9 | Steelwork procurement and manufacture | 06 Jun 22 | 24 Jun 22 | 3w |
| 10 | Steel frame erection | 01 Aug 22 | 12 Aug 22 | 2w |
| 11 | Roof safety netting | 15 Aug 22 | 17 Aug 22 | 3d |
| 12 | Profiled roof sheet and skylights | 18 Aug 22 | 01 Sep 22 | 2w |
| 13 | Wall cladding | 25 Aug 22 | 08 Sep 22 | 2w |
| 14 | Pre-cast concrete mezzanine floor | 15 Aug 22 | 19 Aug 22 | 1w |
| 15 | Pre-cast concrete stairs | 22 Aug 22 | 24 Aug 22 | 3d |
| 16 | Ground floor slab | 25 Aug 22 | 05 Sep 22 | 1w 2d |
| 17 | Internal blockwork walls | 06 Sep 22 | 19 Sep 22 | 2w |
| 18 | External windows and doors | 09 Sep 22 | 15 Sep 22 | 1w |
| 19 | 1st fix M&E | 20 Sep 22 | 26 Sep 22 | 1w |
| 20 | Plastering to offices | 27 Sep 22 | 03 Oct 22 | 1w |
| 21 | Ceilings | 04 Oct 22 | 06 Oct 22 | 3d |
| 22 | 2nd fix M&E | 07 Oct 22 | 13 Oct 22 | 1w |
| 23 | Decorations | 14 Oct 22 | 20 Oct 22 | 1w |
| 24 | Form external roads and car park | 09 Sep 22 | 29 Sep 22 | 3w |
| 25 | Surfacing and lining | 30 Sep 22 | 06 Oct 22 | 1w |
| 26 | Approval under s278 Highways Act 1980 | 21 Oct 22 | 21 Oct 22 | |
| 27 | Planned completion (on site) | 21 Oct 22 | 21 Oct 22 | |
| 28 | Testing, commissioning & snagging | 21 Oct 22 | 27 Oct 22 | 1w |
| 29 | Date for completion | 28 Oct 22 | 28 Oct 22 | |

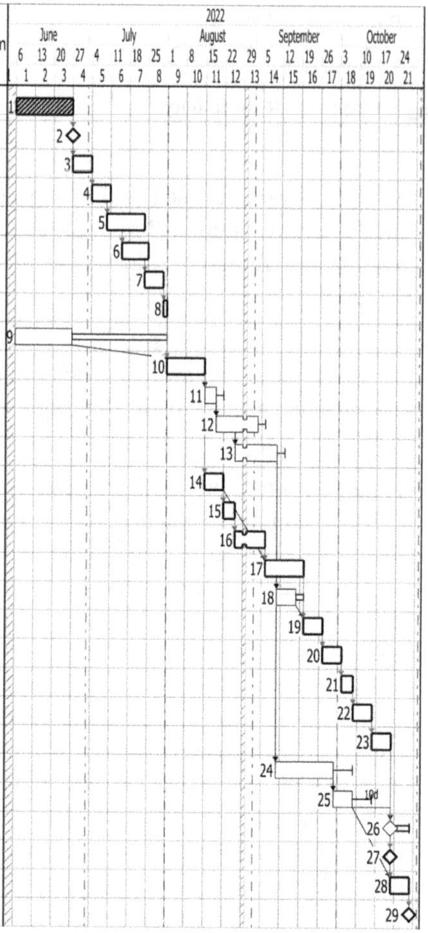

**Figure 3.2** Warehouse impacted as-planned example.

| Line | Name | Start | Finish | Duration |
|---|---|---|---|---|
| 1 | Date of Possession / starting date | 06 Jun 22 | 06 Jun 22 | |
| 2 | Site set up | 06 Jun 22 | 10 Jun 22 | 1w |
| 3 | Strip topsoil / remove trees | 13 Jun 22 | 17 Jun 22 | 1w |
| 4 | Excavate to reduce levels | 17 Jun 22 | 30 Jun 22 | 2w |
| 5 | Excavate foundations | 23 Jun 22 | 01 Jul 22 | 1w 2d |
| 6 | Concrete foundations | 01 Jul 22 | 07 Jul 22 | 1w |

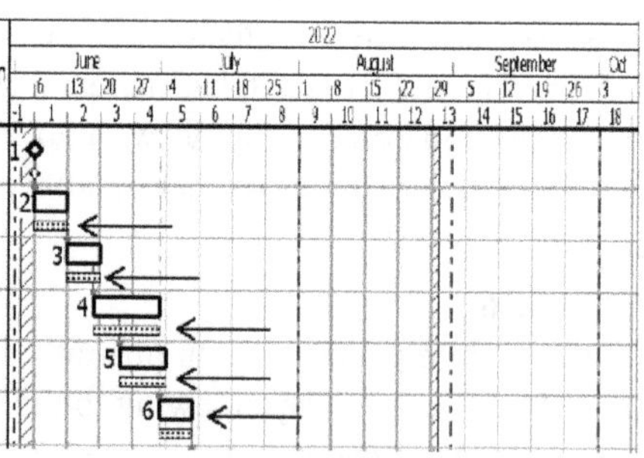

**Figure 3.3** Warehouse example – shown with baseline activities (no progress).

- NEC4:
  - Notify the Project Manager under clause 61.3 because the employer's delay in giving access (a compensation event under clause 60.1(2)) is *'expected to happen'*;
  - If the project manager instructs the contractor to submit a quotation, the contractor can provide the 'impacted as-built' programme – showing the *'proposed changes to the…completion date'* under clause 62.2; and
  - May submit the 'impacted as-planned' programme as the 'revised programme' to the project manager under clause 32.

If (in accordance with recent guidance[116]) the employer's representative (under either JCT or NEC4) when assessing the likely delay does not adopt a 'wait and see approach', then if the contractor's 'impacted as-planned' programme provides the best indication of how the works' progress will be affected, then in those circumstances, the contractor has probably done enough to demonstrate entitlement to an extension of time, albeit in a prospective manner.

However, as soon as the contractor obtains factual information, he should use this when providing the following:

- JCT SBC/Q 2016 – providing the architect/contract administrator with updated particulars and the estimate of delay under clause 2.27.3; or
- NEC4 – revising the programme under clause 32.2.

### 3.5.3 Time Impact Analysis

Time impact analysis is a natural extension of the impacted as-planned method; as very importantly, it introduces actual data by inputting the activities' progress, initially into the 'baseline' programme and subsequently into 'revised' or rescheduled programmes (i.e. those that have been updated to show either the position of the project's current progress or the position that existed when the programme was last revised).

Time impact analysis is an enormously useful method of delay analysis for projects let under NEC4 contracts. Its use of current progress data allows it to not only show an accurate representation of the project's progress[117] but also to forecast the effects of delays caused by compensation events which are either likely to occur or those for which the delaying effects may be continuing.

Some planning software (such as Elecosoft's Powerproject) displays the 'baseline' programme as smaller activity bars underneath what will become the 'actual' activity bars, once progress is applied to them.

Figure 3.3 shows an excerpt of the 'warehouse' programme example (in Figure 3.1) which shows the 'baseline'.

The 'baseline' activities 2, 3, 4 and 5 are shown in Figure 3.3, highlighted with arrows, and the key for baseline activities (for time impact analysis examples) is shown in Table 3.9:

Once the 'baseline' has been set, the programme is ready for progress to be added to the current activities. This is usually added during what are called 'progress periods', which are

116 SCL Delay and Disruption Protocol – (2nd Ed) 2017, Core Principle 4, p5.
117 As required by NEC4 clause 32.1 (first bullet point).

**Table 3.9** Key for 'baseline' activities (time impact analysis).

| Symbol | Description |
|---|---|
| | Baseline activity |

| Line | Name | Start | Finish | Duration |
|---|---|---|---|---|
| 1 | Date of Possession / starting date | 06 Jun 22 | 06 Jun 22 | |
| 2 | Site set up | 06 Jun 22 | 10 Jun 22 | 1w |
| 3 | Strip topsoil / remove trees | 13 Jun 22 | 17 Jun 22 | 1w |
| 4 | Excavate to reduce levels | 17 Jun 22 | 30 Jun 22 | 2w |
| 5 | Excavate foundations | 23 Jun 22 | 01 Jul 22 | 1w 2d |
| 6 | Concrete foundations | 01 Jul 22 | 07 Jul 22 | 1w |
| 7 | Bolt boxes | 08 Jul 22 | 08 Jul 22 | 1d |
| 8 | Steelwork procurement and manufacture | 06 Jun 22 | 24 Jun 22 | 3w |
| 9 | Steel frame erection | 11 Jul 22 | 22 Jul 22 | 2w |

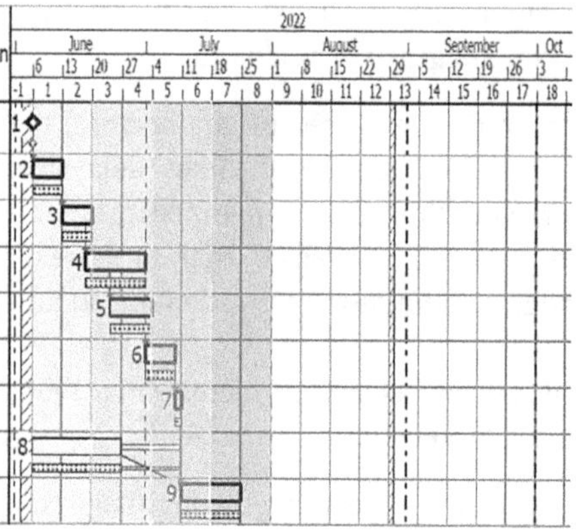


**Figure 3.4** Progress periods shown (against Figure 3.3).

**Table 3.10** Progress periods (for Figure 3.4).

| Key | Description of 'Progress Period' |
|---|---|
| | Period 1: 6 to 20 June, 2022<br>Period 2: 21 June to 4 July, 2022<br>Period 3: 5 to 18 July, 2022<br>Period 4: 19 July to 1 August, 2022 |

often at monthly or more frequent intervals such as weekly or fortnightly. Figure 3.4 shows an excerpt of the 'warehouse' programme example (from Figure 3.3), marked up to indicate the following fortnightly (every 2 weeks) progress periods (Table 3.10):

It should be noted that no progress has been added to the programme in Figure 3.4 so the programme's activities are each in line with their 'baseline'.

Continuing with the example from Figure 3.2, where the employer did not give access to the site until 27 June 2022 (i.e. a 3-week delay from 6 June 2022), the progress for each activity is then added for each progress period (see Table 3.11):

Applying the progress for Period 1 (6 to 20 June 2022) shows that no progress on site was possible for the first few weeks because of the employer's delay in granting possession or access. The only activity that proceeded during the first 3 weeks was the off-site 'Activity 8 – Steelwork procurement and manufacture', which was 100% complete after the first two weeks (i.e. 1 week ahead of its baseline). This is shown in Figure 3.5, with the progress to activity 8 shown as follows (also highlighted win an arrow):

**Table 3.11** Cumulative progress applied over Progress Periods 1, 2, 3 and 4.

| | **Cumulative % progress achieved for each activity in each progress period:** | | | |
|---|---|---|---|---|
| **Activity (number and description)** | **1** | **2** | **3** | **4** |
| 1 – Date of possession/starting date | 0% | 100% | 100% | 100% |
| 2 – Site set-up | 0% | 100% | 100% | 100% |
| 3 – Strip topsoil/remove trees | 0% | 100% | 100% | 100% |
| 4 – Excavate to reduce levels | 0% | 25% | 75% | 100% |
| 5 – Excavate foundations | 0% | 0% | 25% | 100% |
| 6 – Concrete foundations | 0% | 0% | 0% | 100% |
| 7 – Bolt boxes | 0% | 0% | 0% | 100% |
| 8 – Steelwork procurement and manufacture | 100% | 100% | 100% | 100% |
| 9 – Steel frame erection | 0% | 0% | 0% | 0% |

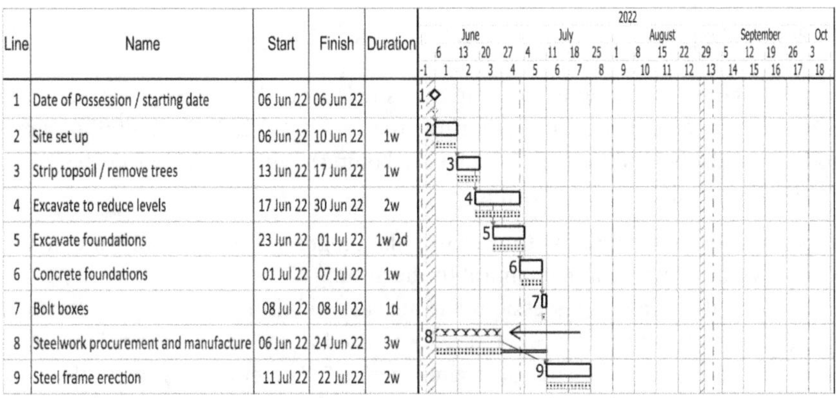


**Figure 3.5** Progress applied to Period 1 (6 to 20 June 2022) but programme not rescheduled.

When the programme is rescheduled after progress period 1 (i.e. after the first two weeks), it is clear that no progress on site was made during weeks 1 and 2. Therefore, when the baseline and the rescheduled bars are compared in Figure 3.6,[118] it can be seen that there is a 2-week delay to the project as activity 9 (steel frame erection) finishes on 5 August 2022.

Applying the progress from all of the progress periods in Table 3.9 and then rescheduling the programme results in Figure 3.7:

118 Figure 3.6 suggests that activity 8 (steelwork procurement and manufacture) actually commences on 26 May 2022 (i.e. 1 week before the planned start date on site (6 June 2022 – see Figure 3.5). This is not correct as in this example, activity 8 is completed 1 week early (i.e. it only takes 2 weeks, and not 3 weeks). Instead of shortening the length of the activity bar for activity 8 from 3 weeks to 2 weeks, in this example, the software has shown activity 8 finishing on the correct date, but starting 1 week earlier than planned. See also Footnotes 118 and 120.

**Table 3.12** Progress applied to activity bar.

| Symbol | Description |
|---|---|
| | Progress applied to activity bar |

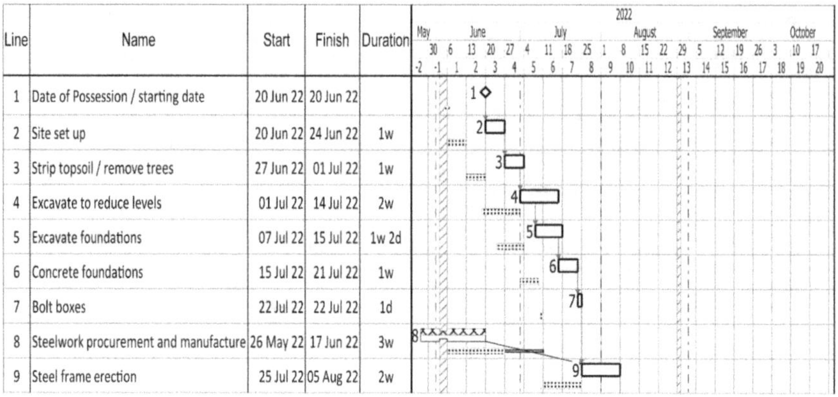


**Figure 3.6** Rescheduled Figure 3.5 after Progress Period 1 (6 to 20 June 2022).

As shown in Figure 3.6, owing to the employer's delay, no site progress was achieved for weeks 1 and 2 (progress period 1[119]). During progress period 2,[120] the overall progress appears to be 2 weeks behind (instead of 3 weeks behind), with no further delays being suffered.[121] This is because of the following:

- If progress is applied to an activity within a particular progress period, the software (in this example) shows the activity commencing at the beginning of the relevant progress period. In the case of activity 1 (date of possession/starting date), whilst this actually occurs on 27 June 2022 (at the end of the first week in progress period 2), the software in this example shows activity 1 occurring at the start of progress period 2. Therefore, it is advisable to use progress periods that are as short as are reasonably practicable (e.g. 1 to 2 weeks). If longer progress periods are used,[122] the programme is less likely to show an accurate representation of when an activity was carried out within that progress period, although the overall percentage of the activity completed within that progress period should be accurate[123];
- Activities 3 (strip topsoil/remove trees) and 4 (excavate to reduce levels) have started earlier than planned; and

119 Progress period 1: 6 to 20 June 2022, shown unshaded in Figure 3.7.
120 Progress period 2: 21 June to 4 July 2022, shown with 20% shading in Figure 3.7.
121 The planning software shows activities (to which progress has been applied) as commencing at the beginning of a progress period. Therefore, even though activity 1 actually commences on 27 June 2022, because it is reported as 100% complete in progress period 2, the software shows it occurring at the beginning of the progress period (i.e. on 20 June 2022). See also Footnotes 115 and 120.
122 For example, a contractor may use progress periods of 1 month, by taking the percentage completed for each activity directly from its monthly applications for interim payments.
123 For example, if the relevant progress period is a period of 4 weeks, and an activity of 1 week's duration commences in week 3 of that progress period and is completed on time, the software used in the Figure 3.7 example will show that activity being completed during week 1. See also Footnotes 115 and 118.

| Line | Name | Start | Finish | Duration |
|---|---|---|---|---|
| 1 | Date of Possession / starting date | 20 Jun 22 | 20 Jun 22 | |
| 2 | Site set up | 20 Jun 22 | 06 Jul 22 | 1w |
| 3 | Strip topsoil / remove trees | 27 Jun 22 | 01 Jul 22 | 1w |
| 4 | Excavate to reduce levels | 30 Jun 22 | 20 Jul 22 | 2w |
| 5 | Excavate foundations | 11 Jul 22 | 25 Jul 22 | 1w 2d |
| 6 | Concrete foundations | 25 Jul 22 | 29 Jul 22 | 1w |
| 7 | Bolt boxes | 29 Jul 22 | 29 Jul 22 | 1d |
| 8 | Steelwork procurement and manufacture | 26 May 22 | 17 Jun 22 | 3w |
| 9 | Steel frame erection | 01 Aug 22 | 12 Aug 22 | 2w |
| 10 | Roof safety netting | 15 Aug 22 | 17 Aug 22 | 3d |
| 11 | Profiled roof sheet and skylights | 18 Aug 22 | 01 Sep 22 | 2w |
| 12 | Wall cladding | 25 Aug 22 | 08 Sep 22 | 2w |
| 13 | Pre-cast concrete mezzanine floor | 15 Aug 22 | 19 Aug 22 | 1w |
| 14 | Pre-cast concrete stairs | 22 Aug 22 | 24 Aug 22 | 3d |
| 15 | Ground floor slab | 25 Aug 22 | 05 Sep 22 | 1w 2d |
| 16 | Internal blockwork walls | 06 Sep 22 | 19 Sep 22 | 2w |
| 17 | External windows and doors | 09 Sep 22 | 15 Sep 22 | 1w |
| 18 | 1st fix M&E | 20 Sep 22 | 26 Sep 22 | 1w |
| 19 | Plastering to offices | 27 Sep 22 | 03 Oct 22 | 1w |
| 20 | Ceilings | 04 Oct 22 | 06 Oct 22 | 3d |
| 21 | 2nd fix M&E | 07 Oct 22 | 13 Oct 22 | 1w |
| 22 | Decorations | 14 Oct 22 | 20 Oct 22 | 1w |
| 23 | Form external roads and car park | 09 Sep 22 | 29 Sep 22 | 3w |
| 24 | Surfacing and lining | 30 Sep 22 | 06 Oct 22 | 1w |
| 25 | Approval under s278 Highways Act 1980 | 21 Oct 22 | 21 Oct 22 | |
| 26 | Planned completion (on site) | 21 Oct 22 | 21 Oct 22 | |
| 27 | Testing, commissioning & snagging | 21 Oct 22 | 27 Oct 22 | 1w |
| 28 | Date for completion | 28 Oct 22 | 28 Oct 22 | |

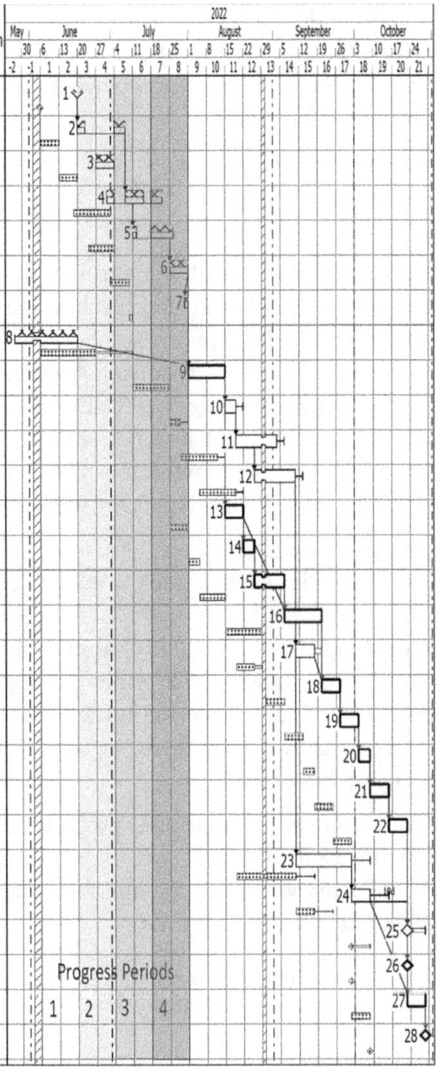


**Figure 3.7** Rescheduled Figure 3.5 after Progress Period 4 (ending on 1 August 2022).

- During progress period 3,[124] the excavation activities 4 and 5 appear to take 1 week longer, and there appears to be 1 day of contractor delay when progressing activity 5 (excavate foundations), but this is recovered in progress period 4, by completing activity 7 (bolt boxes) 1 day earlier than planned.

As at the end of progress period 4[125] (on 1 August 2022), activity 9 (steel frame erection) is ready to commence, and by comparing this to its baseline, it can be seen that progress is 3 weeks behind the original programme. On the basis that (i) the contractor promptly notified the employer (or his representative) under both the JCT standard forms of

124 Progress period 3: 5 to 18 July 2022, shown with 40% shading in Figure 3.7.
125 Progress period 4: 19 July to 1 August 2022, shown with 60% shading in Figure 3.7.

contract,[126] and under NEC4,[127] and (ii) the employer (or his representative) promptly[128] deals with the contractor's notification,[129] the contractor should be entitled to an extension of time[130] of 3 weeks, owing to the employer's delay in giving possession of the site for this period, causing the following:

- JCT contracts: completion is likely to be later than the 'completion date'[131] or under
- NEC4, 'planned completion' will be later than shown on the relevant 'accepted programme'.[132]

As well as tracking the overall delays to a project (refer to Figures 3.4 to 3.7), time impact analysis can act as a useful tool to forecast how the programme may be affected in the event of a change. This method is appropriate for producing the 'revised programme' under clause 32 of NEC4 as this must show *'the actual progress achieved on each operation and its effect upon the timing of the remaining work'* (NEC4 clause 32.1).

Reviewing Figures 3.1, 3.2 and 3.7 shows that the planned critical path remains unchanged[133] for (what NEC4 refers to as) *'the remaining work'*. Using the example of the warehouse programme, let us assume the following:

- After progress period 4 ends on 1 August 2022 (see Figure 3.7), progress continues to be added to activities every 2 weeks until 26 September 2022 (which would be progress period 8);
- After completing 25% of activity 12 (wall cladding) by 29 August 2022 (the end of progress period 6), there is a 4-week delay until work resumes on 26 September 2022; and
- Between the end of progress period 4 (1 August 2022) and progress period 8 (26 September 2022), all other activities (no's 9 to 11, and 13 to 16) are carried out to their intended sequences and durations.

The effects on the programme after progress period 8 ended on 26 September 2022 are shown in Figure 3.8 below.

The delay to activity 12 (wall cladding) has affected the critical path for the remaining work as follows (circled in Figure 3.8).

---

126 Providing the notice and particulars under JCT SBC/Q 2016 – clauses 2.27.1 and 2.27.2, JCT DB 2016 – clauses 2.24.1 and 2,24.2, and the notices under JCT IC 2016 – clause 2.19.1, and JCT MW 2016 – clause 2.7.

127 NEC4, clause 61.3.

128 SCL Protocol (2017) – core principle 4, the employer (or his representative) should not "wait and see" regarding the impact of delay events, but so far as possible, should deal with extensions of time as close to the delaying event as possible – refer to Sections 3.4.1 and 9.1.5 for more detail.

129 Under NEC4, the Project Manager has 6 weeks from the contractor's clause 61.3 notice to request the contractor's quotation (clause 61.4), wait up to 3 weeks for the quotation (clause 62.3) and 2 weeks to accept the contractor's quotation (clause 62.3) whereupon the compensation event is "implemented" (clause 66.1), and the Completion Date (and any Key Date) is changed.

130 A "Relevant Event" under JCT SBC/Q 2016 – clause 2.29.3 / 2.29.7, JCT DB 2016 – clause 2.26.3/2.26.6, JCT IC 2016 – clause 2.20.3/2.20.6, and a *"reason beyond the control of the Contractor"* – JCT MW 2016 – clause 2.7. A compensation event under NEC4, clause 60.1(2).

131 JCT SBC/Q 2016 – clause 2.28.1.2, JCT DB 2016 – clause 2.25.1.2, JCT IC 2016 – clause 2.19.1, JCT MW 2016 – clause 2.7.

132 NEC4 clause 63.5.

133 The planned critical path for activities including and following (i) "Steel frame erection" remains as (ii) "Pre-cast concrete mezzanine floor", (iii) "Pre-Cast concrete stairs", (iv) "Ground floor slab", (v) "Internal blockwork walls", (vi) "1st fix M&E", (vii) "Plastering to offices", (viii) "Ceilings", (ix) "2nd fix M&E", (x) "Decorations", (xi) "Planned completion (on site)", (xii) "Testing, commissioning and snagging" and (xiii) "Date for Completion".

| Line | Name | Start | Finish | Duration |
|---|---|---|---|---|
| 1 | Date of Possession / starting date | 20 Jun 22 | 20 Jun 22 | |
| 2 | Site set up | 20 Jun 22 | 24 Jun 22 | 1w |
| 3 | Strip topsoil / remove trees | 27 Jun 22 | 01 Jul 22 | 1w |
| 4 | Excavate to reduce levels | 30 Jun 22 | 20 Jul 22 | 2w |
| 5 | Excavate foundations | 06 Jul 22 | 25 Jul 22 | 1w 2d |
| 6 | Concrete foundations | 25 Jul 22 | 29 Jul 22 | 1w |
| 7 | Bolt boxes | 29 Jul 22 | 29 Jul 22 | 1d |
| 8 | Steelwork procurement and manufacture | 26 May 22 | 17 Jun 22 | 3w |
| 9 | Steel frame erection | 01 Aug 22 | 12 Aug 22 | 2w |
| 10 | Roof safety netting | 15 Aug 22 | 17 Aug 22 | 3d |
| 11 | Profiled roof sheet and skylights | 18 Aug 22 | 05 Sep 22 | 2w |
| 12 | Wall cladding | 25 Aug 22 | 05 Oct 22 | 2w |
| 13 | Pre-cast concrete mezzanine floor | 30 Aug 22 | 05 Sep 22 | 1w |
| 14 | Pre-cast concrete stairs | 06 Sep 22 | 08 Sep 22 | 3d |
| 15 | Ground floor slab | 09 Sep 22 | 19 Sep 22 | 1w 2d |
| 16 | Internal blockwork walls | 19 Sep 22 | 30 Sep 22 | 2w |
| 17 | External windows and doors | 06 Oct 22 | 12 Oct 22 | 1w |
| 18 | 1st fix M&E | 13 Oct 22 | 19 Oct 22 | 1w |
| 19 | Plastering to offices | 20 Oct 22 | 26 Oct 22 | 1w |
| 20 | Ceilings | 27 Oct 22 | 31 Oct 22 | 3d |
| 21 | 2nd fix M&E | 01 Nov 22 | 07 Nov 22 | 1w |
| 22 | Decorations | 08 Nov 22 | 14 Nov 22 | 1w |
| 23 | Form external roads and car park | 06 Oct 22 | 26 Oct 22 | 3w |
| 24 | Surfacing and lining | 27 Oct 22 | 02 Nov 22 | 1w |
| 25 | Approval under s278 Highways Act 1980 | 17 Nov 22 | 17 Nov 22 | |
| 26 | Planned completion (on site) | 15 Nov 22 | 15 Nov 22 | |
| 27 | Testing, commissioning & snagging | 15 Nov 22 | 21 Nov 22 | 1w |
| 28 | Date for completion | 22 Nov 22 | 22 Nov 22 | |

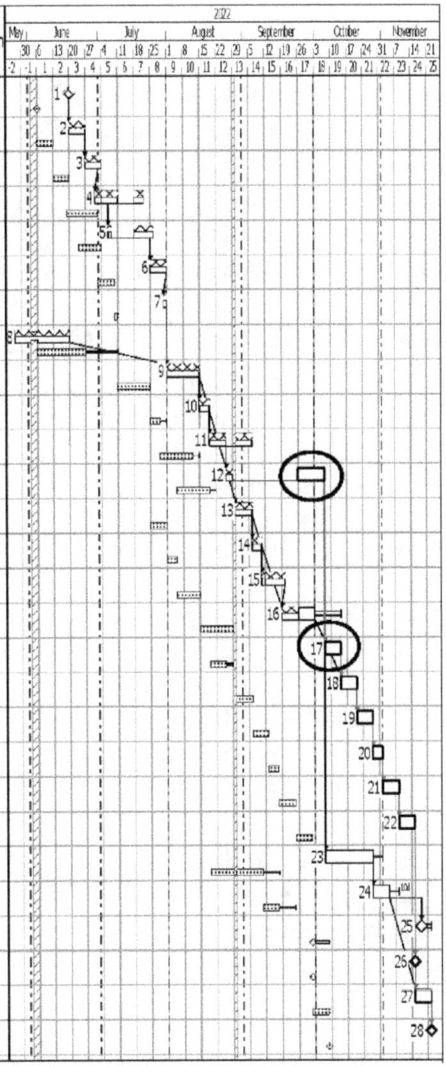


**Figure 3.8** Change in critical path.

- The remaining element of activity 12 now becomes part of the critical path and
- Activity 17 (external windows and doors) now becomes part of the critical path.

Without any contractual justification[134] for the delay in completing activity 12 (wall cladding), the contractor will be responsible for any ensuing period of critical delay caused by pausing this activity.

Another way that planning software can show progress against the baseline programme is to show a 'drop line' on the programme. Figure 3.9 shows the application of the 'drop line' on

134 For example, the employer may have instructed a change in the wall cladding's specification, causing the works to be delayed – which would then be a Relevant Event under JCT SBC/Q 2016 clause 2.29.1 (variation), JCT DB 2016 clause 2.26.1 (Change), JCT IC 2016 clause 2.20.1 (variation), JCT MW 2016 clause 2.7 (following a Variation under clause 3.6.1), or a compensation event under NEC4 clause 60.1(1).

| Line | Name | Start | Finish | Duration |
|---|---|---|---|---|
| 1 | Date of Possession / starting date | 20 Jun 22 | 20 Jun 22 | |
| 2 | Site set up | 20 Jun 22 | 24 Jun 22 | 1w |
| 3 | Strip topsoil / remove trees | 27 Jun 22 | 01 Jul 22 | 1w |
| 4 | Excavate to reduce levels | 30 Jun 22 | 20 Jul 22 | 2w |
| 5 | Excavate foundations | 06 Jul 22 | 25 Jul 22 | 1w 2d |
| 6 | Concrete foundations | 25 Jul 22 | 29 Jul 22 | 1w |
| 7 | Bolt boxes | 29 Jul 22 | 29 Jul 22 | 1d |
| 8 | Steelwork procurement and manufacture | 26 May 22 | 17 Jun 22 | 3w |
| 9 | Steel frame erection | 01 Aug 22 | 12 Aug 22 | 2w |
| 10 | Roof safety netting | 15 Aug 22 | 17 Aug 22 | 3d |
| 11 | Profiled roof sheet and skylights | 18 Aug 22 | 05 Sep 22 | 2w |
| 12 | Wall cladding | 25 Aug 22 | 05 Oct 22 | 2w |
| 13 | Pre-cast concrete mezzanine floor | 30 Aug 22 | 05 Sep 22 | 1w |
| 14 | Pre-cast concrete stairs | 06 Sep 22 | 08 Sep 22 | 3d |
| 15 | Ground floor slab | 09 Sep 22 | 19 Sep 22 | 1w 2d |
| 16 | Internal blockwork walls | 19 Sep 22 | 30 Sep 22 | 2w |
| 17 | External windows and doors | 06 Oct 22 | 12 Oct 22 | 1w |
| 18 | 1st fix M&E | 13 Oct 22 | 19 Oct 22 | 1w |
| 19 | Plastering to offices | 20 Oct 22 | 26 Oct 22 | 1w |
| 20 | Ceilings | 27 Oct 22 | 31 Oct 22 | 3d |
| 21 | 2nd fix M&E | 01 Nov 22 | 07 Nov 22 | 1w |
| 22 | Decorations | 08 Nov 22 | 14 Nov 22 | 1w |
| 23 | Form external roads and car park | 05 Oct 22 | 26 Oct 22 | 3w |
| 24 | Surfacing and lining | 27 Oct 22 | 02 Nov 22 | 1w |
| 25 | Approval under s278 Highways Act 1980 | 17 Nov 22 | 17 Nov 22 | |
| 26 | Planned completion (on site) | 15 Nov 22 | 15 Nov 22 | |
| 27 | Testing, commissioning & snagging | 15 Nov 22 | 21 Nov 22 | 1w |
| 28 | Date for completion | 22 Nov 22 | 22 Nov 22 | |

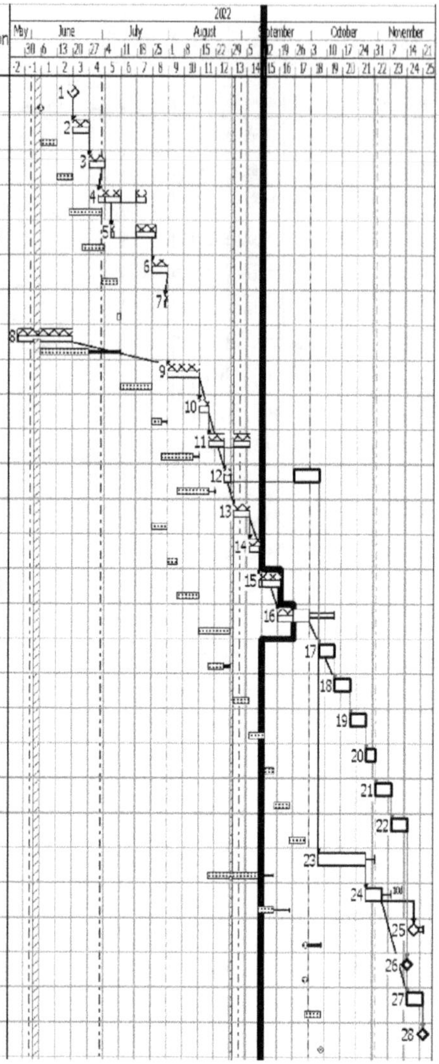


**Figure 3.9** Example showing 'drop line' progress (applied to Figure 3.8).

the rescheduled programme (shown in Figure 3.8). The thick, black 'drop line' shows that as at 12 September 2022, activity 15 (ground floor slab) is 1 week ahead of planned progress, and activity 16 (internal blockwork walls) is 2 weeks ahead of planned progress. All subsequent activities (from activity 17 onwards) are expected to commence in accordance with the rescheduled programme.

### 3.5.4 As-built v As-planned

The 'as-built v as-planned' method is essentially a comparison between the sequence and timing of activities that the contractor originally intended (the 'as-planned') with the actual

**Table 3.13** 'As built' activity.

| Symbol | Description |
|---|---|
| [grey shaded bar] | 'As built' activity (with *italicised* 'Name' descriptions) |

('as-built') sequence and timing of the works. A continuation of the 'warehouse' example is shown in Figure 3.9. The same keys as for Tables 3.7 and 3.8 are used to show the categories of activity, but with the addition of the 'as-built' activities as shown in Table 3.13.

Figure 3.9 uses the following information:

- The 'as-planned' programme from Figure 3.1;
- The employer delay in giving possession from Figure 3.2; and
- The same progress information used in the 'time impact analysis' examples (in Figures 3.6–3.8 and 3.9) has been used, with further progress information (between 26 September and 12 October 2022) being used in order to compare the 'as-built v as-planned' durations for subsequent activities.[135]

Activity 25 (wall cladding 'as-built') shows the same 4-week delay as in Figure 3.8.[136] This delay can either be shown as a gap in progress, as it is in Figure 3.10 (i.e. indicating a pause between the first part of activity 25 'wall cladding' that ended on 26 August 2022, and its resumption on 26 September 2022), or as a separate activity to represent the 4-week delay. If the employer is responsible[136] for the delay to the wall cladding, the relevant activities could be shown as per the example in Figure 3.11.

However, if the delay (3 weeks and 4 days) was caused by the contractor failing to obtain the cladding panels on time, it could be shown as in Figure 3.12:

Table 3.14 shows the activity key for a contractor delay:

Returning to Figure 3.10, once activity 35 (external windows and doors) is completed it appears that the overall delay to the works is 6 weeks and 4 days (activity 36). This is measured from the earliest date that the 'planned' activity 34 ('external windows and doors') could finish – 24 August 2022. Table 3.15 summarises the actual progress made (i.e. both the delays and mitigation of delays) between the planned start date and the actual completion of activity 35 ('External windows and doors'):

**Table 3.14** Contractor delay activity.

| Symbol | Description |
|---|---|
| [hatched bar] | Contractor delay activity |

135 Figure 3.8 applied progress information up to 'wall cladding' (activity 12 in Figure 3.8, now activity 35 in Figure 3.9). Figure 3.9 now shows the 'as-built' sequence and timing for Pre-cast concrete mezzanine floor (activity 27), Pre-cast concrete stairs (activity 29), Ground floor slab (activity 31), Internal blockwork walls (activity 33) and External windows and doors (activity 35).

136 In Figure 3.8, 'wall cladding' was activity 12.

| Line | Name | Start | Finish | Duration |
|---|---|---|---|---|
| 1 | *Employer delay in giving possession* | 06 Jun 22 | 24 Jun 22 | 3w |
| 2 | Date of Possession / starting date | 06 Jun 22 | 06 Jun 22 | |
| 3 | *Date of Possession / starting date* | 27 Jun 22 | 27 Jun 22 | |
| 4 | Site set up | 06 Jun 22 | 10 Jun 22 | 1w |
| 5 | *Site set up* | 27 Jun 22 | 01 Jul 22 | 1w |
| 6 | Strip topsoil / remove trees | 13 Jun 22 | 17 Jun 22 | 1w |
| 7 | *Strip topsoil / remove trees* | 27 Jun 22 | 01 Jul 22 | 1w |
| 8 | Excavate to reduce levels | 17 Jun 22 | 30 Jun 22 | 2w |
| 9 | *Excavate to reduce levels* | 01 Jul 22 | 14 Jul 22 | 2w |
| 10 | Excavate foundations | 23 Jun 22 | 01 Jul 22 | 1w 2d |
| 11 | *Excavate foundations* | 14 Jul 22 | 22 Jul 22 | 1w 2d |
| 12 | Concrete foundations | 01 Jul 22 | 07 Jul 22 | 1w |
| 13 | *Concrete foundations* | 25 Jul 22 | 29 Jul 22 | 1w |
| 14 | Bolt boxes | 08 Jul 22 | 08 Jul 22 | 1d |
| 15 | *Bolt boxes* | 29 Jul 22 | 29 Jul 22 | 1d |
| 16 | Steelwork procurement and manufacture | 06 Jun 22 | 24 Jun 22 | 3w |
| 17 | *Steelwork procurement and manufacture* | 06 Jun 22 | 17 Jun 22 | 2w |
| 18 | Steel frame erection | 11 Jul 22 | 22 Jul 22 | 2w |
| 19 | *Steel frame erection* | 01 Aug 22 | 12 Aug 22 | 2w |
| 20 | Roof safety netting | 25 Jul 22 | 27 Jul 22 | 3d |
| 21 | *Roof safety netting* | 15 Aug 22 | 17 Aug 22 | 3d |
| 22 | Profiled roof sheet and skylights | 28 Jul 22 | 10 Aug 22 | 2w |
| 23 | *Profiled roof sheet and skylights* | 18 Aug 22 | 01 Sep 22 | 2w |
| 24 | Wall cladding | 04 Aug 22 | 17 Aug 22 | 2w |
| 25 | *Wall cladding* | 25 Aug 22 | 05 Oct 22 | 2w |
| 26 | Pre-cast concrete mezzanine floor | 25 Jul 22 | 29 Jul 22 | 1w |
| 27 | *Pre-cast concrete mezzanine floor* | 30 Aug 22 | 05 Sep 22 | 1w |
| 28 | Pre-cast concrete stairs | 01 Aug 22 | 03 Aug 22 | 3d |
| 29 | *Pre-cast concrete stairs* | 06 Sep 22 | 08 Sep 22 | 3d |
| 30 | Ground floor slab | 04 Aug 22 | 12 Aug 22 | 1w 2d |
| 31 | *Ground floor slab* | 09 Sep 22 | 19 Sep 22 | 1w 2d |
| 32 | Internal blockwork walls | 15 Aug 22 | 26 Aug 22 | 2w |
| 33 | *Internal blockwork walls* | 19 Sep 22 | 30 Sep 22 | 2w |
| 34 | External windows and doors | 18 Aug 22 | 24 Aug 22 | 1w |
| 35 | *External windows and doors* | 06 Oct 22 | 12 Oct 22 | 1w |
| 36 | **Overall project delay after 'External windows and doors'** | 25 Aug 22 | 12 Oct 22 | 6w 4d |

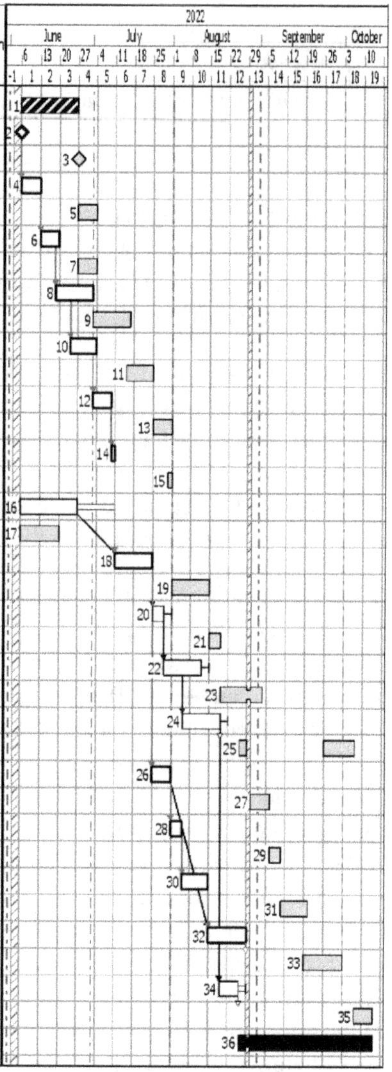

**Figure 3.10** Warehouse As-built v As-planned example.

| Line | Name | Start | Finish | Duration |
|---|---|---|---|---|
| 24 | Wall cladding | 04 Aug 22 | 17 Aug 22 | 2w |
| 25 | *Wall cladding (secondary steelwork)* | 25 Aug 22 | 26 Aug 22 | 2d |
| 26 | *Wall cladding (delay obtaining changed panel specification)* | 30 Aug 22 | 23 Sep 22 | 3w 4d |
| 27 | *Wall cladding (installing changed panel specification)* | 26 Sep 22 | 05 Oct 22 | 1w 3d |

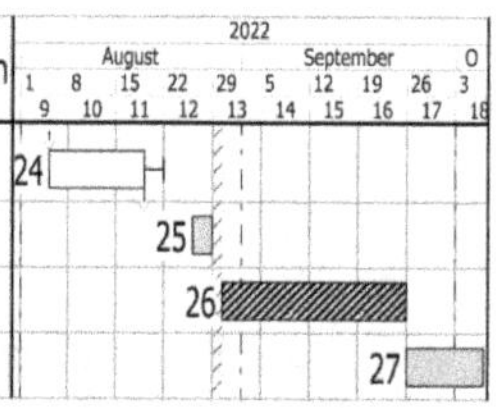

**Figure 3.11** Employer delay to wall cladding.

**Figure 3.12** Contractor delay to wall cladding.

Figure 3.10's activity 36 and the activity-by-activity reconciliation of progress in Table 3.13, show that after when completing activity 35 ('External windows and doors'), progress has been delayed by 6 weeks and 4 days.

Figure 3.9 clearly shows that the initial delay of 3 weeks is the employer's responsibility, so on the basis that the contractor properly notified the employer (or his representative),[137] the contractor is entitled to an extension of time of 3 weeks. However, the following alternatives were considered as causing the delay of 3 weeks and 4 days to activity 25 ('Wall cladding'):

- Alterative 1: The employer instructed a change in the wall cladding panels' specification, and it took the contractor another 3 weeks and 4 days to obtain and fix the panels (refer to Figure 3.10); or
- Alternative 2: The wall cladding panels' specification remained unchanged, but the contractor simply could not obtain the panels so they could be installed on time (refer to Figure 3.13 and Table 3.12).

In Alternative 1, and again, provided it fulfilled the contractual notice requirements, the contractor should be entitled to an extension of time of 6 weeks and 4 days as the completion date is likely to be delayed by that period, and the delay was caused by reasons for which the employer is responsible.

However, in Alternative 2, the contractor is clearly at fault for the delay to the 'Wall cladding' of 3 weeks and 4 days. On the same basis as the first alternative, he would be entitled to a 3 week extension of time for the employer's failure to give possession, but unless he could recover the 'Wall cladding' delay over the course of the following activities, he would be liable to the employer to pay delay damages[138] for the 3-week and 4-day delay.

### 3.5.5 Collapsed As-built

The 'collapsed as-built' method relies upon providing an accurate account of the sequence and timing of (i) all the relevant work operations and (ii) the effects that any events may have had on the progress of the contractor's work. It's sole basis is factual data, and it does not rely upon the contractor's anticipated, or 'baseline' programmes. In Figure 3.9, we saw the 'as-built v as-planned' progress for the warehouse example. This clearly showed the 'as-planned' activities on the baseline programme, against which, the 'as-built' (or actual)

137 Refer to Footnotes 123 to 127.
138 Liquidated damages or unliquidated damages – refer to section 8.1.

**Table 3.15** Reconciliation of progress for Figure 3.10.

| Activity No | Activity Name | Actual Start | Actual Finish | Notes | Cumulative Delay |
|---|---|---|---|---|---|
| 1 | Employer delay in giving possession | 6 June 2022 | 24 June 2022 | Employer delays giving possession of the site for 3 weeks | 3w |
| 5 | Site set up | 27 June 2022 | 1 July 2022 | Planned to commence on 6 June 2022, but owing to activity 1, starts 3 weeks later, and last for 1 week (as planned) | 3w |
| 7 | Strip topsoil / remove trees | 27 June 2022 | 1 July 2022 | Planned to commence the week after 'Site set up' but carried out at the same time, recovering 1 week | 2w |
| 9 | Excavate to reduce levels | 1 July 2022 | 14 July 2022 | Planned to commence 1 day before 'Strip topsoil etc' finishes, and last for 2 weeks and does so in relation to activity 7. | 2w |
| 11 | Excavate foundations | 14 July 2022 | 22 July 2022 | Planned to commence 4 days after 'Excavate to reduce levels' commences and last for 1 week 2 days<br>Actually commences 9 days after activity 9 (1-week delay) and is completed in 1 week 2 days | 3w |
| 13 | Concrete foundations | 25 July 2022 | 29 July 2022 | Planned to commence 1 day before 'Excavate foundation' is completed and last for 1 week<br>Actually starts after activity 11 finishes (1 day delay) and is completed in 1 week. | 3w 1d |
| 15 | Bolt boxes | 29 July 2022 | 29 July 2022 | Planned to commence after 'Concrete foundations' are completed and last for 1 day.<br>Actually completed on the last day of activity 13 (recovery of 1 day) | 3w |
| 17 | Steelwork procurement and manufacture | 6 June 2022 | 17 June 2022 | Not a critical activity | 3w |
| 19 | Steel frame erection | 1 Aug 2022 | 12 Aug 2022 | Planned to commence after 'Bolt boxes' finishes and last for 2 weeks and does so in relation to activity 15 | 3w |
| 21 | Roof safety netting | 14 Aug 2022 | 17 Aug 2022 | Planned to commence after 'Steel frame erection' and last for 3 days and does so in relation to activity 19. | 3w |

| | | | | | |
|---|---|---|---|---|---|
| 23 | Profiled roof sheet and skylights | 18 Aug 2022 | 1 Sept 2022 | Planned to commence after 'Roof safety netting' and last for 2 weeks and does so in relation to activity 21 | 3w |
| 25 | Wall cladding | 25 Aug 2022 | 5 Oct 2022 | Planned to commence 1 week after 'Profiled roof sheet etc' and last for 2 weeks.<br>Actually commences 1 week after activity 23, but after only 2 days' work, there is a delay of 3w 4d until work resumes for a further 1w 3d. | 6w 4d |
| 27 | Pre-cast concrete mezzanine floor | 30 Aug 2022 | 5 Sept 2022 | Owing to the delay to activity 25 – no longer on the critical path (refer to Figure 3.8) | 6w 4d |
| 29 | Pre-cast concrete stairs | 6 Sept 2022 | 8 Sept 2022 | Owing to the delay to activity 25 – no longer on the critical path (refer to Figure 3.8) | 6w 4d |
| 31 | Ground floor slab | 9 Sept 2022 | 19 Sept 2022 | Owing to the delay to activity 25 – no longer on the critical path (refer to Figure 3.8) | 6w 4d |
| 33 | Internal blockwork walls | 19 Sept 2022 | 30 Sept 2022 | Owing to the delay to activity 25 – no longer on the critical path (refer to Figure 3.8) | 6w 4d |
| 35 | External windows and doors | 6 Oct 2022 | 12 Oct 2022 | Planned to commence after 'Wall cladding' and last for 1 week, but was not planned to form part of the critical path.<br>Owing to delays to activity 25, activity 35 now sits on the critical path, and actually commenced after activity 25 and lasted for 1 week. | 6w 4d |

**Table 3.16** Activity requiring crane hire.

| Symbol | Description |
|---|---|
| | Activity requiring crane hire |

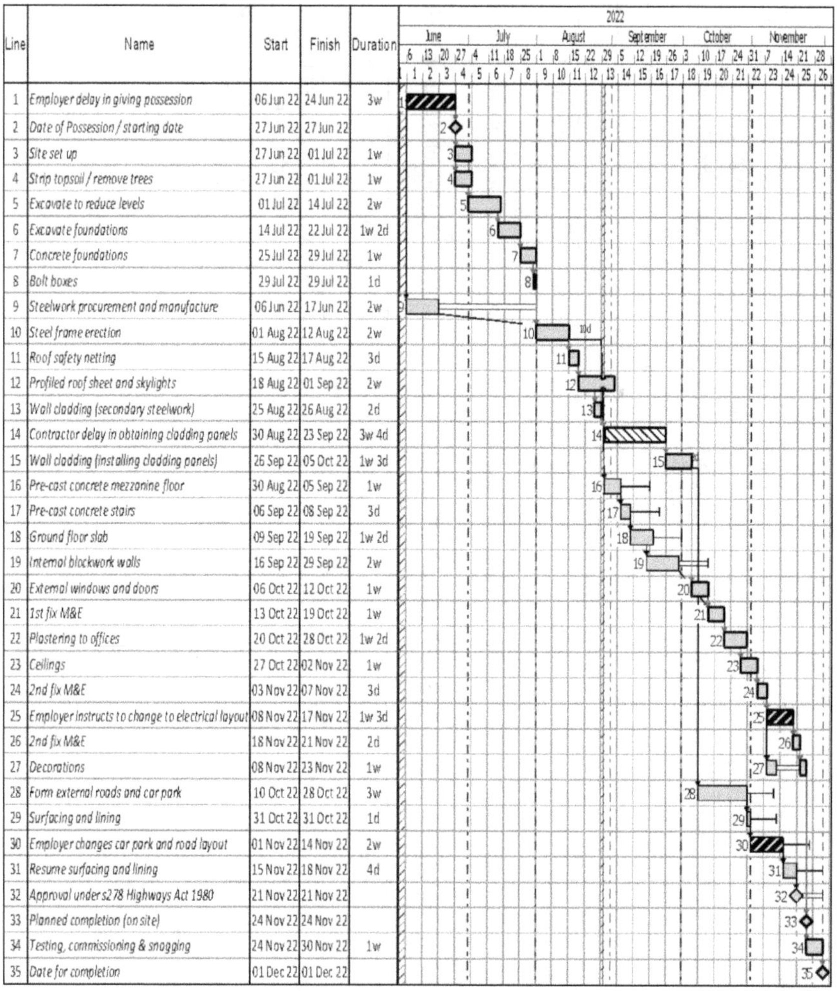


**Figure 3.13** Warehouse example – 'as-built' programme.

durations of those activities were plotted, so a comparison between the two could be made. The progress and delays up to activity 35 (external windows and doors) could then be tracked and ascertained as summarised in Table 3.15. Applying the 'as-built' sequence and durations for all activities and then removing the 'as-planned' activities results in Figure 3.13:

The 'as-built' activities have been logic-linked so the 'as-built' critical path can be shown. The critical path remains in the sequence shown in Figure 3.8.[139] The delays recorded in Figure 3.13 are as follows:

- Employer delays:
  - Activity 1–3-week delay in giving possession of the site
  - Activity 25–1-week 3-day delay from changing the warehouse's electrical layout; and
  - Activity 30–2-week delay from changing the car park and road layout (although, this does not sit on the 'as-built' critical path).
- Contractor delay:
  - Activity 14–3-week 4-day delay in obtaining cladding panels.

To 'collapse' the logic-linked 'as-built' programme in Figure 3.8, the critical delays caused by the employer must be removed.

Figure 3.14 now excludes the 3-week delay caused by the employer giving possession.[140] The 'date for completion' (activity 34) in Figure 3.13 has moved back by 3 weeks, from 1 December 2022 (in Figure 3.12) to 10 November 2022. This indicates that that the contractor would be entitled to extensions of time as follows:

- Under JCT contracts, the contractor would be entitled to have a later date fixed as the completion date that would occur 3 weeks after the original 'date for completion' because:
  - The delay is a Relevant Event[130] and
  - Completion has been delayed beyond the 'Completion Date'.[131]
- Under NEC4, and providing that the programme has been revised correctly under clause 32, the compensation event[132] caused 'planned Completion' to be later than 'planned Completion' on the 'Accepted Programme,[141] the delay would be assessed as 3 weeks.

The programme must be 'collapsed' again to remove the second employer delay – 'Employer instructs change to electrical layout' (activity 24 in Figure 3.13) from the critical path. This is shown in Figure 3.15:

The effect of removing critical activity 24 from Figure 3.14 'collapses' the Figure 3.14 programme to show the 'date for completion' now occurring on 1 November 2022, 1 week and 3 days earlier than in Figure 3.14. There is no need to 'collapse' the programme again to remove activity 28 (Employer changes car park and road layout), as this activity is not on the as-built critical path, and therefore its removal will have no effect on the critical path, or when the 'Date for Completion' (activity 33 in Figure 3.14) occurs. On the same basis as described above,[142] the contractor would be entitled to an extension of time of 1 week and 3 days under both JCT[143] and NEC4 contracts.[144]

139 A time-impact analysis was undertaken; applying the recorded progress for each fortnightly progress period until the work was completed. This verified the 'as-built' critical path, and showed it comprised the same activities as in Figure 3.8.

140 See Figure 3.12, activity 1 'Employer delay in giving possession'.

141 See Footnote 129. "Planned completion" is activity 32 in Figure 3.13.

142 Refer to Footnotes 128 and 129.

143 Activity 34 in Figure 3.13 is an instruction to vary the electrical layout. This is a 'Relevant Event' under JCT SBC/Q 2016 – clause 2.29.1, JCT IC 2016 – clause 2.20.1, and JCT MW 2016 – clauses 3.4/3.6 (a 'variation') or a JCT DB 2016 – clause 2.26.1 (a 'Change').

144 NEC4, a compensation event under clause 60.1(1) an instruction changing the Scope.

| Line | Name | Start | Finish | Duration |
|---|---|---|---|---|
| 1 | *Date of Possession / starting date* | 06 Jun 22 | 06 Jun 22 | |
| 2 | *Site set up* | 06 Jun 22 | 10 Jun 22 | 1w |
| 3 | *Strip topsoil / remove trees* | 06 Jun 22 | 10 Jun 22 | 1w |
| 4 | *Excavate to reduce levels* | 10 Jun 22 | 23 Jun 22 | 2w |
| 5 | *Excavate foundations* | 23 Jun 22 | 01 Jul 22 | 1w 2d |
| 6 | *Concrete foundations* | 04 Jul 22 | 08 Jul 22 | 1w |
| 7 | *Bolt boxes* | 08 Jul 22 | 08 Jul 22 | 1d |
| 8 | *Steelwork procurement and manufacture* | 06 Jun 22 | 17 Jun 22 | 2w |
| 9 | *Steel frame erection* | 11 Jul 22 | 22 Jul 22 | 2w |
| 10 | *Roof safety netting* | 25 Jul 22 | 27 Jul 22 | 3d |
| 11 | *Profiled roof sheet and skylights* | 28 Jul 22 | 10 Aug 22 | 2w |
| 12 | *Wall cladding (secondary steelwork)* | 04 Aug 22 | 05 Aug 22 | 2d |
| 13 | *Contractor delay in obtaining cladding panels* | 08 Aug 22 | 02 Sep 22 | 3w 4d |
| 14 | *Wall cladding (installing cladding panels)* | 05 Sep 22 | 14 Sep 22 | 1w 3d |
| 15 | *Pre-cast concrete mezzanine floor* | 08 Aug 22 | 12 Aug 22 | 1w |
| 16 | *Pre-cast concrete stairs* | 15 Aug 22 | 17 Aug 22 | 3d |
| 17 | *Ground floor slab* | 18 Aug 22 | 26 Aug 22 | 1w 2d |
| 18 | *Internal blockwork walls* | 25 Aug 22 | 08 Sep 22 | 2w |
| 19 | *External windows and doors* | 15 Sep 22 | 21 Sep 22 | 1w |
| 20 | *1st fix M&E* | 22 Sep 22 | 28 Sep 22 | 1w |
| 21 | *Plastering to offices* | 29 Sep 22 | 07 Oct 22 | 1w 2d |
| 22 | *Ceilings* | 06 Oct 22 | 12 Oct 22 | 1w |
| 23 | *2nd fix M&E* | 13 Oct 22 | 17 Oct 22 | 3d |
| 24 | *Employer instructs to change to electrical layout* | 18 Oct 22 | 27 Oct 22 | 1w 3d |
| 25 | *2nd fix M&E* | 28 Oct 22 | 31 Oct 22 | 2d |
| 26 | *Decorations* | 18 Oct 22 | 02 Nov 22 | 1w |
| 27 | *Form external roads and car park* | 19 Sep 22 | 07 Oct 22 | 3w |
| 28 | *Surfacing and lining* | 10 Oct 22 | 10 Oct 22 | 1d |
| 29 | *Employer changes car park and road layout* | 11 Oct 22 | 24 Oct 22 | 2w |
| 30 | *Resume surfacing and lining* | 25 Oct 22 | 28 Oct 22 | 4d |
| 31 | *Approval under s278 Highways Act 1980* | 31 Oct 22 | 31 Oct 22 | |
| 32 | *Planned completion (on site)* | 03 Nov 22 | 03 Nov 22 | |
| 33 | *Testing, commissioning & snagging* | 03 Nov 22 | 09 Nov 22 | 1w |
| 34 | *Date for completion* | 10 Nov 22 | 10 Nov 22 | |

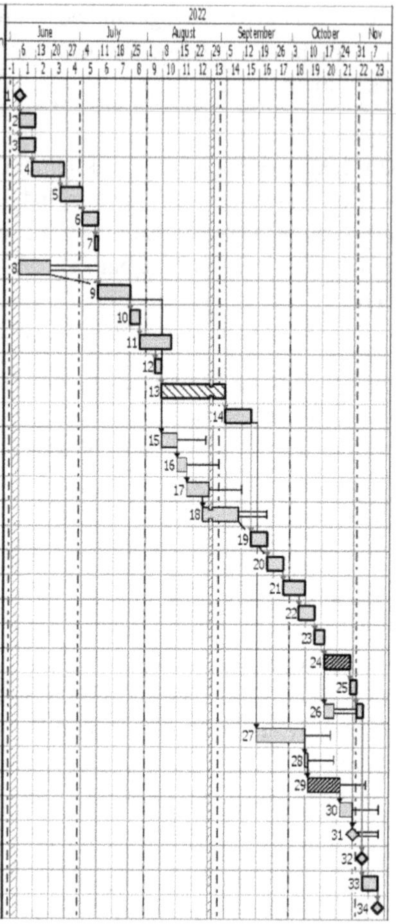

**Figure 3.14** Warehouse example – collapsed 'as-built', 1st critical delay removed.

#### 3.5.5.1 Summary of Delay Analysis Methods

The four methods of delay analysis describe in Table 3.3 have been set out in order of increasing reliance upon factual progress information. It should not be assumed that each method's reliance on factual data increases proportionately from one method to another, as this is not the case. For example, the 'impacted as-planned' method (described in Section 3.5.2) does not rely upon any factual progress information because the period of delay is yet to occur, so 'impacted as-planned' analysis may simply be based upon the contractor's estimates (or forecast) of:

- The intended sequence and timing of the works (i.e. the 'as-planned') and
- The timing and duration of the relevant delay, and its effect on the 'as-planned' programme (i.e. the 'impacted').

In the 'impacted as-planned' example in Section 3.5.2 (refer to Figure 3.2) is effectively a forecast of how the contractor's progress is likely to be affected by the employer's failure to give possession of the site for 3 weeks because it takes place before the date when

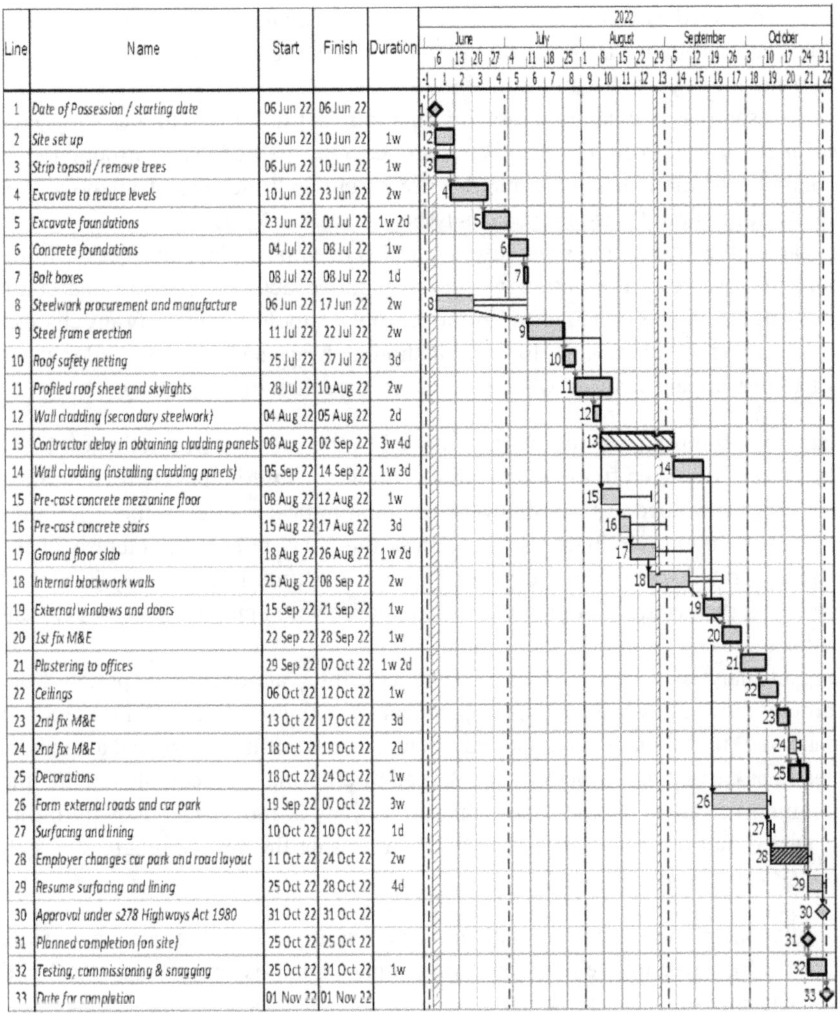


**Figure 3.15** Warehouse example – collapsed 'as-built', 2nd critical delay removed.

the contractor is due to start on site. Therefore, the earliest any factual information can be obtained will be from after the date when the contractor is due to start, and this is only likely to become useful once the contractor is able to actually start on site. Nevertheless, in the 'impacted as-planned' example, the contractor is obliged to notify the employer that either the actual completion date (JCT) or planned completion (NEC4) is likely to take place later than planned – i.e. the relevant date in the contract (JCT) or shown on the 'Accepted Programme' (NEC4). The 'impacted as-planned' method may be the only means that the contractor has at his disposal to provide an estimate of the delay (JCT) or giving his forecast for when subsequent activities may be carried out (NEC4), either as part of revising the 'Accepted Programme' (clause 31) or when altering the 'Accepted Programme' when submitting a quotation for a compensation event under clause 62.2. It is therefore wrong to dismiss the 'impacted as-planned' as an unreliable delay analysis method as a matter of principle, particularly when it is used to provide a forecast or estimate of delay when factual progress information is yet to become available.

However, as soon as factual progress information is introduced to the 'impacted as-planned' method, it's likely to cease being an accurate method. The introduction of factual progress information will permit the use of either 'time impact analysis' (see Section 3.5.3) or an 'as-built v as-planned' analysis (see Section 3.5.4).

'Time impact analysis' is clearly a very useful method to employ. Not only does the introduction of factual progress information into the 'as-planned' activities provide an account of the work's actual progress (see Figures 3.6 and 3.7) but it should also identify any changes in the critical path (see Figure 3.8). However, it should always be borne in mind that the accuracy of a 'time impact analysis' will be limited by the following:

- The accuracy of the progress information applied to the programme's activities. If the activities' progress is taken from either (i) the contractor's interim payment application or (ii) the progress valued in the employer's payment notice, neither of those accounts of progress may be entirely inaccurate – because they may reflect the interests of the party who generated them. The party carrying out the 'time impact analysis' should therefore check the accuracy of progress information before applying it to the relevant programme and
- 'Time impact analysis' will be most accurate when the intervals, or progress periods from which progress information is taken, are practicably, at their shortest. In Figure 3.7, it was explained[145] that if the progress period is 4 weeks and within that period an activity is planned to be carried out in weeks 2 and 3, then by entering 100% completion into the programming software is likely to show that the activity is 100% complete within that progress period, but not necessary precisely when (during the 4-week progress period) the activity is carried out.

The 'as-built v as-planned' method will similarly rely upon accurate accounts of the works' progress. However, unlike with some 'time impact analysis' exercises, the 'as-built' information is ideally shown at the actual time when the activities are carried out. An easy way to visualise this is to think of watching time-lapse video of the works being carried out, and the activity bars being drawn on the programme as the relevant activity progresses. The 'as-built' critical path can either be determined during the course of the works (e.g. by successive 'time impact analyses') or retrospectively, after the work is completed.

The 'collapsed as-built' is an entirely retrospective method of analysis. It solely relies on accurate, factual information in order to show when the work activities were undertaken and how their progress was affected. The as-built critical path is established retrospectively, although in practice, the retrospective actions may include 'fine-tuning' or a final validation of the as-built critical path generated by preceding 'time impact analyses'. Removing the critical employer delay events should indicate the contractor's entitlement to an extension of time, and conversely, removing the critical contractor delay events should indicate the period for which the employer is entitled to delay damages.

145 Refer to Footnotes 118 and 120.

The collection of factual information throughout the relevant stages of a project,[146] particularly during construction[147] should result in the evolution of the contract programmes (mandatory in the case of NEC4 contracts), so the methods of delay analysis change accordingly. Once factual information begins to replace forecast information (such as a 'tender programme') the project programme will evolve and a method of delay analysis with increasing reliance upon factual information will replace those with a lesser reliance.

## 3.6 Float

'Float' is a term used in delay analysis and is helpfully described by the CIOB as:

> '*The amount of time by which an activity can be delayed*[148] *without causing a delay to subsequent activities or the completion date*'[149]

Therefore, if an activity has five days of float, it means that completion of that activity could be prolonged or delayed by up to five days without affecting the following activities or, if on the critical path, the completion date of the project. 'Float' in its widest sense is simply a time contingency that either a contractor builds into his programme or is created by the sequence and timing of other programmed activities. For example, an activity may take 8 weeks to carry out, but the contractor's programme includes a period of 12 weeks; therefore, it includes 4 weeks of 'float'. To identify 'float', a 'gap' is often shown between the relevant activity and its successor (refer to Figure 3.1[150]). It is rare for 'float' to be identified as a separate activity (for example, immediately following its 'parent' activity), as most programming software will not show the 'float' being used up in the event that the 'parent' activity is delayed or prolonged. NEC4[151] specifically requires the programme (and the 'Accepted Programme) to show *'provisions for...float'*.

Most contractors will include float for several reasons, such as to give themselves more time to complete an activity, to reduce the risk of delays to that activity's progress so as not to delay the start of the succeeding activity, and causing a critical delay to the project's overall progress. In this sense, float is often described as a 'time contingency' – providing the contractor with additional time for some activities within the overall programme. In principle, this is no different to a financial contingency, where the contractor may include an additional sum within the price for an activity or the contract sum to account for risks.

There are two 'sub-species' of float that the CIOB define[152] as

- 'Free float' – '*The period by which an activity may be delayed without delaying the start of any of its successor activities*' and

146 The procurement stage of a project (i.e. prior to physical work commencing) involving procurement and / or design may also be relevant.
147 Stage 5 of the RIBA Plan of Work 2020 – "Manufacturing and Construction".
148 Or the activity begins on time but its duration is prolonged.
149 Guide to Good Practice in the Management of Time in Major Projects (2nd Ed) 2018, CIOB.
150 Figure 3.1. 'Total float' is shown after activity 8 (10 days), activity 17 (2 days) and activity 25 (5 days), and 'Free float' is shown for activities 10, 11 and 12 (2 days overall) and activities 23 and 24 (10 days overall).
151 NEC4, clause 31.2 (first sub-heading under the 6th 'bullet' point).
152 See Table 3.7, Footnotes 92, 148, 149 and 150.

| Line | Name | Start | Finish | Duration |
|---|---|---|---|---|
| 1 | Date of Possession / starting date | 06 Jun 22 | 06 Jun 22 | |
| 2 | Site set up | 06 Jun 22 | 10 Jun 22 | 1w |
| 3 | Strip topsoil / remove trees | 13 Jun 22 | 17 Jun 22 | 1w |
| 4 | Excavate to reduce levels | 17 Jun 22 | 30 Jun 22 | 2w |
| 5 | Excavate foundations | 23 Jun 22 | 01 Jul 22 | 1w 2d |
| 6 | Concrete foundations | 01 Jul 22 | 07 Jul 22 | 1w |
| 7 | Bolt boxes | 08 Jul 22 | 08 Jul 22 | 1d |
| 8 | Steelwork procurement and manufacture | 06 Jun 22 | 24 Jun 22 | 3w |
| **9** | **Crane Hire period (inc float)** | **11 Jul 22** | **17 Aug 22** | **5w 3d** |
| 10 | Steel frame erection | 11 Jul 22 | 22 Jul 22 | 2w |
| 11 | Roof safety netting | 25 Jul 22 | 27 Jul 22 | 3d |
| 12 | Profiled roof sheet and skylights | 28 Jul 22 | 10 Aug 22 | 2w |
| 13 | Wall cladding | 04 Aug 22 | 17 Aug 22 | 2w |

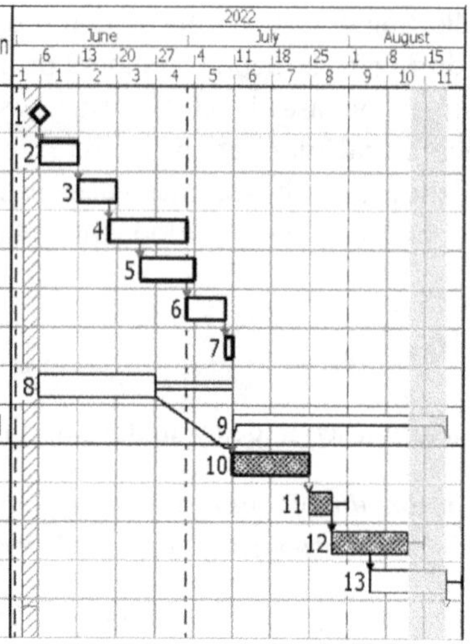


**Figure 3.16** Float (crane hire) example.

- 'Total float' – '*The amount of time an activity can be delayed without delaying a key date, sectional completion date, or the completion date without contravening a constraint.*'

The SCL Protocol includes similar definitions for 'free float' and 'total float' but, unlike the CIOB definitions, do not consider 'Key Dates' constraints or completing the work in sections. A simple example of float is shown in Figure 3.16. The contractor has three activities (no's 10, 11 and 12) that require crane hire for a total period of 4w and 3d. The following activity (no. 13 – Wall cladding) can be completed without the use of the crane, but the contractor has allowed a total period of 5w and 3d for crane hire in his programme (see 'summary' activity no. 9). This leaves a 'float' period of 1w (free float), shown in the grey, shaded section.[153]

It is important to consider the characteristics of float as follows:

- The financial effect (if any) that using up float may have on the cost that the contractor incurs for carrying out the work and its entitlement to recover those costs and
- The effect on the progress of the contractor's work and any entitlement to adjust the period or periods for completing the works (i.e. if the delay postpones the 'Completion Date' (JCT 2016) or 'planned Completion' (NEC4) as summarised in Table 3.17 - 'Extension of time criteria for standard building contracts').

### 3.6.1 'Financial' Effects of Using up Float

The financial effects of using up float are straightforward. If an event occurs for which the employer is responsible and this causes the work operations to use part or all of the float period, then if the contractor can demonstrate his additional costs, he ought to be able to claim the additional costs from the employer. One financial effect may be that the contractor incurs additional costs within the float period that he may not have otherwise incurred, had

153 Figure 3.15. The grey, shaded 'float' period for crane hire is between 11 and 17 August 2022.

**Table 3.17** Extension of time criteria for standard building contracts.

| Contract | Criteria for extension of time |
|---|---|
| JCT SBC/Q 2016[a)]<br>JCT DB 2016[b)]<br>JCT IC 2016[c)] | The 'Relevant Event' is likely to delay the works' (or a section's) completion beyond the 'Completion Date'. |
| JCT MW 2016[d)] | If a delay is beyond the contractor's control and it's apparent that the works will not be completed by the 'date for completion'. |
| NEC4[e)] | Extension of time is 'assessed' as the period of delay (caused by the 'compensation event' that 'planned Completion' occurs later than 'planned Completion' shown on the 'Accepted Programme'. |

a) JCT SBC/Q 2016 – clause 28.1
b) JCT DB 2016 – clause 2.25.1
c) JCT IC 2016 – clause 2.19.1
d) JCT MW 2016
e) NEC4 clause 63.5 (Assessing compensation events)

none the float period been used up. In Section 2.4 (Categories of claim), a simple example of a 'disruption claim' was given – where the contractor plans to use a tower crane for the first 8 weeks of a 12-week period, leaving 'free float' of 4 weeks. Owing to a variation, the tower crane is actually required for 11 weeks of the programmed 12-week period, so 3 weeks of free float are used up. There is no effect on the critical path as the variation is completed within the planned 12-week period (i.e. with 1 week to spare), but the financial effect is that the contractor incurs a further 3 weeks of hire costs for the tower crane. In this particular example, the additional 3 weeks' hire cost ought to be recoverable from the employer as it was caused by a variation.

Clearly, if the event causing the float to be used up is one for which the contractor is responsible, there should be no entitlement to recover any additional costs from the employer. Indeed, and on the basis that the contractor's time-related facilities costs (or 'preliminaries') are covered by the contract sum between the contractual start and finish dates, he is likely to receive some payment for the relevant float period in any event. However, if the contractor can demonstrate that a sub-contractor caused the float to be used up, then provided that the sub-contractor was obliged to complete the preceding activity (or activities) before the float period began, then he ought to be able to recover any loss or additional cost from the sub-contractor. However, contractors should not claim float exclusively for themselves – to the disadvantage of sub-contractors. In one case,[154] the progress of a groundwork sub-contractor's work was delayed by two matters; the main contractor's failure to remove groundwater and the sub-contractor's insufficient resources. The sub-contractor claimed that it did not cause delay to the main contract works because its delays only used up the float period in the main contractor's programme. However, the main contractor argued that only it was entitled to the benefit of the float, and not the sub-contractor. The Judge rejected the contractor's argument:

154 *Ascon Contracting Ltd v Alfred McAlpine Construction Isle of Man Ltd* (2000) 16 Const LJ 316.

*'The float is certainly of value to the main contractor in the sense that delays of up to that total amount, however caused, can be accommodated without involving him in liability for liquidated damages to the employer or, if he calculates his own prolongation costs from the contractual completion date (as McAlpine has here) rather than from the earlier date which might have been achieved, in any such costs. He cannot, however, while accepting that benefit as against the employer, claim against the sub-contractor as if it did not exist. That is self-evident if total delays as against sub-programmes do not exceed the float. The main contractor, not having suffered any loss of the above kinds, cannot recover from sub-contractors the hypothetical loss he would have suffered had the float not existed and that will be so whether the delay is wholly the fault of one sub-contractor, or wholly that of the main contractor himself, or spread in varying degrees between several sub-contractors and the main contractor.'*[155]

### 3.6.2 'Time' Effects of Using up Float

It is self-evident that a delay to an activity that is equal or less than its available float will not directly affect the critical path activities, and therefore, a project's duration will not be prolonged beyond its completion date. Therefore, it is simply wrong for a contractor to argue that it is entitled to an extension of time if a non-critical activity is delayed, and float (or part of it) is used up. This approach is reflected in both the JCT 2016 and NEC4 forms of contract, when considering one of the essential criteria for a contractor's entitlement to an extension of time (see Section 3.2.1) – as shown in Table 3.17:

If the contractor intends to complete his works before the 'completion date', he may do so. This entitlement is reflected in both the JCT 2016 and NEC4 contracts regarding completion[156] insofar as the contractor may complete *'on or before'* the 'completion date'. However, there is no corresponding general obligation for the employer to facilitate the contractor completing the works early,[157] unless:

- The contractor and employer agree to accelerate the work[158] (see Section 3.8) or
- Under NEC4, the contractor revises the programme to show that the 'Client' or 'Others' must now work to earlier dates than were previously shown[159] and this programme then becomes the 'Accepted programme'[160].

Much has been written regarding who 'owns' the float or who is entitled to take benefit of it. Often, examples from other jurisdictions are given to advance a position that the contractor or the employer 'owns' the float. Unless the contract contains a float-sharing

155 See Footnote 154 at para 92.
156 For JCT 2016 contracts, refer to Table 3.1 and for NEC4 (clause 30.1) refer to section 3.1.3.
157 *Glenlion Construction Ltd v The Guinness Trust* (1987) 39 BLR 89.
158 Acceleration provisions in JCT contracts: JCT SBC/Q 2016 – Schedule 2, paragraph 2, JCT DB 2016 – Schedule 2, paragraph 4. Acceleration provisions in NEC4 – clause 36 (Acceleration).
159 NEC4 – otherwise a 'compensation event' under clause 60.1(5).
160 NEC4 – clause 31.3.

clause (and neither of the JCT 2016 nor NEC4 contracts do), then the float is simply expended by whatever delays occur (be they employer delays, contractor delays or a neutral delays or a combination of all three).

## 3.7 Concurrency

There has been considerable discussion over the last 20 years covering the topic of 'concurrency' (or concurrent delays).[161] This volume of discussion is perhaps somewhat surprising as even leading construction law commentators acknowledge that *'true concurrency…will only arise in exceptional factual situations.'*[162]

- Concurrent delay has been defined and described as
  - SCL: *'True concurrent delay is the occurrence of two or more delay events at the same time, one an employer risk event, and the other a contractor risk event, and the effects of which are felt at the same time.'*[163]
  - John Marrin KC: '*…the expression "concurrent delay" is used to denote a period of project overrun which is caused by two or more effective causes of delay which are of approximately equal potency.'*[161]

The current position regarding 'concurrency' in English law has also developed over this period, beginning with what has been called the *'Malmaison approach'*,[164] which has been supported by a number of English cases.[165] This support culminated in a 2012 case *Walter Lilly,*[166] in which the judge (Mr Justice Akenhead) said:

> *'In any event, I am clearly of the view that, where there is an extension of time clause such as that agreed upon in this case and where delay is caused by two or more effective causes, one of which entitles the contractor to an extension of time as being a relevant event, the contractor is entitled to a full extension of time. Part of the logic of this is that many of the relevant events would otherwise amount to acts of prevention and that it would be wrong in principle to construe clause 25*[167] *on the basis that the contractor should be denied a full extension of time in those circumstances. More importantly however, there is a straight contractual interpretation of clause 25 which points very*

---

161 In particular, refer to the Society of Construction Law (SCL) papers by John Marrin KC: 'Concurrent Delay' paper 100 (February 2002) and 'Concurrent Delay Revisited' paper 173 (February 2013), and the SCL's Delay and Disruption Protocol 2nd edition (2017) – see Section 3.3.
162 John Marrin QC, 'Concurrent Delay Revisited' paper 173 (February 2013) and "*…those relatively rare cases where the fact-finding exercise leads to the conclusion that, as a matter of fact, two events are to be regarded as having independently caused delay to the contractor's progress during the same period.*"
163 SCL Delay and Disruption Protocol (2nd Ed) 2017.
164 Following *Henry Boot Construction Ltd v Malmaison Hotel (Manchester) Ltd [1999] 70 Con LR 32.*
165 See *Royal Brompton NHS Trust v Hammond & Ors* (2002) *EWHC 2037 (TCC), Steria Ltd v Sigma Wireless Communications Ltd [2007] EWHC 3454 (TCC), De Beers UK Ltd v Atos Origin IT Services UK Ltd [2010]* EWHC 3276 (TCC).
166 *Walter Lilly & Company Ltd v Giles Patrick Cyril Mackay & Anor (2012)* EWHC 1773 (TCC). The contractor (Walter Lilly) was engaged under a JCT Standard Form of Building Contract 1998 Private Without Quantities (including amendments).
167 JCT SBC/XQ 1998 clause 25 (Extensions of time) is similar to clauses 2.27 and 2.28 of JCT SBC/Q 2016.

*strongly in favour of the view that, provided the relevant events can be shown to have delayed the Works, the contractor is entitled to an extension of time for the whole period of delay caused by the relevant events in question.*'[168]

From *Walter Lilly*, the question of entitlement for extensions of time is resolved in favour of the contractor, provided that delays caused by the employer are those that entitle the contractor to an extension of time under the terms of the building contract. The question of the contractor's financial compensation for concurrent delays also appears to have been resolved. It is suggested that owing to the contractor delay, he cannot satisfy the 'but for' causation test, so he should not be entitled to his costs for the concurrent delay period. The positions regarding time and financial compensation for concurrent delays are now set out in the SCL Protocol[163] under 'Core Principles' 10 and 14:

> '*10. Concurrent delay – effect on entitlement to EOT*
> *True concurrent delay is the occurrence of two or more delay events at the same time, one an employer risk event, the other a contractor risk event, and the effects of which are felt at the same time. For concurrent delay to exist, each of the employer risk event and the contractor risk event must be an effective cause of delay to completion (i.e. the delays must both affect the critical path). Where contractor delay to completion occurs or has an effect concurrently with employer delay to completion, the contractor's concurrent delay should not reduce any EOT due.*
> *14. Concurrent delay – effect on entitlement to compensation for prolongation*
> *Where employer delay to completion and contractor delay to completion are concurrent and, as a result of that delay the contractor incurs additional costs, then the contractor should only recover compensation if it is able to separate the additional costs caused by the employer delay from those caused by the contractor delay. If it would have incurred the additional costs in any event as a result of contractor delay, the contractor will not be entitled to recover those additional costs.*'

It is not uncommon for employers to include amendments to the contract conditions that restrict or remove any entitlement the contractor may have in the event of concurrent delays. The Courts have upheld such amendments.[169]

In summary, in the unlikely event that true concurrency occurs, the contractor ought to be awarded an extension of time (removing his liability for delay damages to the employer) but should not be entitled to his additional prolongation costs for the period of concurrency.

## 3.8 Acceleration

### 3.8.1 Introduction

Construction contracts include more provisions for managing the effects of a decrease in the rate of progress (or managing and avoiding delays), than for managing the rate of progress increasing. In this context, an increase in the rate of progress is referred to

168 *Walter Lilly v Mackay (2012)* at para 370.
169 *Multiplex Constructions (UK) Ltd v Honeywell Control Systems Ltd (No 2)* (2007) EWHC 447 (TCC) at 56, and North Midland Building Ltd v Cyden Homes Ltd [2018] EWCA Civ 1744 at 39.

as 'acceleration', which has the aim of completing the project earlier than its original completion date. Achieving an earlier completion date in this section refers to the employer instructing the contractor to accelerate to either finish earlier, or to overcome a delay for which the employer is culpable. This section does not consider acceleration in order to overcome a delay caused by the contractor, as this is deemed to be 'mitigation' (see Sections 2.7.1 and 3.1.3).

The general nature of construction contracts is to penalise or punish a failure to perform, rather than reward performance which exceeds the contract's requirements. This is understandable from the employer's perspective in two aspects:

- First, the employer may suffer significant inconvenience and financial losses if he cannot use the works from the agreed completion date. In such cases, the contract usually provides for the employer to claim delay damages and an extension of time mechanism to preserve the employer's entitlement to delay damages, if he (and not the contractor) is responsible for part of that delay.
- Second, the employer is normally content to pay a certain price if the contractor satisfactorily completes the works by a definite date. In some instances, the employer may not want the contractor to complete the works before the agreed completion date, as the employer usually becomes responsible for the works at this stage, so depending upon the particular contract's terms, he may become liable for the costs of insurance, maintenance, operation, and security and may have to re-pay any deducted retention to the contractor earlier than he planned.

'Acceleration' has been usefully defined as follows:

- *'The application of additional resources or alternative construction sequences or methodologies seeking to achieve the planned scope of works in a shorter time than planned or execution of additional scope of work within the original planned duration.'*[170]
- *'Action to make up time lost or otherwise complete work at a rate quicker than agreed in the approved schedule [or programme], usually at the employer's expense.'*[171]

The JCT SBC/Q 2016, JCT DB 2016 and NEC4 contracts deal with acceleration in slightly different ways and in varying degrees of detail, but in each case, the contractor provides the employer with a quotation, setting out the costs of accelerating, along with the earlier completion date that may be achieved. For example, none of the three contracts provide any procedure for how the contractor arrives at his price for acceleration (as they otherwise do for valuing variations), so other than perhaps adjudication, there is no default procedure that the parties can rely upon if they can't reach agreement.

The reasons for acceleration usually fall into one the following categories:

- By agreement between the parties or, if the contract so provides, on the instruction of the employer's representative;
- Unilaterally on the initiative of the contractor, often categorised as 'mitigation' by the contractor or as 'using best endeavours' by the employer; or
- Constructive acceleration where the contractor argues that it has no real alternative in the circumstances.

---

170 SCL Delay and Disruption Protocol 2nd Ed (2017).
171 CIOB Guide to Good Practice in the Management of Time in Major Projects 2nd Ed (2018).

### 3.8.2 Acceleration by Agreement or Instruction

Ordinarily, the employer (or his representative) has no power to instruct the contractor to accelerate work because contractor's principal obligation is to complete the work within the time specified, or where no particular contract period is specified – within a reasonable time. The contractor cannot be compelled to complete earlier than the agreed date unless there is an express contract term authorising the architect to require acceleration. With the exception of its Management Contract, JCT contracts did not provide for acceleration as a means of maintaining the completion date (or finishing earlier) until the 2011 editions of its contracts.

However, it is not unknown for the employer or his representative to instruct acceleration where the contract does not provide for it. In that situation, if the contractor simply refuses to carry out the instruction, there is no difficulty. However, problems may arise if the contractor obeys the instruction, as it is doubtful that the contractor has any entitlement to payment. If it is clear that the employer's representative has no power to instruct acceleration, the contractor should be aware and cannot subsequently plead lack of knowledge. Otherwise everything will depend upon the authority, whether ostensible or implied, of the employer's representative to give the instruction as agent for the employer (see Chapter 9). In most cases, the employer's representative will not have such authority. Obviously, there should be no difficulty in the contractor obtaining payment where the employer's representative, in the exercise of powers under a contract, orders acceleration of the works, or the employer and the contractor otherwise agree to implement acceleration measures.

In practice, instructions to, or agreements with, the contractor to accelerate often occur towards the end of a contract period – when it becomes more obvious that the completion date will be exceeded and the employer is desperate to have the project finished by a specific date. The best time to accelerate is as soon as it seems likely that the contract period will be exceeded because that gives the maximum time to the contractor to put the acceleration measures in place.

The only sensible acceleration instructions or agreements are those which require completion by a specific date, as it is essential that the agreement specifies the date for completion whether that is the original completion date or a different date. Inevitably, the contractor will require additional payment so a quotation is essential so that it can be accepted by the employer and both parties know exactly where they stand.

If the acceleration simply amounts to the contractor agreeing to add specific resources and work during a specified number of extra hours per week, that is very unsatisfactory, because the contractor does not undertake to finish by a specific date and, after all the extra resources and overtime working have been applied, there may be no difference whatever in the projected date for completion, because the contractor may have worked inefficiently. If the parties intend to enter into an 'ad-hoc' acceleration agreement, then it would be advisable to follow the structure found in either the JCT 2016 or NEC4 contracts.

There are four basic situations which may result in an acceleration agreement:

- Where it is unlikely that the contractor will complete by the contractual completion date, because of delays for which the contractor would be entitled to an extension of time;
- Where it is unlikely that the contractor will complete by the contractual completion date, because of delays caused by the contractor;

- Where it is unlikely that the contractor will complete by the contractual completion date, because of delays caused by a mixture of the two previous reasons, or if 'neutral' delays occur (such as those caused by adverse weather, or industrial action); and
- Where there is no delay, but the employer wishes the contractor to complete before the original completion date.

The situations in the first and third bullet points may be considered together. In each case, without acceleration the employer's representative would be obliged to give an extension of time. An acceleration agreement would effectively be a variation to the contract. It provides that despite the fact that the contractor is likely to finish after the completion date for reasons which would entitle him to an extension of time, he undertakes to complete by the original completion date or at least a date which will be earlier than any extended date. The situation in the fourth bullet point is similar, because in each of bullet points no's 1, 3 and 4, the contractor is undertaking to complete earlier than required under the terms of the contract. Therefore, in each of these situations, there is 'consideration' on the part of the contractor in that he is agreeing to do something[172] additional, and not required in the original contract.

The somewhat different scenario presented by the situation in the second bullet point raises a problem, because in an agreement to accelerate, the contractor is simply undertaking to do that which it is already obliged to do under the terms of the contracts (i.e. use his 'best endeavours' and having the incentive to mitigate his losses). It is an old rule that simply agreeing to carry out a duty already imposed by a contract between the same parties is not consideration, and does not warrant a further payment.[173] Therefore, the question is: what is the 'consideration' for which the contractor is entitled to be paid, for seemingly doing what it is already obliged to do? The answer may be found in the 1990 Court of Appeal case – *Williams v Roffey Brothers & Nicholls (Contractors) Ltd.*[174]

In that case Williams, a carpenter, entered into a contract with Roffey Brothers, a contractor, to carry out carpentry work on 27 flats for the sum of £20,000. Williams ran into financial difficulties, because the price as agreed was too low and he had failed to supervise his workmen properly. Roffey was concerned that Williams should continue the work to completion, because it would take time to engage another carpenter and they risked being late and incurring liquidated damages. Therefore, it was agreed that the contractor would pay Williams an additional £10,300 to be paid at the rate of £575 for each flat completed. The matter progressed to be heard in the Court of Appeal. The Court held that there was consideration for the agreement and the Roffey accepted that the agreement provided it with a benefit (in the form of avoiding or preventing a dis-benefit), because it ensured that Williams would continue work, that the contractor would not suffer liquidated damages and that it would avoid the expense and trouble of engaging others to finish the work.

The Court set out what it described as 'the present state of the law on this subject' as follows:

> *'(i) if A has entered into a contract with B to do work or, or to supply goods or services to, B in return for payment by B; and*

172 Chitty on Contracts – (30th Ed) 2008at 3-004 defines 'consideration' as "*...the requirement that "something of value" must be given and accordingly...that considering is either some detriment to the promise (in that he may give value) or some benefit to the promisor (in that he may receive value).*"

173 *Stilk v Myrick* (1809) 2 Camp 317.

174 [1990] 1 All ER 512.

*(ii) at some stage before A has completely performed his obligations under the contract B has reason to doubt whether A will, or will be able to, complete his side of the bargain; and*
*(iii) B thereupon promises A an additional payment in return for A's promise to perform his contractual obligations on time; and*
*(iv) as a result of giving his promise, B obtains in practice a benefit, or obviates a disbenefit; and*
*(v) B's promise is not given as a result of economic duress or fraud on the part of A; then*
*(vi) the benefit to B is capable of being consideration for B's promise, so that the promise will be legally binding.'*[175]

Although the court did not purport to overrule *Stilk v Myrick*, but merely to distinguish it, it is not clear how the two can be reconciled.[176]

Any question of the absence of consideration can be resolved quite easily if the parties take the simple step of executing the agreement as a deed, so consideration (in its legal sense) is not required. In each agreement, the grounds which would normally entitle the contractor to an extension of time are subsumed into the agreement so that the completion date as stated in the agreement becomes the new completion date for the contract. The agreement should provide that the employer is entitled to recover liquidated damages in the usual way if the contractor fails to complete by the newly agreed completion date.

There is a further very important question. What is the employer's position where the contractor fails to complete by the contract completion date and, in fact, fails to deal with any of the delay or alternatively deals with only part of the delay? Clearly, the employer would be entitled to liquidated damages for the period of delay, but in some of the situations, the employer would have been entitled to such damages even if there was no acceleration agreement. Moreover, liquidated damages have been held to be exhaustive of damages for delay.[177]

Plainly, a contractor failing to comply with an acceleration agreement would be in breach of that agreement, so is it simply the case that the employer would be entitled to damages in addition to any liquidated damages which may be recoverable for failure to comply with the building contract's requirement to complete by the completion date? It may be argued that damages for breach of the agreement would not be attributable to the same breach for which liquidated damages are recoverable, and, therefore, they would not be caught by the *dicta* in *Temloc v Errill*. Breach of the acceleration agreement could perhaps be distinguished from breach of the completion obligation in building contract in the same way as the court of appeal isolated an entirely separate *benefit* to the contractor in *Williams v Roffey Brothers*[176] even though, on its face, it seemed as if Williams was simply undertaking to do what he had agreed to do under the existing contract. In this instance, the contractor would be in breach of an agreement to provide a separate benefit – i.e. achieving an earlier completion date or maintaining the completion date despite delays. It would be advisable for the acceleration agreement to include clear words to distinguish between any employer's damages for the contractor's breach of the acceleration agreement,

175 [1990] 1 All ER 512 at 522 per Glidewell LJ.
176 See the discussion on this case in *Chitty on Contracts* (2008) 30th edition Sweet & Maxwell, London at paragraphs 3-068 and 3-069.
177 *Temloc Ltd v Errill Properties Ltd* (1987) 39 BLR 30.

and his recovery of delay damages in the event that the contractor is in breach of his completion obligations under the building contract.

Alternatively, it has been suggested that where an employer and a contractor enter into an acceleration agreement they are simply varying the building contract by setting an earlier completion date in return for an additional payment to the contractor. In doing so, they will take into account, the amount (if any) by which the contractor is in culpable delay, the amount of any extension of time which would be applicable and the rate of liquidated damages. The additional payment made by the employer will take account of all these factors including that in the event of a failure by the contractor to achieve the completion date, the only remedy will be liquidated damages. In other words, the parties will simply negotiate in full knowledge of the circumstances. That appears to be the better view and one which has clear similarities to the acceleration provisions in the JCT 2016 and NEC4 contracts.

If further delays occur after entering into the acceleration agreement which would entitle the contractor to an extension of time under the particular contract in use, the employer or his representative must deal with them in accordance with the relevant building contract, with reference to the newly agreed completion date.

### 3.8.3 Unilateral Acceleration

This is the situation where a contractor accelerates its progress with neither any agreement with the employer nor instruction from the employer's representative. No pressure has been placed on it by the refusal of an extension of time, and indeed in this situation, it may be that the contractor is reasonably confident of getting an extension to the contract period or completion date. The contractor may nevertheless decide to increase the resources allocated to the project. One reason for doing this may be in order to find work for operatives from another site which is drawing to a close. The result may be that some time is recovered and an extension of time is not required.

In most such cases, the contractor will find it difficult to contend that it was doing other than using its best endeavours (see Sections 3.1.3 and 11.3.4) to reduce delay.

### 3.8.4 Constructive Acceleration

Claims for 'constructive acceleration' have been recognised by courts in the United States and Canada,[178] but it seems that until recently, their UK counterparts[179] were considerably less likely to do so. Contractors sometimes advance an argument to justify 'constructive acceleration' if they consider that the employer's representative has failed to give an extension of time to which the contractor believes he is entitled. In doing so, a contractor contends that it was forced to increase its resources in order to finish by the date for completion, fearing that it would be charged delay damages because the employer's representative had failed to give an extension of time, to which the contractor felt it is clearly entitled.

178 Burr, A – Delay and Disruption in Construction Contracts, (5th Ed) 2016 Routledge at 13-093.
179 Bailey, J – Construction Law, (3rd Ed) [2020], London Publishing Partnership at 11.198, adding that other jurisdictions such as Australia, Hong Kong and Singapore have not recognised 'constructive acceleration'.

The SCL Protocol[180] defines 'constructive acceleration' as:

> *'Acceleration following failure by the [the employer's representative] to recognise that the Contractor has encountered Employer Delay for which it is entitled to an EOT and which failure required the Contractor to accelerate its progress in order to complete the works by the prevailing contract completion date. This situation may be brought about by the Employer's denial of a valid request for an EOT or by the [employer's representative's] late granting of an EOT. This is rarely recognised under English law.'*

An important question to be asked before a 'constructive acceleration' argument can be advanced is to consider the commercial pressure that the contractor faces. In the absence of (what ought to be a clear entitlement to) an extension of time, the contractor faces the risk of the employer deducting significant delay damages.

The contractor's problem is one of causation. Where the employer's representative wrongfully fails or refuses to make an extension of time, the appropriate contractual (or statutory) route for the contractor to take will be one of the methods of dispute resolution (most likely, adjudication). If the contractor is entitled to an extension of time, it may be said that he should confidently continue the work, without increasing resources, secure in the knowledge that it will be able to recover its prolongation loss and/or expense, and any wrongfully deducted delay damages wrongfully deducted via the contract's dispute resolution provisions. Therefore, if the contractor increases its resources, that is not a direct result of the architect's breach, but is the contractor's decision.

In practice, it must be acknowledged that a contractor in this position may not be so confident. The facts may be complex and the level of delay damages may be high. Faith in an adjudicator may not be total and it may be cheaper, even without recovering acceleration costs, for the contractor to simply accelerate rather than face liquidated damages with no guarantee that an extension of time will ultimately be made. From the viewpoint of simple, commercial realism, the contractor may feel he has no choice other than to accelerate and take a chance as to recovery. Unless the contractor can show that the employer's representative has not given an adequate explanation why the contract period will not be extended, and in those circumstances the amount of delay damages has a real prospect of causing economic hardship (even insolvency), then a 'constructive acceleration' claim probably has little chance of success.[181]

Constructive acceleration was considered by the court in a 2002 case.[182] In the course of an extremely long judgment, the court concluded that the contractor was entitled to recover the cost of acceleration if an extension of time was justified, but refused, and the liquidated damages were 'significant':

> *'[The contractors] say that since they incurred these costs in attempting to comply with [the employer's] wish for the contract to be kept to time and against the background of [the employer's] refusal to grant appropriate extensions of time, they are entitled to be paid for the work which they did in trying to accelerate the work to keep up with the*

180 Society of Construction Law – Delay and Disruption Protocol (2nd Ed) (2017) - 'EOT' means 'extension of time'. No definition is given in the CIOB's 'Guide to Good Practice in the Management of Time in Major Projects' 2nd edition (2018).

181 *Perini Corporation v Commonwealth of Australia* (1969) 12 BLR 82.

182 *Motherwell Bridge Construction Ltd v Micafil Vakuumtechnik and Another* (2002) 81 Con LR 44.

> *schedule ... I am satisfied that they were incurred by [the contractors] in an attempt to recover time lost in completing the work in circumstances where [the contractors] were subject to significant penalties for delay if they failed to complete the work on time. The causes were, in particular, the restrictions which [the contractor's] encountered when they entered on site and the very substantially increased scope of the work ...*'[183]

The authors of *Hudson*[184] consider that if any 'constructive acceleration' claim is to succeed, the contractor must show that the claim is one for breach of contract. In a 2012 case,[185] the court considered acceleration measures that took the form of a 'recovery programme' to mitigate delays:

> *'142 ...I have no doubt and I find that it was sensible and reasonable for SRS to institute its recovery programme by way of mitigation to try to recover the delays totalling 42 days which had been caused by CBUK's breach of contract. It was reasonable because SRS was faced with a very substantial liquidated damages liability to Mace (£500,000 a week) and a six week delay (attributable to CBUK) could potentially have cost SRS some £3 million. To that one would have to add an additional six weeks' worth of site related and overhead costs which SRS would have had to incur. Therefore a concerted effort to work additional hours to try to recover the delay was at every level reasonable. Ultimately, it was not seriously argued otherwise on behalf of CBUK. I have regard to the authorities relating to mitigation of damage, one of the more well-known being Lord MacMillan's dictum in Banco De Portugal v Waterlow & Sons Ltd [1932] AC 452 at page 506:*
>
> *'Where the sufferer from a breach of contact finds himself in consequence of that breach placed in the position of embarrassment the measures which he may be driven to adopt in order to extricate himself ought not be weighed in nice scales at the instance of the party whose breach of contract has occasioned the difficulty. It is often easy after an emergency has passed to criticise the steps which have been taken to meet it, but such criticism does not come well from those who themselves created the emergency. The law is satisfied if the party placed in a difficult situation by reason of the breach of a duty owed to him has acted reasonably in the adoption of remedial measures, and he will not be held disentitled to recover the costs of such measures merely because the party in breach can suggest that other measures less burdensome to him might have been taken.'*
> *143 It follows that the reasonable costs incurred by SRS in implementing acceleration measures to overcome the 42 days delay caused to the steel erection programme by CBUK's failures to deliver steel in accordance with the December Programme should be recoverable. The fact that the acceleration was not ultimately successful due to factors which SRS can not establish are attributable to those failures can not and should not prevent or limit recovery of such costs...'*

183 [2002] 81 Con LR 44 at paragraphs 544 and 548 per Judge Toulmin.

184 Hudson's Building and Engineering Contracts, (14th Ed) 2020 at 6-077.

185 *Cleveland Bridge UK Ltd v Severfield-Rowen Structures Ltd* (2012) EWHC 3652 (TCC) at paragraphs 142 and 143 per Akenhead J.

## 3.9 Sectional Completion

### 3.9.1 Where There Is a Single Date for Completion

Employers and contractors often run into difficulties where the employer chooses to incorporate sectional completion into the building contract. It should be noted that, where there is just one date for completion in the contract, completion of separate sections cannot usually be achieved by simply inserting intermediate dates in the specification or bills of quantities[186] or by a mere instruction to the contractor. Moreover, although a court may be prepared to imply a date for completion of the Works in the absence of agreement, dates for sectional completion will not be implied,[187] and certainly not in JCT 2016 contracts which have a clause[188] giving priority to the printed form over other contract documents.

In such cases, the single completion date will take precedence despite more than one completion dates for particular parts of the works being set out in a subordinate document in the contract. Where there is no 'priority of documents' clause, the ordinary rule will apply that the 'written terms' prevail over the 'printed terms' so any intermediate or sectional completion dates stated in a secondary or subservient document (i.e. the 'printed terms') are unlikely to take precedence over a single completion date set out in the contract itself (the 'written terms'). The 'written terms' in JCT contracts are the 'Contract Particulars' and in NEC4 contracts, they are found in the 'Contract Data'.[189] Attempting to impose sectional completion dates after entering into a building contract that only contains a single completion date would immediately give rise to a number of contractual difficulties. For example, most standard form contracts provide for only one document to set out the contractor's obligations as to completion, rate of delay damages, and for completion of defects. Furthermore, most building contracts refer to extensions of time in relation to the *completion date* or fixing a new *completion date* (singular dates in both respects). Such problems can only be resolved by goodwill on both parties by agreeing to formally[190] vary the building contract so it includes sectional completion dates. The relevant entries for sectional completion in JCT and NEC4 contracts are shown in Table 3.18.

### 3.9.2 Dependent Sections

A common problem occurs where sectional completion has been properly incorporated into the building contract, but there is an interrelationship between two or more of the sections, causing one or more sections to be dependent upon progress or completion of another. This is not a strictly contractual issue, but one of logistics. For example, a school project may be divided into four sections, but section 3 may be dependent on section 1 in the sense that the contractor cannot be given possession of section 3 until section 1 has reached practical completion because the occupants using section 3 have to be moved to section 1.

---

186 *M J Gleeson (Contractors) Ltd v Hillingdon Borough Council* (1970) 215 EG 165.
187 *Bruno Zornow (Builders) v Beechcroft Developments* (1990) 6 Const LJ 132.
188 JCT SBC/Q 2016, JCT DB 2016, JCT IC 2016 – clause 1.3, JCT MW 2016 – clause 1.2 ('Agreement etc, to be read as a whole) – also referred to as a 'priority of documents' clause.
189 In this example, the completion, or sectional completion obligations are set out in 'Contract Data – Part One'.
190 By a deed of variation.

**Table 3.18** Sectional completion: JCT and NEC4 contracts.

| Information required for each Section | JCT 2016Contract Particulars | | | NEC4 Contract Data[a)] |
|---|---|---|---|---|
| | SBC/Q | DB | IC | |
| • Statement that Works are divided into Sections | 6th Recital | 5th Recital | 8th Recital | Option X5 |
| • A clear description of each Section | 6th Recital | 5th Recital | 8th Recital | Option X5 |
| • The date of possession | Cl 2.4 | Cl 2.3 | Cl 2.4 | 3 |
| • The dates for completion of Sections | Cl 1.1 | Cl 1.1 | Cl 1.1 | Option X5 |
| • Maximum periods of deferment of possession | Cl 2.5 | Cl 2.4 | Cl 2.5 | — |
| • Rates of liquidated damages | Cl 2.32.2 | Cl 2.29.2 | Cl 2.23.2 | Option X7 |
| • 'Section Sum' – the value of each section (for apportioning liquidated damages where the employer has taken partial possession of a Section) | Cl 2.37 | Cl 2.34 | Cl 2.29 | — |
| • The rectification period for each Section | Cl 2.38 | Cl 2.35 | Cl 2.30 | 4 |

a) Contract Data Part One.

Suppose in the above example, that the dates for possession and completion of each section are inserted into the contract and expressed as actual dates. The date for completion of section 1 may be 25 March and the date for possession of section 3 may be 30 March to allow the transfer of pupils and staff from one section to another. A problem will arise if section 1 is not finished by the completion date (25 March). This could be because the contractor has been inefficient or it may be because events have occurred which entitle the contractor to an extension of time. Where a building contract splits the works into separate sections, an extension of time must be given in respect of the particular section affected by the delaying event in the section on which it is dependent. Therefore, if the contractor is entitled to an extension of time for a delay to section 1, then the extension of time is given to section 1 only. Whether it is so entitled is irrelevant, because in any event, when the date to start section 3 arrives (30 March), the contractor will still be working on section 1 and the occupants of section 3 cannot be transferred, and the employer cannot give the contractor possession of section 3 on the due date. If the contractor is not entitled to an extension of time for section 1, the employer's representative will rightly say that owing to the contractor's delay in completing section 1; he has prevented the employer from giving possession of section 3 on the due date (30 March).

This approach is to misunderstand the situation entirely – the principles of causation must be applied. The factual cause of the delay to possession of section 3 is not the contractor's delay to section 1, but the fact that the two sections are linked. If they were not, the contractor's delay to section 1 would not affect section 3 at all. However, where the dates for possession and completion are simply expressed as actual dates, the contract terms[191] do

191 Whilst any interrelationship between sections 1 and 3 in this example may not be immediately apparent from the contract terms (e.g. the 'Contract Particulars' under JCT or the 'Contract Data Part One' under NEC4), it is highly unlikely from the other contract documents (such as a contractor's method

not put the contractor on notice of the likely problem – instead, his programming and operational staff must alert him to the problem caused by the project's logistics. The contractor is likely to argue that the employer is in breach of contract for failing to give possession of section 3 and he would be correct. There are two aspects to this problem:

- What can the employer or the employer's representative do to retrieve the situation?
- What could have been done to avoid it in the first place?

If there is provision for the employer to defer possession[192] of any of the sections by a period of time, the employer may take that route and the contractor will be entitled to an extension of time and probably whatever amount of loss it has suffered as a result of the deferment of possession. If there is no deferment provision, the correct analysis of the situation appears to be that there is a breach of contract which, dependent upon its duration, may become repudiatory in nature if the period of possession is significant. In any event, the contractor would be entitled to recover as damages the amount of loss it has suffered as a result. That is fairly straightforward and the amount payable to the contractor, whether by virtue of a loss and/or expense clause in the contract or as damages for the breach may not be substantial. The contractor would have to demonstrate its loss and, essentially, the situation is simply that section 3 has been pushed back in time.

To prevent this situation from occurring, the parties should ensure that the contract contains clear links between completion of one (preceding) section, and commencement of the next, dependent section. Therefore, in the example above, section 1 would remain as having actual dates expressed for the dates of possession and completion, but the date for possession of section 3 would not be expressed as a 'date', but rather as occurring on, or within a specific period of time after completion of section 1. Similarly, there would be no 'completion' date expressed for section 3; instead a period for completion would be written into the contract. The contractor could make any financial claim against the employer in this situation for delays to section 1 which cause a delay to the possession of Section 3, as the employer would not be in breach.

The courts have considered extensions of time in relation to completion of the Works in sections.[193] In that case, the parties were in contract under JCT 98. The date for possession of the first section was a specific date, but the date for possession of each of the remaining sections was dependent on the date of practical completion of the previous section. That seems to be eminently sensible and in accordance with the suggestion in the previous paragraph. However, the completion dates for each section were expressed as specific dates. The contractor was in delay for 8 weeks in the first section and the architect gave 4 weeks extension of time and 4 weeks extension for each of the following sections. That resulted in a period of culpable delay to the first section amounting to 4 weeks for which

statement) that the contractor would be unaware that sections 1 and 3 are linked. However, the issue in this example is the contractor's entitlement to an extension of time under the contract, which will be reliant upon the entries in the contract accurately reflecting the parties' intentions of how the works are to be carried out.

192 Such as under JCT SBC/Q 2016 and IC 2016 – clause 2.5 and JCT DB 2016 – clause 2.4.

193 *Liberty Mercian Ltd v Dean & Dyball Construction Ltd* (2008) EWHC 2617 (TCC).

the employer deducted liquidated damages of £48,000. The contractor's difficulty was that it then found itself in a period of culpable delay for each of the succeeding sections, because in the case of each section, the date for possession was delayed by up to 8 weeks each time, but only 4 weeks extension of time had been given. It was the contractor's case that for each section after the first section, the extension of time should reflect the initial 8-week delay to the first section which irretrievably fixed the date of possession of the following sections. Indeed, the contractor argued that the liquidated damages had become a penalty and that they were unenforceable.

However, the court concluded that the liquidated damages were not a penalty and they were recoverable for all the weeks of culpable delay in each section. The court said:

> *'Attractively though this point was argued, I do not accept [counsel's] submission. It seems to me that it shies away from the critical feature of this contract, namely that the building works were always going to be carried out sequentially, and that the work on one section could not start until the work on the previous section had reached practical completion or (in certain instances) the stage of completion identified in the sectional completion schedule. It is plain from that schedule, and from the sectional completion agreement as a whole, that both sides were aware that culpable delay of 4 weeks on section 1 would automatically mean that work on sections 2, 3, 4 and 5 would start 4 weeks late.*
>
> *... I consider that this is the only sensible construction of the sectional completion agreement ... and the only construction which gives effect to the words used. What is more, such a result cannot be regarded as unfair. On the contrary, if the contractor is in culpable delay for 4 weeks in relation to section 1, which inevitably means that section 2 is also going to start 4 weeks late, so that the contractor's default has caused the delay to section 2, he should therefore be liable for the liquidated damages that will flow in consequence.'*[194]

A rather different situation occurred in *Trollope & Colls Ltd v North-West Metropolitan Regional Hospital Board*.[195] The contract was divided into three sections. Each section had a separate contract sum and set of conditions. The contract had a provision that the third section should commence six months after the issue of the certificate of practical completion for section one, but that it must be completed by a specified date. However, a delay occurred to section one and it delayed the issue of the certificate of practical completion and, therefore, delayed the commencement of section three, but because the completion date for section three was fixed, the period available to the contractor for carrying out the section three work was reduced from 30 months to 16 months.

Matters came to a head, because there was provision in the contract for the nomination of sub-contractors and the contractor required the employer to nominate sub-contractors who could carry out their work within the reduced period. The employer could not find sub-contractors who could do so, therefore the employer said that there must be a term

---

194 *Liberty Mercian Ltd v Dean & Dyball Construction Ltd* (2008) EWHC 2617 (TCC) at paragraphs 23-24 per Coulson J.

195 [1973] 9 BLR 60.

implied in the contract to the effect that the section three completion date must be extended by the amount of any extension of time given to the contractor for delay in section one. The court declined to imply such a term. The result was that the parties would be obliged to negotiate an extension to the completion date of section three with a resulting cost increase. If the delay had been shorter, the consequence might have been that the contractor was faced with a shortened work period for section three. This case does not appear to have been referred to in the *Liberty Mercian*[193] case.

## 3.10 'Milestone' or Key Dates

A building contract may require the contractor to complete a particular 'task' by a specific date. The use of the word 'task' is deliberate in order to distinguish from the contractor completing a work activity or section of the works (described in Section 3.9). The 'task' often has a very short duration or can be almost instantaneous, as the following examples illustrate:

- 'Physical' tasks:
  - Activating part of the works (e.g. switching over from temporary to mains power);
  - Opening an access route (e.g. such as a new entrance road or a bridge)
- 'Administrative' tasks
  - Certification or approval
  - Permissions

JCT 2016 contracts do not provide for 'milestone' dates, whereas NEC4 provides for completing discrete elements of work, or achieving degrees or states of completion as 'Key Dates' in clause 30.3. Key Dates, and the conditions for meeting them, are identified in contract data part One (against '2 The Contractor's Main Responsibilities'), and must also be set out in the programme, and any revision to it (clause 31.2). Clause 25.3 provides that the project manager decides if the contractor has satisfied the particular 'condition' for meeting a key date by the date stated in contract data part one. If, in the project manager's opinion, the contractor has failed to satisfy the key date, and the client incurs (or is likely to incur) the 'additional costs' of either carrying out the work himself, or paying others to carry it out, then the Contractor pays these additional costs to the Client. Clause 25.3 expressly restricts the Client's losses to the two classes of 'additional cost' set out in the clause, so he will be unable to recover any other losses or damages he may incur as a result of the Contractor's failure to hit the Key Date. In the event that achieving a Key Date is affected by the Client, the contractor would most likely be entitled to a compensation event under clauses 60.1(4),[196] 60.1(5),[197] 60.1(16)[198] or 60.1(18)[199] (Table 3.19).

196 NEC4, clause 60.1(4) a Project Manager's instruction to change a Key Date (including to stop or not to start the relevant work).

197 NEC4, clause 60.1(5) The 'Client' or 'Others' do not work in line with the 'Accepted Programme' (clause 31.2 provides that each programme submitted for acceptance must show 'Key 'Dates').

198 NEC4, clause 60.1(16) the 'Client' fails to provide materials, goods, facilities, testing samples or makes inspections (as stated in the 'Scope') which are necessary for a Key Date to be achieved.

199 NEC4, clause 60.1(18) a 'Client' breach of contract (not one of the other 'compensation events').

**Table 3.19** NEC4: Summary of Key Date procedures.

| Description | NEC4 |
|---|---|
| Definition | Clause 11.2(11) |
| Description of Milestone / Key Date (including the criteria for achieving it) | Contract data part one |
| Programme to show Milestone / Key Date | Yes, clause 31.2 |
| Contractor to notify that Milestone / Key Date is complete | No |
| Event signalling the Milestone / Key Date has been met | Project Manager's decision (clause 25.3) |
| Time limit for the Engineer to notify completion | No |
| Employer's / Client's remedy for contractor's failure to achieve the Milestone / Key Date | Additional costs of either (1) Client completing himself or (2) employing others to complete |
| Milestone or key date deemed completed by default | No |
| Extension of time for milestone or key date | Yes, see clause 60.1 |

## 3.11 Foreshortened Programme

### 3.11.1 Contractor's Obligation to Complete

JCT 2016 and NEC4 contracts require the contractor to complete the works 'on or before' the completion date[200] (see Sections 3.1.3 and 3.1.4). The completion date is the date for completion stated in the contract or any later date fixed by the employer's representative when giving an extension of time. Importantly, the contractor may, if it so chooses, complete the works before the date fixed under the contract (see Section 3.6.2). Many employers, and some employer's representatives, believe that they can require the contractor to remain on site until the contract date for completion or any extension of that date. Indeed, it is sometimes said that under the terms of the contract the contractor has undertaken to stay on site until the completion date stated in the contract. That view is clearly the wrong view because JCT 2016 and NEC4 contracts provide that the employer's representative must issue a document[201] that states when, in his opinion, completion has been achieved – whether that occurs before or after the contractual date for completion. Therefore, the contractor does not undertake to stay on site until the contract completion

200 JCT SBC/Q 2016 – clause 2.4, JCT DB 2016 – clause 2.3, JCT IC 2016 – clause 2.4. JCT MW 2016 - clause 2.2 requires the contractor to complete *"by the [Date for Completion] in the Contract Particulars."* NEC4 – clause 30.1.

201 JCT SBC/Q 2016 – under clause 2.30.1 the 'Architect/Contract Administrator' issues the 'Practical Completion Certificate' (or 'Section Completion Certificate' under clause 2.30.2 as appropriate), JCT DB 2016 – under clause 2.27.1 the 'employer' issues the 'Practical Completion Statement' (or 'Section Completion Statement' under clause 2.27.2 as appropriate), JCT IC 2016 – under clause 2.21.1 the 'Architect/Contract Administrator' issues the 'Practical Completion Certificate' (or 'Section Completion Certificate' under clause 2.21.2 as appropriate), and NEC4 clause 30.2, the 'Project Manager' decides when 'Completion' occurs and not later than 1 week after, must 'certify' the same.

date, but only until the works have reached the required state of completion, which may be earlier. The employer's position is understandable because it may be inconvenient for the employer to take possession of the building before the expected date (e.g. staff to run the premises may not have been engaged, rentals may not have started the employer may not have the necessary funding to pay the relevant interim certificate).

### 3.11.2 If the Contractor Could Have Completed Earlier

The point becomes even more important in the ascertainment of loss and/or expense. Where a contractor is claiming for prolongation of the contract period, he may argue that the period of time on which the ascertainment will be based should not be measured from the contract completion date, but from an earlier date, i.e. an earlier date when the contractor had planned to, or would otherwise have been able to complete. Therefore, the question is simply whether the contractor suffers loss and/or expense as a direct result of being kept on site longer than it needed to be on site. On that basis, and in theory, it is conceivable that if a contractor can show that, but for the event relied on,[202] it would have finished two months before the contract date for completion, it would be able to claim its costs for the full length of the prolongation period, including the two months prior to the 'completion' date. This example solely concerns the contractor's claim for additional costs and not for an extension of time (refer to Section 3.6.2). In practice, the contractor would be put to proof that it had actually suffered the losses it claimed.[203]

The arguments in favour of that contention are that, if it was not for the relevant matter or compensation event, the contractor would have completed the works by its earlier, intended date and possibly would have made a certain profit. It was denied that profit by an event which under the terms of the contract entitles the contractor to reimbursement of loss and/or expense or his additional costs. Moreover, neither a 'Relevant Matter' (JCT 2016 contracts) or a 'compensation event' (NEC4) are tied to a prolongation of the contract period beyond the contract completion date. The only criteria are:

- Whether regular progress has been materially affected: there can be no doubt that it has been because if it was not for the event the contractor would have finished by its earlier date;
- whether the contractor has suffered direct loss and/or expense, additional costs, or a loss of opportunity: it must have suffered these by being kept on site longer than would otherwise have been the case; and
- whether it has been reimbursed under some other provision of the contract: it has not been reimbursed under any other provision for the profit it would have made if it had finished at the earlier date.

The arguments against the contention are that, under the terms of JCT 2016 contracts, it is reasonable to assume that the contractor's 'preliminaries' within the contract sum financially covers the contractor for being on site until the contractual 'completion' date.

202 A 'Relevant Matter' under JCT SBC/Q 2016, JCT DB 2016 and JCT IC 2016, or a 'compensation event' under NEC4.

203 For example, if the contract sum contained time-related 'preliminaries' costs for the relevant period before the contractual 'completion' date, then the contractor's claim would be limited to proving other losses.

Therefore, the contractor may not have suffered any loss and/or expense by staying on site between actual completion until the original (or extended) completion date.

The position under NEC4 may be significantly different because the assessment of 'compensation events' under clause 63.1, includes not only the 'actual defined cost'[204] but also the 'forecast defined cost',[205] the 'Fee',[206] and any allowance for risk matters under clause 63.8.[207] It may be that in this example, the only part of assessing a the financial effect of a 'compensation event' under NEC4 that is analogous to ascertaining 'direct loss and/or expense' under JCT 2016 contracts, is assessing the 'actual defined cost' that the contractor has incurred, as the remaining elements to be assessed are either an agreed addition (the Fee), a forecast (forecast defined cost) or a risk allowance (under clause 63.8). Of course, clause 63.5 of NEC4 assesses the delay to the 'Completion Date' by reference to the delay to the 'planned Completion' date. Therefore, if the contractor is intending to complete by the earlier 'planned Completion' date, and his progress is delayed by a compensation event, then he is entitled to additional cost (i.e. the 'change to the Prices) if he completes after the 'planned Completion' date'.

In *J F Finnegan Ltd v Sheffield City Council,*[208] the contractor argued that under a JCT contract,[209] prolongation costs should be calculated from a date calculated by reference to release of batches of houses rather than the contract completion date. The court held that the contract completion date prevailed. The date arrived at by reference to the release of houses was purely a programming date, and it was not a date with which the contractor was obliged to comply.

In JCT 2016 contracts, this position is unaffected by the employer's representative 'accepting' or 'approving' a programme from the contractor showing an earlier date for completion than that set out in the contract particulars because he has no power to vary the terms of the contract.[210] The situation may be complicated if the contractor has proposed a foreshortened programme at the commencement of the project and the employer, rather than his representative, accepted or agreed that programme. In such a case, it may be convincingly argued that the parties have effectively varied the contract by agreeing a new completion date to which all are bound to work instead of the original date. Of course, the position may be very different under NEC4 contracts where a subsequent 'accepted programme' shows an earlier date for 'planned Completion' and earlier dates for the 'Client' (and/or 'Others') to provide access,[211] or carry out work.[212]

---

204 NEC4, clause 63.1, effectively the 'actual Defined Cost' is the cost of work that the contractor has already incurred by the 'dividing date' (which is either the date of Project Manager's or Supervisor's instruction, or the date of the notification of the compensation event under clause 61.3).

205 NEC4, clause 61.3 – 'forecast Defined Cost' is the cost of the work that the contractor must carry out after the 'dividing date' (see Footnote 204).

206 NEC4, the 'Fee' is intended to cover the contractor's ancillary costs such as overheads (such as administration, marketing, recruitment and training) and profit, along with any applicable taxation and bonds or guarantees (NEC4 – Managing Reality Book Four (Managing Change) 3rd Ed (2017) p49) and its calculation is explained in clause 11.2(10).

207 NEC4, clause 63.8, the allowance for risk includes the cost (and time, but that is not relevant to this example) *for matters which have a significant chance of occurring and are not compensation events."*

208 [1988] 43 BLR 124.

209 The JCT 1963 contract.

210 See the commentary to *Glenlion Construction Ltd v The Guinness Trust* at 39 BLR 93.

211 NEC4, failure of the Client to provide access is a 'compensation event' under clause 60.1(2).

212 NEC4, the failure of the 'Client' or 'Others' to carry out work as shown on the 'Accepted Programme' is a compensation event under clause 60.1(5). Also see the 'compensation event' under clause 60.1(3).

### 3.11.3 Late Information

Under JCT 2016 contracts, the position in respect of delay caused by late information from the 'architect/contract administrator' is worthy of separate consideration. Under previous suites of JCT contracts, the late provision of instructions and information was listed as both a 'relevant event' and a 'relevant matter'.

The JCT provisions appear to have been drafted in response to the judgment in the case of *Glenlion Construction Ltd v The Guiness Trust* [213] which considered a term for the provision of information in a previous version of the JCT SBC contract (JCT 63). One of the questions the court had to consider was:

> *'whether there was an implied term of the contract...that, if and in so far as the programme showed a completion date before the date for completion the employer by himself, his servants or his agents should so perform the said agreement as to enable the contractor to carry out the works in accordance with the programme and to complete the works on the said completion date.'*

The court had no hesitation in answering this question in the negative:

> *'It is not suggested by Glenlion that they were both entitled and obliged to finish by the earlier completion date. If there is such an implied term it imposed an obligation on the Trust but none on Glenlion. It is unclear how the variation provisions would have applied. [The extension of time clause] operates, if at all, in relation to the date for completion stated in the appendix. A fair and reasonable extension of time for completion of the works beyond the date for completion stated in the appendix might be an unfair and unreasonable extension from an earlier date.'*[214]

There was, therefore, no obligation on the Trust or its architect to provide information at any times earlier than those necessary to enable the contractor to complete by the contract date. It is, therefore, clear that if the contractor intended to finish earlier than the contract date for completion, it would not have a claim for loss and/or expense associated with being kept on site longer than anticipated if the reason it could not finish on the earlier date was simply that the architect had provided information at such time as to allow completion by the contractual completion date. However, it is sometimes contended that if the contractor, as a matter of fact, had received all the information to allow it to finish at its earlier intended date, it may have a valid claim if it is prevented from finishing on that earlier date by one of the 'relevant matters', even though it still finished before or on the contract completion date.

The JCT 2016 suite of contracts has replaced these clauses (and some others) by a 'catch all' clause referring to acts of impediment, prevention or default on the part of the employer, his representative, or a person for whom the employer is responsible.[215] However, the principle remains the same and the contractor is entitled to an extension of time and loss and/or

213 [1987] 39 BLR 89.

214 [1987] 39 BLR 89 at 103 per Judge Fox-Andrews.

215 A "Relevant Event" under JCT SBC/Q 2016 – clause 2.29.7, under JCT DB 2016 – clause 2.26.6, and JCT IC 2016 – clause 2.20.6, and a "Relevant Matter" under JCT SBC/Q 2016 – clause 4.22.5, JCT DB 2016 – clause 4.21.5, and JCT IC 2016 – clause 4.17.4.

expense if the architect causes delay by not providing the following in accordance with the timescales set out in the contract:

- Under JCT SBC/Q 2016 and JCT IC 2016:
  - Information set out in the 'information release schedule'[216];
  - Further drawings, details and instructions, not given when the contractor reasonably required them (having regard to the progress of works, but expressly not an earlier date than the 'Completion Date',[217] or as notified by the contractor[218]; or
- Under the Design Review Procedure in the contract[219].

The 'Information Release Schedule[218] should not be confused with what may be called an 'information required schedule', often submitted by a contractor at the beginning of a project. Such 'information required' schedules have no particular contractual standing. Under JCT SBC/Q 2016 and JCT IC 2016, the employer's representative (i.e. the 'Architect/Contract Administrator') must provide the contractor with such further details and drawings as are reasonably necessary to explain the contract drawings and to issue such instructions as are necessary enable the contractor to complete the Works in accordance with the contract. Essentially, this means that the employer's representative must provide the instructions in time to enable the contractor to complete the Works by the contract completion date. However, in this context, there are two important qualifications to that obligation which are often overlooked. They amount to this:

- The employer's representative must provide the information and instructions when reasonably necessary for the contractor to receive them, having regard to the progress of the Works.[219] This means that if the contractor is making slow progress, the employer's representative is no longer obliged to provide the instructions in time to enable the contractor to complete the Works by the contract completion date.
- However, if it seems that the contractor is likely to achieve practical completion before the contract date for completion, the employer's representative may revert to providing instructions in time to enable the contractor to complete the Works by the contract completion date. Therefore, although the employer or his representative must not actively prevent the contractor from meeting an earlier date, which it would otherwise have been able to meet, they are not obliged make any special effort to assist the contractor to do so.

216 JCT SBC/Q 2016 – clause 2.11 (Information Release Schedule) and the Fifth Recital. JCT IC 2016 – clause 2.10 (Construction Information) and the Sixth Recital.
217 JCT SBC/Q 2016 – clause 2.12.2 and JCT IC 2016 – clause 2.11.2.
218 JCT SBC/Q 2016 – clause 2.12.3 and JCT IC 2016 – clause 2.11.3.
219 Such as the 'Design Submission Procedure' under the Contractor's Designed Portion in JCT SBC/Q 2016 (Schedule 1), or JCT DB 2016 (Schedule 1).

# 4

# Financial Compensation: Recovery of Direct Loss and/or Expense or Additional Costs

## 4.1 Definition

The legal principles involved in formulating claims for financial compensation are well settled, but often, applying those principles in practice is far from easy. There are many principles which are common to all claims and it seems convenient to gather them together in one place for ease of reference. This chapter is concerned with 'compensation'[1] or reimbursement of the actual costs incurred – akin to common law damages (see Section 4.1). It does not examine 'value-based' claims, as these are explored in Chapter 5.

In most building contracts, the clauses which set out the procedures for claiming financial compensation use a number of phrases. The JCT 2016 contracts all refer to 'direct loss and/or expense'[2], whereas NEC4 refers to 'a change to the Prices'[3]. It is important to understand the meaning of these phrases, otherwise neither the party claiming nor the party claimed against may understand their correct rights and obligations. In general, this is the principle of common law damages, which was illustrated in the case of *Robinson v Harman*:

> '*The rule of common law is, that where a party sustains a loss by reason of a breach of contract, he is, so far as money can do it, be placed in the same situation, with respect to damages, as if the contract had been performed.*'[4]

Therefore, in financial terms, the claimant is reinstated into the financial position he would have enjoyed if the defendant had not been in breach. In short, the claimant is to be reimbursed, but not enriched, so he should not profit from the defendant's breach. Under most building contracts, the events entitling a party to claim 'financial compensation' may not be limited to one party committing a 'breach' of the contract but may also

1 Oxford Dictionary of English (3rd Ed) 2010 *"something, typically money, awarded to someone in recognition of loss, suffering or injury"*.
2 'direct loss and/or expense': JCT SBC/Q 2016 – clause 4.20.1, JCT DB 2016 – clause 4.19.1, JCT IC 2016 – clause 4.15.1, JCT MW 2016 – clause 3.6.3, and used in the corresponding JCT 2016 sub-contracts.
3 'change in the Prices' – NEC4 – clause 63.1.
4 *Robinson v Harman* (1848) 1 Ex 850 at 855.

*Building Contract Claims*, Sixth Edition. Guy Higginbottom and David Chappell.

include 'variations' (changing the scope of works and/or the manner in which it is carried out) or other events such as postponing the works.

### 4.1.1 JCT 2016 v NEC4 Contracts – Different Approaches

When dealing with financial compensation, the approaches taken by JCT 2016 and NEC4 contracts share a number of characteristics with each other, but aim to achieve their conclusion in very different ways. The 'ethos' of NEC contracts has been to swiftly conclude any matter where there is 'difference'. NEC contracts employ a procedure, where the valuation of compensation events[5] is to be agreed (or 'implemented'[6]) not later than 6 weeks after the contractor first notifies the Project Manager. Once it is 'implemented', the compensation event is 'agreed' and concluded – and, save for very limited circumstances, cannot be revised.[7] This should be contrasted to the general approach[8] in JCT 2016 contracts where the financial elements are not concluded until the 'Final Certificate' (or its equivalent)[9] is issued, and this may not occur until around 14 months after the works are completed (i.e. reach 'practical completion' or its equivalent - see Section 3.1.4).

It therefore appears that the JCT 2016 contracts aim to provide an 'accurate answer' to the question of financial compensation, whereas NEC4's aim is to provide a 'swift answer', even if it's not the most accurate.

### 4.1.2 JCT 2016 Contracts – 'Direct Loss and/or Expense'

For JCT contracts, 'loss' and 'expense' may be thought of as two sides of the same coin as the following definitions[10] indicate:

- 'Loss' – '*an amount of money lost by a business or organisation*'; and
- 'Expense' – '*the cost incurred in or required for something*'.

The phrase 'loss and/or expenses' is regarded as 'direct' in contrast with, and excluding, losses and/or expenses regarded as being indirect or too remote (refer to Section 4.2 'Direct v Indirect'). A similar phrase 'direct loss and/or damage' has been judicially considered.[11]

> '*what is its usual, ordinary and proper meaning in the law: one has to ask whether any particular matter or items of loss or damage claimed have been caused by the particular matter.... if it has been caused by it, then one has to go on to see whether*

5 Refer to Section 2.1.1 and Table 2.1.

6 NEC4, clause 66.1 provides that a compensation event is 'implemented' when the Project Manager accepts the contractor's quotation (given under clause 62.3).

7 NEC4 - Clause 66.3 provides that once a compensation event is 'implemented', it is not revised (unless in accordance with limited provisions in the contract).

8 This 'general approach' does not include agreeing "Variation Quotations", or "Acceleration Quotations" (JCT SBC/Q 2016 – Schedule 2), or 'Contractor's Estimates' for 'Valuation of Changes' or 'Loss and Expense' or an 'Acceleration Quotation' (JCT DB 2016 – Schedule 2 'Supplemental Provisions').

9 'Final Certificate' under JCT SBC/Q 2016 – clause 4.26.1, JCT IC 2016 – clause 4.21.1, JCT MW 2016 – clause 4.8.2, or 'Final Statement' under JCT DB 2016 – clause 4.24.1.

10 Oxford Dictionary of English (3rd Ed) 2010. However, in regard to the definition of 'expense', this may also mean actual disbursement – *Chandris v Union of India* (1956) 1 All ER 358 at 363 per Lord Justice Hodson.

11 In *Wraight Ltd v P H & T (Holdings) Ltd* (1968) 13 BLR 27.

> *there has been some intervention or some other cause which prevents the loss or damage from being properly described as being the direct consequence of the [matter].*'[12]

Reference to 'direct loss and/or expense' or similar phrases is clearly the same as a reference to common law damages and should be thought of as equivalent in terms of the measure of damages (i.e. in order to put the claimant '*...in the same position as he would have been in had he not sustained the wrong for which he is now getting compensation or reparation*[13]') and the standard of proof required.[14]

### 4.1.3 NEC4 – 'Change in the Prices'

NEC4's approach to claiming 'financial compensation' retains many of the characteristics found in assessing 'common law damages' and JCT's 'direct loss and/ or expense', but there is a fundamental difference. To facilitate the agreement and implementation of compensation events within the 6 week period,[6] NEC4 not only requires the contractor to give an account of the costs he's already incurred, but also to give a 'forecast' of any costs that he's subsequently likely to incur. By providing his 'forecast' costs for agreement, NEC4 aims to avoid what could otherwise be a lengthy cost-verification process.

The procedure for assessing a 'change in the Prices' (i.e. the financial element of a compensation event) separates the 'actual' financial element from the 'forecast' element at the date when the compensation event was notified[15] (see Section 5.5.1 for more details). This is called the 'dividing date'. The financial elements to be assessed are as follows:

- Separated at the 'dividing date':
  - The 'actual Defined Cost[16]'. In principle, this element has obvious similarities to 'common law damages' or 'direct loss and/or expense' under the JCT 2016 contracts, as it represents the costs incurred up to the 'dividing date', whereas
  - The 'forecast Defined Cost'. This is clearly a forecast[16] of the financial effects of the compensation event after the 'dividing date'; and
- Other elements:
  - The Fee[17] and
  - Other 'risk allowances for cost'.[18]

NEC4's 'compensation event procedure' is explored in detail in Sections 5.3.2 (Compensation Event Procedure) and 13.2 (Assessing Changes to the Prices).

---

12 [1968] 13 BLR 27 at 33 per Megaw J.
13 *Livingstone v Rawyards Coal Company* (1880) 5 App Cas 25 at 39 HL, also see *Keating* (10th Ed) at 9-003.
14 See *F G Minter Ltd v Welsh Health Technical Services Organisation* (1980) 13 BLR 7 CA.
15 NEC4 – clause 63.1.
16 NEC4 does not provide a singular definition of 'Defined Cost'. Firstly, depending upon which Main Contract Options are selected, it may be partially defined in clause 11.2(23) for Options A or B, or in clause 11.2(24) for Options C, D or E in terms of which 'Schedule of Cost Components' is to be used. A further partial definition is provided in clause 52.1 to allocate the contractor's recoverable costs to either 'Defined Cost' or to the 'Fee'. Further explanations of 'actual' and 'forecast' Defined Costs (from clause 63.1) are given in NEC4's Managing Reality Book Four (Managing Change) 3rd Ed (017) pp24–25.
17 NEC4, the 'Fee' is defined in clause 11.2(10).
18 NEC4 clause 63.8.

## 4.2 Direct v Indirect

It is important to understand the difference between direct and indirect (or consequential) loss or damage because a party in breach may not be liable to pay all the damage suffered by the injured party. The courts limit the damages to what is reasonable in the circumstances by applying a rule to exclude all damage that is too remote. The rule has been stated in the case of *Hadley v Baxendale* as follows:

> *'Where two parties have made a contract which one of them has broken, the damages which the other party ought to receive in respect of such breach of contract should be such as may fairly and reasonably be considered either arising naturally, i.e. according to the usual course of things from such breach of contract itself, or such as may reasonably be supposed to have been in the contemplation of both parties at the time they made the contract, as the probable result of the breach of it. Now, if the special circumstances under which the contract was actually made were communicated by the plaintiffs to the defendants, and thus known to both parties, the damages resulting from the breach of such a contract, which they would reasonably contemplate, would be the amount of injury which would ordinarily follow from a breach of contract under these special circumstances so known and communicated.'*[19]

The rule in *Hadley v Baxendale* is said to have two 'limbs', which can be described as follows:

- The first limb refers to damages 'arising naturally'. Such damages are often referred to as 'general damages' – i.e. the kind of damages that anyone would expect to be the result of the breach.
- The second limb refers to damages 'in the contemplation of both parties at the time they made the contract'. These kind of damages depend upon the knowledge of the parties of special circumstances and they are often referred to as 'special damages'.

For example, Ms A buys a car from a dealership and drives it away intending to use it immediately to drive to Southampton; where she is due to start a pre-booked cruising holiday. Suppose then, that the car breaks down on the way to Southampton so that Ms A does not reach the ship in time before its departure. Ms A can certainly claim the cost of necessary repairs to the car, but she cannot claim the cost of the lost holiday, because the dealership knew nothing of the projected holiday or the consequences of a mechanical breakdown. If all those facts had been made known to the dealership before or at the time of the sale contract for the car, Ms A might also have been able to claim the cost of the holiday. However, in practice, it is unlikely that the dealership would accept such a liability by including an exclusion clause in its sale contract to deal with that eventuality,[20] but the principle remains good.

---

19 *Hadley v Baxendale* (1854) 9 Ex 341 at 354 per Alderson B.
20 Refer to Section 4.3.

A useful explanation of this decision and of the authorities generally was given in *Victoria Laundry (Windsor) v Newman Industries, Coulson & Co.*[21] It took the form of propositions which may be summarised as follows:

- The purpose of damages is to put the injured party in the same position, so far as money can, as if its rights had been observed, but to pursue that purpose would provide the party with a complete indemnity and it is considered to be too harsh;
- The injured party may only recover loss reasonably foreseeable at the time of the contract;
- Foreseeability depends on the knowledge of the party committing the breach;
- Knowledge is of two kinds: (i) all reasonable people are assumed to know the kind of loss which is liable to result from a breach in the ordinary course of things and (ii) actual knowledge of special circumstances which may cause greater loss;
- The contract breaker will be liable provided that, if it had asked itself, it would have concluded, as a reasonable person, that the loss was liable to result from that breach; and
- It is enough if the loss could be seen as likely to result.

These propositions were considered by the House of Lords in *The Heron II*.[22] Although they rejected proposition 6 as being too broad, they did not agree the test which should be substituted. Taking their Lordships opinions together, *it is enough if the loss was a serious possibility* appears to be a reasonable consensus. The crucial point which lies at the foundation of the rule in *Hadley v Baxendale*, is the knowledge possessed by the contract breaker at the time the contract was entered into. The point is whether the party had just the ordinary knowledge of an average person in that situation or whether, if the party had considered the matter, its special knowledge would have led it to the conclusion that greater (or perhaps less) than the ordinary loss would result from the breach.

It seems that it is enough if the type of loss is within the reasonable contemplation of the parties even though the *extent* of such loss is far greater than they could have contemplated.[23]

## 4.3 Exclusion of Indirect or Consequential Loss

In theory, the differences between direct and indirect (or consequential) losses are easy to understand; unfortunately, in practice, there may be real difficulties interpretation which is which. In *Saintline Ltd v Richardsons, Westgarth & Co Ltd*,[24] the manufacturers of engines for a ship were in breach of contract and the ship's owners brought an action for damages including loss of profit, wages and storage costs and other fees. The contract contained a clause providing that a manufacturer's liability should not 'extend to any indirect or consequential damages whatsoever'. The court held that all these items were recoverable as

21 [1949] 1 All ER 997 at 1002 per Asquith LJ.
22 *Koufos v Czarnikow Ltd (The Heron II)* (1969) 1 AC 350.
23 *Parsons (Livestock) Ltd v Utley Ingham & Co Ltd* (1977) 2 Lloyds Rep 522.
24 [1940] 2 KB 99.

damages because they were a direct and natural consequence of the breach of contract. While giving judgement, the judge said:

> '*What does one mean by "direct damage"? Direct damage is that which flows naturally from the breach without other intervening cause and independently of special circumstances while indirect damage does not so flow. The words "indirect or consequential" do not exclude liability for damages which are the direct and natural result of the breaches complained of … What the clause does do is to protect the respondents from claims for special damages which would be recoverable only on proof of special circumstances and for damages contributed to by some supervening cause.*'[25]

The same point was considered in *Croudace Construction Ltd v Cawoods Concrete Products Ltd.*[26] Croudace was engaged as the contractor to construct a school and it contracted with Cawoods for the supply and delivery of masonry blocks. Cawoods attempted to restrict its liability by including a term in the contract that stated:

> '*We are not under any circumstances to be liable for any consequential loss or damage caused or arising by reason of late supply or any fault, failure or defect in any materials or goods supplied by us or by reason of the same not being of the quality or specification ordered or by reason of any other matter whatsoever*'

Such clauses usually only exclude liability for consequential damages under the second limb of *Hadley v Baxendale* and so it proved in this case.[27] Croudace made a claim against Cawoods for breach of contract alleging late delivery and defects in materials. They sought to recover loss of productivity and various costs arising out of delay. The question which went before the Court of Appeal was whether such damages were to be regarded as 'consequential' loss. The Court held that the exclusion of consequential loss or damage did not apply to loss resulting in the ordinary course from late delivery of, and defects in, materials – which were heads of direct loss (neither indirect nor consequential). The Court quoted with approval: 'on the question of damages, the word 'consequential' has come to mean 'not direct'…'.[28]

A case where the damage was held to be too remote is *Balfour Beatty Construction (Scotland) Ltd v Scottish Power plc.*[29] Scottish Power contracted to supply temporary power to Balfour Beatty's concrete batching plant, but were held to be in breach of contract after an interruption in the power supply. The facts were that Balfour Beatty were in the process of constructing an aqueduct by employing a continuous concrete pour. Therefore, when the temporary power supply was interrupted, it was impossible to continue with the 'continuous pour' method, so Balfour Beatty had to demolish and work it had carried out, and then re-start. A claim for the cost of the demolition and re-building failed because the House of

25 [1940] 2 KB 99 at 103 per Atkinson J.
26 [1978] 8 BLR 20.
27 *British Sugar plc v NEI Power Projects Ltd* (1987) 87 BLR 42.
28 *Millars Machinery Co Ltd v David Way & Son* (1934) 40 Comm Cas 204 at 210 per Maugham LJ.
29 [1994] 71 BLR 20.

Lords held that Scottish Power could not be presumed to know of the practice of continuous pour construction. It was said:

> *'It must always be a question of circumstances what one contracting party is presumed to know about the business activities of the other. No doubt the simpler the activity of the one, the more readily can it be inferred that the other would have reasonable knowledge thereof. However, when the activity of A involves complicated construction or manufacturing techniques, I see no reason why B who supplies a commodity that A intends to use in the course of those techniques should be assumed, merely because of the order for the commodity, to be aware of the details of all the techniques undertaken by A and the effect thereupon of any failure of or deficiency in that commodity. Even if the Lord Ordinary had made a positive finding that continuous pour was a regular part of industrial practice it would not follow that in the absence of any other evidence suppliers of electricity such as the Board should have been aware of that practice.'*[30]

*Croudace Construction Ltd*, *Saintline Ltd* and *Wraight Ltd*[11] were considered with approval by the Court of Appeal in *F G Minter Ltd v Welsh Health Technical Services Organisation*[31], which confirmed that, in JCT contracts, for 'direct loss and/or expense':

> *'The distinction to be drawn is a distinction between losses and expenses that the law regards as "direct" and those regarded as being "indirect" or "consequential". It is the same distinction as that which has to be made whenever common law damages are being considered and a particular item of loss or expense or damage is to be regarded as "direct" if it can fairly be said to fall within the first rule of Hadley v Baxendale...as being something which arises naturally and in the ordinary course of things...'.*[32]

Although some of these cases are concern JCT contracts, it is likely that many of the principles also apply to claims under other contracts where phrases similar to 'loss and/or expense' are used. In *Robertson Group (Construction) Ltd v Amey Miller (Edinburgh) Joint Venture*[33] the court, after listening to many submissions regarding remoteness of damage, considered that the phrase 'all direct costs and directly incurred losses' permitted the recovery of reasonable sums by way of general corporate overheads and profit. In that instance, the phrase occurred in a letter which was intended to form a temporary contract while the terms of a contract under JCT conditions were finalised. However, it should be noted that, because the recovery of direct loss and/or expense is a procedure under a specific term of the contract, the recovery of certain heads of claim may be permitted under the express contract terms which might not be recoverable as a claim for damages at common law.[34]

30 *Balfour Beatty Construction (Scotland) Ltd v Scottish Power plc* (1994) 71 BLR 20 at 29 per Lord Jauncey.
31 [1980] 13 BLR 7.
32 *F. G. Minter Ltd v Welsh Health Technical Services Organisation* (1980) 13 BLR 1.
33 [2005] ScotCS CSOH 60.
34 Some of the grounds for such contractual claims are not breaches of contract as an examination of the 'Relevant Matters' in the JCT 2016 contracts will confirm (refer to Section 12.1.1).

## 4.4 The More Common Heads of Loss

### 4.4.1 General Principles

The following heads of loss are not intended to be an exhaustive list, but simply those that are most commonly encountered. The basic principle is that, subject to the restrictions of indirect or consequential loss, the contractor should be put into the financial position which it would have been in, had the prolongation or disruption not occurred. If this general principle is observed, a claimant should be better placed in judging or advancing other heads of loss where the particular circumstances permit. There are few limits to the possible heads of loss other than the ingenuity of the claimant in putting together a valid claim.

The heads of loss suggested in the following sections are split into the following categories:

- Prolongation;
- Disruption; and
- Other.

The separation into these categories is not intended to be strict or suggesting that particular heads of loss may only be found in a particular category. This is not the case, as the circumstances under which a suggested head of loss can be incurred may vary – the loss may have been caused by prolongation, disruption or for another reason altogether. Table 4.1 summarises the suggested heads of loss into the categories where they're significantly or most-often encountered:

### 4.4.2 Prolongation Costs

If the works are prolonged beyond the contractual completion date by one of the delay events that entitles the contractor to an extension of time, then provided that the contractor

**Table 4.1** Significant categories of loss.

| | **Significant category of loss** | | |
|---|---|---|---|
| **Head of Loss** | **Prolongation** | **Disruption** | **Other** |
| On-site establishment costs | ✓ | | |
| Management personnel costs | ✓ | ✓ | |
| Plant | ✓ | ✓ | |
| Head office overheads | ✓ | | |
| Loss of profit | ✓ | | |
| Winter working | ✓ | ✓ | |
| Uneconomic working | | ✓ | |
| Increased costs | ✓ | ✓ | ✓ |
| Interest | | | ✓ |
| Financing charges | | | ✓ |
| Sub-Contractor costs | ✓ | ✓ | ✓ |

has complied with the building contract's terms, can demonstrate the causal link, and the building contract also entitles the contractor to additional payment in those circumstances, then the contractor should be able to claim for 'financial compensation'. The latter point is particularly important, as many 'events' may entitle the contractor to an extension of time (refer to Section 3.1.1), but only certain events will entitle it to claim prolongation costs. As a general observation, only those events for which the employer is responsible (such as breaches of the building contract, or instructing variations or postponements of the works) will create the entitlement to 'financial compensation'.

The most common heads of loss included with a prolongation claim are those costs that the contractor incurs on a 'time related' basis.[35] These are often referred to as contractor's 'preliminaries'. The contractor is likely to have included sums of money within the contract sum (or price) for the various resources included in his 'preliminaries' – usually, for the period between the contractual start and finish dates.

These resources, commonly found within the contractor's 'preliminaries', are as follows:

#### 4.4.2.1 On-site Establishment Costs

On-site establishment costs are perhaps the easiest head of claim to establish because all the relevant data should be readily accessible by the contractor in respect of the particular periods of delay. They often consist of the site accommodation (such as temporary buildings), health and safety facilities, some plant (although see Section 4.4.2.3), equipment, tools, communications, costs of temporary electricity and water supplies, costs of welfare and sanitary facilities, along with supervisory and administrative staff engaged for the project, although not necessarily all of them may be based on site (see Section 4.4.2.2).

However, not all of these costs could be 'time related'. Some may be dependent on work or value and care must be taken that inappropriate items and cost duplication are excluded. For example, a crane or scaffolding may be held on site during a delay period; alternatively there may be equipment that is off site during a delay period. Care must also be taken that the on-site staff are really necessary and not simply transferred to site during a period of delay because there is no work for them elsewhere. Rather, the contractor should take steps to move unnecessary people and plant off-site as mitigation of his losses (refer to Section 2.7.2). Relevant labour returns should be provided by the contractor to show what people were doing and why the particular resources were on site. It is always important that a contractor can demonstrate its entitlement to costs under a particular form of contract:

- Under JCT 2016 contracts (and in general – applying the principles of common law damages) provide that the direct loss and/or expense is
  - Direct and have been incurred, or that the contractor is likely to incur them; and
  - The contractor has mitigated his losses.
- Under NEC4 contracts:
  - The 'actual' and 'forecast' Defined Costs generally follow the principles for common law damages and mitigation of loss; and

35 Refer to the RICS' New Rules of Measurement – NRM2: Detailed measurement for building works – (1st ed) 2012. Appendix B contains a 'Template for preliminaries (main contract) pricing schedule (condensed)' that includes numerous 'preliminaries' costs that could be 'time-related'.

- Are permissible under the terms of the contract. For example, 'Defined Cost'[16] under Main Option Clauses C, D (both target cost) or E (cost reimbursable) excludes 'Disallowed Cost'[36] (for example, sums that were neither properly incurred nor substantiated by the contractor's records).

The pattern of resourcing on a project is that there are some days or weeks at the commencement when the resources are being built up, and there will probably a longer period towards the end of the project when the site establishment will be running down – as indicated by the 'S curve' in Figure 2.2 (Chapter 2). It would be inaccurate to simply to take the costs from the date when the Works should have been completed to the date of practical completion or to apply the rates or prices from for the relevant 'preliminaries' resources that form part of the contract sum.

#### 4.4.2.2 Management Personnel Costs

Management personnel may range from a single supervisor (such as a working foreman) to several managers, each overseeing a team of staff on site, and supported with additional office-based personnel. Additional costs of management personnel may be incurred during periods of prolongation and/or when the progress of work is suffering disruption:

- Prolongation: The contractor may simply be claiming for the same level of management personnel as he included in the 'preliminaries' part of the contract sum,[37] but only for a longer period (i.e. for the period of critical delay).
- Disruption: The contractor may be claiming for an increased level of management personnel resources during periods of prolongation. For example, a change in site logistics may require additional management personnel to manage access routes, traffic management, craneage or manual handling of materials and plant.

Matters become complicated when periods of prolongation and disruption overlap, as the management personnel costs for each may be caused by different 'Relevant Matters' (JCT 2016 contracts) or 'compensation events' (NEC4). The contractor will have to separate the relevant costs and allocate them to each cause. This may sound a laborious task, but it may prove to be relatively straightforward. For example, the prolongation costs are likely to include the management staff that were always allocated to the project, whereas the disruption costs are may comprise either (i) additional personnel, often of different disciplines, or (ii) the costs for the existing staff working longer periods (such as weekday overtime or at weekends). Modern construction accounting software usually identifies management personnel on an individual rather than a 'by discipline' basis and shows their time and costs allocated to a project.

Some contractors allocate the costs of their management personnel between the costs of the project and their office overhead costs. Typically, site-based staff that are solely dedicated to a project will be costed to that that project – such as site agents, project-based

36 NEC4, 'Disallowed Cost' is defined clause 11.2(26) of Main Option Clauses C, D and E.

37 It is important to establish the number and quality of supervisors the contractor envisaged at the time the contract was executed and, most important, whether the contractor's assumptions in this regard were well founded and reasonable at the time they were made. This would usually be evident from the contractor's priced breakdown of his 'preliminaries' within the 'contract sum' – refer to Footnote 35.

project managers and contracts managers, engineers, foremen and attendant labour. Management personnel that could be shared between projects are more likely to be allocated to the contractor's general overhead – such as directors, project or contracts managers responsible for more than one project, quantity surveyors, planners/programmers, document controllers and health and safety personnel. If so, the contractor may be showing a higher overhead percentage than would be expected if shared management personnel were costed between projects. The contractor should therefore ensure that any claims for head office overheads (see Section 4.4.2.4) account for the costs of management personnel claimed or recovered elsewhere (such as under prolongation, disruption or within the cost or value of variations).

If management personnel would have been allocated to the project in any event, there can be no claim. Moreover, the contractor should explain the purpose of management's presence on site at the relevant time – particularly if their cost is forming part of a disruption claim. A contractor's method statement with appropriate references to relevant rules or guidelines should be a useful starting point – particularly, if additional costs have been incurred due to a change in the contractor's working method. It is not unknown for a contractor to allocate additional management resources to a project when the contractor has no other work for at that particular time. This is a fairly common tactic for contractors working under the NEC4 target cost contracts (Options C and D[38]), as the costs of management personnel would ordinarily fall within the 'actual Defined Cost'[16], unless those costs fell within the definition of 'Disallowed Cost'[36]. Many 'Clients' expand the definition of 'Disallowed Cost' in NEC contracts to allow the Project Manager to carry out some subjective analysis of the contractor's claim.

Where smaller projects are concerned, it is common for the supervision to be in the hands of a working person-in-charge. Care must be taken that additional supervisory costs are not duplicated within the contractor's claims for measured works or variations.

Under NEC4 contracts, the cost attributable to management personnel are set out in either the 'Shorter Schedule of Cost Components' (for Main Contract Options A and B[39]) or the 'Schedule of Cost Components' (for Main Contract Options C, D or E[40]) under the category entitled 'People'. The 'schedules' clearly set out what costs fall within the Defined Cost for 'People'.

#### 4.4.2.3 Plant

Plant costs may be incurred during periods of prolongation (i.e. the same plant is necessarily on site for a longer period) and/or during periods of disruption (additional plant is required). There are certain important factors which must be taken into account when considering costs related to prolongation. It is necessary to identify which items of plant the contractor has allocated to project. This is likely to be less straightforward than identifying the relevant management resources because not all items of plant may be

38 NEC4 – Option C (Target Contract with Activity Schedule) or Option D (Target Contract with Bill of Quantities).
39 NEC4 – Option A (Priced Contract with Activity Schedule) and Option B (Priced Contract with Bill of Quantities).
40 NEC4 – Option C (Target Contract with Activity Schedule), Option D (Target Contract with Bill of Quantities) and Option E (Cost Reimbursable Contract).

individually identifiable. Unlike allocating management resources, modern constructing accounting software, on its own, is unlikely to provide an acceptable level of detail (i.e. the accounting system may only show the total of a plant hire invoice), so the following additional identification tasks may be needed:

- Plant invoices for 'hired in' plant often include every item of plant that the hirer has provided to the project. The contractor will need to identify the relevant items of plant within each invoice;
- In order to allocate the relevant plant to a disruption claim, the contractor will require evidence (such as daily or weekly plant returns) of what plant was used for a particular activity;
- Similarly, evidence will be needed for shared plant (such as a tower crane or a generator), in order to allocate a period of its usage to the disruption claim;
- The contractor has a duty to mitigate its loss by either trying to use the plant elsewhere else or by terminating the plant hire contract and returning the plant to its owner in accordance with the terms of the hire;[41] and
- The contractor will have to account for the cost of plant already contained within the 'contract sum' – either within its priced 'preliminaries'[42] or included within the rates and prices for carrying out the physical work (including for variations).

The contractor will need to distinguish between the 'hired in' plant and plant it owns. This distinction is necessary if the contractor is a company within a group, and one of the group companies hires out plant and equipment to the others. Once it is established that there is a claim in principle, the question to be answered refers to the amount actually lost or expended by the contractor in hiring the group company's plant. Does the plant hire group company hire out plant to the contractor at the same rates as it would apply to other contractors, or is there a discount? It is for the contractor to prove that it actually has to pay the hire charge. If the contractor is a separate limited company (in other words a separate legal entity), then whatever charge has been paid will be claimable unless there is some arrangement between the companies which allows the contractor to recover its plant expenditure another way. If the contractor and the hirer are actually separate divisions of the same company (i.e. not separate legal entities), it may be difficult for the contractor to immediately demonstrate the value of the plant it has used.

Sometimes a contractor will argue that it is entitled to claim a loss of opportunity for lost plant hire charges because it also operates a plant hire business, hiring out spare plant to other contractors. In such circumstances, the contractor would have to show that if he was not using the particular plant, it would have been able to hire it to another contractor. The (plant hiring) contractor must prove that it had an opportunity to do so.

If the contractor is using its own plant, it may attempt to claim based on a notional hiring charge. That is not a valid basis for the simple reason that the money claimed is not money actually expended. Arriving at the true cost of its plant standing idle is more difficult, but

41 If the contractor is thereby in breach of the plant hire terms, that is entirely a matter for the contractor unless perhaps it can be shown that the damages for the breach are less than the contractor would claim for leaving the plant idle on site. In such circumstances, the contractor probably has a claim for the amount of damages it has had to pay on the grounds that it amounts to mitigation of loss.

42 Refer to Footnote 35.

the case of *B Sunley & Co Ltd v Cunard White Star Ltd*[43] provides assistance. There, an excavating machine had to be transported to Guernsey to be used on a building contract, but it was delayed by 1 week at port. During the delay, it worked for 1 day and earned £16. The contractor claimed £577 for loss of profit, but no evidence was produced. The cost of the machine was £4,500 and its life was said to be three years. The Court of Appeal held that the measure of damage included (i) depreciation, (ii) interest on the money invested, (iii) maintenance costs and (iv) the value of any wages 'thrown away'. The contractor was awarded £30 less the £16 earned while in port. In arriving at that figure, the Court took account of the fact that the machine would not depreciate as much while standing idle as it would when working. At first instance,[44] the court held that the proper measure of damages was the amount the contractor would have made by the use of the machine during the period it was idle, but the Court of Appeal took the view that, without proof of special damages, the contractor could only recover nominal damages based mainly upon a calculation of the rate of depreciation of the machine: There is 'no authority for the proposition that if the owner of a profit-earning chattel does not prove the loss he has sustained the judge may make a fortuitous guess and award him some arbitrary sum'.[45]

In a later case (under JCT 80, but the principle is applicable to the JCT 2016 contracts), the position in *Sunley* was largely upheld where the court has held that ascertainment should take into account the substantiated cost of capital and depreciation, but not the elements normally included in hire rates on the basis that plant will only be profitable for some of the time:

> '*... in ascertaining direct loss or expense under clause 26 of the JCT conditions in respect of plant owned by the contractor the actual loss or expense incurred by the contractor must be ascertained and not any hypothetical loss or expense that might have been incurred whether by way of assumed or typical hire charges or otherwise.*'[46]

A contractor may be able to demonstrate a 'loss of opportunity' (see above) that its own plant could have earned but for it being kept idle on site by factors which would normally entitle the contractor to recover loss and/or expense. The contractor would have to show that there was other work it could have done with the plant. To that extent, such a claim is analogous to a contractor's claim for overheads and profit (see Section 4.4.2.4). It appears that if a contractor can prove its actual loss by submitting detailed calculations based on cost records, it is entitled to recover the proven amount of loss in the normal way.

However, if a contractor cannot prove or evidence actual loss, it is entitled to recover only an amount which is normally limited to depreciation. The absence of real evidence in *Sunley* led to the Court of Appeal taking the view that the depreciation for the claim period would be £29 for the week. However, because it had been said that the working life of the machine was only three years, the Court said that it would not depreciate as much when idle as it did when working and they thought that £20 for the week was all that should be allowed for depreciation. As noted earlier, there are intermediate positions between the two

43 [1940] 2 All ER 97. See also the Canadian case *Shore & Horwitz Construction Co Ltd v Franki of Canada Ltd* (1964) SCR 589 which followed the same principles.
44 [1939] 3 All ER 641.
45 [1940] 2 All ER 97 at 101 per Clauson LJ.
46 *Alfred McAlpine Homes North Ltd v Property and Land Contractors Ltd* (1995) 76 BLR 59 at 93 per Judge Lloyd.

extremes of hired in and contractor's own plant. Each of these positions must be examined carefully, keeping in mind that the key point is that a contractor can only recover, as direct loss and/or expense, what has actually been lost or spent.

Identifying plant costs is likely to pose less of a problem under NEC4 contracts because the rates for particular items plant will be agreed in advance. The 'Shorter Schedule of Cost Components' (for Main Contract Options A and B[39]), or the 'Schedule of Cost Components' (for Main Contract Options C, D, or E[40]) under the categories entitled 'Equipment' and 'Plant and Materials', set out what costs fall within the Defined Cost for plant.

#### 4.4.2.4 Winter Working

Another factor that can lead to a claim which is effectively one for loss of productivity is the carrying out of work in less favourable circumstances, such as excavation work carried out in winter rather than in summer. In such circumstances, the contractor may claim for the additional costs caused by working in a period of less favourable weather.

A 'winter working' claim may fall into both the 'prolongation' and/or 'disruption' categories in the following examples:

- The 'knock on' effect of an earlier delay may push the relevant work into a period of less favourable weather (see Section 2.5). The affected work may become less economical to carry out, requiring more labour, plant equipment and power, and it may require a prolonged duration;
- The employer may instruct the contractor to carry out work in a different sequence – being affected in the same adverse ways as the above example.

Clearly there will be no, or at least little, chance of a claim on this basis where work scheduled to be carried out in winter is pushed into spring or summer.

#### 4.4.2.5 Head Office Overheads

The recovery of head office overheads is one of the more contentious heads of loss. The claim arises when a project has been prolonged, and the contractor's resources remain engaged on that project, and by doing so, they are prevented from earning a contribution to the contractor's overheads, as they are unable to move onto another project. It is important to note that many contractors, of a medium size and larger, will retain many resources that are funded by the overhead contributions generated by their live projects. In addition to the office-based management personnel (see Section 4.4.2.2) that are engaged on the project, the contractor's overheads are likely to cover such expenditure as

- Office rental (e.g. for the head office or regional offices)
- Office-based staff:
  - Human resources
  - Estimating department
  - Accounts department
  - Purchasing department
  - Business development (including marketing)
- Employer's liability/public liability insurance.

It is important to understand that the contractor is not claiming that it has actually *lost* overheads; it is claiming the 'loss of the opportunity' to contribute to head office overheads.

The English courts have considered a number of cases[47] concerning claims for head office overheads and the methods used to advance them. The court's position was definitively set in the case of *Walter Lilly & Company Ltd v MacKay & DMW Developments Ltd*,[48] in which Akenhead J helpfully confirmed the basis of such claims, giving guidance for the method of their calculation and the evidence required:

> '*…the claim is in truth more analogous to a loss of opportunity claim, namely that because [the contractor] could not deploy its [management] team elsewhere and thus is lost the opportunity to tender for any number of projects on the basis of deploying that management team with the probability being that it would have secured sufficient profitable work for that team to produce returns during the period of delay; it has thus lost profit and overhead recovery which it would otherwise probably have secured if the delay had not happened.*'[49]

The judge set out the following principles[50] for making a head office overheads claim:

(a) '*A contractor can recover head office overheads and profit lost as a result of delay on a construction project caused by the factors which entitled it to loss and expense;*
(b) *It is necessary to prove on the balance of probabilities that if the delay had not occurred it would have secured work or projects which would have a return (over and above costs) representing a profit and/or contribution to head office overheads;*
(c) *The use of a formula, such as Emden or Hudson, is a legitimate and indeed helpful way of ascertaining, on the balance of probabilities, what that return can be calculated to be;*
(d) *The 'ascertainment' process under does not mean that the Architect/Quantity Surveyor or indeed the ultimate dispute resolution tribunal must be certain (that is sure beyond reasonable doubt) that the overheads and profit must have been lost. HHJ Lloyd QC was not saying that assessment could not be part of the ascertainment process. What one has to do is to be able to be confident that the loss and expense being allowed had actually been incurred as the result of the…delay or disrupting factors.*'

In *Lilly*, the contractor's claim for head office overheads succeeded because of the following:

47 *J F Finnegan Ltd v Sheffield City Council* (1988) 43 BLR 124, *Alfred McAlpine Homes North Ltd v Property and Land Contractors Ltd* (1995) 76 BLR 59, *AMEC Building Ltd v Cadmus Investment Co Ltd* (1997) 13 Const LJ 50, *City Axis Ltd v Daniel P Jackson* (1998) 64 Con LR 84, *Norwest Holst Construction Ltd v Co-operative Wholesale Society Ltd* (No 2) (1998) EWHC Technology 339, *CFW Architects (a firm) v Cowlin Construction Ltd* (2006) 105 Con LR 116, *Aerospace Publishing v Thames Water Utilities* (2007) 110 Con LR 1 and *Babcock Energy Ltd v Lodge Sturtevant Ltd* (1994) 41 Con LR 45.
48 *Walter Lilly & Company Ltd v MacKay & DMW Developments Ltd* (2012) EWHC 17773 (TCC).
49 See *Lilly* – footnote 48 at paragraph 549.
50 See *Lilly* – footnote 48 at paragraph 541, citing *Alfred McAlpine v Property and Land Contractors* (see footnote 47).

- The contractor provided good evidence of the following during the period of delay (a 99-week delay to a 78-week project):
  - That the prevailing construction market was buoyant;
  - That the contractor's success rate for projects won to opportunities tendered was in the order of 1 in 4;
  - That the opportunities that the contractor had received and then declined were '*precisely detailed on a comprehensive schedule*'; and
  - There was little difference between the quantity surveying experts as how to calculated the claimed amount.
- Oddly, the defendant (employer) did not provide any convincing counter-argument. In particular, he did not put a number of his questions to the contractor's witnesses and generally appeared to miss the point that the contactor's claim was one of lost opportunity.

The court in *Lilly* confirmed that the use of formulae to calculate head office overheads is 'legitimate' and indeed 'helpful' and cited the 'Hudson' and 'Emden' formulae as suitable examples. Prior to *Lilly*, a number of decisions[51] had courted with the idea of using formulae, but none had endorsed their used as definitively as Akenhead J.

The following formulae are often used for calculating head office overheard claims:

- The Hudson Formula

The formula is set out on page 1076 of 11th edition of *Hudson's Building and Engineering Contracts.*[52] The formula's basis is to take the allowance (expressed as a percentage) which the contractor has made in respect of head-office overheads and profit in its original tender and then to divide this figure by the length of original contract period. The result is then multiplied by the period of prolongation of the contract to produce the claimed overheads and profit. Loss of profit is dealt with as a separate head of claim (see Section 4.4.2.6). The 'Hudson formula' is shown in Figure 4.1:

$$\frac{\textbf{Head office overhead}}{\textbf{100}} \times \frac{\textbf{Contract Sum}}{\textbf{Contract Period}} \times \textbf{Period of delay (weeks)}$$

**Figure 4.1** Hudson formula.

The formula combines head-office overheads and profit together on the reasonable basis that contractors normally (and often do) apply a single percentage to their prices to cover both. However, the author says of this calculation that '*in the case of a delayed contract, where the concern is to ascertain the 'profit' which the delayed contract organisation might have expected to earn elsewhere in the market on other contracts, it is this necessary combined operating margin of profit and fixed overhead which, in appropriate market conditions, the contractor's enterprise will have lost as a consequence of the period of owner caused delay...*'.

51 *Tate & Lyle Food and Distribution Co. Ltd v Greater London Council* (1982) 1 WLR 149, *Allied Maples Group Ltd v Simmons & Simmons* (1996) 46 Con LR 134, *Norwest Holst Construction Ltd v Co-operative Wholesale Society Ltd (No.2)* (1998) EWHC 339 (TCC), *How Engineering Services Ltd v Lindner Ceilings Partitions plc* (1999) 2 All ER (Comm) 374.

52 I N Duncan Wallace, *Hudson's Building and Civil Engineering Contracts* (1995) 11th edition, London, Sweet & Maxwell, where the strengths and some weaknesses of this and the Eichleay formula are discussed.

This is not necessarily correct. Because if a contract has overrun, the contractor has not actually lost overheads or, indeed, profits. What it has 'lost' is the opportunity to earn these two elements on other work during the overrun period.

The 'Hudson formula' calculation may not always be convincing, particularly if the percentages are based upon the information that is 'out of date' (such as the contractor's annual accounts prior to the contract). Furthermore, it may never have been feasible for the contractor to earn that level of overhead and profit contribution on that project. In addition, there is a serious inaccuracy because it allows the overheads and profit to be based on a figure including the amount of overheads and profit already included in the contract sum. If it is proposed to use the formula, the amount of overheads and profit already included should be excluded from the contract sum. Other criticisms have been levelled at the formula:

- It requires the total delay period to be reduced if appropriate to take account of various factors for which financial recovery is not permitted under most standard form contracts – such as the contractor's own inefficiency or extensions of time on grounds which do not also permit recovery of loss and/or expense;
- It ignores the contractor's obligation to prove that it had to turn down the opportunity to earn a contribution to overheads and profits during any period of delay;
- It ignores the fact that the contractor should make realistic efforts to deploy its resources elsewhere during a period of delay (mitigation)[53];
- The value of the final account may exceed the contract sum and any proper valuation for variations is likely to have included an element of reimbursement for overheads and profit while fluctuations may have the effect of reducing the percentage of overhead recovery on actual cost; and
- The formula can also produce under-recovery for the contractor where inflation during the period of delay increases the overhead costs envisaged at the time of tender.

Despite *Lilly* seemingly providing approval of the 'Hudson formula', it must be borne in mind that defendant did not provide any convincing counter-arguments. Therefore, the 'Hudson formula' criticisms set out above are likely to remain relevant, although most of these criticisms can be easily addressed by the following steps:

  - Entering the period of critical delay (which entitles the contractor to an extension of time) into the formula instead of the total delay to the project (which might include some contractor's culpable delay);
  - Giving credit for the overheads and profit recovered from variations, or fluctuations and inflation.

- The Emden Formula
  This is a formula which was published in a respected legal textbook, *Emden's Building Contracts and Practice,*[54] which considers the contractor's annual turnover as the basis for the percentage of overheads and profit. Emden formula is shown in Figure 4.2.

53 *Peak Construction (Liverpool) Ltd v McKinney Foundations Ltd* (1970) 1 BLR 114.
54 *Emden's Building Contracts and Practice*, 8th edition, vol 2, at page N/46.

$$\frac{h}{100} \times \frac{c}{cp} \times pd$$

**Figure 4.2** Emden formula '*h*' = the head-office percentage arrived at by dividing the total overhead cost and profit of the contractor's organisation by the total turnover; '*c*' = the contract sum; '*cp*' = the contract period; '*pd*' = the period of delay; '*cp*' and '*pd*' must be calculated using the same units, e.g. weeks.

Essentially, 'Emden' is the same as the 'Hudson' formula, save the clarification on how to calculate the head office overhead percentage (represented by '*h*'). In practice, this is likely to be the same method as would be used to calculate the 'Head office overhead' figure in the 'Hudson formula'.

This approach is open to some of the same criticisms as the Hudson Formula. However, it is more useful than Hudson's Formula if the actual costs of head-office personnel who were directly engaged in dealing with an individual contract are not obtainable. Sometimes, the formula is changed slightly to substitute, for *h,* the proportion of the contractor's overall overhead costs which can be shown from its accounts to be spent on staff who are directly engaged on contracts. This gives an approximation of the cost of staff engaged on the particular contract during the period of delay. However, clearly this approach does not make an allowance for the cost of greater numbers of staff involved during the original contract period due to disruption.

- The Eichleay Formula
  This formula originates in Canada[55] and is a three-stage calculation applying daily rates. Its three stages are shown in Figure 4.3 as follows:

All such formulae are subject to criticisms and the 'Eichleay formula' is no exception. Among the possible drawbacks are:

$$(1)\quad \frac{\textbf{Contract billings}}{\textbf{Total contractor billings for contract period}} \times \textbf{Total head office overheads for contract period} = \textbf{Allowable overhead}$$

$$(2)\quad \frac{\textbf{Allowable overhead}}{\textbf{Days of performance}} = \textbf{Daily contract head office overhead}$$

$$(3)\quad \textbf{Daily contract head office overhead} \times \textbf{Days of compensable delay} = \textbf{Amount of recoverable overheads}$$

**Figure 4.3** Eichleay formula.

55 From the case *Ellis-Don Ltd v The Parking Authority of Toronto* (1978) 28 BLR 98.

- The formula gives only a rough approximation of the recoverable amount for overheads. One notable inaccuracy is that as a financial model, the delayed project cannot be identical to the declined project;
- It does not require proof from the contractor of its actual increased overhead costs resulting from the delay; and
- There is the possibility of double-recovery. To deal with this possibility, it is necessary to deduct any head-office overhead recovery which can be achieved in the valuation of variations.

It is widely used in US Federal Government Contracts and has been adopted in some other non-government cases, but it has been subject to criticism in the courts and it is not universally accepted.[56] However, criticism of the formula is not criticism of the proposition that in a period of reduced activity on site a contractor will incur off-site overheads for which payment is not being recovered from revenue generated at site. It is simply that unintelligent use of the formula will demonstrate its inherent weakness.[55]

#### 4.4.2.6 Loss of Profit

Loss of profit, or to be more precise, loss of the 'opportunity' to earn profit, which the contractor would otherwise have earned had it not been for some delay or disruption, is an allowable head of claim under the financial claims clauses of most standard form contracts. Loss of profit is also recoverable in principle as a head of damage for breach of contract at common law under the first part of the rule in *Hadley v Baxendale*[57]. The contractor is entitled to recover only the profit normally expected to be earned. If a delay prevented the contractor from earning an extremely high profit on another contract, the extremely high profit element would not be recoverable unless, in accordance with the normal rules (the 'second limb'), the employer was aware of the exceptional profit at the time the delayed or disrupted contract has been executed.

The foundation for one type of claim for loss of profit is similar to a claim for recovery of overheads based on lost opportunity.[50] Whether the contractor is prevented from earning a profit on the contract on which it is engaged or whether it is prevented from earning a profit on another contract, it is the loss of opportunity to earn a profit which is important. It does not automatically follow that a contractor can recover a percentage in respect of profit every time it can prove delay or disruption. Such a claim is permissible only in situations where the contractor can show that, as a direct result of the disruption or prolongation, he has been prevented from earning a profit elsewhere in the normal course of his business. The position is similar to that discussed above in Section 4.4.2.5 regarding overheads. Indeed, for convenience, claims for loss of profit are often grouped together with loss of overheads. However, despite that a claim for overheads and a claim for profit are founded

56 See the comments of the USA Court in *E Berley Industries v City of New York* (1978) 385 NE (2d) 281.
57 [1854] 9 Ex 341.

upon a 'loss of opportunity', there can be distinct difference in how these claims for these heads of loss arise:

- A 'loss of opportunity' claim for loss of overheads is based upon the contractor's site resources being prevented from earning an overhead contribution during a period of prolongation, whereas
- A 'loss of opportunity' claim for loss of profit may arise in circumstances such as:
  - The contractor being denied the opportunity to earn a profit by his resources being tied up on a delayed project; or
  - The contractor is denied the opportunity to earn a profit because of some other breach of contract by the employer – such as wrongful termination.

In terms of loss of profit claims, it is suggested that in a situation analogous to a loss of overheads claim on a delayed project (e.g. one calculated on a formula basis), the contractor ought to follow the guidelines in *Lilly*.[50] However, if the loss of profit claim arises for some other reason, such as the employer wrongfully terminating the contractor's employment under the building contract, the contractor's claim rests on (i) the balance of probabilities and (ii) the quantification of his loss may be assessed on this basis. In relation to the burden of proof and quantum of damages, Bailey writes[58]:

> *'It is enough for the innocent party to show, on the balance of probabilities*[59]*, that he missed out on the 'opportunity' of making a profit (or avoiding a loss). If a person is able to show that another's breach of contract deprived him of the opportunity to make a profit, the court will do its best on the available evidence to award compensation to that party for his loss*[60]*. The quantum of damages awarded will reflect the probability of the claimant making a profit (or avoiding a loss)*[61]*.'*

### 4.4.3 Disruption Costs – Uneconomic Working

The basic principle of disruption claims is that the contractor incurs additional costs in carrying out the works in circumstances where the overall duration of the works (or a section) may not have been prolonged (see Section 2.4.2).

Delay and disruption can lead to loss of productivity in two different ways. It may be necessary to employ additional labour and plant or the existing labour and plant may stand idle or be under-utilised. The latter situation is often referred to as 'loss of productivity'. Although this is a permissible head of claim, it can often be difficult to establish the amount

---

58 Construction Law – 3rd Edition (2020) Julian Bailey at 13.19.

59 Citing *Perry v Raleys Solicitors* (2019) UKSC 5, at paragraph 23 per Lord Briggs JSC.

60 Citing *Kellie v Wheatley & Lloyd Architects Ltd* (2014) EWHC 2212 (TCC) at paragraph 94, per HHJ Keyser QC.

61 Citing *John Barker Construction Ltd v London Portman Hotel Ltd* (1996) 83 BLR 31, at page 64 per Mr Recorder Toulson QC in relation to an 'performance bonus' agreement to pay the contractor £20,000 if it completed by 26 August 1994: *"It is impossible to tell whether, as a matter of probability, that the [contractor] would have finished by 26 August 1994, but for those changes. They would have had a reasonable opportunity of doing so, but they could easily have failed for all manner of reasons. In those circumstances I would hold that the plaintiffs are entitled to damages for the loss of that chance equal to 50 per cent of the agreed performance bonus, or £10,000."*

of the actual additional expenditure involved, unless the contractor keeps accurate, detailed and contemporaneous records. Contractors commonly attempt to overcome this problem by presenting the claim on a 'global' or 'total cost' basis (refer to Sections 2.4.2 and Chapter 7). They often attempt to demonstrate loss of productivity by reference to original tender figures to establish the anticipated percentage productivity and then apply actual labour figures to indicate a fall in productivity. Usually, a new percentage is calculated to form the basis of calculating the claim. A contractor should be able to establish the actual costs incurred, but it will clearly be impossible to prove beyond an educated guess, what the costs would have been had the delay or disruption not occurred.

A contractor's tender breakdown is likely to have little relevance. The contractor's intended labour and plant resources (by reference to the original programme of work[62]) may be inaccurate and possibly inadequate. Some of the additional labour and plant time may be the difference between the contractor's wrongly estimated proposed resources (that may be insufficient) and what it would have used even if the contract had proceeded without delay or disruption. In appropriate cases, it is possible to demonstrate the true loss by ignoring the tender breakdown showing intention and simply comparing a period of normal working with a period when disruption is present. This is known as the 'measured mile' approach. Provided that the relevant building work is of a fairly repetitive nature, the 'measured mile' method produces a fairly convincing ratio to apply to work of a similar nature.[63] The 'measured mile' method must be used appropriately. For example, if a particular type of excavation work (using specific plant and resources) proceeded unencumbered for 3 out of its 6 weeks' duration, but then suffered disruption for the remaining 3 weeks, then the contractor has a directly applicable and consistent 'mile' against which to 'measure' the effects of disruption. However, the 'measured mile' is of limited use in the following circumstances:

- If the relevant works are disrupted from the start; or
- If there is no comparable work (i.e. the 'mile'), against which the 'measure' can be made.

A further difficulty is that of relating particular items of additional expenditure under these or indeed other heads to particular events. Contractors may not always keep records in such a detailed form as to enable this to be done, particularly where there may be several concurrent causes of delay and disruption, some but not all of which may entitle the contractor to make a financial claim.[64] Provided that the contractor submits meaningful evidence to isolate the disruption elements (from the costs arising from the concurrent delay[65]), it is a matter for the employer's representative to determine the amount due.

In Chapter 2, at Section 2.4 (Categories of claim), two simple examples of disruption claims were given, which arose from variations instructed by the employer. The method of evaluation in Chapter 2's examples is 'value based' (i.e. it includes an element of profit),

62 Refer to Section 3.5.2 – Impacted as-planned delay analysis.

63 The method was approved by Mr Recorder Percival in the 1985 *Whittal Builders v Chester-Le-Street District Council* (the 1985 case) (1996) 12 Const LJ 356. There were two cases by this name (one in 1987).

64 Concurrency and the inherent difficulties are discussed in Chapter 3, Section 3.7.

65 Refer to Section 3.7 – the general rule during periods of 'concurrent delay' is that the contractor is not entitled to financial compensation, and the employer is not entitled to delay damages.

rather than seeking compensation (or reimbursement) for a loss, or additional expense – on the usual 'common law damages' basis for breaches of contract, as are considered in the following example:

- After a competitive tender, a contractor is appointed to build a new three-storey office building for the employer. The contractor is experienced in carrying out this kind of project and finds this kind of work profitable. Progress is on time, and the contractor is able to securely store his materials and plant in the partially completed building, ready for the internal fitting-out to begin. However, the employer then introduces a new contractor to carry out other parts of the internal fitting out works and tells the original contractor he will have to accommodate the new contractor on site. However, the new contractor is obstructive, using up the contractor's materials and storage areas and gets in the way of the contractor's tradesman – blocking stairwells and corridors with scaffold towers, etc. The disruption caused by the employer's new contractor causes the original contractor's part of the fit out works to take twice as long as planned, but they are completed defect-free. The original contractor keeps daily records of the disruption suffered. The employer accepts that the contractor is entitled to an extension of time and prolongation costs, but refuses to pay the additional (disruption) costs for the fit-out works. In order to pursue a claim for disruption, the original contractor has to demonstrate the following:
  - The employer has duty to provide the contractor with sole possession of the site, so by introducing the new fit-out contractor, he is in breach of that duty;
  - The new fit-out contractor clearly obstructs and disrupts the progress of the original contractor's work, so the original contractor can prove the causal link between the employer's breach and the alleged disruption;
  - The contractor incurs significant costs for additional labour (around £75k more) and for access plant (approximately £5k more) when carrying out his fit out works. However, the contractor faces the following difficulties in demonstrating the extent of his losses that are directly attributable to the disruption:
    1. As the disruption began at the start of the fit-out works, the contractor is unable to employ the 'measured mile' approach – i.e. to compare[66] the fit out works' productivity during a period of 'normal' progress, against a period of disruption;
    2. The contractor's tendered prices for the fit out works may not be a reliable source of comparison. However, as the contractor is experienced, and he usually earns a profit carrying out this type of work, it is reasonable to assume that overall his contract price ought to be sufficient to carry out the work profitably. Nevertheless, the contractor examines his tender and discovers that the fit-out works contain a pricing error, so the contractor would have lost £10k;
    3. As a parallel exercise, the contractor checks the output man-hours that would usually be required to a unit (such as a linear metre) of fit out works on time, from an accepted source (such as a recognised pricing book[67]). It compares this output against the actual man hours achieved in completing the required linear meterage of fit-out works and applies the hourly rate paid to its fit out operatives. This indicates that the disrupted or reduced output has cost the contractor around £68k.

66 The 'measured mile' approach.

67 For example, Spons' Architects and Builders Price Books include the 'Labour Hours' for completing a unit of works (such as lm, $m^2$, $m^3$ etc).

- So how much is the contractor likely to recover?
  - Out of the contractor's total loss (or additional expense) of £75k for the fit-out works, its pricing error means that it would have always lost £10k when carrying out the fit out works. Therefore, the contractor's loss caused by the disruption should be account for this amount (reducing from £75k to £65k);
  - The contractor has sensibly carried out a verification exercise (in point (3) above), by comparing published output data against the actual output data for the fit-out works. This indicates that the contractor would have had to spend an additional £68k to complete the fit-out works at this disrupted level of output. This provides good evidence that, save for the £10k pricing error, the contractor's tendered output for the fit-out works was reasonably accurate.
  - On the face of it, the contractor in this example has provided a logical assessment of its loss and has accounted for an obvious error in its original pricing. The example also tells us that the contract kept daily records of the disruption caused and completed the fit out work with no defects. If the matter proceeded to be decided by a court, or an adjudicator, then despite the contractor's best efforts in seemingly completing the works as efficiently as possible in the circumstances, the assessment of its claim is likely to include a further reduction,[68] that on the balance of probabilities, the contractor suffered from inefficiencies[69] to some extent (albeit small), that the employer may raise or that were not captured in the contractor's records.

Accounting for points (i), (ii) and (iii), it seems highly unlikely that the contractor would recover an amount in excess of £65k. A more reasonable amount, accounting for any contractor inefficiencies, would be in the region of £55k to £60k.

### 4.4.4 Other Costs

The costs considered under Section 4.4.4 are those which may arise from causes other than prolongation or disruption of the contractor's works. The contractor may be entitled to claim for increased costs of resources (such as labour and materials or goods), or interest or financing costs accruing from a late payment.

#### 4.4.4.1 Increased Costs

Additional expenditure on labour, materials or plant due to increases in cost is usually an allowable head of claim in the following circumstances:

- The cost increase occurred during a period of delay (for which the contractor is entitled to an extension of time and loss and/or expense) and was unavoidable; or

---

68 *Bridge UK.COM Ltd (t/a Bridge Communications) v Abbey Pynford plc* (2007) EWHC 728 (TCC). Ramsey J, applied a 20% reduction "*...to allow for the inherent uncertainty in [what was in that case] this retrospective method*" in order to account for resources that were wrongly included in the claim. See also *Harvey Shopfitters Ltd v ADI Ltd* (2003) EWCA 1757, where only 25% of a claim's value was allowed, as the defendant provided evidence of an *"unproductive and wasteful use of labour."*.

69 Common allegations of contractor inefficiency include non-productive overtime, weather conditions (although not relevant to the above example as the works are internal), poor management and direction of labour, rectifying defective work, and lack of continuity for labour and management.

- Increases in the cost of labour, materials or plant occur during the contract period, but are unconnected to any entitlement to loss and/or expense may be claimable if the contract in question includes 'fluctuation' clauses.

The circumstances summarised in the first bullet point would apply if the contractor was unable to avoid an increase in labour, material or plant prices that occurred during a period of prolongation (for which the contractor is entitled to an extension of time and loss and/or expense). A fairly recent example (post Covid-19) could involve material suppliers fixing their prices at the delivery date, so that the contractor is unable to be sure of a fixed price, once the materials are delivered. If the delivery takes place within the prolonged period, the contractor should be able to recover the additional costs.

A fluctuations claim may be sustainable in situations where disruption has resulted in labour-intensive work being delayed and carried out during a period after an increased wage award. It should be noted that, where a claim of this kind is being made in respect of a delay to completion, it is not only the work carried out during the period of delay that should be considered. The correct measure would be the difference between what the contractor would have spent on labour, materials and plant and what it has actually had to spend over the whole period of the work as a direct result of the delay and disruption concerned (see Section 4.4.3). However, detailed proof is necessary and, in making this calculation, proper allowance must be made for any recovery of increased costs under any applicable fluctuations clauses in the contract (see below). Moreover, it is what the contractor 'should' have spent excluding expenditure due to its own defaults, which must be considered.

Contractors may seek to make such calculations easier by using an arbitrary formula or a notional percentage. Such approaches are not acceptable, as the contractor must show that the increases in costs have been the inevitable consequence of the cited occurrence. This can be quite complicated in the case of materials, because the contractor must demonstrate that it could not reasonably have placed its order earlier to avoid the price increases. Above all, it must not be assumed that all work and all materials after the period of delay or during a prolongation period after the contract completion date will automatically suffer a price increase, so this kind of claim must be made on an item by item basis and each price increase must be properly evidenced. This head of claim, whether for increases in labour, goods or materials costs, is itself very laborious to prepare, which is probably why the convenient use of arbitrary formulae or notional figures is so attractive, albeit wrong.

The second bullet point concerns the contractor's entitlement to recover price fluctuations in any event, irrespective of whether the works are delayed or varied. The fluctuations may not simply involve changes in material prices. JCT 2016 contracts also provides for price adjustments for 'Fluctuations' if the option is selected in the Contract Particulars.[70] The parties can select either that no Fluctuations Provisions applies or one of the following:

- JCT Fluctuations Option A (Contribution, levy and tax fluctuations[71]);

70 JCT SBC/Q 2016 – Contract Particulars, entry against clause 4.3 and 4.14, JCT DB 2016 – Contract Particulars, entry against clause 4.2, 4.12 and 4.13 (Fluctuations Provisions), JCT IC 2016 – Contract Particulars, entry against clauses 4.3 and 4.9, and JCT MW 2016 – Contract Particulars, entry against clauses 4.3 and 4.8.

71 JCT Fluctuations Options A is set out in Schedule 7 to the JCT SBC/Q 2016 and JCT DB 2016 contracts, Schedule 4 of JCT IC 2016, and Schedule 2 of JCT MW 2016.

- JCT Fluctuations Option B (Labour and materials cost and tax fluctuations)[72];
- JCT Fluctuations Option C (Formula adjustment)[72]; or
- Another fluctuations option selected by the parties (which must be set out in the Contract Particulars).

NEC4, at Option X1 (Price Adjustment for Inflation), is used to change the prices for Options A, B, C and D on a similar basis to the Fluctuations Options in the JCT 2016 contracts.

#### 4.4.4.2 Interest

The law relating to a payee's entitlement to interest has taken some time to develop,[73] but a creditor now enjoys a statutory right to simple interest and compensation in the form of the Late Payment of Commercial Debts (Interest) Act 1998 (as amended[74]) ('LPCDA'), which first came into law in November 2002. As the name implies, it only deals with a 'qualifying debt[75]' (consumers are excluded), but this will encompass the vast majority of sums overdue under a building contract. Broadly, the creditor is entitled to claim interest at 8% above the Bank of England Base Rate. In addition, the creditor may recover a contribution towards its debt recovery costs under s5A of the LPCDA.[76]

JCT 2016 contracts include similar provisions to the LPCDA, entitling the payee to interest on late payments – both interim and final payments, whether to be paid by employer or by the contractor. The JCT 2016 provisions[77] are less punitive than those in the LPCDA as simple interest is calculated at 5% above the prevailing base rate.[78]

NEC4 also entitles the payee to interest under clause 51.2 if the payment is late, or late because the Project Manager has failed to issue a 'certificate'. Unlike in the JCT 2016 contracts, the interest rate[79] in NEC4 can be varied by the parties as follows:

- NEC4 – Interest rate entries in Contract Data Part One (Section 5 – payment):
  - The additional percentage above the applicable base rate (but should not be less than 2%);
  - The particular base rate to apply, and
  - The relevant bank applying the base rate.

---

72 JCT Fluctuations Options B and C are no longer included in the JCT DB 2016 contract, but can be downloaded from the JCT website.

73 See cases such as *London, Chatham and Dover Railway Co v South-Eastern Railway Co* (1893) AC 429, *President of India v La Pintada Cia Navegacion SA* (1984) 2 All ER 773, and the Law Commission's 1978 Report on Interest Cmnd 7229. Eventually, this was put into effect, at least in part, by the LPCDA.

74 By the Late Payment of Commercial Debts Regulations 2013 (SI 2013 395).

75 LPCDA, s3.

76 LPCDA s5A – the creditor may recover an additional £40 (if the debt is less than £1,000), £70 if the debt is between £1,000 and £10,000, and £100 if the debt is over £10,000.

77 JCT SBC/Q 2016 – clauses 4.11.6 and 4.16.7, JCT DB 2016 – clauses 4.9.6 and 4.9.7, JCT IC 2016 – clauses 4.12.6 and 4.12.7, and JCT MW 2016 – clauses 4.6.1 and 4.6.2. In each case, any unpaid interest is recoverable as a debt.

78 The 'Interest Rate' is defined in clause 1.1 of the JCT 2016 contracts.

79 NEC4, Contract Data – Part One, Section 5 (Payment).

A successful claimant may also recover interest in adjudication,[80] litigation[81] or in arbitration.[82]

#### 4.4.4.3 Financing Charges

Prior to the LPCDA introducing a statutory right to interest for creditors, parties making financial claims under building contracts would often include 'financing charges' as a head of loss.[83] Financing charges would be claimed on the following bases:

- As a direct expense[84] – the creditor's bank charging a sum (usually fixed fee and simple or compound interest[85]) to the creditor for borrowing money to make good the financial deficit caused by funding the other elements of direct loss and/or expense or damages it had suffered; or
- As a 'loss of opportunity' (see Section 4.4.2.5 and footnote 50) – that the creditor would have earned on the money that, but for the breaches of contract causing it to incur loss and/or expense or damages, it would have retained in its bank account (usually a percentage of interest).

Given the relatively punitive nature of the LPCDA and the interest clauses in JCT 2016 and NEC4 (i.e. entitling the creditor to an additional percentage of between 2% and 8% above the relevant interest base rate), claiming interest under the aforementioned has largely replaced the practice of claiming financing charges. Simply put – interest claimed under

---

80 The Scheme for Construction Contracts (England and Wales) Regulations 1998 (as amended), Part I – Adjudication, paragraph 20(c) – the adjudicator having regard to the contract's terms regarding simple or compound interest.

81 Senior Courts Act 1981, s35A.

82 Arbitration Act 1996, s49.

83 See *F. G. Minter Ltd v Welsh Health Technical Services Organisation* (1980) 13 BLR 7 – a case a contract under the JCT 63 terms.

84 See *F. G. Minter* Footnote 83: *'[In] the building and construction industry the cash flow is vital to the contractor and delay in paying him for the work he does naturally results in the ordinary course of things in his being short of working capital, having to borrow capital to pay wages and hire charges and locking up in plant, labour and materials capital which he would have invested elsewhere. The loss of the interest which he has to pay on the capital he is forced to borrow and on the capital which he is not free to invest would be recoverable for the employer's breach of contract within the first rule in Hadley v Baxendale without resorting to the second, and would accordingly be a direct loss, if an authorised variation of the works, or the regular progress of the works having been materially affected by an event specified… has involved the contractor in that loss.'* at para 15 per Stephenson LJ.

85 *Rees & Kirby Ltd v Swansea City Council* (1985) 5 Con LR 34 at 51 per Robert Goff LJ: "*'There remains to be considered the question whether [they] are entitled to recover their financing charges only on the basis of simple interest, or whether they are entitled to assess their claim on the basis of compound interest, calculated at quarterly rests, as they have done. Now here, it seems to me, we must adopt a realistic approach. We must bear in mind, moreover that what we are considering is a debt due under a contract; this is not a claim for interest as such, … but a claim in respect of loss or expense in which a contractor has been involved by reason of certain specified events. [the contractor] like (I imagine) most building contractors, operated over the relevant period on the basis of a substantial overdraft at their bank, and their claim in respect of financing charges consists of a claim in respect of interest paid by them to the bank on the relevant amount during that period. It is notorious that banks do themselves, when calculating interest on overdrafts, operate on the basis of periodic rests: on the basis of the principle stated by the Court of Appeal in Minter's case, which we here have to apply, I for my part can see no reason why that fact should not be taken into account when calculating the [contractor's] claim for loss or expense …*".

these methods is far simpler and, in the majority of cases, provides a greater commercial benefit for the creditor than claiming financing charges.

### 4.4.5 Costs of a Claim

It is common for a contractor to seek payment for the additional resources used to prepare (or assist in the preparation of) a claim to the employer. These additional resources may consist of members of the contractor's organisation who would not normally become involved in a project (such as senior staff, e.g. heads of commercial or planning departments, or company directors), or external consultants that the contractor has appointed. It is necessary to examine the circumstances that give rise to the 'claim':

- Does the 'claim' arise out of the ordinary course of administering the contract; or
- Does the claim arise from a breach of contract by the employer?

For the first circumstance, it is suggested that the contractor is not entitled to be reimbursed for any costs incurred in preparing the claim for the simple reason submitting a 'claim' for an extension of time, and/or loss and/or expense under the JCT 2016 contracts, or a quotation for a 'compensation event' under NEC4 is part of the contractor's obligations under those contracts and does not entitle the contractor to additional payment. However, it is worth exploring the positions under these contracts in more detail:

- Under JCT 2016 contracts:
  - When claiming an extension of time, the contractor is obliged to notify[86] the employer or his representative (identifying the applicable 'Relevant Event(s)'), provide particulars of the delaying effects, an estimate of the delay beyond the Completion Date,[87] provide updates of any changes to the expected delay and provide '*such further information*' that may be '*reasonably*' required[88]; and
  - When claiming loss and/or expense, the contractor is not only obliged to notify[89] the employer or his representative, and provide an initial assessment[90], but also to provide 'such information as is reasonably necessary' for his claim to be ascertained, and updates[91] on a monthly basis.

  Therefore, unless the employer or his representative are not acting 'reasonably', it highly unlikely that the additional staff or external consultants would fall outside the scope of those obligations (particularly for a loss and expense claim). The position for an extension of time claim may be slightly different. JCT 2016 contracts do not place much emphasis on the contractor providing a programme (see Section 3.5.1), and none of the extension of time clauses expressly require a critical path analysis (CPA) method to be employed. Indeed, there is also support for the position that the employer or his representative does not necessarily have to use CPA methods or planning software when assessing extensions of time.[92] It may therefore be arguable that the costs for providing a CPA method in

86 JCT SBC/Q 2016 – clause 2.27.1, JCT DB 2016 – clause 2.24.1, and JCT IC 2016 – clause 2.19.1.
87 JCT SBC/Q 2016 – clause 2.27.2, JCT DB 2016 – clause 2.24.2, and JCT IC 2016 – clause 2.19.1.
88 JCT SBC/Q 2016 – clause 2.27.3, JCT DB 2016 – clause 2.24.3, and JCT IC 2016 – clause 2.19.4.2.
89 JCT SBC/Q 2016 – clause 4.21.1, JCT DB 2016 – clause 4.20.1, and JCT IC 2016 – clause 4.16.1.
90 JCT SBC/Q 2016 – clause 4.21.2, JCT DB 2016 – clause 4.20.2, and JCT IC 2016 – clause 4.16.2.
91 JCT SBC/Q 2016 – clause 4.21.3, JCT DB 2016 – clause 4.20.3, and JCT IC 2016 – clause 4.16.3.
92 *City Inn Limited v Shepherd Construction Ltd* [2008] BLR 269, paragraphs 27 to 29.

support of an extension of time claim fall outside out the contractor's existing obligations and may be recoverable.

- Under NEC4,

  Unlike JCT 2016 contracts, NEC4's 'programme' obligations in clauses 31 and 32, effectively operate as CPA method, so it is difficult to envisage how this, and the process for quoting[93] for compensation events (in other words, NEC4's 'claim' procedure) could fall outside the contractor's duties, and entitle him to recover the costs of doing so. However, the contractor's entitlement to the costs of preparing quotations for compensation events depend upon the following:
  - For Main Option A (Priced Contract with Activity Schedule) and B (Priced Contract with Bill of Quantities), it now appears that the contractor may recover the cost for providing quotations[94] if the cost of 'People' assisting with the clause 62 quotation fall within the definition in the Short Schedule of Cost Components;[95]
  - For Main Options C (Target Cost with Activity Schedule), D (Target Cost with Bill of Quantities) and E (Cost Reimbursable Contract), it is suggested that the contractor may recover the costs of the 'claim' or clause 62 quotation as 'Defined Cost[96]', provided that:
    - The cost of 'People' providing or assisting with the clause 62 quotation fall within the definition in the 'Schedule of Cost Components';[97] but
    - The costs of preparing a 'claim' for proceedings such as Adjudication, or for the tribunal, are excluded, and treated as 'Disallowed Cost'.[36]

The second situation (arising from a breach by the employer) is different from the first. In this example, the employer (or his representative) have either not acted reasonably (in the case of JCT 2016 contracts' examples) or have committed some other breach. On the face of it, the contractor's costs of the claim could fall under the first limb of *Hadley v Baxendale.* Some confusion was been caused by the case of *James Longley & Co Ltd v South West Regional Health Authority*[98], as some considered it authority that a claims consultant is entitled to be paid for work in preparing the claim. The case decided no such thing. The facts are that, following an arbitration which was settled during the hearing, taxation (or assessent) of the parties legal costs were reviewed and the fees of a claims consultant were allowed in respect of work done in preparing the contractor's case for arbitration (the preparation of three schedules annexed to the Points of Claim). His fees for preparing the contractor's claim for presentation to the architect were not allowed. The fees which were allowed were fees of a potential expert witness in the arbitration.

The principle is that if the claim is prepared as part of arbitration or litigation, the costs can be recovered as part of the costs of the action. The employment of a claims consultant

---

93 NEC4, clause 62.

94 NEC3: clauses 11.2(22) of Options A and B expressly excluded the costs for preparing clause 62 quotations for compensation event. However, NEC4 contains no such exclusions (refer to clauses 11.2(21) in Option A, and 11.2(22) in Option B).

95 NEC4 – definitions of 'Defined Cost' in clause 11.2(23) of Options A and B as the costs in the Short Schedule of Cost Components.

96 NEC4 – definitions of 'Defined Cost' in clause 11.2(24) of Options C, D, and E as the costs in the Schedule of Cost Components, less 'Disallowed Cost' (defined in clause 11.2(26) of Options C, D and E).

97 For Main Contract Options C, D or E, the 'Schedule of Cost Components' is used. Point 1 (People) provides that the relevant personnel must be working in the 'Working Area' (defined in clause 11.2(20)).

98 [1984] 25 BLR 56.

by a contractor is, at least in theory, unnecessary. In practice, of course, many contractors do need assistance in presenting their documents in the best possible way. Evidencing a claim is not something that every contractor knows how to do without such assistance. But whether a contractor employs a claims consultant or a solicitor to give advice or even takes counsel's opinion on the matter, the fees involved are not recoverable unless incurred as part of arbitration or legal proceedings. It has been held that managerial time spent in dealing with a problem may be claimable as a claim for 'special damages' in an action at common law.[99] In a 2007 case,[100] Mr Justice Ramsey adopted the following approach to the recovery of management time:

> *'as a matter of principle, such head of loss (i.e. the costs of wasted staff time spent on the investigation and/or mitigation of the tort) is recoverable, notwithstanding that no additional expenditure "loss", or loss of revenue or profit can be shown. However, this is subject to the proviso that it has to be demonstrated with sufficient certainty that the wasted time was indeed spent on investigating and/or mitigating the relevant tort; i.e. that the expenditure was directly attributable to the tort - see per Roxburgh LJ in British Motor Trades Association at 569. This is perhaps simply another way of putting what Potter LJ said in Standard Chartered namely that to be able to recover one has to show some significant disruption to the business; in other words that staff have been significantly diverted form their usual activities. Otherwise the alleged wasted expenditure on wages cannot be said to be "directly attributable" to the tort.'*

### 4.4.6 Sub-Contractor Claims

As the majority, if not all of the contractor's work will be sub-contracted, the contractor is likely to receive claims from its sub-contractors if the progress of their work is affected. Clearly, the contractor will not be entitled to include sub-contractor claims in his claim to the employer, when the cause of those sub-contractor claims is the contractor's mismanagement, or if he is in breach of his obligations under those sub-contracts. This section only considers claims from sub-contractors that arise from the equivalent causes (i.e. either breaches of contract caused, or variations instructed by the employer). Sub-contractor claims, including their entitlements under the JCT 2016 and NEC4 sub-contracts, are considered in Chapter 14.

A sub-contractor will be susceptible to the same nature of costs and expenses that a main contractor will incur – as described in Sections 4.4.2 (prolongation costs), 4.4.3 (disruption costs), 4.4.4 (other costs) and 4.4.5 (claim costs). Many sub-contractors will also sub-let portions of their works (to sub-sub-contractors), so they may also face 'stepped down' versions of the costs and expenses (described in this section) from sub-sub-contractors.

It is often the case that the information provided by a sub-contractor contains significantly less detail than the equivalent information if it was provided by a main contractor. This can

---

99 See *Tate & Lyle Food and Distribution Co Ltd v Greater London Council* (1982) 1 WLR.
100 149 *Bridge UK.COM Ltd (t/a Bridge Communications) v Abbey Pynford plc* [2007[ EWHC 728 (TCC) at paragraph 125, citing *R+V Verischerung AG v Risk Insurance and Reininsurance* SA (2006) EWHC (Comm), per Mrs Justice Gloster DBE, at paragraph 77.

cause difficulties for the main contractor when advancing (or including) sub-contractor claims 'up the line' to the employer. It is quite normal for sub-contractors to provide fewer details than the main contractor would provide, because the nature of some sub-contract works, along with the sub-contract conditions under which they're engaged, may be far simpler and straightforward than the main contract works and the main contract conditions. Therefore, a sub-contractor may not have the same extensive obligations to substantiate its claims for loss and expense as the contractor may have under the main contract. Employers and their consultants should be aware of this and make the appropriate allowances when assessing the elements of a main contractor's claim that originates with his sub-contractors.

## 4.5 Quantum Meruit Claims

A general introduction to 'quantum meruit' claims and four examples were given in section 1.2.3.

The circumstances giving rise to 'quantum meruit' claims are often misunderstood. It now seems clear that an entitlement to be paid 'a reasonable sum[101]' is not the same as an entitlement to be paid on a 'quanum meriuit' basis.[102] Quantum meruit is strictly appliable on the concept of 'unjust enrichment'[103] which has been defined[104] as:

> *'A field of law covering cases in which one person has been unduly enriched at the expense of another and is under an obligation 'quasi ex contractu' (as if from a contract) to make restitution to him. In many cases of quasi contract, the defendant has received the benefit from the claimant himself. The claimant may have paid money to him under a mistake of fact, or under a void contract, or may have supplied services under the mistaken belief that he was contractually bound to do so…'*

Clearly, a party cannot make a quantum meruit claim if there is an agreement to either pay an agreed sum[105] or upon the basis of calculating such a sum.

Examples of the circumstances in which 'quantum meruit' claims arise, the appropriate methods of valuing the resulting, are considered separately below.

### 4.5.1 Circumstances Giving Rise to Quantum Meruit Claims

The following circumstances in building contracts may give rise to quantum meruit claims:

- The 'letter of intent' scenario – where the parties are negotiating over a contract's essential terms, but do not reach agreement, and work was carried out in the expectation that a contract would be agreed;[106]

101 For example, under s15(1) of the Supply of Goods and Services Act 1982.
102 Hudson (14th Ed) 2023 at 5-015.
103 *Benedetti v Sawaris* (2013) UKSC 50.
104 Oxford Dictionary of Law – (9th Ed) 2018 defines "unjust enrichment" by reference to 'quasi-contract'.
105 Keating (10th Ed) 2019 at 4-031.
106 *British Steel Corporation v Cleveland Bridge & Engineering Co Ltd* (1981) 24 BLR 94, *Turriff Construction Ltd and Turriff Ltd v Regalia Knitting Mills Ltd* (1971) 9 BLR 20.but may not apply in the case

- If the contract under which work was carried out is void or unenforceable;[107]
- For undertaking work that is 'outside' the scope of an existing contract[108] or varied work in excess of a previously agreed financial limit;[109] and
- Where a party carries out work on a speculative basis (in the expectation that its costs will be covered by being awarded a contract), but that work is made use of by the prospective employer.[110] However, if the work (such as an architect's design) is purely speculative – i.e. dependent upon some external factor which does not materialise, then the claimant has no entitlement.[111]

### 4.5.2 Valuation of Quantum Meruit Claims

The valuation of construction work on a quantum meruit basis should consider a fair commercial rate[112] for work carried out, and (arguably to a lesser extent for building work) the market value of the services at the time that the enrichment was received[103].

Elements to consider when valuing work on a quantum meruit basis are as follows:

- The net cost of labour[113] and materials, plus an element for overheads and profit;[114]
- Any unnecessary additional costs incurred by the claimant carrying out the work[115] (such as deductions for defective work[112]);
- The prevailing circumstances, including:
  - The site conditions[115]; and
  - The other party's conduct[115] (e.g. if the works were prolonged[112]).

However, contractors should be cautious when attempting to argue that it is entitled to recover on the basis of a complete re-rating of the bills of quantities or on a *quantum meruit* because the whole scope and character of the work has changed.

---

of all failed negotiations – for example, if the parties were able to withdraw at any time, and were 'subject to contract' – see *Regalian Properties v London Docklands Development Corporation* (1995) 1 WLR 212. See also *Hall & Tawse South Ltd v Ivory Gate Ltd* (1997) 62 Con LR 117 to be distinguished from *Regalian* because (i) whilst the claimant could withdraw for negotiations at any time, once it commenced work, it was obliged to continue, and (ii) the valuation of work was not on a 'quantum meruit' basis, but based upon the rates and prices in the bills of quantities that would have been incorporated into a finalised contract – similarly, see also *ERDC Group Ltd v Brunel University* (2006) 109 Con LR 114.

107 *A L Barnes Ltd v Time Talk (UK) Ltd* (2003) EWCA Civ 402. However, it is suggested that a claimant may not recover on a quantum meruit basis for an illegal contract - see *Taylor v Bhail* (1995) 50 Con LR 70.

108 *Thorn v London Corp* (1876) 1 App Cas 120. *Amantilla Ltd v Telefusion PLC* (1987) 9 Con LR 139 at 145 per Judge Davis.

109 *Sir Lindsay Parkinson & Co Ltd v Commissioners of Works and Public Buildings* (1950) 1 All ER 208.

110 *William Lacey (Hounslow) v Davis* (1957) 1 WLR 932. *Marston Construction Co Ltd v Kigass Ltd* (1989) 15 Con LR 116.

111 *Stephen Donald Architects Ltd v Christopher King* (2003) 94 Con LR 1.

112 *ERDC Group Ltd v Brunel University* (2006) 109 Con LR 114, *Laserbore v Morrison Biggs Wall* (1993) CILL 896, and *Costain Civil Engineering Ltd and Tarmac Construction Ltd v Zanin Dredging and Contracting Company Ltd* (1996) 85 BLR 85 (it can be valued by reference to any profit on the work made by the other party and to any competitive edge which the provider of the service enjoys – for example, already being on site and, therefore, avoiding the need for further mobilisation costs).

113 In relation to a quantum meruit claim for professional services, obtaining reliable evidence of the reasonable time spent and a reasonable hourly rate – *Tallington Lakes Ltd (No. 01595671) v Larking Gardens (a firm)* (2014) EWCA Civ 959.

114 See footnote 112 and *Weldon Plant Ltd v Commissioner for the New Towns* (2000) EWHC 76.

115 *Serck Controls Ltd v Drake & Scull Engineering Ltd* (2000) 73 Con LR 100.

## 4.6 Exclusive Remedy Clauses

It is not unusual for building contracts to contain exclusive remedy clauses which limit a contracting party to a particular remedy. Indeed, liquidated damages clauses usually limit the employer with an exclusive remedy[116] for the contractor's failure to complete the works on time – so the employer is unable to seek common law damages as an additional avenue of recovery for the contractor's breach.

NEC4's clause 63.6 states that the only rights that the employer ('Client') and the contractor have to change the Prices, Completion Date or a Key Date are in respect of a compensation event. However, the restrictiveness of this clause is tempered by compensation event 60.1(18), which effectively allows 'common law' damages to be claimed for any Client breach as a compensation event – provided, of course, that the contractor complies with the relevant notification requirements (see Section 5.3.2 and Figure 5.1). Under a JCT 2016 contract, an 'Employer' may attempt to impose an exclusive remedy clause on the contractor by a contract amendment to remove the contractor's right to claim common law damages and restrict his rights of compensation to the matters[117] in the contract that entitle the contractor to financial compensation.[118] A simple exclusion clause is introduced to restrict the contractor to the Relevant Matters (JCT), then the contractor should not suffer much of a disadvantage, as JCT 2016 contracts also contain 'catch all' clauses,[119] that entitles the contractor to claim financial compensation for acts of prevention, impediments or defaults by the employer. However, as with NEC4, the contractor will have to comply with the relevant provisions that entitle it to claim loss and expense (see chapter 12).

A claim in tort (such as for negligence or nuisance) is another remedy which a wronged party may be able to rely on outside of a contract – although in building contract disputes, such claims are relatively rare.

The same requirements apply to exclusive remedies clauses as they apply to any other type of exclusion clause, including the 'reasonableness' test under the Unfair Contract Terms Act 1977's ('UCTA').

It is important to check that any exclusive remedies provisions in the main contract are reflected in the sub-contracts; otherwise, the contractor could be caught out by any shortfall in his ability to pass risks and liabilities down the contractual chain.

Exclusive remedy clauses are types of exclusion clauses, and in certain circumstances (such as the parties entering into a building contract on one party's standard terms of business[120]), they will be subject to UCTA's test of reasonableness.[121] There is nothing inherently unreasonable in adding an exclusive remedy clause that removes the contractor's right to claim common law damages, thereby limiting his remedies to the Relevant Matters or 'compensation events' expressly included in the contract. However, if the employer deletes

116 *Temloc Ltd v Errill Properties Ltd* (1987) 39 BLR 30 and *Piggott Foundations Ltd v Shepherd Construction Ltd* (1993) 67 BLR 48.
117 That is, 'Relevant Matters' under the JCT 2016 contracts.
118 *Scottish Power UK Ltd v BP Exploration Operating Company Ltd & Ors* (2016) EWCA 1043.
119 JCT SBC/Q 2016 – clauses 2.29.7 and 4.22.5, JCT DB 2016 – clauses 2.26.6 and 4.21.5, JCT IC 2016 – clauses 2.20.6 and 4.17.4, JCT MW 2016 – clause 2.7 ('*...for reasons beyond the control of the Contractor*'), and NEC4, clause 60.1(18).
120 Unfair Contract Terms Act 1977 s3.
121 Unfair Contract Terms Act 1977 s11(1), and see Schedule 2 in regard to 'time bar' clauses.

the 'catch all' clauses and any other clauses from the contract, whilst insisting upon the exclusive remedy clause, then the contractor may be put to a significant disadvantage, possibly unable to recover compensation even in the event of breaches by the employer. In such cases, the exclusive remedy clause may fall foul of UCTA.

## 4.7 Claims at Termination of Contractor's Employment

Termination (also referred to as 'determination') is the means for one party to bring the contract to a premature end. He can do this if the other party has become insolvent, commits a serious breach of that contract, or either party may terminate following some 'neutral' event resulting in it being impossible or impracticable to continue with the project. A party's right to terminate will normally arise after it has notified the other party, allowing both parties to begin to get their affairs in order before the termination takes effect and then providing a procedure for concluding other aspects of the project (such as finishing any remaining work) once termination has occurred. Termination is usually a two-stage process: the first stage being a 'warning' to the other party and the second stage usually concludes or confirms that termination has taken place. In the case of one party committing a serious breach, the 'first stage' notification period provides the innocent party with a 'self help' remedy. It informs the other party to stop committing the breach (or warns them not to repeat it) and lets them know that if they fail to do so, the breach is sufficiently serious to allow the innocent party to terminate, usually by issuing a further 'second stage' notice.

As well as the 'contractual' termination provisions (summarised in Table 4.2 and dealt with in more detail below), a party may have a common law right to terminate the contract. This is often referred to as repudiation or recission. If the other party commits a repudiatory breach (i.e. one that is sufficiently serious; often going right to the heart of his obligations under the contract – such as the contractor abandoning the works and leaving site without justification), then the other party can accept that repudiatory breach and bring the contract to an end. However, this section is only concerned with the procedures following contractual termination.

Once termination has taken place, JCT 2016 and NEC4 building contracts contain procedures for concluding contract, such as resolving the contractor's final account and dealing with any further payments from the employer to the contractor or from the contractor to the employer as the case may be.

### 4.7.1 Termination Claims Under JCT 2016 Contracts

The calculation of payments to the contractor will produce starkly differing results depending upon whether termination was by the employer or by the contractor, as the other (non-terminating) party will have committed a 'specified default'. The exception being that in the case of termination by either party,[122] the employer bears the risk of termination.

If the employer terminates, he is able to recover the additional costs[123] of completing the Works, following the principle of common law damages (see Section 1.2.2 and Chapter 6).

---

122 JCT SBC/Q 2016, JCT DB 2016, JCT IC 2016 – clause 8.11, JCT MW 2016 – clause 6.10.1.
123 JCT SC/Q 2016, JCT DB 2016, JCT IC 2016 – clause 8.7.4, JCT MW 2016 – clause 6.7.3.

**Table 4.2** Grounds for termination.

| **Ground for Termination (E) Employer, (C) Contractor** | **JCT 2016 clauses** | **NEC4 clauses** |
|---|---|---|
| *Insolvency* | | |
| Of the contractor | (E) SBC/Q, DB, IC 8.5.1, MW 6.5.1 | (E) & (C) 91.1 (R1) to (R10) |
| Of the employer | (C) SBC/Q, DB. IC 8.10.1, MW 6.9.1 | |
| *Breach/Default by contractor* | | |
| Failure to provide performance security | — | (E) 91.2 (R12) |
| Wrongfully ceases or suspends work | (E) SBC/Q, DB, IC 8.4.1.1, MW 6.4.1.1 | (E) 91.2 (R11) |
| Fails to proceed regularly and diligently | (E) SBC/Q, DB, IC 8.4.1.2, MW6.4.1.2 | (E) 91.3 (R14) |
| Fails to comply with a rectification instruction | (E) SBC/Q, DB, IC 8.4.1.3 | (E) 916 (R18) |
| Fails to comply with sub-contracting provisions[a)] | (E) SBC/Q, DB, IC 8.4.1.4 | (E) 91.1 (R13) |
| Fails to comply with assignment provisions | (E) SBC/Q, DB, IC 8.4.1.4 | — |
| Fails to comply with CDM provisions | (E) SBC/Q, DB, IC 8.4.1.5, MW 6.4.1.3 | (E) 91.3 (R15) |
| Engages in corrupt/fraudulent activity | (E) SBC/Q, DB. IC, 8.6, MW 6.6 | (E) 91.8 (R22) |
| *Breach/default by employer* | | |
| Fails to pay the amount due by the final date for payment | (C) SBC/Q, DB, IC 8.9.1.1, MW 6.8.1.1 | (C) 91.4 (R16) |
| Interferes with, or obstructs a certificate | (C) SBC/Q, IC 8.9.1.2, MW 6.8.1.2 | |
| Fails to comply with assignment provisions | (C) SBC/Q, IC 8.9.1.3, DB 8.9.1.2 | — |
| Prolonged suspension of the works | (C) SBC/Q, DB, IC 8.9.2,[b)]MW 6.8.2 | (C) 91.6 (R19) |
| Fails to comply with CDM provisions | (C) SBC/Q, IC 8.9.1.4, DB 8.9.1.3, MW 6.8.1.3 | — |
| *Event not the fault of either party* | | |
| If released from the contract by law | (E&C) SBC/Q, DB, IC 8.11.1.5, MW 6.10.1.5 | (E&C) 91.5 (R17) |
| Employer's representative doesn't instruct a re-start/not to complete the works | (E&C) SBC/Q, DB, IC 8.8.1 | (E) & (C) 91.6 (R20) |

**Table 4.2** (Continued)

| **Ground for Termination (E) Employer, (C) Contractor** | **JCT 2016 clauses** | **NEC4 clauses** |
|---|---|---|
| Employer prevents the contractor from completing the whole of the works | (E&C) SBC/Q, DB 6.14,[c)] IC 5.7<br>SBC/Q, DB, IC 8.11.1[d)] | (E) 91.7 (R21)[e)] |
| Grounds under s73 of the PC Regulations | (E&C) SBC/Q, DB, IC 8.11.3, MW 6.10.3 | — |

a) JCT SBC/Q DB and IC refer to failure to comply with the 'Consent to sub-contracting' clause (JCT SBC/Q – clause 3.7, JCT DB 2016 – clause 3.3, JCT IC 2016 – clause 3.5), but JCT IC 2016 contains an additional ground for termination for failing to comply with clause 3.7 'Named Sub-Contractors'
b) JCT DB 2016, the period of suspension under clause 8.9.2 is a "specified suspension event", and must exceed the period of suspension set out against the entry for clause 8.9.2 in the Contract Particulars (the default period is 2 months)
c) JCT DB 2016, clause 6.14 (Loss or damage to Existing Structures – right of termination)
d) JCT DB 2016, clauses 8.11.1.1 (force majeure), 8.11.1.2 (employer's instructions under clauses 2.13, 3.9 or 3.10 arising from Statutory Undertaker's default), 8.11.1.3 (loss or damages caused by an insurable risk, or an Excepted Risk – see clause 6.8), 8.11.1.4 (civil commotion, terrorism), 8.11.1.6 (delay in permission/approval of Development Control Requirements).
e) NEC4, clause 91.7, the event also prevents the Contractor from completing the whole of the works within 13 weeks after planned Completion (as shown on the Accepted Programme).

It does so by requiring the employer to prepare a statement[124] within 3 months of completing the Works (or the making good of defects if any are present) to calculate the difference between:

1) The actual amount that the employer had paid for the Works, including:
   - The amount already paid by the employer to the (original) contractor;[125]
   - Amounts paid to a replacement contractor (or contractors) to complete the Works;[126] and
   - Any further 'expenses' incurred by the employer from any other breaches of contract by the contractor (including those expenses arising from the termination);[127] and
2) The amount that the employer would have paid to the original contractor for completing the Works.[128]

124 JCT SBQ/Q 2016, JCT DB 2016,JCT IC 2016 – clause 8.7.4, JCT MW 2016 – clause 6.7.3.
The employer's statement must be prepared within completion of the Works, or making good of defects.
125 JCT SBC/Q 2016, JCT DB 2016, JCT IC 2016 – clause 8.7.4.2, JCT MW 2016 – clause 6.7.3.2.
126 JCT SBC/Q 2016, JCT DB 2016, JCT IC 2016 – clause 8.7.4.1, JCT MW 2016 – clause 6.7.3.1.
127 JCT SBC/Q 2016, JCT DB 2016, JCT IC 2016 – clause 8.5.3.2, JCT MW 2016 – clause 6.5.2.2 (termination by employer following contractor's insolvency) and 8.10.3, or JCT MW 2016 – clause 6.9.2 (termination by contractor following employer's insolvency) expressly state that the Contractor's obligations under Article 1 are suspended. JCT SBC/Q 2016, JCT DB 2016, JCT IC 2016 – clauses 8.7.1 (termination by employer), 8.12 (Termination by Contractor) or clause 8.11 (by either party) imply that the Contractor's obligation and entitlement to complete the Works ceases (JCT MW 2016 – clauses 6.7.1, 6.10 and 6.11). The "expenses" incurred prior to the date of termination may include liquidated damages.
128 JCT SBC/Q 2016, JCT DB 2016, JCT IC 2016 - clause 8.7.4.3, JCT MW 2016 – clause 6.7.3.2.

If the amount in (1) exceeds the amount in (2), then the difference shown on the employer's statement[124] is a debt due from the contractor to the employer.[129] This is likely to occur in the majority of cases as the employer will want to resume the Works as quickly as possible in order to avoid the works suffering damage and to limit any additional financial effects (such as delayed sale or use of the project). Consequently, the employer may not have time to competitively obtain a price for completing the works, and a replacement contractor may not want to be held to a fixed price for completing another contractor's work.

However, in the (unlikely) event that the sum in (2) exceeds the sum in (1), and the employer would always have had to pay the original contractor the higher sum for completing the Works,[130] then this produces a perverse result – as the difference is paid to the contractor. Whilst the employer may not have to pay any more for the Works that he always would have done, it cannot be right that a contractor who commits a 'specified default' can be rewarded by being paid a sum greater than the value of work he has carried out, when he has committed a 'specified default'. If the employer is able to use his commercial acumen to obtain a lower price for completing the work (i.e. mitigating his losses), then he should be able to gain its benefit in financial terms.

JCT 2016 contracts do not expressly state when the 'due date', or the 'final date for payment' occur for the 'difference' set out in the employer's statement. They merely state that the difference becomes a 'debt due'.[129] It may be that the Scheme[131] applies, and the 'due date' is likely to be either the date that the employer sends his statement, or the date when the contractor receives it. The final date for payment is likely to be 17 days after the due date[132]. This is important if the contractor disputes the 'difference' in the employer's statement, then he can issue a 'pay less notice'. JCT 2016's final account procedures provide that the contractor may issue a 'pay less notice' not later than 5 days before the final date for payment under the 'final account'.[133]

If the contractor has terminated[134] (or either party has done so[135]), then the contractor's termination account[136] is comparatively more straightforward. His 'termination account' should be submitted either as soon as is '*reasonably practicable*' after termination, or if

129 JCT SBC/Q 20J6, CT DB 2016, JCT IC 2016 - clause 8.7.5, JCT MW 2016 – clause 6.7.4.

130 For example, if the economic conditions were such that it was a "contractor's market" when the contract was concluded, but subsequently and significantly changed to an "employer's market" by the time the employer terminated the contractor's employment.

131 The Scheme for Construction Contracts (England & Wales) Regulations 1998 SI 649, as amended.

132 The Scheme, Part II (Payment), paragraph 8(2).

133 JCT SBC/Q 2016 – clause 4.115.2, JCT IC 2016 – clause 4.12.5.2, JCT MW 2016 – clause 4.5.4.2 (paying less than stated in the 'Final Certificate') or JCT DB 2016 – clause 4.9.5.2 (paying less than stated in the 'relevant statement').

134 JCT SBC/Q 2016, JCT DB 2016, JCT IC 2016 – clause 8.9, or JCT MW 2016 – clause 6.8.

135 JCT SBC/Q 2016, JCT DB 2016, JCT IC 2016 – clause 8.11, or JCT MW 2016 – clause 6.10 (also see Footnotes 124 and 125).

136 JCT SBC/Q 2016. JCT DB 2016, JCT IC 2016 – clause 8.12.3, or JCT MW 2016 – clause 6.11.2.

either party has terminated, then not later two months thereafter[137]. It should show the following:

- The total of work properly carried out up to the date of termination (valued in accordance with the contract),[138] including goods or materials that the contractor has ordered and must pay for;[139]
- Any loss and expense, either occurring before[140] or after[141] termination;
- The costs of removing the contractor's facilities;[142] and
- Other amounts due under the contract.[143]

The contractor's 'termination account' does not need to show the amounts that the employer has paid to date, but the employer must account for such payments when making payment in respect of the contractor's termination account. The final date for the payment of the contractor's 'termination account' is 28 days after the employer receives it, and no retention can be deducted.[143] It appears that the 'due date' for payment of the contractor's 'termination account' is the date it submitted to the employer, so again, it is advisable for the contractor to retain proof of the date when it was sent. In practical terms, and particularly where the 'termination account' is accompanied by significant documentation, it is unlikely that the employer has sufficient time to issue a 'payment notice'[144] within 5 days of the account's receipt. It is more likely that the employer will issue a 'pay less notice'[145] as he will have more time to do so. It appears that a pay-less notice ought to be issued not later than 5 days before the final date for payment,[146] but as the JCT 2016 contracts do not stipulate such a period for the 'termination account', a cautious employer may issue a pay-less notice in respect of the 'termination account' not later than 7 days before the final date for payment, so it complies with paragraph 10 of Part II of the Scheme.[131]

The employer has the option to not to complete the Works,[147] but this only seems to apply if the employer has terminated the contractor's employment under the clauses

---

137 JCT SBC/Q 2016, JCT DB 2016, JCT IC 2016 – clause 8.12.3.
138 JCT SBC/Q 2016, JCT DB 2016, JCT IC 2016 – clause 8.12.3.1, or JCT MW 2016 – clause 6.11.2.1.
139 JCT SBC/Q 2016, JCT DB 2016, JCT IC 2016 – clause 8.12.3.4, or JCT MW 2016 – clause 6.11.2.2.
140 JCT SBC/Q 2016, JCT DB 2016, JCT IC 2016 – clause 8.12.3.2.
141 JCT SBC/Q 2016, JCT DB 2016, JCT IC 2016 – clause 8.12.3.5, or JCT MW 2016 – clause 6.11.2.3 and in the case of termination by either party under (where the Works are damaged by a an event that is an Excepted Risk, or covered by the Works Insurance), only in respect of negligence or a default on the part of the employer (or any employer's Person) – see Footnote 135.
142 JCT SBC/Q 2016, JCT DB 2016, JCT IC 2016 – clause 8.12.3.3.
143 JCT SBC/Q 2016, JCT DB 2016, JCT IC 2016 – clause 8.12.5, or JCT MW 2016 – clause 6.11.4.
144 HGCRA 1996, s110A(1)(a) and s110A(2).
145 HGCRA 1996, s111(3).
146 As is the case for interim payments or for the final payment under JCT SBC/Q clause 4.11.5, JCT DB 2016 clause 4.9.5, JCT IC 2016 – clause 4.12.5, or JCT MW 2016 – clause 4.5.4.
147 JCT SBC/Q 2016, JCT DB 2016, JCT IC 2016 – clause 8.8.

for 'Default by Contractor'[148] or 'Insolvency of Contractor'.[149] The contracts set out a procedure whereby the employer must notify the contractor within 6 months (from the date on which the contractor's employment is terminated) that he does not intend to complete the Works.[150] The employer must then provide a statement to the contractor[151] showing:

- The total value of work that the contractor properly carried out at either the date of termination or on the date when the contractor became insolvent, along with any other amounts due under the contract ;[152] and
- The total (or 'aggregate') of expenses that the employer has properly incurred, including those arising from the contractor's defaults or breaches (either before termination or after).[153]

Somewhat imprecisely, the relevant clause[154] states that the amounts that the employer has already paid to the contractor must be taken into account but doesn't say exactly how. However, first, taking the example of the employer's statement, and second, the 'difference'[154] it seems that the amounts previously paid to the contractor fall under the 'expenses' described in clause 8.8.1.2. The 'difference' (which seemingly means the total after subtracting the amount in clause 8.8.1.2 from the amount in clause 8.8.1.1) is paid by the employer to the contractor (if a positive sum), or by the contractor to the employer (if a negative sum).

### 4.7.2 Termination Claims Under NEC4

Clause 93 (Payment on Termination) sets out four stages of calculating the amount due (A1,[155] and A2 to A4[156]) on termination which are available to each party as follows:

- The amount due under method A1 is due for any reason for terminating[157] and includes[158]:

148 JCT SBC/Q 2016, JCT DB 2016, JCT IC 2016 – clause 8.4.
149 JCT SBC/Q 2016, JCT DB 2016, JCT IC 2016 – clause 8.5.
150 JCT SBC/Q 2016, JCT DB 2016, JCT IC 2016 – clause 8.8.1.
151 JCT SBC/Q 2016. JCT DB 2016, JCT IC 2016 – clause 8.8.1 provides that the employer provides the clause 8.8 statement either within a "reasonable time" after notifying the contractor (in accordance with clause 8.8.1), or if no such notification is given, then within 2 months after the expiry of the 6 month period in clause 8.8.1.
152 JCT DB 2016, clause 8.8.1.1.
153 JCT DB 2016, clause 8.8.1.2.
154 JCT SBC/Q 2016, JCT DB 2016, JCT IC 2016 – clause 8.8.2.
155 NEC4, Clause 93.1 – A1.
156 NEC3, Clause 93.2 – A2 to A4.
157 NEC4, User Guide (2020), p80.
158 NEC4, clause 93.1.

| | |
|---|---|
| Amount assessed for 'normal payments' | £A |
| Excluding amounts to be permanently paid by the contractor:<br>• Failing to meet a Key Date under clause 25.3<br>• Client's cost of rectifying an uncorrected defect under clause 46.1 | |
| Defined Cost for Plant and Materials:<br>In the Working Areas; or<br>Where title has passed to the Client and the contractor will accept delivery | £B |
| Any other Defined Cost that the contractor 'reasonably incurred'[a)] | £C |
| Amounts that the Client has retained[b)] | £D |
| Less any unpaid instalment or balance of an advanced payment[c)] | (£E) |
| **Total amount due under A1:** | **£A+£B+£C+£D-£E** |

a) For example, in NEC4 – Option A (Priced Contract with Activity Schedule) – including the value of uncompleted activities as at the date of termination.
b) For example, retention under Option X16, or amount retained under clause 50.5 if the contractor has not issued its first programme (under clause 31.1).
c) NEC4, Option X14 (Advanced Payment to the Contractor), clause X14.3.

- A1 is used by the parties as follows:
  - By either party in the event of the other's insolvency (reasons R1 to R10), if the law releases either of them from further obligations (R17) or if an instruction to stop work arises from no fault of either party (R20);
  - By the Client only for reasons R11 to R15 or R22 (all contractor default) or if an unforeseeable event (which is neither party's fault) prevents completion (R22); or
  - By the contractor only in the event of the Client's failure to pay (R16) or from a Client default (R19);

**Table 4.3** NEC4 – Clause 92 (A4), application of 'fee percentage' on termination.

| Main Contract Options | Fee percentage applied to: |
|---|---|
| Option A: Priced Contract with Activity Schedule<br>Option B: Priced Contract with Bill of Quantities<br>Option C: Target Cost Contract with Activity Schedule<br>Option D: Target Cost Contract with Bill of Quantities | 'fee percentage' is applied to £X<br>Where £X = [total of the Prices at the Contract Date] – [Price for Work Done to Date] |
| Option E: Cost Reimbursable Contract | 'fee percentage' is applied to £Y<br>Where £Y = [first forecast of Defined Cost for the works] – [(Price for Work Done to Date) – (the Fee)] |

**Table 4.4** NEC4 – Methods of calculating the termination account.

| Reason (clause 91.1) 'E' = Client, 'C' = contractor | | Procedure (clause 92) | | | | Calculation of Amount Due (clause 93) | | | |
|---|---|---|---|---|---|---|---|---|---|
| | | P1 | P2 | P3 | P4 | A1 | A2 | A3 | A4 |
| Individual Insolvency | (R1) Other party presents bankruptcy application | **E**<br>**C** | E | E | C | E<br>C | C | E | C |
| | (R2) Other party has bankruptcy order against it | **E**<br>**C** | E | E | C | E<br>C | C | E | C |
| | (R3) Receiver appointed (over assets) | **E**<br>**C** | E | E | C | E<br>C | C | E | C |
| | (R4) Made arrangement with creditors | **E**<br>**C** | E | E | C | E<br>C | C | E | C |
| Company Insolvency | (R5) Winding up order made against other party | **E**<br>**C** | E | E | C | E<br>C | C | E | C |
| | (R6) Provisional liquidator appointed | **E**<br>**C** | E | E | C | E<br>C | C | E | C |
| | (R7) Resolution for winding up passed[a)] | **E**<br>C | E | E | C | E<br>C | C | E | C |
| | (R8) Administrator appointed/administration order | **E**<br>C | E | E | C | E<br>C | C | E | C |
| | (R9) Receiver appointed over substantial assets | **E**<br>C | E | E | C | E<br>C | C | E | C |
| | (R10) Made arrangement with creditors | **E**<br>C | E | E | C | E<br>C | C | E | C |
| (R11) Contractor substantially fails to satisfy his obligations | | E | **E** | E | — | E | — | E | — |
| (R12) Contractor fails to provide bond or guarantee | | E | **E** | E | — | E | — | E | — |
| (R13) Contractor sub-contracts a substantial part of the Works without Project Manager's approval | | E | **E** | E | — | E | — | E | — |

| | | | | | | | | | |
|---|---|---|---|---|---|---|---|---|---|
| (R14) Contractor substantially hinders the Client | | E | **E** | E | — | E | — | E | — |
| (R15) Contractor commits a serious health and safety breach | | E | **E** | E | — | E | — | E | — |
| (R16) Client fails to pay contractor within 13 weeks | | C | — | — | C | C | C | — | C |
| (R17) Law releases either party from further obligations | | E<br>C | — | — | E<br>C | E<br>C | E<br>C | — | — |
| Instruction to stop work | (R18) Instruction arises from contractor's default | E | E | E | — | E | — | E | — |
| | (R19) Instruction arises from Client's default | C | — | — | C | C | C | — | C |
| | (R20) Instruction does not arise from a party's default | E<br>C | — | — | E<br>C | E<br>C | E<br>C | — | — |
| (R21) Unforeseeable, non-fault[b] event prevents completion | | E<br>C | — | — | E<br>C | E | E | — | — |
| (R22) Contractor commits a 'corrupt act' | | E | E | E | — | E | — | E | — |

a) NEC4, clause 91.1 (R7) – resolution for winding up passed exception for the purposes of amalgamation or reconstruction.
b) NEC4, clause 91.7.

- A2 – comprises the forecast Defined Cost[159] of removing Equipment and is used by the parties as follows:
  - By the contractor when the Client is insolvent (reasons R1 to R10), fails to pay (R16), or following an instruction to stop work arising from the Client's default (R19);
  - By either party for reasons R17 (released from obligations by law) or R20 (instruction to stop work not arising from any party's default); or
  - By the Client in the event an unforeseeable event prevents completion (R20).
- A3 – The additional cost to the Client (as forecast) of completing the Works. A3 is solely available to the Client.
- A4 – This is the application of the 'fee percentage[160]' which is solely available to the contractor when terminating for the Client's insolvency (R1 to R10), when the Client fails to pay (R16), and an instruction to stop work arising from the Client's default (R19). Under A4, the fee percentage is applied depending upon which Main Option Clause applies – as shown in Table 4.3.

There are additional sub-clauses applying to Option A (Priced Contract with Activity Schedule),[161] Option C (Target Cost Contract with Activity Schedule)[162] and Option D (Target Cost Contract with Bill of Quantities),[163] concerning the calculation of the amount due on termination.

NEC4's reasons for termination, the applicable procedures and methods for calculating the amount due on termination are shown in Table 4.4.

159 See Section 5.5.1 and Table 5.15.
160 NEC4, clause 11.2(10) defines the "Fee".
161 NEC4, Option A, clause 93.3 – Grouping of activities is not taken into account when assessing the amount due on termination.
162 NEC4, Option C, clause 93.4 – The Project Manager assesses the contractor's share after issuing the termination certificate. Clause 93.6 If the Project Manager's assessment of the contractor's share is positive (in the event of a saving) it is added to the amount due to the contractor on termination. Conversely, if the assessment is negative (an excess) it is deducted from the amount due to the contractor on termination.
163 NEC4, Option D, clauses 93.5 and 93.6 (having similar effect as clauses 93.4 and 93.6 under Option C).

# 5

# Claims for Variations Under the Contract, Acceleration, Bonus Clauses and Contractor's Share

## 5.1 Definitions and Scope

There are clauses permitting variations in all standard forms of building contract. If there was no such clause in lump sum contracts, the contractor could not be compelled to vary the works, and it could insist upon completing only the work originally specified in the contract. Although building contracts contain a clause permitting variations, the power is not unfettered, and no power to order variations should be implied.[1]

A variation may not be ordered if it changes the whole scope and character of the works. To determine whether this has been done in any particular case; first, reference must be made to the part of the contract (usually a 'recital') that describes the work to be done. As a broad rule of thumb, if the variation is not wholly outside the scope of the works' description, it is unlikely to be a variation which changes the whole scope and character of the works, so it is likely to be valid. To a large extent, the point is academic, as a contractor will usually welcome the opportunity to carry out additional work and thereby earn money in the valuation of the variation and possibly in the formation of a claim for disruption or prolongation.

Variations usually happen in most building projects, and the chance of any contract being completed without any variation to the works is relatively low. The problem is the innate complication of the building process and the fact that it is almost impossible for any employer to finally decide on every part of a building before it is substantially constructed. There are, of course, other instances where variations are virtually imposed upon the employer in complex, prestigious or state of the art building with a high technical content (such as hospitals, media studios and concert venues). No employer wants to feel that the finished building is already out of date and the instructing of often substantial variations will continue throughout the life of the project. In addition, there are many employers who are just incapable of visualising the finished building until it is actually or very nearly completed. The building – as tendered – will bear a general resemblance to the finished product in these instances, but no more. As the building begins to take shape, the employer may have second thoughts about certain parts of it. Other reasons for variations include questions of timing of the project. Often the employer's representative and the

1 *Stockport MBC v O'Reilly* (1978) 1 Lloyd's Rep 595.

*Building Contract Claims*, Sixth Edition. Guy Higginbottom and David Chappell.

contractor are asked to start on site long before it is realistic to do so. Unless the employer's representative and the contractor are firm in those circumstances, the project may be set for disputes at a later date.

This chapter also provides guidance regarding the changes under the relevant building contracts for 'Acceleration' (Section 5.6) and 'Bonus clauses' (Section 5.7).

### 5.1.1 Definition of a 'Variation'

The word 'variation' in building contracts refers to a change in the works – most commonly a change in the physical works (such as an addition or a change in the works' specification), or it can mean a change in the way that the work is carried out. This is known as a 'variation under the contract' and should be distinguished from a 'variation of the contract itself' which is quite different (briefly considered in Section 5.1 above, and Chapter 6, Section 6.3). A 'variation' can be generally defined as follows:

- '*a change or slight difference in the condition, amount, or level, typically within certain limits*'[2]; and
- '*(change) a modification or alteration to that agreed under the terms of the contract, or that described in the bill of quantities*'.[3]

The second bullet point definition refers to the 'bill of quantities', but a reference to any of the principal documents incorporated into a building contract describing the scope of works (such as the drawings, specification, scope, schedules of works or priced documents[4]) is equally applicable. For JCT DB 2016 (or the CDP element of JCT SBC/Q 2016), it is only a change in the 'Employer's Requirements'[5] that will create a variation, but not a change in either the 'Contractor's Proposals' or the 'Contract Sum Analysis' (or CDP Analysis). Consequently, the employer can only instruct a 'Change', including the expenditure of a 'Provisional Sum'[6] only if it is in the Employer's Requirements.[7] Reference is sometimes made to the Recitals[8] which, it is argued, suggests that the employer accepts the Contractor's Proposals. A close reading of the Recital indicates that it is of little practical or legal effect, as it is a principle of contract interpretation that if words in the main part of the contract are ambiguous, only then may one turn to the recitals to assist finding the true meaning of the words. However, if the words in the main part of the contract are clear, the recitals

---

2 Oxford Dictionary of English (3rd Ed) 2010.

3 Oxford Dictionary of Construction, Surveying & Civil Engineering – (2nd Ed) 2020.

4 Refer to Sections 5.1.3 and Table 5.2 (Bills of Quantities – for JCT and NEC4), 5.1.4 (Other pricing documents), 5.4.1 and Table 5.11 (JCT Valuation Rules) and 5.5.1 and Table 5.11 (NEC4).

5 In practice, the Employer's Requirements (in JCT DB 2016, or in the CDP for JCT SBC/Q 2016) often contain drawings, specification and a scope of works that the contractor is to design). The employer can only alter the requirements on which the contractor has based its design. The contractor is free to respond to that change in any way it wishes and there is no provision for the employer to be consulted other than when construction drawings are to be submitted. For example, if the employer requires an auditorium to be capable seating fifteen hundred instead of one thousand people, the contractor may satisfy that change, if not precluded by some other requirement, by making the auditorium longer, wider, a combination of the two or by introducing a gallery.

6 Refer to Section 5.2.3.2.

7 JCT DB 2016 – clause 3.11, and JCT SBC/Q 2016 including a CDP – clause 3.16.

8 JCT DB 2016 – Third Recital, and JCT SBC/Q 2016 including a CDP – Twelfth Recital.

cannot change them.[9] In any event, the words in the relevant clauses[10] are unambiguous, and the Third Recital of JCT DB 2016 simply records the employer's general satisfaction with the Contractor's Proposals. The employer, even if receiving professional advice, cannot be expected to check the Contractor's Proposals in detail. Any approval or acceptance which the employer gives must be understood in that context.[11]

The method in which the works may be undertaken may also change – and become a 'variation'. For example, if access routes to the site or access to the workfaces changes, or works' operations change from being carried out by machine to being carried out by hand.

The scope and nature of variations differ under each particular contract. For example:

- Under JCT 2016 contracts:
  - More comprehensive contracts (such as JCT SBC/Q 2016) are likely to contain a more comprehensive definition of variations, compared to simpler contracts (such as JCT MW 2016)[12];
  - If the contractor is responsible for the works' design (or an element of it), there may be fewer grounds for a variation; and
  - Depending upon the format of the pricing document, variations may arise if the pricing document's entries have not been set out correctly (e.g. correcting and entry in a bill of quantities).
- Under NEC4 contracts:
  - NEC4's 'compensation events' are 'umbrella' terms for all matters under that contract that would entitle the contractor to additional money or time in which to complete the works. Unlike JCT 2016 contracts (which deal with such matters separately), 'compensation events' also include procedural matters and breaches of contract by the employer.
- Table 5.10 in Section 5.3.1 identifies which 'compensation events' are a 'variation' to the scope of work.

In Table 2.1 (in Chapter 2), a comparison of the matters entitling the contractor to additional payment under JCT SBC/Q 2016 and NEC4 is shown. In the following sections, the contractor's entitlement to 'variations' is discussed.[13]

Sometimes, a contractor will argue that it is not prepared to carry out an instruction requiring a variation unless the price is agreed first. Although most building contracts provide for the cost of a variation to be agreed first if the employer or his representative so desires, the contractor is usually required to comply with any variation instruction forthwith, subject only to the right of reasonable objection in certain circumstances. It would not be classed as a reasonable objection if the contractor refused to carrying out the instruction until the price had been agreed. Most standard forms of building contracts provide that, unless the employer (or his representative) wishes to agree the price first, the

9 *Leggott v Barrett* (1880) Ch D 306; *Royal Insurance Co. Ltd v G & S Assured Investments Co. Ltd* [1972] 1 Lloyds Rep 267.

10 JCT DB 2016, clauses 2.2, 3.11 and 5.1, and JCT SBC/Q 2016 including a CDP – clauses 2.2, 3.16, 5.1.1 and 5.1.2.

11 *Hampshire County Council v Stanley Hugh Leach Ltd* (1991) 8-CLD-07-12.

12 Summarised in Table 5.1.

13 Refer to Tables 5.3, 5.5, 5.6 and 5.8 (for JCT 2016 contracts) and 5.10 for NEC4.

contractor must comply with the instruction and the variation is valued in accordance with the building contract's terms. In some cases, this may be the valuation that the employer's quantity surveyor decides. If the contractor disputes the valuation, his remedy lies in the contract's dispute resolution procedures (such as adjudication).

The parties may agree that the value of variations is to be agreed (along with any other effects on programme) before the contractor carries out the varied work. This is obviously the mandatory approach taken by the NEC4 building contract, but JCT SBC/Q 2016[14] and JCT DB 2016[15] also include options for contractor's 'quotations' to be agreed before the works are undertaken.

It is not uncommon for a contractor to try encourage or persuade the employer (or his representative) to accept the contractor's valuation of a variation. Clearly, it is a good practice to agree the value of a variation prior to carrying out the varied work as this provides all parties with certainty and reduces the likelihood of disputes. However, and much in the same way as some employers (or their representatives) may attempt to coerce a contractor into signing a contract before he is fully satisfied over his price, design or programme, a contractor may attempt to coerce the employer into agreeing the value of a variation by refusing to carry out the work unless an agreement is reached. Such an approach by the contractor is rarely successful. Contracts that oblige the contractor to comply with an instruction 'forthwith',[16] and if the contractor fails to comply within the allotted time, they provide the employer with a right to employ others to carry out the instructed work and deduct any addition cost from the contractor.[17]

### 5.1.2 Scope of a 'Variation' – Contract Documents

Before there can be a variation, it is essential to know 'what' is being varied. Although this sounds trite, it is of fundamental importance. Given that building contracts allow the employer (or his representative) to instruct variations, the original (or unvaried) scope of work must be firstly be known or established – i.e. the total amount of work (including its relevant design) which the contractor agreed to do when executing the contract. Although it is obvious that a variation to the scope cannot arise unless that scope is being varied, it is surprisingly how often that point is overlooked because it may not be clear if an instruction amounts to a change in scope, or is simply confirming the scope of work that the contractor already agreed to provide under the contract.[18] This point is very clearly made in NEC4[19] confirming that a 'compensation event' arises from an instruction '*changing the Scope*'.

---

14 JCT SBC/Q 2016 – clause 5.3 (Variation Quotation) – agreed in accordance with Schedule 2 (Variation and Acceleration Quotation Procedure), provided that.

15 JCT DB 2016 – Supplemental Provisions, Schedule 2, paragraph 2 (Valuation of Changes – Contractor's estimates), paragraph 3 (Loss and expense – contractor's estimates) and paragraph 4 (Acceleration Quotation).

16 JCT SBC/Q 2016 – clause 3.10, JCT DB 2016 – clause 3.5, JCT IC 2016 – clause 3.8, and JCT MW 2016 – clause 3.4 – but see Section 5.2.3.2 (contractor's reasonable right of objection). NEC4 – clause 27.3 provides that the contractor 'obeys' an instruction that's properly given by the Project Manager.

17 JCT SBC/Q 2016 – clause 3.11, JCT DB 2016 – clause 3.6, JCT IC 3.9, and JCT MW 2016 – clause 3.5. In each case, the employer (or his representative) notifies the contractor of his failure to comply with the instruction, and the employer can employ and pay others to give effect to that instruction (and deduct the cost from the contractor) if the contractor fails to comply within 7 days of the aforementioned notice.

18 *Sharpe v San Paulo Railway* (1873) 8 Ch App 597.

19 NEC4, clause 60.1(1) 'Compensation events'.

The position can be uncertain where the contract documents do not provide a detailed account of the intended scope for the works – for example, if the scope of work falls short of achieving Stage 4 (Technical Design) of the RIBA Plan of Work 2020. An absence of detailed information can be encountered in contracts where the contractor is responsible for all, or part of the works' design (e.g. JCT DB 2016 or the CDP element of JCT SBC/Q 2016), or in the 'simpler' forms of contract such as JCT MW 2016. Careful examination of the 'contract documents' is required in order to, firstly, ascertain the 'scope' of works that the contractor is obliged to carry out, and secondly, to ascertain if there is a 'variation' to that scope. Table 5.1 summarises the contract documents in which the 'scope' of works for each contract are found.

**Table 5.1** Contract Documents Containing the 'Scope' of Works.

| **Contract document** | **JCT 2016 contracts:** | | | | |
|---|---|---|---|---|---|
| | **SBC/Q** | **DB** | **IC** | **MW** | **NEC4** |
| Contract bills | Second Recital | — | *Third Recital*[a)] | — | B and D |
| Contract drawings | Third Recital[b)] | — | Second Recital | Second Recital | — |
| Employer's requirements (JCT SBC/Q 2016 – CDP) | Tenth Recital | First Recital[c)] | — | — | — |
| Priced Document | — | — | Fourth Recital | — | — |
| Specification | — | — | *Third Recital*[a)] | Third Recital | — |
| Work schedules | — | — | *Third Recital*[a)] | *Third Recital*[d)] | — |
| Schedule of rates | — | — | — | *Second Recital* | — |
| The scope | — | — | — | — | Scope[e)] |

a) JCT IC 2016, in the Third Recital, identifies alternative documents in the 'Contract Bills', 'Specification' or 'Work Schedules', to be deleted as appropriate.
b) JCT SBC/Q 2016 does not separately identify a 'specification' in the same way as in the Third Recital of JCT IC 2016 and JCT MW 2016, however, specification documents are ordinarily 'listed' within the 'Contract Drawings'.
c) JCT DB 2016 – the 'Employer's Requirements' are likely to contain any drawings and specification from which the contractor is to produce the 'Contractor's Proposals'.
d) JCT MW 2016, Third Recital identifies alternative documents in the 'Works Schedules' or 'Schedule of Rates' (to be deleted as appropriate).
e) NEC4 – the 'Scope' is likely to contain all the information from which the contractor is to construct the works (including drawings, specifications and other documents – such as those identifying which elements of the works the contractor is to design).

The 'scope' of works usually comprises many of the following classes of documents:

- Drawings (e.g. architectural, structural and building services engineering).
- Specifications (often in the 'NBS[20]' format) for all architectural, structural and building services elements. The building services' specification is often a 'performance' specification, leaving the contractor (via its building services sub-contractor) to design these works under the CDP.

  The contract's drawings and specifications may be read in conjunction with Bills of Quantities (JCT SBC/Q 2016, JCT IC 2016, and NEC4 Options B[21] and D) or the Employer's Requirements (JCT DB 2016 and JCT SBC/Q 2016 with the Contractor's Designed Portion – 'CDP').
- Pre-construction information provided by the employer as part of the Construction (Design & Management) Regulations 2015.
- Reports and test results (such as ground investigation reports and surveys).

  Provided that the 'contract documents' containing the scope are well-defined, then it should be relatively simple to establish if a 'variation' has been instructed, by reference to the following (see Section 5.2):
  - Additional work – either an additional quantity of work that is contained in the 'scope', or new work that is different from that contained in the scope;
  - Omitted work – work omitted from the 'scope'; or
  - Substituted work – the same quantity of work found in the 'scope' but of a different specification.

### 5.1.3 Scope of a Variation – Bills of Quantities

Not all building contracts contain 'bills of quantities' – as is shown in Table 5.1. Bills of quantities are required for JCT SBC/Q 2016 and NEC4, Option B (Priced Contract with Bill of Quantities) and Option D (Target Contract with Bill of Quantities), and are one alternative (see Footnote *a* in Table 5.1) for JCT IC 2016. JCT DB 2016 no longer includes an option for using bills of quantities.[22]

There are perceived disadvantages in using bills of quantities. The perception probably originates in the old legal principle that, unless there are express provisions to the contrary, a contractor will be expected to do everything necessary in order to complete the works, irrespective of whether the necessary work is expressly referred to in the contract documents and even if it is not possible to foresee the necessity for the work at the time of tender.[23] Many older cases deal with the contractor's duty under certain circumstances to carry out work which is not shown in documents provided at the time of tender.[24]

20 National Building Specification (RIBA).

21 NEC4: Option B (Priced Contract with Bill of Quantities), and Option D (Target Contract with Bill of Quantities).

22 A 'bill of quantities' option was available under a previous version of JCT's design and build contract (JCT DB 2011), under paragraph 3 of the 'Supplemental Provisions' in Schedule 2 of JCT DB 2011.

23 *Thorn v London Corporation* (1876) 1 App Cas 120.

24 See for example *Tharsis Sulphur & Copper Company v McElroy & Sons* (1878) 3 App Cas 1040; *Jackson v Eastbourne Local Board* (1886) Hudson, Building Contracts, (4th ed) Vol 2, p 81; *Re Nuttall and Lynton and Barnstaple Ry* (1899) 82 LT 17.

A further case[25] concerned a contract for housing, where the specification did not mention floorboards, but did say:

> '*the whole of the materials mentioned or otherwise in the foregoing particulars, necessary for the completion of the work, must be provided by the contractor*'.

It was held that the provision and laying of the floor boards was included in the contract, and the contractor was not entitled to recover for them as an extra.

Most of these cases date from the nineteenth century. The common law has moved on, and it may well be that many of these cases would now have a different outcome, even on the same facts. Generally, if a document is provided to a contractor in the knowledge that it is obliged to rely on it in the preparation of its tender (such as a modern contract with bills of quantities[26]), the courts will require very clear words in a contract or in the document itself, before it will hold that an employer should not be responsible for errors or omissions in it.[27] However, where there are no bills of quantities, or they are not guaranteed to be accurate, the decisions in some of the old cases may still have relevance. In the *Williams* case,[25] the general clause noted above appears to have been the key factor in the decision. Moreover, following developments in the reliance doctrine,[28] a contractor tendering on the basis of documents that state that errors will not be adjusted and paid for, may be able to bring an action in the courts directly against the person who prepared the quantities if it can be shown that they were negligently prepared.[29]

In one relevant case,[30] the bills contained the following paragraph:

> '*where and to the extent that materials, products and workmanship are not fully specified they are to be:*
>
> (1) *suitable for the purpose of the Works stated in or reasonably to be inferred from the Contract Documents.*
> (2) *in accordance with good building practice, including the relevant provisions of current BSI documents*'.

The court concluded that the paragraph meant if the materials and workmanship were fully specified, the contractor had carried out its obligation by doing what was specified. If they were not fully specified, the contractor had the duties set out in sub-paragraphs (1) and (2). This is clearly a valuable paragraph to include in the bills of quantities from the employer's point of view.

Bills of quantities are intended to give a very precise correlation between the amount of work and materials and the price to be paid for the Works. Therefore, many of the risks borne by contractors in some of the older cases no longer apply. Risks such as the accidental mis-measurement or omission of work are generally borne by the employer. For example, if 20 $m^2$ of paving is measured in the bills but 40 $m^2$ is actually required, it is clear that,

25 *Williams v Fitzmaurice* (1858) 3H & N 844.
26 Such as JCT SBC/Q 2016, JCT IC 2016 (where 'Contract Bills' are selected), and NEC 4 Option B (Priced Contract with Bill of Quantities) and Option D (Target Contract with Bill of Quantities).
27 *Bacal Construction (Midlands) Ltd v Northampton Development Corporation* (1976) 8 BLR 88.
28 Hedley Byrne *& Co Ltd v Heller & Partners Ltd* (1964) AC 465.
29 *Auto Concrete Curb Ltd v South Nation River Conservation Authority* (1994) Const LJ 39; *Henderson v Merritt Syndicates* [1995] 2 AC 145; (1994) 69 BLR 26.
30 *Bowmer & Kirkland Ltd v Wilson Bowden Properties Ltd* 11 January (1996), unreported.

under JCT SBC/Q 2016, JCT IC 2016 and NEC4 (Options B and D), the contractor will be automatically entitled to payment for the extra 20 $m^2$ of paving. However, the same outcome would be unlikely under a contract where drawings and specification were the only contract documents providing the 'scope' of work, and 40 $m^2$ of paving was accurately shown or could be measured from the drawing. This point is often misunderstood by contractors engaged on smaller projects (often referred to 'plan and spec' jobs).

It perhaps inaccurate to suggest that the use of bills of quantities has many disadvantages. There is often little doubt regarding the extent of the work which the contractor has undertaken (along with its value), and the certainty that a bill of quantities provides is a very clear benefit to both the employer and the contractor, as it removes a substantial area for possible disputes. Indeed, it thought likely that if a modern civil engineering contract been used, *Sharpe's* case[18] would probably never have reached the courts at all; the employer would have had to pay for the extra two million cubic yards of earthworks, but that would have been, on the facts, the correct price for the work. If the employer objected to the court's decision, he may have had a remedy against the engineer. Although the operation of bills of quantities is usually straightforward, it may contain some entries such as 'Approximate' or 'Provisional' quantities that complicate matters. These are explored in more detail towards the end of Section 5.1.3.

Under the JCT SBC/Q 2016, JCT IC 2016 and NEC4 contracts, establishing the 'scope' of work from the bills of quantities (or 'contract bills') is a relatively straightforward task because:

a) The bills of quantities' accuracy is effectively guaranteed by the employer:
   - If different quantities of work (to those set out in the bills of quantities) are required, the bills should be corrected and any difference between the 'actual' quantities and the quantities stated in the bills[31] is a 'variation'; and
   - The employer retains a lesser degree of responsibility for work that is described as an 'Approximate Quantities' in the bills of quantities but only to the extent that the quantities are not a reasonably accurate forecast. If they are not, then the contractor may not only be entitled to an increased value for the relevant quantities of work[32] but also an extension of time,[33] loss and expense[34] or a compensation event[35];

---

31 JCT SBC/Q 2016 (clause 2.14.1) and JCT IC 2016 (clause 2.12.2) provide that *'If in the Contract Bills there is…any error in quantity …the…error shall be corrected'* and such correction is a 'Variation' (JCT SBC/Q 2016 – clause 2.14.3 and JCT IC 2016 clause 2.13.1). Note that JCT SBC/Q 2016 provides in clause 2.14.3 that any such correction is to be 'treated' as a Variation, whereas JCT IC 2016 does not include any such 'treating' provision. Instead, clauses 2.12.2 of JCT IC 2016 states that an error 'shall be corrected', and clause 2.13.1 states that the Architect/Contract Administrator 'shall issue instructions in this regard. Clearly, the errors described do not include errors in the pricing of the contract bills as clause 4.2 of both JCT SBC/Q 2016 and JCT IC 2016 state that any errors in calculating the Contract Sum is accepted by the employer and the contractor.

NEC4 contains similar provisions so that any differences between the final quantity and the quantity stated in the Bill of Quantities is a compensation event under clause 60.4 (Options B and D).

32 JCT SBC/Q 2016 – clause 5.6.1.5, JCT IC 2016 – clause 5.3.1.5.

33 JCT SBC/Q 2016 – clause 2.29.5, JCT IC 2016 – clause 2.20.4.

34 JCT SBC/Q 2016 – clause 4.22.4, JCT IC 2016 – clause 4.17.3.

35 NEC4 – clause 60.4 (Options B and D). The third bullet point refers to the final value of an item of work (i.e. the actual quantity of work multiplied by the Bill of Quantities' rate) exceeding 0.5% of the 'total of the Prices at the Contract Date'.

b) The bills of quantities are prepared according to a set of measurement rules.[36] In the event that a description in the bill of quantities is either not given in accordance with the measurement rules, or a departure from the measurement rules is not stated, then correcting the bills of quantities entitles the contractor to a variation.[37]

The difference between how the 'quantities' and 'Approximate Quantities' are treated (particularly in the JCT 2016 contracts) is often not immediately apparent. Table 5.2 summarises these differences.

It may therefore be the case, that even before the employer decides to change the scope, the contract itself may generate an instruction requiring a variation in order to correct the contract bills (either in terms of correcting their quantities or descriptions). This means that the contractor has contracted (for the contract sum) to carry out only what is shown in the contract bills and not what is shown on the contract drawings (if they differ from what's described in the contract bills). However, the wording of JCT SBC/Q 2016, JCT IC 2016 and NEC4 makes it clear that the contractor has contracted to construct what is specified in the contract documents;[38] therefore, the contractor cannot refuse to carry out any work which is different from that shown in the bills if it is shown on the drawings, and if it is to be constructed in accordance with the drawings rather than the bills, it will require the correction of a discrepancy and be a variation.

The bills of quantities may contain items of work, where the quantity is stated to be 'provisional'. Entries for 'provisional quantities' are now comparatively rare and were sometimes found in local authority contracts (nowadays, public sector procurement often employs NEC4 'target cost' contracts instead). The effect of a 'provisional quantity' is suggested to be that the contractor's rate or price for a unit of that work is not subject to adjustment,[39] irrespective of whether the actual quantity of work significantly differs from the 'provisional quantity' set out in the contract bills.[40]

Where there are no bills of quantities, the contractor may be required to allow for something which is difficult to estimate in its tender. For example, the contractor may be asked to include for all foundation work whatever the sub-soil or sub-surface conditions may be. Unless the contractor has had the opportunity to carry out extensive site investigations in putting together the price, he may price for working in more severe ground conditions than he may actually encounter. The employer will then be in a position of having to pay a higher price even if the contractor encounters far more favourable ground conditions. In practice, most contractors will make a sensible allowance and price for conditions between the two extremes, but the principle remains the same. The employer may pay for the worst conditions even if the best are encountered. The clear merit of bills of quantities

36 Typical methods of measurement include the RICS' 'New Rules of Measurement' (or pre 2009 – the Standard Method of Measurement – 7th Ed), or the ICE's Civil Engineering Standard Method of Measurement – 4th Ed).

37 JCT SBC/Q 2016 (clause 2.14.1) and JCT IC 2016 (clause 2.12.2) and NEC4 clause 60.6 (Options B and D).

38 Clause 2.1 of both JCT SBC/Q 2016 and JCT IC 2016. Clause 20.1 of NEC4.

39 For example, as would otherwise be the case in JCT SBC/Q 2016 (clause 5.6.1.2), JCT DB 2016 (clause 5.4.2), and JCT IC 2016 (clause 5.3.1.2) – refer to Section 5.4.1 and Table 5.9.

40 Bailey, J. Construction Law (Third Ed) 2020, at 3.32.

**Table 5.2** 'Quantities' and 'Approximate Quantities'.

<table>
<tr><th>Description</th><th>'Quantities'</th><th>'Approximate Quantities'</th></tr>
<tr><td>Definition:</td><td>JCT SBC/Q & IC 2016:<br>None<br>NEC4 – none</td><td>JCT SBC/Q & IC 2016:<br>Clause 1.1</td></tr>
<tr><td rowspan="2">Method of valuation:<br>If the quantities in the contract bills are accurate</td><td>JCT SBC/Q & IC 2016:<br>Valued using the rates and prices in the contract bills.[a)]</td><td>JCT SBC/Q & IC 2016:<br>Valued at the rate for the 'Approximate Quantity' in the contract bills.[b)]</td></tr>
<tr><td colspan="2">NEC4 (Option B): If the value (i.e. rate × final quantity) of an item of work <0.5% of the 'total of the Prices', this is not a 'compensation event' so valued as the quantity of work completed in the bill of quantities, multiplied by the rate.[c)]</td></tr>
<tr><td rowspan="2">If the quantities in the contract bills are not accurate</td><td>JCT SBC/Q & IC 2016:<br>Valued using the rates and prices in the contract bills, unless the quantity 'significantly changes' then the rates or price in the contract bills are the 'basis' for valuation, plus a 'fair allowance' for the difference in quantity.[d)]</td><td>JCT SBC/Q & IC 2016:<br>If the 'Approximate Quantity' is not a 'reasonably accurate forecast' of the actual quantity, then the rates or price in the contract bills are the 'basis' for valuation, plus a 'fair allowance' for the difference in quantity.[e)]</td></tr>
<tr><td colspan="2">NEC4 (Options B and D): If either:<br>a) The final total quantity of work is different from the quantity in the Bills of Quantities[f)]; or<br>b) the value (i.e. rate x final quantity) of an item of work >0.5% of the 'total of the Prices',[f)]<br>this is a 'compensation event' so the valuation is as the assessment of the 'change to the Prices' for a compensation event.[g)]</td></tr>
<tr><td rowspan="2">Affecting the time for completion</td><td>JCT SBC/Q & IC 2016:<br>Yes, if correcting the contract bills is a Variation (which should be instructed),[h)] which could either result in an extension of time for additional quantities,[i)] or (for JCT SBC/Q 2016) a reduction in an extension of time (for a 'Relevant Omission')[j)] for a reduction in quantities.</td><td>JCT SBC/Q & IC 2016<br>Yes, if the 'Approximate Quantity' is not a 'reasonably accurate forecast' then this is a 'Relevant Event' for an extension of time under both contracts, but does not have to be instructed.<br>If the 'Approximately Quantity' is an unreasonably excessive quantity, then it appears that any lesser, actual quantity cannot be a 'Relevant Omission' because this must be instructed.</td></tr>
<tr><td colspan="2">NEC4:<br>1. The final total quantity of work is different from the quantity in the Bills of Quantities[f)]; or<br>2. the value (i.e. rate × final quantity) of an item of work >0.5% of the 'total of the Prices',[f)]<br>this is a 'compensation event' so is as per the assessment of a 'delay to the Completion Date' for a compensation event.[k)]</td></tr>
</table>

**Table 5.2** (Continued)

| Description | 'Quantities' | 'Approximate Quantities' |
|---|---|---|
| *Additional cost:* | JCT SBC/Q & IC 2016:<br>Yes, if correcting the contract bills is a Variation (which should be instructed),[h] which results in direct loss and/or expense under both contracts.[l] | JCT SBC/Q & IC 2016:<br>Yes, if the 'Approximate Quantities' are not a 'reasonably accurate forecast' under both contracts.[m] |
| | NEC4: If either:<br>• The final total quantity of work is different from the quantity in the Bill of Quantities[f]; or<br>• the value (i.e. rate × final quantity) of an item of work >0.5% of the 'total of the Prices',[f]<br>this is a 'compensation event' so the valuation is as the assessment<br>of the 'change to the Prices' for a compensation event.[g] | |

a) JCT SBC/Q 2016 – clause 4.14.1.1 and JCT IC 2016 – clause 4.9.1.1.
b) JCT SBC/Q 2016 – clause 5.6.1.4, and JCT IC 2016 – clause 5.3.1.4.
c) NEC4 – clause 50.3 (assessing the amount due) the 'Price for Work Done to Date' – Option B definition in clause 11.2(3).
d) JCT SBC/Q 2016 – clause 5.6.1.2 and JCT IC 2016 – clause 5.3.1.2.
e) JCT SBC/Q 2016 – clause 5.6.1.5 and JCT IC 2016 – clause 5.3.1.5.
f) NEC4 (Options B and D), a compensation event under clause 60.4.
g) NE4, clause 63.1.
h) JCT SBC/Q 2016 – clauses 2.14.1 and 2.14.3, JCT IC 2016 – clauses 2.12.2 and 2.13.1.
i) A 'Relevant Event' under JCT SBC/Q 2016 – clause 2.29.1, or JCT IC 2016 – clause 2.20.1.
j) JCT SBC/Q 2016 – clauses 2.26.3, 2.28.5.2 and 2.28.6.4.
k) NEC4 – clause 63.5.
l) A 'Relevant Matter' under JCT SBC/Q 2016 – clause 4.22.1, and JCT IC 2016 – clause 4.17.1.
m) A 'Relevant Matter' under JCT SBC/Q 2016 – clause 4.22.4, and JCT IC 2016 – clause 4.17.3.

is that the employer simply pays for the work that the contractor actually does. A provisional sum can be included for difficult excavation in order to obtain a 'budget price', so if rock or other obstructions are discovered, they can be measured and valued in the appropriate way. The use of bills of quantities allows a very precise allocation of risk. The applicable methods of measurement[36] clearly show where those risks lie – sometimes with the employer and sometimes with the contractor, but generally on the basis of a specified amount of precisely described work, against which the contractor can give a realistic price.

Bills of quantities are not always suitable where the employer is seeking a firm cost for the project. An employer may be better off using a design and build contract that adopts a less detailed pricing document (such as a Contract Sum Analysis[41]) – where the risk for the works' quantities falls on the contractor, but where the contractor effectively controls the design for the works. This approach is suitable if it is most important to the employer that a stated price for the works will not be exceeded – prioritising cost certainty over the lowest cost.

41 As found in JCT DB 2016, or for the CDP element of JCT SBC/Q 2016 (the 'CDP Analysis').

### 5.1.4 Other Pricing Documents

In this section, 'other' pricing documents are those that are not prepared in accordance with a method of measurement. Unlike bills of quantities (for which the employer is contractually obliged to ensure is prepared in accordance with a method of measurement), the contractor will be responsible for ensuring that his pricing allows for all work that is indispensably necessary[42] to fulfil his obligations under the contract.

Some 'pricing documents', such as the Contract Sum Analysis in JCT DB 2016, or the CDP Analysis for the CDP work in JCT SBC/Q 2016 are clearly not intended to provide the information related to the works' 'scope', as this information is set out in the 'Employer's Requirements' in those contracts.[43] However, other pricing documents under JCT IC 2016 or JCT MW 2016 may include 'scope' information, such as:

- JCT IC 2016: 'Priced Document', Specification or Work Schedules; and
- JCT MW 2016: Specification, Work Schedules and Schedule of Rates.

## 5.2 JCT 2016 Contracts – Definitions of Variations

A 'Variation' (described as a 'Change' under JCT DB 2016) can arise in many situations under the JCT 2016 contracts. These variations (or changes) range from the more obvious changes to the works[44] and the conditions imposed upon them being carried out,[45] to more procedural matters, such as resolving discrepancies or inadequacies in the contract documents, insurance matters and dealing with design and some sub-contract matters. These situations are summarised in Table 5.3.

### 5.2.1 Alteration of Design, Quality or Quantity of the Work

The relevant clauses that provide for 'Variations' (or 'Changes') that arise from altering the design, quality or quantity of work are set out in clause 5.1 of JCT SBC/Q 2016, JCT DB 2016 (see Footnote *a* in Table 5.3) and JCT IC 2016. A more general description is given in clause 3.6.1 of JCT MW 2016. References to clause 5.1.1 in this section should also be taken to refer to clause 3.6.1 of JCT MW 2016.

It may appear that clause 5.1.1 defines a 'variation' in almost unlimited ways, so that anything and everything is included. For example, clause 5.1.1.1 refers to an addition, omission or substitution of any work. If that was strictly interpreted, the employer or his representative would be able to instruct virtually anything. However, a single phrase in the contract

42 Keating on Construction Contracts 10th Ed (2016 and 2019) at 4–045. Also refer to the cases in section 5.1.3 in Footnote no's 23, 24 and 25.

43 Hence the exclusion of the 'Contract Sum Analysis' and the 'CDP Analysis' from Table 5.1. The 'Contractor's Proposals' (applicable under both JCT DB 2017 and CDP element of JCT SBC/Q 2016) are also excluded from Table 5.1 as save for resolving any discrepancies, ambiguities or inadequacies in the Employer's Requirements (see Section 5.3.2.1), they do not provide the 'scope' of works, merely an account of how the contractor intends to satisfy the Employer's Requirements.

44 Such as the 'Variations' (or 'Changes') to the scope of works in clause 5.1.1 (JCT SBC/Q 2016, JCT DB 2016, and JCT IC 2016), and the simpler, more general description in JCT MW 2016, clause 3.6.1.

45 See Footnote 44, save under clause 5.1.2 in JCT SBC/Q 2016, JCT DB 2016, and JCT IC 2016.

**Table 5.3** 'Variations' or 'Changes' Under the JCT 2016 contracts.

| Description of Variation (or Change[a]) | SBC/Q | DB[a] | IC | MW |
|---|---|---|---|---|
| *Alteration of design, quality or quantity:* | | | | |
| Addition, omission or substitution of work | 5.1.1.1 | 5.1.1.1 | 5.1.1.1 | 3.6.1 |
| Altering kinds or standards of materials or goods | 5.1.1.2 | 5.1.1.2 | 5.1.1.2 | |
| Removal of compliant work, materials or goods | 5.1.1.3 | 5.1.1.3 | 5.1.1.3 | |
| Changes in 'other works' conditions | 5.9 | 5.6 | 5.5 | |
| *Changes or restrictions imposed by the employer:* | | | | |
| Access and use to the site or parts of it | 5.1.2.1 | 5.1.2.1 | 5.1.2.1 | 3.6.1 |
| Working space limitations | 5.1.2.2 | 5.1.2.2 | 5.1.2.2 | |
| Working hours limitations | 5.1.2.3 | 5.1.2.3 | 5.1.2.3 | |
| Undertaking or completing work in a particular order | 5.1.2.4 | 5.1.2.4 | 5.1.2.4 | |
| *Other circumstances giving rise to a Variation:* | | | | |
| Correction of divergence in the Employer's Requirements ('ER's') with site boundary | — | 2.10.1 | — | — |
| Correction, alteration, modification of BoQ[b] and ER's | 2.14.3 | 2.12.2 | 2.14 | 2.4 |
| Correcting 'Contractor's Proposals' or CDP analysis | 2.14.4 | — | — | — |
| Correction of discrepancy in ERs' | 2.16.2 | 2.14.2 | — | — |
| Correcting divergence between Contract Documents (in ER's) and Statutory Requirements[c] | 2.17.2.1<br>2.17.2.2 | 2.15.2.1<br>2.15.2.2<br>2.15.2.3 | 2.15.2 | — |
| Emergency compliance with Statutory Requirements | 2.18.3 | — | 2.16.3 | — |
| *Instructions:* | | | | |
| Instructions issued | 3.14 | 3.9 | 3.11 | 3.4 |
| Postponing work | 3.15 | 3.10 | 3.12 | 3.4 |
| Expenditure of 'Provisional Sums' | 3.16 | 3.11 | 3.13 | 3.7 |
| Non-compliant work, including goods, materials | 3.18.3[d] | 3.13.2[d] | — | — |
| Non-compliant workmanship | 3.19[d] | 3.14[d] | — | — |
| Instructions to deal with antiquities | 3.22.2[e] | 3.15.2[e] | — | — |

(Continued)

**Table 5.3** (Continued)

| Description of Variation (or Change[a)]) | SBC/Q | DB[a)] | IC | MW |
|---|---|---|---|---|
| Design Submission procedure, compliance with 'Architect'/ Contract Administrator' comments | Scheme 1, para 7 | Scheme 1, para 8.1 | — | — |
| Insurance claims and reinstatement (Terrorism cover, and Options B and C) | 6.13.5.3<br>6.13.6 | 6.13.5.3<br>6.13.6 | 6.13.5.3<br>6.13.6 | 5.6.6 |
| Remedial measures for Joint Fire Code breach | 6.19.2 | — | 6.17.2 | — |
| *Named Sub-Contractors ('NAM/SC')* | | | | |
| Removal of impediment for NAM/SC | — | — | Scheme 2, para 3 | — |
| Instructions for NAM/SC work if the NAM/SC is terminated | — | — | Scheme 2, para 8.2 | — |

a) A Change" under JCT DB 2016 is the equivalent of a Variation (under JCT SBC/Q 2016, JCT IC 2016 and JCT MW 2016).
b) 'BoQ' means the bills of quantities, or the 'Contract Bills' in JCT SBC/Q 2016 (clause 1.1 and Second Recital).
c) Refer to Sections 5.2.3.1 and 5.2.3.2 and Table 5.5 for more details.
d) A 'Variation' can be instructed, but does not entitle the contractor to any addition to the Contract Sum, or an extension of time.
e) The 'Architect/Contract Administrator' (JCT SBC/Q 2016) or 'Employer' (JCT DB 2016) may issue instructions to deal with antiquities, and which this clause does not expressly mention the instruction being a 'Variation', it is difficult to conceive how it could not be if the work fell within the 'Variation' definition (clause 5.1 of JCT SBC/Q 2016, JCT DB 2016 and JCT IC 2016).

must not be taken out of context. JCT 2016 contracts must be read as a whole,[46] and in this instance, the key provision governing clause 5.1.1 is the clear reference to the 'Works' in the context of any alterations or modifications of design, quality or quantity, and also states that the sub-clause referring to 'any' work is included within the governing meaning. Therefore, it is only the 'Works' described in the contract documents[47] that are subject to variation. For example, the employer (or his representative) has no power to instruct work that is outside the site boundary, as this would not fall within the definition of the 'Works'. For example, an employer may own another property near the site and may think it more economical to get the contractor carrying out the 'Works' on the first project, to do some other work at the additional property. The employer's representative has no power to instruct such a

46 *'Agreement etc to be read as a whole'* – JCT SBC/Q 2016, JCT DB 2016, JCT IC 2016 – clause 1.3, JCT MW 2016 – clause 1.2.
47 'Works' is described in the 'First Recital' in JCT SBC/Q 2016 (including the 'Contractor Designed Portion'), JCT DB 2016, JCT IC 2016 and JCT MW 2016.

variation (whether by way of additions, omissions or changes), requiring the contractor to execute work clearly outside the site and not contemplated by the original contract.[48] Of course, the employer would be able to agree a separate contract with the contractor for him to carry out work on the other property, but ordinarily, the employer's representative would have no authority to do so.

If the employer's representative issues an instruction adding work, it immediately becomes part of the Works, which he may vary by further instruction. The employer (or his representative) cannot substantially alter the 'nature' of the Works – for example, by changing a traditionally constructed building into a building wholly constructed from prefabricated elements, as the contractor may successfully argue that the varied building is now substantially different from the building for which it originally tendered.[49] It is suggested that the employer (or his representative) cannot validly instruct the contractor to build something substantially in excess of (and certainly something substantially less than[50]) what was provided under the contract. In essence, the guiding principle is that the Works as completed must still be capable of being identified as the Works set out in the building contract (e.g. in the first recital of a JCT 2016 contract).

On a casual reading, the wording of clause 5.1.1.3 might appear to be an instruction issued by the employer's representative under the relevant clause[51] is a 'Variation'. However, it requires careful reading to see if the removal from site of work, materials or goods constitutes a variation except where they are removed because they are not in accordance with the contract. It could be confused in practice with the employer's (or his representative's) power to issue instructions to the contractor to remove from site work and materials not in accordance with the contract[52] so the contractor is not entitled to extra money (see Footnote *d* in Table 5.3). However, this clause is at pains to highlight this difference and relates to specific instructions for work or materials that *are* in accordance with the contract, but which for some reasons, the employer (or his representative) wishes to remove from the site of the Works either temporarily or permanently.

### 5.2.2 Changes or Restrictions Imposed by the Employer

Clause 5.1.2[53] allows the employer to impose obligations and restrictions upon the contractor and to change those already imposed by the relevant contract documents[54] in respect to the following five matters:

- access to the site[55];
- use to be made of specific parts of the site[55];

---

48 *Sir Lindsay Parkinson & Co Ltd v Commissioners of Works* (1950) 1 All ER 208.
49 *Blue Circle Industries plc v Holland Dredging Company (UK) Ltd* (1987) 37 BLR 40.
50 *Abbey Developments Ltd v PP Brickwork Ltd* (2003) CiLL 2033.
51 JCT SBC/Q 2016 – clause 3.18.3, JCT DB 2016 – clause 3.13.2.
52 JCT SBC/Q 2016 – clause 3.18.3, and 3.19 (in the case of workmanship), and JCT DB 2016 – clause 3.13.2 (or clause 3.14 in the case of workmanship).
53 As in section 5.2.1, clause 5.1.2 of JCT SBC/Q 2016, JCT DB 2016 and JCT 2016 should also be considered the equivalent within clause 3.6.1 of JCT MW 2016.
54 JCT SBC/Q 2016 – the 'Contract Bills' or the 'Employer's Requirements' (under the CDP), JCT DB 2016 – the 'Employer's Requirements', and JCT IC 2016 – the 'Contract Documents'.
55 JCT SBC/Q 2016, JCT DB 2016, JCT IC 2016 – clause 5.1.2.1.

- limitations of working space available to the contractor[56];
- limitations of hours to be worked by the contractor[57]; and
- requirements that the Works be carried out or completed in any specific order.[58]

There is a curious feature about this clause, because at first sight, it appears to be the only variation which does not originate from an 'instruction' (either from the employer or his representative).

Clauses 5.1.1 and 5.1.2 are clear in these circumstances - 'Variation' means the 'imposition by the employer' of any changes or restrictions of the type set out in clauses 5.1.2.1 to 5.1.2.4. This is relatively unusual because in most circumstances the 'employer's representative'[59] (and not the 'employer' - save for JCT DB 2016) issues instructions requiring a variation.[60] There is no qualification on that power in the rest of clause 5.1.2 or, indeed, in the rest of the contract. Whilst the contractor may raise a reasonable objection[61] to such an instruction (i.e. one creating a variation under clause 5.1.2), this does not give the employer's representative the power to impose such changes or restrictions himself. Therefore, in addition to the unilateral imposition of such changes or restrictions by the employer, the employer's representative may also issue instructions in regard to clause 5.1.2 matters, effectively confirming the changes that the employer has imposed. The conclusion to be drawn from this is that if the employer imposes obligations or restrictions or changes them (as set out in clause 5.1.2), it clearly constitutes as a 'Variation' under the terms of the contract and is to be valued accordingly, irrespective of whether the employer's representative has confirmed this in an instruction. Furthermore, under each JCT 2016 contract, a 'Variation' automatically creates an entitlement[62] to both an extension of time[63] and loss and/or expense.[64] In view of the particular wording of clause 5.1.2, it is suggested that, in giving an instruction requiring a variation under this clause, the employer's representative should clearly refer to the imposition of obligations or restrictions by the employer.

Employers rarely avail themselves of the powers apparently given to them under this clause, which is fortunate in view of the confusion that could result. If an employer did

56 JCT SBC/Q 2016, JCT DB 2016, JCT IC 2016 – clause 5.1.2.2.
57 JCT SBC/Q 2016, JCT DB 2016, JCT IC 2016 – clause 5.1.2.3.
58 JCT SBC/Q 2016, JCT DB 2016, JCT IC 2016 – clause 5.1.2.4.
59 In JCT 2016 contracts, the 'employer's representative' is the 'Architect/Contract Administrator' (JCT SBC/Q 2016, JCT IC 2016 and JCT MW 2016). In JCT DB 2016, whilst the 'employer' carries out most of the functions of the 'Architect/Contract Administrator' under the other JCT 2016 contracts, in practice, they are carried out on the employer's behalf by an a third party; as an 'agent' of the employer.
60 JCT SBC/Q 2016 – clause 3.14, JCT DB 2016 – clause 3.9, JCT IC 2016 – clause 3.11 and JCT MW 2016 – clause 3.4.
61 JCT SBC/Q 2016 – clause 3.10.1, JCT DB 2016 – clause 3.9.2, and JCT IC 2016 – clause 3.11.2. The contractor has no right to 'reasonably object' to an instruction under JCT MW 2016.
62 Except under JCT SBC/Q 2016 – clauses 3.18.3 and 3.19, and JCT DB 2016 – clauses 3.13.2 and 3.14. See Table 5.1 and Footnote *d* in Table 5.3.
63 JCT SBC/Q 2016 – clause 2.29.1, JCT DB 2016 – clause 2.26.1, JCT IC 2016 – clause 2.20.1, and JCT MW 2016 – clause 2.7.
64 JCT SBC/Q 2016 – clause 4.22.1, JCT DB 2016 – clause 4.21.1, JCT IC 2016–4.17.2, and JCT MW 2016 – clause 3.6.3.

so, and either he or his representative informed the contractor other than in writing, the contractor could always confirm that instruction in writing, within seven days.[65]

The power of the employer (or his representative) is limited to only those specific obligations and restrictions in clause 5.1.2 as there is no similar powers in respect of imposing further obligations or restrictions of any other kind. It is clear that the employer's power is not limited to varying obligations and restrictions already imposed through the medium of the relevant contract documents[54] but extends to imposing fresh obligations or restrictions – provided that they are of the kinds listed in clauses 5.1.2.1 to 5.1.2.4 and subject to the contractor's right of reasonable objection[61] (to instructions imposing an obligation or restriction - refer to Section 5.2.3.2).

The five matters in clause 5.1.2 often attract little attention in practice. Imposing any of them will almost certainly give rise to additional costs to the contractor; some of the restrictions may be far reaching, and it is surprising that more claims have not been founded on them.[66] One reason is possibly because employers have adopted a sensible approach and declined to impose obligations in any of these categories unless they are unavoidable. Restricting access to the site, or the times of such access could be quite catastrophic and where there is a need for such restriction, it should be included in the contract documents (if the restriction is known, or reasonably expected at the time they are produced). It is more likely that any instructions issued under this head will be concerned with the relaxation of previously imposed restrictions.

A similar comment can be made for the other categories in clauses 5.1.2.2 and 5.1.2.3. Where the employer or his representative issues instructions regarding the use of various parts of the site, it must not be confused with failure to give possession of the whole site on the date of possession in the contract. Possession must be given on that date, but the employer (or his representative) can restrict this use. It is doubtful that this allows the employer (or his representative) to postpone work because the contracts have already given them that power.[67] The employer or his representative may wish to restrict the contractor from storing certain materials, erecting cranes, siting cabins and the like. Limiting working space probably falls into the same category, whilst limiting hours might be necessary in order to allow third parties to have access to the employer's premises, or be in response to complaints and visits from local authority inspectors when the project is in progress.

The final matter in clause 5.1.2.4 can cause the most difficulty. An obvious method for the employer to exert more control over the sequence in which the Works are carried out is to divide the Works into 'sections' (refer to Section 3.9). A strict reading of the clause suggests that although the order of completion and of carrying out the work may be varied, there is no power under this clause to require the contractor to complete parts (or sections) of the Works by different start and completion dates if there is just one completion date in the contract (see Section 3.9). In any event, this power must be exercised with great care where

---

65 JCT SBC/Q 2016 – clause 3.12.1, JCT DB 2016 – clause 3.7.1, JCT MW 2016 – clause 3.4. No such express right of confirmation exists under JCT IC 2016.

66 Claims for the effects of the matters in clause 5.1.2 may appear 'initially intangible' – see section 1.1.1.

67 JCT SBC/Q 2016 – clause 3.15, JCT DB 2016 – clause 3.10, JCT IC 2016 – clause 3.12, and JCT MW 2016 – clause 3.4.

the employer has also chosen to divide the Works into sections and set out individual dates for possession and completion in the contract.

The employer (or his representative) should not attempt to alter the content of any of the sections by using clause 5.1.2.4 because this may invalidate the liquidated damages clause and possibly cause the employer further difficulties (such as rendering 'time at large' – see Section 3.1.5). There is no mechanism in the contract to amend liquidated damages in this way,[68] and it may be suggested that a substantial change in the sequence of work under clause 5.1.2.4 may invalidate the liquidated damages provisions. It is thought that a restriction or change under clause 5.1.2.4 can only relate to work that is within either:

- the 'Works' as a whole; or
- a 'Section'.

The JCT 2016 contracts[69] only provide the contractor with limited grounds to object to instructions (see Section 5.2.3.2 and Table 5.9). Therefore, if the obligations or restrictions have just been imposed by the employer,[70] then the contractor can avoid complying with them, by objecting under the terms of the contract. That makes sense, because the contractor would not be aware of the restrictions at the time of tender (or, at the latest, when it executed the contract) and will then be entitled to:

- recover any costs of the 'Variation' via the appropriate method of valuation (see Section 5.4),
- recover any loss and/or expense (either in the form of 'prolongation'[71] or 'disruption'[72] costs); and
- avoid liquidated damages for late completion by being entitled to an extension of time[73]; or
- In circumstances where the clause 5.1.2.4 change or restriction creates a clear breach of the employer's obligations under the CDM Regulations,[74] the contractor may be able to terminate his employment under the contract.[75]

### 5.2.3 Instructions and Other Circumstances Giving Rise to a 'Variation'

As summarised in Table 5.1, JCT 2016 contracts provide a number of circumstances in which a 'Variation' (or 'Change') may arise. Each of these will be covered in the following sub-sections:

---

68 As opposed to a proportionate reduction in liquidated damages to reflect the employer taking 'partial possession' of areas of the Works, or of a Section: JCT SBC/Q 2016 – clause 2.37, JCT DB 2016 – clause 2.34, and JCT IC 2016 – clause 2.29.
69 Save for JCT MW 2016 – refer to footnote 61.
70 For example, if the contractor came to site one Monday morning and found that over the weekend, the employer had changed the access routes into and out of the site.
71 Refer to sections 2.4.1 and 4.4.2.
72 Refer to sections 2.4.2 and 4.4.3.
73 Refer to section 3.2 and Chapter 11.
74 A breach by the employer under: JCT SBC/Q 2016 – clause 3.23, JCT DB 2016 – clause 3.16, JCT IC 2016 – clause 3.18, or JCT MW 2016 – clause 3.9.
75 Employer's 'specified default' under JCT SBC/Q 2016 and JCT IC 2016 – clause 8.9.1.4, JCT DB 2016 – clause 8.9.1.3, and JCT MW 2016 – clause 6.8.1.3.

- Corrections to contract documents, resolving discrepancies, ambiguities, inadequacies, etc.;
- Instructions (including for postponing the works and the expenditure of provisional sums); and
- Design Submission Procedure and Contractor's Designed Portion ('CDP')

#### 5.2.3.1 Corrections to Contract Documents (Resolving Discrepancies, Ambiguities, Inadequacies, etc.)

The contractor's entitlement to a variation will depend upon the nature of the contract document which may need correcting. This, in turn, will depend upon the type of JCT 2016 contract. As a general rule, if the contractor has prepared a 'design' document (such as the 'Contractor's Proposals' under the JCT SBC/Q 2016 CDP, or under JCT DB 2016), there will be no entitlement to a 'Variation' if it contains deficiencies because of the following reasons:

- The contractor is responsible for preparing his 'Contractor's Proposals';
- He should have prepared them in accordance with, and following a detailed review of the 'Employer's Requirements'; and
- It is assumed that the 'Contractor's Proposals' will be correct and satisfy the Employer's Requirements.[76]

However, it is often the case that documents prepared by, or on the employer's behalf contain discrepancies, ambiguities, inadequacies, or may not have been prepared in the way in which they should have been.

Table 5.4 summarises the 'Contract Documents' under each of the JCT 2016 Contracts.

The contractor's entitlement to a 'Variation' (or 'Change') arising from correcting or resolving discrepancies, divergences or inadequacies in the contract documents is summarised in Table 5.5.

The various discrepancies between the documents described in Table 5.3 giving rise to 'Variations' (or 'Changes') are explored below:

- Correction of the 'Contract Bills'. This 'Variation' will only occur under JCT SBC/Q 2016 and JCT IC 2016 (see Footnote g in Table 5.4), where bills of quantities are used. Under both contracts, the bills are to be prepared in accordance with the 'Measurement Rules' (NRM2[77]). If an item in the Contract Bills is not described in accordance with NRM2, and this difference in description has not been identified or explained in the bills, then the description shall be corrected. Such correction is a 'Variation'. In *C Bryant & Son Ltd v Birmingham Hospital Saturday Fund*,[78] the contractor undertook to erect a convalescent home under an early version of the JCT standard form, which provided:

76 JCT DB 2016, it is clear that the 'Contractor's Proposals' are subservient to the 'Employer's Requirements' by virtue of (i) the Third Recital, (ii) clauses 2.2.1 and 2.2.2 providing that materials, goods and workmanship must comply with specific descriptions in the 'Employer's Requirements', (iii) a 'Change' can only be instructed to the 'Employer's Requirements' (not to the 'Contractor's Proposals', because a deficiency in the 'Contractor's Proposals' will simply be a breach), and (iv) only resolving 'discrepancies' or 'inadequacies' in the 'Employer's Requirements' constitutes a 'Change'.

77 RICS New Rules of Measurement – Detailed Measurement for Building Works (NRM2) as published at the Base Date. NRM2 replaced SMM7, as an amendment to the 2011 JCT contracts.

78 [1938] 1 All ER 503.

**Table 5.4** JCT 2016 Contracts – 'Contract Documents'.

| JCT 2016 contract | Employer Document | Contractor Document | Pricing Document |
|---|---|---|---|
| SBC/Q | Contract Bills[a)]<br>Contract Drawings[b)] | — | Contract Bills[a)]<br>Activity Schedule[c)] |
| (if the 'Contractor's Designed Portion is used): | Employer's Requirements[d)] | Contractor's Proposals[e)] | CDP Analysis[e)] |
| DB | Employer's Requirements[f)] | Contractor's Proposals[g)] | Contract Sum Analysis[g)] |
| IC | Contract Drawings[h)]<br>Bills of Quantities[i)]<br>Specification[i)]<br>Work Schedules[i)] | — | Bills of Quantities[j)]<br>Specification[j)]<br>Work Schedules[j)]<br>Contract Sum Analysis[k)]<br>Schedule of Rates[k)] |
| MW | Contract Drawings[l)]<br>Contract Specification[l)]<br>Work Schedules[l)] | — | Contract Specification[m)]<br>Work Schedules[m)]<br>Schedule of Rates[m)] |

a) JCT SBC/Q 2016 – Second Recital.
b) JCT SBC/Q 2016 – Third Recital.
c) Refer to Footnote *a*. The 'Activity Schedule' is an optional document, including a series of priced activities, so that the sum of those prices equates to the 'Contract Sum' (excluding 'Provisional Sums' and any 'Approximate Quantities' included in the Contract Bills).
d) JCT SBC/Q 2016 – Tenth Recital.
e) JCT SBC/Q 2016 – Eleventh Recital.
f) JCT DB 2016 – First Recital.
g) JCT DB 2016 – Second Recital. There is now no provision for a 'Bill of Quantities' to be included in JCT DB 2016. Previously, JCT DB 2011 included an option to use a 'Bill of Quantities' in the Schedule 2 of the 'Supplemental Provisions'.
h) JCT IC 2016 – Second Recital.
i) JCT IC 2016 – Third Recital (only one of the 'Bills of Quantities', the 'Specification' or the 'Work Schedules' should be provided by the employer.
j) JCT IC 2016 – Fourth Recital – 'Pricing Option A' (stating which of 'Bill of Quantities', the 'Specification' or the 'Work Schedules' have been priced by the contractor).
k) JCT IC 2016 – Fourth Recital – 'Pricing Option B' (providing a 'Contract Sum Analysis' or a 'Schedule of Rates' – the 'Priced Document' or 'Pricing Option B').
l) JCT MW 2016 – Second Recital (delete as appropriate).
m) JCT MW 2016 – Third Recital (delete as appropriate).

> *'The quality and quantity of the work included in the contract sum shall be deemed to be that which is set out in the bills of quantities, which bills, unless otherwise stated, shall be deemed to have been prepared in accordance with the standard method of measurement …'*

The 'method of measurement' in question provided that, where practicable, the nature of the soil must be described, attention must be drawn to any existing trial holes, and that excavation in rock must be given separately. Although the bills of quantities referred to drawings and the site conditions about which the contractor was to satisfy itself, excavation in rock was not shown separately. The court held that the contractor could treat the rock excavation as an extra.

**Table 5.5** Corrections and Resolution of Errors in Contract Documents.

| Description of 'error' in Contract Document | Entitlement to a 'Variation' (or 'Change') | | |
|---|---|---|---|
| | **JCT SBC/Q 2016** | **JCT DB 2016** | **JCT IC 2016** |
| Correction of 'Contract Bills' | Yes – clauses 2.14.1, 2.14.3 | — | Yes, clauses 2.12.2, 2.13.1, 2.14 |
| Inadequacy in Employer's Requirements[a)] | Yes – clauses 2.14.2, 2.14.3 | Yes, clause 2.12.2 | — |
| Discrepancy in Employer's Requirements | Yes – clause 2.16.2 | Yes, clause 2.14.2 | — |
| Discrepancies in contract documents | Yes, clauses 2.15, 2.17 | Yes, clauses 2.10,[b)]2.14.2 | Yes, clauses 2.13, 2.14 |
| Divergence from Statutory Requirements | Yes clauses 2.17.2.1,[c)]2.17.2.2 | Yes, clauses 2.15.2.1,[c)]2.15.2.2,[d)] 2.15.2.3[e)] | Yes, clause 2.15.2 |
| Emergency compliance with Statutory Requirements | Yes, clause 2.18.3 | — | Yes, clause 2.16.3 |

a) JCT SBC/Q 2016 – clause 2.14.2 only applies where the contractor is not obliged to verify the Employer's Requirements' accuracy under clause 2.13.2 (relating to clause 2.17 'Divergences from Statutory Requirements').
b) JCT DB 2016 – clause 2.10 (Divergence in Employer's Requirements and definition of site boundary). However, it is clear from the wording of the clause that the contractor has no obligation to look for or find divergences, merely to give notice if it finds any - London Borough of Merton v Stanley Hugh Leach Ltd (1985) 32 BLR 51.
c) The divergence is only treated as a 'Variation' if the change in Statutory Requirements occurs after the Base Date.
d) If a change in the 'Contractor's Proposals' is required to confirm with any permission or approval by a Statutory Authority after the Base Date (and not precluded in the 'Employer's Requirements'), then complying with the permission or approval is a 'Change'.
e) If the Employer's Requirements need amending to conform with the Statutory Requirements.

- Resolving errors in the Employer's Requirements ('ER's'). This 'Variation' or 'Change' will only occur under the JCT SBC/Q 2016 contract where the 'Contractor's Designed Portion' is used[79] or in JCT DB 2016.
  - Resolving an inadequacy. An 'inadequacy' means that '*there is a lack of quality or quantity*'[80] or that ER's are insufficient in some way. For example, the ER's for a student accommodation project require rooms to be partly cooled by having opening windows, and partly by mechanical ventilation but expressly prohibit any air cooling methods (such as air-conditioning). The employer subsequently instructs a change – omitting the opening windows, in the incorrect belief that the mechanical ventilation will be sufficient to cool the rooms. The rooms overheat during the summer months, because the ER's are inadequate in providing a method of cooling. Resolving the inadequacy means the introduction of air conditioning, which is a 'Variation' or a 'Change'.
  - Resolving a 'discrepancy' in the ER's. A 'discrepancy' means a 'lack of compatibility between two or more facts'.[80] A simple example where the contract drawings and the specification contradicted one another (e.g. if the drawings showed internal

79 JCT SBC/Q 2016 – see the Ninth Recital.
80 Oxford Dictionary of English 3rd Ed (2010).

stud partitions constructed from a single layer of 12 mm plasterboard either side, whereas the specification required two layers of 15 mm fire-resistant plasterboard each side).

- Divergence from Statutory Requirements.[81] The contractor may be entitled to a variation under all of the JCT 2016 contracts if the contract documents diverge from the Statutory Requirements. This entitlement to a variation (or Change) arises if the relevant contract documents do not permit the works to be undertaken in accordance with the Statutory Requirements (which, as can be seen in the definition in footnote[81], are wide-ranging). The general rules are:
  - Any documents prepared before the 'Base Date'[82] should be compliant with the Statutory Requirements;
  - If the document does not comply and is prepared by the employer, then provided that the contractor is not responsible for verifying the document's accuracy (see Footnote *a* in Table 5.5), then correcting the document is a 'Variation' or 'Change' (see Footnote *e* in Table 5.5); and
  - If the contractor prepares the document, then only changes in the statutory requirements after the Base Date that require corrections are a 'Variation' or 'Change' (see Footnote *c* in Table 5.5).
- Emergency compliance with Statutory Requirements. The entitlement to a 'Variation' for emergency compliance with statutory requirements is only found in JCT SBC/Q 2016 and JCT IC 2016. In any event, and irrespective of the JCT 2016 contract,[83] the contractor must notify the employer (or his representative) of any of the following discrepancies:

---

81 JCT 2016 contracts define 'Statutory Requirements' as *'any statute, statutory instrument, regulation, rule or order made under any statute or directive having the force of law which affects the Works or performance of any obligations under this Contract and any regulation or bye-law of any local authority or statutory undertaker which has any jurisdiction with regard to the Works or with whose systems the Works are, or are to be, connected....'*

82 The 'Base Date' is the date after which changes in the following may affect the employer or contractor (depending upon the contract terms): (i) Joint Fire Code, (ii) Statutory Requirements, (iii) exercise by the UK Government, Local or Public Authority of its statutory powers, (iv) VAT exemption, (v) the RICS' 'Definition of Prime Cost of Daywork carried out under a Building Contract', and (vi) any applicable Fluctuations. A JCT definition (in relation to fluctuations) is *'In JCT contracts, the Contract Sum is deemed to have been calculated at the agreed Base Date. The Base Date is usually stated to be the date of the tender or priced offer, which means that the risk of inflation between the tender and contact execution lies with the supplying party. However, if the date of execution of the contract or commencement of the works is used as the Base Date, then the risk of inflation over this period rests with the Employer.'* www.corporate.jctltd.co.uk (Dec 2022).

83 JCT DB 2016 obliges the contractor to notify the employer of discrepancies under clause 2.13.1 (Employer's Requirements), clause 2.13.2 (Contractor's Proposals or other Contractor's Design Documents), or clause 2.13.3 (any instruction issued by the employer), but unlike JCT SBC/Q 2016 or JCT IC 2016. However, none of the clause 2.13 'documents' in JCT DB 2016 entitle the contractor to a 'Change' under the 'Emergency compliance with Statutory Requirements' clause (clause 2.16).

- JCT SBC/Q 2016[84]: '*departure, error, omission or inadequacy*' or '*discrepancy or divergence*'; or
- JCT IC 2016[85]: '*departure, error or omission*', '*error in description or quantity, any omission or any inconsistency*'.

between any of the relevant documents[86] – including those listed in Table 5.6.

In the event that an 'emergency' arises from one of the stated discrepancies between the documents listed in Table 5.4 and the Statutory Requirements, and provided that the contractor has notified the employer's representative under the relevant clause,[87] the contractor is entitled to a Variation.[88]

#### 5.2.3.2 Instructions

Under each of the JCT 2016 contracts, the employer (or his representative) may issue instructions for a number of reasons. The instructions could be to provide the contractor with further information, resolve discrepancies etc. (see Section 5.2.3.1), instruct the contractor to do something – such as varying the works, or the instructions may be procedural in nature. Under JCT SBC/Q 2016 and JCT IC 2016, the employer's representative (the 'Architect/Contract Administrator') may sanction any variation made by the contractor even if the variation is not the result of an instruction.[89] The sanction must be in writing and this power can be used to ratify instructions which have been given but never confirmed by either the Architect/Contract Administrator or contractor,[90] or it may be used to sanction instructions given on site by the employer or by a consultant without the employer's representative's knowledge. Whether it is wise for the employer's

**Table 5.6** JCT SBC/Q 2016 and JCT IC 2016 Contracts – Notices of Discrepancies in Documents.

| Discrepancy between the following documents: | JCT SBC/Q 2016 | JCT IC 2016 |
|---|---|---|
| Contract Drawings | 2.15.1 | 2.13.1.1[a)] |
| Contract Bills | 2.15.2 | 2.13.1.1[a)] |
| Issued by the employer/representative: | | |
| – Instructions | 2.15.3 | 2.13.1.2 |
| – Drawings or documents | 2.15.4 | 2.13.1.3 |

a) JCT IC 2016 – clause 2.13.1.1 refers to discrepancies in the 'Contract Documents' which include 'the Contract Drawings' and either the 'Priced Document' (which may include priced Bills of Quantities, Specification of Work Schedules – see Footnote *j* in Table 5.4), or the priced 'Contract Sum Analysis' or 'Schedule of Rates' (see Footnote *k* in Table 5.4).

---

84 JCT SBC/Q 2016 – clause 2.15 (Notice of discrepancies etc).
85 JCT IC 2016 – clause 2.13 (Instructions on errors, omissions and inconsistencies).
86 JCT SBC/Q 2016 – in addition to the documents noted in Table 5.4, the contractor must notify the employer's representative of discrepancies in the Contractor's Proposals or other Contractor's Design Documents (under clause 2.15.5).
87 JCT SBC/Q 2016 – clause 2.15 (Notice of discrepancies etc).
88 JCT SBC/Q 2016 – clause 2.17.3. JCT IC 2016 – clause 2.13.3.
89 JCT SBC/Q 2016 – clause 3.14.4 and JCT IC 2016 – clause 3.11.3.
90 JCT SBC/Q 2016 – clause 3.12.3.

representative to sanction instructions of that kind given by other parties is another question. The employer's representative cannot be obliged by the employer to ratify the employer's instructions and he is only authorised to deal only with instructions properly issued under the terms of the contract. It has already been seen that, with the possible exception of imposing restrictions and obligations (see Section 5.2.2), the only person authorised by the contract to issue instructions is the employer's representative.

Table 5.7 summarises the nature of instructions that may be given to the contractor under the JCT 2016 contracts.

However, not all 'instructions' constitute a 'Variation' or 'Change' under the terms of each of the JCT 2016 contracts (refer to Table 5.2).

- General points:
  - The employer (or his representative[91]), is empowered to issue instructions,[92] and they must be issued in writing.[93] If they are received by the contractor's 'site manager', they are deemed to be received by the contractor.[94]
  - The contractor must comply with the instruction forthwith,[95] and if he fails to do so, then the employer (or his representative) may notify him of such failure.[96] If the contractor fails to comply after seven days after receiving the notice, the employer may engage another contractor to carry out the work in the relevant instruction and deduct the costs from monies due to the contractor.
  - If any instruction is not issued in writing, then it may not be in accordance with the contract, and may be of no effect. However, JCT SBC/Q 2016 and JCT DB 2016 provide that any instructions issued otherwise than in writing can be confirmed in writing by the contractor to the employer (or his representative) within seven days,[97] and if not dissented from in writing within a further seven days from receipt (not despatch) is to take effect from the expiry of the second seven days. As the time periods are measured from when the contractor's confirmation is received, a prudent contractor or will send its confirmation by recorded, or special delivery post – where proof of receipt can be obtained. If the confirmation is one that reports a significant variation, then the contractor should not solely rely on emails. Even if the emailed confirmation does not meet with an 'undeliverable' message, this may not be evidence that his confirmation was received. If the employer (or his representative) confirms the instruction

91 JCT SBC/Q 2016 – clause 3.4 provides that an instruction by a 'clerk of works' is no effect unless (1) the same matter also allows the Architect/Contract Administrator to issue instructions, and (2) that the 'Architect/Contract Administrator' confirms the clerk of works' 'direction' by issuing an instruction within 2 Business Days after the 'direction' was given.

92 JCT SBC/Q 2016 – clause 3.13, JCT DB 2016 – clause 3.8, JCT IC 2016 – clause 3.10, JCT MW 2016 – clause 3.6, including those for 'Variations' (JCT SBC/Q 2016 – clause 3.14, JCT IC 2016 – clause 3.11, JCT MW 2016 – clause 3.6) or 'Changes' (JCT DB 2016 – clause 3.9).

93 JCT SBC/Q 2016, JCT DB 2016, and JCT IC 2016 – clause 1.7.1. JCT MW 2016 – clause 1.6.1.

94 JCT SBC/Q 2016, JCT DB 2016 and JCT IC 2016 – clause 3.2.

95 JCT SBC/Q 2016 – clause 3.10, JCT DB 2016, clause 3.5, JCT IC 2916 – clause 3.8, JCT MW 2016 – clause 3.4. 'Forthwith' in this context does not necessarily mean 'immediately', because the instruction may vary work not yet done, but it imposes an obligation upon the contractor to carry out the work as soon as it reasonably can do so. *London Borough of Hillingdon v Cutler* (1967) 2 All ER 361.

96 JCT SBC/Q 2016 – clause 3.11, JCT DB 2016 – clause 3.6, JCT IC 2016 – clause 3.8, JCT MW 2016 – clause 3.5.

97 JCT SBC/Q 2016 – clause 3.12. JCT DB 2016 – clause 3.7.

**Table 5.7** JCT 2016 contracts – Nature of Instructions.

| Nature of 'Instruction': | SBC/Q | DB | IC | MW |
|---|---|---|---|---|
| *General:* | | | | |
| Instructions to be in writing | 1.7.1 | 1.7.1 | 1.7.1 | 1.6.1 |
| Correcting a divergence in the Employer's Requirements and definition of site boundary | — | 2.10.1 | — | |
| Issuing further drawings, details, instructions | 2.12 | — | 2.11 | 2.3 |
| Instructions for discrepancies/inconsistencies | 2.15 | 2.13 | 2.13.1 | 2.4[a)] |
| Discrepancies in design (in CDP) documents | 2.16.1 | 2.14[b)] | — | — |
| Divergencies from Statutory Requirements | 2.17.2 | 2.15.2 | 2.15.2 | 2.5 |
| Emergency compliance with Statutory Requirements | 2.18.3 | — | 2.16.1 | — |
| Adjusting the completion date (related definitions) 'Relevant Omission' | 2.26.3 | 2.23.3 | — | — |
| Schedules of defects and instructions | 2.38.1<br>2.38.2 | 2.35.1<br>2.35.2 | 2.30[c)] | 2.10[c)] |
| Site manager is the 'Contractor' for instructions | 3.2 | 3.2 | 3.2 | — |
| Clerk of works' instructions | 3.4 | — | — | — |
| *Instructions:* | | | | |
| Compliance with instructions | 3.10 | 3.5 | 3.8 | 3.4 |
| Non-compliance with instructions | 3.11 | 3.6 | 3.9 | 3.5 |
| Instructions other than in writing | 3.12 | 3.7 | — | — |
| Provisions empowering instructions | 3.13 | 3.8 | 3.10 | 3.6 |
| Instructions requiring Variations (or Changes) | 3.14 | 3.9 | 3.11 | |
| Postponement of work | 3.15 | 3.10 | 3.12 | — |
| Provisional Sums | 3.16 | 3.11 | 3.13 | 3.7 |
| Inspections and tests | 3.17 | 3.12 | 3.14 | — |
| Removal of work, materials or goods that are not in accordance with the Contract | 3.18.1 | 3.13.1 | 3.16.1 | — |
| Instructions dealing with non-compliant workmanship | 3.19 | 3.14 | 3.16.2 | — |

*(Continued)*

**Table 5.7** (Continued)

| **Nature of 'Instruction':** | **SBC/Q** | **DB** | **IC** | **MW** |
|---|---|---|---|---|
| Exclusion of persons | 3.21 | — | 3.17 | 3.8 |
| Antiquities | 3.22.2 | 3.15.2 | — | — |
| *Procedures:* | | | | |
| Variation quotation | 5.3.1<br>5.3.2<br>5.3.3 | — | — | — |
| Breach of Joint Fire Code | 6.19.12 | — | 6.17.2 | — |
| Instruction for 'Named Sub-contractor' | — | Scheme 2 Para 1.1.2 | Scheme 2 Paras 2, 7 | — |
| Inviting a Variation quotation | Scheme 2 Para 1 | Scheme 2 Para 2 | — | — |
| Accepting a Variation or acceleration quotation | Scheme 2 Para 4 | Scheme 2 Para 4.3 | — | — |
| Not accepting a Variation or acceleration quotation | Scheme 2 Para 5 | — | — | — |
| Opening up for testing – Code of Practice | Scheme 4 | Scheme 4 | — | — |
| Cost savings and value improvements – implementing a change | Scheme 8 Para 3.3 | Scheme 2 Para 7.3 | Scheme 5 Para 3.3 | Scheme 3 Para 3.3 |

a) JCT MW 2016, clause 2.4 refers to the 'correction of inconsistencies' but unlike the other JCT 2016 contracts, does not stipulate the form of the correction.
b) JCT DB 2016 – clauses 2.14.1 only refers to the 'decision' of the employer, but in practice, this decision is likely to be in the form of an instruction.
c) JCT IC 2016 – clause 2.30, and JCT MW 2016 – clause 2.10: the employer's representative 'notifies' the contractor of defects, which is seemingly the equivalent of an 'instruction' under clause 2.38 (JCT SBC/Q 2016) or clause 2.35 (JCT DB 2016).

in writing within the first seven days, the contractor need not confirm it itself, but it will take effect from the date of the confirmation. If neither the contractor nor the employer's representative confirms the instruction within the time limit, but the contractor nevertheless complies with it, then the employer's representative may confirm it himself at any time prior to the issue of the final certificate, and the instruction will then, strangely, be deemed to have taken effect from the date it was originally given. That provision could result in interesting problems of interpretation.

- Correcting errors, divergencies and discrepancies (refer to Section 5.2.3.1. The contractor may be entitled to an extension of time and direct loss and/or expense in the circumstances shown in Table 5.8.

**Table 5.8** Extension of Time/Loss and Expense for Instructions Resolving Discrepancies.

| Compliance with an instruction issue to resolve discrepancy in or between: | JCT SBC/Q 2016 | JCT DB 2016 |
|---|---|---|
| • The Contract Drawings | EoT 2.29.2.1<br>L&E 4.22.2.3 | — |
| • The Contract Bills | EoT 2.29.2.1<br>L&E 4.22.2.3 | — |
| • The Employer's Requirements | — | EoT 2.26.2.1 |
| - An Instruction | EoT 2.29.2.1<br>L&E 4.22.2.3 | — |
| - Any drawings subsequently issued | EoT 2.29.2.1<br>L&E 4.22.2.3 | — |
| • Contractor's Proposals and other Contractor's Design Documents | EoT 2.29.2.1<br>L&E 4.22.2.3 | — |

- 'Relevant Omissions'. The employer (or his representative) can instruct reductions in the quantity of work, or omissions to be made from the works (refer to Section 5.2.1). JCT SBC/Q 2016 and JCT DB 2016 expressly treat such 'Relevant Omissions'[98] as the only grounds for the employer (or his representative) to reduce an extension of time that has been already granted[99] (see Sections 11.1.1 and 11.3).
- Instructions for rectifying defects after 'practical completion'. All of the JCT 2016 contracts[100] provide for the employer (or his representative) to issue instructions (in the form of a schedule) for the rectification of defects after 'practical completion' – i.e. during the rectification period.
- Instructions for 'Variations' or 'Changes' – refer to Sections 5.2.1, 5.2.2 and 5.2.3.
- Postponement of work. The JCT 2016[101] contracts permit the employer (or his representative) to instruct[102] the work (including its design) to be postponed. The maximum period of postponement is set out in the 'Contract Particulars'.[103] Postponement entitles the contractor to the following:
  - An extension of time[104];
  - Direct loss and/or expense[105]; and

98 JCT SBC/Q 2016 – clause 2.26.3, JCT DB 2016 – clause 2.23.3.
99 JCT SBC/Q 2016, clauses 2.28.4, 2.28.5.2 and 2.28.6.4. JCT DB 2016, clauses 2.25.4, 2.25.5.2 and 2.25.6.4.
100 Refer to Table 5.5 and Footnote *c* in Table 5.7.
101 Save for JCT MW 2016.
102 JCT SBC/Q 2016 – clause 3.15, JCT DB 2016 – clause 3.10, and JCT IC 2016 – clause 3.12.
103 JCT SBC/Q 2016, JCT DB 2016 and JCT IC 2016 – entry in the 'Contract Particulars' against clause 8.9.2 'Period of Suspension' (for which the default period is 2 months).
104 'Relevant Event' under JCT SBC/Q 2016 – clause 2.29.2.1, JCT DB 2016 – clause 2.26.2.2, and JCT IC 2016 – clause 2.20.2.1.
105 'Relevant Matter' under JCT SBC/Q 2016 – clause 4.22.2.1, JCT DB 2016 – clause 4.21.2.1, and JCT IC 2016 – clause 4.17.2.1.

  - If the period of postponement exceeds the period set out in the 'Contract Particulars',[103] that becomes a 'specified default'[106] of the employer, and if the contractor complies with the contract's termination provisions, he may terminate his employment under the contract.
- Provisional sums. All of the JCT 2016 contracts allow the employer (or his representative) to instruct[107] the expenditure of 'provisional sums'.

  A provisional sum is often used to provide a budget within the contract sum for an uncertain element of work. The 'uncertainty' could arise for any of the following reasons where the work in question:
  - has not been fully designed (e.g. the employer may not have decided upon the precise specification for an item of work);
  - may not ultimately be required (e.g. stabilisation of ground or a structure);
  - its extent (or quantity) is unknown (e.g. whilst the employer may definitely require some refurbishment work to be carried out, its extent may depend upon the results of a subsequent survey or report).
  - The contractor does not carry out any provisional sum work (including its design) unless the employer instructs the provisional sum to be expended. The employer has no obligation to instruct the provisional sum's expenditure, and it is not uncommon for provisional sums not to be instructed. If the employer does instruct its expenditure, then the amount of money expressed as the 'provisional sum' is omitted from the contract sum and replaced with the sum of money calculated in accordance with the contract. Depending upon the 'Valuation Rules' in a particular contract (see Section 5.4.1.3), the calculation may be as simply as totting up the actual costs of the work and adding the contractor's overheads and profit.

  By including a provisional sum, the employer accepts all, or a significant cost risk for that work, as he cannot hold the contractor to a fixed price, and consequently, the contractor faces considerably less commercial risk. The use of provisional sums can therefore benefit both parties, as the employer won't face the risk of paying a high (or 'worst case scenario') price for work for which the actual cost may be much less (see Section 5.1.3), and the contractor isn't held to a fixed price for work, the scope and extent of which cannot be priced with certainty. It is generally accepted that because of provisional sums' temporary nature, the contractor does not include any time allowance within his programme for carrying out provisional sum work. Neither should there be any allowance included within the contractual period for carrying out provisional sum work (calculated from the contracts' start and completion dates). The absence of any time allowance (in either the programme or period for completion) for provisional sum work can often be contentious. It is fairly common for the Employer's Requirements to try and distinguish between types

---

106 'Specified default' – JCT SBC/Q 2016 and JCT IC 2016 – clause 8.9.2.1 (specifically referring to an instruction under the relevant clause – see Footnote 102). However, JCT DB 2016 – clause 8.9.2 does not refer to an instruction under clause 3.10 (postponement of work), instead referring to *'by reason of any impediment…by act…by the Employer'* and such 'act' could clearly be an instruction under clause 3.10.

107 JCT SBC/Q 2016 – clause 3.16, JCT DB 2016 – clause 3.11, JCT IC 2016 – clause 3.13, and JCT MW 2016 – clause 3.7.

of provisional sum, often (incorrectly) describing them as either 'defined' and 'undefined'. If these distinctions are set out in the Employer's requirements, the employer will deem that the contractor has included sufficient allowances for carrying out the defined provisional sum work in both its programme and in pricing its preliminaries.

The difference between 'defined' and 'undefined' provisional sums has been set out in various methods of measurement for building works - the correct description are a 'provisional sum for defined work' or a 'provisional sum for undefined work' (previously described in Rule 10.3 of SMM7,[108] and from 2012, in Section 2.9 of NRM2).[77] If bills of quantities are adopted as the pricing document (see Section 5.2.3.1), and NRM2 is the stated method of measurement, then the Employer's Requirements' description for 'defined' provisional sums (i.e. a 'provisional sum for defined work') will have to contain sufficient information for the contractor to price the work, as if he was offering a fixed price. This information should include the following:

- Descriptions of the type of work, how it is to be constructed, and from what it is to be constructed;
- The location(s) where it is to be fixed, how it is to be fixed (including any other work that is to be fixed to it), and any details of any specific limitations must be given; and
- Sufficiently accurate quantities.

If the provisional sum's description is lacking in respect of any of the above, then the provisional sum will be not be for 'defined work' (it will be 'undefined'), and the contractor will not be deemed to have made any allowance for its execution, either within his programme or in pricing its preliminaries.

- Managing the quality of work. The employer (or his representative) may instruct[109] the contractor to deal with quality issues as follows:
  - Ordering inspections and tests to be carried out[110];
  - Work, materials or goods not in accordance with the contract[111]; and
  - Workmanship not in accordance with the contract.[112]

  The general rule is that the instructions must be 'reasonable'. If they are, the contractor is not entitled to an extension of time of any addition to the contract, sum, but if the instructions are not 'reasonable', or the test or inspections shows the relevant work complies with the contract, then the contractor may be entitled to additional costs[113] and an extension of time.[114]
- Antiquities. Under JCT SBC/Q 2016[115] and JCT DB 2016,[116] the employer (or his representative) may issue instructions regarding the discovery of antiquities. The contractor is

108 Standard Method of Measurement for Building Works – 7th Ed (2000).
109 All JCT 2016 contract, except for JCT MW 2016. Under JCT SBC/Q 2016 and JCT DB 2016, the employer (or his representative) ought to comply with the 'Code of Practice' (Schedule 4 in both contracts).
110 JCT SBC/Q 2016 – clause 3.17, JCT DB 2016 – clause 3.12 and JCT IC 2016–3.14.
111 JCT SBC/Q 2016 – clause 3.18.1, JCT DB 2016 – clause 3.13.1, and JCT IC 2016 – clause 3.16.1.
112 JCT SBC/Q 2016 – clause 3.19, JCT DB 2016 – clause 3.14, and JCT IC 2016 – clause 3.16.2.
113 'Relevant Matter' under JCT SBC/Q 2016 – clause 4.22.2.1, JCT DB 2016 – clause 4.21.2.2, and JCT IC 2016 – clause 4.17.2.3.
114 'Relevant Event' – JCT SBC/Q 2016 – clause 2.29.2.2, JCT DB 2016 – clause 2.26.2.3, and JCT IC 2016 – clause 2.20.2.3.
115 JCT SBC/Q 2016 – clause 3.22.2.
116 JCT DB 2016 – clause 3.15.2.

entitled to both an extension of time[117] and direct loss and/or expense[118] for complying with such instructions.
- Variation and acceleration quotations, cost savings. The JCT 2016 contracts provide various mechanisms for the employer to seek or obtain savings and agreed prices for variations and acceleration. These can range from:
  - The contractor proposing specification or design changes that could benefit the employer – changes in the prices to be instructed[119];
  - Instructing the contractor to provide a quotation for Variations[120] or Changes[121];
  - Instructing the contractor to provide a quotation to accelerate the works[122] (see Section 5.6);
  - Instructions for 'Named Sub-Contractors'.[123]
- If there is a breach of the Joint Fire Code.[124]

The contractor has limited rights to reasonably object to certain instructions given by, or on behalf of the employer. These are summarised in Table 5.9.

**Table 5.9** JCT 2016 Contracts: Contractor's Reasonable Right of Objection.

| Ground for objection | JCT SBC/Q 2016 | JCT DB 2016 | JCT IC 2016 |
|---|---|---|---|
| If an instruction will infringe a copyright or patent | — | 2.20.2 | — |
| Variation (or Change) imposing obligations or restrictions | 3.10.1, 3.14.2 | 3.5.1, 3.9.2 | 3.8, 3.11.2 |
| Instruction changing the CDP design | 3.10.3 | 3.9.1[a)] | — |
| Expending a Provisional Sum – for CDM reasons | — | 3.9.4 | — |
| An instruction to rectify work not in accordance with the Contract | — | — | 3.15.2 |

a) JCT DB 2016 – clause 3.9.1 prohibits the employer from instructing a 'Change' to the Works' design without the contractor's consent. The word 'design' of course does not simply mean the drawings, but also the written part of the Contractor's Proposals such as specifications and schedules of work – *John Mowlem & Co Ltd v British Insulated Callenders Pension Trust Ltd* (1977) 3 Con LR 63.

117 'Relevant Event' under JCT SBC/Q 2016 – clause 2.29.4 or JCT DB 2016 – clause 2.26.4.
118 'Relevant Matter' under JCT SBC/Q 2016 – clause 4.22.4 or JCT DB 2016 – clause 4.21.3.
119 'Supplemental Provisions' – 'Cost savings and value improvements' (which apply by default unless de-selected): JCT SBC/Q 2016 – Schedule 8, para 3.3, JCT DB 2016 – Schedule 2, para 7.3, JCT IC 2016 – Schedule 5, para 3.3 and JCT MW 2016 – Schedule 3, para 3.3.
120 'Supplemental Provisions' – 'Variation Quotation': JCT SBC/Q 2016 – clause 5.3 (Variation Quotation) and Schedule 2, paragraph 1, paragraph 4 to accept the quotation, and paragraph 5 to reject the quotation.
121 'Supplemental Provisions' – 'Variation Quotation': JCT DB 2016 – Schedule 2, paragraph 2.
122 'Supplemental Provisions' – 'Acceleration Quotation': JCT SBC/Q 2016 – Schedule 2, paragraph2 (for acceptance and rejection, see Footnote 120), JCT DB 2016 – Schedule 2, paragraph 4.
123 'Supplemental Provisions': JCT DB 2016 – Schedule 2, paragraph 1.1.2, JCT IC 2016 – Schedule 2, paragraphs 2 and 7.
124 JCT SBC/Q 2016 – clause 6.19.2 and JCT IC 2016 – clause 6.17.2.

#### 5.2.3.3 Design Submission Procedure and Contractor's Designed Portion ('CDP')

Under JCT DB 2016 (and JCT SBC/Q 2016 where the CDP applies[125]), the employer is able to review the contractor's design by using the 'Design Submission Procedure'.[126] This procedure allows the employer to review each of the 'Contractor's Design Documents'[127] and return copies to the contractor marked 'A', 'B' or 'C', which have the following meanings[128]:

- 'A' – The contractor proceeds with the work in accordance with the submitted document;
- 'B' – As 'A' but the contractor must also proceed in accordance with any comments from the employer (or his representative); or
- 'C' – The contractor accounts for the comments, and either resubmits his document for approval, or notifies the employer (or his representative) within seven days that he considers the comments constitute a 'Change' or 'Variation. The contractor does not proceed with any work which has received a 'C' category comment[128].

If the employer (or his representative) does not withdraw the status 'C' comment that was the subject of the contractor's notification, and confirms it, then the contractor should resubmit his document and, subject to the contract, the incorporation of the status 'C' comments may be a 'Change' or a 'Variation'.[129] This sounds a little vague, and it is suggested that if the employer (or his representative) confirms the status 'C' comment, then the deciding factor of whether a 'Change' or 'Variation' occurs will be if the comments fall into clauses 5.1.1 or 5.1.2 of the contract (refer to Sections 5.2.1 and 5.2.2).

The CDP element in JCT SBC/Q 2016 is clear that an instruction with regard to the CDP must be issued as an alteration or modification to the 'Employer's Requirements'[130] (see Section 5.1.1), so it is important that the Employer's Requirements contain the appropriate level of detail.

Clause 2.2.2 of JCT SBC/Q 2016 allows the employer's representative to issue what are referred to as 'directions' for the 'integration' of the CDP's design within the design of the rest of the Works.[131] The word 'directions' is not defined.[132] Since JCT SBC/Q 2016 uses

125 JCT SBC/Q 2016 – see the Ninth Recital.

126 JCT SBC/Q 2016 and JCT DB 2016 – Schedule 1 'Design Submission Procedure'.

127 JCT DB 2016 and JCT SBC/Q 2016 (inc the CDP) at clause 1.1 define 'Contractor's Design Documents' as *'the drawings, details and specifications of materials, goods and workmanship and other related documents and information prepared by or for the Contractor in relation to the design of the Works (including such as are contained in the Contractor's Proposals), together, where applicable, with any other design documents or information to be provided by him under the BIM Protocol.'*

128 JCT DB 2016 and JCT SBC/Q 2016 – Design Submission Procedure, paragraph 5.

129 The 'Design Submission Procedure' was introduced for JCT DB 2005, as a simpler procedure was included in JCT WCD98 (in Supplemental Provision S2). JCT's 'Design and Build Contract Guide (DB/G)' for the 2005 edition gave the following guidance (at paragraph 134); '*...Unless the Contractor follows this procedure then compliance with a comment will not be treated as giving rise to a Change. Where the Employer confirms a comment, whether or not it actually gives rise to a Change will remain to be determined by reference to the provisions of the Contract.*'

130 JCT SBC/Q 2016 – clause 3.14.3.

131 JCT SBC/Q 2016 – see also clause 3.10.3.

132 'Direction' is not a defined term in JCT SBC/Q 2016. The ordinary meaning of 'direction' is *'guidance of someone or something'* (Oxford Dictionary of English, 3rd Ed 2010), and appears to have this meaning in the context of the CDP works (and in respect of clerk of works 'directions' under clause 3.4 – refer to Footnote 91), but may not constitute a 'Variation' unless (a) it constitutes a 'variation' in the manner

both 'instruction' and 'direction', it is appropriate to believe that a somewhat different meaning is intended for 'direction', otherwise only the word 'instruction' could have been used.[133] The distinction is not clear in ordinary usage, but one definition[132] of 'direction' is 'guidance' which probably approximates to the intended meaning. Under clause 3.4, the employer's representative may (within 2 business days) confirm directions from the clerk of works to the contractor, otherwise the clerk of works' direction will have no effect. It therefore appears that something less than 'instructions' is intended by the use of 'directions' here. Therefore, the meaning of 'direction', wherever it occurs in the contract should be the same, and the use of 'direction' instead of 'instruction' must be assumed to convey a different meaning.

Integration of the CDP design with the rest of the Works' design can be a fruitful area for claims. The contractor may contend that the employer's representative's directions for integration unavoidably result in additional work and, therefore, entitle the contractor to additional cost and/or an extension of time. The situation tends to be misunderstood. The principles are straightforward even though particular circumstances may need careful consideration, but in essence, there are four basic situations:

- If the Works' design includes interfaces with the CDP work, it is a matter for the contractor to allow in the Contractor's Proposals for the proper integration of the CDP with the rest of the Works' design. However, even if the Works' design does not 'integrate' the CDP design, and if the employer's representative does not instruct a variation under clause 3.14.3 (alteration or modification of the CDP's Employer's Requirements), the contractor has no claim for any additional cost;
- If the invitation to tender is not supported by sufficient information to enable the contractor to properly design the interface between the CDP and other work and the contract documents are executed without the inadequacy being resolved, the contractor, ought to be entitled to a 'variation' under clauses 2.14.2 and 2.14.3 (refer to Table 5.3);
- If the employer's representative subsequently issues instructions to change the rest of the Works, which affects the CDP, or concern the CDP by instructing through the Employer's Requirements, then irrespective of whether the employer's representative issues a clause 2.2.2 'direction' for the integration of the CDP design within the design for the (now varied) Works, the contractor's entitlement to a 'variation' arises from the employer's representative's instruction; but
- If the employer's representative issues a clause 2.2.2 'direction' to the contractor to alter the CDP design in order to correct its own error, the contractor must bear those costs itself.

In the second and third circumstances outlined above, it is said that the contractor ought to be entitled to a 'Variation'. It is suggested that where the contractor suffers adverse effects (such as incurring additional costs or the progress of its work is delayed) by clause 2.2.2 'directions' issued by the employer's representative – which do not constitute 'Variations' as defined in the contract, then the contractor may claim the 'direction' is an '*impediment,*

described in Table 5.1, or (b) changes the CDP's 'employer's Requirements' in the same manner, and (c) the direction originates from, or is given by the Architect/Contract Administrator.

133 *John Jarvis Ltd v Rockdale Housing Association* (1986) 36 BLR 48. A different word will be presumed to convey a different meaning.

*prevention or default*' and seek an extension of time under clause 2.29.7, and/or loss and expense under clause 4.22.5.

All instructions issued in regard to contractor's designed portion work are subject to the contractor's right of reasonable objection under clause 3.10.3 (refer to the Section 5.2.3.2 and Table 5.9).

## 5.3 NEC4 Contracts – Definition of 'Compensation events' (Including Variations)

NEC4, like its predecessors (NEC to NEC3), deals with 'variations' in a novel way compared to other construction or engineering contracts published by the JCT, FIDIC or ICE. It groups together all events and matters for which the contractor would be entitled to changes in the price for the works, or the period in which he must complete the works, and calls them 'Compensation Events'. Whilst appearing slightly unusual to practitioners who may be more used to JCT contracts that include separate sections dealing with (i) 'Variations' (i.e. scope changes and procedures) and (ii) breaches of the contract by the Client (i.e. entitling the contractor to extensions of time and loss and expense, NEC's concept of the 'compensation event' is simple and clear.

The operation of NEC4's compensation event procedure retains many of the characteristics of JCT's Variation (or Change), extension of time and loss and expense procedures, but includes some important and fundamental differences that go to the heart of its operation. For the majority of this book, how the JCT 2016 building contracts deal with each type of claim (such as variations, extensions of time and loss and expense) is captured in individual sections (see Sections 5.2 and 5.4 and Chapters 11 and 12), but for clarity, and to highlight the relatively unique nature of NEC4's compensation event procedure, these are covered in Sections 5.3.1 (NEC4 – Compensation events) and 5.3.2 (NEC4 – compensation event procedure) below. The single heading of 'compensation events' reflects NEC's ethos in dealing with project changes.

Book Four (Managing Change) of NEC's 'Managing Reality' series summarises[134] the overriding principles for compensation events as:

- The contractor should not be '…out of pocket for things that happen that are outwith its control'; and
- 'Compensation events are events that are at the Client's risk in the contract.'

### 5.3.1 NEC4 – Compensation events

A simple way to understand NEC4's compensation events is to compare them to similar, or equivalent matters under a JCT 2016 contract (as shown in Table 2.1 in Chapter 2). Table 5.10 repeats the various compensation events from clause 60 of NEC4 and identifies which compensation event is a 'variation' in scope, which arises because of a particular 'procedure', and which is a 'Client breach'.

134 Managing Reality – Book Four (Managing Change) (3rd Edition) 2017, "1.1. Introduction" p4.

**Table 5.10** NEC4 Compensation events Cross-Reference Table.

| **NEC4 compensation events – clause 60.1 (for all Main Option Clauses B and D and Secondary Options)** | **Variation** | **Procedure** | **Client Breach** |
|---|---|---|---|
| (1) Project Manager's Instruction ('PMI') to the Contractor to change the Scope[a)] | ✓ | | |
| (2) The Client prevents access to parts of the site by the 'access date' or relevant dates on the Accepted Programme | | | ✓ |
| (3) The Client fails to provide something by the Accepted Programme's date | | | ✓ |
| (4) PMI not to start, or stop work | | ✓ | |
| (4) PMI to change a Key Date | ✓ | | |
| (5) The Client (or other contractors) do not do their work in accordance with the Accepted Programme | | | ✓ |
| (5) The Client (or other contractors) do not work in accordance with the Scope | | | ✓ |
| (5) The Client (or other contractors) undertake work that is not set out in the Scope. | ✓ | | |
| (6) The PM[b)] fails to respond to a communication within the period for reply | | ✓ | |
| (7) PMI dealing with an object of archaeological value or that is of interest found on the site | ✓ | | |
| (8) The PM changes a previous decision | ✓ | | |
| (9) The PM's reason(s) for not accepting something is/are not set out in the contract | | ✓ | |
| (10) The contractor is instructed to search for a defect but there isn't one | | ✓ | |
| (11) Testing causes an otherwise avoidable delay | | ✓ | |
| (12) The Contractor encounters unforeseen physical conditions within the site, which are not weather conditions | ✓ | | |
| (13) The contractor encounters adverse weather conditions within the meaning of the contract | ✓ | | |
| (14) The Client causes an event for which he is liable | | | ✓ |
| (15) PM certifies 'taking over' before 'Completion' and the 'Completion Date' | | ✓ | |
| (16) Client fails to provide necessary facilities, equipment and materials for tests | | | ✓ |
| (17) The PM corrects an assumption he made about a previous Compensation Event | ✓ | | |
| (18) Client commits a breach of the contract which is not a Compensation Event | | | ✓ |

| | | | |
|---|---|---|---|
| (19) An event occurs (which is not another Compensation Event), which prevents the contractor from completing the works[c] that is outside of each party's control, or that the contractor could not have reasonably foreseen | | ✓ | |
| (20) The PM does not accept a contractor's quotation for a proposed instruction | | ✓ | |
| (21) Any additional Compensation Events set out in the Contract Data Part One occur | | ✓ | |
| Options B and D – differences between the final quantities of work and the quantities in the Bill of Quantities: | | | |
| • If it does not arise from a change in the Scope; | | ✓ | |
| • If it causes the Defined Cost to change (see Tables 5.12 and 5.13 in Section 5.5.1); or | | ✓ | |
| • If the Bill of Quantities for that rate exceeds 0.5% of the total of the Prices. | | ✓ | |
| (all Clause 60.4) | | | |
| • If the difference in quantity delays Completion or a Key Date (clause 60.5) | | ✓ | |
| • If the Project Manager corrects a mistake in the Bills of Quantities arising from a departure from the method of measurement, ambiguity or inconsistency (clause 60.6), including between the Bills of Quantities and another document (clause 60.7) | ✓ | | |
| Secondary Option X2 (Changes in the law) clause X2.1 | ✓ | | |
| Secondary Option X12 (Multiparty Collaboration): | | | |
| • Instruction to change the Partnering Information: reduced prices – clause X12.3(6) | ✓ | | |
| • Change of programme to match the revised timetable: reduced prices – clause X12.3(7) | ✓ | | |
| Secondary Option X14 (Advanced Payment) clause X14.2 | | | ✓ |
| Secondary Option X15 (Contractor's Design) clause X15.2 | | ✓ | |
| Y2.4 Contractor suspension | | | ✓ |

a) NEC4 – including an instruction to change (a) the Scope to resolve an ambiguity or inconsistency under clause 63.10, or (b) to correct a change in the Condition for a Key Date under clause 63.11.
b) Including those working on the Project Manager's behalf – such as the Supervisor.
c) NEC4 clause 60.1(19) – the event can also prevent the contractor from completing by the Accepted Programme's planned completion date.

In addition to the compensation events under clauses 60.1 and Y2.4, Main Options B (Priced Contract with Bill of Quantities) and D (Target Contract with Bill of Quantities) include additional compensation events dealing with the differences in the final quantities of work,[135] and Project Manager's instruction to correct the Bills of Quantities.[136]

In terms of 'variations', the most obvious similarity to a Variation (or Change) under JCT 2016 contracts would be a Project Manager's instruction to change the Scope under clause 60.1(1). As with JCT's 2016 contracts, the contractor is obliged to carry out that instruction immediately.[137]

Given the 'Accepted Programme's' importance as a management tool (see Section 3.5.1[138]), it is unsurprising that NEC4 emphasises the need for the Client to comply with it. Compensation Events under clauses 60.1(2), 60.1(3) and 60.1(5) all encourage the Client, along with anyone working on his behalf (referred to as 'Others') to comply with the Accepted Programme. Similarly, and in keeping with the NEC ethos of dealing with matters contemporaneously, the Compensation Event under clause 60.1(6) encourages the Project Manager to promptly respond to communications.

Compared to JCT 2016 contracts, NEC4 may seem a little simplistic when dealing with matters such as 'force majeure' and those covered by insurance, such as the Compensation Event in clause 60.1(19). The larger JCT 2016 contracts set out such matters in more detail and in separate clauses. However, the same point could be applied to JCT DB 2016, which uses the 'impediment, prevention, default' Relevant Event (clause 2.26.6) to the same effect as many of NEC4's compensation events (for example, clauses 60.1(2), 60.1(3), 60.1(5), 60.1(9), 60.1(14), 60.1(16) and 60.1(18)).

### 5.3.2 NEC4 – Compensation Event Procedure

NEC4's compensation event procedure shares many of the characteristics found in JCT 2016 contracts, in terms of the contractor giving prior notifications and providing quotations for the varied work (refer to Section 5.4.3). The contractor provides some prior warning or notification to the Project Manager if an event may occur that could affect the project. 'Affect' in this context means an event which is likely to adversely affect the Client in terms of:

- Something that may disrupt, delay or prolong the progress of the works – resulting in the project (or a section of it) being completed later than was either planned or agreed; or
- An event that may result in the Client having to make additional payments to the contractor – either for carrying out a variation, or paying financial compensation to cover the contractor's costs of dealing with disruption, delays or prolongation of its works.

Most Clients, and particularly those who are public bodies embarking on infrastructure projects will have a project budget, which ideally, should not be exceeded. The purpose of a contractor giving a prior warning or notification to the Project Manager is not (as some contractors often believe) intended to prevent a contractor becoming entitled to additional time

135 NEC4 – Options B and D, clauses 60.4 and 60.5.
136 NEC4 – Options B and D, clause 60.6, to either change how the method of measurement is applied, or resolve an inconsistency or ambiguity.
137 NEC4 – clause 27.3 provides that the contractor 'obeys' a valid instruction given by the Project Manager or the Supervisor. See also section 5.1.1 and Footnote 16.
138 Section 3.5.1 (Initial steps: the programme) and Tables 3.4,3.5 and 3.6.

or money, but to give the employer (or the Project Manager on his behalf) an opportunity to either avoid or mitigate any adverse effects of the event.

For JCT 2016 contracts, the requirement for the contractor giving prior warning or notice is used for events which may adversely affect the works' progress and entitle the contractor to loss and expense or additional costs. However, with NEC4, the prior warning requirement also applies to variations because, as Table 5.10 shows,[139] NEC4's compensation events include variations, along with matters giving rise to extensions of time and loss and expense (or additional costs).

In keeping with its core principles of mutual trust and co-operation,[140] and its emphasis on monitoring and managing the project,[141] NEC4 places the obligation to notify a compensation event on both the Project Manager and the contractor. Similar in procedure to the JCT 2016 contracts' optional 'quotation' procedures (refer to Section 5.4.3), NEC4 requires the contractor to submit 'quotations' for the Project Manager to assess, although of course, providing quotations in NEC4 is mandatory. In the context of 'variations', notifying a compensation event will occur as follows:

- Clause 61.1 provides that the Project Manager must notify the Contractor of a compensation event when issuing an instruction[142]; or
- Clause 61.3 obliges the Contractor to notify the Project Manager of a compensation event if firstly, the Contractor 'believes' the event is a compensation event, and secondly, if the Project Manager has not already notified the contractor in this regard.

The notification obligations in clauses 61.1 and 61.3 compliment each other in both the practical and contractual senses. It is fairly obvious that an instruction to increase the quantity of work will constitute a compensation event (as the Prices will increase), so it makes practical sense for the Project Manager (or Supervisor) to notify the Contractor accordingly. If the effects of a Project Manager's (or Supervisor's) instruction are less obvious and may not have been immediately apparent to the person issuing the instruction, then clause 61.3 permits the Contractor to notify the compensation event. The Contractor's obligation to notify under clause 61.3 would arise if the Supervisor had issued an instruction, but had not informed the Project Manager, who consequently failed to notify a compensation event under clause 61.1. If the Project Manager notifies the contractor of a compensation event,[143] clause 61.2 provides that he also instructs the contractor to submit quotations.

Clause 61.3 contains perhaps one of the most 'infamous' provisions found in any of the modern building contracts – namely, the 'time bar' clause that unless the contractor notifies the Project Manager that an event (or expected event) is a compensation event within eight weeks of becoming aware of the event, then he will not be entitled to any changes in the Prices, a Key Date or the Completion Date. Clause 61.3 is considered to be an exclusion clause, intended to limit the Client's liability to the contractor, if the contractor fails to comply. In order to be effective, exclusion clauses must be very clearly worded, contain no

139 Also refer to Table 2.1 in Chapter 2.
140 NEC4 – clauses 10.1 and 10.2.
141 The importance of the 'Accepted Programme' is set out in sections 8.6.2 and 8.6.2 above.
142 NEC4 – clause 61.1. The "instruction" can be given by the Project Manager or the Supervisor, but only the Project Manager notifies the contractor of a compensation event.
143 NEC4, clause 61.1.

ambiguities, and if such an ambiguity is present, then the clause will be interpreted against the party who drafted it, as Lord Bingham of Cornhill said in *Dairy Containers Ltd v Tasman Orient Line CV*[144]:

> *'The general rule should be applied that if a party, otherwise liable, is to exclude or limit his liability or to rely on an exemption, he must do so in clear words; unclear words do not suffice; any ambiguity or lack of clarity must be resolved against that party.'*

The wording of clause 61.3 seems clear when read alongside the NEC4 contract as a whole, and in particular, considering the way in which NEC4 is written. Clause 61.3 sets out:

- What the contractor must do[145];
- The period of time in which he must do it[146]; and
- The sanction imposed upon him if he fails to comply.[147]

The operation of the 'time bar' clause is contentious and has a number of perfectly valid arguments for and against. A contractor falling foul of the time bar provision will naturally feel aggrieved if this prevents him from recovering additional payment and time, for what would otherwise be a perfectly legitimate compensation event under the contract (for example, a Project Manager's instruction under 60.1(1) that adds significant work to the Scope). Conversely, a Client may feel aggrieved if he was prevented from effectively managing a compensation event because the contractor failed to issue either an early warning[148] or failed to notify the Project Manager. These circumstances may arise where the Client is a public body and faces deadlines or specific timescales for obtaining additional funding for the project.

Time bar clauses have a number of advantages and disadvantages as follows:

- Advantages:
  - Encourages the parties to deal with matters as they happen (or relatively soon after), instead of leaving them to the later stages or the end of the project;
  - Promotes the resolution of issues or conflicts during the course of the project; and
  - Contractor's notification obligations assist the Client with managing the compensation event (such as taking steps to mitigate delay or further time to allocate additional funding to the project).
- Disadvantages:
  - Perceived unfairness of the risk of losing a legitimate entitlement on a 'technicality' (failure to comply with an administrative procedure);
  - Can result in a lack of clarity whereby the contractor formats all communications as 'notices' to reduce the risk of failing to comply with the time-bar clause; and

144 [2005] 1 WLR 215.
145 NEC4 – clause 61.3. The contractor must notify the Project Manager that an event has occurred that has resulted, or may result in a compensation event.
146 NEC4 – clause 61.3. The contractor must notify the Project Manager within 8 weeks after becoming aware that the event has happened.
147 NEC4 – clause 61.3. The sanction is clear in that the contractor will not be entitled to any change in the Prices, the Completion Date or a Key Date.
148 NEC4 – clause 15 (Early warning).

- Can be require additional administrative resources for the contractor, the Project Manager and the Client.

The exceptions to the Contractor having to comply with the eight-week 'time bar' are as follows:

- The Project Manager (or Supervisor) issued an instruction, a certificate or has changed a previous decision which caused the event[149]; or
- It was reasonable for the Contractor not to 'believe' that the event is a compensation event.

Whilst not avoiding compliance with clause 61.3, the Contractor may claim that he did not reasonably become 'aware' of the event until later but must still notify within eight weeks of when he did. Clause 61.3's requirements for the contractor to 'believe' and become 'aware' that an event is a compensation event are subjective. It is thought that the Contractor would have to genuinely 'believe' that an event was not a compensation event and had genuinely not become aware of the event, in order to justify a failure to notify under clause 61.3. There may be some discrete arguments when a contractor could genuinely become aware of an event, or genuinely believe that an event is a compensation event. These may centre on such factors as the contractor's belief or awareness that may be triggered by the later consequences of the event, instead of the occurrence (or liklihood) of the event itself. Any deliberate avoidance by the contractor of either believing that the event is a compensation event or becoming aware of the event (or even its subsequent consequences) is unlikely to assist the contractor in justifying any failure to comply with clause 61.3.

Upon receipt of the contractor's clause 61.3 notification (of a compensation event), the Project Manager has one week (or such longer period as the contractor agrees) to issue one of the following replies[150]:

- The Project Manager confirms that the event (in the contractor's clause 61.3 notification) is a compensation event and instructs the contractor to submit quotations; or
- He may reject the contractor's clause 61.3 notification, either because it was late[151] or the notified event:
  - Is not one of the compensation events listed in clause 60.1;
  - Is either unlikely to, or is not going to occur, or will not affect the Defined Cost, Completion Date or a Key Date; or
  - Was the contractor's fault.

NEC's 'Managing Reality' refers to the above as the 'Five Point Test'.[152]

Clause 61.4 also places a 'time bar' on the Project Manager. If he fails to respond to the contractor's clause 61.3 notification within either one week (or a longer period if agreed with the contractor), the contractor notifies the Project Manager, and if a response isn't

149 NEC4 – clause 61.3. The NEC4 version of clause 61.3 is less subjective than its NEC3 predecessor. Clause 61.3 of NEC3 provided that the Contractor must notify a Compensation Event within 8 weeks, or would not be entitled to any changed Prices, Key Dates or changes to the Completion date, *'unless the Project Manager should have notified the event to the Contractor but did not.'*

150 NEC4 – clause 61.4.

151 NEC4 – The contractor failed to provide the clause 61.3 notification within 8 weeks.

152 NEC4 'Managing Reality' – Book 4 'Managing Change' (2018) 1.7.4 on page 22.

forthcoming for another two weeks,[153] then the Project Manager's failure to respond in time is firstly considered to be his acceptance that the notified event is a compensation event, and secondly, is treated as his instruction to the contractor to submit quotations.

If an event is either yet to occur, or has occurred and in both instances, it is impossible for the contractor to accurately forecast the effects of the event, then the Project Manager's clause 61.2 instruction[154] includes his assumptions,[155] upon which his assessment of the compensation event (under clause 63) will be based. If the Project Manager's assumptions prove to be incorrect, he corrects them,[156] which then becomes a compensation event under clause 60.1(17).

The contractor submits his quotation (or a revised quotation) within three weeks[157] of the Project Manager's instruction. The three-week period can be extended with the Project Manager's and contractor's agreement before the Project Manager's reply is due.[158] Clause 62.2 of NEC4 requires the contractor's quotation to include[159]:

- His proposed changes to the Prices (this can be an addition or a reduction in the case of an omission); and
- Any changes to the 'Accepted Programme' to show alterations to the Completion Date, and any Key Dates.

Unless the duration is extended (see above), the Project Manager must reply to the contractor's quotation within two weeks of receiving it, and his clause 62.3 reply must set out:

- That he accepts the contractor's quotation[160];
- That a revised quotation is to be submitted by the contractor, provided that the Project Manager gives his reasons why[161]; or
- The Project Manager will assess any changes to the Prices, Completion Date or any Key Dates.

As with the Project Manager's response to the contractor's clause 61.3 notification, another 'time bar' applies to his reply to the contractor's quotation; however, the Project Manager is given a greater opportunity to issue his clause 62.3 reply. If he fails to provide his clause 62.3 reply within two weeks (or such extended period as is agreed), then the contractor notifies the Project Manager under clause 62.6 of that failure. The contractor's clause 62.2 quotation[162] is treated as being accepted by the Project Manager, if he fails to reply within a further two weeks following the contractor's clause 62.6 notification.

153 NEC4 – clause 61.4. The 2 week period after the contractor's notification under clause 61.4 (i.e. of the Project Manager's failure to respond to the contractor's clause 61.3 notice) is not adjustable by agreement.
154 NEC4 – clause 61.2, the Project Manager's instruction to the contractor to submit quotations.
155 NEC4 – clause 61.6.
156 NEC4 – via a notification under clause 61.6.
157 NEC4 – clause 62.3.
158 NEC4 – clause 62.5.
159 NEC4 – clause 62.2.
160 NEC4 – The Project Manager's acceptance of the contractor's quotation is a 'notification' (refer to clause 13.7).
161 NEC4 – clause 62.4.
162 NEC4 – clause 62.6 or if the contractor has provided more than one quotation, the quotation nominated by the contractor in his clause 62.6 notification.

It appears that NEC4 maintains a similar position to JCT 2016 contracts, insofar as the Project Manager cannot instruct any compensation events to vary the Scope after the Completion Date. Clause 11.2(16) defines the 'Scope' as including valid instructions, and clause 11.2(2) defines 'Completion' as when the contractor has completed the Scope, implying that once the Scope of work is complete, no further instructions to vary can be issued. Clause 61.7 also provides that no compensation event is notified after the Defects Certificate is issued.[163] This should not be interpreted as permitting the Project Manager or Supervisor to issue instructions to vary the Scope after Completion, but only allowing the contractor to notify compensation events occurring during the 'defect correction period',[164] for example, those occurring under clauses 60.1(10),[165] 60.1(11)[166] and 60.1(16).[167]

Clause 63.6 sets out that the only rights that the contractor and client possess in terms of changing the Prices, the Completion Date or any Key dates are provided by the compensation event procedure. An initial reading may appear as if NEC4 intended to impose an 'exclusive remedies' clause by including clause 63.6. This would mean that the parties' rights to claim any common law damages for breach of contract by the other would be excluded, and their sole rights would be confined to the compensation event procedure. However, there are a number of points to consider regarding clause 63.6:

- In legal terms[168]:
  - In order to effectively take away any common law rights or remedies that a party may have, the clause in question should contain clear words to that effect; and
  - A lack of clarity or ambiguity in the clause is resolved against the party seeking to rely on it.

Comparing the wording of NEC4 clause 63.6 with, say, clause 44.4 of the MF/1 (1988) contract,[169] which in addition to restricting the parties' 'rights' restricted their 'obligations and liabilities', it appears that a party's liability to the other for common law damages may not be affected.[170]

- Clause 63.6 obviously refers to compensation events, which include any breach by the Client at clause 60.1(18). Therefore, the compensation event procedure would appear to permit the contractor to make a claim that he could otherwise make as a common law damages for the Client's breach, under the compensation event procedure in any event.
- If Secondary Option X18 (Limitation of liability) is selected, then arguably, clause 63.6 may not restrict the contractor's liability to the Client for matters such as consequential losses.[171]

163 NEC4 – issued by the Supervisor (clause 44.3).
164 NEC4 – Refer to Contract Data Part One (4 – Quality Management).
165 NEC4 – clause 60.1(10) – Supervisor instructs the contractor to search for a defect but none are found.
166 NEC4 – clause 60.1(11) – unnecessary delays are caused by a test or an inspection.
167 NEC4 – clause 60.1(16) – Client's failure to provide necessary facilities etc., for tests and inspections.
168 *Diary Containers Ltd v Tasman Orient Line CV* (2005) 1 WLR 215.
169 *Strachan & Henshaw Ltd v Stein Industrie (UK) Ltd and GEC Alsthom Ltd* (1997) 87 BLR 52.
170 However, a number of NEC4 commentaries suggest that similar wording in clause 25.3 effectively excludes the Client's rights to recover amounts, other than the 'additional cost' (as stated in that clause) if the contractor's work fails to achieve the condition to meet a Key Date.
171 NEC4 – Option X18 (Limitation of liability), clause X18.2.

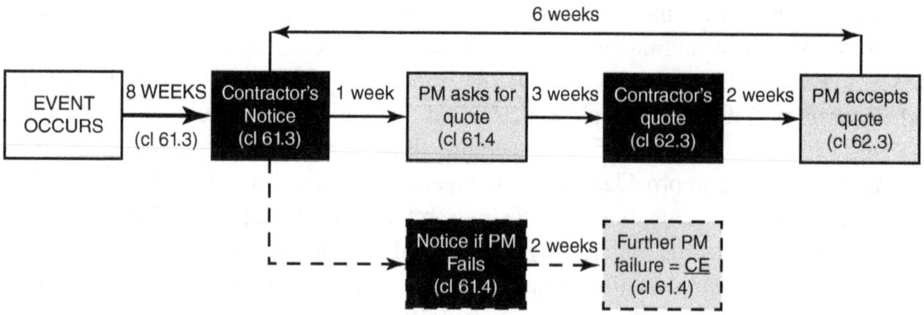


**Figure 5.1** NEC4 Compensation Event Procedure (contractor's submission).

Figure 5.1 summarises the Compensation Event procedure[172] which includes the contractor's quotation and valuation of the same (covered in Sections 5.3.2 and 5.5).

The Project Manager and the contractor may agree to extend[173] the usual periods for the contractor to provide his quotation[174] and/or for the Project Manager to provide his reply.[175]

Once the contractor has submitted his quotation, the Project Manager must respond. He may simply accept the contractor's quotation by notification,[176] instruct the contractor to provide an alternative quotation,[177] or that he will assess the compensation event himself.[178] NEC4 contains default provisions in the event that the Project Manager fails to respond to the contractor's quotation. Firstly, the contractor must notify the Project Manager of his failure to reply,[179] and if this failure continues for two weeks after the contractor's notification, then the contractor's quotation (or if an alternative quotation was submitted, the quotation specified in the contractor's notification) is deemed to be accepted.[179] The 'assessment procedure' is summarised in Figure 5.2. For a reconciliation of the compensation event timetables found in the NEC4 main contract, the NEC4 Subcontract and the NEC4 Short Subcontract, refer to Section 14.2.2 and Figure 14.3.

At first glance, clause 64 appears to echo the provisions of clause 62.3 (the Project Manager's reply), but it is an extension of that clause where the Project Manager is carrying out his own assessment. If the Project Manager does not provide his assessment within a period equivalent[180] to that in which the contractor was to provide his quotation, then the contractor may notify him of this failure.[181] If the Project Manager's failure to provide his

172 NEC4 – Figure 3 also shows the procedure for 'assessment' of Compensation Events – discussed further in section 8.9 (Valuation (or assessment) of variations).
173 NEC4 – clause 62.5.
174 NEC4 – clause 62.3 – the contractor ordinarily provides his quotation within 3 weeks after the Project Manager's instruction to submit a quotation (under clause 61.4).
175 NEC4 – clause 62.3, the Project Manager ordinarily replies to the contractor's quotation within 2 weeks.
176 NEC4 – clause 62.3 (first bullet point).
177 NEC4 – clause 62.3 (second bullet point). The Project Manager must provide the contractor with his reasons for instructing him to provide an alternative quotation (clause 62.4).
178 NEC4 – clause 62.3 (third bullet point).
179 NEC4 – clause 62.6.
180 NEC4 clause 64.3. The 'default' period for the contractor to submit his quotation is 3 weeks (clause 62.3) unless either (i) the Project Manager extends the time, or (ii) the Project Manager and the Contractor agree (clause 62.5).
181 NEC4 clause 64.4.

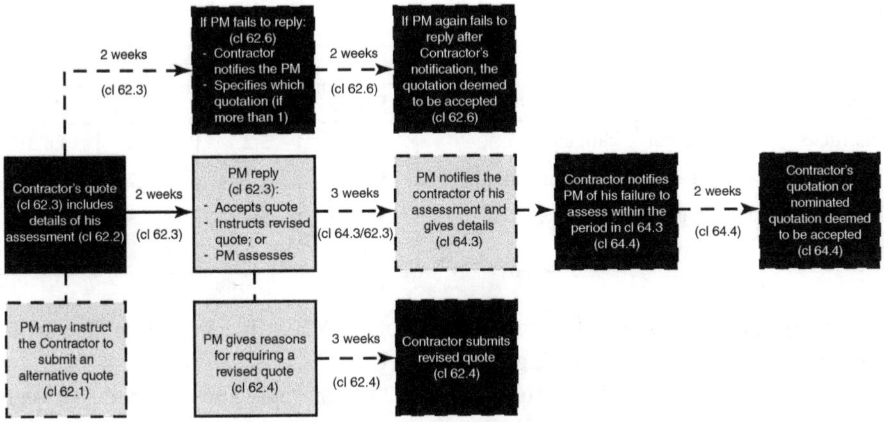


**Figure 5.2** NEC4 Compensation Event Assessment Procedure.

assessment continues for another two weeks following the contractor's notification, then the contractor's quotation (or nominated quotation if there have been more than one) is deemed to be accepted.[181]

Clause 66.1 provides that the compensation event is 'implemented' if the Project Manager:

- Accepts the contractor's clause 62.3 quotation (or it is deemed to have been accepted); or
- Notifies the contractor of his assessment under clause 64.3.

Once a compensation event is implemented, unless the contract provides otherwise, it cannot be revised (refer to Sections 5.3.2, 5.5 and 6.1.2).

## 5.4 Valuation Under JCT 2016 Contracts

Once a 'Variation' (or 'Change') has been established (see Section 5.2), the contractor must quantify its effects. He can do this in essentially two ways:

- Set out his entitlements when applying for an interim payment; or
- If particular procedures are to be followed – provide a 'quotation' to the employer or his representative.

The effects of a 'variation' upon the contractor's works usually fall into the following categories:

- Financial:
  - The value of carrying out a different quantity of work (see Sections 5.2.1 and 5.2.3); and/or
  - Any direct loss and/or expense arising from carrying out the variation (see Section 5.2.2); and
- Time: Any extension of time that the contractor requires (see Chapters 3 and 11), or if the progress of works is to be accelerated (refer to Sections 3.8 and 5.6).

There is a very important distinction to make between the ways that JCT 2016 contracts and NEC4 contracts approach a 'Valuation' of 'variations'.

**Table 5.11** JCT 2016: 'Valuation' Matters.

| 'Valuation' matter | JCT 2016 contract | | | |
|---|---|---|---|---|
| | **SBC/Q** | **DB** | **IC** | **MW** |
| All instructed 'Variations' (or 'Changes') | 5.2.1.1 | 5.2.1 | 5.2.1 | 3.6.3 |
| All work to be treated as a 'Variation' (or 'Change') | 5.2.1.2 | 5.2.2 | 5.2.2 | — |
| All instructed 'Provisional Sum' work | 5.2.1.3 | 5.2.3 | 5.2.3 | 3.6.3 |
| All 'Approximate Quantity' work | 5.2.1.4 | — | 5.2.4 | — |

The primary approach for a 'Valuation' under all of the JCT 2016 contracts is the parties to 'agree' the amount[182] of the following in Table 5.11.

Under JCT SBC/Q 2016, JCT DB 2016 and JCT IC 2016, the 'agreement' is made between the 'employer' and the 'contractor'. On the face of it, this encourages the contractor to by-pass the employer's representative[183] and agree the 'Valuation' directly with the employer. Given the perceived relative simplicity of the works under a JCT MW 2016 contract, it is perhaps surprising that 'agreement' of a valuation cannot be made directly between the contractor and employer, but only with the employer's representative (the Architect/Contract Administrator).

In the event that an 'agreement' over the matters in Table 5.11 cannot be reached, a 'Valuation' must be made using the contract's 'Valuation Rules' (shown in Table 5.12). Under the express terms of the particular JCT 2016 contract, the 'Valuation' is carried out by:

- The 'Quantity Surveyor' – JCT SBC/Q 2016 and JCT IC 2016;
- Not stated in JCT DB 2016 – but in practice, the contractor ought to apply the Valuation Rules when valuing the matters described in Table 5.11 when applying for payment,[184] and the employer (or the 'Employer's Agent, or a quantity surveyor acting on the employer's behalf) should also apply the Valuation Rules when stating the sum due at the due date[185] in a 'Payment Notice', or in a 'Pay Less Notice'[186]; or
- The 'Architect/Contract Administrator' – JCT MW 2016.

Under the 'Valuation Rules', the contractor's pricing (i.e. that comprises the 'Contract Sum') will be the primary reference for valuing the works, unless, the 'variation' is for something entirely new. This means that the contractor takes responsibility for the sufficiency of his price when pricing the original work, and this responsibility continues when valuing variations. This can sometimes lead to some extreme scenarios under JCT 2016 contracts, such as:

182 The 'Employer' and the 'Contractor' are to 'agree' the 'Valuation' under JCT SBC/Q 2016 (clause 5.2.1), JCT DB 2016 and JCT IC 2016 (clause 5.2), whereas clause 3.6.2 of JCT MW 2016 requires the 'agreement' to be between the Architect/Contract Administrator and the contractor.
183 The employer's representative will be the 'Architect/Contract Administrator' under JCT SBC/Q 2016 and JCT IC 2016. Whilst JCT DB 2016 refers to an 'Employer's Agent' (see Article 3) being able to act on the employer's behalf, there are relatively few express duties for him, but in practice, the Employer's Agent effectively assumes most of the employer's administrative duties under the contract.
184 JCT DB 2016 – the contractor submits its 'Interim Payment Application' under clause 4.7.3.
185 JCT DB 2016 – clause 4.7.5.
186 JCT DB 2016 – clause 4.9.5.

**Table 5.12** The JCT 2016 'Valuation Rules'.

| **'Valuation Rule'** | **JCT 2016 Contract ('pricing document' type) and clause** | | |
|---|---|---|---|
| | **SBC/Q[a)]** | **DB[b)]** | **IC** |
| *Allowance (addition or omission) to be made in Valuations for the relevant design work* | — | 5.4.1 | — |
| *If the 'additional' or 'substituted' work can be valued by measurement,[c)] and:* | | | |
| • Is of similar character<br>• Executed under similar conditions<br>• Quantity does not significantly change; then the rates and prices in the 'pricing document' are used for its valuation.[d)] | Contract Bills<br>5.6.1.1[e)] | CSA<br>5.4.2[d)] | Priced Document<br>5.3.1.1 |
| • Is of similar character, but is not<br>• Executed under similar conditions; and/or<br>• Significantly changes in quantity, then the rates and prices in the 'pricing document' shall be the 'basis' for valuation, but shall include a 'fair allowance' for differences in conditions, or quantities. | Contract Bills<br>5.6.1.2[e)] | CSA<br>5.4.2 | Priced Document<br>5.3.1.2 |
| • Is not of 'similar character' to that in the 'pricing document'<br>It shall be valued using 'fair rates and prices' | Contract Bills<br>5.6.1.3, 5.10 | CSA<br>5.4.2, 5.7.1 | Priced Document<br>5.3.1.3 |
| • The 'Approximate Quantity' is a reasonably accurate forecast of the actual quantity;<br>The rate or price for the 'Approximate Quantity' in the 'pricing document' shall be used for its valuation. | Contract Bills<br>5.6.1.4[f)] | — | Priced Document<br>5.3.1.4 |
| • The 'Approximate Quantity' is not a reasonably accurate forecast of the actual quantity;<br>The rate of price for the 'Approximate Quantity' in the 'pricing document' shall be the 'basis' for the valuation, but shall include a 'fair allowance' for the difference in quantity. | Contract Bills<br>5.6.1.5[f)] | — | Priced Document<br>5.3.1.5[f)] |
| *If work is 'omitted':* | | | |
| • The rates and prices in the 'pricing document' are used to value the omitted work | Contract Bills<br>5.6.2 | CSA<br>5.4.3 | Priced Document<br>5.3.2 |

*(Continued)*

**Table 5.12** (Continued)

| 'Valuation Rule' | JCT 2016 Contract ('pricing document' type) and clause | | |
|---|---|---|---|
| | SBC/Q[a)] | DB[b)] | IC |
| *If the 'additional' or 'substitute' work cannot be valued by measurement; then* • The valuation of work will be caried out on a 'daywork' basis. | 5.7 | 5.5 | 5.4 |

a) Refer to SBC/Q clause 5.9 – Change of conditions for other work.
b) An appropriate allowance for any addition or reduction in the relevant design. JCT DB 2016 – clause 5.4.1. Also see JCT SBC/Q – clause 5.8.1 for a variation in work in the Contractor's Designed Portion ('CDP').
c) When valuing any additional, substituted or omitted work, measurement shall be carried out in accordance with the 'Measurement Rules' applicable to the 'pricing document', allowances shall be made for (a) any 'percentages or lump sum adjustments' in the pricing document, and allowances (addition or reduction) in 'preliminaries' (save for expending a Provisional Sum for defined work) – JCT SBC/Q 2016 (clause 5.6.3), or JCT IC 2016 (clause 5.3.3). Under JCT DB 2016, clause 5.4.4 provides that the valuation shall allow for additions or reductions in provision of site administration or site facilities, including temporary works.
d) JCT DB 2016 – clause 5.4.2. It should be noted that the valuation must be 'consistent' with work of similar character in the CSA (whereas in JCT SBC/Q 2016 and JCT IC 2016, the rates and prices for work of similar character must 'determine' the valuation).
e) JCT SBC/Q 2016 – for a variation in the CDP work, refer to clauses 5.8.2 to 5.8.4 and clause 5.8.1 (Footnote *b* in Table 5.12 for the design allowance).
f) JCT SBC/Q 2016 – clause 5.6.1 provides that clauses 5.6.1.4 and/or 5.6.1.5 apply only if the sole difference is the quantity of the work subject to a 'Variation'. See also JCT IC 2016, clause 5.3.1 (in relation to valuation under clauses 5.3.1.4 and 5.3.1.5).

- Having priced a single item of work with a healthy profit margin, a variation increasing the quantity of this item would result in a very high profit being earned, whereas
- If the contractor priced a particular item of work at a loss, a variation increasing the quantity of that item would compound the contractor's losses.

The JCT 2016 contracts' approach should be contrasted with that of NEC4, which clearly intends that where a variation is instructed, it cannot be the contractor's 'fault', so consequently, the contractor should not lose financially – even if it priced the original work at a loss (refer to Section 5.5).

JCT 2016 contracts operate what they call the 'Valuation Rules' for valuing the physical work and any change in conditions under which that work is undertaken that arise from a 'variation'.

### 5.4.1 The 'Valuation Rules' and Their Application

In principle, the 'Valuation Rules' in JCT 2016 contracts remain fairly consistent across the SBC/Q, DB and IC versions. The provision for valuing work in MW is found in clause 3.6.3 and is effectively a condensed version of the 'Valuation Rules' found in the JCT SBC/Q, DB and IC 2016 contracts. JCT MW 2016 clause 3.6.2 firstly encourages the contractor and the employer's representative to agree a price for the instruction before the contractor carries it out. Failing agreement under clause 3.6.2, JCT MW 2016 clause 3.6.3 requires the work to be valued on a *'fair and reasonable basis using the any relevant rates and prices in the [priced document*[187]*]'*.

Table 5.12 summarises the valuation rules across the JCT SBC/Q, DB and IC 2016 contracts.

#### 5.4.1.1 Valuation by Measurement

If the contractor and the employer (or where applicable, his representative) do not agree an amount as the 'Valuation' of the matters set out in Table 5.11, then the 'Valuation Rules' apply (see Table 5.12). JCT SBC/Q 2016 and JCT IC 2016 both require any measurement of work to be carried out in accordance with the 'Measurement Rules'.[188]

In respect of each individual item of Work comprising the Valuation, the first question to be asked is whether the varied work it is capable of being measured? If so, then, the following three criteria must be considered which dictate the method of valuing the measurable work, namely, whether:

1) The work is of similar character to that work,
2) It is executed under similar conditions; and
3) If it significantly changes the quantity of work.

The Valuation Rules provide that

- If all three criteria are the same as those related to an item of work already set out in the relevant pricing document (see Section 5.2.3.1 and Table 5.4) – i.e. if the character and conditions are similar and the quantity is not significantly changed by the variation – the rate or price set out in the pricing document against that item must be used for the valuation of the variation.
- If the character is similar to that of an item of work set out in the relevant pricing document, but the conditions are not similar and/or the quantity is significantly changed, the rate or price in the relevant pricing document against that item must still be used as the basis of the valuation of the item but must be adapted so as to make a 'fair allowance' to allow for the changed conditions and/or quantity.
- If the work is not of similar character, the 'Valuation' must be made using fair rates and prices (the rates and prices in the relevant pricing document are no longer relevant).

It is clear that, if the character of the work to be measured is similar to that of an item which is already in the relevant pricing document, and to which a rate or price is fixed the Valuation must use those the rates and prices. A change to those rates and prices can be made only if there is a change in the conditions in which the work is carried out, or if there is a significant change in its quantity. It is only when the character of the measured work is not similar to anything in the relevant pricing document, that the Valuation is made using 'fair rates and prices'.

---

187 JCT MW 2016 – the 'priced document' is the 'Contract Specification/Work Schedules/Schedule of Rates' (clause 3.6.3).

188 JCT SBC/Q 2016 – clause 5.6.3.1 and JCT IC 2016 – clause 5.3.3.1. See Footnote 77 referring to the RICS' New Rules of Measurement.

Therefore, the key word in these rules (in points (1) and (2) above) is 'similar', and it is important that the meaning of the phrases 'similar character' and 'similar conditions' should be properly understood. The JCT 2016 contracts do not define 'similar', so presumably they rely upon its plain and ordinary meaning:

> *'Similar – adjective. Having a resemblance in appearance, character or quality, without being identical'.*[189]

Unfortunately, these phrases have caused some difficulty in practice, with various industry commentators interpreting them in different ways.[190] It is clear from the dictionary definition, that 'similar' does not mean 'identical'.

If the item is of 'similar character', it is also necessary to examine the relevant meaning of 'character':

> *'character – noun. The distinctive nature of something'.*[189]

In determining whether the relevant work is of similar character to work included in the relevant pricing document, it may be necessary to consider the differences in the nature of its constituent elements, namely, the material and workmanship required to construct it. If the material can readily be purchased from the same supplier and requires the same workmanship (i.e. using the same tools, discipline of tradesmen and methods), then there is a greater likelihood that the work is of 'similar character'.

The word 'similar' is also used to qualify the 'conditions' under which the relevant work is undertaken. Again, it seems clear that it does not mean 'identical' but has the same meaning as used to qualify 'character'. However, there are different considerations. It is possible to precisely define the 'character' of an item by its description in the relevant pricing document, but it is rarely possible to closely define anything but the basic conditions under which the work is to be carried out. Therefore, whether or not the conditions are 'similar' is likely to be influenced by considering what conditions the contractor ought to have reasonably anticipated in light of the available information when the contract was executed. The 'conditions' referred to in the valuation rules are the conditions to be derived from, and described in the relevant pricing document, along with the drawings, specification and any other relevant contract documents.[191] Therefore, the Valuation should not account of the actual conditions that prevailed when the contract was made, unless those conditions were sufficiently described in the contract documents. Many employers amend the terms of their contracts so the contractor is deemed to have inspected, and be aware of

189 Oxford Dictionary of English (3rd Ed) 2010.

190 *'If the instruction requiring a variation alters in any way [the] description of the work, it must therefore become different and may well in fairness be deserving of a different rate. To interpret the words in any other way will be to prevent the quantity surveyor from applying a fair valuation where the varied description of the item, although arguably remaining of a similar character to the original, justifies a different rate. Furthermore, the last few lines of clause 3.7.4 [similar to the last few lines of SBC clause 5.6.1.2] makes it clear that due allowance for any change in conditions is made only where the work has not been modified other than in quantity so that the character of the work itself must remain unchanged* Neil F Jones & Simon Baylis, *Jones & Bergman's JCT Intermediate Form of Contract*, 3rd edition, 1999, Blackwell Science at p. 172.

191 *Wates Construction (South) Ltd v Bredero Fleet Ltd* (1993) 63 BLR 128.

the conditions under which the work will be carried out, in order to avoid claims under this heading.

The word 'similar' is not used to describe a change in 'quantity'; the criterion used here being that quantity has been 'significantly[192] changed' by the variation. The change in quantity is likely to be relative, so a small change in quantity may be significant for items if their original quantity was small, but a larger change in quantity may not be significant in other circumstances. There are no precise rules, and it is a matter of fact and degree in each particular case. It is normal to assume that relatively large increases in quantity require reductions in the rate (for example, using economies of scale and/or if the work can be carried out in an easier way – by machine) and vice versa. However, that may not always follow.

It is not always appreciated that the rate or price in the relevant pricing document must be used as the basis of calculation of price, and it can be adjusted only to take account of changed conditions and/or significantly changed quantity. If a contractor's rate or price is far too low or far too high, there is no means of altering it, as JCT 2016 contracts state that the contract sum can be adjusted only in accordance with the express provisions of the contract.[193] Therefore, if the contractor has made a mistake and no one notices until after the contract is executed, the parties are left with the consequences. It has been suggested that, if a rate or price is clearly erroneous, the rate or price may be corrected, and changed to one that is reasonable before using it as a basis for the valuation. That is plainly incorrect.[194] The contractor has agreed under the terms of the contract to carry out variations to the Works, and the employer has agreed to pay for them on that basis. The Valuation Rules do not say that the rates and prices shall form the basis of valuation after the parties subsequently decide that the rates and prices are reasonable. The parties can only work with the rates and prices in the relevant pricing document and have no express power to change them. Neither the employer nor the contractor can avoid the consequences of rates and prices being too high or too low.

On occasion, a contractor may speculatively use a deliberately high rate or price for an single item, or an item of work for which a relatively low quantity is included in the relevant pricing document. In doing so, he may gamble on the quantity of that item significantly increasing, so he can earn additional profit. However, employers (or their quantity surveyors) are likely to check for anomalous rates and prices in the relevant priced document at the tender stage, so they should be alive to such pricing tactics. However, unless the employer is in a strong bargaining position, he can do little about it. It can simply be a part of a contractor's commercial strategy.[195]

Where a Valuation is to be made for the work that is not of 'similar character', a 'fair rates and prices' should be used. This seemingly precludes the common (and basic) quantity surveying practice of using a pro rata adjustment of the rates and prices in the relevant

192 *'Significantly – adverb. In a sufficiently great or important way as to be worthy of attention'* Oxford Dictionary of English (3rd Ed) 2010.

193 JCT SBC/Q 2016 and JCT IC 2016 – clause 4.2, and JCT DB 2016 – clause 4.1.

194 *Henry Boot Ltd v Alstom Combined Cycles Ltd* (1999) BLR 123. See also *Dudley Corporation v Parsons & Morrin Ltd*, 8 April (1959) CA unreported.

195 *Convent Hospital v Eberlin & Partners* (1988) 14 Con LR 1. The case went to appeal, (1989) 23 Con LR 112, but not on this point.

pricing document.[196] The Court considered the meaning of a 'fair valuation' in the case of *Weldon Plant v Commission for the New Towns*[197] as follows:

> '...*a fair valuation had to include each of the elements which are ordinarily to be found in a contract rate or price: elements for the cost of labour, the cost of plant, cost of materials, cost of overheads and profit. In my judgement a fair valuation has not only ordinarily to include something on account of each of these elements, but also it would not be a fair valuation within the meaning of the contract*[198] *if it did not do so.*'[199]

With regard to the matter of and addition for overheads, and the principle of the contractor not being penalised by a 'fair valuation', HH Judge Humphrey Lloyd QC continued:

> '...*the arbitrator did not deal with the addition which in my judgement has to be made in order to ensure that the contractor obtains a contribution from the costs of the business it undertakes towards its fixed or running overheads. For reasons that I have already given it would not be fair if the valuation did not include an element on account of such contribution. It would mean that such contribution would have to be found elsewhere, presumably from the contractor's margin for profit or risk. In my view a valuation which in effect required the contractor to bear that contribution itself would not be a fair valuation, in accordance with the principles of clause 52(1)*[200] *which are intended to secure that the contractor should not lose as a result of having to execute a variation (except, as I have stated, to the extent that its costs etc are of its making).*'

Where the Valuation includes the contractor's design, the relevant contracts also provide that allowances should be made for the following additions or omissions:

- The value of any design work[201]; and
- Site facilities (including administrative functions) and any temporary works.[202]

---

196 A simple example would be changing the thickness of a brick wall from 1 brick, to 2 bricks thick. It's unlikely that this change would be of similar character (as the wall's character and appearance would be changed), so in the absence of a provision requiring a valuation by 'fair rates and prices,' a pro rata valuation is made, doubling the original rate (or price) to reflect the doubling of the wall's thickness.
197 *Weldon Plant Ltd v Commission for the New Towns* (2000) EWHC 76 in which the contractor (Weldon) was appealing against the effects of an Arbitrator's award.
198 *Weldon Plant Ltd v Commission for the New Towns*. In this case, the contract was the ICE Conditions (6th Edition).
199 *Weldon Plant Ltd v Commission for the New Towns* (2000) EWHC 76 at paragraph 16.
200 JCT DB 2016, clause 5.4.2 is similar to clause 52(1) of ICE 6th Edition which states: '*Value of ordered variation: (1) The value of all variations ordered by the Engineer in accordance with clause 51, shall be ascertained by the Engineer after consultation with the Contractor in accordance with the following principles. (a) Where work is of similar character and executed under similar conditions to work priced in the Bill of Quantities it shall be valued at such rates and prices contained therein as may be applicable; (b) Where work is not of a similar character or is not executed under similar conditions...the rates and prices int eh Bill of Quantities shall be used as the basis for valuation so far as may be reasonable failing which a fair valuation shall be made....*'
201 JCT DB 2016 – clause 5.4.1, and JCT SBC/Q 2016 (for CDP work) – clause 5.8.1.
202 JCT DB 2016 – clause 5.4.4, JCT SBC/Q 2016 (for CDP work) – clause 5.8.4 (referring to clause 5.6.3.3).

A fair valuation should also be made for the 'Additional provisions' if any of the work (be that additions, substitutions or omissions of work) or other liabilities directly arising from a Change cannot reasonably be valued by measurement.

The rule for the valuation of omissions from the contract Works could hardly be simpler, as the Valuation Rules state that they are to be valued at the rates and prices set out in the relevant pricing document. However, there are certain circumstances where this may not appear to be a fair way of valuing omissions – particularly the case where the employer's representative (with the consent of the employer) has instructed the contractor not to make good any defects[203] in return for making an 'appropriate deduction' the contract sum. However, the operation of the Valuation Rules regarding omissions must not be overlooked and, if the omissions substantially change the conditions under which other work is executed, the resulting changed conditions for carrying out the relevant remaining work must be valued accordingly.

The Valuation Rules under JCT SBC/Q 2016 and JCT IC 2016 are also clear that, in carrying out valuation, the quantity surveyor must take into account several factors other than the rates and prices in the relevant pricing document for individual items, or valuation at fair rates and prices as the case may be.[204] When considering the valuation of additional or substituted work, or the omission of any work, or an instruction for the expenditure of a provisional sum for undefined work, the quantity surveyor must make allowance for any percentage or lump sum adjustments in the 'Contract Bills'.[205] The sums referred to are any such percentages or lump sums which are usually to be found in the general summary at the end of the bills. They must be applied pro rata to all prices for measured work and, therefore, to all variations.

JCT SBC/Q 2016's and JCT IC 2016's Valuation Rules also provide that the quantity surveyor is also required to make allowance,[206] when valuing and where appropriate for any addition to or reduction of preliminary items of the type referred to in the 'Measurement Rules'.[77] The clause does not specifically oblige the quantity surveyor to use the rates and prices set out in the bills against such items, but simply to make allowance for any addition to or reduction of such items. However, the quantity surveyor must be able to justify the method of calculating the allowance and, in practice, most quantity surveyors will use the rates and prices in the contract bills.

The Valuation Rules and clauses also include sections for the following:

- 'Change of conditions for other work'; and
- 'Additional provisions'.

The 'Change of conditions…' clauses[207] provide that a variation to one part of the Works can have another effect on the way in which other work (including where applicable, the contractor's design) must be carried out. It states that, where the introduction of a variation changes the conditions under which other work (which is not varied) is executed, the quantity surveyor must value that other work as if it had been varied. In practice, this

203 JCT SBC/Q 2016 – clause 2.38, JCT DB 2016 – clause 2.35, and JCT IC 2016 – clause 2.30.
204 JCT SBC/Q 2016 – clause 5.6.3, and JCT IC 2016 – clause 5.3.3.
205 JCT SBC/Q 2016 – clause 5.6.3.2, and JCT IC 2016–5.3.3.2.
206 JCT SBC/Q 2016 – clause 5.6.3.1, and JCT IC 2016 – clause 5.3.3.1.
207 JCT SBC/Q 2016 – clause 5.9, JCT DB 2016 – clause 5.6 and JCT IC 2016 – clause 5.5.

is likely to mean that the Variation of other work must be re-valued on the basis of the rates and prices in the relevant pricing document, against the appropriate items adjusted in respect of the changed conditions. It may also be necessary for the employer (or his quantity surveyor) to make allowance for other factors such as consequential changes in preliminary items or lump sum adjustments.

The 'Additional provisions' clearly exclude direct loss and/or expense from being included in the valuation of Variations or Changes.[208] It follows that, in making any allowances in respect of preliminary items, the employer (or his quantity surveyor) must not make allowance for the material effect of the particular variation on the regular progress of the Works. This permits the employer (or his quantity surveyor) to make allowance for the less than material effect upon regular progress. It should be noted that, in stating that an allowance must be made, the contract is deliberately giving the employer (or quantity surveyor) scope to do rather more, or less, than would be the case if the word 'ascertainment' had been used.

The 'Additional provisions' also require the employer (or his quantity surveyor) to make a fair valuation of any liabilities directly associated with a variation,[209] if the valuation cannot reasonably be carried out by the application of the other Valuation Rules or clauses.[210] There is no other restriction and the clause appears to oblige the quantity surveyor to make a fair valuation in such cases. Such liabilities might include the loss to the contractor where a variation to the work results in materials already properly ordered for the Works (i.e. as originally specified in the contract) becoming redundant. It will also include the valuation of the effect of any instruction which does not require the addition, omission or substitution of work, i.e. obligations or restrictions (see Section 5.2.2).

#### 5.4.1.2 If the Work Cannot be Measured – Valuation on a Daywork Basis

If the Variation (or Change) consists of additional or substituted work, but this cannot be valued by measurement, then it must be valued on a 'daywork' basis.[211] This essentially means that the value comprises the actual:

- Time spent by the personnel carrying out the work;
- Plant and equipment; and
- Materials or goods used;

on the proviso that the contractor shall provide a daily record of the above (commonly provided by a 'daywork sheet' or voucher[212]) for the employer, or his representative[213] to verify not later than seven 'Business Days'[214] after the work was carried out.

208 JCT SBC/Q 2016 – clause 5.10.2, JCT DB 2016 – clause 5.7.2 and JCT IC 2016 – clause 5.6.2.

209 JCT SBC/Q 2016 – clause 5.10.1, JCT DB 2016–5.7.1 and JCT IC 2016 – clause 5.6.1.

210 JCT SBC/Q 2016 – clauses 5.6 to 5.9, JCT DB 2016 – clauses 5.2 to 5.6, and JCT IC 2016 – clauses 5.3 to 5.5.

211 JCT SBC/Q 2016 – clause 5.7, JCT DB 2016 – clause 5.5 and JCT IC 2016 – clause 5.4.

212 The commonly-used description is a 'daywork sheet' although the all of JCT 2016 contracts refer to 'vouchers'.

213 JCT DB 2016 requires the daywork sheet (voucher) to be submitted to the 'Employer', whereas JCT SBC/Q 2016 and JCT IC 2016 refer to the 'Architect/Contract Administrator'.

214 In all JCT 2016 contract, a 'Business Day' is defined in clause 1.1 (Definitions and interpretation) as a day which is not a Saturday, Sunday, or a Public Holiday (itself defined in clause 1.1 as Christmas Day, Good Friday, or a day which under the Banking and Financial Dealings Act 1971 is a public holiday).

Typical examples of work falling into the 'daywork' category could be investigative works (e.g. to find buried services) or work to repair or commission a piece of equipment.

The valuation rules[211] provide that the basis is as follows:

- The 'prime cost' of the work (expressed as hourly rates) calculated in accordance with either of the following current at the Base Date[82]:
  - The 'Definition of Prime Cost of Daywork carried out under a Building Contract' by the RICS[215]; or
  - If the work is carried out by a specialist trade, definition of daywork given the body representing that trade[216]; plus
- The 'Percentage Additions' on the prime cost of those rates as set out in the contract; or
- The 'All-Inclusive Rates' for daywork.

The calculation of daywork rates from the first two bullet points (i.e. the rate in the RICS' or specialist body's daywork definition, plus the percentage addition), seems particularly long-winded and unnecessary. What is effectively a 'base rate' for daywork is provided in the relevant bodies' definition, and a percentage addition (expressed in either the 'Contractor Particulars'[217] or the relevant pricing document[218]) is then applied, so the calculated rate equates to the hourly rate that the contractor wants to be receive for carrying out 'daywork'.

It seems far easier to simply insert an 'All-Inclusive Rate' for daywork into the relevant part of the contract. This largely reflects the approach taken by professional services contracts, and more recently, by NEC contracts where a single rate for a professional or operative of a particular discipline or grade of seniority is set out in the contract.

The JCT 2016 contracts require the submission of daywork sheets (the contract refers to 'vouchers') and the associated timescale is not ideal. The relevant clauses states that the daywork sheets must be delivered to the employer or his representative for verification not later than seven 'Business Days[214]' after the work has been executed. Therefore, if the work is carried out on a Monday, the last date for submitting the daywork sheet will be Wednesday of the following week. The relevant clauses' intent is seemingly that the seven-day timescale refers to the contractor submitting (i.e. the 'delivery') of the daywork sheets, and not a period for 'verification' by the employer or his representative. Clearly, the employer or his representative must verify the properly submitted daywork sheets for each interim payment or stage payment.

The method of daywork 'verification'[219] is not stated in any of the JCT 2016 contracts. It is often common practice that a daywork sheet is signed by the employer, and this is taken to signify agreement that the stated time spent, and plant and materials used are correct. There is some force to this position, insofar as if the person to whom the sheet was submitted was unsure that the daywork sheet's information was correct, then he should not

215 Currently the 3rd Ed. JCT SBC/Q 2016 – clause 5.7.1, JCT DB 2016 – clause 5.5.1, and JCT IC 2016 – clause 5.4.1.

216 JCT SBC/Q 2016 – clause 5.7.2, JCT DB 2016 – clause 5.5.2 and JCT IC 2016 – clause 5.4.2.

217 JCT DB 2016 – the 'Contract Particulars' entry against clause 5.5 (Daywork).

218 JCT SBC/Q 2016 – the 'Percentage Additions' are set out in the 'Contract Bills', and for JCT IC 2016, the 'Percentage Additions' are set out in the 'Priced Document' (Fourth Recital) which can be Bills of Quantities, Specification, Works Schedules, or Schedule of Rates.

219 'Verification' meaning 'the process of establishing the truth, accuracy or validity of something'. Oxford Dictionary of English (3rd Ed) 2010.

sign it. The signing of daywork sheets is therefore a matter of agreeing evidence, and not (as some contractors misunderstand), an agreement that the work is to be valued on a 'daywork basis'. The often used phrase 'for record purposes only' provides the signatory with little assistance. Firstly, it is a matter of fact and application of the contract's valuation rules whether the work will be valued on a daywork basis. Secondly, of course the signature is for 'record purposes only' as it provides a form of evidence (although perhaps not irrefutable evidence) that the daywork sheet is correct[220]! In these circumstances, it appears that the employer's representative (including his quantity surveyor) has no right to substitute his own opinion for the hours and other resources described on the daywork sheets.[221] It has been held that where the employer has set out a system of verification by signing, but has neglected to do so, the sheets will stand without further proof as evidence of the work done unless they can be shown to be inaccurate.[222]

There are now many means for the employer (or his representative) to 'verify' a daywork sheet. If the site is using an electronic, or fingerprint entry system, each operative will (often part of his site induction) have to register so he can enter and exit the site. The entry system will provide evidence of how long each operative was present on site, so in the event that the operative was only on site for 8 hours, but a daywork sheet says he worked for 12 hours, then the entry systems' evidence can be used for verifying the hours worked for him, or any other operatives. In a similar but less precise way, plant return and material delivery evidence can also be relied upon to a degree, when verifying dayworks.

It is has previously been suggested that the contractor ought to give 'advance warning' to the employer and his representative that he intends to submit daywork sheets. However, in practice, this is unlikely to happen because (a) the contractor has no obligation to do so, and (b) it may be a sub-contractor submitting daywork sheets that alerts the contractor that a Variation (or Change), or it is reasonable to expect that the work can be valued by measurement.

Given that (1) the employer, or his representative are afforded some protection from a 'daywork ambush' by the seven-day submission rule, (2) that only 'Variations' (or Changes) will give rise to dayworks and (3) they will have been instructed by the employer or his representative, then they should already be alerted to the likelihood that the varied work cannot be valued by measurement, there is little room for complaint if the contractor submits daywork sheets for verification.

#### 5.4.1.3 Valuation of the Contractor's Design (Including the 'Contractor's Designed Portion') and 'Provisional Sums'

In any 'Valuation' carried out under JCT DB 2016, clause 5.4.1 clearly provides that an appropriate allowance (an addition or an omission) should be made for the relevant design work. It appears that only design work related to a 'Change' can be valued. This emphasises the importance, not only of making sure that there is a rate or price for design work in the Contract Sum Analysis, but also that there is a change with which the design work can be associated. The contractor should also be aware of carrying out design work for changes which may be abortive. The contractor may be unable to claim the cost of such work, because it will not be 'relevant' or connected to any change.

220 *Inserco v Honeywell* 19 April 1996, unreported.
221 *Clusky (trading as Damian Construction) v Chamberlain*, Building Law Monthly, April 1995, p. 6.
222 *JDM Accord Ltd v Secretary of State for the Environment, Food and Rural Affairs* (2004) CILL 2067.

Valuation of the contractor's designed portion ('CDP') is only found in JCT SBC/Q 2016, where the CDP has been selected. Clause 5.8 deals with valuation of the CDP work. There is an overall stipulation[223] that clauses 5.6.3.2 (percentage or lump sum adjustments), 5.6.3.3 (adjustment of preliminary items), 5.7 (daywork) and 5.9 (change in conditions for other work) will apply to the CDP work if relevant. However, the nature of this work makes the inclusion of specific provisions necessary. Therefore, clause 5.8.1 requires an allowance to be made for the addition or omission of design work. A prudent contractor will include a basis for valuing design work in its CDP analysis (e.g. providing hourly rates for particular design personnel). In addition, the valuation of any variations to the CDP work must be consistent with similar character of work in the CDP analysis. Echoing clause 5.6.1.2, allowance is to be made for any change in conditions or significant change in quantity. The comments already made with regard to 'similar' and significant change in quantity also apply here (refer to Section 5.4.1.1). If there is no work of a similar character, the quantity surveyor is to make a fair valuation. Echoing clause 5.6.2, the valuation of omissions is to use the values in the CDP analysis.

The differences between 'Provisional Sums' for 'defined work' and 'undefined work' are summarised in Section 5.2.3.2.

Provisional sums for defined work are only found in contracts operating a bill of quantities that have been prepared in accordance with the contract's 'Measurement Rules' such as JCT SBC/Q 2016 and JCT IC 2016, which now operate the RICS' 'New Rules of Measurement – Detailed Measurement for Building Works (NRM2)'[224] ('NRM2'). JCT DB 2016 no longer includes a provision for 'bills of quantities' so any provisional sums under this contract will,[225] by definition, be for undefined work. In order to qualify as a provisional sum for defined work, the bills of quantities must provide very specific information in relation to the provisional sum – effectively, the same level of information that would otherwise be provided in the description of any other item in the bills of quantities. This information (listed below) is to enable the contractor to make '*due allowance in his…programming, planning and pricing preliminaries.*'[226] Three very obvious effects of instructing the expenditure of a provisional sum for defined work are as follows:

- that the contractor is not entitled to an extension of time for carrying it out,[227] as his programme is deemed to allow for its duration;
- The contractor is not entitled to a direct loss and/or expense[228]; and

223 JCT SBC/Q 2016 – clause 5.8.4.

224 'Measurement Rules' are defined as the 'RICS New Rules of Measurement – Detailed Measurement for Building Works (NRM2)' in clause 1.1 (Definitions) of JCT SBC/Q 2016 and JCT IC 2016.

225 JCT DB 2016 – clause 1.1 (Definitions and interpretation) provides that 'Provisional Sums' are to be included in the Employer's Requirements. Therefore, there is no provision for the inclusion of provisional sums in the Contractor's Proposals or the CSA, and any such sum set out in those documents, must also be set out in the Employer's Requirements, otherwise (1) the employer will be powerless to deal with them and (2) it is arguable that the contractor will be entitled to payment of the sum (which is likely to be deemed a fixed price) without carrying out the relevant work.

226 RICS NRM 2 – Rule 2.9.1.3.

227 JCT SBC/Q 2016 – clause 2.29.2.1 and JCT IC 2016 – clause 2.20.2.1. In both cases, the 'Relevant Event' confirms that instructions for the expenditure of a 'Provisional Sum for defined work' is excluded.

228 JCT SBC/Q 2016 – clause 4.22.2.1, and JCT IC 2016 – clause 4.17.2.1. In both cases the 'Relevant Matter' confirms that instructions for the expenditure of a 'Provisional Sum for defined work' is excluded.

- No allowance for adjusting preliminary items is made when valuing the work required to comply with an instruction to expend a Provisional Sum for defined work.[229]

Rule 2.9.1.2 of NRM2 sets out that the following information shall be provided for every Provisional Sum for defined work:

- The nature and construction of the work;
- A statement of how and where the work is fixed to the building and what other work is to be fixed thereto;
- A quantity or quantities which indicate the scope and extent of the work; and
- Any specific limitations identified.

However, if the 'provisional sum' work does not contain the information set out in NRM2's rule 2.9.1.2, it is for undefined work, and the contractor is deemed not to have made any allowance for it in programming and pricing preliminaries. This addresses a difficulty which contractors and employer's representatives have often misunderstood. The fact is that a contractor will be unable to make any firm attempt to programme or price preliminaries to deal with a provisional sum for undefined work – one which may be little more than a title and a figure. For example, a provisional sum described simply as 'mechanical installation – £10,000' contains none of the information required by NRM2 and is almost useless as a description.

Where work can be described in items in accordance with the Measurement Rules, but accurate quantities cannot be given, an estimate, called an 'Approximate Quantity' must be given (refer to Section 5.1.3 and Table 5.2). The valuation of work for which an approximate quantity is included in the contract bills is covered by the Valuation Rules.[230] Where the approximate quantity is a reasonably accurate forecast, the valuation must be in accordance with the rates and prices for the approximate quantity. If it is not a reasonably accurate forecast, the rate or price forms the basis for the valuation, but the employer (or his representative or quantity surveyor) is to make a fair allowance for the difference in quantity.

Under the Valuation Rules and their associated clauses,[231] if the contractor's compliance with an instruction to expend a provisional sum

- for undefined work; or
- for defined work to the extent that the instruction is different from the description in the contract bills; or
- for the execution of work for which an inaccurate approximate quantity is in the contract bill

substantially changes the 'Conditions for other work', the other work is to be treated as though it was the subject of an instruction for a variation. This clause simply brings this kind of item under the same rules as variations to measured work. The differences highlight the extent of the contractor's knowledge about the kind of work and the amount at the time the contract was made.

---

229 JCT SBC/Q 2016 – clause 5.6.3.3, and JCT IC 2016 – clause 5.3.3.3.
230 JCT SBC/Q 2016 – clauses 5.6.1.4 and 5.6.1.5, and JCT IC 2016 – clauses 5.3.1.4 and 5.3.1.5.
231 JCT SBC/Q 2016 – clauses 5.9.1 to 5.9.4, and JCT IC 2016 – clauses 5.5.1 to 5.5.4.

### 5.4.2 Variation and Acceleration Quotations and Contractor's Estimates

JCT SBC/Q 2016 and JCT DB 2016 contain provisions for the employer (or his representative) to request the contractor to provide 'quotations' for the following matters:

- JCT SBC/Q 2016:
  - 'Variation Quotation' under clause 5.3 and Schedule 2;
  - 'Acceleration Quotation' under Schedule 2; and
- If the relevant 'Supplemental Provisions' are selected in JCT DB 2016's 'Contract Particulars',[232] the 'Contractor's estimates' for the following must be provided:
  - Valuation of Changes[233];
  - Loss and Expense[234]; and
  - Acceleration.[235]

The contractor may include sums for loss and expense (refer to Section 12.2.7) in the Variation Quotation[236] and Acceleration Quotation[237] in JCT SBC/Q 2016, and in the contractor's estimates for 'Valuation of Changes[238]' and the Acceleration Quotation[239] (as well as the estimate for 'Loss and Expense') under JCT DB 2016.

The procedures and timescales governing the submission and acceptance of 'Variation Quotations' (under JCT SBC/Q 2016) and 'Acceleration Quotations' (under JCT SBC/Q 2016 and JCT DB 2016) are similar and are illustrated in Figure 5.3.

The process for either a 'Variation' or 'Acceleration' quotation begins with the contractor either being instructed,[240] or invited[241] to provide the Variation Quotation or Acceleration Quotation, respectively. The contractor may respond as follows, in the event that he considers that:

- In the case of a 'Variation Quotation', insufficient information has been provided to him, he may notify the Architect/Contract Administrator within seven days of the instruction[242]; or

---

232 JCT DB 2016 – refer to the entries in the Contract Particulars for the Seventh Recital (Schedule 2). If no option ('applies/does not apply') is selected for Supplemental Provisions 1 to 3, then they do not apply. Conversely, if neither of those option is selected for Supplemental Provisions 4 to 10, then they apply.

233 JCT DB 2016 – entry for Seventh Recital – Part 1 of Schedule 2 – Supplemental Provision 2 (Valuation of Changes – Contractor's estimates).

234 JCT DB 2016 – entry for Seventh Recital – Part 1 of Schedule 2 – Supplemental Provision 3 (Loss and expense – Contractor's estimates).

235 JCT DB 2016 – entry for Seventh Recital – Part 2 of Schedule 2 – Supplemental Provision 4 (Acceleration Quotation).

236 JCT SBC/Q 2016 – see clause 5.3, and Schedule 2 (Variation and Acceleration Quotation Procedures) paragraph 1.2.3.

237 JCT SBC/Q 2016 – see clause 5.3, and Schedule 2, paragraph 2.1.1.

238 JCT DB 2016 – see Schedule 2 (Supplemental Provisions) – paragraph 2.3.1.

239 JCT DB 2016 – see Schedule 2, paragraph 4.1.2.

240 Variation Quotation – instruction by the Architect/Contract administrator under JCT SBC/Q 2016 – clause 5.3.1 and paragraph 1.1 of Schedule 2.

241 Acceleration Quotation – invitation by the Architect/Contract Administrator under JCT SBC/Q 2016 – paragraph 2.1 of Schedule 2, or by the employer under JCT DB 2016 – paragraph 4.1.1 of Part 2 of Schedule 2.

242 JCT SBC/Q 2016 – paragraph 1.1. of Schedule 2.

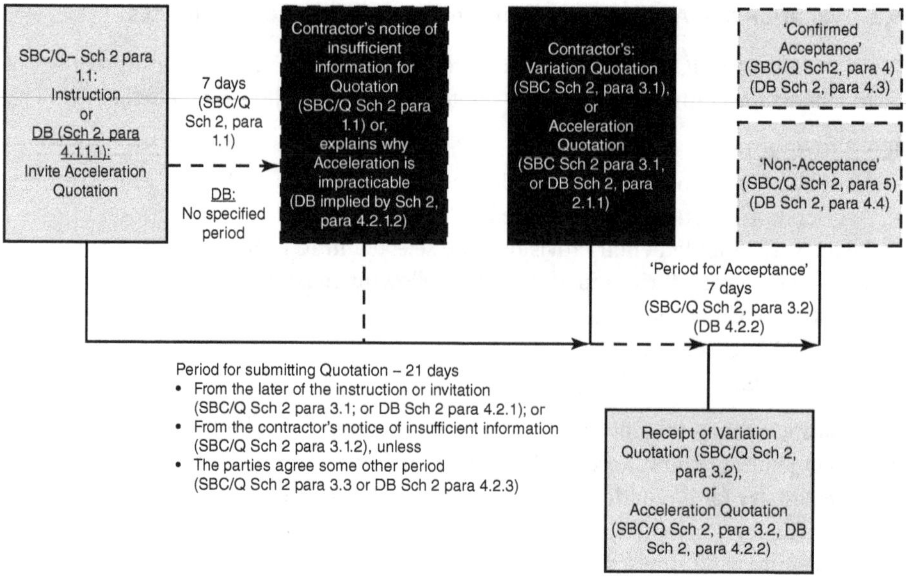


**Figure 5.3** JCT Variation and Acceleration Quotation Procedures.

- In the case of an 'Acceleration Quotation', the contractor explains why it's impractical to accelerate progress.[243]

In each case the contractor should then be provided with sufficient information from which to compile the relevant quotation,[244] and the period for submitting the relevant quotation will be 21 days from either (1) receipt of that further information[245] or (2) if the contractor gave no notification of insufficient information, or explanation why accelerating progress is impractical - 21 days from the date he received the instruction or invitation to provide the relevant quotation.[246]

It appears that the contractor cannot withdraw the relevant quotation before its acceptance (as it could otherwise do in the course of ordinary negotiations) because in this instance, the contractor is bound by the contract terms to keep its offer open. The relevant quotation must contain the information set out in Table 5.13.

The procedure under JCT DB 2016 for the 'Valuation of Changes – Contractor's estimates' is similar – as shown in Figure 5.4.

The information that the contractor must include in his estimate is summarised in Table 5.14.

If the contractor fails to comply with the 14-day time period for submitting his estimate, the clause 3.9 instruction shall be assessed in accordance with clauses 2.23 to 2.25

243 JCT SBC/Q 2016 – paragraph 2.1.2 of Schedule 2, and JCT DB 2016 – paragraph 4.1.1.2. Unlike the 7 day period specified in Footnote 242, no period is specified for the contractor to provide its explanation.

244 Variation Quotation – JCT SBC/Q 2016 – paragraphs 1.1 and 3.1.2 of Schedule 2.

245 Variation Quotation – JCT SBC/Q 2016 – paragraph 3.1.2 of Schedule 2, or Acceleration Quotation – JCT SBC/Q 2016 – implied by the following paragraph: (i) JCT SBC/Q 2016 – paragraph 3.1.2 of Schedule 2, or (ii) JCT DB 2016 – paragraph 4.2.1.2 of Part 2 of Schedule 2.

246 JCT SBC/Q 2016 – paragraph 3.1.1 of Schedule 2 for the Variation Quotation or Acceleration Quotation, or JCT DB 2016 – paragraph 4.2.1.1 of Part 2 of Schedule 2.

**Table 5.13** Variation and Acceleration Quotation Information.

| Information to be included | Variation Quotation | Acceleration Quotation |
|---|---|---|
| The amount by with the Contract Sum is adjusted (including calculations[a)]) | SBC/Q – Para 1.2.1 | SBC/Q – Para 2.1.1<br>DB – Paras 4.1.1.1, 4.1.2 |
| Any adjustment to time[b)] | SBC/Q – Para 1.2.2 | SBC/Q – Para 2.1.1<br>DB – 4.1.1.1 |
| Amount of loss and expense | SBC/Q – Para 1.2.3 | SBC/Q – Para 2.1.1<br>DB – Para 4.1.2 |
| Quotation's reasonable preparation costs | SBC/Q – Para 1.2.4 | SBC/Q – Para 2.1.1<br>DB – Para 4.1.2 |
| Indicative resources | SBC/Q – Para 1.2.5.1 | — |
| Method of working | SBC/Q – Para 1.2.5.2 | — |
| Other conditions | — | SBC/Q – Para 2.1.1<br>DB – Para 4.1.1.1 |

a) JCT SBC/Q 2016 – Calculations shall include reference to the Contract Bills' rates and prices, and allowances for adjusting preliminary items, respectively where relevant and appropriate.
b) For the Acceleration Quotation – both JCT SBC/Q 2016 and JCT DB 2016 must show the 'time that can be saved'.

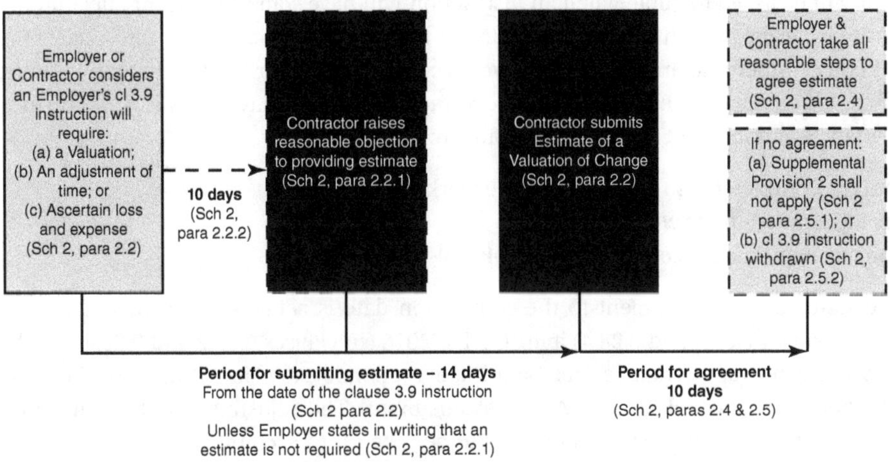


**Figure 5.4** JCT DB 2016 – 'Valuation of Changes – Contractor's Estimates.'

(extension of time), 3.9 (a Change[247]) and 4.20 (loss and expense), but any amounts that would otherwise be additions to the Contract Sum shall not be included in any Interim Valuations, and will only be included with the relevant Final Statement[248] setting out the adjustments to the Contract Sum under clause 4.2.[249] This is perhaps one of the most

247 JCT DB 2016- a clause 3.9 instruction is valued in accordance with clause 5.2.
248 JCT DB 2016 – the 'relevant Final Statement' is either the 'Final Statement' provided by the contractor under clause 4.24.1, or in default (as provided in clause 4.24.3) the 'Employer's Final Statement' under clause 4.24.4.
249 JCT DB 2016 – Schedule 2, para 2.6.

**Table 5.14** JCT DB 2016 – Contractor's Estimate Information.

| Information to be included | Valuation of Change Contractor's Estimate |
|---|---|
| (1) The amount by with the Contract Sum is adjusted (including calculations) | Schedule 2, para 2.3.1 |
| (2) Any additional resources | Schedule 2, para 2.3.2 |
| (3) Method of working (statement) | Schedule 2, para 2.3.3 |
| (4) Any extension of time | Schedule 2, para 2.3.4 |
| (5) Amount of loss and expense | Schedule 2, para 2.3.5 |

significant uses of a 'time bar' clause in JCT 2016 contracts. The contractor's failure to submit the estimate within the 14-day period is likely to have serious consequences for him, although unlike the time bar clause 61.3 in NEC4 (refer to Section 5.3.2), the contractor does not permanently lose his entitlements; instead, he faces a delay in receiving them. In the event that the contractor's estimate contains significant values for carrying out the clause 3.9 instruction, and/or for loss and expense, he may have to wait for over a year[250] after practical completion, to be paid. This is likely to have a seriously detrimental effect on his cashflow.

The employer has an important role to play in this process, probably because the contractor will be asked to quote when an instruction will have some significant effect on the contract in terms of additional expenditure or time. It is for the employer to accept the quotation or otherwise and, if the employer accepts, then he (or his representative) must confirm his agreement in writing to the contractor within 10 days. This is a 'Confirmed Acceptance' and its effect is to formally confirm:

- that the contractor is to proceed with the work;
- the adjustment to the contract sum;
- and any new date for completion (if applicable) to be fixed.

The status of the adjustment to the completion date is acknowledged in JCT SBC/Q 2016 (in clauses 2.26.2 and 2.28.4) and JCT DB 2016 (in clauses 2.23.2 and 2.26.4), so the employer (or his representative) need not issue a separate extension of time for the same instruction as it is treated as a 'Pre-Agreed Adjustment'. The adjustment to the completion date noted in the confirmed acceptance is to be taken into account by the employer (or his representative) in the normal way when considering further extensions of the contract period.

For a 'Variation Quotation' under JCT SBC/Q 2016, if work carried out under a Confirmed Acceptance is subsequently varied by the employer or his representative, clause 5.3.3 requires it to be valued on a fair and reasonable basis, but having regard to the figures in the contractor's quotation. Clause 5.3.3 also includes a requirement to include any direct loss and/or expense resulting from compliance with such instruction. It appears that

250 JCT DB 2016 – it is likely that the earliest that final payment (as set out in the relevant Final Statement – see Footnote 248) becomes due will be 1 month after the expiry of the Rectification Period (the 'default' period is 6 months after practical completion, but more commonly, it is 12 months).

this is just for clarity, as this type of variation would seem to be a relevant matter under clause 4.22.

If no agreement over the contractor's quotation is reached within 10 days (i.e. no 'Confirmed Acceptance' is issued), the employer (or his representative) must either instruct that:

- the variation is to be carried out and valued under the Valuation Rules[251]; or
- instruct that the variation is not to be carried out (JCT SBC/Q 2016), or is withdrawn (JCT DB 2016).

There seems to be no provision for the employer or the quantity surveyor on behalf of the employer to negotiate on the quoted price. It is either to be accepted or rejected within the 10-day time limit. If no Confirmed Acceptance is issued, and the employer (or his representative) either instructs that work is not to be carried out (JCT SBC/Q 2016) or withdraws the instruction (JCT DB 2016), the contractor is entitled to be paid for the preparation of his quotation or estimate:

- JCT SBC/Q 2016 (schedule 2, paragraph 5.2) provides that a fair and reasonable amount must be added to the contract sum to represent the cost of preparation, whereas
- JCT DB 2016 (schedule 2, part 1, paragraph 2.5) includes a similar provision (which includes the cost of any additional design work) but clarifies that other than paying for the costs of preparation, withdrawing the instruction will be at no cost to the employer.

Both JCT SBC/Q 2016 and JCT DB 2016 clearly state that if the employer does not accept the quotation, the quotation cannot be used for any purpose at all.[252] Therefore, the quotation or estimate must be treated as though it had never been submitted. This is to avoid the employer (or his representative, including the quantity surveyor) using the submitted quotation or estimate to value other variations or Changes, or to assess any extension of time or loss and expense.

## 5.5 Valuation Under NEC4 Contracts (Including 'Contractor's Share')

As with JCT 2016 contracts (where variations are valued by the employer or his representative), the Project Manager assesses the compensation event under NEC4. The Project Manager's assessment can be carried out in the following ways:

- Under clause 63.2, the Project Manager and the contractor can effectively 'agree' the change in the Prices arising from the compensation event by agreeing lump sums or rates; however, there is no provision for them to 'agree' the length of time by which the Completion Date or any Key Dates are delayed[253];

251 JCT SBC/Q 2016 Schedule 2, paragraph 5 states that the variation is '*...to be valued under the Valuation Rules (clauses 5.6 to 5.10)*' whereas JCT DB 2016, Schedule 2, Part 1, paragraph 2.5.1, states that '*...Supplemental Provision 2 shall not apply in respect of this instruction*' – i.e. applying the Valuation Rules for valuation.

252 JCT SBC/Q 2016 – Schedule 2, paragraph 5.3, and JCT DB 2016, Schedule 2, part 1, paragraph 2.5.

253 NEC4 – clause 63.5 sets out the method for assessing delays to the Completion Date.

- If the contractor provides a clause 62.1 quotation within the allotted period from the Project Manager's request,[254] the Project Manager assesses this in accordance with clause 63, unless the following occur (as provided in clause 64.1):
  - He decides that the contractor's clause 62.1 quotation does not properly assess the compensation event, and no revised clause 62.1 quotation has been instructed; or
  - If the contractor fails to include a programme (showing the relevant changes) with his clause 62.1 quotation; or
  - The Project Manager has not accepted the contractor's programme,[255] when he provides the clause 62.1 quotation.

  in which case, the Project Manager assesses the compensation event himself.

- Clause 64 provides that the Project Manager makes his own assessment of a compensation event if the contractor;
  - Fails to provide his clause 62.1 quotation within the time allowed[256]; or
  - If the contractor has defaulted regarding the programme,[257] by either failing to provide an Accepted Programme,[258] failing to submit a revised, or altered programme for acceptance[259]; or the contractor's submitted programme falls foul of one (or more) of the reasons that the Project Manager may refuse to accept the programme.[260]

  The Project Manager assesses the effects of a compensation event as follows:

- Changes to the Prices (where not agreed under clause 63.2):
  - Additions (or increases) are dealt with under clauses 63.1;
  - Omissions (expressed as reductions in the Prices) are covered under clauses 62.3 and 62.4; and
- Delays to the Completion Date and/or to Key Dates are covered under clause 63.5.
- The Project Manager must also consider applying '*risk allowances for cost and time matters*' under clause 63.8.

The Project Manager must notify the contractor of his assessment within the same period as the contractor was given for providing his quotation.[261] The period for his assessment is stated to begin as soon as the '*...Project Manager's assessment becomes apparent*', so in practice, this is likely to start when one of the four events in clause 64.1 occurs, or where he

254 NEC4–3 week sunder clause 62.3, or within any period extended by the Project Manager under clause 62.5.

255 NEC4 – The Project Manager does not accept the contractor's programme for one of the reasons in clause 31.1.

256 NEC4 – clause 64.1.

257 NEC4 – clause 64.2.

258 NEC4 – clause 11.2(1) definition of 'Accepted Programme.'

259 NEC4 – Some examples: the contractor may have failed to provide a first programme for acceptance within the period set out in the Contract Data (clause 31.1), or within the 'period for reply,' or a period longer than the interval stated in the Contract Data (clause 32.2).

260 NEC4 – The Project Manager does not accept the contractor's programme for one of the reasons in clause 31.3 – see section 8.6.3 and Table 26.

261 NEC4 – clause 64.3. Usually, the contractor provides his quotation within 3 weeks of the Project Manager's instruction to do so (clause 62.3) unless the Project Manager extends the period under clause 62.5.

notifies the contractor that he will make his own assessment.[262] It appears that the Project Manager is obliged to provide details of his assessment,[263] although they may not necessary to accompany his clause 64.3 notification, but must be provided within the relevant period. If he fails to do so, the contractor notifies the Project Manager, giving him a further two weeks in which to assess the compensation event, failing which the contractor's quotation (or if more than one quotation was submitted, the contractor notifies the Project Manager which one should be used) is deemed to be accepted.[264]

Clause 66.1 provides that once the contractor's quotation is accepted (either by the Project Manager notifying the contractor of his acceptance,[265] or because he failed to reply on time,[266] or after the Project Manager notified the contractor of his own assessment[267]), then the compensation event is 'implemented'. Implementing a compensation event almost irreversibly[268] changes the Prices, the Completion Date (and any Key Date),[269] and clause 66.3 provides that a revised assessment of these changes cannot be made unless the following occur:

- The Project Manager corrects an assumption[270] on which the compensation event's assessment was based; or
- The Project Manager's assessment of the final amount (under clause 53) does not become 'conclusive' evidence for the reasons set out in that clause, although the scope for altering any implemented compensation event appears limited (see Section 9.6.3).

The applicable methods of assessment changes to the Prices are described in Sections 5.5.1 and 13.3. The method for assessing changes to the Completion Date or to a Key Date are set out in Section 13.3.

### 5.5.1 Assessing Changes to the Prices

NEC4's principles for compensation events area set out in Section 5.3 (above) and can be summarised as

- No compensation event is the fault of the contractor because they all arise from employer risk events; and
- The contractor should not suffer a disadvantage (either commercially or in terms of time for completion) as a result of a compensation event.

262 NEC4 – clause 64.3. Also see 'Managing Reality – Book 5 Managing Procedures' – 'Pro Forma 21: Project Manager's Reply to a Quotation for a Compensation Event' p59.
263 NEC4 – clause 64.3.
264 NEC4 – clause 64.4.
265 NEC4 – clause 62.3 the Project Manager's acceptance of the contractor's clause 62.2 quotation.
266 NEC4 – clause 62.6.
267 NEC4 – clause 64.3 provides that the Project Manager notifies the contractor of his assessment within a period commencing from when it was apparent that the Project Manager's assessment was required, and lasts for the same period of time as the contractor had to provide his clause 62.2 quotation (either 2 weeks under clause 62.2, or such extended period under clause 62.5).
268 Akin to the effect of the Final Statement in JCT DB 2017 (clause 1.8 – Effect of Final Statement).
269 NEC4 – clause 66.2.
270 NEC4 – clause 61.6 Project Manager's correction of an assumption, which is a compensation event under clause 60.1(17).

In the event that the Project Manager and the contractor have not agreed lump sums or rates for assessing changes to the Prices, the Project Manager assesses the increases or decreases (the changes) in the Prices by the extent to which the compensation event has affected the Defined Cost and the resulting Fee adjustment (clause 63.1). This method applies to the assessment of compensation events for all of NEC4's Main Option Clauses. The assessment comprises the following stages set out in clause 63.1:

- Establishing the actual 'Defined Cost' of any work carried out up to the 'dividing date'; then
- Forecasting the Defined Cost of any work not done by the 'dividing date'; and
- Applying the resulting Fee.

NEC4's companion books (Managing Reality) explain the basis for assessing compensation events by using Defined Cost. The theory being that the contractor is '*no better or worse off*'[271] if the Client instructs a change for which he bears the risk. It further explains that the contractor's risk position (when providing quotations for compensation events that change the prices) is the same as when he was tendering for the project.

For such an important and widely applicable term, the definitions of 'Defined Cost' are a little scattered throughout NEC4, and some degree of common-sense interpretation is required. First, the correct definition for the applicable Main Option Clause should be identified. These are summarised in Table 5.15.

Clause 52.1 provides another definition of 'Defined Cost', but this focusses more on allocating the contractor's recoverable costs into either 'Defined Cost' and the 'Fee'.[272] The categories of Defined Cost in clause 63.1 are as follows:

- actual Defined Cost; and
- forecast Defined Cost.

Unfortunately, NEC4 does not define what is meant by the 'actual' or 'forecast' categories of Defined Cost, but it is not difficult to arrive at a common-sense interpretation, when

**Table 5.15** Applicable Definitions of 'Defined Cost'.

| Definition Clause | Defined Cost Definition | Applicable Main Option Clauses[a)] |
|---|---|---|
| 11.2(23) | The cost of the components in the Short Schedule of Cost Components is the Defined Cost | A and B |
| 11.2(24) | The cost of the components in the Schedule of Cost Components excluding the Disallowed Cost.[b)] | C, D and E |

a) NEC4 – Main Option Clause F (Cost reimbursable contract) is not considered in this book.
b) NEC4 – Main Option Clauses C, E and E – 'Disallowed Cost' is defined in clause 11.2(26). For further details, see section 10.2.4.

271 NEC4 – Managing Reality Book Four (Managing Change) 3rd Ed (2017) p45.

272 NEC4 – clause 52.1 gives a definition of 'Defined Cost' as essentially the costs incurred by the contractor, which are not included in the Fee (which is defined in clause 11.2(10) and is the 'fee percentage' stated in Contract Data Part Two (under '1 General') which is applied to the 'Defined Cost).

**Table 5.16** Explanation of Actual and Forecast Defined Cost.

| Category of Defined Cost (clause 63.1) | Explanation/Interpretation |
|---|---|
| 'actual Defined Cost' | This is the Defined Cost of work that the contractor has already undertaken ('*the work done*') by the 'dividing date'. It is not the value of work carried out to date – as set out in the rates or prices contained in the Short Schedule of Cost Components, the Schedule of Cost Components or any Bills of Quantities. |
| 'forecast Defined Cost' | This is a forecast of the Defined Cost of the work that the contractor must carry out after the 'dividing date'. It is not the value of the work yet to be carried out – as calculated from rates and prices contained the Short Schedule of Cost Components, the Schedule of Cost Components or the Bills of Quantities, but the forecast Defined Cost of such work as is necessary to give effect to the compensation event. |

clause 63.1 is read alongside the explanation provided in NEC4's companion – Managing Reality[273] – as Table 5.16 summarises.

In very basic terms, the two categories of Defined Cost are either the actual or forecast costs (such as the costs for labour, plant, materials and sub-contractors) of the physical work that the contractor has carried out, or will carry out. When assessing the 'forecast Defined Cost' element, the Project Manager should consider any 'risk allowances for cost and time' as required by clause 63.8.

The 'Fee' is intended to cover the contractor's ancillary costs such as overheads (such as administration, marketing, recruitment and training) and profit, along with any applicable taxation and bonds or guarantees.[274]

The first step for the Project Manager is to establish when the 'dividing date' occurred, which will either be (see clause 63.1):

- The date of an instruction from the Project Manager or Supervisor;
- The date when the contractor issued the Project Manager with his notification of a compensation event under clause 61.3; or
- The date of any other communication from the Project Manager or Supervisor that comprises any other notification, issuing a certificate or changing an earlier decision.

In practical terms, the date of the Project Manager's or Supervisor's instruction is likely to be an instruction to change the Scope[275] or change a previous decision.[276] As this will often be the earliest date that the contractor hears about the change, it is highly unlikely that he will have started to do any of the necessary work resulting from the instructed change in the Scope or a previous decision. Conversely, if the contractor has notified a compensation event under clause 61.3, there is a greater likelihood that some of the compensation event's

273 NEC4 – Managing Reality Book Four (Managing Change) 3rd Ed (2017) pp24–25.
274 NEC4 – Managing Reality Book Four (Managing Change) 3rd Ed (2017) p49.
275 NEC4 – A compensation event under clause 60.1(1).
276 NEC4 – A compensation event under clause 60.1(8).

effects have already taken place (for example, the contractor may have already incurred delay cost – perhaps for a period of up to eight weeks[277]).

Once the Project Manager has established the 'dividing date', he can then assess the change in the Prices for the compensation event by

- Assessing the actual Defined Cost of any work carried out up to the 'dividing date'; and
- Assessing the forecast Defined Cost of any work not carried out by the 'dividing date'.

It seems clear that no 'risk allowances' for either cost or time will be considered when assessing the actual Defined Cost because this work has already been carried out. However, when assessing the forecast Defined Cost of any work not carried out, the Project Manger should consider any such 'risk allowances' for time and cost – as provided by clause 63.8. This clause qualifies the 'risk allowances' (for cost and time) by stating they should be for '...*matters which have a significant chance of occurring and are not compensation events*'.

NEC4's Managing Reality and the NEC4 User Guide provide the following guidance:

- '*The conceptual idea is that the Contractor includes in quotations for compensation events for the "risks" they carry under the contract in the same way as when they are tendering for the work.*'[278]
- '*Allowances for risk must be included in forecasts of Defined Cost and Completion in the same way that the Contractor should allow for risks when pricing its tender...*'[279]; however
- '*It is not the intention....for a blanket percentage to be added to each compensation event as has been the temptation in some contracts; risk for each compensation event should be considered for each event....*'[280]

The last stage of assessing the change in the Prices will be calculating and including any Fee.

Drawing on NEC4's overriding principles for compensation events, along with the categories of Defined Cost together, an illustration of how the change to the Prices are assessed can be given:

Examples:

1) For a highways project let under NEC4 Main Option Clause B (Priced Contract With Bill of Quantities), the Project Manager instructs a change in the Scope, adding 100 pre-cast concrete bollards – 300 mm Ø and 750 mm high.
2) In the Bills of Quantities, the contractor has priced the same bollard at £205 each, yet the actual cost of installing each bollard is £240. On this basis, and if there are no compensation events, the contractor will lose £35 for each bollard that he installs. This would be the case if the contract was let under the JCT SBC/Q 2016 form, where the contractor

277 NEC4 – The contractor is only entitled to a change in the Prices or Completion Date/Key Date if he notifies a compensation event within 8 weeks after becoming aware that the event has happened.

278 NEC4 – Managing Reality Book 4 (Managing Change) (2017) p90.

279 NEC4 – User Guide volume 4 (2017) p65.

280 NEC4 – Managing Reality Book 4 (Managing Change) (2017) p93. Managing Reality gives a thorough risk analysis example on pages 91 to 93 for all the elements of cost in the Shorter Schedule and Schedule of Cost Components (such as wage increases, availability and performance (poor) of labour and sub-contractors, equipment reliability and availability, shortages and delays of materials or goods, design liability for temporary and permanent works, site access and logistical restrictions, weather conditions, changes in the law, and unforeseen risks.

takes the risk for the sufficiency of the rates and prices, even in the event of a variation – so would only be entitled to recover £205 for installing each bollard.

3) However, applying clause 61.3, it is clear that the 'dividing date' is the date of the Project Manager's instruction, so at that date, the contractor has not started the compensation event work. Therefore the assessment of the change in the Prices will only use 'forecast Defined Costs'. NEC4's principle provides that the contractor should not suffer a commercial disadvantage when undertaking compensation events, so the compensation event would be assessed as the forecast Defined Cost of £240 per bollard plus the Fee. The work is straightforward and no 'risk allowances' for time or cost (under clause 63.8) have been assessed as having a significant chance of occurring.

It is vital that the Project Manager consistently applies the concept of 'Defined Cost' (either actual or forecast) to his assessments. Assessing the compensation event on a basis that is not consistent with clause 61.3, even if intended to be for the parties' convenience, is likely to result in an imbalance between the risk positions that the Client and the contractor have accepted under NEC4. For example, if the Project Manager and the contractor agree rates or lump sums[281] to assess the change in prices for one compensation event, but revert to the 'Defined Cost' approach when assessing another, their risk positions may vary as follows:

- For the compensation event assessed on the basis of agreed rates and prices:
  - The contractor may benefit from the Project Manager agreeing a rate or price which particularly lucrative, providing a greater profit margin than he would obtain via the 'Defined Cost' method of assessment; or
  - The Contractor may agree rates or prices that provide less of a profit margin or even result in a loss.
  - The party believing that they have 'lost out' may seek to recover their losses (in the case of the contractor) or additional monies paid (in the case of the Client) when assessing subsequent compensation events.
  - The Project Manager (or a cost consultant assisting him) may misuse or misapply the agreed rates and prices when assessing subsequent compensation events. He may wrongly try to compare the price assessed using the 'Defined Cost' method with an agreed rate or price, even when the two methods are based upon entirely different bases and concepts of risk.

Managing Reality claims[282] that the 'Defined Cost' approach dispenses with arguments that often occur under the JCT 2016 contracts when valuing works using unit rates or prices. In particular, arguments over the applicability or appropriateness of adjusting a rate or price (in the Contract Bills, a priced schedule of work or Contract Sum Analysis) to give a new rate that reflects the changes in quantities, changes in specification or changes in conditions under which the variation is (or will be) carried out.

In the author's view, and from a quantity surveying perspective, the opposite of Managing Reality's claim is often true. The reasons for this are as follows:

281 NEC4 – clause 63.2.
282 NEC4 – Managing Reality Book Four (Managing Change) 3rd Ed (2017) p45.

- Unlike JCT 2016 contracts, the rates and prices for work in NEC4's activity schedule[283] or bills of quantities[284] are not used as the basis for assessing the change in the Prices, as NEC4 clearly assesses the change in the Prices using the 'Defined Cost' method. Therefore, under NEC4, the contractor only bears the pricing risk of the rates or prices in the applicable activity schedule or bill of quantities, and that risk isn't transferred when assessing compensation events. Contrasting this with the contractor's position under the JCT 2016 contracts, where the contractor almost always bears the pricing risk for changes where his rates and prices form the basis of valuing the works.
- Where a rate or a price is to be adjusted, the basis of that rate or price remains, so ordinarily, any argument is confined to the rate 'adjustment' only. The rate adjustment is usually assessed as additional costs required to carry out the work. For example, if the work is carried out at height, then the additional elements are likely to include the costs of providing access, and any labour for extra handling. The amounts in respect of physically carrying out the item of work itself (i.e. the elemental costs for labour, materials, plant or sub-contractor work) are likely to be unchanged.

However, NEC4's 'Defined Cost' method is likely to create more potential matters for dispute for the parties to argue over, as following elements can be disputed:

- The forecast 'Defined Cost' of the original work. For example, the contractor may claim that the relevant price in his activity schedule or bills of quantities represents the forecast defined cost. However, this may not be the case, as the contractor may have under-priced the work, in order to provide a competitive tender. As a matter of course, the Project Manager should investigate this in order to ascertain an accurate value for the original work's forecast defined cost.
- The forecast 'Defined Cost' for the changed work;
- The Project Manager and contractor may be in dispute over the appropriate rates or prices in the Schedule of Cost Components or the Short Schedule of Cost Components (which are often expressed as hourly rates) to use for the resources required (items of plant and personnel) in order to carry out both the original and the changed work; and
- The durations for which the resources in the Schedule of Cost Components or Short Schedule of Cost Components (plant and people[285]) are required to carry out the original and the changed work. A quantity surveyor will be able to measure the quantities of both the original and changed work, and then apply the appropriate rates of output per unit of the original and changed work. For example, Spon's pricing books include the outputs (or durations of required resource per unit of work) for most items of work,[286] to which the rates and prices in the Schedule of Cost Components or Short Schedule of Cost Components can be applied.

---

283 NEC4 – Option A (Priced Contract with Activity Schedule).

284 NEC4 – Option B (Priced Contract with Bill of Quantities).

285 NEC4 – Option B, clause 63.13, the Project Manager and contractor can agree a new rate for 'People' but not for any other category of resource.

286 Spons' 'Architects and Builders' Price book includes 'Labour hours' for most items of work, and its 'Civil Engineering and Highway Price Book' includes 'Gang hours.'

It is often far less complicated to adopt what is a more 'traditional approach' which is:

- One source of cost data[287] from which to value variations; and
- The contractor takes the pricing risk for those rates and prices.

This ensures that any disputes over valuing work using new, or derived rates, are usually confined to the adjustments to that rate or price, rather than a wholesale dispute over all elements of it. The rate or price can be directly applied to the quantity of work, so arguments over the duration, or output for that work can similarly be avoided.

This 'traditional approach' neither fits with NEC4's ethos for the Client taking the risk for all compensation events nor its 'Defined Cost' method. This is not to say that the methods of valuation for the JCT 2016 contracts are better or worse, but parties to NEC4 contracts should be aware that Managing Reality's claim that seemingly, NEC4 inherently reduces the disputes over valuation of variations (when compared to JCT), may not be entirely correct, and actually require a greater quantity surveying involvement than would otherwise be required under the JCT 2016 contracts.

The methods of assessing the change to the Prices (or Defined Cost) for each of NEC4's Main Option Clauses A to D are set out in Table 5.17.

**Table 5.17** Summary of NEC4 Main Option Clauses' Assessment of Compensation Events.

| Main Option Clauses for assessing compensation events | Option A | Option B | Option C | Option D |
|---|---|---|---|---|
| Are the Prices (Options A and B) or Defined Cost (Options C and D) reduced from changes to the Client's scope proposed by the contractor. | Yes 63.12 | Yes 63.12 | No 63.13 | No 63.13 |
| Prices are reduced by multiplying the compensation event's effect by the 'value engineering percentage'[a)] | 63.12 | 63.12 | — | — |
| Changed Price assessments are in the form of changes to: | Activity Schedule 63.14 | Bill of Quantities 63.15 | Activity Schedule 63.13 | Bill of Quantities 63.15 |
| • For work 'not yet done' | — | Changed rate, quantity or lump sum, or new rate[b)] | — | Changed rate, quantity or lump sum, or new rate[c)] |
| • For 'work done' | — | New lump sum item | — | New lump sum item |
| New rate for 'People Rates' | 63.16 | 63.16 | — | — |

a) NEC4 – 'Value engineering percentage' – see Contract Data Part One (6 – Compensation events) for Options A or B.
b) NEC4 – Option B, clause 63.15 – if there is no rate in the Bill of Quantities, a new rate is calculated by reference to the 'method of measurement' (set out in Contract Data Part One, 6 – Compensation Events), unless the Project Manager and Contractor otherwise agree.
c) See Footnote 286.

287 Such as the rates and prices in a bill of quantities or a schedule of rates.

Option E (Cost Reimbursable Contract) assesses a change to the Prices in accordance with clauses 63.1 to 63.4 of NEC4's Core Clauses.

If selected, Option X1 (Price Adjustment for Inflation) is used to change the prices for Options A, B, C and D on a similar 'fluctuations' basis as the Fluctuations Option in JCT 2016 contracts.

Under Option C (Target Contract with Activity Schedule) or Option D (Target Contract with Bill of Quantities), the Project Manager must assess the 'Contractor's share'.[288] He does this on two occasions; firstly providing a 'preliminary assessment' at Completion, and then a 'final assessment' with assessing the 'final amount due'. As both assessments are essentially tasks that take place on and after completion, they are covered in Section 5.5.2, along with the incentive payment under Option X22.

### 5.5.2 Contractor's Share (Options C and D) and Incentive Payment (Option X22)

The three other financial matters that NEC4 expressly includes in the 'final amount due' in clause 53 are:

- The Contractor's share under Option C (Target Contract with Activity Schedule) in clauses 54.1 to 54.4;
- The Contractor's share under Option D (Target Contract with Bill of Quantities) in clauses 54.5 to 54.8; and
- The Incentive payment under Option X22 (Early Contractor Involvement), clause X22.7(3) – see Section 5.4.3.

The clauses governing the Contractor's share under Options C and D are identical. The Contractor's share is assessed by the Project Manager on the following two occasions, albeit on a slightly differing basis:

- A 'preliminary assessment' is assessed at Completion[289] using
  - A forecast of the Price for Work Done to date; and
  - The final total of the Prices, whereas
- The 'final assessment' is made and included in the 'final amount due'[290], using
  - The final Price for Work Done to Date; and
  - The final total of the Prices.

The calculation for the 'Contractor's share', first, uses the difference[291] between the following elements, as set out in Table 5.18.

The calculation of the 'difference' is as follows:

| **'Total of the Prices'** | **£A** |
|---|---|
| Less: the 'Price for Work Done to Date' | (£B) |
| **Difference** | **£C** |

288 NEC4, refer to Option C, clauses 54.1 to 54.4, and Option D, clauses 54.5 to 54.8 for the 'Contractor's share.'
289 NEC4, Option C, clause 54.3, or Option D, clause 54.7.
290 NEC4, Option C, clause 54.4, or Option D, clause 54.8.
291 NEC4, Option C, clause 54.1 or Option D, clause 54.5.

**Table 5.18** NEC4 Elements of the 'Contractor's share' Calculation.

| Element of the 'difference' for the 'Contractor's share' calculation | Description |
|---|---|
| The 'total of the Prices'; and | The 'Prices' are essentially the 'value' of the Works, or the 'target' – i.e. the total of the prices from the relevant pricing document, as adjusted (increased or decreased) by compensation events.<br>• Option C, clause 11.2(32)[a] the total of the activities in the Activity Schedule[b] as adjusted in accordance with the contract<br>• Option D, clause 11.2(35)[c] means the total of:<br>– The quantity of work completed multiplied by the rate in the Bills of Quantities[d]; and<br>– A portion of each lump sum of work in the Bills of Quantities that the contractor has completed |
| The 'Price for Work Done to Date' | The 'Price for Work Done to Date' is calculated as the 'Defined Cost' (i.e. the costs[e] used for interim payments) plus the Fee.<br>• Option C, clause 11.2(32); and<br>• Option D, clause 11.2(35) means the total Defined Cost plus the Fee. |

a) NEC4, Option D, 'the total of the Prices' is not defined in Option C, but is defined in clause 11.2(35) of Option D.
b) NEC4, Option C, 'Activity Schedule' is defined in clause 11.2(21).
c) See Footnote 289.
d) NEC4, Option D, 'Bill of Quantities' is defined in clause 11.2(22).
e) NEC4, see clause 52.1 and refer to 8.9.3.1.

The 'Difference' is then expressed as follows[292]:

- A 'saving' if the Price for Work Done to Date is less than the total of the Prices, in which case the Client pays the 'Contractor's share' to the contractor; or
- An 'excess' if the Price for Work Done to Date is greater than the total of the Prices, in which the Contractor pays the 'Contractor's share to the Client.

The final stage of calculating the Contractor's share is to apply the 'Contractor's share percentages' to the 'share ranges' set out in Contract Data Part One.[293] The 'share ranges' are expressed a percentages, applying to the 'difference' (i.e. both the 'saving' and the 'excess'), and are divided into four 'increments' or bands. A separate 'Contractor's share percentage' is then set against each of these increments (or bands). The following two examples show how the Contractor's share range is calculated against a 'saving' and against an 'excess':

Saving example:

- The 'difference' between the 'total of the Prices' (which is £2 M) and the 'Price for Work Done to Date' (which is £1.8 M) is £200 000; and

292 NEC4, Option C, clause 54.2, or Option D, clause 54.6.
293 NEC4, Option C, clause 54.1, or Option D, clause 54.5.

- The 'share range' and the corresponding 'Contractor's share percentage' are as follows:

| Share range: | | Contractor's share percentage: |
|---|---|---|
| Less than | 1% | 10% |
| From | 1% to 5% | 20% |
| From | 6% to 10% | 40% |
| Greater than | 10% | 50% |

- The calculation of the 'Contractor's share' is as follows:

| Share range: | | £ |
|---|---|---|
| Less than 1% | Up to 1% (£2,000), the contractor is paid 10%: | £200 |
| From 1% to 5% | For the next 1% to 5% (£8,000), the contractor is paid 20%: | £1,600 |
| From 6% to 10% | For the next 5 to 10% (£10,000), the contractor is paid 40% | £4,000 |
| Greater than 10% | The saving does not exceed 10% | £0 |
| | **Total 'Contractor's share' paid by the Client** | **£5,800** |

Excess example:

- The 'difference' between the 'total of the Prices' (which is £4.5 M) and the 'Price for Work Done to Date' (which is £4.8 M) is (£300 000); and
- Applying the same share ranges as the 'Saving example', shows an excess of 6.67% the calculation of the 'Contractor's share', is as follows:

| Share range: | | £ |
|---|---|---|
| Less than 1% | Up to 1% (£3,000), the Client is paid 10%: | (£300) |
| From 1% to 5% | For the next 1% to 5% (£12,000), the Client is paid 20%: | (£2,400) |
| From 6% to 10% | For the next 5 to 10% (£5,100)), the Client is paid 40% | (£2,040) |
| Greater than 10% | The excess does not exceed 10% | £0 |
| | **Total 'Contractor's share' paid to the Client** | **(£4,740)** |

If Option X22 – Early Contractor Involvement[294] applies (see Section 5.7.3), then the Project Manager will also include his assessment of the contractor's 'budget incentive' in the 'final amount due' under clause 53.1. The 'budget incentive' will only be payable if the 'final Project Cost[295]' is less than the Budget.[296] The Project Manager calculates the 'budget incentive' as follows:

294 NEC4, Option X22 Early Contractor Involvement is only used with Option C (Target Contract with Activity Schedule) or Option E (Cost Reimbursable Contract).
295 NEC4, Option X22 (Early Contractor Involvement), clause X22.1(2) defines 'Project Cost' as the total paid by the Client to the contractor (and Others – as defined in clause 11.2(12)) for the items included in the 'Budget.'
296 NEC4, Option X22 (Early Contractor Involvement), 'Budget' is defined in clause X22.1(1).

([Budget] – [final Project Cost]) × % in Contract Data Part One[297] = 'budget incentive'

Once the final assessment is provided (either by the Project Manager under clause 53.1, or by the contractor under 53.2 – hereafter referred to as the 'relevant assessment'), it becomes 'conclusive evidence' of the final amount due, unless the party disagreeing with the relevant assessment of the final amount takes one the dispute resolution options under clause 53.3. Clause 53.3 sets out the appropriate procedure under whichever of dispute resolution options W1, W2 or W3 have been selected to apply in Contract Data Part One.[298] The selection of Options W1, W2 or W3 will depend upon whether the HGCRA 1996 applies, and what the parties select as the 'tribunal' (in Contract Data Part One).

Option W2 will be selected if the HGCRA 1996[299] applies – which will be in the majority of cases if the project is in the United Kingdom and subject to the UK law, and the construction contract is a contract for 'construction operations' within the meaning of s105 of the HGCRA 1996.

## 5.6 Acceleration Provisions

A general description of how acceleration may be dealt with under building contracts is given in Section 3.8 of Chapter 3.

### 5.6.1 Acceleration Under JCT SBC/Q 2016 and JCT DB 2016

Both JCT SBC/Q 2016 and JCT DB 2016 include provisions for acceleration within the Supplemental Provisions of those contracts,[300] the procedures for which were set out in detail in Section 5.4.2. However, there are no acceleration provisions found in the JCT IC 2016 or JCT MW 2016 contracts.

Under JCT SBC/Q 2016 and JCT DB 2016, the option to accelerate progress can only be exercised by the employer. This may partly be due to the contractor's obligation to continuously use his 'best endeavours'[301] to avoid delays (see Section 3.1.3 above). The employer firstly invites the contractor to confirm either:

- If it's practicable to achieve an earlier completion date; and if so,
- The time can be saved and the contractor's costs for accelerating (the 'Acceleration Quotation'). The contractor is entitled to the reasonable costs or providing an Acceleration Quotation, irrespective of whether the employer accepts it and instructs him to accelerate.

297 NEC4, Contract Data Part One – entry under Option X22 for the applicable percentage of the saving as calculating the 'budget incentive' as.
298 NEC4, Contract Data Part One, Section '1 General' entry against *'Option for resolving and avoiding disputes.'*
299 Housing Grants Construction and Regeneration Act 1996 (as amended) – refer to 8.7.
300 JCT SBC/Q 2016 – Supplemental Provisions, Part 2, paragraph 2 (Acceleration Quotation) and JCT DB 2016, Schedule 2 – Supplemental Provisions, Part 2, paragraph 4 (Acceleration Quotation).
301 JCT SBC/Q 2016 – clause 2.28.6.1 and JCT DB 2016 – clause 2.25.6.1.

Unless the parties agree to longer or shorter periods, the contractor has 21 days from the employer's invitation in which to provide the Acceleration Quotation,[302] and the employer must have at least 7 days to accept it. The employer must provide a 'Confirmed Acceptance' of the contractor's Acceleration Quotation in order to implement the proposed acceleration measures.[303] This must confirm the amount by which the Contract Sum is adjusted,[304] the adjustment to the Completion Date[305] (or Completion Dates of any relevant Section) and any other conditions set out in the employer's invitation.[306] The Acceleration Quotation may not be used for any other purpose[307] if a Confirmed Acceptance is not issued.

### 5.6.2 Acceleration Under NEC4

In NEC4, the purpose of the acceleration clause 36 could not be clearer – it is to complete the works before the Completion Date. Whilst the contractor and the Project Manager may discuss accelerating the works, the acceleration quotation process must be commenced by the Project Manager, instructing the contractor to provide a quotation (which includes details of any Key Dates to be changed). The contractor provides the clause 36 quotation within three weeks thereafter and it must include[308]:

- A revised programme showing the earlier Completion Date and any Key Dates that are to be changed; and
- The changes to the Prices.

The Project Manager replies to the clause 36 quotation within three weeks, and it's accepted, and he must change the Completion Date (and any Key Dates) and the Prices accordingly, along with accepting the revised programme.

### 5.6.3 Summary of Acceleration Provisions

Table 5.19 summarises the contracts' acceleration provisions.

Refer to Section 5.4.2 (Variation and Acceleration Quotations) for the relevant procedures under JCT SBC/Q 2016 and JCT DB 2016.

302 The employer may ask for amended proposals during the 21 day (or extended) period for the contractor to provide the Acceleration Quotation (JCT SBC/Q 2016, Schedule 2, paragraph 3.3. JCT DB 2016 Schedule 2 Supplemental Provisions, Part 2 paragraph 4.2.3).
303 JCT SBC/Q 2016 Schedule 2, paragraph 4, JCT DB 2016 Schedule 2 Supplemental Provisions, Part 2 paragraph 4.3.
304 JCT SBC/Q 2016 Schedule 2, paragraph 4.1, JCT DB 2016 Schedule 2 Supplemental Provisions, Part 2 paragraph 4.3.1.
305 JCT SBC/Q 2016 Schedule 2, paragraph 4.2, JCT DB 2016 Schedule 2 Supplemental Provisions, Part 2 paragraph 4.3.2.
306 JCT SBC/Q 2016 Schedule 2, paragraph 4.3, JCT DB 2016 Schedule 2 Supplemental Provisions, Part 2 paragraph 4.3.3.
307 JCT SBC/Q 2016 Schedule 2, paragraph 5.3, JCT DB 2016 Schedule 2 paragraph 4.4.2.
308 NEC4, clause 36.2. The quotation must also provide 'details of the assessment' but it is not clear what this means.

**Table 5.19** Summary of Acceleration Provisions.

| Description | JCT SBC/Q 2016 | JCT DB 2016 | NEC4 |
|---|---|---|---|
| Acceleration provisions (clause or section) | Schedule 2 Supplementary Provisions, Paragraphs 2–4 | Schedule 2 Supplementary Provisions, Part 2, para 4 | Clause 36 |
| Acceleration process initiated by the employer | ✓ | ✓ | ✓ |
| Contractor may initiate acceleration process | — | — | — |
| Contractor provides a quotation | ✓ | ✓ | ✓ |
| Period for contractor's quotation | 21 days | 21 days | 3 weeks |
| Quotation includes a programme | | | ✓ |
| Quote includes earlier completion date | ✓ | ✓ | ✓[a)] |
| Quote includes costs of acceleration | ✓ | ✓ | ✓ |
| Quote acceptance period | 7 days | 7 days | 3 weeks |
| Means of acceptance | 'Confirmed Acceptance' | 'Confirmed Acceptance' | Acceptance |
| Prohibition of using rejected quotation | ✓ | ✓ | – |

a) NEC4 – Also includes changes to any Key Dates.

## 5.7 Bonus Clauses

Few standard forms of building contract include provision for bonus clauses as a standard option. Bonus clauses are usually written into contracts if the employer wishes to provide an incentive for the contractor to complete a particular task in a way that exceeds his ordinary obligation under that particular contract.

Some notable examples are included in NEC4 contracts:

- The contractor's 'gain share' for completing the works for less than the 'target price' in Options C and D;
- Early completion under Option X6 (Bonus for early completion); and
- Payment of the 'budget incentive' under Option X22 (Early contractor involvement).

### 5.7.1 Contractor's Share

The JCT 2016 contracts covered in this book do not contain any specific provisions for a 'bonus' in the form of NEC4's 'gain share' to be paid to the contractor. The 'contractor's share' under NEC 4 contracts is covered at Section 5.5.2.

### 5.7.2 Early Completion Bonus

This is set out in Option X6 of NEC4 and provides that the contractor is paid the 'early completion bonus' for completing the works, or a section earlier than the Completion Date. If the works are to be completed in sections, a different sum can be expressed as the bonus for each section. The Client pays to the contractor a sum of money per day,[309] for each day between the Completion Date[310] and the earlier of the following:

- Completion[311]; or
- The date that the Client takes over[312] the works.

In effect, the early completion bonus is the reverse of a liquidated damages provision – the 'liquidated' sum being 'added to' rather than 'deducted from' payments due to the contractor.

### 5.7.3 'Budget Incentive' for Early Contractor Involvement (Option X22)

NEC4 introduced a new secondary option (X22) that sets out a procedure for early contractor involvement. It is only intended for use with main option clauses 'Option C' (Target Contract with Activity Schedule) and 'Option E' (Cost Reimbursable Contract).

Option X22 divides the work into either two or three sections as follows, which the Client must clearly describe in the Scope:

- The entirety of the works is divided into Stage 1 and Stage 2,[313] so the whole of the works is subject to the early contractor involvement procedure; or
- Only parts of the works are designated as Stage 1 and Stage 2, with the remaining parts of the works (i.e. the 'third' stage) being described in the Scope in the normal way and not subject to the early contractor involvement procedure in X22.

The Client sets the 'Budget[314]' for the works in Stage 1 an Stage 2, and the incentive mechanism[315] (similar to the 'Contractor's share[316]') encourages the Contractor to complete the works for less than the Budget, by offering a percentage of the saving (the 'budget incentive'). The Contractor provides the following:

---

309 The sum is expressed as the 'bonus' for either (1) the 'whole of the works' – as a single figure, or (2) a 'bonus' for each section of the works, in Contract Data Part One – against the entries for Option X6.

310 NEC4: 'Completion Date' is defined in clause 11.2(3).

311 NEC4: 'Completion' is defined in clause 11.2(2).

312 NEC4 – clause 35 (Take over).

313 X22.1(3), and the corresponding entries in Contract Data Part One against Option X22.

314 X22.1(1) and the corresponding entries in Contract Data Part One against Option X22 for 'Budget' and the 'Budget incentive' percentage – clause X22.7(1).

315 X22.7(1).

316 Option C – clause 54 (The *Contractor's* share) – see sections 5.5.2 and 5.7.1.

- Stage 1 – detailed forecasts of the Defined Cost[317] for Stage 1 at the intervals set out in the Contract Data;
- Stage 2 – submits design proposals which include forecasts of their effects on the Project Cost[318] and on the Accepted Programme[319];
- Stages 1 and 2 – forecasts of the Project Costs[320] (in conjunction with the Project Manager) at the intervals stated in the Contract Data.
- Approvals and consents from 'Others'.[321]

The Project Manager considers the Contractor's Stage 1 forecasts and Stage 2 design proposals, and either accepts them or rejects them.[322] A typical calculation for the 'budget incentive' under Option X22 is shown in Section 5.5.2.

### 5.7.4 Bonus Clauses and Acts of Prevention

The operation of the bonus clauses described in Sections 5.7.1 to 5.7.3 appears to be relatively straightforward. However, their relatively simplicity may be unhelpful in circumstances where the employer hinders or prevents the contractor from satisfying the conditions for becoming entitled to the 'bonus'. In these cases, the employer is likely to face a 'loss of opportunity' claim[323] from the contractor. Under the standard forms of contract (assuming for this example, that the employer and contractor have agreed to amend a JCT 2016 contractor to include a 'bonus' clause), the contractor becomes entitled to additional time and costs if the employer hinders or prevents him in the following circumstances:

- Under JCT 2016 contracts, if the employer (or someone acting on his behalf) causes an '*impediment, prevention or default*' then the contractor is entitled to:
  - An extension of time[324]; and
  - Direct loss and/or expense.[325]
- Under NEC4, a 'compensation event' under clauses:
  - If the Client (or others on the client's behalf) fail to provide access, or something (including testing facilities, etc), or work in accordance with the 'Programme' or 'Scope' under clauses 60.1(2), 60.1(3), 60.1(5) and 60.1(16); or

317 X22.2(1), Option C – clause 11.2(24).
318 X22.1(2), utilising the clause X22.1(4) 'Pricing Information' provided by the Contractor and set out in Contract Data Part Two against Option X22.
319 11.2(1).
320 X22.2(5).
321 X22.3(6), although Option X22 does not specify at what stage the Contractor applies for or obtains 'approvals and consents.'
322 The reasons for rejecting (not accepting) forecasts or design submission proposals are that (a) they do not comply with the Scope, (b) the forecast includes unnecessary work, (c) the design proposals will cause the Client to incur unnecessary cost, and (d) that the Project Manager is not satisfied that the Contractor has properly assessed the Prices (or changes to them) – clauses X22.2(2) and X22.3(3).
323 See section 4.4.2.5 for the principles of a 'loss of opportunity' claim – *Walter Lilly & Company Ltd v MacKay & DMW Developments Ltd* (2012) EWHC 17773 (TCC).
324 A 'Relevant Event' under JCT SBC/Q 2017 – clause 2.29.7, JCT DB 2016 – clause 2.26.6, and JCT IC 2016 – clause 2.20.6. For JCT MW 2016, the contractor may have to claim that the employer has breached one of the 'implied terms' of the contract (refer to section 6.2).
325 A 'Relevant Matter' under JCT SBC/Q 2016 – clause 4.22.5, JCT DB 2016 – clause 4.21.5, and JCT IC 2016 – clause 4.17.4. For JCT MW 2016, the contractor may have to seek common law damages for breaches of one of the implied terms (refer to section 6.2).

- If the Client commits another breach of the contract under clauses 60.1(14), 60.1(18) or Y2.4.

What amounts to a bonus clause may perhaps arise without the parties being entirely aware of it. In a 1996 case,[326] the parties entered into an acceleration agreement, stipulating that the contractor would be paid additional sums of firstly £50 000 in July 1994, then three payments of £20 000 in August 1994 if the contractor completed the relevant work by 26 August 1994. Considering the final (third) payment of £20 000, Mr. Recorder Toulson said:

> *'I conclude that the £20,000 was agreed to be a performance related payment if the plaintiffs [contractor] completed by the 26 August 1994… It was also an express term of the acceleration agreement that the defendants [the hotel] would supply the plaintiffs with all outstanding information by the end of 12 July 1994, and I accept that there were implied non-hindrance terms as pleaded in paragraph 7 of the re-amended statement of claim.*
> *By reason of the numerous changes made after the acceleration agreement the defendants were in breach of those implied terms, if not also of the express term. The latter point turns on whether the express duty was conditional upon receipt of a specific request for information from the plaintiffs. I doubt that it was, but the point is academic. It is impossible to tell whether, as a matter of probability, the plaintiffs would or would not have finished by 26 August 1994, but for those changes. They would have had a reasonable opportunity of doing so, but they could easily have failed for all manner of reasons. In those circumstances I would hold that the plaintiffs are entitled to damages for loss of that chance equal to 50 per cent of the agreed performance bonus, or £10,000'.*[327]

Although based on a specially worded acceleration agreement, nevertheless, this part of the judgment gives useful guidance on the way in which the courts may decide whether and to what extent a contractor has been deprived of the opportunity to earn a bonus by the actions or defaults of the employer and architect. In this instance, the judge took a robust, if somewhat rough and ready, approach. The matter of implied terms is covered in Section 6.2.[328]

326 *John Barker Construction Ltd v London Portman Hotels Ltd* (1996) 50 Con LR 43.
327 [1996] 50 Con LR 43 at 69 per Mr. Recorder Toulson.
328 Implied terms including those requiring co-operation between the employer and the contractor may be particularly relevant to building contracts that do not contain express provisions that deal with extensions of time, or loss and expense (such as JCT MW 2016 – refer to Footnotes 321 and 322).

# 6

# Basis for Common Law Claims

## 6.1 General

Most of the claims which are considered in this book concern the contractor seeking additional time or financial reimbursement under the express terms of the particular building contract. Notable examples being, claims for 'variations', the contractor's right to claim for an extension of time, or direct loss and/or expense under the JCT 2016 contracts, or a 'compensation event' under NEC4.

However, the contractor may also claim damages at common law for breach of contract (refer to Section 2.6), or in limited circumstances, under the law of tort. Common law claims are predominantly financial in nature, but can include claims for extensions of time. The common circumstances under which a contractor may raise a common law claim are set out in Section 1.2.2 (Common Law Claims) and may arise in the following circumstances:

- The parties have contracted under a very simple, bespoke building contract that does not contain any provisions for 'claims' matters; or
- A building contract that only contains general conditions for extensions of time and/or financial compensation (such as the JCT MW 2016 contract).

It had been suggested that claims based in common law are only available under JCT 2016 contract and not under NEC4, for the following reasons:

- JCT 2016 contracts[1] expressly confirm that the contractor's right to claim direct loss and/or expense at common law is are preserved, whereas
- NEC4 expressly states that the 'Client' and the 'Contractor' only have rights to change the 'Prices' (along with changes to the completion date and any key dates) in accordance with the contract – i.e. only via the 'compensation event' procedure.[2]

1 JCT SBC/Q 2016 – clause 4.24, JCT DB 2016 – clause 4.23, JCT IC – clause 4.19. JCT MW 2016 – whilst no equivalent clause is included, Guidance Note 18 states *'Statutory and common law rights are not restricted by the terms of the Contract'*.
2 NEC4 – clause 63.6. At first glance, JCT 2016 contracts contain a similar provision (JCT SBC/Q 2016 and JCT IC 2016– clause 4.2, and JCT DB 2016 – clause 4.1) but the 'loss and expense' clauses all preserve the contractor's right to common law damages (see Footnote 1).

*Building Contract Claims*, Sixth Edition. Guy Higginbottom and David Chappell.

However, both the JCT 2016 and NEC4 contracts include 'catch all' provisions that allow the contractor to claim extensions of time[3] and additional costs[4] in the event that the employer (or someone for whom the employer is responsible) commits an act of prevention.[5] It appears that the position for common law claims under JCT 2016 contracts is clear, but less clear in respect of NEC4 because it appears to contain a conflict between (i) attempting to restrict the parties' rights to the 'compensation event' procedure but (ii) their rights to consequential losses may be unaffected[6] (see also Section 5.3.2).

In practical terms, JCT 2016 contracts may allow the contractor to make additional or alternative claims[7] for damages based on the same facts as those which it has put forward as part of an application for loss and/or expense under the terms of the contract. This means that the contractor can look for a remedy under a contract term without thereby prejudicing its right to claim at common law on the same grounds. It is suggested that for JCT 2016 contracts, this may present the contractor with the following options:

- Where the contractor has failed to comply with particular requirements of the contract terms, such as an obligation to give notice within a reasonable time of becoming aware that regular progress is being or is likely to be affected, a common law claim may avoid the contractual restrictions imposed upon the contractual claim.[8]
- There is nothing to prevent a contractor from pursuing both claims in tandem provided that it can make good such claims.[9]
- Moreover, it should be noted that settlement of a claim made under contractual terms will not preclude the contractor from pursuing a claim for damage arising from the same facts provided only that additional damages (e.g. for consequential loss) are claimed and that there is no element of double recovery.[10]

In this context, it is important to remember that not all the matters which (under a standard form contract) entitle the contractor to direct loss and/or expense are breaches of contract. For example, the issuing of an instruction may entitle a contractor to loss and/or expense, but the giving of that instruction is not a breach, because it is expressly empowered by the contract. Therefore, a failure on the part of the contractor to properly operate the provisions of the clause giving a contractual remedy will sometimes leave it without any remedy at all.

---

3 'Relevant Events' under JCT SBC/Q 2016 – clause 2.29.7, JCT DB 2016 – clause 2.26.6 and JCT IC 2016 – clause 2.20.6.

4 'Relevant Matters' under JCT SBC/Q 2016 – clause 4.22.5, JCT DB 2016 – clause 4.21.5 and JCT IC 2016–4.17.4.

5 NEC4 – clause 60.1(18) and 60.1(19) – an event which hinders the contractor's progress and *'is not one of the other compensation events listed in the contact'*.

6 Keating on NEC (2nd Ed) 2022 at 7–182. See also *Saintline Ltd v Richardsons, Westgarth & Co Ltd* (1940) 2 KB 99.

7 It is important to understand that such a remedy may fall short of what the contractor might obtain by claiming at common law for breach of contract.

8 *London Borough of Merton v Stanley Hugh Leach Ltd* (1985) 32 BLR 51.

9 *Fairclough Building Ltd v Vale of Belvoir Superstore Ltd* (1990) 28 Con LR 1.

10 *Whittal Builders Ltd v Chester-Le-Street District Council* (1996) 12 Const LJ 356 (reporting the 1985 *Whittal* case).

In the event that the employer intends to restrict or prevent the contractor from claiming common law damages he may insert one, or both of the following clauses:

- Excluding indirect or consequential losses (refer to Section 4.3); and/or
- An 'exclusive remedy' clause (so as to restrict the contractor to the 'Relevant Matters' or 'compensation events' expressly included in the contract – refer to Section 4.6).[11]

Some of the most frequent common law claims are considered below. No standard form contract is a wholly self-contained document; it must be read as a whole, against the background of the general law, and common law claims can arise under many of the standard form contracts in current use. A claim made under the terms of the contract must be in accordance with those terms. Where the contractor makes a claim under a specific contract provision, such as for loss and expense under a JCT 2016 contract[12] or a 'compensation event' under NEC4,[13] it must comply with any conditions precedent relating to the timing of notifications, applications or the information to be provided. If compliant (and assuming the facts support his claim), the contractor will be entitled to the remedy prescribed by the contract.

### 6.1.1 Authority of Employer or His Representative

The extent of the employer's authority, and particularly the authority of his representative, is particularly important when dealing with common law claims. This is explored further in Chapter 9 (Duties of the Employer's Representative). This is because the conditions of the relevant building contract may only permit the employer (or his representative) to ascertain or assess claims that are made under the terms of that contract (such as for 'loss and expense' under JCT 2016 contracts or for 'compensation events' under NEC4). Ordinarily, and unless the employer takes one of the following steps to vary (i.e. extend) the contract's conditions to either:

- Permit the agreement of a claim directly with the contractor (i.e. waiving the contract's provisions for notification and assessment of claims); or
- Expressly confirm that his representative now has authority to do the same on his behalf;

then the employer (or his representative) has power only to ascertain and certify (or assess) those claims which have been raised under the relevant contract clauses. The employer's representative has no power to ascertain, much less certify or assess, amounts due to the contractor in respect of common law claims.

There is often an expectation for the employer's representative to certify or assess sums agreed by the employer in response to a contractor's common law claim, but he should decline to do so. Not only does the employer's representative have no authority to ascertain, certify or assess such claims – that in doing so may render any payment notice or certificate as invalid – but also his issuing such a document implies that he played a major part in ascertaining (or assessing) the particular sum. In the event that such sum was later

11 Under NEC4, an exclusive remedy clause is likely to require clause 60.1(18) to be deleted.
12 JCT SBC/Q 2016 – clause 4.21, JCT DB 2016 – clause 4.20.1, and JCT IC 2016 – clause 4.16.1 (particularly, in respect of the Relevant Matters described in Footnote 4).
13 NEC4 – clause 61.3 (see Footnote 5).

challenged by the employer, his representative could be placed in a difficult position and face a claim for negligence from the employer. For example, under JCT 2016 contracts, it seems clear that only an arbitrator (and probably an adjudicator) may decide claims based on tort (as well as breach of contract) on the proviso that the tort in question must arise out of the transaction which is the subject-matter of the contract.[14] Sometimes a contractor will submit a claim which relies partly on the contractual machinery and partly on damages for breach of contract. If the contractor has properly separated the contractual and common law parts of the claim, the employer's representative will be able to deal with the former but decline the latter. However, very often the two elements will be mixed together. In that case, the employer's representative must either request that the contractor disentangle its claim and submit only the contractual part, or if possible, only assess the parts of the claim that are made under the terms of the contract.

The use of 'catch all' clauses JCT 2016 contracts[4] and NEC4[5] allows the employer's representative to deal (as additional costs) with many, if not all situations of employer default which previously would have had to have been dealt with as common law claims for a breach of contract. Of course, as noted earlier, the employer's representative only has such authority, if the contractor makes the claim under the terms of the contract and not as a common law claim for breach of contract (e.g. without complying with the contract in terms of notification).

### 6.1.2 Effect of 'Final Certificates'

The final certificates[15] (or final statement[16]) issued under the JCT 2016 contracts are conclusive evidence[17] that any reimbursement of loss and/or expense is in final settlement of all the contractor's claims arising out of any of the 'Relevant Matters', thus precluding any common law claims not raised under the 'loss and expense' provisions under the relevant contract. Because these clauses are quite comprehensive, the issue of the final certificate (or its equivalent) under the JCT 2016 contracts will effectively put an end even to common law claims unless presented with exceptional ingenuity. In the absence of a final certificate or in situations where the final certificate is not conclusive (such as JCT MW 2016), common law claims will only be subject to the Limitation Act 1980. Therefore, under JCT MW 2016, the contractor may pursue such common law claims at any time within the period of limitation, i.e. 6 years for actions based on simple contract or tort or 12 years if the contract is a deed. The limitation period is usually taken to run from practical completion so far as the contractor is concerned.[18]

NEC4 now contains a 'Final Assessment' procedure in clause 53, but this provides little in the way of 'conclusive evidence', compared to the final certificate or final statement under JCT 2016 contracts. There is effectively no need for NEC4 to include a procedure

14 *Re Polemis and Furness, Withy & Co* (1921) 2 KB 560; The Arbitration Act 1996.
15 JCT SBC/Q 2016 – clause 4.26 and JCT IC 2016 – clause 4.21.
16 JCT DB 2016 – clause 4.24.
17 JCT SBC/Q 2016 and JCT IC 2016 – clause 1.9.1.4, and JCT DB 2016 – clause 1.8.1.3.
18 *Tameside Metropolitan Borough Council v Barlow Securities Group Services Ltd* (2001) BLR 113 CA. This part of the first instance judgment was not overturned on appeal.

to 'conclude' any claims matters because they should have already been concluded by the 'compensation event' procedure (refer to Section 5.3.2):

- The only way that the Prices can be changed under NEC4 is via the 'compensation event' procedure[2];
- Once 'implemented', the compensation event cannot usually be revised[19]; and
- Compensation events cannot be notified after the Defects Certificate[20] has been issued.[21]

## 6.2 Implied Terms

### 6.2.1 General Principles

The written terms found in contracts to which the parties are deemed to have applied their minds and agreed are referred to as 'express terms'. When one refers to 'implied terms', it is understood that one is referring to terms which are not written terms found in the particular contract (i.e. express terms) that a court would imply into a particular contract. It is a somewhat complex topic which is comprehensively treated in the standard texts dealing with the law of contract. In relation to building contracts, implied terms usually fall into one of the following categories:

- Implied (or imposed) by statutes:
  - Rights
  - Standards
- At common law
- Other bases for implied terms in building contracts

In practice, and in the context of building contract claims, the concern is with those terms which will be implied into the contract by the courts, in order to make the contract commercially effective.[22]

A term will not be implied into a contract simply because one of the parties thinks it would have been reasonable to insert it. There can never be an implied term to give business efficacy to a contract if there is already an express term dealing with the same matter.[23]

It has been erroneously suggested that this principle applies to all implied terms; however, it does not apply to those terms which are already implied by law (e.g. by statute or at common law). Further, the courts will not imply a term into a contract which is otherwise perfectly clear, simply to sort out a problem for one or both parties. That is the case even where it is clear that the parties had not sufficiently applied their minds to the problem

19 NEC4 – clause 60.1(17) provides that a compensation event can be revised – which in itself, is a compensation event.
20 NEC4 – clause 44.3.
21 NEC4 – clause 61.7.
22 *The Moorcock* (1889) 14 PD 64.
23 *Les Affréteurs Réunis v Leopold Walford* (1919) AC 801.

before experiencing the problem, as was the case where a delay in a phased contract had unexpected results. The position has been aptly put in a well-known passage:

> *'An unexpressed term can be implied if and only if the court finds that the parties must have intended that term to form part of their contract: ...'*[24]

### 6.2.2 Terms Implied (or Imposed) by Statute

Statutes are grouped into the primary and secondary legislation. The primary legislation is usually an 'Act of Parliament', and the secondary legislation, called a Statutory Instrument, is usually referred to as 'Regulations'. In simple terms, the primary legislation (the 'Act') provides the basis for the right or standard, and the secondary legislation (the 'Regulations') provides the necessary detail.

- Rights:
  - Right to interim payments and payment notifications. The primary legislation is the Housing Grants Construction and Regeneration Act 1996 (as amended[25]) ('HGCRA 1996') which provides contractors and sub-contractors (as the payees) with the right to interim payments if their contract's duration exceeded 45 days, and that employers (or payers) were obliged to issue payment and pay-less notices, failing which they could be liable to pay the amount claimed (or subsequently notified) by the payee. The details of which are provided by the secondary legislation – in the Scheme for Construction Contracts Regulations 1998[26] (as amended[25]) ('the Scheme');
  - Right to adjudication. The HGCRA 1996 provides that parties to a 'construction contract' have the right to have a dispute (which can be any difference) decided by adjudication, and the decision is temporarily binding until finally resolved by the Court, agreement or by arbitration;
  - Time for performance. In the event that no programme of works or completion date is agreed between the parties, s14 of the Supply of Goods and Services Act 1982 provides that the contractor must complete the works within 'a reasonable time' (see Section 3.1.1);
  - Under s.12 of the Sale of Goods Act 1979, a condition is implied into the sales of chattels that the seller has a right to sell the goods, that the buyer shall enjoy quiet possession of them and that they are free from any charge or encumbrance to a third party.
- Standards
  - That materials must be 'satisfactory quality' under s14(2) of the Sale of Goods Act 1979[27];

24 *Trollope & Colls Ltd v North West Metropolitan Regional Hospital Board* (1973) 9 BLR 60 at 70 per Lord Pearson.

25 Amended by the Local Democracy, Economic Development and Construction Act 2009.

26 The 'Scheme' appropriate to the particular UK regional jurisdiction will apply – The Scheme for Construction Contracts (England and Wales) Regulations 1998 (as amended), or the Scheme for Construction Contracts (Scotland) Regulations 1998 (as amended).

27 It is generally implied that a contractor will supply good and proper materials, construct the work in a good and proper manner and that the finished structure will be reasonably fit for its intended purpose so far as that purpose has been made known to the contractor and provided that there is no independent

- The primary legislation is the Building Act 1984, and this imposes a duty to comply with the secondary legislation in the form of The Building Regulations 2010 (and 2018) and the Building (Amendment) (England) Regulations 2024. The Building Regulations provide that the work should comply with the appropriate 'Approved Documents[28]';
- Imposing duties when constructing dwellings that the work is done in a workman-like or professional manner, using proper materials so that the dwelling will be fit for human habitation under s1 of the Defective Premises Act 1972[29]; and
- Services must been performed with 'reasonable skill and care' under s13 of the Supply of Goods and Services Act 1982.[30]

### 6.2.3 Common Law (Case Law)

Terms implied by common law (or by case law) often clarify terms implied by statute[31] or confirm principles that are implied into building contracts. Three of the most common implied terms for which parties often rely upon to formulate their claims are:

- Prevention (by a party)
- Co-operation
- Reliance

First, one of the most common implied terms from which a contractor may base a claim for breach of contract is the employer's duty not to carry out any acts of prevention, and to do everything reasonably necessary to allow the contract to be properly performed. Examples of the employer preventing the contractor carrying out the works are more obvious (such as restricting access to, or movement around the site) and failing to provide the contractor with the necessary construction information in a timely manner. However, employers and their consultants are often guilty of preventing and obstructing the contractor's work, by less obvious means. After entering into the building contract, they may impose new procedures upon the contractor, which are ill-thought out, administratively

designer. *Young & Marten Ltd v McManus Childs Ltd* (1969) 9 BLR 77. *Hancock v B W Brazier (Anerley) Ltd* (1966) 2 All ER 901; *Test Valley Borough Council v Greater London Council* (1979) 13 BLR 63; *Viking Grain Storage Ltd v T H White Installations Ltd* (1985) 3 Con LR 52.

28 The 'Approved Documents' (A to S and 7) are published by the UK Government: https://www.gov.uk/government/collections/approved-documents

29 See also *Test Valley Borough Council v Greater London Council* (1979) 13 BLR 63, it was held by the High Court that the question: '*whether there were implied terms of the said agreement that the respondents would provide completed dwellings which were constructed (a) in a good and workmanlike manner, (b) of materials which were of good quality and reasonably fit for their purpose and (c) so as to be fit for human habitation.*

30 See also *IBA v EMI and BICC* (1980) 14 BLR 1 at 44 per Lord Fraser: *It is now well recognised that in a building contract for work and materials, a term is normally implied that the main contractor will accept responsibility to his Employer for material provided by nominated sub-contractors. The reason for the presumption is the practical convenience of having a chain of contractual liability from the Employer to the main contractor and from the main contractor to the sub-contractor....* Accordingly, *the principle that was applied in Young and Marten in respect of materials ought in my opinion to be applied here in respect of the complete structure, including its design'.*

31 For example, refer to Footnotes 27, 30 and 30.

cumbersome and time-consuming, and often none of them are found in the contract! Some examples are as follows:

- Introducing additional or different procedures for 'approving' the contractor's work or design;
- Unduly complicated and long-winded cost-verification procedures, where the agreed or established value of a variation must be approved by several tiers of management in the employer's organisation;
- Refusal to instruct necessary variations until the contractor provides a quotation that the employer (or his consultants) finds acceptable; and
- Refusing to pay or delaying interim payments until the contractor has provided a VAT invoice, when amounts due under the contract are exclusive of VAT.[32]

In these (and other) cases, the employer may be relatively unsophisticated so far as building contracts are concerned, and may require a great deal of support and guidance from his consultants. However, the employer should not let his consultants 'run riot' and attempt to 'reinvent the wheel' where the building contract already contains perfectly sensible and practical procedures.

The implied 'prevention' term was considered in some detail in the case of *London Borough of Merton v Stanley Hugh Leach Ltd*,[33] where one of the many points at issue concerned the implication of certain implied terms. The contractor was asking the court to agree that the following terms should be implied into the building contract (then the JCT 63 version):

- That the employer would not hinder or prevent the contractor from carrying out its obligations in accordance with the terms of the contract and from executing the Works in a regular and orderly manner; and
- That the employer would take all steps reasonably necessary to enable the contractor to discharge its obligations and to execute the Works in a regular and orderly manner.

The court was quite clear that these terms ought to be implied:

> *'The implied undertaking not to do anything to hinder the other party from performing his part of the contract may, of course, be qualified by a term express or to be implied from the contract and the surrounding circumstances. But the general duty remains so far as qualified. It is difficult to conceive of a case in which this duty could be wholly excluded'*[34]

A somewhat older case is to the same effect:

> *'There is an implied term by each party that he will not do anything to prevent the other party from performing the contract or to delay him in performing it. I agree that generally such a term is by law imported into every contract.'*[35]

---

32 Article 2 (Contract sum) of the JCT 2016 contracts provides that the contract sum is 'VAT-exclusive' and clause 5.15 of NEC4 provides that *'Any tax which the law requires a Party to pay to the other Party is added to any payment made under the contract'*.

33 *London Borough of Merton v Stanley Hugh Leach Ltd* (1985) 32 BLR 51.

34 [1985] 32 BLR 51 at 80 per Vinelott J.

35 *Barque Quilpé Ltd v Brown* (1904) 2 KB 264 at 274 per Vaughan Williams LJ. See also *Cory Ltd v City of London Corporation* (1951) 2 All ER 85.

The second implied term is one of co-operation between the parties. In *Merton v Leach*, the contractor wanted the court to imply a term of co-operation. The court said:

> *'As regards the second of these two terms it is well settled that the courts will imply a duty to do whatever is necessary in order to enable a contract to be carried out'.*[33]

Other courts have reached much the same conclusion. In *Mackay v Dick* it was said:

> *'Where in a written contract it appears that both parties have agreed that something should be done which cannot effectively be done unless both concur in doing it, the construction of the contract is that each agrees to do all that is necessary to be done on his part for the carrying out of that thing though there may be no express words to that effect'.*[36]

which is echoed in *Luxor (Eastbourne) Ltd v Cooper*:

> *'If A employs B for reward to do a piece of work for him which requires outlay and effort on B's part ... generally speaking, where B is employed to do a piece of work which requires A's co-operation ... it is implied that the necessary co-operation will be forthcoming'.*[37]

Therefore, it appears to be well settled that both employer and contractor must co-operate to the extent necessary to enable the contract to be performed. The scope of co-operation between the employer and contractor can be wide and varied in modern building contracts. At one end of the scale, it may amount to little more than the employer refraining from hindering the contractor in the execution of the Works. Conversely, the progress of the contractor's work may significantly depend upon the employer's co-operation – for example, in obtaining permits or approvals[38] to allow particular work operations to proceed.

However, there are exceptions to these principles. More correctly, the principles should be applied with a degree of caution and with a view to what is necessary to make the contract work. The guidance set out in *Mona Oil Equipment Co v Rhodesia Railway Co* is useful:

> *'I can think of no term that can properly be implied other than one based on the necessity for co-operation. It is, no doubt, true that every business contract depends for its smooth working on co-operation, but in the ordinary business contract, and apart, of course, from express terms, the law can enforce co-operation only in a limited degree - to the extent that it is necessary to make the contract workable. For any higher degree of co-operation the parties must rely on the desire that both of them usually have that the business should be done'.*[39]

---

36 *Mackay v Dick* (1881) 6 App Cas 251 at 263 per Lord Blackburn.

37 *Luxor (Eastbourne) Ltd v Cooper* [1941] 1 All ER 33.

38 The employer may have to provide information to the contractor in support of applications under s278 of the Highways Act 1980, or under s106 of the Town and Country Planning Act 1990.

39 *Mona Oil Equipment Co v Rhodesia Railway Co* (1949) 2 All ER 1014 at 1018 per Devlin J.

None of the JCT 2016 contracts displace these two implied terms, and thus, the employer is liable at common law for breach of them. This can have important practical implications, and consideration must be given to one of the most common grounds of common law claim arising under a building contract. NEC4, of course, expressly includes an obligation for not just the parties, but the Project Manager and the Supervisor to *'act in a spirit of mutual trust and co-operation'*.[40]

In *Holland Hannen & Cubitts v WHTSO*,[41] a term was implied that *'the Employer would do all things necessary to enable the contractor to carry out and complete the works expeditiously, economically and in accordance with the contract'*. In *Thomas Bates & Son Ltd v Thurrock Borough Council*[42], a further term was implied that where the employer is to provide goods for the works, they will be supplied in time to enable the contractor to carry out the works expeditiously and in accordance with the contractor's planned progress.

Every building contract contains an implied term that the employer will give possession of the site to the contractor in sufficient time to enable it to complete the Works by the due date[43] – and most JCT 2016 contracts[44] and NEC4[45] contain express terms to the same effect. The employer's failure to give possession – as provided in the contract, is a breach for which the contractor will be entitled to damages in respect of any resultant loss.

A third implied term is one where a party seeks to rely upon the accuracy of the documents provided by the other party (see also Section 6.7). Claims often arise when the contractor has incurred higher costs than he either expected (or allowed for within his price) for carrying out an element of the works – groundworks in particular. In *Bacal Construction (Midlands) Ltd v Northampton Development Corporation*,[46] the contractors recovered damages for the employer's breach of an implied term or warranty that the ground conditions would accord with the basis on which they were instructed to design. The contract was substantially in JCT 63 form, and the Court of Appeal held that the necessary re-designing of the foundations and the additional work caused by the discovery of porous, calcium carbonate rock did not rank as variations for the purposes of the contract. The court, in an analysis which was accepted by the Court of Appeal, said:

> *'Bacal have submitted that there are strong commercial reasons for implying such a term or warranty in the contract as they have suggested. First, before designing the foundations of any building, it is essential to know the nature of the site conditions. Secondly, where the contract is for a comprehensive development of the kind here in question, the contractor must know the soil conditions at the site of each projected block in order to be able to plan his timetable and estimate his requirements for materials. These are matters which relate directly to the contract price. Thirdly, if the work is interrupted or delayed by unforeseen complications, the contractor is unlikely to be able to complete his contract in time'*.[48]

---

40 NEC4 – clause 10.2.

41 *Holland Hannen & Cubitts v WHTSO* (1981) 18 BLR 80 - concerning the JCT 1963 contract.

42 *Thomas Bates & Son Ltd v Thurrock Borough Council* (1976) JPL 34.

43 *Freeman & Son v Hensler* (1900) 64 JP 260. Possession is dealt with in more detail in Section 4.6 of this chapter.

44 JCT SBC/Q 2016 and JCT IC 2016 – clause 2.4, and JCT DB 2016 – clause 2.3. There is no equivalent express term in JCT MW 2016, so this term will be implied.

45 NEC4 – refer to clauses 30.1 and 60.1(2).

46 *Bacal Construction (Midlands) Ltd v Northampton Development Corporation* (1976) 8 BLR 88.

When providing surveys or reports, many consultants include disclaimers that they are not responsible for the accuracy of any information that they provide, and employers avoid any liability by removing this imply term by including a 'no reliance' clauses in their building contracts.

### 6.2.4 Other Bases for Implied Terms in Building Contracts

Terms can be implied into building contracts in other ways, which are described as follows:

- For reasons of 'business efficacy'. Without implying such a term, the building contract would not achieve the result intended by the parties, and possibly to the extent that it may not operate at all;
- By convention, including:
  - By custom (including local custom), provided that it is not inconsistent with the contract as a whole[47]; or
  - If there is a 'course of dealing' between the parties, similar terms (including the parties' own interpretation) may be implied into a new contract made on the same basis.[48] This is a fairly rare occurrence although one which is argued by contractors where the parties have either not executed a contract, or had only agreed some of the contract's intended terms. In order for a term to be implied on this basis, the parties must have '*consistently on former and similar occasions adopted a particular course of dealing*'.[49] It may not be sufficient to found a course of dealing if there are only two or three previous occasions or if previous contracts were different in nature.
  - Particular trade usage in a trade or profession where there is a long-standing practice in support. The usage must be certain, long-standing, well known in that trade or profession and it must be a practice which fair minded people would adopt.[50]
- If a term 'goes without saying'. Seemingly closely related to 'business efficacy' principle, this is often referred to as the 'officious bystander test'. The idea is that if, when the parties were agreeing their terms, an officious bystander had been asked if a particular term was included, the bystander would have testily replied 'Yes, of course'.

An 'entire agreement' clause (i.e. one which restricts the governing contract terms to those expressly written in the contract) will usually defeat a claim made on the basis of a breach of any implied terms.

## 6.3 Variation of the Contract Itself (Not a Variation Under the Terms of the Contract)

Variation of the contract itself is often mistaken with a variation of the works under that contract – probably because many standard forms of building contract expressly refer to variations and variation clauses. Often, the parties have little, or no understanding of the

47 *London Export Co Ltd v Jubilee Coffee Roasting Co Ltd* (1958) 1 WLR 661.
48 *McCutcheon v David McBrayne Ltd* (1964) 1 WLR 125.
49 Chitty on Contracts (30th Ed) at 13-022.
50 *Symonds v Lloyd* (1859) 141 ER 622.

meaning of a variation of the contract and may never have encountered that concept. A 'Variation' (often as a defined term in standard forms of building contracts) usually refers to the employer (or his representative), instructing the contractor to change in the works in some way (refer to Sections 5.1.1 to 5.1.4). However, when one refers to 'variation of the contract', reference is being made to a change in the terms of the building contract itself.

The parties to a building contract may agree to change any of its terms, provided only that such changes are not actually unlawful, or that do not result in the contract becoming void or voidable. Unless a representative or an agent of one of the parties is clearly given authority to agree such a change to the contract terms on their behalf,[51] any agreements must be made by the parties themselves – i.e. between the employer and contractor, or between a contractor and sub-contractor. Where the parties agree to a change in the terms of the original building contract, they must do so by forming another contract which varies the original and is made on the same basis. Therefore, if the original building contract was executed as a deed, the new contract (to vary the original) must also be a deed (commonly referred to as a 'deed of variation').

If the original contract was not executed as a deed, then 'consideration' must be present. To put this simply, each party must do something 'extra' for the other. For example, the contractor may agree to the employer making interim payment at less frequent intervals (e.g. every 2 months), in return for the employer making an additional payment (either a lump sum or a percentage addition to all amounts due to the contractor). Usually, there is sufficient evidence of consideration in such cases, but to avoid any doubt, it is advisable for the contract varying the first to be executed as a deed.

Some common examples of varying the original building contract include:

- The addition of a substantial quantity work that is of a completely different nature to the work contained in the original building contract, for which the original contract's variation (and valuation) procedure may be inadequate.[52] For example, if the original building contract required the contractor to build a warehouse for the employer, and subsequently, the employer wanted the contractor to build some housing on an adjoining plot of land;
- Subsequently agreeing to sub-divide the works so 'sectional completion' will apply instead of a single completion date; and
- An agreement to accelerate where the contract does not provide for it.[53]

## 6.4 Omission of Work to Give it to Others

It is widely understood that a contractor is entitled to carry out the whole of the works in the building contract. If the employer attempted to prevent the contractor carrying out any part of the Works, it would be a breach of contract. Most standard form building contracts provide that the employer (or his representative) may instruct the omission of parts of the

51 For example, a solicitor instructed by one of the parties. Ordinarily, the Architect/Contract Administrator (JCT) or Project Manager (NEC4) has no authority to vary the terms of the building contract. See *Sharpe v San Paulo Railway* (1873) 8 Ch App 597.

52 Most standard form building contracts have 'Variation' clauses, but not all may be sufficiently comprehensive (such as JCT MW 2016).

53 See section 3.8.

Works. However, such clauses do not permit the omission of work so that it can be carried out by another contractor (for example, if the second contractor was cheaper). Ordinarily, most building contracts allow work to be omitted from the overall scope of works, but not omitted in order to give it to someone else.[54]

In a 1997 case,[55] the court held that the architect's power to omit a provisional sum cannot be exercised so as to omit work if the employer intends to have the omitted work carried out by another contractor.[56] The authorities were reviewed in a 2003 case[57] where it was said:

> *'The justification for these decisions is in my judgment to be found in fundamental principles. A contract for the execution of work confers on the contractor not only the duty to carry out the work but the corresponding right to be able to complete the work which it contracted to carry out. To take away or vary the work is an intrusion into and an infringement of that right and is in breach of contract... reasonably clear words are needed in order to remove work from the contractor simply to have it done by somebody else; whether because the prospect of having it completed by the contractor will be more expensive for the employer than having it done by somebody else, although there can well be other reasons such as timing and confidence in the original contractor. The basic bargain struck between the employer and the contractor has to be honoured, and an employer who finds that it has entered into what he might regard as a bad bargain is not allowed to escape from it by the use of the omissions clause so as to enable it then to try and get a better bargain by having the work done by somebody else at a lower cost once the contractor is out of the way (or at the same time if the contract permits others to work alongside the contractor)'.*

This principle may also apply where the employer subsequently intends to change who provides work within the contractor's scope of works – for example, from a 'domestic' sub-contractor of the contractor's choosing to a nominated sub-contractor or supplier. On the other hand, if work is stipulated to be carried out by a named (or specified) sub-contractor, the main contractor cannot be forced to do it, nor can he decide to do it itself.[58]

The situation is less clear where an employer wishes to omit work, in order to have it carried out by another contractor, but after the first contractor has completed the contracted scope of work. For example, the employer feels that the first contractor is not performing well and wants to give the work to another, more capable, contractor. The employer may omit work and perhaps intends to never to have it done at all. Subsequently, after completion (by the first contractor, and perhaps after the rectification period has ended), the employer may decide that the work should be carried out after all, and instructs the second

54 See the American case of *Gallagher v Hirsch* (1899) NY 45, where the appeal court held that the 'omission' meant work which was not to be done at all and did not include work which was taken from the contractor and given to another contractor to carry out.

55 *AMEC Building Ltd v Cadmus Investment Company Ltd* (1997) 13 Const LJ 50. See also *Vonlynn Holdings Ltd v Patrick Flaherty Contracts Ltd* 26 January 1988, unreported.

56 Ibid at 64 to 67.

57 *Abbey Developments Ltd v PP Brickwork Ltd* (2003) EWHC 1987 (TCC) at paragraphs 45 and 47 per Judge Lloyd.

58 *T A Bickerton & Son Ltd v North West Metropolitan Regional Hospital Board* (1970) 1 All ER 1039.

contractor to do it. Although these possibilities have been suggested as being within the employer's discretion,[59] a court thought it unlikely:

> *'It remains to be decided (but it is very doubtful) that work could be omitted simply because the owner is dissatisfied with the performance of the contractor, since the contract itself could and should, and in many cases does, make provision for what is to happen if the contractor's performance is so poor that the employer, having lost confidence in the contractor's ability to complete the work in accordance with the contract, is entitled then to take the whole or part of the work then outstanding away from the contractor in order that it can be done by others more satisfactorily. But such a provision for termination, or partial termination, is something which must be the subject of clear words, because otherwise it would be an intrusion into the contractor's right to finish the work. … The editor of Hudson refers to the possibility that there may be "sound technical or commercial reasons for omitting the work" which would justify an otherwise unlawful omission. It is as difficult to see how that can be imported legitimately into a contract as it is to see how to give effect to the policy that you may not omit work but to have it done by someone else. Could it be implied - if so, does this mean that an employer would be liable to be interrogated as to its motives every time there was a variation by way of omission or which was seen as a prelude to or paving the way to an omission? … The test must therefore be whether the variations clause is or is not wide enough to permit the change that was made. If, with the advantage of hindsight, it turns out that the variation was not ordered for the purpose for which the power to vary was intended then there will be a breach of contract. So the motive or reason is irrelevant…'.*[60]

In practice, it may be quite difficult to decide whether an employer requires an omission to be a permanent omission of work, or whether it is simply a way of having the work done by others at a cheaper cost – particularly if the omitted work is actually carried out after (i) practical completion or (ii) even when the original contractor's obligations (including rectifying defects) are complete, and he has left site. The correct view seems to be that if the employer has engaged others to carry out previously omitted work, at any time before the parties' obligations to each other are at an end (e.g. before the 'final certificate' or its equivalent is issued), the original contractor ought to be able to mount a claim against the employer for any damages suffered as a result of the breach. This situation is clearer under the JCT 2016 contracts where the 'final certificate' (or its equivalent) is 'conclusive evidence' in many fundamental aspects of the contract;[61] however, it remains to be seen how the same principle could apply to contracts such as NEC4, where there is no single concluding 'certificate'[62] or where the financial aspects of the contract are concluded progressively.[63]

---

59 I N Duncan Wallace, 'Hudson's Building and Engineering Contracts', 11th edition (1995) Sweet & Maxwell at paragraph 4.202.

60 *Abbey Developments Ltd v PP Brickwork Ltd* (2003) EWHC 1987 (TCC) at paragraphs 48, 49 and 50 per Judge Lloyd.

61 For example, refer to clause 1.9.1 of JCT SBC 2016 or JCT IC 2016, clause 1.8.1 of JCT DB 2016.

62 Under JCT MW 2016.

63 Following implemented 'compensation events' under NEC4 clause 66.3.

Where a main contractor brings others onto the site to supplement the labour of a sub-contractor, already lawfully on site, without that sub-contractor's consent, the main contractor is in breach of contract which may be repudiatory in nature so as to entitle the sub-contractor to treat the sub-contract as at an end, and leave site.[64]

## 6.5 Extra Work

The contractor is not entitled to payment if (of his own volition – without instruction or agreement) he carries out 'extra' work which either (i) is an additional quantity to that included in the contract or (ii) exceeds the specification. In both instances, the contractor may be in breach of contract, because he has done something different from that which he undertook to do under the contract.[65] However, if an employer gains benefit from work which he instructed the contractor to carry out then ordinarily, he is obliged to pay for it, but the contractor may not be entitled to payment for work which has not been instructed. Disputes often arise because the employer insists that certain work is included in contract's scope of works, but the contractor maintains that it should be treated as an 'extra' (see Section 5.1). For example, if the employer refuses to issue an instruction requiring a 'variation' and the work is substantial, the contractor may be tempted to refuse performance unless an instruction is issued and, in the absence of such instruction, to treat the contract as repudiated and sue for damages.[66] Clearly, this is a very risky strategy for the contractor, and one that should rarely be attempted.

Standard forms of building contract oblige the contractor to comply with 'instructions' forthwith,[67] even if the instruction is not for a 'variation' (or 'compensation event'). Therefore, the employer can often instruct the contractor to carry out the disputed work without treating it as a 'variation'. If the contractor fails to comply with such an instruction, then the employer may (after notifying the contractor) instruct and pay others to carry out the relevant work[68] and deduct costs from the contractor. Furthermore, if the contractor's refusal persists, the employer may be able to terminate the contractor's employment under the contract.[69]

In most circumstances, it would be advisable for the contractor to proceed with the work and resolve the dispute or difference (of whether or not the disputed work constitutes a variation) via the contract's dispute resolution procedures (e.g. by adjudication). In doing

64 *Sweatfield Ltd v Hathaway Roofing Ltd* (1997) CILL 1235.

65 *Holland Hannen & Cubitts v Welsh Health Technical Services Organisation* (1981) 18 BLR 80.

66 *Peter Kiewit Sons' Company of Canada Ltd v Eakins Construction Ltd* (1960) 22 DLR (2d) 465.

67 JCT SBC/Q 2016 – clause 3.10, JCT DB 2016 – clause 3.5, JCT IC 2016 – clause 3.8, JCT MW 2016 – clause 3.4 (provided the instruction is in writing), and the obligation to comply 'forthwith' is implied in NEC4 – clause 27.3.

68 JCT SBC/Q 2016 – clause 3.11, JCT DB 2016 – clause 3.6, JCT IC 2016 – clause 3.8, JCT MW 2016 – clause 3.5, and NEC4 clause 27.3. NEC4, clause 27.3 does not set out a timescale for the contractor's compliance with a valid instruction from the Project Manager or Supervisor, but the scope of the clause appears wide enough to permit the relevant instruction to include a timescale, with which the contractor must comply.

69 Provided that the disputed works is a 'substantial' part of the works – see JCT SBC/Q 2016, JCT DB 2016 and JCT IC 2016 – clause 8.4.1.1, and JCT MW 2016 – clause 6.4.1.1, and NEC4 clause 91.2 (reason R11).

so, the contractor should clearly inform the employer that it is proceeding with the disputed work, without prejudice to its assertion that the work is a 'variation'.

The employer (or his representative) may only instruct such variations as the contract expressly provides[70] (refer to Section 5.1). Although an employer's representative acts as an agent for the employer, he can only do so with limited authority, which is either provided by specific provisions in his contract with the employer, or in the absence of such, is usually limited to what is stated in the contract.[71] Where the scope of the instructed variation is not expressed in precise terms (such as JCT MW 2016), it is normally implied that the employer's representative can only issue instructions which are within the scope of the contract. Therefore, if the employer's representative issues instructions which are not empowered by the contract, the contractor should firstly seek advice, and secondly not comply, otherwise he may be in breach of contract. Where the employer's representative appears to have authority (see Section 9.1) to instruct a variation, even if unauthorised by the employer, the employer will normally be liable to the contractor for the price, and the employer may in turn look to his representative for reimbursement.

Where the contract (such as JCT MW 2016[72]) does not comprehensively list 'variations' to the Works which the employer (or his representative) has power to instruct, the contractor may have a real problem if he's instructed to do something which is not clearly provided for under the terms of the contract. Some contracts have clauses which permit the contractor to require the architect to specify the empowering clause[73] and this protects the contractor, even if the employer or his representative is wrong. Despite JCT MW 2016 not containing an equivalent clause, it would nevertheless be advisable for the contractor to enquire under what provision the instruction was issued.

If the employer gives a direct instruction to the contractor, then save for under JCT DB 2016, it would not be authorised under most standard forms (which clarify the scope of instructions that can be issued and who can issue them). It is likely that the giving of such an instruction could create a separate contract[74] (i.e. outside the terms of the building contract between the contractor and employer), by which the contractor would most likely be entitled to payment on a *quantum meruit* basis.

There may be some exceptional circumstances which entitle the contractor to additional payment:

- If the employer has implicitly promised to pay if the work is, as a matter of law, additional to the contract;[75] or

70 The employer's representative may not have an automatic right to instruct 'variations' – *Cooper v Langdon* (1841) 9 M & W 60.

71 *Sir Lindsay Parkinson & Co Ltd v Commissioners of Works and Public Buildings* (1950) 1 All ER 208.

72 Contract clause 3.6.1 of JCT MW 2016, in particular *'or other change to the Works'* with clause 5.1 of JCT SBC/Q 2016, JCT DB 2016, and JCT IC 2016, and NEC4 clause 60.1 (compensation events).

73 'Provisions empowering instructions', see JCT SBC/Q 2016 – clause 3.13, JCT DB 2016 – clause 3.8, and JCT IC 2016 – clause.

74 Unless the parties agreed to 'vary' the building contract's terms to include the directly instructed variation (refer to Section 6.4).

75 *Molloy v Liebe* (1910) 102 LT 616.

- If the employer agrees to pay, and if it is subsequently found that the work was included in the original contract, but the employer will has received some additional consideration in return for the promise to pay.[76]

## 6.6 Possession of the Site

### 6.6.1 General Position

There is an implied term in every building contract that the employer will give possession of the site to the contractor within a reasonable time to enable the contractor to complete the Works by the contractual date stipulated for completion.[77] The employer must give the contractor possession of the site on the due date and an employer who is in breach of that obligation is liable in damages.[78] Under the terms of the contracts considered in this book, express provisions are set out for employer to give possession to the contractor (refer to Section 3.1.2).

### 6.6.2 Sufficient Possession

The question often arises whether the contractor must have possession of the whole site or whether it is enough if he has possession of sufficient of the site to begin to carry out the Works. In this context, the meaning of 'sufficient' is twofold:

- Has the employer given sufficient areas, or parts of the site so the contractor can carry out the works without any significant encumbrances or impediments; and
- Is the contractor entitled to exclusive possession of those areas or parts, or does he have to share possession with other parties?

The first question can be answered in general terms. Excluding any separate sectional completion or 'start' dates,[79] the Courts considered the meaning of 'possession' and held that it cannot be given 'piecemeal', and possession of the site must be given by the employer to the contractor.[80] Clearly, where sectional completion applies, the employer will be obliged to give the contractor possession of an entire section by the relevant date for possession (see Section 3.1.2). However, it is suggested that the extent of possession must be significant[81] and sufficient so the contractor '*...must clearly be allowed such use of the site as it requires for the purpose of carrying out the work in the way it thinks best*',[82] without the employer committing any acts of 'prevention' (see Section 6.2.3). Possession must also include access.[83]

76 *Sharpe v San Paulo Railway* (1873) 8 Ch App 597.
77 Hudson 14th Ed (2020) at 3-089.
78 *Rapid Building Group v Ealing Family Housing Association Ltd* (1984) 1 Con LR 1.
79 For example, any 'Dates of Possession of Sections' under JCT SBC/Q 2016 and JCT IC 2016 - clause 2.4, JCT DB 2016 – clause 2.3, or NEC4's *'access dates'* – clause 30.1.
80 *Whittal Builders v Chester-Le-Street District Council* (1987) 40 BLR 82.
81 *The Rapid Building Group Ltd v Ealing Family Housing Association Ltd* (1984) 1 Con LR 1.
82 Hudson 14th Ed (2020) 3-089.
83 See *LRE Engineering Services Ltd v Otto Simon Carves Ltd* (1981) 24 BLR 127, a case concerning a main contractor not providing access to a sub-contractor, where (1) the sub-contract stated that *'access...shall be afforded'* and (2) access was denied by unlawful picketing during a strike.

**Table 6.1** Contractor's remedies for the employer's failure to give possession.

| Breach by failing to give possession | JCT SBC/Q 2016 | JCT DB 2016 | JCT IC 2016 | NEC4 |
|---|---|---|---|---|
| Extension of time/loss and expense<br>*'impediment, prevention or default'* | EoT 2.29.7<br>L&E 4.22.5 | EoT 2.26.6<br>L&E 4.21.5 | EoT 2.20.6<br>L&E 4.17.4 | — |
| Compensation Event | — | — | — | 60.1(2) |
| Possible Termination[a)]<br>The employer prevents the contractor from completing the whole (or a substantial) part of the works (default periods of suspension are in brackets) | 8.9.2.2<br>(2 months) | 8.9.2<br>(2 months) | 8.9.2.2<br>(2 months) | 91.7<br>(13 weeks) |

a) JCT MW 2016 allows 'suspension' for a continuous period of 1 month (clause 6.8.2.2).

The answer to the second question, particularly when considering the standard forms of building contract considered by this book, appears to be 'no'. Most contain express provisions that remove the contractor's right to exclusive possession, as the contractor is obliged to allow the following to have access to the works:

- The employer and his representative[84]; and
- Other contractors[85]

### 6.6.3 Failure to Give Possession

If the employer fails to give possession on the date stated in the contract, he is in breach of contract, and if the failure continues for an inordinate period of time, it may be a breach that is sufficiently serious as to entitle the contractor to either terminate his employment under the contract or accept a repudiatory breach by the employer. It is clear that without possession of the site, the contractor cannot execute the Works, so such a breach is likely to be fundamental and repudiatory. It is a breach not only of the express terms of the JCT and NEC4 contracts but also a breach of the term that would be implied at common law in the absence of an express term.

When possession of the site (or a section) is deferred (refer to Section 6.6.4), the contractor will be entitled to an extension of time, but not always additional costs (no entitlement to loss and expense under JCT 2016 contracts, but deferment is a 'compensation event' under NEC4). The contractor may also have the right to terminate his employment under the contract as shown in Table 6.1.

Given that the contractor would be entitled to an extension of time and (in the case of NEC4) additional costs (see Table 6.1), it is unlikely that a contractor would elect to treat the breach as an opportunity to bring its obligations to an end, either by the contract's

84 JCT SBC/Q 2016, JCT DB 2016, and JCT IC 2016 – clause 3.1 (access for the employer or his representative). NEC4, clause 27.2.
85 See 'Work not forming part of the Contract': JCT SBC/Q 2016 and JCT IC 2016– clause 2.7 and JCT DB 2016 – clause 2.6. NEC4 clause 25.1 where the contractor must share the site (Working Areas) with 'Others' (as defined in clause 11.2(12)).

**Table 6.2** Contractor's remedies for prolonged deferment and suspension.

| Breach for prolonged deferment or suspension (postponement) | JCT SBC/Q 2016 | JCT DB 2016 | JCT IC 2016 | NEC4 |
|---|---|---|---|---|
| Default period of deferment | 2.5 (6 weeks) | 2.4 (6 weeks) | 2.5 (6 weeks) | — |
| Default period of suspension | 2 months | 2 months | 2 months | 13 weeks |
| Instruction to suspend work | 3.15 | 3.10 | 3.12 | 34.1 |
| Extension of time (JCT) | | | | |
| • Deferment | 2.29.3 | 2.26.3 | 2.20.3 | — |
| • Suspension (or postponement) | 2.29.2.1 | 2.26.2.2 | 2.20.2.1 | — |
| Loss and expense (JCT) | | | | |
| • Deferment | — | — | — | — |
| • Suspension (or postponement) | 4.22.2.1 | 4.21.2.1 | 4.17.2.1 | — |
| Compensation Event (NEC4) | — | — | — | 60.1(2)<br>60.1(4) |
| Possible Termination[a)]<br>The employer prevents the contractor from completing the whole (or a substantial) part of the works (default periods of deferment and/or suspension are shown above) | 8.9.2.2 | 8.9.2<br>(2 months) | 8.9.2.2<br>(2 months) | 91.7<br>(13 weeks) |

a) Refer to Section 4.4.2.3 (Plant).

termination provisions or by repudiation. Any consideration of termination or repudiation is likely to include whether the duration for failing to be given possession will exceed the period of deferment (or suspension) stated in the contract (refer to Table 6.1) and the commercial benefit[86] to the contractor for continuing with the project. It is more likely that a contractor in this position will rely on its common law right to claim damages, while keeping its employment under the contract alive.

### 6.6.4 Deferring Possession and Suspension by the Employer

Deferring possession and suspending (or postponing) the works are closely related and most modern building contracts provide a default periods of deferment and suspension that should not be exceeded. Deferment occurs at the very start of the works, or of a section, and suspension (again, of the works, or of a section), can occur at any time before completion; and is the postponement of the contractor's operations. The contractor's remedies for prolonged deferment and suspension, beyond the default periods, are similar to those set out in Table 6.1, and are shown in Table 6.2.

86 The contractor is entitled to damages for the breach as the least of its remedies – see *London Borough of Hounslow v Twickenham Garden Developments Ltd* (1970) 7 BLR 81.

Under the three JCT contracts in Table 6.2, the contractor has no entitlement to additional costs (loss and/or expense) in the event that the employer defers possession for up to the default (6 week) or other agreed period. It seems that the basis for removing the contractor's entitlement in this manner is that because he cannot begin the works until the deferment period is over, it is unlikely that he will incur significant costs for the period or deferment. Whilst a larger contracting organisation may be able to mitigate its losses by allocating site resources (such as staff[87]) to other projects on a temporary basis, he may have had to reserve significant items of plant and equipment (see Footnote a in Table 6.2), along with his site accommodation[88] prior to commencing the works. Furthermore, the contractor will not be earning a contribution towards its head office overheads[89] during a period of deferment. A prudent contractor working under these JCT contracts should allow a risk or contingency sum within its contract price to cover it for losses during a deferment period.

## 6.7 Incorrect Information

### 6.7.1 Basis of Claim

Claims under this head commonly arise where the employer has provided the contractor with information concerning the works (either information about the site, or describing the work to be undertaken), and in reliance upon the accuracy of such information, the contractor's allowances in terms of its price and/or programme turn out to be insufficient. The contractor's reliance upon information, which is subsequently found to be inaccurate, is partly covered in Section 6.2.3.

### 6.7.2 Provision of Correct Information

It is suggested that there is an implied term that where the employer is to provide the contractor with information concerning the works and then that information should be correct. These are the following two bases for implying this term:

- From case law and
- As a claim for 'misrepresentation'.

There are three cases, each of which required the issue of correct information to be resolved by the court, albeit, two concern the provision of information by the contractor to a sub-contractor:

- In *Merton v Leach*,[90] the court accepted an implied term that the 'The Architect would provide [the contractor] with ... correct...information concerning the works';
- A similar clause was implied in *J&J Fee v Express Lift Co*[91] – that the contractor would '...provide the [sub-contractor] with correct information concerning the works...'; and

87 Refer to Section 4.4.2.2 (Management costs).
88 Refer to Section 4.4.2.1 (On-site establishment costs).
89 Refer to Section 4.4.2.5 (Head office overheads).
90 *London Borough of Merton v Stanley Hugh Leach Ltd* (1985) 32 BLR 51 at 81 to 83.
91 *J&J Fee Ltd v The Express Lift Company* (1994) 10 Const LJ 151 QBD.

- The Scottish case of *Scottish Power v Kvaerner*[92] implied a wider term that the contractor had 'a duty to provide or arrange for the provision to the [sub-contractor] of such full, correct and co-ordinated information concerning the sub-contract works as was or ought reasonably to have been known by the [contractor]...to enable the [sub-contractor] to fulfil their obligations under the sub-contract'.

The extent to which such an implied term may apply may be affected by the express terms of the building contract insofar as they apply to the accuracy of the relevant information. For example, in arriving at his decision in *Merton*, the judge considered the contractor's obligation to notify the architect if any '*discrepancy or divergence between the contract drawings and the contract bills*' was discovered.[93]

The other ground for a claim is misrepresentation. A statement, supposedly of fact, made by the employer[94] and upon which the contractor is intended to rely is a 'representation'. If the statement is untrue, it is a 'misrepresentation'. A misrepresentation may be the basis of a claim, if it is incorporated into the contract, or if it was an inducement to the contractor to enter into the contract (e.g. made during tender discussions, or set out in tender documents to persuade the contractor to reasonably believe that the work could be profitable). A misrepresentation can be fraudulent, negligent, innocent or under statute and the significance lies in the remedies available. Misrepresentations made by the employer may entitle a contractor to a claim for negligent misrepresentation and/or breach of warranty and/or under the Misrepresentation Act 1967 (as amended) ('the 1967 Act').

Under the 1967 Act, the remedies which were formerly restricted to cases of fraud or recklessness now apply to all kinds of misrepresentations unless the party who made the representation can prove '*that he had reasonable ground to believe and did believe up to the time the contract was made that the facts represented were true*'.[95] Ordinarily, the general rule is that the employer does not warrant that (i) the drawings, bills of quantities or specification are accurate or (ii) that the site is fit for the works[96] or (iii) that the contractor will be able to construct on the site[97] is thought not to excuse the employer from liability for misrepresentation.[98] Most of the cases on which such common law principles are founded are based on nineteenth century cases which must be treated with caution in the light of the

92 *Scottish Power plc v Kvaerner Construction (Regions) Ltd* 1999 SLT 721. Lord Macfadyen considered *Merton* and *J&J Fee Ltd*: *'In Merton the arbitrator upheld a term similar to that proposed...but under deletion of the words 'full' and 'co-ordinated' on the basis that they were unnecessary. That decision was not the subject of appeal [from the arbitrator's decision]. In J&J Fee Ltd, the term was agreed in the same modified form. It seems to me, however, that it is the word 'correct' which adds nothing. It would be nonsensical to imply a term that there was a duty to provide information but no duty that the information would be correct. Nevertheless, no harm is done by stating in the formulation of the implied term the requirement that the information be correct. It seems to me, however, that there is a practical point in stating that the information is to be full and co-ordinated. I cannot envisage the hypothetical reasonable main contractor balking at the inclusion of those words. It therefore seems to me to be appropriate to include them in the formulation of the implied term'*.

93 Refer to Footnote 90, e.g. JCT SBC/Q 2016 – clause 2.15, JCT DB 2016 – clause 2.13, JCT IC 2016 – clause 2.13.3, or NEC4 – clause 17.1.

94 In Section 6.7.2, references to the 'employer' also include his representative, or other person acting on the employer's behalf.

95 S.2(1) Misrepresentation Act 1967.

96 Hudson – 14th Edition (2020) at 3-086.

97 *Thorn v The Mayor and Commonalty of London* (1876) LR 1 HL 120.

98 *Appleby v Myers* (1867) LR 2 CP 651; *Thorn v London Corporation* (1876) 1 App Cases 120.

1967 Act. The employer's representative may be liable at common law for any fraudulent or negligent misstatement or representation[99] and also liable under the 1967 Act and it is thought that the scope of such liability is increasing.[100] Section 3 of the 1967 Act restricts the employer's power to exclude liability for misrepresentation.

> *'If any agreement (whether made before or after the commencement of this Act) contains a provision which would exclude or restrict –*
> *(a) any liability to which a party to a contract may be subject by reason of any misrepresentation made by him before the contract was made; or*
> *(b) any remedy available to another party to the contract by reason of such a misrepresentation;*
> *that provision shall be of no effect except to the extent (if any) that, in any proceedings arising out of the contract, the court or arbitrator may allow reliance on it as being fair and reasonable in the circumstances of the case'.*[101]

A representation followed by a warning that the given information may not be accurate will not usually be sufficient to protect the employer, because it is a clear intention to circumvent section 3 of the 1967 Act, and indeed, such a statement may convert the representation into a misrepresentation.[102] In many instances, a representation is made in the 'preliminaries' section of the contract documents,[103] as a statement about the subsoil, underground services or aspects of the site. Other statements made in the preliminaries regarding the sequence of operations, statements in letters written by the architect and possibly by the quantity surveyor, and certainly statements made by the employer at meetings held prior to the contract being executed may all be called in aid in appropriate circumstances.[104] When the contract is executed, such representations become terms of the contract and if they are incorrect, the contractor may have a claim for damages for breach of contract or possibly for breach of warranty where it was not expressly made part of the contract.[105]

If the representation is fraudulent (i.e. '*knowingly, without belief in its truth, or recklessly, careless whether it is true or false*'[106]), the contractor may be able to recover more substantial damages under section 2(1) of the 1967 Act, which probably entitle it to recover all losses, provided that they are not too remote. A very clear example of fraudulent misrepresentation is to be found in the old case of *Pearson v Dublin Corporation*.[107] This

99 *Hedley Byrne & Co v Heller and Partners Ltd* (1964) AC 465.
100 See the fuller discussion of this point in Chapter 1, Section 1.4.
101 Substituted by s. 8(1) of the Unfair Contract Terms Act 1977.
102 *Cremdean Properties Ltd v Nash* (1977) 244 EG 547.
103 Found in the introductory sections of the 'Contract Bills' (JCT SBC/Q 2016), the 'Employer's Requirements' (JCT DB 2016), 'Bills of Quantities', 'Specification' or 'Work Schedules' (JCT IC 2016 and JCT MW 2016) or the 'Site Information' or 'Scope' (NEC4), and usually following the format of 'Preliminaries (main contract) – Part A: Information and requirements' in NRM2 (or its previous equivalent in SMM7).
104 *Holland Hannen & Cubitts (Northern) Ltd v Welsh Health Technical Organisation* (1981) 18 BLR 80.
105 *Bacal Construction (Midlands) Ltd v Northampton Development Corporation* (1976) 8 BLR 88.
106 *Derry v Peek* (1889) 14 App Cases 337.

concerned a complex project for the construction of an outfall sewer and associated works for a lump sum price. A key point was that an existing wall continued to a depth of 9 ft below a datum, as shown on information provided by the employer's engineers. The true situation was that the wall scarcely reached 3 ft deep. This caused the contractor considerably more expense than expected. It emerged that the engineers had carried out no proper survey and, indeed, doubted the accuracy of the very information they provided. The court held that this amounted to fraud. The contract required the contractor to satisfy itself regarding the dimensions of the existing work and stated that the employer was not responsible for the accuracy of statements it had made about the existing structure. However, none of this was enough to provide a defence to the allegation of fraud. Had it not been fraud, it might have then succeeded. Whether such a defence would be enough now is doubtful.

Depending upon the precise circumstances, it seems clear that the contractor may be able to found a claim against the employer on the ground of misrepresentations made during pre-contractual negotiations on the part of the employer with regard to the site and its conditions. In *Morrison-Knudsen International Co Inc v Commonwealth of Australia*,[108] the contractor claimed that the basic information provided at pre-tender stage regarding the soil at the site was false, inaccurate and misleading. He also alleged that the clays at the site, contrary to the information provided, contained large quantities of cobbles. The court had to consider a preliminary issue and concluded:

> *'The basic information in the site document appears to have been the result of much highly technical effort on the part of a department of the [employer]. It was information which the [contractor] had neither the time nor the opportunity to obtain for themselves. It might even be doubted whether they could be expected to obtain it by their own efforts as a potential or actual bidder [tenderer]. But it was indispensable information if a judgment were to be formed as to the extent of the work to be done…'.*

The period leading up to the signing of a building contract can take a long time, involves a number of persons acting on the employer's behalf and requires the employer to rely (to a varying extent) upon the accuracy of design information prepared by consultants. Not unexpectedly, employers may be concerned over the risks that (i) representations may have been made by various persons with appropriate authority or (ii) that design information may not be completely accurate. The simplest measures available to the employer are to introduce express terms in the form of contract amendments which have the following effects:

- 'No reliance' clauses (see Section 6.2.3) – clearly stating:
  - that the contractor is not entitled to rely upon the accuracy of any statement or design provided to him, and that he promises (warrants) that he is satisfied with that information; and

107 [1907] AC 351.
108 *Morrison-Knudson International Co Inc v Commonwealth of Australia* (1972) 13 BLR 114 at 121 per Barwick CJ.

- That the contractor is responsible for, and is satisfied that the conditions of the site (including any existing buildings and sub-surface conditions[109]) are adequate for it to carry out the works for the contract price and by the completion date.
- An 'entire agreement' clause[110] – ensuring that no statements or representations that are not expressly included in the contract (i.e. in writing) have any effect.

---

109 *Co-operative Insurance Society Ltd v Henry Boot Scotland Ltd* [2002] EWHC 1270 (TCC). The contractor was not entitled to rely on a ground investigation report if it was merely referred to on a drawing which was not incorporated into the contract, but merely noted to identify a source of relevant information for the contractor. This was insufficient to override a clause in the contract which placed on the contractor the obligation to satisfy itself about the nature of the site and the sub-soil.
110 NEC4, clause 12.4.

# 7

# Global Claims

## 7.1 General Points

Global claims have been a popular topic of discussion in the construction industry for some time. Industry commentators have extensively charted the Courts' decisions over the years: some siding with the 'employer's team' (as the recipient of a global claim) and other siding with the 'contractor's team' (the party issuing the global claim). Those on the 'employer's team' sought decisions from the Court that sounded the 'death knell' for global claims, whilst those on the 'contractor's team' looked for any decisions which supported making claims on a 'global' basis.

Thankfully, what felt like an endless 'debate' over 'global claims' has largely been brought to a neat conclusion by a fairly recent and very practicable decision of Court, one which has been subsequently endorsed. However, global claims will remain contentious – particularly if you're the party receiving one.

## 7.2 Basic Principles of Global Claims

The SCL Delay and Disruption Protocol describes a 'global claim' as

> *'A global claim is one in which the Contractor seeks compensation for a group of Employer Risk Events but does not, or cannot demonstrate a direct link between the loss incurred and the individual Employer Risk Event.'*[1]

1 SCL Delay and Disruption Protocol – 2nd Edition (2017) Society of Construction Law. The SCL Protocol defines an 'Employer Risk Event' as 'An event or cause of delay or disruption which under the contract is at the risk and responsibility of the Employer' such as a 'Relevant Event' or 'Relevant Matter' under JCT 2016 contracts or a 'compensation event' under NEC4.

*Building Contract Claims*, Sixth Edition. Guy Higginbottom and David Chappell.

Another, and perhaps slightly 'softer', definition is taken from the 1995 edition of 'Hudson':

> '…*those where a global or composite sum, however computed, is put forward as the measure of damages or of contractual compensation where there are two or more separate matters of claim or complaint, and where it is said to be impractical or impossible to provide a breakdown or sub-division of the sum claimed between those matters.*'[2]

Whilst both definitions seemingly refer to 'loss' or a 'sum' in the financial sense, a global claim can also be made for an extension of time.

In general, it is necessary for the contractor to establish each and every head of claim, by means of supporting documentation and other evidence. However, often a contractor will attempt to form its claim on a global basis. A simple example of a global claim is shown below:

| **Total Costs incurred** | | **£2,500,000** |
|---|---|---|
| Less | (£2,100,000) | |
| Value recovered through contract (including the value of work for 250 variations) | | |
| Amount Claimed for Prolongation | | |
| (20 weeks @ £10,000 per week) | (£200,000) | |
| Claim total | (£2,300,000) | (£2,300,000) |
| Amount of 'overspend' remaining | | £200,000 |
| (Disruption, disturbance due to 250 variations) | | |

The above example assumes that all 250 variations caused the contractor to lose £200,000 after accounting for (i) the value of works and variations and (ii) the amount recovered for prolongation.

The 'debate' over global claims began in 1967.[3] A case arose because an arbitrator proposed to award a lump sum where the contractor made a broad claim for delay and disruption because a large number of disparate matters had delayed completion and caused financial loss, but some of these matters did not entitle the contractor to time or money. The employer contended that the contractor should only recover in relation to specific claims where losses could be separately proved. The court upheld the opinion of the arbitrator and set out guidelines for the 'global' approach which have formed their basis ever since. Essentially, the global approach is only justified where a claim depends 'on an extremely complex interaction in the consequences of various denials, suspensions and variations' and where '*it may well be difficult or even impossible to make an accurate apportionment of the total extra cost between the several causative events*'. In those clearly defined circumstances, it seems that there is no reason why an architect, engineer or arbitrator '*should not recognise the realities of the situation and make individual awards in respect of those parts of individual*

2 Duncan-Wallace, I N. Hudson's Building and Engineering Contracts Vol 11 (1995) at 8-200.
3 *J Crosby & Sons Ltd v Portland UDC* (1967) 5 BLR 121.

*items of the claim which can be dealt with in isolation and a supplementary award in respect of the remainder of those claims as a composite whole'*.[3]

Further cases involving global claims followed in 1991[4] and 1993,[5] the first of which was seized upon by some commentators as ringing the 'death knell' on global claims. However, instead of the court deciding that the concept of 'global claims' was not permissible, the global claims submitted in both cases were criticised for lack of evidence and particularity. There were more cases[6] throughout the remainder of the 1990's, none of which appeared to set out a definitive case for or against the principle of global claims. Instead, they provided pleading guidelines, stating that the link between 'cause' and 'effect' (the 'nexus') should be shown (also see Section 2.1.2).

Nevertheless, the courts allowed claims to be made on a global basis,[7] on the basis set out in *Merton v Leach*:

> *'It is implicit [in Crosby] that a rolled-up award can only be made in a case where the loss or expense attributable to each head of claim cannot in reality be separated and secondly that a rolled-up award can only be made where apart from that practical impossibility the conditions which have to be satisfied before an award can be made have been satisfied in relation to each head of claim'*.[8]

It therefore seemed that the contractor was entitled to put its claim in any rational way albeit that putting forward a claim on a global basis may present particular evidential difficulties.[9] A particular difficulty is the establishment of a definable connection (nexus) between the alleged wrong and the consequent delay and damage.[10]

## 7.3 The Current Position

The 2012 case of *Walter Lilly v Mackay*[11] considered many claims from the contractor (Lilly), including a 'global' claim. In the context of the 'practical impossibility' referred

4 *Wharf Properties Ltd v Eric Cumine Associates* (1991) 52 BLR 1.

5 *Mid Glamorgan County Council v J Devonald Williams and Partners* (1993) 8 Const LJ 61, where Oliver LJ stated '...*the pleading is hopelessly embarrassing as it stands*...' and provided '*no agenda for trial*'. See also *Imperial Chemical Industries PLC v Bovis Construction Ltd, GMW Partnership and Oscar Faber Consulting Engineers* (1992) 8 Const LJ 293.

6 See *John Holland Construction & Engineering Pty Ltd v Kvaerner R J Brown Pty Ltd* (1996) 82 BLR 83, and *London Underground Ltd v Kenchington Ford plc & Others* (1998) 63 Con LR 1, and *GAB Robins Holdings Ltd v Specialist Computer Centres Ltd* (1999) 15 Const LJ 43.

7 See for example *Nauru Phosphate Royalties Trust v Matthew Hall Mechanical and Electrical Engineers Pty Ltd and Another* (1992) 10 BCL 178; *British Airways Pension Trustees Ltd v Sir Robert McAlpine & Sons Ltd & Others* (1994) 72 BLR 102; *Bernard's Rugby Landscapes Ltd v Stockley Park Consortium Ltd* (1997) 82 BLR 39; *John Doyle Construction Ltd v Laing Management (Scotland) Ltd* (2002) BLR 393.

8 *London Borough of Merton v Stanley Hugh Leach Ltd* (1985) 32 BLR 51 at 102 per Vinelott J.

9 *GMTC Tools and Equipment Ltd v Yuasa Warwick Machinery Ltd* (1994) 73 BLR 102.

10 *Ralph M Lee Pty Ltd v Gardner and Naylor Industries Pty Ltd* (1997) 12 Const LJ 125.

11 *Walter Lilly & Company Ltd v Giles Patrick Cyril Mackay and DMW Developments Ltd* (2012) EWHC 1773 (TCC).

to in *Merton*[9], Akenhead J described the circumstances in which the global claim was submitted:

> *'Even if a global cost claim cannot be allowed unless it was impracticable or very difficult for the Contractor to relate every penny of loss to each established and pleaded event which entitled it to loss and/expense, I am satisfied that it was impracticable or very difficult for [Lilly] in this case. This project was, essentially, a complete mess from the administrative side on the part of [the employer] and its professional team. The job started with remarkably little design, there were hundreds of variations, there was throughout the project hopelessly late provision of information and instructions to [Lilly], there was a substantial level of discord between, principally, [the employer] and most of his professional team most of the time and there was a strategy evolved by [the employer] which involved aggressively supervising the professional team and omitting a substantial amount of work whilst leaving [Lilly] with the work which he thought [Lilly] would have difficulty in dealing with in time. By early 2007 until the end of the project, it was virtually impossible sensibly to programme all the works because it was not known what work and with what detail it would be ordered; this period was also confused by unsubstantiated hints that [the employer] might omit yet further work.'*[12]

The judge drew 'the relevant threads together' from previous cases[13] and set out the following seven principles for global claims:

1) the contractor must prove its case on the facts and the balance of probabilities;
2) the contractor must satisfy any contractual requirements for making the claim (e.g. complying with conditions precedent for notifying claims), or the claim will be disqualified;
3) a party is at liberty to prove its claim with whatever evidence it feels will satisfy the tribunal by meeting the required standard of proof;
4) although there is nothing 'wrong' with a global claim, there are evidential difficulties to overcome – to establish that the loss incurred would not have been incurred in any event (i.e. causation);
5) the presence of one or a series of events or factors, which are either not pleaded or a 'contractor risk event',[14] caused or contributed to the total or global loss, does not necessarily mean that the contractor can recover nothing, as it depends on the impact that those events had – i.e. that the global claim would simply be reduced by the loss resulting from that event;
6) it is unnecessary for the court to go down the global claim route if the actual cost attributable to an event can be readily determined (although the court will be more sceptical about the global cost claim); and
7) a global claim should not be rejected just because the claimant (the contractor in most cases) has caused the "practical impossibility" of disentanglement.

12 Ibid at 487.

13 *Crosby* (footnote 3), *Merton* (footnote 9), *Wharf* (footnote 5), *John Holland* (footnote 7), *Bernhard's* and *John Doyle* (footnote 8), also see *Petramec Inc v Brasilerio SA Petrobas* (2007) EWCA Civ 1371.

14 SCL Delay and Disruption Protocol – (2nd Ed) 2017 defines 'Contractor Risk Event' as *'An event or cause of delay or disruption which under the contract is at the risk and responsibility of the contractor'*.

## 7.4 Summary

What seems clear is that the global approach is the 'when all else fails' approach and it should only be adopted when the contractor faces a 'practical impossibility' of disentangling the various circumstances giving rise to the claim. A contractor should not be misled into believing that the current view from *Lilly v Mackay* allows a claim to be made when a contractor cannot actually demonstrate a valid claim but can only show a shortfall in its income. The evidential difficulties (described above) that a contractor faces when pursuing a 'global claim' are very clear.[15]

15 Refer to *John Sisk & Son Ltd v Carmel Building Services Ltd (in administration)* (2016) EWHC 806 (TCC) at 55 and 57.

# 8

# Employer Claims Against the Contractor

## 8.1 General

The majority of claims made under building contracts are made by the contracting party (that being a main contractor or a sub-contractor) against their employer, and this is mainly for the following reasons:

- The contracting party will have far more obligations to their employer, with which the employer can interfere or breach; and
- Instead of the contracting party having to formally claim and recover additional payment, the employer can simply deduct any sums for 'claims' from amounts that would otherwise be paid to the contracting party.

The most common claim that an employer will raise against a contractor is to recover its additional costs or losses because the contractor has failed to complete the works by the 'completion date', in other words 'damages for late completion', or 'delay damages'. In most building contracts, the amount that the employer can claim is a pre-agreed, or 'liquidated' amount of damages, applied to whatever period of time has passed after the 'completion date' until the contractor has completed the works to the standard required by the contract. Of course, the employer can claim 'damages' in the ordinary way (refer to Section 2.6 and Chapter 6) by claiming his actual costs, but the majority of employers find it far more convenient to recover liquidated damages (see Section 8.2) and accordingly, adopt the liquidated damages provisions in their contracts.

Under most building contracts, the employer's ability to claim 'damages for late completion' provides additional protection[1] against defective, or non-compliant work. As described in Section 3.1.4 (Completion obligations), in order for the works to be considered 'complete', the contractor's work must be free of patent defects, save for any *de-minimis* items that (in his discretion), the employer or his representative consider will not prevent completion. Therefore, even if the contractor has completed the quantity of work required, the works will not be considered 'complete' until all patent defects have been rectified. In such cases,

1 In addition to the employer's remedies of (a) instructing the contractor to rectify non-compliant work under JCT SBC/Q 2016 – clause 3.19, JCT DB 2016 – clause 3.14, JCT IC 2016 – clause 3.15, JCT MW 2016 – clauses 2.3 and 3.4 (although not specifically addressing non-compliant work), and NEC4 – clauses 43 and 44 and (b) employing others to rectify non-compliant work, in the event that the contractor fails to do so: JCT SBC/Q 2016 – clause 3.11, JCT DB 2016 – clause 3.6, JCT IC 2016 – clause 3.9, JCT MW 2016 – clause 3.5, and NEC4 – clause 46.1.

*Building Contract Claims*, Sixth Edition. Guy Higginbottom and David Chappell.

there is no need for the employer to claim for the costs of rectification because the contractor is obliged to do this at his own cost, and the employer may also be recovering damages for late completion while he does so. The employer may claim the costs of rectifying defective work from the contractor in the following circumstances:

- If, as part of concluding the building contract, he wishes to accept a defect in return for a reduction (abatement) of the contract sum; and
- To claim for rectifying latent defects that have been discovered after the works are 'complete' and the contract's defects rectification period has expired.

Contract permitting, the employer may be able to make other claims against the contractor. Two examples include the contractor's failure to

- Construct the works to or within specific tolerances or achieve a performance specification; and
- Meet a 'Key Date' under NEC4 (see Sections 3.1.3 and 3.10).

## 8.2 Delay Damages (Liquidated Damages) for Late Completion

### 8.2.1 The Meaning and Purpose of Liquidated Damages

Liquidated damages means a fixed and agreed sum as opposed to unliquidated damages (e.g. actual costs) which are neither fixed nor agreed and must be proven. Liquidated damages[2] have been defined as follows:

- *'A fixed sum, usually per week or per day, written into the contract as being payable by the Contractor in the event that the works are not completed by the contract completion date (original or extended).'*[3]
- *'A Liquidated damages clause is a clause in a contract which stipulates what amount of money will be payable as damages for loss caused by a breach of the contract irrespective of what loss may actually be suffered if a breach of the relevant kind (typically, delay in performance of the contract) occurs. Liquidated damages clauses are a standard feature of major construction and engineering contracts and commonly provide for damages to be payable at a specified rate for each week or day of delay in the completion of work by the contractor after the contractual completion date has passed. Such a clause serves two useful purposes. First, establishing what financial loss delay has caused the employer would often be an intractable task capable of giving rise to costly disputes. Fixing in advance the damages payable for such delay avoids such difficulty and cost. Second, such a clause limits the contractor's exposure to liability of an otherwise unknown and open-ended kind, while at the same time giving the employer certainty about the amount that it will be entitled to recover as compensation. Each party is therefore better able to manage the risk of delay in the completion of the project.'*[4]

2 'Liquidated damages' are also referred to as 'liquidated and ascertained damages', or abbreviated to 'LAD's' or 'LD's'.
3 Delay and Disruption Protocol – (2nd Ed) 2017. Society of Construction Law.
4 *Triple Point Technology Inc v PTT Public Company Ltd* (2021) UKSC 29, at para 74.

Ordinarily, and in order to recover general (or common law) damages in matters involving breaches of contract (e.g. to complete the works by the 'completion date'), it is necessary to first, prove that the defendant had a contractual obligation to the claimant; second, that there was a failure to partly or wholly fulfil that obligation; and finally, that the claimant suffered loss or damage thereby. Very often it is clear that there has been a breach, and indeed, a contractor's failure to complete the works on time is likely to be self-evident, but it can be difficult, time-consuming and often expensive to prove the actual costs or losses (unliquidated damages)[5] that were incurred. To avoid what may be an expensive and painstaking process of proving actual costs (or losses), the parties may decide, when they enter into a contract, that for convenience' sake, in the event of a breach of a particular kind the party in default will pay an agreed sum to the other. This sum is termed liquidated damages.

Liquidated damages clauses rely upon the parties' clear agreement that the party breaching a particular contract condition will pay the agreed sum or a sum calculated by an agreed method (e.g. utilising agreed rates to be applied to particular periods of time). Provided that the liquidated damages clause uses a genuine pre-estimate of the likely loss, and its method of operation is clear, the defaulting party will have to pay liquidated damages to the injured party, even in the event that the injured party suffers no actual loss.[6]

In the building industry and elsewhere, the terms 'liquidated damages' and 'penalty' are commonly used as though they were interchangeable; however, they are totally different in concept. Whereas liquidated damages are compensatory and should be a genuine pre-estimate of the damages likely to flow from a particular breach, a penalty is a sum which is not related to probable damages, but rather as a punishment, stipulated *in terrorem*.[7] The courts will usually enforce the former, but not the latter.[8]

Liquidated damages clauses must be clear, unambiguous and able to be operated simply. This not only means that the liquidated damages clauses in the contract's conditions have to be clearly drafted but also the parties must enter the correct information into the contract as follows:

- JCT 2016 contracts – in the 'Contract Particulars' and in
- NEC4 contracts – in 'Contract Data – Part One' for Option X7.

5 *Clydebank Engineering Co v Don Jose Yzquierdo y Castenada* (1905) AC 6.
6 *BFI Group of Companies Ltd v DCB Integration Systems Ltd* (1987) CILL 348 – see also Section 8.2.4 and Footnote 69. For example, completion of a residential development may be delayed, and the contractor must pay liquidated damages to the employer. However, changes in the property market during the period of delay resulted in the properties' sale prices increasing to the extent that the employer may have suffered no loss, or may even have benefitted.
7 *Cellulose Acetate Silk Co Ltd v Widnes Foundry* (1933) AC 20. See also (1) *United International Pictures v Cine Bes Filmcilik ve Yapimcilik AS* (2003) EWCA Civ 1669 at 13 *'whether a provision is to be treated as a penalty is a matter of construction to be resolved by asking whether at the time the contract was entered into the predominant contractual function of the provision was to deter a party from breaking the contract or to compensate the innocent party for breach. That the contractual function is deterrent rather than compensatory can be deduced by comparing the amount that would be payable on breach with the loss that might be sustained if breach occurred'*. Per Mance LJ and (2) *Murray v Leisureplay plc* (2005) EWCA Civ 963.
8 See *Cavendish Square Holding BV v Talal El Makdessi* (2015) UKSC 67.

These entries must provide the following information as a minimum:

- A date for commencing work[9] or a mechanism for determining when the contractor starts;
- A date when the contractor must complete the works or period of time for him to do so[10]; and
- A rate for liquidated damages (usually to be applied, per day, or per week).

Complications can arise if the whole of the works is sub-divided into sections or the contract allows for the employer to take possession of areas of the work (or of a section).

### 8.2.2 Liquidated Damages Clauses Must be Clear and Unambiguous

Liquidated damages clauses can have a significant effect on the contractor's finances – particularly, in instances when the rate of liquidated damages is high or where sectional completion applies, and the contractor is late in completing more than one of those sections.

Liquidated damages clauses must be simple and clear to operate, as is discussed below:

- Owing to their relatively serious, and potentially draconian nature, liquidated damages clauses are interpreted *contra-proferentum*[11] which means:

  > *'against the profferer,' i.e. against the person who drafted or tendered the document. If there is an ambiguity in a document to which all other methods of construction have failed to resolve, so that there are two alternative meanings to certain words, the court may construe the words against the party who put forward the document and give effect to the meaning more favorable to the other party.*[12]

- When considering a JCT building contract with contractor's design,[13] Dyson J illustrated the relative simplicity of calculating liquidated damages:

  > *'The reality is that there are only 3 elements to a claim for [liquidated damages]. The first is the date when the Works or Section of the Works ought to have been completed. For the purposes of a deduction of [liquidated damages], this is fixed, for the time being at least, by the giving of a notice under clause 24.1. The second is the date of practical completion of the Works or the relevant Section. The third is the application of the amount or amounts included in the contract for LADs to the period between the two dates just mentioned, a matter of simple arithmetic.'*

- The calculation of liquidated damages can be complicated by matters such as
  - The employer taking possession of parts of the Works or a Section[14]; or

9 See Section 3.1.2.
10 See Section 3.1.4.
11 *Peak Construction (Liverpool) Ltd. v McKinney Foundations Ltd.* (1970) 1 BLR 114.
12 Keating on Construction Contracts (10th Ed) 2016 at 3-044.
13 *Bovis Lend Lease Ltd v Braehead Glasgow Ltd* (2000) EWHC Technology 118 at para 23. Whilst the judgement states that the contract was for *'the design and construction of a shopping center'*.
14 For the employer taking 'partial possession' see Section 3.1.4 and Section 8.2.6.

- Inconsistent, incomplete or ambiguous entries in the contract for (i) the rates of liquidated damages in the contract[15] or (ii) the sections of the works to which the relevant rates of liquidated damages are to be applied.[16]

Therefore, it is essential that the particular building contract contains provisions for adjusting the calculations for liquidated damages when the employer takes over a part and that the relevant entries in the 'Contract Particulars' (JCT 2016 contracts) or 'Contract Data' (NEC4) are fully and correctly completed.

Whilst the courts have thoroughly examined the incorporation of liquidated damages clauses into contracts,[17] they are reluctant to find a contractual provision to be void for uncertainty, if they can find an interpretation that gives effect to the parties' intentions.[18]

### 8.2.3 'Liquidated Damages' or a 'Penalty'

It is extremely important that the sum entered into a contract as liquidated damages is a genuine pre-estimate of the likely loss and not a penalty. The courts often pay little attention to the parties' terminology in liquidated damages clauses,[19] and modern forms of contract do not use the word 'penalty'. The words 'liquidated damages' and 'delay damages' are used in the JCT 2016[20] and NEC4 contracts,[21] respectively.

The rules for deciding whether a sum is to be considered liquidated damages or a penalty were formulated by Lord Dunedin in *Dunlop Pneumatic Tyre Co Ltd v New Garage & Motor Co Ltd*[22] ('Lord Dunedin's four tests'[23]). These are set out below:

- *'(i) Though the parties to a contract who use the words penalty or liquidated damages may prima facie be supposed to mean what they say, yet the expression used is not conclusive.*

15 Refer to Section 8.2.4 and 8.2.5 (below).
16 *Vinci Construction UK Ltd. v Beumer Group UK Ltd.* (2017) EWHC 2196 (TCC).
17 *Blu-Sky Solutions Ltd v Be Caring Limited* (2021) EWHC 2619 (Comm) at para 110. *Blu-Sky* was not a construction case, and it is highly unlikely that a court would find that any standard forms of building contract (such as JCT or NEC4) fail to meet the requirements for bringing an onerous clause (for liquidated damages in this case) to a party's attention. However, *Blu-Sky* sets out important guidelines for drafting bespoke building contracts.
18 See Footnote 16, at Paragraphs 47 and 48.
19 *Kemble v Farren* (1829) All ER 641. In that case, not only was the sum expressed by the parties as liquidated damages, it was clearly stated that it was 'not a penalty or penal sum'. Notwithstanding the clear words, the court had little hesitation in finding that the sum was a penalty. See also *Ranger v Great Western Rail Co* (1854) All ER 321 at 332 per Lord Cranworth LC: *All the circumstances which have been relied on in the different reported cases, as distinguishing liquidated damages from penalty, are to be found here. The injury to be guarded against was one incapable of exact calculation. The sum to be paid is not the same for every default, for that which should occasion small as for that which should occasion great inconvenience, but one increasing as the inconvenience would become more and more pressing, and finally, the payments are themselves secured by the penalty of a bond'.*
20 JCT SBC/Q 2016 – clause 2.32 and JCT DB 2016 – clause 2.29 (Payment or allowance of liquidated damages), JCT IC 2016 – clause 2.23 (Liquidated damages for non-completion) and JCT MW 2016 – clause 2.8.1 (Damages for non-completion).
21 NEC4: Option X7 (Delay Damages). '*The essence of a penalty is a payment of money stipulated as in terrorem of the offending party; the essence of liquidated damages is a genuine covenanted pre-estimate of damage*' *Clydebank Engineering Co v Don Jose Yzquierdo y Castenada* (1905) AC 6, was cited as authority for this proposition.
22 [1915] All ER 739.
23 Referred to Footnote no's 26, 27 and 28 (below).

*The court must find out whether the payment stipulated is in truth a penalty or liquidated damages…'*

- *'(ii) The essence of a penalty is a payment of money stipulated as in terrorem of the offending party; the essence of liquidated damages is a genuine…pre-estimate of damage*[24]'
- *'(iii) The question whether a sum stipulated is a penalty or liquidated damages is a question of construction to be decided upon the terms and inherence circumstances of each particular contract, judged of as at the time of the making of contract, not as at the time of the breach*[25].' and
- *'(iv) To assist this task of construction various tests have been suggested, which if applicable to the case under consideration, may prove helpful or event conclusive. Such tests are:*
  - *It will be held to be a penalty if the sum stipulated for is extravagant or unconscionable in amount in comparison with the greatest loss which could conceivably be proved to have followed from the breach…*
  - *It will be held to be a penalty if the breach only in not paying a sum of money, the sum stipulated is a sum greater than the sum which ought to have been paid.'*

The test of whether a clause should be struck out as a penalty was recently considered by the Supreme Court,[26] where it restated the test as follows:

- *…in so far as it refers to 'punishment' and 'an additional or different liability' as opposed to 'in terrorem' and 'genuine pre-estimate of loss,' this definition seems to us to get closer to the concept of a penalty than any other definition we have seen. The real question when a contractual provision is challenged as a penalty is whether it is penal, not whether it is a pre-estimate of loss. These are not natural opposites or mutually exclusive categories. A damages clause may be neither or both. The fact that the clause is not a pre-estimate of loss does not therefore, at any rate without more, mean that it is penal. To describe it as a deterrent (or, to use the Latin equivalent, in terrorem) does not add anything. A deterrent provision in a contract is simply one species of provision designed to influence the conduct of the party potentially affected. It is no different in this respect from a contractual inducement. Neither is it inherently penal or contrary to the policy of the law. The question whether it is enforceable should depend on whether the means by which the contracting party's conduct is to be influenced are 'unconscionable' or (which will usually amount to the same thing) 'extravagant' by reference to some norm.*[27]' and

---

24 See also *Robophone Facilities Ltd v Blank* (1966) 3 All ER 128 at 142 per Diplock LJ: 'The onus of showing such a stipulation is a 'penalty clause' lies upon the party who is sued upon it. The terms of the clause may themselves be sufficient to give rise to the inference that it is not a genuine estimate of damage likely to be suffered but is a penalty. Terms which give rise to such an inference are discussed in Lord Dunedin's speech in *Dunlop Pneumatic Tyre Co. v New Garage & Motor Co* (1915) AC 79 at 87. But it is an inference only and may be rebutted. Thus it may seem at first sight that the stipulated sum is extravagantly greater than any loss which is liable to result from the breach in the ordinary course of things, i.e. the so-called 'first rule' in *Hadley v Baxendale* (1854) 9 Exch 341. This would give rise to the prima facie inference that the stipulated sum was a penalty. But the plaintiff may be able to show that owing to special circumstances outside 'the ordinary course of things' a breach in those special circumstances would be liable to cause him a greater loss of which the stipulated sum does represent a genuine estimate'.

25 See also *Public Works Commissioner v Hills* (1906) AC 368, [1906] All ER 919, and *Webster v Bosanquet* (1912) AC 394, were cited as authorities for this proposition.

26 *Cavendish Square Holding BV v Talal El Makdessi* (2015) UKSC 67.

27 *Makdessi* (see Footnote 26) at para 31 per Neuberger LJ and Sumption LJ. See also *Imperial Tobacco Co v Parsley* (1936) 2 All ER 515 at 521 per Lord Wright MR: '*I do not think the word 'unconscionable' there has*

- *'The true test is whether the impugned provision is a secondary obligation which imposes a detriment on the contract-breaker out of all proportion to any legitimate interest of the innocent party in the enforcement of the primary obligation. The innocent party can have no proper interest in simply punishing the defaulter. His interest is in performance or in some appropriate alternative to performance. In the case of a straightforward damages clause, that interest will rarely extend beyond compensation for the breach, and we therefore expect that Lord Dunedin's four tests would usually be perfectly adequate to determine its validity. But compensation is not necessarily the only legitimate interest that the innocent party may have in the performance of the defaulter's primary obligations…'*[28]

Two important considerations are (i) the extent to which an accurate pre-estimate of loss can be carried out and (ii) the existence of different events, each of which are said to give rise to liquidated damages. The House of Lords held that hypothetical situations cannot be used to defeat a liquidated damages clause. The court will take a pragmatic approach:

> *'Whatever the degree of care exercised by the draftsman, it will still be almost inevitable that an ingenious argument can be developed for saying that in a particular hypothetical situation a substantially higher sum will be recovered than would be recoverable if the plaintiff was required to prove his actual loss in that situation. Such a result would undermine the whole purpose of the parties to a contract being able to agree beforehand what damages are to be recoverable in the event of a breach of contract. This would not be in the interest of either of the parties to the contract since it is to their advantage that they should be able to know with a reasonable degree of certainty the extent of their liability and the risks which they run as a result of entering into the contract.'*[29]

Other, fairly recent decisions of the courts have confirmed that they favour a pragmatic approach, and they are unlikely to decide that a sum is a penalty based purely on hypothetical rather than factual grounds:

> *'Because the rule about penalties is an anomaly within the law of contract, the courts are predisposed, where possible, to uphold contractual terms which fix the level of damages for breach. This predisposition is even stronger in the case of commercial contracts freely entered into between parties of comparable bargaining power.'*[30]

In that case, important factors were that the parties had comparable bargaining power, it was a commercial contract and the level of damages survived scrutiny by the parties' legal advisors. Perhaps, most important was that the court was following the lead set by higher courts and was predisposed to uphold terms which fix the level of damages.

---

*any reference to the fact that the parties were on an unequal footing. It does not bring in at all the idea of an unconscionable bargain. It is merely a synonym for something which is extravagant and exorbitant'.*

28 *Makdessi* (see Footnote 26) at para 32 per Neuberger LJ and Sumption LJ.

29 *Philips Hong Kong Ltd v Attorney General of Hong Kong* (1993) 61 BLR 41 at 54 per Lord Woolf.

30 *Alfred McAlpine Capital Projects Ltd v Tilebox Ltd* (2005) 104 Con LR 39 at 50 per Jackson J. See also *Lordsvale Finance v Bank of Gambia* (1996) QB 752 and *Indian Airlines v GIA International* (2002) EWHC 2361 (Comm).

Further guidance regarding whether liquidated damages clauses are extravagant, exorbitant or unconscionable that applies the test in *Makdessi* is as follows:

- Each liquidated damages provision must be considered in the context of the contract as a whole[31];
- The court should be cautious to interfere with the parties' agreement to a liquidated damages provision, particularly if the parties negotiated with the benefit of legal advice[32];
- The party benefitting from the liquidated damages provision should have a legitimate interest in enforcing it – for example, if the employing party is likely to suffer losses[33] caused by the contractor's failure to complete on time[34];
- The difficulty in quantifying general damages that the employing party could incur[35]; and
- Evidence put before the court that the level of liquidated damages would be unconscionable or disproportionate.[36]

### 8.2.4 The Sum Expressed as Liquidated Damages and its Calculation

Building contracts express the amount liquidated damages as a specific sum of money that the employer can deduct for a particular interval (or period of time), in the following ways:

- If the entire works are not divided into sections, so there is only one completion date[37]:
  - The sum of money for each period or unit of time. In JCT contracts, the rate of liquidated damages is expressed as *'at the rate of £[insert sum] per [insert period]'*[38] – with the 'period' usually being a 'week' or a 'day'
  - NEC4 expresses delay (liquidated) damages as *'Delay damages for Completion of the whole of the works are [insert sum] per day'*.[39]
- If the entire works are divided into sections (see Section 3.9):
  - In addition to a rate of liquidated damages applicable for the relevant period and applicable to each Section, the contract must also contain:
  - A separate completion date for each Section; and
  - A comprehensive description of the each Section, so that the contractor can clearly identify the work that needs to be completed.[40]

Other contracts (particularly bespoke contracts) have expressed liquidated damages as a sum of money *'per week, or part thereof'*. That is probably because at one time that form of

---

31 *Eco World – Ballymore Embassy Gardens Company Ltd v Dobler UK Ltd* (2021) EWHC 2207 (TCC) at 77.
32 *Eco World* (see Footnote 31) at 79.
33 Where the employing party is a management contractor, his losses may include those caused by a trade contractor failure to complete having an adverse effect (including disruption) on following trade contractors.
34 *Eco World* (see Footnote 31) at 80.
35 *Eco World* (see Footnote 31) at 81.
36 *Eco World* (see Footnote 31) at 82.
37 The completion date must (i) be clearly stated in the particular building contract, and (ii) be extendable in the event of the employer's default.
38 Entry in the 'Contract Particulars' – JCT SBC/Q 2016–2.32.2, JCT DB 2016 – clause 2.29.2, JCT IC 2016 – clause 2.23.2 and JCT MW 2016 - clause 2.8.
39 NEC4, Contract Data Part One – Option X7: Delay damages.
40 See *Taylor Woodrow Holdings Ltd. v Barnes & Elliott Ltd.* (2004) EWHC 3319 (TCC) and *Vinci Construction UK Ltd. v Beumer Group UK Ltd.* (2017) EWHC 2196 (TCC).

words was printed in a standard form. It is likely that the intention behind the inclusion of that phrase is to indicate that the rate will be reduced pro rata the proportion of the week excluded from consideration. Of course, the phrase actually means, in simple terms, '*at the rate of £X for each whole week and at £X for any part of a week*' which is why standard forms do not use the expression. There is a presumption that the provision amounts to a penalty because it purports to levy the same rate of liquidated damages whether the delay is a whole week or only one day (or indeed one hour) of that week. If the rate is correct for a week, it must be excessive for a lesser unit of time (such as a day or an hour). It is a presumption which can be rebutted, for example by showing that the employer's damages begin to accrue at the very beginning of the week.

In terms of the actual sum expressed as liquidated damages, the following points can also be considered:

- The fact that the sum stipulated as liquidated damages bears no relation to the contract sum is not relevant[41];
- There is a strong inference that a sum is liquidated damages where the parties have agreed a sum or sums as liquidated damages and the sum claimed is not excessive in relation to the actual loss suffered.[42]
- Where there is a single event:
  - If the pre-estimate of likely loss is relatively straightforward, a sum may be a penalty if it is greater than such loss, but otherwise is liquidated damages[43]; but
  - If the pre-estimate of likely loss less straightforward, it may be more likely that the sum is liquidated damages as the difficulty of pre-estimation increases.[44]
- Where there are several events:
  - If the pre-estimate of likely loss in respect of any one of them is relatively easy, a sum will be a penalty if it is greater than such loss, but liquidated damages otherwise; but
  - If the pre-estimate of likely loss is difficult, the sum is likely to be liquidated damages;
- If one sum is payable in respect of several events:
  - But which result in different kinds and amounts of loss, it is likely to be a penalty[45]; however
  - If the damage is the same in kind, but giving rise to differing amounts of loss, it may be liquidated damages.[46]

---

41 *Imperial Tobacco Co v Parsley* (1936) 2 All ER 515 at 524 per Lord Slesser LJ. Also see *Alfred McAlpine Capital Projects Ltd v Tilebox Ltd* (2005) 104 Con LR 39 at 50 per Jackson J: *'In my view, a pre-estimate of damages does not have to be right in order to be reasonable. There must be a substantial discrepancy between the level of damages stipulated in the contract and the level of damages which is likely to be suffered before it can be said that the agreed pre-estimate is unreasonable'*.

42 See *Philips Hong Kong Ltd v The Attorney General of Hong Kong* (1993) 61 BLR 41 at 59 per Lord Woolf repeated in *Ballast Wiltshire PLC v Thomas Barnes & Sons Ltd*, unreported, 29 July 1998 at Paragraph 50 per Judge Bowsher: 'The fact that the issue has to be determined objectively, judged at the date the contract was made, does not mean what actually happens subsequently is irrelevant. On the contrary it can provide valuable evidence as to what could reasonably be expected to be the loss at the time the contract was made'.

43 *Kemble v Farren* (1829) All ER 641. See also *Eco World* in Footnotes 31, 32, 34 to 36.

44 *Dunlop Pneumatic Tyre Co Ltd v New Garage & Motor Co Ltd* (1915) All ER 739.

45 *Ford Motor Co v Armstrong* (1915) 31 TLR 267.

46 *Dunlop Pneumatic Tyre Co Ltd v New Garage & Motor Co Ltd* (1915) All ER 739.

- If one sum is expressly stated to be an average of the pre-estimated loss resulting from each of several events, it is likely to be liquidated damages.[47]
- Where different sums are payable in respect of different events, they are likely to be liquidated damages.[48]

The relative ease or difficulty of calculating of a genuine pre-estimate of loss for liquidated damages is likely to depend upon the nature of building project itself. The employer may have little idea how much loss he may suffer if the building is not completed by the due date. Whilst liquidated damages are especially suited to situations where precise estimation is difficult,[49] the employer should try to calculate as accurate a figure as possible. He should include every item of additional cost which can reasonably be said to flow directly from the contractor's failure to complete by the completion date, that is, the damages recoverable under the first limb of the rule in *Hadley v Baxendale* (refer to Section 4.2). It seems that the sum can be increased to include amounts which would normally only be recoverable under the second limb if the employer can show that special circumstances were involved.[50] It remains unclear whether, in the case of liquidated damages, the special circumstances must be known to the contractor when the contract is made. It seems appropriate to reveal such circumstances at tender stage although it could be argued that the higher figure, agreed as liquidated damages, is itself a sufficient prior notification.

For a project where the employer will use the building for commercial purposes, such as to earn revenue (such as property sales, or rent), the calculation is likely to be straightforward. Providing that the building contract contains a 'partial possession' clause (see Section 8.2.6) that enables liquidated damages to be reduced in proportion with any partial possession taken, the pre-estimate of loss is likely to consider the following on a time-related (e.g. weekly or daily) basis:

- Loss of rental income;
- Financing charges for any loan repayments that would otherwise have been repaid from rental income;
- The cost of retaining the building contract's professional team during a period of contractor-delay;
- The costs of maintaining an empty building (e.g. maintenance fees, energy costs, business rates, etc);
- The costs of retaining sales staff for the period of contractor-delay;
- Costs of temporary accommodation (including alternative office or storage facilities); and
- Ancillary costs such as transport.

However, if the building does not have a primary commercial purpose perhaps because it's intended for public use (such as a library or museum), then the significant causes of loss may be more difficult to ascertain. Some local authorities and other public bodies

---

47 *English Hop Growers v Dering* (1928) 2 KB 174.
48 *Imperial Tobacco Co v Parsley* (1936) 2 All ER 515.
49 See *Eco World – Ballymore Embassy Gardens Company Ltd v Dobler UK Ltd* (2021) EWHC 2207 (TCC) – Footnote 35.
50 *Philips Hong Kong Ltd v Attorney General of Hong Kong* (1993) 61 BLR 41. A Format/checklist for calculating liquidated damages is to be found in Chappell, Cowlin and Dunn's *Building Law Encyclopaedia* (2009) Wiley-Blackwell.

make use of a formula calculation which basically depends upon a percentage of the capital sum. Whether that would constitute liquidated damages will depend on the precise circumstances and particularly the difficulty with which a precise calculation could be made. Use of a formula is a perfectly sensible approach where it is obvious that substantial loss will be suffered in the event of a delay, but where it is virtually impossible to calculate precisely in advance what that loss would be.[51]

### 8.2.5 Liquidated Damages as the Only Remedy

A question often asked is whether a party to a building contract containing a liquidated damages clause can sue for the actual loss suffered or whether that party is restricted to the sum expressed as liquidated damages. This question often arises where the actual losses (or damage) exceed the amount recoverable under the liquidated damages clause.

The sum expressed as liquidated damages was held to be exhaustive of the remedies available to the claimant for late completion in the case of *Temloc Ltd v Errill Properties Ltd.* [52] This case concerned works carried out by the contractor (Temloc) for a property developer (Errill) under a JCT 1980 Edition Private Without Quantities building contract ('JCT 80'). The Appendix[53] to the JCT 80 expressed liquidated damages *'at the rate of £____ per_____'.* However, the sum of liquidated damages was stated to be '£nil' and period of time for which that rate was intended to apply was left blank. It was held that the parties had agreed that, in the event of late completion, no (or £Nil) liquidated damages should be applied. Importantly, the Court of Appeal considered that even if a rate (i.e. greater than £0 per week) had been stated, the court considered that the rate would have represented an exhaustive agreement as to damages which were or were not to be payable by the contractor in event of his failure to complete on time:

> *'I think it is clear, both as a matter of construction and as one of common sense that if (1) clause 24 is incorporated into the contract and (2) the parties complete the relevant part of the appendix, either by stating a rate at which the sum is to be nil, then that constitutes an exhaustive agreement as to the damage which are or are not to be payable by the contractor in the event of his failure to complete the works on time.'*[54]

That, of course, does not preclude the employer from recovering as unliquidated damages other losses not directly caused by the breach of obligation to complete, but which may be connected to such breach. In *Piggott Foundations Ltd v Shepherd Construction Ltd,*[55] the court followed *Temloc* and held that liquidated damages were exhaustive of the damages which could be recovered for failure to complete the Works on time. Indeed, NEC4 provides

51 *Philips Hong Kong Ltd v The Attorney General of Hong Kong* (1993) 61 BLR 41 PC.
52 *Temloc Ltd v Errill Properties Ltd* (1987) 39 BLR 30, referred with approval in *Biffa Waste Services Ltd and Another v Maschinenfabrik Ernst Hese GmbH and Another* (2008) 118 Con LR 104.
53 The 'Appendix' under JCT 1980 was the forerunner of the 'Contract Particulars' found in the JCT 2016 contracts. The Appendix entry for clause 24.2 of JCT 80 dealt with 'liquidated and ascertained damages' (as they were then described).
54 *Temloc Ltd v Errill Properties Ltd* – see Footnote 52, as per Nourse LJ at 39.
55 [1993] 42 Con LR 98.

separate remedies in the event the contractor fails to *'meet a Key Date'* and fails to complete the works by the 'Completion Date' as follows:

- The contractor pays unliquidated damages (described as 'additional costs' incurred by the Client) if he fails to meet a Key Date[56]; and
- Provided that Option X7 (Delay damages) is selected, the contractor pays delay damages (liquidated damages) if he fails to complete the works by the 'Completion Date'.

Further questions arise under traditional forms of contract where the entries in the contract have not been completed as expected (see Section 8.2.3 above). For example, if there is space to insert a rate but no rate is inserted and the space is left blank. JCT 2016[57] and NEC4 contracts do not provide for unliquidated damages to apply by default in such cases. Clearly, unliquidated damages are recoverable if Option X7 (Delay damages) was not selected under NEC4, but if it has been, and the relevant entries for Option X7 are not completed in contract data part one, then as with JCT 2016 contracts, there is no default provision to re-apply unliquidated damages. For JCT 2016 contracts, if the liquidated damages entry in the contract particulars[57] is crossed out, then this may be good evidence of the parties' intention to not to apply liquidated damages. As a matter of contract construction, as the 'written terms' (i.e. the 'contract particulars' in JCT 2016 contracts, or the contract data – part one in NEC4) ordinarily take precedence over the 'printed terms' (i.e. the clauses themselves), so deleting all of the liquidated damages entries in the contract may suffice in disapplying liquidated damages, rather than just deleting the liquidated damages clause from the 'conditions'. It is suggested that deleting both the relevant contract entry for liquidated damages and the liquidated damages clause would be sufficient to re-apply common law or unliquidated damages for the contractor's failure to complete the works on time.

### 8.2.6 Where There Is Partial Possession

Provisions for the deduction of liquidated damages often run into trouble where the contract also permits the employer to take partial possession of the Works. Ordinarily, and unless possession is taken by the employer strictly in accordance with the contract terms, possession by the employer is not a trigger for the end of liquidated damages.[58]

In *M J Gleeson (Contractors) Ltd v Hillingdon Borough Council*,[59] the contract for the construction of blocks of houses provided for liquidated damages at the rate of £5 per dwelling per week. Only one date in the contract was stipulated as the completion date, but the bills of quantities which formed part of the contract provided for the Works to be completed in sections. The employer attempted to deduct liquidated damages with reference to the sectional dates for completion set out in the bills. It was held that the bills

56 NEC4 – clause 25.3.

57 Entries for liquidated damages in Contract Particulars: JCT SBC/Q 2016 – clause 2.32.2, JCT DB 2016 – clause 2.29.2, JCT IC 2016 – clause 2.23.2 and JCT MW 2016 – clause 2.8. Entries for NEC4 (Option X7) in Contract Data Part One.

58 In *Impresa Castelli SpA v Cola Holdings Ltd* (2002) 87 Con LR 123, the employer had occupied part of the Works, but not under the partial possession clause of WCD 98. It was held that such occupation did not amount to partial possession, there was no mechanism to reduce liquidated damages and, therefore, the full amount of liquidated damages could be recovered.

59 *M J Gleeson (Contractors) Ltd. v Hillingdon Borough Council* (1970) 215 EG 165.

did not override the printed contract[60] and, therefore, the single contract completion date prevailed. Consequently, as soon as the employer started to take possession of the houses, liquidated damages were no longer deductible.

In *Bramall & Ogden v Sheffield City Council*,[61] the council required the construction of 123 houses. However, there was no sectional completion in the contract being used. There was provision for partial possession in clause 6, and the appendix stated the rate of liquidated damages as £20 per week for each uncompleted dwelling. However, the partial possession clause also provided that liquidated damages were to be calculated by proportioning the liquidated damages in the same ratio as the value of the part taken into possession bore to the contract sum, but the liquidated damages sum had already been split into a rate per dwelling. The court summarised the contractor's argument as follows:

> *'The 'Works' cover not only the houses but the other items above referred to. Condition 22 refers to a failure 'to complete the Works' by the extended date. As from that date the employer becomes entitled to liquidated damages until the works are completed. Condition 16 deals with the consensual taking of possession of part of the works. Condition 16(e) provides for the sum payable after taking possession in respect of the period during which the works remain incomplete. The way in which the liquidated damages are dealt with is set out in the Appendix. This does not allow of the calculation to be made which is required by Condition 16(e), and one cannot operate the Appendix and Condition 16(e) in the circumstances of this case. The inconsistency can only be reconciled if provision is made in the contract for sectional completion of those parts which are taken over and to which specific liquidated damages provisions are applied.'*[62]

*Bramall & Ogden* was followed by *Avoncroft Construction Ltd v Sharba Homes (CN) Ltd*.[63] Although the case was mainly concerned with an adjudicator's decision and a party's attempt to set-off against it, the court expressly stated that the liquidated damages clause failed according to the principle in the *Bramall & Ogden*, because partial possession was taken.[64] The building contract (JCT Private With Quantities 1998 Edition) had no provision for sectional completion so the liquidated damages could not apply.

Current editions of the standard building contracts that contain partial possession clauses provide that the liquidated damages will be proportioned in the same way as the value of the part taken into possession bears to the contract sum.[65] It has already been seen that this formula will work only if the rate of liquidated damages in the contract particulars

60 The building contract provided that the bills did not override the contract's terms and conditions similar to clause 1.3 of JCT SBC/2016, JCT DB 2016, JCT IC 2016 and clause 1.2 of JCT MW 2016.

61 *Bramall & Ogden Ltd. v Sheffield City Council* (1983) 29 BLR 73.

62 *Bramall & Ogden Ltd v Sheffield City Council* (1983) 29 BLR 73 at 85 per Judge Hawser.

63 *Avoncroft Construction Ltd v Sharba Homes (CN) Ltd* (2008) 119 Con LR 130.

64 See also the court's decision in *Stanor Electric Ltd v R Mansell Ltd* (1998) CILL 399. A single sum for liquidated damages for the works (consisting of two houses) was not capable of division, because there was no provision in the contract. The employer tried to deduct half the sum to represent one of the two houses. If both houses had been delayed by an equal amount, the sum would not have been held to be penal.

65 JCT SBC/Q 2016 – clause 2.37, JCT DB 2016 – clause 2.34, JCT IC 2016 – clause 2.29 and NEC4 (Option X7) – clause X7.3.

is expressed as a single sum in respect of the whole of the Works. Any attempt to express it as a figure for various parts of the Works (as in *Bramall & Ogden*) is likely to render the liquidated damages clause unworkable.

There are some important practical considerations for the employer taking partial possession if the liquidated damages rate is expressed as a single (and therefore divisible) amount. Since the rate of liquidated damages is for the whole of the Works, proportionally reducing the rate to reflect the portion of Works taken into possession should be fairly straightforward. However, a situation may arise where reducing the liquidated damages rate by the same proportion the value of that partially possessed area may not properly represent a genuine pre-estimate of the losses that the employer may suffer. To avoid this situation, the employer should carefully consider the project's logistics and whether sectional completion and separate rates for liquidated damages for each section would be more appropriate.

Generally, the courts are less inclined to interfere with liquidated damages provisions.[66] Where the differences are trivial, or where the mechanism for calculating liquidated damages is 'cumbersome',[67] it is unlikely that the courts will throw out the liquidated damages provisions.[68]

### 8.2.7 Maximum Recovery – Including a Damages 'cap'

An important question is whether a claimant is entitled to recover the amount specified as liquidated damages if the damage actually suffered is less than the amount or perhaps nothing at all? Further, is a claimant able to recover liquidated damages though it can be demonstrated that it has actually gained a benefit from the breach? In *BFI Group of Companies Ltd v DCB Integration Systems Ltd*,[69] it was found that, although the claimants had suffered no actual loss as a result of being unable to use part of the works, they remained entitled to liquidated damages.

An often-encountered question concerns the employer's position if liquidated damages are held to be void or a penalty. Is the employer restricted to recovery of such amount as can be proved up to, but not greater than, the amount of the liquidated damages sum that was held to be a penalty? In other words, does the amount of liquidated damages that would otherwise be recoverable, operate as a limit of the contractor's liability? There have been

66 *Eco World – Ballymore Embassy Gardens Company Ltd v Dobler UK Ltd* (2021) EWHC 2207 (TCC) at paras 74 to 83.

67 *Mansion Place Limited v Fox Industrial Services Ltd (No.2)* (2021) EWHC 2972 at Paragraphs 97 and 98.

68 *Buckingham Group Contracting Limited v Peel L&P Investments and Property Ltd* (2022) EWHC 1842 (TCC) at.

69 *BFI Group of Companies Ltd v DCB Integrations Systems Ltd* (1987) CILL 348. The form of contract was based upon the JCT Minor Works 1980 terms and provided for the payment of liquidated damages if completion was delayed beyond the completion date. The claimants were given possession by the contractor on the extended date for completion although the arbitrator found that practical completion had not taken place. Had they not been given possession, they would have been obliged to wait until practical completion was certified before being able to execute the fit-out works. Unlike JCT contracts, such as JCT SBC/Q 2016, JCT DB 2016, or JCT IC 2016, there is no provision for possession of part of the Works before practical completion and the possession granted to the claimants in this case was a concession. Therefore, the claimants were able to carry out work during the period within which they were receiving liquidated damages. This represented a considerable advantage to the claimant. See also *Golden Bay Realty Pte Ltd v Orchard Twelve Investments Pte Ltd* (1991) 1 WLR 981 at 986.

some competing views by leading authorities, comparing the position in charterparty cases with that in construction. The court has decided as follows:

- That as a matter of general principle, and following the Supreme Court's decision in *Makdessi*,[70] *'if a liquidated damages provision is void for uncertainty or as a penalty, it is wholly unenforceable and the Employer's entitlement to general damages will not be subject to a cap.'*[71]
- However, the court will consider the words used in the liquidated damages clause, and each clause *'…must be construed in accordance with the established principles of contractual interpretation…'*[72] and *'…even where a liquidated damages clause is found to be wholly unenforceable as a penalty, it may on a true construction be found to operate as a limitation of liability provision.'*[73]

In *Buckingham*, the court considered whether a provision stating *'Cap on Maximum LAD's 7.5% £1,928.253.77'* would limit the contractor's liability to general damages in the event that liquidated damages were unenforceable. The meaning of 'Maximum LAD's' was considered to be plain, and not referring to *'…anything other than LAD's'* such as general damages, and the court concluded that *'Given the clarity of the works used by the parties in this case, I see no basis for concluding that the clause should serve the additional function of operating as a cap on Buckingham's overall liability for delay arising from breach of that or some other provision.'*[74]

JCT 2016 contracts[75] contain clauses limiting of the contractor's liability for loss of use, loss of profit or other consequential loss arising from its design, can be limited to the amount set out in the contract particulars. In each case, the relevant limitation clearly states that the limitation does not apply to liquidated damages.[76]

NEC4 provides an option X18 (limitation of liability) and similarly to its JCT counterpart, this does not limit the contractor's liability for delay damages. If selected, Option X18 operates as a limitation of the contractor's liability for the following; capping (or limiting) liability to the amounts set out in contract data part one:

- *indirect or consequential loss*;[77]
- for any one event *'loss of or damage to the Client's property'*;[78]
- *'liability for defects' due to [the Contractor's] design'*;[79] and

70 See Footnote no's 8, and 26 to 28.

71 *Eco World – Ballymore Embassy Gardens Company Ltd v Dobler UK Ltd* (2021) EWHC 2207 (TCC) at 110.

72 *Eco World* (see Footnote 71) at para 116, and *Buckingham* (see Footnote 68) at 96.

73 *Eco World* (see Footnote 71) at para 111.

74 *Buckingham* (see Footnote 68) at para 98.

75 JCT SBC/Q 2016 – clause 2.19.3 (in relation to CDP Design Work), JCT DB 2016 – clause 2.17.3 and JCT ICD 2016 – clause 2.34.3.

76 Liquidated damages under JCT SBC/Q 2016 – clause 2.32, JCT DB 2016 – clause 2.29 and JCT ICD 2016 – clause 2.23.

77 NEC4 – clause X18.2.

78 NEC4 – clause X18.3. Seemingly, further loss of or damage to the Client's property is excluded under clause X18.5.

79 NEC4 – clause X18.4.

- *'all matters arising out of or in connection with the contract',*[80] except for the following matters which are excluded:
  - Delay damages deductible under Option X7;
  - Low performance damages under Option X17;
  - Contractor's share option (under Main Options C or D).

## 8.3 Procedures for Deducting Liquidated Damages for Late Completion

### 8.3.1 General Points

In Section 8.2.2, the requirement for liquidated damages clauses to be clear and unambiguous was highlighted. The general position is that there are '3 elements' to a claim for liquidated damage, as described the court in *Bovis Lend Lease Ltd v Braehead Glasgow Limited*[81]:

> *'The reality is that there are only 3 elements to a claim for LAD's. The first is the date when the Works or Section of the Works ought to have been completed. For the purposes of a deduction of LAD's, this is fixed, for the time being at least, by the giving of a notice… The second is the date of practical completion of the Works or the relevant Section. The third is the application of the amount or amounts included in the contract for LAD's to the period between the two dates just mentioned, a matter of simple arithmetic.'*

Most standard forms of building contract generally follow these three 'elements.'

### 8.3.2 Liquidated Damages Under JCT 2016 Contracts

The procedures for the employer to deduct or withhold liquidated damages under JCT 2016 contracts are more extensive than the '3 elements' from *Bovis v Braehead*[81]. They include several mandatory notifications from the employer that are intended to verify the contractor's failure to complete on time and to notify him that the employer may seek payment, or deduction of liquidated damages. In terms of the 'three elements,' the relevant Completion Date for the Works (or a Section) will either be as set out in the contract particulars[82] or will be such later date (or dates) as are fixed by the employer, following the operation of the extension of time procedures (see Section 3.2). The second and third elements (practical completion and the rate of liquidated damages in the contract) are self-explanatory.

Where the Works are divided into Sections, it is essential that the contract particulars[83] accurately identify each Section. In the unfortunate event that the Sections are not

---

80 NEC4 – clause X18.5.

81 *Bovis Lend Lease Ltd v Braehead Glasgow Ltd* (2000) EWHC Technology 118 at para 23.

82 JCT DB 2016, the entry (Date for Completion of the Works) or entries (Dates for Completion of Sections) against clause 1.1 in the Contract Particulars.

83 JCT DB 2016, the entry (Descriptions of Sections) against the Fifth Recital.

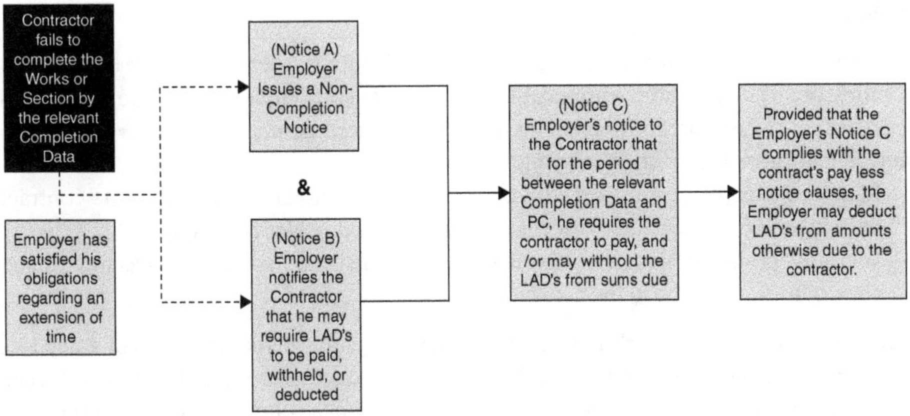


**Figure 8.1** JCT 2016 contracts – procedure for the employer to deduct liquidated damages.

accurately described, or an element of the Works falls outside the Sections' descriptions, then the employer may not be able to deduct liquidated damages because of the uncertainty surrounding the parts of the relevant Sections that remain uncompleted.[84]

Assuming that the extension of time provisions have been operated correctly and firstly, there has either been no employer delay, or any employer delay has already been properly accounted for when fixing a later date as the Completion Date(s) and secondly, the works (or a Section) have not reached practical completion[85] by the relevant completion date(s), then under clause 2.27, the employer issues a non-completion notice to the contractor.

If the employer intends to deduct liquidated damages from sums that are otherwise due to the contractor under JCT 2016 contracts,[86] then he must carry out the procedure summarised below in Figure 8.1:

The procedure requires employer to issue the following three essential documents to the contractor:

- Both of the following:
  - (Notice A) – the non-completion certificate[87]; and
  - (Notice B) – notification[88] that the employer may require the contractor to pay liquidated damages to the employer, or the employer may withhold or deduct liquidated damages;

84 Taylor Woodrow Holdings Ltd v Barnes & Elliott Ltd (2004) EWHC 3319 (TCC).

85 See Section 3.1.4.

86 Figure 8.1 only shows the procedures for the employer deducting liquidated damages under JCT SBC/Q 2016, JCT DB 2016 and JCT IC 2016. The procedure for deducting liquidated damages under JCT MW 2016 (in clause 2.8) is much simpler.

87 A 'Non-Completion Certificate' under JCT SBC/Q 2016 - clause 2.31, or a 'certificate of non-completion' under JCT IC 2016 – clause 2.22 A 'Non-Completion Notice' under JCT DB 2016 – clause 2.28. There is no provision for a 'non-completion notice' under JCT MW 2016.

88 JCT SBC/Q – clause 2.32.1.2, JCT DB 2016 - clause 2.29.1.2, and JCT IC 2016 – clause 2.23.1.2: 'Notice B' must be issued before the due date for the final payment.

then followed by:

- (Notice C) – notification to the contractor, stating that for the period between the relevant Completion Date(s), and the applicable date of practical completion (for the Works or a Section(s) – when the entitlement to deduct liquidated damages for the Works or the relevant Section will cease), the employer will:
  - Require the contractor to pay liquidated damages[89] (either at the rate in the contract particulars or a lesser rate[90]) for the period described in Notice C; and/or[91]
  - Withhold or deduct liquidated damages (at the rates and for the period as mentioned above) from sums due to the contractor.[92]

Provided that the employer's notice either complies with the requirements of a pay-less notice or any separate payment notice[93] or pay-less notice that refers to the employer's Notice C, then the employer may deduct liquidated damages from sums due to the contractor.

In the event that the employer has taken partial possession (see Section 8.2.6) of parts of the Works or a Section(s), then the rate of liquidated damages applicable to Works or a Section, other than the 'Relevant Part' shall be reduced in proportion. The relevant proportion of liquidated damages is calculated as the value of the relevant part, compared to either

- For the works as a whole, the contract sum[94]; or
- For a section, the section sums[95] in the contract particulars.

It is therefore essential, that the contract sum and section sums are properly set out in the contract particulars, otherwise the employer may not be able to deduct liquidated damages in the event of partial possession.

### 8.3.3 Liquidated Damages Under NEC4

Unlike JCT 2016 contracts, NEC4 does not provide an automatic entitlement for the Client to deduct liquidated damages if the contractor fails to complete by the completion date, or the relevant completion date for a section if Option X5 (sectional completion) is selected. The client's right to deduct liquidated sums as delay damages is provide by selecting secondary Option X7 (delay damages).[96] In the absence of selecting Option X7, it is thought that the Client would be able to deduct unliquidated damages if the contractor

89 JCT SBC/Q – clause 2.32.2, JCT DB 2016 - clause 2.29.2 and JCT IC 2016 – clause 2.23.2.
90 Lesser rates of liquidated damages may be payable if the employer has taken 'partial possession'.
91 JCT DB 2016. Clause 2.29.2 refers to the employer requiring the contractor to pay liquidated damages 'and/or' the employer withholding or deducting liquidated damages from sums due to the Contractor, whereas the employer's clause 29.1.2 notice must state one or other (i.e. payment or withholding/deduction).
92 JCT SBC/Q 2016 – clause 2.32.2.2, JCT DB 2016, clause 2.29.2.2 and JCT IC 2016 – clause 2.23.2.2.
93 JCT 2016 contracts refer to both a payment notice for interim payments, and a payment notice for the final payment.
94 JCT 2016 contracts - the Contract Sum is set out in Article 2.
95 The Section Sums are set out in the Contract Particulars: JCT SBC/Q 2016 – 2.32.2, JCT DB 2016 – clause 2.29.2, and JCT IC 2016 – clause 2.29.
96 NEC4, Contract Data Part One, entry for 'Secondary Options' in section '1 General'.

was in breach of his completion obligations. Unlike the JCT 2016 contracts, NEC4 does not require the Project Manager to notify the contractor of a failure to complete the works by the Completion Date[97]

Clause X7.1 provides that delay damages are as follows:

- Calculated on a daily basis;
- At the rate (or rates if Option X5 is selected) stated in the contract data[98];
- Between the completion date[99] and the earlier of:
  - Completion[100]; or
  - When the Client takes over[101] the works (or section); and
- Are paid by the contractor.[102]

As the Project Manager assesses[103] the 'amount due' for payment to the contractor under clause 50, it appears that unlike JCT 2016 contracts, under NEC4 it is the employer's representative (the Project Manager), and not the Client who calculates the delay damages that the contractor must pay.

Clause X7.2 deals with repayment of delay damages to the contractor in the event of a change in the Completion Date (extension of time) being awarded. In addition to repaying the delay damages to the contractor, the Client must also pay interest,[104] calculated from the date that delay damages were either deducted from (or paid by) the contractor, to the date when they're repaid by the Client.

Clause X7.3 reduces the delay damages if the Client takes over part of the works (or a Section) before Completion. The damages are reduced in proportion to the benefit (as assessed by the Project Manager) that the Client gains for taking over a part, compared to the benefit the Client would obtain if he took over the whole of the Works. The 'benefit' may be subjective and unclear. For example, if the Client took over all of the works, with exception of the power supply, it could be argued that he has obtained no benefit at all, because without power, the works cannot be operated. NEC4's 'User Guide' gives what is seemingly meant to be a simple assessment of the proportion of benefit the Client may obtain and how this is applied to the daily rate of delay damages.[105] However, this example considers many variables for a manufacturing Client, including:

- The Client's forecast profit for each of two sections of the Works (assuming Option X5 applies); and
- The actual profit that the Client would obtain for using each section as at the Completion Date.

97 See Figure 8.1 – 'Notice A' being the 'Non-Completion Certificate' (or its equivalent) under the particular JCT 2016 contract. See Footnote 87.
98 NEC4, Contract Data Part One – the relevant entries for 'delay damages' are found under 'X7: Delay Damages', including when Option X5 (Sectional Completion) is also selected.
99 NEC4, clause 11.2(3).
100 NEC4, clause 11.2(2).
101 NEC4, clause 35 (Take over).
102 NEC4, clauses 50.3 and 50.4 provide that the 'amount due' must account for '*…amounts to be paid by or retained from the Contractor*'.
103 NEC4, clause 50.2.
104 NEC4, clause 51.4 and the relevant entry for the 'interest rate' is set out in Contract Data Part One, in '5 Payment.'
105 NEC4 User Guide (2020), p88.

Given that liquidated sums for delay damages are intended to simplify matters, by providing a pre-agreed sum for each period of delay (per day in the case of NEC4), that for convenience, is accepted by the contractor as the amount he must pay, it seems peculiar to apply a subjective and significantly more complicated calculation of what may be an 'intangible' benefit to the Client, in order to proportionally reduce the delay damages to reflect the work taken over by the Client. Considering the variable elements of NEC4's example (above), there is simply more for the parties to argue over in terms of disputing the Project Manager's assessment under clause X7.3. It would be far more sensible to adopt a similar and far less subjective approach based upon the value of the works taken over.

### 8.3.4 Damages for Late Completion Summary

The delay (or liquidated) damages provisions for the three design and build contracts are summarised in Table 8.1.

## 8.4 Defences to Liquidated Damages for Late Completion

Contractors are naturally concerned by having to pay liquidated damages, as the sums claimed by the employer can be ruinously expensive and have catastrophic effects on a contractor's financial position. Clearly, the contractor will be aware of the rate or rates (if sectional completion applies) of liquidated damages for failing to complete by the contractual date.

### 8.4.1 By Agreement

The parties can agree that the employer will limit or not exercise his right to claim liquidated damages the following ways:

- By agreeing (either orally or expressly in writing) that liquidated damages shall not be deducted or
- By the operation of a 'waiver' or forbearance.

Agreements to limit or prohibit the deduction of liquidated damages are usually recorded in writing. Such agreements usually occur at the beginning of a project (when the parties are agreeing the building contract's terms) or much later in the project – usually when the employer is able to, or likely to deduct liquidated damages.

Firstly, it is becoming common for contractors to negotiate a 'LAD free' period when agreeing the building contract's terms. This usually takes the form of an agreed period (usually expressed in weeks) starting at the contractual completion date and during which the employer is prevented from deducting liquidated damages. The drafting required to give effect to such an agreement is usually straightforward and ought to be effective as long as it is sufficiently clear. For example, writing 'LAD-free period of 4 weeks' against the relevant entries in the contract[106] may suffice. Alternatively, the parties may decide

106 Against the relevant entry in the 'Contract Particulars' in JCT 2016 contracts, or against Option X7 (Delay Damages) in NEC4's Contract Data Part One.

**Table 8.1** Summary of damages for late completion.

| Delay (liquidated) damages provision | JCT SBC/Q | JCT DB | JCT IC | JCT MW | NEC4 |
|---|---|---|---|---|---|
| Liquidated (delay) damages apply | Clause 2.32 | Clause 2.29 | Clause 2.23 | Clause 2.8 | Option X7 |
| Unliquidated (common law) damages apply | No | No | No | No | Not selecting Option X7 |
| Employer has to notify failure to complete | Yes, clause 2.31 | Yes, clause 2.28 | Yes, clause 2.22 | No | No |
| Employer must notify contractor (in addition to a payment/pay-less notice) | Clause 2.32.1.2 Clause 2.32.2 | Clause 2.29.1.2 Clause 2.29.2 | Clause 2.23.1.2 Clause 2.23.2 | No | No |
| Liquidated damages are an exclusive remedy | Yes | Yes | Yes | Yes | Yes |
| Separate rates of damages for Sections | Yes | Yes | Yes | No | Yes Options X5 and X7 |
| Unit of time applying to delay damages | Per week | Per week | Per week | Per day or per week | Per day |
| Damages reduced for 'partial possession' or 'taking over parts' | Clause 2.37 | Clause 2.34 | Clause 2.29 | No | Yes – X7.3 |
| Employer repays liquidated damages for EoT | Yes, clause 2.32.3 | Yes, clause 2.29.3 | Yes clause 2.34 | Not stated | Yes – X7.2 |
| Limitation of delay (liquidated) damages | No | No | N/A[a)] | No | No – X18.5 |

a) JCT IC 2016 does not contain the 'limitation' of liability for damages caused by contractor's design, as this is found in JCT ICD 2016 (refer to Footnote 76).

to amend the contract's relevant liquidated damages clauses and/or add a specific entry into to the written part of the contract.[106] Of course, this type of agreement only provides a limitation to the employer's ability to claim liquidated damages. After the 'LAD-free period' has expired, it is then open to the employer to deduct liquidated damages in the usual way (refer to Section 8.3).

The parties may reach an agreement that the employer will not deduct liquidated damages. Such agreements are rarely reached without the contractor making some concession in return, and indeed without such 'consideration' (in the legal sense), an agreement will not be binding unless it is set out and executed as a deed. The contractor may have competing claims for an extension of time, along with claims for loss and expense, so in return for the contractor withdrawing such claims, the employer may agree to not to deduct liquidated damages. The precise details of reaching such an agreement may be contentious,[107]

107 *Mansion Place Limited v Fox Industrial Services Ltd* (2021) EWHC 2972 (TCC).

so it is advisable for the parties to promptly confirm their understanding of the agreement in each case.

Secondly, the employer may agree to waive or forbear his right to claim liquidated damages. The important distinction between (i) a 'waiver' or 'forbearance' and (ii) the ad hoc agreement referred to above is that the former does not require any consideration. If one party indicates to the other, either by plain words or by conduct, that it intends to forego a right, then subsequently, it may not exercise that right if circumstances change, as it would be inequitable to do so. The requirements for a 'wavier' or forbearance to operate are as follows[108]:

- A legal relationship between the parties that governs their rights and obligations;
- One party must promise (or represent) to the other that he will not enforce his strict legal rights arising from that legal relationship;
- The promisor must intend that the promisee will rely upon that promise (or representation); and
- The promisee relies upon the promise (or representation).[109]

Two useful illustrations are:

- The contractor is, say, ten weeks from contract completion date when the employer promises not to enforce liquidated damages. The contractor, who was on target to complete on time using an optimum amount of labour, relies upon the employer's promise, and relaxes his progress, but the employer withdraws his promise five weeks before the original completion date. The contractor still has work to carry out, but because he is working at a reduced rate of progress, he will not complete the work for another 7 weeks. If the contractor can complete on time (i.e. in 5 weeks), he will have expended more money than he originally expected because it is less efficient to increase labour beyond an optimum point. If, despite his efforts, he finishes after the completion date, he will be liable to pay liquidated damages which, but for reliance upon the promise, would not have been incurred.
- Applying the same facts as above, save that the employer withdraws his promise two weeks after the contract completion date, but when the contractor still has three more weeks work to carry out working at the reduced rate. The contractor can do nothing but attempt to reduce the time to complete the work. Even if, the contractor is not liable to pay liquidated damages for the two weeks between the due date and the employer's withdrawal, there is nothing he can do to prevent a liability for up to three weeks liquidated damages from the time of withdrawal until actual completion.

It is doubtful that the employer would be entitled to terminate the arrangement once the contractor has taken any significant steps in reliance on it, such as paying its sub-contractors in full without deducting any damages for their failure to complete in time.[110]

---

108 Chitty on Contracts, (30th Ed) 2008 at para 3–086.

109 See *Charles Rickards Ltd v Oppenhaim* (1950) KB 616 and *Hughes v Metropolitan Railway Co* (1877) 2 App Cas 439.

110 *London Borough of Lewisham v Shepherd Hill Civil Engineering* 30 July 2001, unreported.

### 8.4.2 Deficiencies in the Building Contract

The law regarding liquidated damages is now well-developed, so it is rare that any deficiencies are found in the liquidated damages clauses in standard forms of building contract. However, not all projects use standard forms of building contract, as bespoke forms of contract may be adopted, or the terms of a standard building contract could be amended. In the event that the liquidated damages clauses differ from those in unamended, standard forms of building contract, then there is a greater likelihood that deficiencies may occur. The following general points will apply:

- Liquidated damages clauses must be clear and unambiguous and are treated as 'contra proferentum' (see Section 8.2.2) – interpreted against the party offering the clause in the event that the clause is ambiguous; however
- The courts will attempt to give effect to liquidated damages clauses, even if their operation is 'cumbersome' (see Section 8.2.6).

The liquidated damages provisions in standard forms of building contract have been tested over many years, so they should operate correctly if the relevant entries in the building contract[111] have been made. The importance of correctly and fully completing those entries in the building contract that relate to liquidated damages cannot be overstated. The following points are relevant:

- Clearly expressing whether 'unliquidated' or common law damages, or liquidated damages will apply in the event that the contractor fails to complete the works (or a section) by the contractual completion date:
  - If 'unliquidated' damages are to apply, then the employer should delete[112] the liquidated damages clause from the contract;
  - The employer should not attempt to dis-apply the liquidated damages clause by simply writing '£Nil' against the relevant entry in the contract for liquidated damages (see Section 8.2.5);
  - Care should be taken if the contract requires the amount of liquidated damages to be reduced in proportion with the part of the works of which the employer has taken partial possession (see Section 8.2.6); and
- If the works are to be completed in sections, include clear descriptions of each section to which liquidated damages apply (see Section 8.2.4). If it is not possible to ascertain the works within each section, it will be difficult to determine if that section has been completed and, depending upon this degree of difficulty, the liquidated damages clause may be inoperable.

### 8.4.3 Employer Failing to Observe the Contract Terms

Section 8.3 summarises and illustrates the procedures with which the employer must comply, in order to be entitled to deduct liquidated damages under the JCT 2016 and NEC4 standard forms of building contract.

---

111 In the 'Contract Particulars' for JCT 2016 contract, or in Contract Data Part One (against Option X7) for NEC4 contracts.

112 In the case of NEC4 contracts, simply not select Option X7 (delay damages) as being applicable. Where a JCT 2016 contract is being used, the employer should strike out the relevant entries in the 'Contract Particulars' along with the text of the liquidated damages clauses.

Under the JCT 2016 contracts, there are a number of conditions precedent to the deduction of liquidated damages by the employer, as shown in Figure 8.1. The view that under JCT contracts, a certificate (or notice) of non-completion is a condition precedent was clearly endorsed by the court[113]:

> *'There can be no doubt that a certificate of failure to complete given under clause 24.1 and a written requirement of payment or allowance under the middle part of 24.2.1 were conditions precedent to the recovery of [liquidated damages] under the latter part of Clause 24.2.1.'*

Doubt was thrown upon whether the employer's requirement in clause 24.2.1 of JCT80 – to demand liquidated damages in writing was a condition precedent was considered by the court.[114] However, this would not apply under modern building contracts. Not only are the procedure and content for written notices concerning the deduction of liquidated damages under the JCT 2016 contracts clear and mandatory[115] (see Figure 8.1) but those contracts must also comply with the current payment legislation for construction contracts, which provide that a payment notice, and/or pay-less notice must firstly be in writing,[116] and secondly specify *'the basis on which that sum is calculated'*.[117]

### 8.4.4 Time Has Been Extended

Liquidated damages are only payable in the event that the contractor fails to complete the works (or a section) by the contractual completion date, because progress has been delayed by one or more grounds for which the contractor is responsible. However, if the employer has caused the works' progress to be delayed, the contractor is usually entitled to an extension of time for completing the works or a section (refer to Section 3.2, and Chapters 11 and 13). In the event that the contractor is entitled to an extension of time, he is excused from completing the works later than the original completion date, and in addition to recovering financial compensation (if the circumstances permit), he will avoid liability for liquidated damages.

### 8.4.5 Time Is at Large

If the contract terms (other than those relating to liquidated damages) that concern extensions of time are deficient in some way; such as failing to adequately deal with an act of

---

113 *A Bell & Son (Paddington) Ltd v CBF Residential Care & Housing Association Ltd* (1989) 46 BLR 102 at 107 per Judge Newey. The form of building contract was JCT Standard of Building Contract 1980 Edition, Private with Quantities ('JCT80' - a predecessor of JCT SBC/Q 2016). Clause 24 of JCT 80 dealt with liquidated damages.

114 *Jarvis Brent Ltd v Rowlinson Constructions Ltd* (1990) 6 Const LJ 292. Clause 24.2.1 of JCT80 stated that *'the Employer may require in writing…'* (underlining added),

115 JCT 2016 contracts all state (1) the written notice requirements are mandatory (i.e. '*Provided:…*'), and (2) that that the employer must state the period between the Completion Date and the date of practical completion (or of a section): JCT SBC/Q 2016 – clause 2.32.2, JCT DB 2016 – clause 2.29.2 and JCT IC 2016 – clause 2.23.2.

116 The Housing Grants Construction and Regeneration Act 1996 (as amended), s115(6).

117 The Housing Grants Construction and Regeneration Act 1996 (as amended), – 'payment notice' complying with s110A(2)(a)(ii) and a pay-less notice complying with s111(4)(b). Also see the Scheme for Construction Contracts (England and Wales) Regulations 1998 (as amended, or their Scottish or Northern Irish counterparts).

prevention (see Section 3.2.2) by the employer, then time may become 'at large' (refer to Section 3.1.5). If time does become 'at large', then the employer cannot insist upon the contractor completing the works by the contractual completion date, as the contractor will become entitled to complete the works within a reasonable time instead. Consequently, and in such cases, the employer will not be able to deduct liquidated damages.

As described in Section 3.1.5, the circumstances in which time may become 'at large' are now few and far between when the parties have opted for a modern building contract such as the JCT 2016 contracts or NEC4. These contracts contain what are often called 'catch all' provisions, that entitle the contractor to an extension of time (and indeed, loss and expense in the qualifying circumstances) if the delay caused by the employer is not covered by one of the more specific grounds that entitle the contractor to an extension of time.

However, the likelihood of time becoming 'at large' may be greater under a bespoke contract or amended version of a JCT 2016 or NEC4 contract.

### 8.4.6 The Contract Is Terminated

The law regarding the operation of liquidated damages clause has been recently and decisively clarified by the Supreme Court in the case of *Triple Point Technology Inc v PTT Public Company Limited.* [118] The case concerned a contract for the provision of specialist software (by Triple Point) to a commodities trading company (PTT). Triple Point was culpably late in completing the software and claimed payment from PTT, who refused to pay. Triple Point then suspended work on the software, and PTT terminated the contract on the ground that Triple Point had wrongfully suspended and ultimately, ceased work. The contract entitled PTT to deduct liquidated damages in the event that Triple Point failed to complete the work by the agreed completion dates. Triple Point began court proceedings in February 2015, which were heard in the Technology and Construction Court (TCC) in late 2016.[119] The TCC judge Mrs. Justice Jefford found Triple Point liable for the delays, dismissed Triple Point's claim and awarded PTT approximately \$4.5 m in damages, comprising circa \$3.4 m in liquidated damages and around \$1 m for general damages.

Triple Point then appealed[120] to the Court of Appeal, where Sir Rupert Jackson's decision (with whom Lord Justice Floyd and Lord Justice Lewison agreed) identified the that PTT's entitlement to liquidated damages could be considered as follows:

- The liquidated damages clause did not apply (following the 1913 case of *British Glanzstoff Manufacturing Company Ltd v General Accident, Fire and Life Assurance Corporation Ltd*[121]);
- PTT could only deduct liquidated damages until the contract was terminated (or determined); or
- PTT could deduct liquidated damages until a replacement contractor (in place of Triple Point) had completed the works.

Prior to the Court of Appeal's decision, the argument that liquidated damages clauses only applied up to termination (the second point above), was generally supported, although

118 *Triple Point Technology Inc v PTT Public Company Limited* (2021) UKSC 29.
119 *Triple Point Technology Inc v PTT Public Company Limited* (2017) EWHC 2178 (TCC).
120 *Triple Point Technology Inc v PTT Public Company Limited* (2019) EWCA Civ 230.
121 *British Glanzstoff Manufacturing Company Ltd v General Accident, Fire and Life Assurance Corporation Ltd* (1913) AC 143.

no clear authority on this point had been established.[122] The Court of Appeal's decision seemingly leaned towards the position that liquidated damages clauses would not apply if the contract was terminated – parties who had previously relied upon the certainty of a known sum per week (or per day) of liquidated damages would have to prove and argue over the actual cost incurred for failing to complete by the completion date. Whilst it stated that each case must be decided on its own facts, the majority of liquidated damages clauses (including those in building contracts) are drafted in a similar way to the liquidated damages clause in the contract between PTT and Triple Point. It was also felt that the Court of Appeal's decision did not clarify the applicable circumstances in which each of the three approaches applied.

The Supreme Court overturned the Court of Appeal's decision, principally for the reason that it went against commercial reality and the aim and purpose of a liquidated damages clause:

> *'The difficulty about this approach is that it is inconsistent with commercial reality and the accepted function of liquidated damages. Parties agree a liquidated damages clause so as to provide a remedy that is predictable and certain for a particular event (here, as often, that event is a delay in completion). The employer does not then have to quantify its loss, which may be difficult and time-consuming for it to do. Parties must be taken to know the general law, namely that the accrual of liquidated damages comes to an end on termination of the contract (see Photo Production Ltd v Securicor Transport Ltd [1980] AC 827, 844 and 849). After that event, the parties' contract is at an end and the parties must seek damages for breach of contract under the general law. That is well-understood: see per Recorder Michael Harvey QC in Gibbs v Tomlinson (1992) 35 Con LR 86, p 116. Parties do not have to provide specifically for the effect of the termination of their contract. They can take that consequence as read. I do not, therefore, agree with Sir Rupert Jackson when he holds in the second sentence of para 110 of his judgment that "If a construction contract is abandoned or terminated, the employer is in new territory for which the liquidated damages clause may not have made provision." The territory is well-trodden, and the liquidated damages clause does not need to provide for it'.*[123]

The Supreme Court has now clarified that unless the contract uses very clear words to the contrary, liquidated damages will only apply up to the point that the contract is terminated.

### 8.4.7 The Stipulated Sum Is Actually a Penalty

The grounds on which a contractor may put forward this contention are noted in Section 8.2.2 of this chapter.

---

122 See *Glanzstoff* (Footnote 121), and *Harbutt's Plasticine Ltd v Wayne Tank and Pump Co Ltd* (1970) 1 All ER 225 CA.

123 *Triple Point Technology Inc v PTT Public Company Limited* (2021) UKSC 29 at para 35 – the decision of Lady Arden with whom Lord Leggatt and Lord Burrows agreed (para 73), and Lord Sales and Lord Hodge agreed (at para 123). The Supreme Court considered three issues, the first of which was *'Are liquidated damages payable...where Triple Point never completes the work and PTT never accepts it?'* with which all five law lords agreed.

## 8.5 Defects: Reduction/Abatement of the Contract Price

Claims for the rectification of defective work currently sit behind extensions of time as the second most common claim against a contractor.[124]

This section is not concerned with the procedural remedies available to the employer (e.g. instructing the contractor to rectify)[125] or claims made against the contractor after the contract has been concluded (such as claims for rectifying latent defects), but the financial remedies available to the employer under the contract.

### 8.5.1 General Points

If the contractor is carrying out or completes the work but elements of the work are defective, the employer has the following remedies under most building contracts:

- If the works are in progress, but not completed, he can:
  - Instruct the contractor to remove the defective (or non-compliant work) and reinstate it to the standard required by the building contract; and/or
  - Abate (reduce) the amount due to the contractor as an interim payment accordingly;
- If the work has been completed[126] and the rectification[127] or defects correction[128] period has not expired:
  - Instruct the contractor to rectify the defective or non-compliant work; or
  - Accept the non-conformity or defect in the work, in return for an abatement (a reduction) of the contract price; or
- If the rectification or defects correction period has expired, but the contract has not been concluded, the employer may:
  - Abate the contract price for the defective work; or
  - Set-off (deduct) the deduct either the damages (losses) arising from the employer correcting the work or sums equating to the diminution in value of the employer's property or loss of amenity.

### 8.5.2 Abatement/Reduction of the Contract Price

Abatement in this context means to 'reduce[129]' the contract price for an item of work, to reflect the work's defective or non-compliant nature. Abatement can never exceed the contract price that would otherwise be paid for the work, if it wasn't defective.

---

124 Refer to the 'Construction Contracts and Law Report' (2012) by the RIBA. In its 2022 survey, claims for defective work were reported by 41% of respondents, and 'extension of time' claims were reported by 50% of respondents.

125 In addition to instructing the contractor to rectify defective work, the employer may adopt further measures (such as engaging others to do so, or termination of the contractor's employment under the contract) if the contractor fails to comply with the employer's rectification instruction.

126 The 'quantity' of work is completed, but elements of that work are defective – i.e. it contains patent defects that would prevent the works reaching 'practical completion' (JCT 2016 contracts) or 'completion' under NEC4.

127 The 'Rectification Period' – JCT 2016 contracts.

128 The 'defects correction period' – NEC4.

129 Oxford Dictionary of English – (3rd Ed) 2010.

Calculating the appropriate value of abatement will vary in complexity, depending upon the nature of the defect or the non-compliance in each case, particularly if the missing or defective element is simple to quantify. Two examples are:

- The building contract includes a lump sum price for constructing a metal stud partition wall, including its finishes which include 3 coats of paint. The contractor may only have applied 2 coats of paint, so the price would be abated (reduced) by the appropriate value for 1 coat of paint; or
- A heating system may only provide 85% of the specified output. In this case, and provided that the 85% output meets or exceeds the output required for the building's normal operation, the lump sum price for the heating system could be reduced accordingly.

Abatement can become complicated or even wholly inappropriate in some cases, usually involving a serious defect that may render the work valueless. Consider the examples involving a building's structural elements where an inadequate or missing component may render the structural element useless and therefore, valueless. If the steel reinforcement in a concrete beam or foundation is either missing, or inadequate, then reducing the value of that component is likely to be an inadequate remedy. The defective structural element will have to be broken up, removed from site and reconstructed correctly, in accordance with the contract. In such a case, the costs of all the work operations required to replace the defective structural element are likely to exceed the contract price for that element, so the appropriate remedies for the employer, will be as follows:

- If the works are in progress and the contractor is rectifying or replacing the defective work, then the employer need only pay for the value of work that the contractor has properly carried out. This usually means that the employer would firstly either not include or omit the value of the original and defective work.[130] Secondly, once the defective work has been removed from site, the employer would pay for the value of the properly completed replacement work; or
- If the works are complete and the rectification, or defects correction period has begun, as an alternative to instructing the contractor to correct the defective work,[131] the employer can accept the defect in return for abating the contract price. JCT 2016 and NEC4 contracts both permit the employer to abate the contract price in this way (see Sections 8.5.3 and 8.5.4 below).

### 8.5.3 Abatement Under JCT 2016 Contracts

JCT 2016 Contracts[132] permit the employer to accept a defect but abate the contract price by instructing the contractor accordingly (not to rectify the defect), and making an appropriate deduction from the Contract Sum. Under JCT SBC/Q, IC and MW 2016, the employer

130 NEC4 Option C (Target Cost with Activity Schedule) or Option D (Target Cost with Bill of Quantities) may permit the 'Defined Cost' of the defective work to be included in payments to the contractor, unless that is 'Disallowed Cost' – see clauses 11.2(24) and 11.2(26) in both Options.
131 NEC4 Target Cost Options C and D both include the 'cost of correcting Defects after Completion' within the definition of 'Disallowed Cost' – clause 11.2(26) in both Options.
132 JCT SBC/Q 2016 – clause 2.38, JCT DB 2016 – clause 2.35, JCT IC 2016 – clause 2.30, and JCT MW 2016 – clause 2.10.

decides to accept the defect, and the Architect/Contract Administrator instructs the contractor accordingly. However, under JCT DB 2016, the employer not only makes the decision to accept a defect but also instructs the contractor. There is less risk for any consultants engaged by the employer, who could otherwise be liable for any consequences of the defect's continued presence.[133] However, it would be advisable for the employer and any consultants to carefully consider the effect that a remaining defect may have, before issuing such an instruction (for example, there should not be any detrimental effects to the building itself, or to its use or maintenance under the CDM Regulations 2015). The contractor should also insist that the employer's instruction reduces the specification accordingly, so the contractor will not be in breach – so the instruction has a similar effect to clause 45 of NEC4 (see Section 8.5.4 below), where the 'Scope' is changed so the defect does not have to be corrected.[134] Additionally, it is advisable for the contractor to insist upon an indemnity from the employer, absolving him of any responsibility if the employer chooses that the defect can remain.

### 8.5.4 Abatement Under NEC4

Under NEC4, when the Client is using the works (i.e. after 'completion'), he can choose if he wants the contractor to return to correct defects. If the Client does want the contractor to correct the remaining notified defects, then the Project Manager arranges access to the relevant part of the works for the contractor to do so.[135] The contractor must then rectify the defect within the defect correction period, failing which the Project Manager assess the cost of the Client engaging others to correct the defect, and the Scope is changed to accept the defect.[136]

If the Client does not want the contractor to correct the defect, he has the following options:

- He can simply deny the contractor access to correct the defect, in which case, the Project Manager assesses the amount it would cost the contractor to correct the defect (which the contractor must pay to the Client)[137]; or
- The Project Manager proposes to the contractor that the defect is accepted without it being corrected, whereupon the contractor submits a quotation to reduce the Prices.[138]

For each of the above Client's options, the Scope is changed so the defect does not have to be corrected.[139]

---

133 Contrast with JCT SBC 2016, JCT IC 2016 and JCT MW 2016 where the Architect/Contract Administrator instructs the contractor not to make good a defect in return for a reduction in the Contract Sum (albeit, with the employer's consent).
134 NEC4, clause 45. In return for the Project Manager (and obviously the Client) agreeing to reduce the Scope in order to accept a defect, the contractor submits a quotation to reduce the 'Prices' and/or achieve an earlier Completion Date for the Project Manager to Accept.
135 NEC4, clause 44.4 – expresses the contractor's access to correct defects as 'use' of the relevant part of the works that have been taken over. Also see Section 8.6.2 in relation to NEC4 Option X17 (low performance damages).
136 NEC4, clause 46.1.
137 NEC4, clause 46.2.
138 NEC4, clauses 45.1 and 45.2.
139 NEC4, clauses 45.1, 45.2 and 46.2.

### 8.5.5 Costs of Rectifying Defective Work

Ordinarily, it is assumed that the claimant (as the injured party) will seek to rectify the defective work.[140] However, the claimant ought to attempt to mitigate its costs by initially, giving the building contractor the opportunity to carry out the rectification/replacement of defective work. In the event that it is unreasonable for the claimant to either (i) fail to give the original contractor the opportunity to make good the defects or (ii) if it unreasonably refuses the original contractor's offer to carry out the same, then the claimant's losses may be limited to what it would have cost the original contractor to rectify/replace the defective work.[141]

It used to be the case that the measure of damages was the diminution (reduction) in value of the property caused by the defects. However, more recently, the measure of damages is taken to be the cost of reinstatement or repair because that is the foreseeable consequence of defective work.[142] In terms of quantifying the damages for reinstatement or repair, the Court has given the following guidelines[143]:

a) *'One needs first of all to determine that the loss in question flows naturally from and is effectively causatively linked to the established breach of contract;*
b) *One then needs to determine, in the case of actual reinstatement or repair as the case may be, whether it involved putting right the defect or physical damage as the case may be. If there has been a breach which caused some physical damage then prima facie it is the cost of reinstatement of that physical damage which is recoverable;*
c) *It then needs to be determined whether and if so to what extent the remedial solution actually adopted was reasonable in all the circumstances. In this regard, the advice of an expert may well be a material factor in the determination of what was a reasonable course of action for an innocent claimant to take;*
d) *In so far as this is a different exercise, one should determine whether the claimant, as a matter of mitigation, acted reasonably in pursuing the remedial solution and in that context the Court should have regard to whether the claimant acted on the advice of experts retained by it;*
e) *The claimant has a duty to mitigate its loss; it is not a heavy duty and in all or most cases the duty is not to act unreasonably in the context of incurring loss. The burden of proving a breach is on the defendant which must show that the claimant acted unreasonably; [and]*
f) *Every case depends upon its facts but in considering reasonableness the Court needs to look at all the facts and in particular the circumstances in which the innocent claimant finds itself at the time when it commits to a particular remedial course of action'.*

---

140 *Ruxley Electronics and Construction v Forsyth* (1995) 73 BLR 1.
141 *Martlet Homes Ltd v Mulalley & Co Ltd* (2022) EWHC 1813 (TCC) at para 527.
142 *Ian McGlinn v Waltham Contractors Ltd & Ors* (2007) EWHC 149 (TCC) at Paragraph 787.
143 *Linklaters Business Services v Sir Robert McAlpine Ltd & Ors* (2010) EWHC 2931 (TCC) at Paragraph 131, after considering *Golden Strait Corporation v Nippon Yusen Kubishika Kaisha* (2007) UKHL 12, *British Westinghouse Electric and Manufacturing Co Ltd v Underground Electric Railways Co of London Ltd* (1912) AC 673, *East Ham BC v Bernard Sunley & Sons Ltd* (1966) AC 406, Ruxley (see Footnote 140), *Banco De Portugal v Warterlow & Sons* (1932) AC 452, *Darlington BC v Wiltshier Northern Ltd* (1995) 1 WLR 68, *Skandia and another v Thames Water* (1999) BLR 338, *Axa Insurance UK Plc v Cunningham Lindsey United Kingdom* (2007) EWHC 3023, *Board of Governors of Hospitals for Sick Children v McLaughlin and Harvey and Others* (1987) 19 Con LR 25 and *McGlinn* (see Footnote 142).

Taking the guidelines from *Linklaters*[143], the Courts have given some useful examples, particularly where the claimants sought to recover the rectification costs for defective works from the defendants in cases that followed the Grenfell Tower tragedy,[144] in which significant time-related costs that are ancillary to the costs of the rectification or replacement works were incurred. Relevant heads of claim are as follows:

- Costs of variations under the building contract for rectification or replacement works, where the defendant could not show that the claimant acted unreasonably:
  - Work that was additional to that specified in the rectification/replacement works contract but nevertheless was always required by the claimant may be allowed, even if this is valued and claimed on a 'daywork' basis'[145] or against a provisional sum;
  - Changing the location of the remedial/replacement works contractor's site set-up.[146]
- Contingency. The Court may consider that a contingency may be appropriate and express this as a percentage (%), to cover matters such as design and price risk, and necessary items of work whose scope is unknown. However, if the circumstances are relatively well-established (e.g. if the building has been inspected or tested by 'experienced construction professionals',[147] then the contingency will be reduced to reflect the knowledge of the building).
- Allowance for adverse weather conditions. The Court may consider some allowance for adverse weather affecting the remedial/replacement works, but it is reasonable to expect that the remedial/replacement works contractor will rely upon weather data, and programme his works accordingly[148];
- Settlement agreements reached between the claimant and the remedial/replacement works contractor. It is not uncommon for claimant to reach a compromise or settlement agreement with the remedial/replacement works contractor – for example, agreeing to waive any claim for delay damages, in return for the contractor agreeing to waive claims for additional costs (such as loss and/or expense for prolongation). In cases where the Courts find that such 'settlement agreements' are commercially sensible in the circumstances, such agreements may have the following effects on the claim:
  - If the settlement agreement includes for the claimant waiving a claim for delay damages, it is unlikely that he will be able to recover any other costs arising from the remedial/replacement contractor failing to complete by his contractual completion date[149];

144 On 17 June 2017, fire spread from the interior of a fourth floor flat around the exterior of the 24 storey tower block via the external cladding, constructed from aluminium composite material (ACM). This led to the Building Safety Act 2022 c30. Also see the resulting public enquiry - www.grenfelltowerinquiry.org.uk/

145 *Martlet Homes Ltd v Mulalley & Co Ltd* (2022) EWHC 1813 (TCC) at 459.

146 *Martlet Homes* (see Footnote 145) at para 438.

147 *Zagora Management & Ors v Zurich Insurance plc & Ors* (2019) EWHC 140 (TCC) at paras 10.10.9 and 10.10.9.

148 *Martlet* Homes (see Footnote 145) at Paragraph 513: 'nowadays contractors are able to obtain reasonably reliable information about expected weather conditions, at least for the next few days, so as to schedule their work accordingly'. When assessing a 'contingency' for adverse weather, the claimant may wish to rely upon available weather data that could indicate what proportion of time (e.g. per year) is lost in the relevant location.

149 *Martlet Homes* (see Footnote 145) at 460 to 465.

- Similarly, the defendant will not be able to argue that the claimant's claim should be reduced because of its '...*failure to exercise a clear contractual right and that either the full amount or at least a substantial proportion of the total amount should be deducted from the amount [that the defendant] is liable...to pay the claimant.*'[150]
- Time-related costs (including the costs of supervision and specialist supervision). Unless the defendant can demonstrate that the claimant's claim is unreasonable,[151] the claimant may be entitled to recover time-related costs of supervision. If particular circumstances apply, such as there being an imminent risk to the building occupants' safety (such as in the case of fire), then the costs of specialist supervision (such as fire wardens or other similarly qualified professionals) may be recoverable.
- Professional fees, including those for an environmental consultant will ordinarily, be permitted.[152]
- A reduction for any 'betterment' will be applied.[153]

## 8.6 Employer Claims for Other Breaches of Contract

The majority of claims by employers are for delay damages and defective work. However, employer may raise other claims against the contractor, but these are usually found where (i) a standard form of building contract has been amended to permit such a claim or (ii) a separate entitlement exists for the employer.

### 8.6.1 Liquidated Damages for Failing to Construct the Work to a Specified Area ($m^2$ or $ft^2$)

Where the building contract requires the contractor to carry out work where the plan area (or areas) within the building have an essential or significant effect upon the employer's ability to earn revenue from that building, then the employer may amend the building contract to recover liquidated damages from the contractor if those areas are not met. Typical examples include:

- Commercial office space advertised at a certain area (usually $ft^2$);

---

150 *Martlet Homes* (see Footnote 145) at Paragraphs 477 to 496.

151 *Martlet Homes* (see Footnote 145) at Paragraph 539, concerning the claimant's claim for £2,987,205.93 for the 'waking watch' costs of fire supervision for 1 year and 9 months: '...*the claimant's initial approach from June 2017 was to investigate, with the benefit of professional advice, whether the appropriate course was removal, replacement or repair, regardless of whether any, and if so what, cost was ultimately recoverable from the defendant*'. '*The defendant has to establish that the claimant unreasonably failed to mitigate its loss by showing that it could and should have acted differently in specified respects and that, had it done so, the overall duration would have been significantly reduced*' and at Paragraph 546 '*In my judgment the defendant has failed to establish that this decision was unreasonable so as to justify reducing the overall period of time over which the waking watch should be allowed. The defendant has adopted what in my view is a fundamentally flawed strategy of advancing a wholly unrealistic case*'.

152 *Martlet Homes* (see Footnote 145) at Paragraphs 522 and 524.

153 *Martlet Homes* (see Footnote 145) at Paragraph 471.

- Commercial retail space:
  - A particular end-user or tenant may have specified that the retail premises be a minimum area (such as a supermarket's sales floor); or
  - Specific minimum areas are required (such as warehousing, or storage);
- Residential dwellings may marketed as being a particular, or minimum area.

The building contract will be amended to show the following:

- The area (often expressed as the 'gross internal area' or 'net internal area') or areas to be achieved by the contractor; and
- The liquidated damages per unit area that the employer may deduct if the constructed area is less than the area stated in the building contract.

The definition and methods of measuring the relevant areas is often stated in the amendments to the building contract.[154]

In the event that such a clause is found in the building contract, it is fairly clear that achieving the correct areas or subareas is important to the employer. If the contractor is engaged under a design and build contract, then he will be responsible for ensuring that the design results in the constructed works meeting the employer's requirements for particular areas. This is far simpler in new-build construction. However, these clauses are often found in building contracts for refurbishing, or working in an existing building, so the contractor should either satisfy himself that the relevant areas can be achieved within the existing building layout or reject the clause.

### 8.6.2 Other Damages and Costs Payable Under NEC4

NEC4 contains other provisions for the client recovering monies from the contractor in the event that:

- The work fails to satisfy the 'Condition' required to achieve a Key Date; and
- If Option X17 (low performance damages) applies, the works do not achieve the performance level set out in contract data part one.

If the project manager decides that contractor has failed to meet the 'condition'[155] for satisfying a key date, by the date stated as the 'key date',[155] then provided that the client incurs either:

- Additional costs of carrying out the relevant work; or
- Pays 'others'[156] to carry out the relevant work,

  then clause 25.3 provides that within 4 weeks the project manager assesses the 'additional cost' within 4 weeks of the condition for the key date being met. The contractor then pays this 'additional cost.'

154 For example, in accordance with the 'Code of Measuring Practice' published by the RICS. The current version is the 6th Edition (May 2015).
155 Contract Data Part One – '2 The Contractor's Main Responsibilities'.
156 NEC4, clause 11.2(12) defines 'Others'.

Clause 25.3 also clarifies that the client is only entitled to recover the 'additional costs' if the contractor fails to meet a Key Date. The contractor's failure to meet a Key Date is seemingly distinguished from his failure to complete the works by the Completion Date, so any delay damages (either as liquidated damages under Option X7) or common law damages are unlikely to apply (see Section 8.2.5).

Option X17 (low performance damages) provides the client with an additional remedy[157] if any defects are present in the work. If a defect is shown in the Defects Certificate,[158] and this causes a particular part of the work to perform at a lower level than that stated in contract data part one, the contractor must pay the stated amount of 'low performance damages' to the client.

157 In addition to either having the contractor correct the defect (NEC4 clause 44.1), or reducing (abating) the prices in return for accepting a defect under NEC4 clause 45 (refer to Section 8.5.4).
158 NEC4, clause 11.2(7) defines the 'Defects Certificate'.

# 9

# Duties of the Employer's Representative

## 9.1 Employer's Representatives (Architects', Contract Administrator's and Project Manager's) Powers and Liability to the Contractor

### 9.1.1 General Points

Modern building contracts require the administrative and valuation functions (for which the employer is responsible) to be carried out by someone acting on the employer's behalf. In this chapter, these persons (although it can also be a firm or company, as well as an individual) will generally be referred to as the 'employer's representative'. The precise name and duties of the particular employer's representative will vary with each building contract, as shown in Table 9.1:

It can be seen from Table 9.1 that each building contract requires a specific person to undertake the contract's administrative functions on the employer's behalf. These are:

- The Architect[1] or Contract Administrator:
  - JCT SBC/Q 2016
  - JCT IC 2016; and
  - JCT MW 2016
- Employer's Agent – JCT DB 2016; and
- Project Manager – NEC4

Aside from the Employer's Agent under JCT DB 2016, the employer's representative usually carries out some design work – so his professional background is usually that of an architect or an engineer. If the employer's representative's professional background is based on management,[2] then the employer's design consultants will often report to him.

1 The Architects Act 1997 provides that a person may only refer to themselves as an 'Architect' if they are registered on the Register of Architects. Consequently, if either the employer's representative (as an individual) or as a firm does not meet the requirements of the Architects Act 1997, then the employer's representative must be described as the 'Contract Administrator'.
2 For example, if acting as the Contract Administrator under JCT SBC/Q 2016, JCT IC 2016 or JCT MW 2016, or as the Project Manager under NEC4, the employer's representative may not be qualified as a designer (such as an architect or engineer) but may be qualified in project management, quantity surveying or building surveying.

*Building Contract Claims*, Sixth Edition. Guy Higginbottom and David Chappell.

**Table 9.1** Employers' representatives for JCT 2016 and NEC4 contracts.

| Contract | Employer's representative(s) | | | | |
|---|---|---|---|---|---|
| | Administrative functions | | | Other functions | |
| | Architect/Contract Administrator | Employer's Agent | Project Manager | Quantity Surveyor | Supervisor |
| JCT SBC/Q 2016 | ✓ | | | ✓ | |
| JCT DB 2016 | | ✓ | | | |
| JCT IC 2016 | ✓ | | | ✓ | |
| JCT MW 2016 | ✓ | | | | |
| NEC4 | | | ✓ | | ✓ |

Depending upon the building contract used, other functions, such as the valuation of works undertaken and assisting the employer's representative with the management of defective work, are respectively carried out by a quantity surveyor (JCT SBC/Q 2016 and JCT IC 2016) and the Supervisor (NEC4) respectively. However, it is not uncommon for a quantity surveyor to assist the Project Manager under NEC4 contracts, albeit unlike the 'Supervisor', a quantity surveyor is not named in the NEC4 contract. Any quantity surveying functions under an un-amended NEC4 contract will be carried out as part of the Project Manager's duties.

The employer's representative's duties mainly focus on administering the building contract between the employer (or Client) and the contractor. This is a very important function and can materially affect the outcome of the project for both parties. Depending upon the form of building contract, the employer's representative's responsibilities may vary, but generally, he must act fairly and professionally.[3] If the conditions of the appointment between him and the employer do not specify the contract administration services that he must carry out, then ordinarily, he must ensure that he satisfies all of his and the employer's express obligations under the particular building contract. Determining the scope of services in this way is relatively straightforward for JCT 2016 contracts and NEC4, as the employer's representative's duties are clearly expressed. It is generally thought that the roles and responsibilities of the 'Contract Administrator', 'Project Manager' and 'Employer's Agent' are closely analogous to those of an architect,[4] insofar as any decision-making must be fair, unbiased and holding the balance between the employer and the contractor. The employer may face serious consequences if he attempts to interfere with the duties of his representative, particularly if those duties concern the issuing of formal certificates, notices or statements under the terms of the building contract. Such interference may entitle the contractor to terminate his employment under the contract.[5]

3 Keating on Construction Contracts (10th Ed), 14-016.
4 Costain v Bechtel [2005] EWHC 1018.
5 JCT SBC/Q 2016 and JCT IC 2016 – termination under clause 8.9.1.2, and JCT MW 2016 – clause 6.8.1.2.

**Table 9.2** Standard forms of engagement for the employer's representative.

| Contract/role | Appointment document |
|---|---|
| JCT SBC/Q 2016, JCT IC 2016, JCT MW 2016: | |
| Architect | RIBA Professional Services Contract 2020 |
| Contract Administrator | RIBA Professional Services Contract 2020<br>RICS[6] Contract Administration Services May 2022 |
| Quantity Surveyor | RICS[6] Quantity Surveyor Services May 2022 |
| JCT DB 2016 | |
| Employer's Agent | RICS[6] Employer's Agent Services May 2022 |
| NEC4 | |
| Project Manager | NEC4 Professional Services Contract<br>RICS[6] Project Manager Services May 2022 |
| Supervisor | NEC4 Professional Services Contract |

### 9.1.2 Employer's Representative(s) – General Duties

The precise duties that the employer's representative carries out for the employer vary from project to project. The employer may require his representative to be engaged under one of the standard forms of appointment (see Table 9.2), or he may engage his representative under a bespoke form of appointment. Typical standard forms of appointment for the employer's representative under building contracts are as follows:

With the exception of the NEC4 Professional Services Contract, all of the other means of appointing the employer's representative include a schedule of services[7] for him to perform. Under NEC4, the 'Client' and the 'Project Manager' will have to agree the services for the Project Manager – for example, they could adopt the RICS' Project Manager Services, if appropriate.

The services for the employer's representative are usually expressed in the order of each stage of the RIBA Plan of Work 2020.[8]

6 The RICS services (e.g. for Contract Administration, Employer's Agent, Project Manager or Quantity Surveyor) are governed by either the RICS' Standard Form of Consultant's Appointment (May 2022) or the RICS Short Form of Consultants Appointment (May 2022).

7 The RIBA Professional Services Contract 2020 has an integral schedule of services, covering the 'Role Specifications' for 'Architect/ Consultant', 'Lead Designer', 'Project Lead', and 'Contract Administrator', along with 'Specialist Roles' that can be agreed between the employer and his representative. The 'schedule of services' includes tasks allied to each stage of the RIBA Plan of Work 2020.

8 The RIBA Plan of Work 2020 Stages are: 0 – Strategic Definition, 1 – Preparation and Briefing, 2 – Concept Design, 3 – Spatial Co-Ordination, 4 – Technical Design, 5 – Manufacturing and Construction, 6 – Handover, and 7 – Use.

Ordinarily, any professional person will owe their client (or employer) an implied duty to carry out their services with reasonable skill and care:

- RIBA Professional Services Contract 2020 – clause 3.1[9];
- RICS Standard Form of Consultant's Appointment (May 2022)[6] – clause 3.1[10]; and
- NEC4 Professional Services Contract – clause 20.2.[11]

Irrespective of whether the employer's representative has been engaged under a formal or written contract (often referred to a 'conditions of engagement', or 'appointment') this duty will be imposed upon the employer's representative (and quantity surveyor – see Section 9.2) by s13 of the Supply of Goods and Services Act 1982. This same standard of performance is included in the collateral warranties that employer's representatives, quantity surveyors and other consultants often enter into, on behalf of funders, purchasers and/or tenants.

### 9.1.3 Employer's Representative(s) – Relevant Duties

This section assumes that the employer's representative has been engaged under either bespoke terms of appointment, or one of the standard forms of appointment (which include the relevant services) described in Section 9.1.2 and Table 9.2. The relevant duties carried out by the employer's representative, as far as they relate to the contractor, will be those carried out during RIBA Stage 5 (Manufacturing and Construction). There may also be some that relate to completion or 'Handover' (in RIBA Stage 6).

Whether the employer's representative[12] is an Architect, Contract Administrator, Employer's Agent, or Project Manager, their relevant duties during RIBA Stage 5 (Manufacturing and Construction) ordinarily fall into the following six categories:

- Design obligations and design coordination
  - Providing other design consultants with architectural design
  - Coordinating other consultants' design
  - Advising the employer
  - Duties under the CDM Regulations 2015
- Manage and report on other design consultants:
  - Instruct the consultants

---

9 RIBA Professional Services Contract 2020 – clause 3.1 '*…the Architect/Consultant will exercise the reasonable skill, care and diligence expected of an Architect/Consultant experienced in the provision of such services for projects of a similar size, nature and complexity …*'.

10 RICS Standard Form of Consultant's Appointment (May 2022) – clause 3.1 '*The Consultant exercises in the performance of the Services the Requisite Standard*'. The 'Requisite Standard' is defined as '*The standard of reasonable skill and care to be expected of an appropriately qualified professional consultant of the discipline stated in the Appendix holding itself out as having the competence and resources to perform the Services, and who is experienced in providing services in connection with works of a similar size, scope, nature, complexity and value…*'. If the RICS Short Form of Consultant's Appointment (May 2022) is used, a slightly shorter version of the 'Requisite Standard' applies (in clause 2): '*The Consultant, when performing the Services, exercises the reasonable skill and care to be expected of an appropriately qualified professional consultant of the same discipline as the Consultant holding itself out as having the competence and resources to perform the Services (the 'Requisite Standard')*.

11 NEC4 Professional Services Contract – clause 20.2: '*The Consultant's obligation is to use the skill and care normally used by professionals providing services similar to the service.*'.

12 Not a quantity surveyor – as their duties fall into two additional categories.

  - Management of consultants
  - Co-ordination of consultants' responses to questions
  - Monitor consultants' performance against the programme[13]
- Administration of the building contract
  - Maintain the building contract's management procedures (contract administration/project management)
  - Issue instructions in accordance with the building contract
  - Liaise with the employer and other consultants (in relation to contractor claims)
  - Issue notifications to the contractor under the building contract
- Quality control
  - Inspect the works at the appropriate times during construction
  - Carry out tests and inspections[14]
  - Certify/notify completion when the works are free of patent defects
- Programme monitoring:
  - Inspect the works, and report on progress against the programme[13] (of both consultants and the contractor)
  - Comment on the programme
- Site meetings

The quantity surveyor and Supervisor ordinarily have the following roles:

- The quantity surveyor – JCT SBC/Q 2016 and JCT IC 2016:
  - Producing regular/monthly cost reports for the employer
  - Valuation of variations and ascertainment of loss and expense for interim valuation purposes;
  - Assessing the contractor's 'final account'; and
  - Payment/pay-less notice obligations.
- The Supervisor – NEC4;
  - Identifying (marking) materials and plants that are intended for the works but are off site;
  - Instructing the contractor to search for and correct a defect
  - Carrying out tests and inspections
  - Issuing the Defects Certificate.

### 9.1.4 The Relevance of the Building Contract's Terms

The previous Sections (9.1.1 to 9.1.3) focused on the employer engaging his representative under one of the standard forms of appointment and obliging them to perform specific tasks – as set out in a schedule of services. However, the employer's representative may not have been engaged under a formal appointment – so how does he (and the contractor) know what duties he is to perform? The answer to this question has the following two aspects:

13 The employer's representative and the consultants are reliant upon a programme (JCT 2016) or the programme (NEC4) produced by the contractor, against which the progress of the consultants' design and the contractor's progress can be monitored.
14 NEC4 – duties of the Supervisor under clauses 41 (tests and inspections) and 44 (issuing the Defects Certificate)

- Duties that the employer's representative ought to perform for the employer; and
- Duties that the employer's representative ought to perform when acting on the employer's behalf when administering the terms of the building contract.

The first aspect concerns the duties that the employer's representative should be expected to perform for the employer, and are likely to include the management of other consultants, and reporting cost, quality and programme matters to the employer (see Section 9.1.3). Unless the contractor has seen the employer's representative's appointment document, he will not be aware of the scope of these duties, other than perhaps, in very general terms. In any event, the employer's representative's duties to the employer should not materially or directly affect the contractor, as his duties under the terms of the building contract will be clear (see below).

The second aspect concerns the employer's representative's duties under the building contract, so the contractor will have a much greater idea of the extent of the duties owed to him. These will be each duty or task that the employer's representative must carry out under the JCT SBC/Q 2016, JCT IC 2016, JCT MW 2016 and NEC4 contracts, and in all likelihood, every duty that the 'Employer' must carry out under JCT DB 2016.

The law presumes that the employer's representative is aware of the whole of the terms of the building contract under which he is acting, and this is fundamental to his responsibilities.[15] Unfortunately, some employers' representatives have limited knowledge of the contracts which they purport to administer. Lack of knowledge in that situation can be negligent. The employer's representative has no intrinsic powers by virtue of being the employer's representative under a particular building contract. Take, for example, JCT SBC/Q 2016 clause 3.14.1, which states that the employer's representative (Architect/Contract Administrator) may issue instructions requiring a variation. If that clause were not in the contract, the employer's representative would have no power to issue such instructions. It should also be noted that the employer's representative is not given general power under the contract to issue any instruction which may seem appropriate, but only such instructions are expressly empowered by the contract.

### 9.1.5 Agency

Employer's representatives sometimes believe that they have general powers of agency which enable them to act on the employer's behalf in every way, provided that the action is connected with the project. Contractors are often under the same misapprehension. The powers of the employer's representative to act as agent for the employer are to be found in their terms of engagement or, if there are no written terms,[16] by necessary implication. In either case, the powers of the employer's representative to act as agent (e.g. in place of the employer) will usually only be those that are absolutely essential for the administration of the contract. Ordinarily, employer's representatives have no power to affect the legal relationship that exists between the employer and the contractor,[17] so they

15 *City Inn Ltd v Shepherd Construction Ltd* [2010] CSIH 68 (upholding [2007] CSOH 190).
16 It should be noted that the Codes of Professional Conduct of both the Royal Institute of British Architects and the Architects Registration Board require architects to conclude written terms of engagement.

are unable to agree on matters that are outside the scope of their authority. If an employer's representative exceeds their authority, the consequences for them can be severe.

Both the contractor and the employer's representative should exercise care when working under the JCT DB 2016 – as the employer's representative is the 'Employer's Agent'. This title should not be taken to imply that the employer's representative is empowered to act as 'agent' for the employer in the purely legal sense. Instead, the extent of the employer's representative's (Employer's Agent's) authority should generally be understood to cover the express duties of:

- the 'Employer's Agent'[18]; and
- The administrative duties of the employer – save for those which could significantly affect the relationship between the employer and the contractor, such as:
  - Notifying the contractor of the intention to deduct liquidated damages[19]; or
  - Notifying the contractor of the employer's intention to terminate the contractor's employment under the contract.[20]

The employer's representative would be well-advised to seek the employer's approval for either of these courses of action.

### 9.1.6 Employer's Representative's Authority, Exercising Discretion, and Making Decisions

With regard to claims, the employer's representative may only act in the way set out in the contract. Therefore, if there is a pre-condition which must be satisfied before he can act, such as the giving of notice by the contractor, an employer's representative who acts without that pre-condition being satisfied is acting without authority and may become liable to the employer. It is important to note that ordinarily, the employer's representative often has no powers to certify payment of *quantum meruit*, *ex gratia* or discretionary claims under most building contracts. Therefore, contractors should be wary of making agreements with the employer's representative that fall outside the apparent scope of his authority as provided by the contract's terms (see Section 9.1.4). It is not uncommon for the employer to rely upon the advice of his representative, and in doing so, may give him the authority to negotiate, or to reach a 'deal' or agreement with the contractor that may sit outside the terms of the building contract. In such circumstances, the contractor should obtain confirmation of the extent of the employer's representative's authority to act on the employer's behalf. In doing so, it would be advisable for the contractor to copy the employer into any correspondence between himself and the employer's representative in this regard.

17 *Charles Rickards Ltd v Oppenheim* [1950] 1 KB 616.

18 JCT DB 2016 hardly mentions the "Employer's Agent". He's identified in Article 3 and clause 1.1, and if communicating with the contractor, must do so in writing (clause 1.7). Under clause 2.7.3, the contractor should keep copies of the building contract and the "Contractor's Design Documents" on site, and allow access for the employer's Agent or any person he so authorises to areas on and off-site, where the works are being prepared (clause 3.1). For interim payments or the final payment, the employer's Agent may give payment notices, pay-less notices (clause 4.10.2), and clause 7.4.3 includes the employer's Agent within the meaning of "consultants" in respect of Third Party Rights and Collateral Warranties. All other duties to the contractor are stated to be performed by the employer.

19 JCT DB 2016, notifications under clauses 2.29.1.2.

20 JCT DB 2016, employer's notices under clauses 8.4.1 and 8.4.2.

**Table 9.3** Employer's representative's exercise of 'opinion' in JCT 2016 contracts.

| Clause for employer's representative to form opinion | JCT SBC/Q 2016 | JCT DB 2016 | JCT IC 2016 | JCT MW 2016 |
|---|---|---|---|---|
| Providing further information to the contractor, if ER's opinion is that completion is before the Completion Date | 2.12.2 | — | 2.11.2 | — |
| Delay event in ER's opinion is a Relevant Event | 2.27.1 | 2.24.1 | 2.19.1 | — |
| After the Completion Date, fixing different completion dates that in the ER's opinion are fair and reasonable: | | | | |
| Fixing a later completion date | 2.28.5.1 | — | — | — |
| Fixing an earlier completion date | 2.28.5.2 | — | — | — |
| ER certifies practical completion | 2.30 | — | 2.21 | — |
| Defects made good (partial possession) | 2.35 | — | 2.27 | — |
| Defects made good | 2.39 | — | 2.31 | — |
| Previous ER's 'opinion' cannot be overruled | 3.5.2 | — | 3.4.2 | Article 2 |
| Arbitrator can review or open up any 'opinion' | 9.5 | 9.5 | 9.5 | Schedule 1 para 3 |

The employer's representative often has some power to consider claims that are more 'common law damages' in nature[21] (see Section 6.1), but the contractor should be encouraged to make those claims under the appropriate clause in the contract[21] where possible.

When exercising his power to consider claims, the employer's representative often has to form his own opinion or exercise discretion. On the face of it, JCT 2016 contracts include more scope for the employer's representative (referred to as 'ER's' in Table 9.3) to form his own opinion or exercise discretion as shown in Table 9.3:

The exercise of discretion or the forming of an opinion must be done reasonably and rationally, and the following criteria[22] should apply to the decision reached by the employer's representative[23];

- Did the employer's representative neglect or refuse to take something into account that should have been taken into account, or conversely, did the employer's representative consider something that was irrelevant?
- Was the decision so perverse that no reasonable person, acting reasonably, could have made it, even though the decision-making process itself could not be faulted?

The Project Manager in NEC4 exercises his discretion, or makes 'decisions' over a number of matters. These 'decisions' of the Project Manager usually take the form of notifying the contractor whether he has complied with the contract so something can be 'accepted'.

21 In respect of certifying or including payment for 'common law' damages, and in practical terms, the employer's representative's does have some power to consider such claims; see JCT SBC/Q 2016 – clause 4.22.5, JCT DB 2016 – 4.21.5, JCT IC 2016 – clause 4.17.4, JCT MW 2016 – see Section 6.1, and NEC4 clauses 60.1(18) and 60.1(19).

22 *Braganza v BP Shipping Ltd and Anor* [2015] UKSC 17.

23 Or the 'Employer' in the case of JCT DB 2016.

A pivotal example of the Project Manager's powers of acceptance will be the contractor submitting his revised programme for under clause 31.3 (refer to Section 3.5.1). If the Project Manager accepts the programme (or revised programme), it will become the 'Accepted Programme'. In this example, the Project Manager's ability to reject (called 'non-acceptance') does not wholly depend upon his 'decision' (or discretion) as clause 31.3 sets out four reasons for him to notify the contractor of his 'non-acceptance'), although he must obviously decide if those reasons apply.

Table 9.4 sets out the 'decisions', 'acceptances' or 'non-acceptances' that the Project Manager and/or the Supervisor have to make under the NEC4 contract. The reasons for the Project Manager not accepting (non-acceptance) a matter are set out in the footnotes.

### 9.1.7 Employer's Representative and the Contractor

Ordinarily, the employer's representative owes no duty to the contractor. This is because of the following:

- There is no direct contractual relationship between the employer's representative and the contractor, so he owes no contractual duty to the contractor (the 'privity of contract' rule); and
- Any action[24] brought by the contractor for negligence on the part of the employer's representative will be limited to pure economic loss.

### 9.1.8 The Employer's Representative's Duties - Extensions of Time

Under the terms of most forms of building contract, the employer's representative[25] has the following duties:

- The precise role and duties of the employer's representative will be determined by the terms of the contract under which he is required to act[26];
- He should apply the relevant provisions of the particular building contract[27];
- To act fairly, lawfully, rationally, logically[27], reasonably and impartially[25] and in an unbiased[28] manner;
- Make a logical and methodical analysis[27] of the effect that any delay event has had, or is likely to have, on the programme (or progress); and
- Calculate the relevant critical delay, instead of making an impressionistic, general assessment.[27]

When determining the appropriate extension of time, the employer's representative must first act fairly and impartially, and in an unbiased manner between the parties. This is an

24 The action will be one brought in 'tort' and not in contract.
25 *Costain Ltd v Bechtel Ltd & Anor* [2005] EWHC 1018 (TCC) at paragraphs 27 to 53. The Project Manager's duties to act impartially (in that case, when certifying) under NEC are similar to those of the equivalent duties of an Architect/contract administrator under JCT contracts.
26 *Scheldebouw BV v St James Homes (Grosvenor Dock)* Ltd [2006] EWHC 89 (TCC) at para 34, citing *Costain Ltd v Bechtel Ltd* [2005] EWHC 1018 (TCC).
27 *John Barker Construction Ltd v London Portman Hotel Ltd* (1996) 83 BLR 31.
28 *Sutcliffe v Thackrah* [1974] 1 All ER 859.

**Table 9.4** NEC4 – Express 'decisions', 'acceptances' and 'non-acceptances' of the Project Manager and/or Supervisor.

| Project Manager's or Supervisor's 'decision' (D) or acceptance/non-acceptance (A) | Category | NEC4 Clause |
|---|---|---|
| Revising the 'Early Warning Register' to record decisions | D | 15.4 |
| Contractor's proposal to reduce Scope/cost[a)] | A | 16.2 |
| Contractor's proposal to add/extend the Working Areas[b)] | A | 16.3 |
| Contractor's particulars of design[c)] | A | 21.2 |
| Contractor's design of equipment[d)] | A | 23.1 |
| Contractor's replacement personnel[e)] | A | 24.1 |
| The condition for a Key Date has not been met | D | 25.3 |
| Contractor's proposed sub-contractors[f)] | A | 26.2 |
| Contractor's sub-contract documents[g)] | A | 26.3 |
| Decides the date of Completion | D | 30.2 |
| Contractor's programme[h)] | A | 31.3 |
| Contractor's revised programme[h)] | A | 32.2 |
| Contractor's acceleration quotation | A | 36.1 |
| Contractor's quality policy statement and quality plan[i)] | A | 40.2 |
| Contractor's quotation to reduce prices for defects | A | 45.2 |
| Changing the Scope to reflect uncorrected and accepted defects | A | 46.2 |
| Decides the first 'assessment date' | D | 50.1 |
| Options C, D, E and F: Accepts or notifies contractor of 'Defined Cost' | A | 50.9 |
| Option A: revision of Activity Schedule[j)] | A | 55.4 |
| Contractor's notice for compensation event deemed to be accepted | A | 61.4 |
| Decides no early warning notice was given | D | 61.5 |
| Contractor's compensation event quotation | A | 62.3 |
| Compensation event quotation deemed to be accepted | A | 62.6 |
| Contractor did not properly assess the compensation event | D | 64.1 |
| Compensation event quotation deemed to be accepted | A | 64.4 |
| Proposed instruction quotation | A | 65.2 |
| Option F: Decides the 'change' in Prices, Completion Date or Key Date if the parties do not agree | D | 63.17 |
| Contractor's certificate of insurance | A | 84.1 |
| Option X4: Ultimate holding company guarantee[k)] | A | X4.2 |
| Option X10: Information Execution Plan (inc revisions)[l)] | A | X10.1<br>X10.4 |
| Option X12[m)]: Core Group Members' decisions | D | X12.1(4)<br>X12.2(3)<br>X12.3(5) |
| Option X13: Performance Bond[n)] | A | X13.1 |
| Option X14: Bank issuing advance payment bond[o)] | A | X14.2 |
| Option X15: Contractor's designer's PI insurance certificate[p)] | A | X15.6 |

Table 9.4 (Continued)

| Project Manager's or Supervisor's 'decision' (D) or acceptance/non-acceptance (A) | Category | NEC4 Clause |
|---|---|---|
| Option X16: Contractor's retention bond[q)] | A | X16.3 |
| Option X21: Whole life costing quotation | A[r)] | X21.3 |
| Option X22: Early contractor involvement forecasts (stages 1 & 2) | A | X22.2[s)]<br>X22.3[t)] |

a) NEC4 – clause 16.2. The Project Manager may give 'any reason' for not accepting the contractor's proposal.
b) NEC4 – clause 16.3. The Project Manager's reasons for not accepting the contractor's proposals are (a) that extending the Working Areas is unnecessary, and (b) that it will not be used for work 'not in the contract'.
c) NEC4 – clause 21.2. The Project Manager's reasons for not accepting the contractor's design are that it does not comply with (a) the Scope, or (b) the applicable law.
d) NEC4 – clause 23.1. The Project Manager's reasons for not accepting the contractor's design for equipment are that it is not in accordance with (a) the Scope, (b) the design already accepted by the Project Manager, or (c) the applicable law.
e) NEC4 – clause 24.1. The Project Manager's reason for not accepting the contractor's replacement personnel is that the replacement's qualifications and experience are not as good.
f) NEC4 – clause 26.2.
g) NEC4 – clause 26.3. The Project Manager's reasons for not accepting the sub-contract documents are (a) they will not allow the contractor to 'Provide the Works', or (b) they do not include a statement of mutual trust and co-operation.
h) NEC4 – clause 31.3. The Project Manager's reasons for not accepting the contractor's programme are (a) the plans shown are not practicable, (b) it does not provide the information required by clause 31.2, (c) it does not realistically show the contractor's plans and (d) it does not comply with the Scope.
i) NEC4 – clause 40.2. The Project Manager's reason for not accepting the quality policy statement and quality plan are that it does not allow the contractor to 'Provide the Works'.
j) NEC4 – Option A, clause 55.4. The Project Manager's reasons for not accepting the revised Activity Schedule are (a) it does not relate to the operations shown on the Accepted Programme, (b) changed Prices are not evenly distributed between the relevant uncompleted activities, and (c) the total of the Prices is changed.
k) NEC4 – Option X4 – clause X4.2. The Project Manager's reason for not accepting the guarantee is that the guarantor's commercial position is not strong enough to carry the guarantee.
l) NEC4 – Option X10 – clause X10.4(2). The Project Manager's reasons for not accepting the Information Execution Plan are (a) it does not comply with the Information Model Requirements, or (b) it does not allow the contractor to 'Provide the Works'.
m) NEC4 – Option X12 Multiparty collaboration. In the event that the Project Manager and/or Supervisor are named either a 'Promotor' or within the 'Schedule of Partners' in Contract Data Part One.
n) NEC4 – Option X13 – clause X13.1. The Project Manager's reason for not accepting the bond is that the bondsman's commercial is not strong enough to carry the bond.
o) NEC4 – Option X14 – clause X14.2. The Project Manager's reason for not accepting the bond is that the bondsman's commercial is not strong enough to carry the bond
p) NEC4 – Option X15 – clause X15.6. The Project Manager's (or the Client's) reason for not accepting the PI (professional indemnity) insurance is that the insurer's commercial position is not strong enough to carry the insurance.
q) NEC4 – Option X16 – clause X16.3. The Project Manager's reason for not accepting the bond is that the bondsman's commercial is not strong enough to carry the bond
r) NEC4 – Option X21 – clause X21.3. The Project Manager may give 'any reason' for not accepting the quotation.
s) NEC4 – Option X22 – clause X22.2(2). The Project Manager's reasons for not accepting the Stage One forecast are (a) it does not comply with the Scope, or (b) it includes unnecessary work.
t) NEC – Option X22 – clause X22.3(3). The Project Manager's reasons for not accepting the Stage Two forecast are (a) it does not comply with the Scope, (b) the Client will incur unnecessary costs to Others, or (c) he is not satisfied that the Prices (or changes thereto) have been properly assessed.

onerous duty, because in many instances, the employer's representative must act as judge of his own behaviour. It is sadly common for an employer's representative to give an extension of time for the wrong reason (such as a neutral reason like exceptionally adverse weather conditions), in order to cover up his own failings (e.g. late issue of information or instructions). At the other end of the spectrum, an honest employer's representative who acts fairly may often find that his employer is less than happy! An employer's representative should be encouraged to obtain professional advice before making a decision – particularly if the effect of that decision is likely to have significant consequences for the employer and/or the contractor.

Although not strictly necessary, an employer's representative often notifies the employer as a matter of courtesy before giving an extension of time. In some instances, the employer may not understand the situation and may attempt to persuade or cajole the employer's representative to decide upon a shorter extension, or that no extension should be given. The employer is not entitled to interfere with his representative's duties. If the employer attempts to effectively substitute himself for his representative, this goes against the whole purpose of engaging a representative in the first place. Such interference may entitle the contractor to terminate his employment under the contract (refer to Section 9.1.1).

In a 1990 case,[29] under the 1980 edition of JCT Minor Works Contract, the employer's representative (an architect) concluded that the contractor was entitled to an extension of time and notified him accordingly. The employer objected and rejected the extension (and a later extension), at one point visiting the site and telling the contractor that the architect had no power to give extensions of time. The employer even went as far as to notify the architect that his approval was required for any extension of time. In due course, the architect, after obtaining legal advice, ceased to act. The resulting case dealt with several issues, among them the extension of the time point in relation to which the judge said:

> *... the [employer] sought to interfere with the [architects'] performance of their duties under [the extension of time clause] which they very properly resisted. Some of [the employer's] letters were also very offensive and indicated a total lack of confidence in the [architects]. [The employer and their] Solicitors also undermined the [architects'] position in relation to the contractors. In my judgment the [employer's] letters, the Solicitors' letters and the [employer's] conduct were in breach of contract and the [architects] were amply justified in treating their engagement as at an end.*[30]

Firstly, the court's view was that the architect acted properly but the employer had not. The architect was entitled to damages from the employer for the unlawful termination.

Second, the employer's representative should correctly apply the extension of time provisions in the particular building contract. If his duties to assess and decide upon an extension of time are triggered by receipt of the contractor's delay notification, he should not proceed with deciding an extension of time to appease the contractor without receiving the delay notice. Similarly, he should not appease a dissatisfied employer by either refusing to make

29 *Argyropoulos & Pappa v Chain Compania Naviera SA* Unreported, 1 February 1990; reported in abridged form in (1990) 7-CLD-05-01.

30 Unreported, 1 February 1990 at 19 per Judge Newey.

a decision over an extension of time, or if granting an extension, ensure that the correct reasons for the extension are given (see above), even if that means the employer will have to pay compensation to the contractor and lose an entitlement to claim delay damages.

Third, the employer's representative should ascertain as many facts as can be reasonably obtained; for example, by asking questions of the employer, other consultants, and even the contractor. In respect of this latter point, many standard forms of building contract oblige the contractor to provide the employer's representative with further information,[31] so those avenues of enquiry should be explored.[32] The employer's representative should seek to establish the facts to enable the length of delays to be established with accuracy, as it is essential he considers all the evidence before making a decision. In keeping with his duty to act impartially, the employer's representative should consider all facts, both those that favour the contractor and those that are unfavourable. However, the employer's representative is encouraged not to adopt a 'wait and see' approach to assessing extension of time (see Section 3.5.2), so he will have to carefully consider the extent of information he needs in order to carry out this duty.

The employer's representative's use of planning software applications is not thought to be essential to him assessing extensions of time,[33] as older methods (implicitly not involving planning software) remain valid if planning software cannot be used accurately. However, given that in the majority of cases,[34] it is only practicable to show many of the programme's express requirements (particularly for NEC4 contracts – refer to Section 3.5.1 above) by using planning software, it would be advisable for the employer's representative to either have access to the relevant software himself, or if not, the services of someone who has. The use of planning software and particularly its application for delay analysis often require the input of specialist consultants. If the employer's representative is unsure that he has the required expertise, he should advise the employer accordingly, given the employer has the option of appointing a more experienced or specialist planner or delay analyst.

### 9.1.9 Extension of Time Decisions and Time Limits

The periods of time in which the employer's representative should provide his extension of time decision vary in importance, depending upon the form of building contract being used. For NEC4 contracts, the periods of time for the employer's representative (the Project Manager) are clearly set out, along with the consequences if he fails to respond to the contractor

31 JCT SBC/Q 2016 – clauses 2.27.3 and 4.21.3, JCT DB 2016 – clauses 2.24.3 and 4.20.3, JCT IC 2016 – clauses 2.19.4.2 and 4.16.3, and NEC4 – Options C, D and E – clause 50.9.

32 For example, the employer's representative must have properly considered a contractor's entitlement to an extension of time before issuing a notice prior to terminating the contractor's employment on the grounds that the contractor failed to work regularly and diligently - *Sindall Ltd v Solland* (2004) 80 Con LR 152.

33 *City Inn Limited v Shepherd Construction Ltd* [2008] BLR 269 paras 27 to 29.

34 SCL Delay and Disruption Protocol – 2nd Ed (2017), Guidance Part B – Guidance on Core Principles (at paragraph 1.43) encourages the contractor to provide its programme to the employer's representative in its 'native electronic form' (e.g. Oracle Primavera, Elecosoft Powerproject, or Microsoft Project) and not just as a.pdf.

within those periods[35] – refer to Chapter 5, Section 5.3.2, Figures 5.1 and 5.2 and Chapter 13, Section 13.3. However, JCT contracts take a different approach. Before 'practical completion' and provided that the contractor has notified the employer's representative[36] (and provided any further information required by the particular contract[37]), the employer's representative must notify his extension of time decision to the contractor within 12 weeks.[38]

This 12-week period has been considered by the courts in a number of cases. In a 1987 case,[39] the 12-week period was considered to be *'directory only'* [40]

However, in a later case, it's the court's view over the 12-week period that appeared a little firmer:

> *'The process of considering and granting extensions of time is to be completed not later than 12 weeks after the date of practical completion and the architect must, within that timescale, either finally fix the completion date or notify the contractor that no further extensions of time are to be granted.'*[41]

When considering the issue of the final certificate and the pre-conditions, which must be fulfilled before it is issued, the court continued:

> *'It would be much more consistent with the mandatory language of the conditions and would give effect to that language if all the 'shalls' are read in this way: the architect must issue the various certificates and the final certificate and in the sequence and with the prescribed time intervals between the successive steps. If the time limits prescribed by the conditions are not kept or maintained, the architect must still issue the certificates in question as soon as it is reasonably possible to issue them subject to the terms of any agreement as to their issue that has been reached or acknowledged by the parties. …The correct starting point is that the power is mandatory. It is then necessary to consider what the minimum relaxation would be that is necessary to give that mandatory requirement business efficacy. In that context, it can be seen that it is necessary to allow*

---

35 For example, clause 60.1(6) creates a compensation event if the Project Manager (or Supervisor) fails to respond to a communication within the period stated in Contract Data Part One, and clause 61.4 where the Project Manager's continued failure to notify the contractor becomes a compensation event and an invitation for the contractor to submit quotations.
36 JCT SBC/Q 2016 – clause 2.27, JCT DB 2016 – clause 2.24, JCT IC 2016 – clause 2.19 JCT MW 2016 does not specify any period for the employer's representative to give an extension of time.
37 Such as the 'particulars' of the delay's expected effects – JCT SBC/Q 2016 – clause 2.27.2 and JCT DB 2016 – clause 2.24.2. JCT IC 2016 and JCT MW 2016 only require the contractor to issue a notice of delay.
38 Unless there are less than 12 weeks until the Completion Date, the employer's representative must notify his extension of time decision to the contractor before the Completion Date. JCT SBC/Q 2016 – clause 2.28.2 (after receiving the contractor's particulars – see Footnote n in Table 9.4), JCT DB 2016 – clause 2.25.2 and JCT IC 2016 – clause 2.19.3.
39 *Temloc Ltd v Errill Properties Ltd* (1987) 39 BLR 30 at page 39. The case principally concerned the rate of liquidated damages payable under a JCT Private Without Quantities 1980 contract.
40 *'…directory only as to time and is not something which would invalidate the calculation and payment of liquidated damage.'* Ibid.
41 *Cantrell and Another v Wright & Fuller Ltd* (2003) 91 Con LR 97 at 147 per Judge Thornton. This case also concerned a 1980 edition of a JCT contract.

*for a relaxation of the prescribed timescales subject to the imposition of a requirement of reasonableness. ...Although the certifier has an implied power to issue certificates out of time, this power is limited since it would be unworkable and contrary to the presumed intentions of the parties for the power to issue certificates not to be subject to the obvious constraint that the parties should be notified of its intended exercise and should not be prejudiced by its exercise. In other words, the power must be exercised reasonably.*'[42]

The court is clearly referring to the giving of certificates, in particular, the final certificate,[43] the effects of which can be far more serious and possibly permanent and unalterable.[44] This must be contrasted with the employer's representative's decision over an extension of time, which can be revised by either:

- Before or after practical completion, reducing the extension of time after part of the remaining works has been omitted;
- After practical completion, fixing a later date, either after reviewing a previous decision or otherwise; or
- Having an adjudicator open up and review any disputed extension of time decisions.

The court seems to hold the door slightly ajar; its view being that the time limits in the JCT contract were 'neither rigid nor immutable'.

### 9.1.10 Loss and Expense Decisions and Time Limits

The employer's representative must also make decisions regarding loss and expense, as the ascertainment, certification, or notification of loss and expense will form an integral part of both interim payments and, in the case of the JCT 2016 contracts, the final payment. Therefore, the time limits imposed on the employer's representative will also be linked to his relevant duties in respect of interim and final payments (such as the provision of payment notices, and/or pay-less notices). JCT 2016 contracts[45] provide that the employer's representative (or the employer in the case of JCT DB 2016) must respond to the contractor's initial estimate of loss and expense, and any 'updates' to that within 28 days and 14 days respectively (refer to Sections 12.3.2 and 12.3.3).

Under NEC4, the Project Manager has to carry out similar duties within particular timescales. As 'loss and expense' or additional cost resulting in a 'change to the Prices' (to use the correct NEC4 language), must be assessed by the Project Manager under the 'compensation event' procedure – as covered in Chapter 5 in Section 5.3.2 and summarised in Figures 5.1 and 5.2. The valuation of those additional costs is described in Section 5.5.

---

42 Cantrell and Another v Wright & Fuller Ltd (2003) 91 Con LR 97 at 136- 140 per Judge Thornton.

43 Or the Final Statement – the equivalent under JCT DB 2016

44 For example, once the Final Certificate (or Final Statement in the case of JCT DB 2016) has become 'conclusive', then this has the effect that all and only extensions of time as are due under the contract have been given – JCT SBC/Q 2016 and JCT IC 2016 – clause 1.9.1.3, and JCT DB 2016 – clause 1.8.1.2.

45 JCT SBC/Q 2016, JCT DB 2016 and JCT IC 2016.

### 9.1.11 Employer's Liability for Breaches by the Employer's Representative

The court in *Merton v Leach* was of the view that the employer's implied undertaking was to do all things necessary to enable the contractor to carry out the work:

> *'...extends to those things which the architect must do to enable the contractor to carry out the work and that the building owner is liable for any breach of this duty on the part of the architect.'*[46]

Most instances of co-operation will be between the employer's representative (and not the employer) and the contractor. The employer will only be liable for the breach of the duty to co-operate (see Section 6.2.3 – so far as those functions performed by the employer's representative acting as the employer's agent are concerned and not usually when his representative is acting as a certifier), because in the latter case the employer may not be vicariously liable. The position of the employer's representative has been set out thus:

- As 'architect' (or 'contract administrator') under a JCT contract:

> *'Under the standard conditions, the architect acts as the servant or agent of the building owner in supplying the contractor with the necessary drawings, instructions, levels and the like and in supervising the progress of the work and ensuring that it is properly carried out... To the extent that the architect performs these duties the building owner contracts with the contractor that the architect will performs them with reasonable diligence and with reasonable skill and care. The contract also confers on the architect discretionary powers which he must exercise with due regard to the interests of the contractor and the building owner. The building owner does not undertake that the architect will exercise his discretionary powers reasonably; he undertakes that although the architect may be engaged or employed by him, he will leave him free to exercise his discretion fairly and without proper interference by him ...*[47]*.'*

- As 'Project Manager' under NEC contracts:

> *'Upon examining these provisions, I am unable to find anything which militates against the existence of a duty upon the project manager to act impartially in matters of assessment and certification. It is quite true...that what the project manager decides is determinative. It is subject to later review by others. But this is also true of all the other conventional construction contracts. The interim certificates of an architect or engineer may be challenged by arbitration. Indeed, since May 1998 every construction contract has also contained, either expressly or by implication, provisions for adjudication which precisely mirror clause 92 of the present contract. For present purposes, I cannot see any relevant matter of distinction between dispute resolution procedure of [this] contract...and the dispute resolution procedures of conventional construction contracts...'*[48]

46 London Borough of Merton v Stanley Hugh Leach Ltd (1985) 32 BLR 51 at 81 per Vinelott J.
47 London Borough of Merton v Stanley Hugh Leach Ltd (1985) 32 BLR 51 at 78 per Vinelott J.
48 Costain Limited & Ors v Bechtel Ltd & Anor [2005] EWHC 1018 (TCC).

- The current position is suggested to be:

> *'In undertaking these (contract administration) functions, the Architect does not act as the agent of the employer but, since he is engaged by the employer, he has a contractual obligation to act fairly, impartially and in accordance with the powers given to him by the conditions. The employer may not interfere in the timing of the issue of any certificate but is not himself in breach of contract if a particular certificate is not issued or is erroneous unless he is directly responsible for that failure. However, if and when it comes to his notice that the Architect has failed to comply with his administrative obligations, by for example failing to issue a certificate required by the contract, the employer has an implied duty to instruct the Architect to perform that function in so far as it remains within the power of the Architect to perform it and the employer is in breach of the contract with the Contractor to the extent that he does not intervene to arrange for the correct or a correcting step to be taken by the Architect.'*[49]

In some instances, the employer will carry out the functions ordinarily fulfilled by his representative. This may be the case where the employer is a local authority or similar public body and has appropriately qualified personnel (such as architects, engineers, project managers) employed 'in house'. However, unless there are very clear words in the contract to indicate the contrary,[50] the employer is obliged to carry out any 'decision-making' functions (such as certification, granting extensions of time, assessing or valuing the amounts due to the contractor) in an independent, impartial, fair and honest manner.[51]

## 9.2 Quantity Surveyor's Role

A quantity surveyor is usually engaged for most building projects in the United Kingdom, despite not always being 'named[52]' in the building contract (see Table 9.1 above). However, as different forms of building contract require tasks for which the quantity surveyor is clearly suited to be undertaken under the umbrella of either the employer or the employer's representative, these tasks are often delegated to a quantity surveyor. Some examples of these quantity surveying tasks are as follows:

- The valuation of works under JCT DB 2016 or under JCT MW 2016, which the employer is to carry out; or
- The assessment of the amount due under NEC4, which is carried out by the Project Manager (the employer's representative under that contract).

The quantity surveyor's scope of services and his responsibilities to the employer are usually set out in his contract (conditions of appointment). Typical quantity surveying services comprise the following:

---

49 *BR and EP Cantrell v Wright and Fuller* [2003] BLR 412.
50 *Balfour Beatty Civil Engineering Ltd v Docklands Light Railway Ltd* [1996] 78 BLR 42.
51 *Scheldebouw BV v St James Homes (Grosvenor Dock)* Ltd [2006] EWHC 89 (TCC).
52 The 'Quantity Surveyor' is only 'named' in JCT SBC/Q 2016 and JCT IC 2016.

- Providing initial cost and budget advice to the employer;
- Procurement services, including advising on forms of contract, producing the pricing document (e.g. bills of quantities, schedules of works, etc.), tender reviews, contractor appointment advice;
- Construction phase services:
  - Interim valuations (payment notices, pay-less notices)
  - Providing regular cost reports/cashflow forecasts to the employer
  - Valuation/agreement of any changes to the scope of works
- Post contract, including agreeing to the 'final account' (JCT), contractor's share (NEC4), valuation of abatement for defects.

It is uncommon for the contractor to be fully aware of all the services that the quantity surveyor is carrying out for the employer, as he would need a copy of the quantity surveyor's conditions of engagement.[53] Where the quantity surveyor is mentioned in the contract (such as in JCT SBC/Q 2016 or JCT IC 2016), the contractor should know the general scope of the quantity surveyor's duties that directly affect them from the relevant clauses in those contracts.

However, for contracts where the typical quantity surveying services are undertaken by either the employer or the employer's representative, the contractor may not be aware of the full extent of the quantity surveyor's services, and importantly, the extent (if any) of the quantity surveyor's authority. This latter point may become an issue if the contractor is corresponding with and submitting documents (such as interim payment applications) to the quantity surveyor, when the quantity surveyor is carrying out duties delegated to him by either the employer or by the employer's representative. This extent of the quantity surveyor's authority is particularly relevant where 'claims' are concerned. This is because an employer may not have chosen to include any 'claims' assessment services in the quantity surveyor's scope of services – as they are often expressed as 'additional' or 'supplementary services'[54] (see also Section 12.3.2 and Table 12.7).

Table 9.5 illustrates the tasks of a quantity surveying nature that are likely to be carried out by a quantity surveyor; either directly as named as such in the contract, or indirectly (shown in *italics*) where he would be assisting the employer's representative; with functions delegated to the quantity surveyor on behalf of the employer (JCT DB 2016) or his representative (such as the Project Manager under NEC4):

53 Whilst a contractor should have a good awareness of the scope of services that an architect or structural or building services engineer may undertake from design and build contracts where such professionals are novated from the employer to the contractor, quantity surveyors are rarely novated.
54 For example, the RICS' Quantity Surveyor Services (May 2022) includes the following as 'Supplemental services': '2.2.3 Advise on the Contractor's extensions of time. Analyse and report on the Contractor's application(s) for extensions of time. Prepare recommendations for the Client's 'approval', '2.2.4 Advise on the cost and contractual consequences arising from an acceleration instruction' and '2.2.5 Advise on the Contractor's entitlement to loss and expense. Analyse and report on the Contractor's loss and expense claim(s). Prepare recommendations for the Client's 'approval'.

**Table 9.5** Quantity surveying functions under each contract.

| Function | JCT SBC/Q 2016 | JCT IC 2016 | JCT DB 2016 | JCT MW 2016 | NEC4 |
|---|---|---|---|---|---|
| Quantity surveyor 'named' in the contract | Article 4 | Article 4 | — | — | — |
| QS included in definitions: | | | | | |
| • 'Contractor's Persons' | 1.1 | 1.1 | — | — | — |
| • 'Employer's Persons' | 1.1 | 1.1 | — | — | — |
| • 'Quantity Surveyor' | 1.1 | 1.1 | — | — | — |
| • 'Valuation' | 1.1 | 1.1 | — | — | — |
| Notices to the QS to be in writing | 1.7.1 | 1.7.1 | — | — | *13.1* |
| QS to use rates and prices only for purposes of contract | 2.8.4 | 2.8.3 | — | — | — |
| Impediment, etc. (inc by QS) is a delay event | 2.29.7 | 2.20.6 | *2.26.6* | *2.7* | *60.1(3), 60.1(6), 60.1(8)* |
| Replacement of QS | 3.5.1 | 3.4.1 | — | — | — |
| Interim valuations made by the QS | 4.9.2 | 4.8.3 | *4.7.1* | *4.3* | *50.1* |
| Contractor submits 'payment application' to the QS | 4.10.1 | 4.11.1 | *4.7.3* | *4.4.1* | *50.2* |
| Pay-less notice may be issued by QS | 4.12.1.1 | 4.13.1.1 | *4.9.5* | *4.4.5* | *Y2.3* |
| QS prepares statement of retention | 4.17.2 | — | — | — | — |
| Retention bond: QS' statement of retention otherwise held | 4.18.1 | — | — | — | — |
| QS may ascertain loss and expense (interim payment) | 4.21.2 | 4.16.2 | *4.20.2* | — | *50.1* |
| QS notifies contractor of ascertained loss and expense | 4.21.4 | 4.16.4 | *4.20.4* | — | — |
| Impediment, etc. by QS is a Relevant Matter | 4.22.5 | 4.17.4 | *4.21.5* | — | *60.1(3), 60.1(6), 60.1(8)* |
| QS receives 'final account' documents from Contractor | 4.25.1 | 4.20.1 | *4.24.1* | *4.8.1* | *53.1, 53.3, 53.4*[a)] |
| QS may ascertain the loss and expense (final account) | 4.25.2.1 | 4.20.2.1 | — | — | — |
| QS prepares statement of adjustments to Contract Sum | 4.25.2.2 | 4.20.2.2 | *4.24.1* | *4.8.2* | — |
| QS carries out a 'Valuation' | 5.2.1 | 5.2 | *5.2* | *3.6.3* | *50.1* |

*(Continued)*

**Table 9.5** (Continued)

| Function | JCT SBC/Q 2016 | JCT IC 2016 | JCT DB 2016 | JCT MW 2016 | NEC4 |
|---|---|---|---|---|---|
| Assessment of the 'Contractor's share' | — | — | — | — | *54.3, 54.4* |
| QS is a 'consultant' in the Rights Particulars | 7.4.3 | 7.3.2 | *7.4.3* | — | — |
| Impediment, etc. by QS is a termination event for Contractor | 8.9.2.2 | 8.9.2.2 | — | — | — |
| Assessing the amount due on termination | 8.7.4, 8.12.3[b)] | 8.7.4, 8.12.3[b)] | 8.7.4, 8.12.3[b)] | 6,7.3, 6.11.2[b)] | *93* |
| 'Variation Quotation' to the QS | Sch 2, 1.1 | — | *Sch 2, 2.2* | — | *61.2, 62.1, 62.4, 63.1, 63.2, 65.1, 63.12,*[c)] *63.13,*[d)] *63.14,*[e)] *63.15,*[f)] *63.16,*[g)] *64.1, 64.3* |
| *Loss and expense – contractor's estimates* | — | — | *Sch 2, 3.2* | — | — |
| 'Acceleration Quotation' to the QS | Sch 2, 3.2 | | *Sch 2, 4.1.1.1* | — | *61.2, 62.1, 62.4, 63.1, 63.2, 65.1, 63.12, 63.13, 63.14, 63.15, 63.16, 64.1, 64.3* |
| Inspection of contractor's cost records | — | — | — | — | *52.4* |
| Agree fluctuations with QS | Sch 7, A5, A7 | Sch 4, 5, 7 | *Sch 7, A5, A7* | — | *X1* |

a) NEC4 clause 53 (Final Assessment) is limited to assessing (i) the 'Contractor's share' under clauses 54.1 to 54.4 of Option C (Target Contract with Activity Schedule), (ii) the 'Contractor's share' under clauses 54.5 to 54.8 of Option D (Target Contract with Bill of Quantities), or (iii) the incentive payment under Option X22 (Early Contractor Involvement)
b) JCT SBC/Q 2016, JCT DB 2016 and JCT IC 2016 – the employer's clause 8.7.4 'statement' is likely to be prepared by the Quantity Surveyor, who is likely to assess the Contractor's clause 8.12.3 'account'. The respective equivalents under JCT MW 2016 are the clause 6.7.3 'account' (in lieu of the clause 8.7.4 statement) and the clause 6.11.2 'account'.
c) NEC4 – Options A and B, clause 63.12 – reducing the Prices
d) NEC4 – Options C and D clause 63.13 – reducing the Defined Cost
e) NEC4 – Option C, clause 63.14 – compensation event assessments for changed Prices are expressed as changes to the Activity Schedule.
f) NEC4 – Option B and D clause 63.15 - compensation event assessments for changed Prices are expressed as changes to the Bill of Quantities
g) NEC4 – Option A and B, clause 63.16 – agreement of a new 'People Rate'

### 9.2.1 Valuation of Work

The most important task of the quantity surveyor under a building contract is to value the works.[55] The valuation of work is required in order to:

- Calculate the amount due to the contractor:
  - For interim payments;
  - For final payments (including where the contractor's employment has been terminated);
  - Calculating the value of any abatement (or reduction) of the contract sum for defective work; and
  - Assisting the Project Manager in calculating any 'share' due to or from the contractor (under NEC4).
- Valuing and agreeing to variations under the contract:
  - The value of 'Variations', 'Changes' or quotations for these under the relevant JCT 2016 contract; or
  - Assessment of 'compensation events' under NEC4 (including to adjust the 'target' under Options C or D) on behalf of the Project Manager.

The nature of the quantity surveyor's powers under JCT 2016 forms has been clearly stated:

> *'His authority and function under the contract are confined to measuring and quantifying. The contract gives him authority, at least in certain instances, to decide quantum. It does not in any instance give him authority to determine any liability, or liability to make any payment or allowance.'*[56]

The position appears to be the same under the JCT SBC/Q and IC 2016 contracts so far as the quantity surveyor's powers are concerned. The terms of the contract, express and implied, give the quantity surveyor no independent authority.

### 9.2.2 Direct Loss and/or Expense

Under the JCT SBC/Q and IC 2016 forms of contract[57] the principal responsibility for ascertaining the amount of 'direct loss and/or expense' incurred by and reimbursable to the contractor initially rests with the employer's representative (the architect or contract administrator). He has the option of involving the quantity surveyor.

- To check that the application made by the contractor is correct in principle; that is to say the employer's representative must be sure that the application satisfies all the conditions and that the relevant matters on which it relies are accurately cited and, most important, that the contractor is entitled to some reimbursement; and

55 *County and District Properties Ltd v John Laing Construction Ltd* (1982) 23 BLR 1 at 14 per Webster J. - '*...function and his authority under the contract are confined to measuring and quantifying. The contract gives him authority, at least in certain instances, to decide quantum. It does not in any instance give him authority to determine any liability, or liability to make any payment or allowance.*'.

56 *Country & District Properties Ltd v John Laing Construction Ltd* (1982) 23 BLR 1, at 14 per Webster J, where this question arose under a JCT 63 contract.

57 JCT SBC/Q 2016 – clause 4.25.2.1, JCT IC 2016 – clause 4.20.2.1.

- To ascertain the actual amount of money which should be paid to the contractor as a result. The employer's representative may decide to instruct the quantity surveyor to carry out this ascertainment.

The quantity surveyor only has the power to carry out that function if expressly instructed by the employer's representative. In practice, the quantity surveyor is usually best suited by training, experience and qualifications to perform that task, and most claims put forward by the contractor will have been calculated in some detail, often before an entitlement to anything is established. The contractor's calculations have usually been carried out by its own quantity surveyor or perhaps by an external claims consultant, who will often be a quantity surveyor. Therefore, it makes perfect sense for any financial discussions to be dealt with by another quantity surveyor, who 'speaks the same language'.

Another reason for the employer's representative instructing the quantity surveyor to ascertain the contractor's loss and expense is to reduce his own liability. Employers often consider contractor's 'claims' to be contentious and are unhappy to receive them. The employer's representative may not want to be exposed to the risk of the employer disagreeing with his ascertainment of the contractor's loss and expense, so he may delegate this duty to the quantity surveyor. However, the employer's representative should not attempt to avoid any involvement or responsibility for ascertaining the contractor's claim. It is advisable for him to work closely with the quantity surveyor, particularly when establishing or verifying the facts upon which the contractor's claim is based, as a significant proportion of the loss and expense claim could be based upon decisions made by the employer's representative.

### 9.2.3 Quantity Surveyor's Duty

In some respects, the quantity surveyor's position is similar to that of the employer's representative (see Section 9.1.8) although, strictly speaking under the JCT SBC/Q 2016 and JCT ICD 2016 contracts, the quantity surveyor has no power to decide liability; his duty is purely concerned with ascertaining the amount (or quantum) of the loss and expense.[58] Matters of liability are usually decided by the employer's representative.[59]

To the contractor, the extent of the quantity surveyor's duties may be wrongly perceived, as he may mistakenly believe that the quantity surveyor's duty is to establish liability. As mentioned in Section 9.2 and in Table 9.4, the clearest indication of the quantity surveyor's duty is provided by the building contract. Under JCT SBC/Q 2016 and JCT IC 2016, it is a duty of ascertainment, if the employer's representative so instructs.

Employers often expect their representatives and quantity surveyors to resist, defend or rebut the contractor's claims. However, such measures form no part of the quantity surveyor's duties, which, so far as the quantity surveyor is concerned, comprise establishing (on the architect's instructions and in strict accordance with the contract), the amount payable to the contractor. The quantity surveyor's duty is a positive one – he must 'ascertain'. He does not have a negative duty (i.e. to resist ascertainment), so he should

58 JCT SBC/Q 2016 – clause 4.21.4, and JCT IC 2016 – clause 4.16.2.

59 The employer's representative under JCT DB 2016 (the 'Employer's Agent') or under NEC4 (the Project Manager) may be a quantity surveyor, but their functions and duties are distinguished from those of the quantity surveyor for the purposes of this chapter.

not look for reasons not to ascertain, particularly if those reasons stray into the realms of deciding liability. Of course, an integral part of the quantity surveyor's ascertainment will be to scrutinise the contractor's claims, and identify any of the following:

- Examples of duplication ('double counting') – e.g. where elements of the contractor's loss and expense claim have already been ascertained and paid (such as in variations); and
- Examples where the contractor's records or other substantiation do not support the sums claimed.

### 9.2.4 Duty to the Employer

The quantity surveyor's primary duty to the employer is to act with reasonable skill and care (see Section 9.1.2). The quantity surveyor must always act in strict conformity with the professional standards of the discipline and maintain the relevant ethical standards. Section 9.1.2 considers the 'Requisite Standard' imposed upon the quantity surveyor by the RICS' consultant's appointments.[60] This is often a higher standard than that imposed by many bespoke conditions of engagement for quantity surveying services.

In respect of claims, the quantity surveyor's duty is to ensure that the employer pays no more than the actual amount of loss or expense directly and properly incurred by the contractor, but the duty extends to ensuring that the contractor recovers no less. It is not part of the quantity surveyor's duty to strive to reduce the amounts properly recoverable under the contract.

The quantity surveyor, like the employer's representative, has a duty to the employer to act fairly between the parties. That duty arises as a result of the nature of the tasks which the building contract requires the quantity surveyor to carry out. The nature of these tasks demand that the quantity surveyor acts with even-handedness when carrying them out. If the quantity surveyor fails to carry out those tasks in accordance with the terms of the building contract, the employer may be liable to the contractor for that failure as a breach of a contractual undertaking, but only, it seems, if the employer was aware of the quantity surveyor's duty and of the breach.[61] However, it must be emphasised that the quantity surveyor is not a party to the building contract, any more than the employer's representative. Therefore, for example, the contractor can neither refer a dispute with the employer's representative, nor with a quantity surveyor, to adjudication, as he can only adjudicate against the employer. A referring contractor may ask the adjudicator to make a declaration or finding in respect of an alleged breach by the quantity surveyor (or the employer's representative) as part of deciding an overall dispute, but adjudicators are often unable to do so, because any allegations of professional negligence can only be made by the employer (i.e. the party engaging the quantity surveyor). However, asking an adjudicator to provide the following redress would provide implied criticism of the quantity surveyor and the employer's representative:

- If the adjudicator's valuation of the matter in dispute (e.g. an application for interim payment) is significantly higher than that made by the employer's quantity surveyor

60 RICS' Standard Form of Consultants Appointment (May 2022) and the Short Form of Consultant's Appointment (May 2022).
61 *Penwith DC v V P Developments Ltd* 1999.

(particularly when valuing loss and/or expense under at JCT SBC/Q 2016 or JCT IC 2016 contract);

- A declaration that the contractor has carried out a function correctly:
  - That an application for interim payment or quotation for a compensation event has been submitted on time, in the correct format and contains the correct substantiation of the amounts claimed;
  - That a notice of delay, or a notification of a compensation event, is valid; or
- A decision that implies that the employer's representative has not carried out its duties correctly:
  - That 'completion' (of a section or of the works) has been achieved; or
  - The 'rectification' or 'defects' period has expired, and no outstanding defects remain.

### 9.2.5 Quantity Surveyor's Liability to Others

For the same reasons as apply to the employer's representative (see Section 9.1.7), it is highly doubtful that the quantity surveyor owes a duty to anyone other than the parties with which he has entered into a contract (such as formal conditions of engagement), or a collateral warranty.

### 9.2.6 Quantity Surveyor Functions Under JCT 2016 Contracts

- JCT SBC/Q 2016 and JCT IC 2016

  The default position is that all variations, and all work executed by the contractor in accordance with the employer's representative's instructions for the expenditure of provisional sums shall be valued by the quantity surveyor named in the contract.[62] Therefore, save for (i) any agreement between the contractor and the employer[63] or (ii) for any variation quotation,[64] it is entirely a matter for the quantity surveyor to determine the price to be paid or allowed in respect of a variation.[65]

  The quantity surveyor is not obliged to consult with the contractor when valuing any work – whether it is a 'Variation' or for the purposes of an interim or final payment. For interim valuations, both the JCT SBC/Q 2016 and JCT IC 2016 contracts contain the default position that the employer's representative carries out the interim valuations and issues an Interim Certificate[66] (payment notice) to the contractor for each. There is an

62 JCT SBC/Q 2016 – clause 5.2.1, JCT IC 2016 – clause 5.2.

63 *'The value of. All Variations [etc.]...shall be such amount as is agreed by the* Employer *and the* Contractor...' – JCT SBC/Q 2016 – clause 5.2, JCT IC 2016 – clause 5.2. Similar provisions for 'a Valuation' exist in JCT DB 2016 – clause 5.2 (but not expressly involving a quantity surveyor), or JCT MW 2016 – clause 3.6.2 (but agreement over the Valuation is between the contractor and the employer's representative, not the employer). See also *County and District Properties Ltd v John Laing Construction Ltd* (1982) 23 BLR 1 at 14 per Webster J: *'In my view the word agreed can only mean "agreed between the parties", although it may well be that on occasion a quantity surveyor may perhaps be given express authority by the Employer to make such an agreement. But the JCT contract does not give him the authority.'*.

64 JCT SBC/Q 2016 – clause 5.3 and Schedule 2.1.1. A similar provision is contained in JCT DB 2016 – in Schedule 2.1 (Valuation of Changes – Contractor's Estimates), but without expressly mentioning a quantity surveyor.

65 See section 5.4 'Valuation under JCT 2016 Contracts'.

66 JCT SBC/Q 2016 – clause 4.9.1, and JCT IC 2016 – clause 4.8.1.

implication that the quantity surveyor will assist the employer's representative, and is obliged to do so if the employer's representative considers it necessary.[67] For the purposes of a valuation, the contractor has the right to be present for the measurement of any work under JCT SBC/Q 2016,[68] but no corresponding right exists under JCT IC 2016. In practical terms, there is nothing to prevent the contractor from observing the quantity surveyor measuring work for a valuation. This right to be present for measurement[68] does not extend to having to agree quantities with the quantity surveyor, as it is merely a right to observe. Both JCT SBC/Q 2016 and JCT IC 2016 provide the contractor with the opportunity to submit his 'Payment Applications' for interim payments,[69] instead of solely relying on the employer's representative and (in all likelihood) the quantity surveyor to calculate the amount due. The implication is that the contractor's 'Payment Application' is sufficiently clear and accurate, and contains all the information required, so any differences between the amount claimed and the amount shown in the Interim Certificate[66] are not significant. In the event of significant differences between either (1) the value that the contractor could reasonably expect the employer's representative and quantity surveyor to show in an Interim Certificate, or the amount claimed in a Payment Application, and (2) the amount due as shown in an Interim Certificate, the likely grounds for the contractor to dispute this difference are:

- If the quantity of the relevant work has not been measured correctly by the quantity surveyor[68]; and/or
- That the quantity surveyor has failed to correctly apply the applicable Valuation Rules (see Section 5.4.1).

For the final payment, the quantity surveyor may proceed without the contractor's input and may (via the employer's representative) provide the contractor with the statement[70] of all the adjustments to the contract sum, which would include a list of the variation valuations. The contractor has the option of accepting it or, if he disputes it, referring the matter to adjudication.

In practice, the situation is usually quite different, and far more collaborative between the quantity surveyor and the contractor. The quantity surveyor will usually consult with the contractor on an ongoing basis (e.g. for each interim valuation) as measurement and valuation take place. Generally, the quantity surveyor will attempt to reach an agreement with the contractor on matters concerning measurement and valuation in order to minimise the areas of dispute. It is obviously helpful if the contractor has agreed the whole of the quantity surveyor's (or relevant) statement[70] of adjustments of the contract sum. Sometimes, the quantity surveyor may send the statement to the contractor with a sheet inviting the contractor to sign, indicating agreement. This practice has probably arisen from the quantity surveyor's duties to provide regular cost reports to the employer,[71]

67 JCT SBC/Q 2016 – clause 4.9.3, and JCT IC 2016 – clause 4.8.3.
68 JCT SBC/Q 2016 – clause 5.4.
69 JCT SBC/Q 2016 – clause 4.10, and JCT IC 2016 – clause 4.11.
70 JCT SBC/Q 2016 – clause 4.25.2.2, and JCT IC 2016 – clause 4.20.2.2. Similar provisions are contained in JCT DB 2016 - clause 4.24.1 and JCT MW 2016 – clause 4.8.2.1, but these 'statements' are provided by the employer and the employer's representative respectively.
71 RICS' Quantity Surveying Services (May 2022) – 1.1.2 (prepare monthly/regular cost reports),.

and possibly 'agree' variations and the final account.[72] This practice can lead to the false assumption by employer's representatives and quantity surveyors alike, that the contractor's agreement to these statements is necessary. The employer's representative does not require the contractor's agreement before issuing the final certificate, but the quantity surveyor producing (and the employer's representative issuing) this statement are conditions precedent to the final certificate having the conclusive effects as required by the contract.[73] The contractor may contest any valuation of work, by adjudication, or by legal proceedings or by arbitration until the expiry of 28 days after the date of issue of the Final Certificate.[74]

Occasionally, an employer's representative will issue an instruction requiring a variation, stating *'the work carried out under this instruction is to be valued at the rates and prices in the contract bills'*. That part of the instruction regarding the valuation method would only have effect if the Valuation Rules (refer to Section 5.4.1) stated as such. If not, the quantity surveyor must follow the Valuation Rules for the correct method of valuation.

It is rare for the employer's representative to certify a different amount from that valued by the quantity surveyor. Financial certification is an onerous and specialist task, and the employer's representative must be reasonably satisfied regarding the quantity surveyor's valuation before certifying. The usual ground for an employer's representative to alter the quantity surveyor's valuation is if defects are present that warrant an abatement (reduction) in the works' value (refer to Section 8.5). Ideally, the employer's representative (if an architect or similarly qualified consultant) would inform the quantity surveyor of the nature and extent of any defective work, so that the quantity surveyor's valuation will account for any appropriate abatements (reductions) in value. However, if defects are discovered either just before the Interim Certificate (or payment notice) is issued, or before a pay-less notice[75] must be issued, then the employer's representative may not have time to consult the quantity surveyor, so he'll have to make the reduction himself. The employer's representative will usually require the quantity surveyor to provide a simple breakdown of the defective work's value if it had been carried out correctly, and apply the appropriate reductions to the quantity surveyor's value. It is no part of the employer's representative's duty to re-value work that has already been valued by the quantity surveyor under the provisions of the contract. Indeed, it has been said above that an architect has no such power. However, it is advisable for the employer's representative to at least carry out some simple checks (such as arithmetic). If, in circumstances when the employer's representative can show that the quantity surveyor's valuation is either too high or too low, or contains arithmetical errors, then the employer's representative's duty is to certify what he believes to be the correct amount.

---

72 RICS' Quantity Surveying Services (May 2022- 1.7.4 (agree the cost of instructions), and 1.7.9 (facilitate agreement of the final account).

73 *Penwith District Council v VP Developments Ltd*, 21 May 1999, unreported.

74 JCT SBC/Q 2016 and JCT IC 2016 – clause 1.9.2.1. JCT DB 2016 – clause 1.8.2.1 contains a similar 28-day time limit that runs from the date of either the 'Final Statement' (issued by the contractor) or the 'Employer's Final Statemen" (clause 4.24.6).

75 A pay-less notice may be given by the employer, or by the employer's representative on the employer's behalf: JCT SBC/Q 2016 clause 4.12.1.1 – and JCT IC 2016 – clause 4.13.1.1.

- JCT DB 2016

  A quantity surveyor is often engaged by the employer under the JCT DB 2016 form of contract. Under JCT DB 2016, only a few of the duties ordinarily carried out by the employer's representative under the JCT SBC/Q 2016 and JCT IC 2016 contract are carried out by the employer. The employer's representative is the 'Employer's Agent', but he is hardly mentioned.[76]

  Usually, and in practical terms, the employer engages his representative to carry out (1) all of the employer's obligations to the contractor,[77] as well as the obligations designated to the 'Employer's Agent'[77]; and (2) the usual quantity surveying functions.[78] Consultants often describe these two aspects of the employer's representative's role as 'EA/QS' (meaning 'employer's agent/quantity surveyor'). Any quantity surveying functions (i.e. interim valuations and valuation of Changes[79] using the Valuation Rules[80]) are likely to be similar to those under a JCT SBC/Q 2016 or JCT IC 2016 contract, with the exception that JCT DB 2016 no longer provides for a bill of quantities and any specified method of measurement.

  The employer's agent is exactly that: an agent of the employer and it is probable that the agent owes no duty to the employer to act fairly between the parties. The notices issued by the employer's agent do not have the status of a certificate of an independent architect.[81] It seems, however, that the agent must demonstrate a very high duty of good faith.[82]
- JCT MW 2016

  Owing to the relatively simple nature of the works envisaged by a JCT MW 2016 contract, it is unlikely that a quantity surveyor will be appointed. When acting as the employer's representative, few architects have a quantity surveying qualification and an architect intending to exercise this function should make sure that he has the appropriate professional indemnity insurance in place to cover the risk. It is more likely that the employer's representative will be a quantity surveyor, carrying out a similar hybrid role to that of 'Employer's Agent' and quantity surveyor under the JCT DB 2016 contract.

---

76 JCT DB 2016: The 'Employer's Agent' is He's identified in Article 3 and clause 1.1, and if communicating with the contractor, he must do so in writing (clause 1.7). Under clause 2.7.3, the contractor should keep copies of the building contract and the "Contractor's Design Documents" on site, and allow access for the Employer's Agent or any person he so authorises to areas on and off-site, where the works are being prepared (clause 3.1). For interim payments or the final payment, the Employer's Agent may give payment notices, pay-less notices (clause 4.10.2), and clause 7.4.3 includes the Employer's Agent within the meaning of "consultants" in respect of Third Party Rights and Collateral Warranties.

77 For example, under the RICS' 'Employer's Agent Services (May 2022).

78 For example, under the RICS' Quantity Surveying Services (May 2022).

79 JCT DB 2016 – clause 5.1 describes 'Changes' (variations under this form of contract).

80 JCT DB 2016 – clause 5.4 sets out the 'Valuation Rules'.

81 *J F Finnegan Ltd v Ford Seller Morris Developments Ltd* (1991) 53 BLR 38. However, those notices must comply with the requirements for a 'payment notice' and a 'pay less notice' as per the terms of the JCT DB 2016 contract (and of course, the HGCRA 1996).

82 *Balfour Beatty Civil Engineering Ltd v Docklands Light Railway Ltd* (1996) 78 BLR 42.

### 9.2.7 Quantity Surveyor Functions Under NEC4

There is no mention of a quantity surveyor in NEC4 (refer to Section 9.2 above) as the employer engages the Project Manager to carry out the financial assessments and valuations of work that the contract requires. In many cases, the Project Manager is selected from an engineering or other appropriate design background, or is chosen for his management abilities (e.g. his profession is as a Project Manager). In either case, it may be unusual for him to carry out any quantity surveying functions himself, unless he is a qualified civil engineer. Instead, these functions are often carried out by a quantity surveyor who is either engaged:

- Directly by the employer; or
- As a sub-consultant to the Project Manager

If the quantity surveyor is directly engaged by the employer, his role will be to carry out particular functions that would otherwise be carried out by the Project Manager (refer to Table 9.4), although it would be the employer who would pay the quantity surveyor's fees. It is likely that the employer would either engage the quantity surveyor under a bespoke appointment or under the NEC4 Professional Services Contract. The quantity surveyor may be engaged as a sub-consultant by the Project Manager. In either case, the quantity surveyor's services would include complying with the Project Manager's instructions and performing all of the relevant valuation and duties on the Project Manager's behalf, such as:

- Receiving the contractor's application for payment,[83] assessing the amount due for interim payments[84] and the final payment[85];
- Advising the Project Manager regarding the contractor's quotations for compensation events[86] or a proposed instruction[87];
- Assessing or assisting the Project Manager with assessing compensation events[88];
- Assessing the 'Contractor's share'[89];
- Inspecting the contractor's records of Defined Cost, subcontractors or other records required by the Scope[90]; and
- Assessing the amount due on termination.[91]

83 NEC4 – clause 50.2.
84 NEC4 – clause 50.1.
85 NEC4 – clauses 53.1, 53.3 and 53.4.
86 NEC4 – advising the Project Manager: if the compensation event has any effect upon the Defined Cost (clause 61.2), for the contractor to submit alternative quotations (clause 62.1), and for the contractor to submit a revised quotation (clause 62.4).
87 NEC4 – advising the Project Manager of the need for a quotation for a proposed instruction (clause 65.1).
88 NEC4 – advising the Project Manager: regarding the change to the Prices (clause 63.1), for agreement of rates or lump sums with the contractor (clause 63.2), if the compensation event reduces the total Defined Cost (clause 63.3 and 63.4, Options A and B – clause 63.12, Options C and D – clause 63.13), regarding changes to the Activity Schedule (Option A – clause 63.14) or Bill of Quantities (Options B and D – clause 63.15), regarding a new 'People Rate' (Options A and B – clause 63.16), regarding 'assumptions' and reasonably incurred Defined Cost (clause 63.9), and assessing the compensation event (clause 64.1), and providing details thereof (clause 64.3) on the Project Manager's behalf.
89 NEC4 – Options C and D (clause 54)
90 NEC4 – Options C and D (clause 52.4)
91 NEC4 – clause 93.

### 9.2.8 Commercial Settlements

In most cases, many of the quantity surveying services like the valuation of work for interim payment, variations and ascertaining loss and expense will involve discussion between the contractor and the quantity surveyor (although see Section 9.2.6). However, in practice, most claims are ultimately settled by agreement. The quantity surveyor, of course, is not normally empowered to 'do a deal' with the contractor, and where some sort of 'broad brush' settlement is clearly to the benefit of the parties, the employer's representative and the quantity surveyor must place the options in front of the employer and obtain his instructions (see Sections 9.1.5 and 9.1.6). Where such a commercial settlement is agreed and incorporated into the final account, the employer's representative will often have difficulty issuing a final certificate that has the effect of applying all the contractual terms governing the calculation of the finally adjusted contract sum, when some of those constituent sums may have been agreed outside the terms of the contract. In such cases, it is advisable for the contractor and employer to set the final account agreement out in deed of settlement, that outlines how the overall deal was agreed, and sets out which sums are due under the terms of the contract (e.g. can be included in the final certificate or relevant final statement) and which sums were agreed outside the contract, between the employer and the contractor.

# 10

# Preparation and Substantiation of Claims

## 10.1 Preparing a Claim

### 10.1.1 Introduction

When a contractor is submitting a claim; be that for the value of a variation, an extension of time, or for reimbursement of loss and expense, he must provide evidence to support that claim. The nature and type of this evidence (or substantiation) is not always specified in every building contract, so it is either the contractor who offers certain information in the hope that it provides sufficient evidence of his claim, or supporting information is requested by the employer or by his representative. The parties must take note of, and observe the precise wording of the contract in order for the contractor to properly evidence and evaluate their claim.

In Chapter 2,[1] demonstrating an entitlement to a claim was divided into the following stages:

- Duty
- Breach
- Causation
- Loss.

Providing evidence of the 'duty' is often relatively straightforward, as it will be found in the form of the express conditions in the contract, or conditions that may be implied. The contractor will have to provide factual evidence that the 'breach' has been committed, explain the 'causation' – causing him to suffer loss – and demonstrate the 'loss' itself. In particular, substantiating 'causation' can be a difficult and lengthy task, and is the one that contractors often overlook to some degree. For example, in the case of providing loss and expense for a delay, a contractor may simply take the difference between the contract date for completion and the actual, later date of completion, but then base detailed calculations of loss on that period, without providing a similar level of detail to prove that (i) the reason for the delay was one of the grounds listed in the contract or (ii) the causation linking the alleged delays with its claim for loss. In a different example, a contractor may wrongly assume that an extension of time entitles him to financial compensation. He may go to a considerable effort to show the causation between the events that caused substantial

1 Refer to Section 2.1.2 (*Duty, breach, causation and loss*).

*Building Contract Claims*, Sixth Edition. Guy Higginbottom and David Chappell.

delays and other problems but fail to show the basis of his entitlement because the events do not fall within the grounds for loss compensation in the contract. Claims for disruption (refer to Section 2.4.2) are often difficult to substantiate and, for that reason, they are often either ignored or treated in a casual fashion. Unfortunately, it will not take an experienced employer (or his representative) long to realise that the contractor is almost acknowledging the unlikelihood that his disruption claim will succeed!

Conversely, when an experienced contractor has submitted a properly evidenced claim, it will not take him long to realise that the employer or his representative is either unwilling or unable to deal with it. Common examples of employer's (or his representative's) unwillingness to deal with a claim include procrastinating and a seemingly never-ending series of requests for further information that descend into pedantry.

Under most building contracts, the employer or the employer's representative has a positive duty to 'ascertain[2]' a contractor's claim or make an 'assessment[3]' of it. They do not have an active obligation to disprove a contractor's claim. Of course, the extent of a contractor's entitlement and the evidence provided may not be sufficient to warrant a significant additional payment or extension of time, or any further entitlement at all.

### 10.1.2 General Principles

The building contracts covered in this book do not provide much in the way of detailed guidance for either the contractor or the employer (or his representative) for the extent or the nature of information to be provided in a claim. Out of all the contracts, the most specific details to be provided are those set out in NEC4's requirements for the 'programme' and 'revised programme' when quoting for compensation events (see Sections 3.5.1 and 13.3) along with the Schedule of Cost Components and the Shorter Schedule of Cost Components (particularly, if set out in the Scope when the target cost main Options C and D are selected).[4]

Some useful guidance into the types, or classes of information, required to support a claim can be found in the rules for civil litigation[5] – in the Pre-Action Protocol for Construction and Engineering Disputes (2nd ed)[6] ('the Protocol'). It is advisable to adopt the Protocol for all construction and engineering disputes,[7] as it has the following objectives:

- *'to exchange sufficient information about the proposed proceedings broadly to allow the parties to understand each other's position and make informed decisions about settlement and how to proceed'*[8]; and
- *'to make appropriate attempts to resolve the matter without starting proceedings and, in particular, consider the use of an appropriate form of ADR in order to do so.'*[9]

---

2 JCT SBC/Q 2016 – clauses 4.21.4 and 4.25.2.1, JCT DB 2016 – clause 4.20.4, JCT IC 2016 clauses 4.16.2 and 4.20.2.1.
3 NEC4 – clauses 50.1, 63.1, 63.5, 64.1 and 64.2.
4 NEC4 – Option C (Target Contract with Activity Schedule) and Option D (Target Contract with Bill of Quantities) – clause 52.2 '...*other records as stated in the Scope*'.
5 www.justice.gov.uk/courts/procedure-rules/civil/rules
6 www.justice.gov.uk/courts/procedure-rules/civil/protocol/prot_ced (17 February 2017).
7 Protocol paragraphs 1 – and should also be adopted for professional negligence claims against architects, engineers and quantity surveyors.
8 Protocol – Paragraph 3.1.1.
9 Protocol – Paragraph 3.1.2.

Substituting 'proposed proceedings' for 'claim' and 'settlement' for 'ascertainment' or 'assessment' shows that in principle, the Protocol's objectives effectively translate into a useful guide for the parties when attempting to resolve a dispute (or contentious matter) themselves, without referring a claim to adjudication, or other formal methods of dispute resolution. Paragraph 7.1.3 of the Protocol provides further useful guidance regarding the content of a claim and its substantiation:

- A brief summary of the claim:
  - That is proportionate to the claim; and
  - Contains the major contractual or statutory provisions relied upon; and
- A summary of what is claimed (e.g. time or money) including:
  - The monetary value of the claim;
  - A proportionate level of breakdown; and
  - Any expert report that is succinct and central to the claim.

The Protocol also provides guidance for the defendant's (or claim recipient's) response, which should contain a similar level of information. This provides a basic description of the kinds of information that should be provided in support of, and in response to, a claim.

Parties may have a tendency to only provide evidence that supports their position – i.e. that substantiates a claim or an argument against a claim. However, a party should be cautious when adopting a selective approach to providing evidence, as at a later date, he may be compelled to provide documents within his control that may not support his case. If a claim proceeds to litigation outside the small claims track,[10] then the rules of 'standard disclosure' are likely to apply. The Court is likely to order each party to provide to the other:

- The documents on which that party relies[11]; but also
- Documents that adversely affect his own case.[12]

Documents falling outside the scope of 'standard disclosure' would be those subject to legal privilege or outside a party's control. Parties are also at liberty to apply to the Court for disclosure of documents at the pre-action (or 'Protocol') stage.[13]

### 10.1.3 Requests for Further Information

Many employers or their representatives (architects, project managers and quantity surveyors) and also contractors may be uncertain about the way in which to approach the reimbursement of loss and/or expense. If the contractor does not submit adequate supporting information when it makes its application for reimbursement, it is helpful if the employer's representative (and a quantity surveyor, if so instructed – see Section 9.2) in exercising their powers to request information, also take the opportunity to correct any inconsistencies or errors in the contractor's approach. Information should be requested as precisely as circumstances allow.

---

10 Civil Procedure Rules (CPR) rule 31.1(2).
11 CPR rule 31.6(a).
12 CPR rule 31.6(b)(i).
13 CPR rule 31.16.

On no account should the contractor simply be requested to provide 'more information' or 'better proof' – often expressed under the guise of 'further and better particulars'. It leaves the contractor unclear as to the kind of information which will satisfy the employer or his representative. It also gives an unscrupulous contractor the opportunity to submit less information than may be necessary on the basis that he has not received any clear indication to the contrary. It is easy for a contractor to say that, in his opinion, the information he's already provided substantiates his claim. Then, unless the employer's representative is prepared to specify exactly what is required, there is often an impasse. It is therefore better for all concerned if the required information is clearly stated at the outset.[4] However, the employer or his representative should consider how the particular building contract may permit the contractor to submit further information on a progressive basis (refer to Section 10.2.1).

The following list, which is not exhaustive, includes some of the points which might be initially raised with the contractor:

- Loss and/or expense (JCT), or a change to the Prices (NEC4), may not follow automatically from an extension of time (JCT) or a delay to the Completion Date (NEC4);
- The contractor's entitlement to claim may be:
  - time-barred from submitting a claim[14]; or
  - possibly be restricted because his written application should have been made as soon as it has become, or should reasonably have become, apparent that the regular progress of the Works has been or is likely to be affected[15]; however, JCT 2016 contracts envisage that claims (for both extensions of time and loss and expense) may be developed progressively and encourage the contractor to subsequently provide further information (see Section 10.2.1);
- As far as it is reasonably practicable (and particularly for interim payments), each cause should be linked to its effect;
- Amounts claimed should be linked to the relevant matters (JCT 2016 contracts) or particular compensation events (NEC4) in the contract;
- Where prolongation is involved, some form of critical path analysis (refer to Chapter 3) should be provided;
- Only actual costs may be acceptable for a final ascertainment under the JCT 2016 contracts. Forecast, provisional (e.g. 'on account') or formulaic claims are unlikely to support a significant valuation, and the contractor should expect any 'on account' valuations for interim payment purposes to reflect the deficiencies in the claim's supporting evidence;
- For NEC4 contracts, assessing the contractor's quotation or the Project Manager's own assessment must be carried out in accordance with the contract;
- For a final ascertainment or assessment, substantiation in the form of invoices, labour returns, pay slips, detailed schedules of plant and equipment and amounts paid to sub-contractors etc. should be provided; and
- In some cases, the employer's representative may not have the authority to ascertain or assess a particular claim (e.g. if the employer has instructed the contractor to carry out work outside the Works or the Site).

---

14 NEC4 – clause 61.3. The contractor must notify the Project Manager within 8 weeks of becoming aware that the event that could lead to a 'compensation event' has happened.

15 JCT SBC/Q 2016 – clause 4.21.1, JCT DB 2016 – clause 4.20.1, and JCT IC 2016 – clause 4.16.1.

The employer's representative should not procrastinate by unreasonably requesting more and more information from the contractor for the following reasons:

- Throughout the construction phase,[16] the employer's representative should maintain his own records of contractor's progress[17] and of the quality of work carried out,[18] so he should not have to rely solely upon contractor-provided information to verify the contractor's claim; and
- The employer and his representative are exposed to the risk of adjudication in the following circumstances:
  - If the contractor feels that the employer's representative already has sufficient information that warrants significantly more being awarded (e.g. a higher sum for the claim in an interim payment, or a greater extension of time) then he may adjudicate (e.g. on the basis that the relevant interim payment is unreasonably low); and
  - If the employer's representative has failed to respond within any time limits imposed by the building contract:
    - Within the 12 week period has not provided any decision regarding the fixing of a later completion date or giving an extension of time[19]; or
    - Failed to respond within two weeks after the contractor notified him of his prior failure to assess a compensation event within the time allowed.[20]
- If the employer is found to be liable for any failure by the employer's representative to ascertain or assess a contractor's claim (including having to foot the bill for legal fees during an adjudication), then the employer's representative may face a claim for professional negligence.

### 10.1.4 Setting Out the Claim

It is quite impossible to set out a detailed sample claim which a contractor can complete by simply filling in the blanks and which is suitable for every occasion. Every claim is different and the way it is set out must reflect that difference. However, it is possible to highlight some important issues and an outline claim is shown in Section 10.1.5 below. The key points are as follows (see also Section 10.1.2):

- The contractor must satisfy any conditions precedent to its claim, such as the timely submission of its application and the provision of any further details of the loss and/or expense incurred for the purpose of deciding the validity in principle of the claim and to

---

16 RIBA Plan of Work 2020 – Stage 5 (Manufacturing and Construction).

17 RIBA Standard Professional Service Contract 2020 – Stage 5 (Manufacturing and Construction) example 'Contract Administrator' services include '*...inspect the construction of the works with respect to general compliance with the Building Contract and Project Programme*' and '*Review the progress of the construction works against the Project Programme*'. The RICS' Quantity Surveyor Services (May 2022) require: the preparation of regular/monthly cost reports (1.1.2), periodic site visits to assess progress for interim payment purposes (1.7.2) and the optional services: advising on extensions of time (2.2.2) and loss and expense (2.2.5).

18 RIBA Good Practice Guide: Inspecting Works (2009) page 94 regarding 'Inspection records'.

19 JCT SBC/Q 2016 – clause 2.28.2, JCT DB 2016 – clause 2.25.2 and JCT IC 2016–2.19.3, see also Section 9.1.9.

20 NEC4 – clause 64.4.

enable the employer or his representative (including the quantity surveyor) to carry out the ascertainment;

- The contractor must set out the facts upon which he relies as showing his entitlement to reimbursement and provide any necessary evidence of those facts; and
- He must reference the particular express contract term(s) and/or any implied terms which give rise to the claimed entitlement and set out:
  - the financial basis of its claim together with evidence to support the amounts claimed; and/or
  - some critical path analysis (refer to Chapter 3) that shows the critical delays for which an extension of time is claimed, along with supporting evidence.

Although standard form contracts do not expressly require the contractor to submit a separate claim document (e.g. setting out in detail and with evidence what the contractor considers to be its entitlement), most contractors adopt this approach or at least they attempt to do so. The sound reason is that a properly set out claim which is correctly argued and supported by persuasive evidence is the best way for the contractor to recover the amounts he claims. It is advisable to draft the claim on the assumption that the reader knows nothing about the project. Ideally, the claim will require little alteration to form part of a referral notice or pleadings in any respective subsequent adjudication or legal or arbitration proceedings.

By way of an example, it may be useful to briefly consider what a contractor is required to do if claiming loss and/or expense for a variation under the JCT SBC/Q 2016 contract, under clause 4.20 (a more in-depth examination is given in Chapter 12). Where there is a claim for direct loss and/or expense arising under the relevant matter in Clause 4.22.1 (Variations), in respect of a 'Variation' (for additional work), the contractor must establish the following:

- An instruction is deemed to have been properly given to the contractor by the employer's representative, either:
  - In writing (e.g. under clause 3.10); or
  - Other than in writing (orally) which (i) the contractor confirmed in writing within seven days and (ii), the employer's representative does not dispute by notifying the contractor within a further seven days (Clause 3.12.1);
- The instruction constitutes a Variation[21] and has been or will be carried out by the contractor;
- The Variation has not been the subject of a 'Confirmed Acceptance' of a 'Variation Quotation' under Clause 5.3, containing loss and/or expense under the provisions of Paragraph 1.2.1 of Schedule 2;
- Any significant period of time between:
  - the instruction's date of issue, the date when the contractor receives the instruction; or moreover
  - the date when the instruction is given other than in writing, and the expiry of the seven day period during which the employer's representative must notify the contractor of his dissent to the contractor's written confirmation of the said instruction;

21 JCT SBC/Q 2016 – e.g. to carry out additional work is a Variation under clause 5.1.1.1.

This is an important point because the length of time that the contractor is given to comply with the instruction[22] may affect his ability accommodate the variation into his overall programme of Works (i.e. the appropriate or optimum time to carry out the variation as efficiently as is reasonably possible. The contractor ought to show that it carried out the variation at a time without incurring any more direct loss and/or expense which in the circumstances would be reasonable. With plenty of notice, the contractor should be able to plan and execute the variation at minimum overall cost, and at the most suitable time. If the contractor was obliged, either by the timing or the nature of the instruction, to carry out the variation at an inconvenient time, then the amount of loss and/or expense might be increased, which the contractor should explain.

- Carrying out of the variation has directly affected regular progress of the contractor's other Works. The contractor must establish that the effect upon regular progress was a direct result of the instruction and not the consequence of some other, or intervening event;
- The effect on the regular progress of works has been 'material' and more than trivial;[23]
- The way in which, and the extent to which, other work was affected as a direct result of the carrying out of the variation. It is obviously essential to show how, and to what extent regular progress was affected by the variation. This is at the very heart of the claim; and
- The date on which the contractor made a proper and timely written application to the employer or his representative in respect of the direct loss and/or expense being claimed is of some importance, because for sound practical as well as legal reasons, the contractor's entitlement to reimbursement and possibly to quantum under the contract provisions may depend upon the extent that the administrative provisions of the contract have been satisfied.

### 10.1.5 Likely Content of a Claim

Previous editions of this book, and indeed, prevailing industry practice, seemed to consider a 'claim' to be a stand-alone document, that comprised a 'narrative' along with many lever arch files of the supporting information. This conjures up images of the contractor's team of quantity surveyors, lawyers and claims consultants working feverishly on the claim for many months after the project has been completed. This now seems more accurate in terms of describing how pleadings or statements of case are assembled for litigation or arbitration, or possibly for a substantial adjudication. The 'General Introduction' in Section 1.1 pointed out that it may simply not be in the contractor's interests to submit claims after a projects' completion. Furthermore, if the contractor waits too long before submitting his 'claim', he may lose some or all of his entitlements because he's time-barred (see Section 2.9.2).

This section attempts to summarise the kinds of information that are appropriate to provide with a 'claim' at the following stages of a project:

- For interim payment purposes during the construction phase;
- For the 'final account' after the works are completed;

22 The contractor is obliged to comply with the instruction 'forthwith' (clause 3.10). However, complying forthwith may not always result in the contractor being able to immediately commence the physical work that 's instructed. For example, he may have to carry out some design work beforehand (e.g. if the Variation is to the 'contractor's designed portion', or there may be a long procurement period for the necessary materials required by the Variation.

23 JCT SBC/Q 2016 – clause 4.20.1 provides that progress must be 'materially affected'.

- For adjudication (at any time, either before or after completion); and
- For litigation or arbitration after the project's completion.

Table 10.1 shows the typical details to provide in support of claims for the 'stages' mentioned above:

## 10.2 Types of Evidence Required to Support a Claim

### 10.2.1 Introduction

In some cases, even if the contractor is unable to provide certain information in support of his claim, this may not necessarily deprive him of the right to some reimbursement for loss and expense, as some estimate may be made[24] by the employer or his representative. In practice, however, without substantial evidence, the contractor is unlikely to receive anywhere near the amount it originally claimed.

The employer (or his representative) may initially perceive that initially, some claims lack adequate supporting evidence, but there may be perfectly understandable reasons for this absence of detail such as:

- The contractor may be obliged to notify the employer as early as is reasonably possible, or within a specific period of time, so he may not be in possession of sufficient (and certainly not 'full') supporting information when the claim is first notified; and
- If a (main) contractor is relying upon his sub-contractors to provide evidence, he may have to either wait for them to provide the supporting information, or need to discuss the nature of information to be provided by them.

It is natural for an employer (or his representative) to become impatient regarding the provision of supporting evidence, as they will want to know the extent of their potential liability (or advise the employer of such) as early as possible. However, the employer (or his representative) should remember how the particular building contract deals with claims:

- JCT 2016 contracts allow claims to be built up progressively, to be finally ascertained at the final account stage after the project has been completed, whereas
- NEC4 requires 'claims' to be dealt with as 'compensation events' and agreed during the course of the works – before completion.

For JCT 2016 contracts, Table 10.2 shows how the contractor may build up his claim progressively until its conclusion:

What is often misunderstood is that when the contractor is making the claim, it is a matter for him to prove that it's valid. It is advisable for a claimant to proceed on the basis that everything he asserts must be authenticated unless it's self-evidently correct, or it is agreed as common ground. The contractor need not prove what are referred as established facts; that being, facts which are accepted as common knowledge and which it would be absurd to prove (refer to Section 2.2.2). However, it is always better to provide proof that's as comprehensive as possible than to omit something, which may later turn

24 *Chaplin v Hicks* [1911] 2 KB 786.

**Table 10.1** Typical Supporting Information for Claims.

| Information | Interim payment | Final account | Adjudication | Litigation or arbitration |
|---|---|---|---|---|
| General Project Information[a]: | — | — | — | — |
| • Parties' names and addresses | — | — | ✓ | ✓ |
| • Project title | — | — | ✓ | ✓ |
| • Project address | — | — | ✓ | ✓ |
| • Brief description of the work | — | — | ✓ | ✓ |
| • Identity and address of the employer's representative(s) | — | — | ✓ | ✓ |
| Contract information[b]: | — | — | — | — |
| • The form of contract (inc amendments) | — | — | ✓ | ✓ |
| • Extent of contractor design | — | — | ✓ | ✓ |
| • Contract sum (or contract price) | — | — | ✓ | ✓ |
| • Start and finish dates (inc for each section) | — | — | ✓ | ✓ |
| • Other dates (such as Key Dates) | — | — | ✓ | ✓ |
| General 'claim' information: | — | — | — | — |
| • A brief summary including the 'nature' of the claim – e.g. variation or compensation (time and/or money) | ✓ | ✓ | ✓ | — |
| • A detailed explanation of the 'claim' | — | ✓ | ✓ | ✓ |
| • The claim's origin – e.g. an instruction or breach of contract (express or implied) | ✓ | ✓ | ✓ | ✓ |
| • Method of submission (e.g. interim payment application, quotation) | ✓ | ✓ | — | — |
| • Date(s) of relevant notifications | — | ✓ | ✓ | ✓ |
| Specific 'claim' information | — | — | — | — |
| • The amount of money claimed | ✓ | ✓ | ✓ | ✓ |
| • The period of time claimed | ✓ | ✓ | ✓ | ✓ |
| • The extent of the dispute (describing any elements accepted by the employer) | — | ✓ | ✓ | ✓ |
| • The contractual basis of entitlement (citing the express terms of the contract and/or any implied terms)[c] | ✓ | ✓ | ✓ | ✓ |
| • A sufficient explanation of the facts | ✓ | — | — | — |
| • A detailed explanation of the facts | — | ✓ | ✓ | |
| • A breakdown of the sum claimed | ✓ | ✓ | ✓ | ✓ |
| • Supporting evidence of the sum claimed[d] | ✓ | ✓ | ✓ | ✓ |
| • Programme(s) showing delay(s) to the completion date (or other dates) | [e] | ✓ | ✓ | ✓ |
| • An explanation of the delay(s) | ✓ | ✓ | ✓ | ✓ |
| • Witness statements from key personnel | — | — | ✓ | ✓ |

*(Continued)*

**Table 10.1** (Continued)

| Information | Interim payment | Final account | Adjudication | Litigation or arbitration |
|---|---|---|---|---|
| Other information: | — | — | — | — |
| • Any previous adjudicator decisions | — | — | ✓ | ✓ |
| • Copies of and court/arbitrator orders | — | — | — | ✓ |
| • Expert quantity surveyor's report | — | — | ✓ | ✓ |
| • Expert delay analyst's report | — | — | ✓ | ✓ |
| • Other expert report (e.g. technical) | — | — | ✓ | ✓ |
| Format: | — | — | — | — |
| • Paragraph numbers for text explanation | — | — | ✓ | ✓ |
| • Paginated 'bundle' of supporting info | — | — | ✓ | ✓ |
| • List of contents | — | — | ✓ | ✓ |

a) Ordinarily and under the standard forms of building contract, there is no requirement to provide the 'General' or 'Contract' information with an interim payment application, as it should be clear that the interim or final application is applicable to the relevant project. However, amended contracts may set out precise requirements for a 'valid' interim or final payment application (i.e. one that the payer is obliged to consider), so information under these headings may have to be provided. The application may also have to be submitted in a specific way, such as being sent by recorded post and email to specified personnel.
b) See Footnote 23.
c) For JCT SBC/Q, DB and IC 2016 contracts, all 'Relevant Events' and 'Relevant Matters' should be described. For NEC4 contracts, all 'compensation events' should be described.
d) See Section 4.4.
e) A programme must be provided under NEC4 projects, but there is no requirement for the contract to provide a programme under JCT 2016 contracts. See Chapter 3.

**Table 10.2** JCT 2016 contracts – provisions to progressively 'build up' a claim.

| Requirement | JCT SBC/Q | JCT DB 2016 | IC 2016 |
|---|---|---|---|
| Extension of time: | — | — | — |
| • Initial notice: 'when reasonably apparent' and 'forthwith' | 2.27.1 | 2.24.1 | 2.19.1 |
| • Updates (notify material changes) | 2.27.3 | 2.24.3 | —[a)] |
| Loss and/or expense: | — | — | — |
| • Initial notice: 'as soon as…reasonably apparent' | 4.21.1 | 4.21.1 | 4.16.1 |
| • Monthly updates 'until all information reasonably necessary…has been supplied' | 4.21.3 | 4.20.3 | 4.16.3 |

a) JCT IC 2016 – the contractor is not expressly required to provide the 'material circumstances', 'particulars of delay' or 'estimate of the expected delay' as required by JCT SBC/Q 2016 or JCT DB 2016. However, there is an implication that the contractor must provide further details to enable the employer's representative (*'as soon as he is able'* – clauses 2.19.1 and 2.19.2) to estimate the length of the delay. The contractor must also provide further information if required (clause 2.19.4.2).

out to be important. The old adage 'he who asserts must prove' is clearly relevant. Some claims consist merely of broad and sweeping assertions (such as delay, disruption and loss) without making any attempt towards proving them. In extreme examples, contractors have been known to simply refer to events which occurred during the progress of the Works, and then submit a stack of contemporary correspondence (refer to Chapter 7). Although most standard form contracts do not specify that the contractor makes a detailed claim (as if it was a submission in litigation or arbitration), they certainly do not require the employer's representative to put the contractor's claim together for him.[25] In any event, despite what the contract may require, only a foolish contractor would expect the employer's representative to assemble a claim from an unexplained pile of papers.

A contractor who is to provide substantiating evidence should have adequate records. The contractor is usually obliged to provide the employer's representative with all the information necessary for them to validate and ascertain the contractor's claim. The extent of information required should be proportionate (and will obviously vary accordingly) to the complexity of project and the claim. It will usually include records of accounts (including overheads, profit and salary details), sub-contractor claims, labour returns and time sheets, delivery notes and receipts, programmes showing a critical path analysis, proof of the intended head office overheads and profit, and sometimes a report from an expert witness. These requirements place a heavy burden upon the contractor, but unless evidence of this kind is provided, it may be difficult for the employer's representative to properly carry out his duty to form an opinion, ascertain the amount due, or make an assessment. That is a practical consideration is reflected as a contractual requirement. The point received comment in *Peak Construction (Liverpool) Ltd v McKinney Foundations Ltd*,[26] when the court was dealing with the kind of evidence necessary to substantiate a claim for head office overheads and loss of profit.[27] The court made the point that not only was the evidence of the contractor's auditor as to profitability relevant, but also

> '*...some evidence as to what the site organisation consisted of, what part of the head-office staff is being referred to, and what they were doing at the material times...*'[28]

The court felt it would be useful to have an analysis of the contractor's yearly turnover for a period of some seven years, in order to establish the level of profitability and the effect of the particular contract disruption and overrun had upon it.

For claims under JCT 2016 contracts, contractors sometimes use the rates for preliminary items (e.g. for site personnel or accommodation), averages, generalisations or even figures plucked out of the air (typically percentages for lack of productivity). There is nothing inherently wrong with such an approach for the purposes of notifying a claim or for the initial stages of producing a claim for interim payment purposes, but the contractor should expect an equally basic ascertainment or assessment (in the form of an 'on account' valuation) from the employer or his representative. However, when finalising a claim, it is

---

25 Although the Project Manager can make his own assessment of a 'compensation event' under clause 64 of NEC4.

26 *Peak Construction (Liverpool) Ltd v McKinney Foundations Ltd* (1970) 1 BLR 114.

27 See Section 4.4.2.6.

28 (1970) 1 BLR 114 at 122 per Salmon LJ.

the *actual* loss and/or *actual* expense which is required and not figures taken from the contract's pricing document (e.g. bills of quantities), or from indices or other sources.

NEC4 takes a different approach. Unless the employer's representative (the Project Manager) and the contractor agree rates or lump sums[29] to assess changes to the Prices for a compensation event, the changes in the Prices are assessed by calculating the 'actual Defined Cost' and 'forecast Defined Cost' (for 'work already done' and 'work not done' by the 'dividing date')[30] – refer to Section 5.5.1. Both the 'actual' and 'forecast' Defined Cost are calculated by using whichever of Short Schedule of Cost Components (Options A and B) or the Schedule of Cost Components (Options C and D) applies (refer to Table 5.14) then adding the Fee.[31] Despite some rates and prices being included in the 'Schedules', the contractor will have to demonstrate the amounts he has paid under the other categories of Defined Cost.

### 10.2.2 The Explanation (or Narrative)

Most claims begin with an explanation (or 'narrative') telling the story of the claim. A clearly worded explanation that accurately describes the facts (often in chronological order) provides valuable assistance to the person who has to understand and assess the claim (usually, the employer or his representative).[32] It is often the case that a contractor will assume that the employer (or his representative) knows all of the facts and circumstances relating to the claim, and may automatically assume that no further explanation is required. This is rarely the case, as unless they have worked for a contracting organisation, some employer's representatives may not be fully aware of the logistical and operational factors that apply to the contractor's work.

It is not uncommon for the employer to engage different personnel to investigate a claim. He may engage a different (often a senior or more experienced) member of the firm providing the employer's representative, or he may engage external consultants to assess a claim. Therefore, if the initial investigation of a claim is likely to be a 'desktop study', carried out by a person with little, or even no prior knowledge of the project, then a clear explanation will undoubtedly help them. Obviously, the explanation should describe the duty, breach (or entitlement), causation and loss (or value) of the claim (refer to Section 2.1.2). Not only is it advisable to clearly explain the claim on the basis that it may be assessed by someone with no prior involvement with the project but also consider that it may subsequently be put before an adjudicator, an arbitrator or the Court.

Contractors should avoid using emotive or general language, and not make 'sweeping statements' when explaining the claim. They should also avoid repetition. These approaches were fairly common several decades ago – and presumably, the authors of those types of claims believed that 'strongly worded' narratives may frighten the employer into settling the claim to the contractor's advantage. Experienced employer's representatives,

29 NEC4 – clause 63.2.

30 NEC4 – clause 63.1.

31 NEC4 – the 'Fee' is defined in clause 11.2(10), and in clause 52.1 as *'All the Contractor's costs which are not included in the Defined Cost are treated as included in the Fee…'*.

32 A senior claims consultant once told the author that *'If your Mum can understand the claim and thinks that the claimant has been 'hard done by', then you've written a clear and understandable narrative'*!

claims consultants and lawyers will immediately see through such poor tactics – as they are usually employed in a vain attempt to mask an absence of facts and evidence. The recipient of a claim is likely to be far more concerned when confronted with a clearly explained claim; one that includes detailed references to dates, facts, entitlements and provides evidence of the same, as he (or his advisers) will need to verify the claims that have been made.

In compiling a claim, the contractor or their advisers often paraphrase parts of correspondence, site meetings, etc, to emphasise their arguments. However, paraphrasing should only be used to highlight the relevant sections of text (i.e. cutting out the irrelevant words) and not be used to paint an inaccurate or false picture. The claim's recipient will refer to the original documents and, so inappropriate paraphrasing may result in the claim being discredited and compromised.

The claim's author should also be discouraged from making general or sweeping statements. Statements such as '*…the employer's representative constantly issued late instructions*' naturally implies that no instructions were issued at the appropriate time, whereas the true circumstances may be very different. The 'lateness' in this example may simply be perceived by the contractor either as a general and unchecked assumption, or by the contractor incorrectly assessing the progress of his work. If the correct circumstances are verified, the following may be found:

- None of the instructions were issued late. They may have been issued later than the contractor would have originally liked, but they were issued at a reasonable time in relation to the progress of the contractor's work;
- Some of the instructions were issued late, but the affected work was not on the critical path, so the completion date remained unaffected; or
- Only isolated instructions were issued late in relation to the progress of the contractor's work.

Another phrase which is often used to argue that the progress of work has been seriously affected is '*The work was becoming increasingly complex*'. That kind of vague statement invites the following questions:

- How complex was the work originally?
  - Are the works either new build or refurbishment?
  - Were the works or a relatively simple nature (e.g. the construction of a warehouse), or did it involve the integration of many different specialist trades (such as a hospital)?
- How do you measure complexity?
  - What has actually changed, so a comparison can be made?
  - Has complex work been added to what was originally a simple project? For example, has the building's footprint been increased, and parts of the larger foundations now require piling or underpinning, or work is now required to adjacent or existing buildings?
- At what stage in the project did the complexity increase?
  - Was the change introduced at the start or part-way through the construction?
  - Did the additional complexity require existing work to be adapted (e.g. dismantled and re-constructed)?
  - Does the increased complexity require design studies and further design to be carried out?

  - Was there a distinct period of time when the complex work was incorporated into the existing work, or did it affect the entire project?

  to say nothing of such questions as

- What made it more complex?
  - Did the employer introduce new and more complex works by instructing a change in the design (e.g. additional work or substituted works)?
  - Or did the contractor underestimate the complexity of the original works?
- Whose responsibility was that?
  - Who was responsible for any design changes?
  - A change in the employer's design or the contractor's design?

Unless the claimant can provide substantiated answers to those questions, it indicates the inadequacies of the original statement.

Other phrases often used are *'Circumstances beyond the contractor's control'* and *'…of no fault of the contractor'*. The implication of such phrases is that the matter complained of is the employer's fault, but this is not always the case.[33]

Unless the contract expressly states that factors beyond the contractor's control will entitle it to an extension of time,[34] or claim loss and/or expense, the use of this expression is generally unhelpful. Many things are outside the contractor's control but not all entitle him to an extension of time or additional payment. The risks that the employer and the contractor each agree are set out in the contract.

Assembling a claim requires many of the disciplines of legal argument or a piece of scholarly research, analytical in nature, leaving nothing out, taking nothing for granted and citing appropriate authority or evidence for every statement.

### 10.2.3 Time Records (Programmes and Similar Documents)

The requirements for the contractor's programme, both in terms of what is set out in the contract[35] and for employing one of the recognised methods of delay analysis, are set out in Sections 3.5.1 to 3.5.6.

However, its usefulness in substantiating a time or monetary claim depends on the way in which the programme was prepared. If the contractor has employed an experienced and competent planner to prepare an initial or 'tender' programme (not when a claim is envisaged but at the beginning of the work), it should provide good evidence of the sequence and timing that the contractor intended to adopt when carrying out the originally specified scope of works. The more detail that can be incorporated, the better. In addition to depicting the sequence and timing of the contractor's works in the form of a Gantt chart, most planning software allows plant and labour resources to be applied, so cashflows and outputs can readily be calculated.

---

33 *Scott Lithgow Ltd v Secretary of State for Defence* (1989) 45 BLR 1 (HL).

34 JCT MW 2016, clause 2.7 (Extensions of time).

35 Refer to Section 3.5.1. The references to a programme in JCT 2016 contracts are relatively few (refer to clauses 2,9.1 to 2.9.3 of JCT SBC/Q 2016), whereas the programme (and 'Accepted Programme') in NEC4 is a pivotal document for assessing extensions of time – delays to the Completion Date and/or Key Dates (see clause 63.5). Table 3.4 summarises the information to be provided in the programme under NEC4, Table 3.5 summarises the requirements for revising the programme, and Table 3.6 sets out the reasons which the employer's representative (Project Manager) may cite to reject the revised programme.

However, it is important to remember that unless the contract provides for the programme to be accepted (by the employer's representative) and used as the only means of assessing delays (as it is under NEC4), then strictly speaking, the programme is only a depiction in the bar-chart form, of how the project was, and/or is forecast to be constructed. In order to substantiate the programme, the contractor will have to demonstrate the accuracy of the activities' durations. He can do this by referring to several sources:

- Programmes submitted to indicate progress throughout the project's construction stage. These may show the '% completion' of each activity as specific intervals and progress shown against a baseline – either by rescheduling the subsequent activities or as indicated by a 'drop line' (see Section 3.5.3 and Figure 3.9);
- Progress reports with a description of progress;
- Progress photographs and time-lapse video (see Section 10.2.6); and
- The progress of activities claimed in the contractor's applications for interim payment. If there is a significant difference between the values of work claimed by the contractor, and the amounts set out in the employer's payment notice or a pay-less notice, then an average could be taken unless other evidence is to be preferred.

### 10.2.4 Cost Records

Cost records are the building blocks of a financial claim, and without them, the contractor will have a significant difficulty to overcome. Most contractors will operate software accounting systems of some description. The sophistication and functions of such software vary and have grown; from recording and filing financial information, to a fully fledged and cloud-based financial ERP[36] database that's dedicated to construction companies. In terms of substantiating claims, such databases are not only likely to contain a financial record of each transaction (e.g. in a spreadsheet) but also copies of invoices, orders, sub-contracts and applications for payment. 'Set price' transactions (such as the purchasing of materials or hire of plant) are usually recorded as the cost claimed on the relevant invoices, but transactions that are usually subject to the contractor's adjustments (such as its sub-contractor's applications for payment) may be recorded as

- The amount applied for by the sub-contractor; or
- The amount set out in the contractor's payment notice or a pay-less notice to the sub-contractor.

In either case, an accrual may have to be made until the end of an account recording period. For example, if the contractor is attempting to demonstrate the value of work that a sub-contractor carried out in a calendar month, but the due date for sub-contractor payments occurs in the middle of a month, an accrual (or adjustment) will have to be made to the amount claimed by the sub-contractor, so it is projected (or forecast) to the end of a month.

For most other transactions, the software will show the date of the claim, so the actual costs incurred by a contractor can simply be extracted by looking up all the actual costs (plus sub-contractor accruals) to a particular date, or for a particular period.

36 ERP – 'Enterprise resource planning', such as 'COINS' provided by Access Technology Group Ltd.

The ERP software may allow the costs to be sorted and allocated, but it is often easier to import the relevant cost information into a spreadsheet like Microsoft Excel. Even if a direct transfer of electronic data cannot be made into a spreadsheet, hard copies of account ledgers can often be digitally scanned, and optical character recognition software can then be employed to read the scanned material and transfer it into a fully working spreadsheet for further analysis (as if it had be created electronically in the first place).

Different contractors may report certain costs in different ways. For example, one contractor may allocate costs for staff who may be engaged on more than one project (such as a project manager, quantity surveyor or planner) to its overheads, rather than record their costs against individual projects. Other contractors may require their staff to complete a weekly timesheet, so their time (and cost) can be accurately allocated to all the projects that they're working on. It will still remain the fact that the adequacy and accuracy of cost records will depend upon the accuracy and completeness of recording and allocating those costs.

When verifying the 'Defined Cost'[37] in NEC4's target cost contracts (Options C and D), the Project Manager must subtract what is called 'Disallowed Cost'. For Options C and D, 'Disallowed Cost' has identical definitions in Clause 11.2(26) which are summarised below:

- Costs which cannot be justified from the contractor's accounts;
- Costs incurred because the contractor did not comply with:
  - Sub-contracts or contracts for supply;
  - An acceptance or procurement procedure set out in the Scope;
  - The contract's 'early warning' procedure;[38] or
  - The requirement to notify the Project Manager of an adjudication between himself and a sub-contractor or supplier;
- Costs of correcting Defects:[39]
  - After Completion;[40]
  - Caused by the contractor failing to comply with a constraint on how to provide the Works that is set out in the Scope;
- Costs of resources (e.g. plant and materials) not used to provide the Works, for either:
  - Allowing for a reasonable amount of waste, availability or utilisation; or
  - Not taken away from the Working Areas when requested by the Project Manager; and
- Dispute resolution costs:
  - Of preparing for, and participating in an Adjudication;
  - Payments to DAB[41] members; or
  - Other proceedings (using the 'tribunal'[42]) between the contractor and the client.

---

37 Refer to Section 5.5.1 and Tables 5.15 and 5.16 for an explanation of 'Defined Cost'.

38 NEC4, clause 15 (Early warning).

39 NEC4, 'Defect' is defined in clause 11.2(6).

40 NEC4, 'Completion' is defined in clause 11.2(2).

41 NEC4: 'DAB' is the 'Dispute Avoidance Board' in Option W3.

42 NEC4: The 'tribunal' is set out in Contract Data – Part One (under 'Resolving and avoiding disputes') and finally resolves disputes (e.g. the 'tribunal' can be the Court, an Arbitrator or an Expert – for 'expert determination').

The numerous elements of 'Disallowed Cost' will have to be identified by the Project Manager carefully, scrutinising the contractor's accounts and records.[43] It is not uncommon for NEC4 contracts to be amended to provide extensive auditing rights for the Project Manager, or others (such as a quantity surveyor) to examine the contractor's accounts and records.

### 10.2.5 Correspondence and Similar Documents

Copies of relevant correspondence, along with notices should be attached to the claim, together with copies of other relevant documents such as payment applications, payment notices, etc. – as required by the terms of the contract. Nowadays, most parties conduct virtually the whole of their business correspondence electronically. This can include issuing documents and correspondence by email or sending it via a project-specific 'portal'. The use of 'portals' is increasingly common on large-scale projects that are let under the NEC forms of contract. The portal will allow the parties to submit early warning notices, notifications of compensation events, quotations and assessments of compensation events, along with other notifications, payment applications and assessments and give each party access to these documents. Additionally, these portals can provide summaries of each type of communication or document – showing the dates when documents were submitted and responded to. The building contract may be amended to ensure mandatory use of the portal so any correspondence issued by other means is rendered invalid.

If a project document portal is not employed, then it is likely that the parties will be using some kind of email software. Even if a party has not set up a central email file, it is not difficult to collate emails sent and received by each member of the contractor's or employer's team, but simply saving all of their emails as '.pst' (personal storage table) files and collating them into a single folder or database. The emails can easily be sorted by various criteria including:

- The names of the sender and recipient;
- The emails' subjects;
- The dates when sent;
- If the emails contain attached files;
- By a category (if allocated to a particular category);
- If the email has been read or remains unread;
- If marked as 'urgent'; and
- By searching for a 'keyword'.

The parties must also be alive to the proper means of notifying the other under the terms of the contract. Despite the almost universal use of electronic correspondence, many forms of building contract (such as JCT 2016) require certain notices to be delivered by a method than can be formally verified. This includes recorded or 'signed for' post or by actual delivery. The contracts often contain default procedures, whereby if a notices' delivery cannot be verified, it is deemed to have been delivered a specified number of business days after posting.

43 NEC4 – clause 11.2(26) first bullet point.

### 10.2.6 Photographs, Time-Lapse Video

A visual record of a project can be extremely valuable, as it can give a clear and factual account of the progress of work and its quality at a particular point in time. Where photographs are concerned, it should be clearly apparent when a photograph was taken and what its subject is. Digital photographs record their time and date automatically, but care should be taken when transferring or copying them to another data storage device so that their recorded date is not lost and replaced with the date of transferral. If progress photographs are being taken periodically, then it is useful to take subsequent photos of particular parts of the works from similar positions and angles, so a simple visual comparison of the progress of work from photo to photo can be made. In terms of selecting the photographs' subject, it is advisable to focus on the critical path activities, so as clear a record of critical activities can be obtained. Mounting digital cameras on drones can produce photographs or video footage of the entire site, including areas that would otherwise be inaccessible, if regular 'fly-bys' are carried out.

Another source of visual evidence is time-lapse video. Many prestigious or notable projects record their progress via time-lapse video – for example, if a developer is keen to show the progress of how his new building is being constructed. It is not uncommon for such videos to be available to the public on a property developer's website – particularly for projects where external progress may be important to show (e.g. construction of a housing estate or section of road or a bridge). For other projects, the building contract may require the contractor to provide time-lapse video of the project's progress that can be accessed by the employer and other consultants.

Time-lapse video can be a very powerful and definitive tool when compiling and assessing claims for prolongation and/or disruption, as frequent photographs (e.g. every 15 minutes) can be taken. If the video footage is taken from several positions across the project, it may provide a very comprehensive account of the building's progress, particularly of its exterior and showing its external operations. By way of example, a large-scale development in the centre of Birmingham required a large, landmark building to be demolished. The contractor claimed that the progress of demolition was interrupted because, amongst other factors, purging of the existing gas main could not be completed. However, the time-lapse video showed that for a significant period of the claimed delay, the contractor had either failed, or was unable to cart-away sufficient spoil arising from the demolition to allow his plant to move around on the site.

External time-lapse video becomes less useful when progress shifts from external work (which can be easily viewed by the camera) to the internal works. It would not be much use to determine the progress of an interior fit-out works (e.g. to an office or residential block). Internal time-lapse video would be relatively more complicated to obtain.

### 10.2.7 Other Records (Site Meetings, Site Diaries, Labour Returns)

For smaller projects, the cost of providing an on-line communications portal, time-lapse or drone video may be disproportionately high. The contractor and the employer may then resort to more 'traditional' records.

Notes, records or 'minutes' of site meetings may provide useful supporting evidence. Usually, such records are prepared by one party (such as the employer's representative or the contractor) without reference to the other, although one of the first agenda items of a subsequent meeting is usually to record agreement of the previous meeting's minutes. Without any consensus as to their accuracy, the value of minutes may be no more than a 'one-sided' set of internal notes taken by one of the participants. Minutes that are not recorded as agreed in later minutes have a lesser evidential value, somewhat similar to personal notes. Conversely, and if they are produced by one party, they can be very helpful to the other party, because the author of the records will have difficulty refuting his or her own statements. Usually, the records will be taken as a good record of what was agreed at the meeting, because on receipt of their copies, all parties usually check and challenge any inaccuracies, ambiguities at that time, and ensure that the relevant corrections are recorded both in correspondence and in the record of the next meeting. If such inaccuracies are not challenged contemporaneously, with the passage of time, it will become increasingly difficult (though not impossible) to establish that the minutes are not an accurate record of what actually happened. Such records often form important parts of claims because matters are recorded there which are not referred to elsewhere. Moreover, the 'minutes' do not just contain a list of the points that were discussed, as they often contain progress reports from not just the contractor, but from the employer's team as well. However, care should be taken to provide an accurate account of matters – in particularly, the progress of work. A contractor who is recorded as regularly reporting a project as being on time, may find it difficult to subsequently argue that a critical delay occurred at that time. If a contractor is unsure whether the progress of his work is being or is likely to become delayed, he should either qualify any progress reports or programmes, or remain silent on the matter. Such a stance may frustrate the employer, as it appears non-committal, but it is exactly the position that JCT 2016 and NEC4 contracts call for. In Section 10.2.1, the JCT 2016 concept of permitting the contractor to 'build up' his claim was highlighted (i.e. by providing the employer's representatives with updates for both the estimate of delay and for loss and/or expense). For NEC4 contracts, the contractor submits revised programmes, and the employer's representative may accept them, or reject them under particular grounds (refer to Section 3.5.1 and Table 3.6).

Other types of written records are:

- Site and other diaries. A contractor's site diary can be a very useful source of information, as it will often record the following information:
  - The weather (e.g. temperature, rainfall);
  - The number of tradesmen of each discipline (or company) on site and the areas in which they are working;
  - Any visits from the employer's representative or other consultant engaged by the employer;
  - Any visits from any other inspectors (e.g. Building Control, the local authority);
  - Materials and plant deliveries; and
  - Information received from employer's representative.

Often the diaries kept by the contractor and an employer's representative (such as the clerk of works) will differ in what they say about the same events. One such situation was considered in *Oldschool v Gleeson (Construction) Ltd*, where there was a conflict between the diaries kept by a consulting engineer and the site agent. The judge said:

> *'I found [the agent's] diary entries unsatisfactory, in the sense that they do not record warnings and complaints when, as I believe, these were given. I cannot believe that the district surveyor could have written in these terms on 21 March if [his assistant] had not in fact given the same warning, and yet not a hint of it appears in [the agent's] diary; which, together with other instances, leads me to think that he was not anxious to record criticisms or complaints when they were made, and where such appear in [the engineer's] diary and not in the diary of [the agent], I am bound to say that I have no hesitation in accepting [the engineer's] contemporaneous record as being the accurate one.'*[44]

Diaries can be very helpful if one of the parties is trying to prove that the employer's representative was on site at a particular time on a certain day. If it can be shown that the site diary regularly and consistently noted when the employer's representative was on site, and the fact that there is no note of his presence on a particular day will tend to substantiate an allegation that he was not on site. It is less likely that the site agent has simply forgotten to log the employer's representative in the diary if the contractor can show a pattern of entries on past occasions. Many sites now operate a fingerprint entry system, which require each entrant to record their details (such as their company) and can provide a very accurate record of their entry and exit times into the site.

- Labour returns. Labour returns or time sheets are invaluable as a record of the amount of labour and other resources used on a project and, often more importantly, on which areas of the site they have been working. This is particularly important when a claim for disruption costs is being made. Even if the site uses a fingerprint entry system, this will only record the tradesmen's entry and exit times but will not provide evidence of where they were working, unless the site is sub-divided into areas that each require such access. Many contractors are in the habit of providing such sheets to the employer's representative on a regular basis and sometimes the preliminaries section of the contract bills or specification requires such sheets to be provided by the contractor every week.
- Plant returns. Like labour returns, plant returns provide a record of what plant was engaged on a project – for how long and on what part. Their relevance is often limited on construction projects, as major items of plant may not be required for a significant duration of the works (e.g. they may only be required during a groundwork stage), although keeping a record of access plant's operations (such as tower crane, hoist or mast climber) can often be useful. The importance of plant returns for civil engineering projects, or projects where a significant proportion of the works is mechanised, will be greater.

### 10.2.8 A Scott Schedule

A Scott schedule is a document which was often used in in dispute resolution and was first used by Mr. G A Scott QC over 100 years ago. It's use in construction has been commonplace since spreadsheets were used to set out contractors' applications for payment,

44 (1976) 4 BLR 103 at 117 per Sir William Stabb J.

and payment notices and pay-less notices on behalf of the employer. They are especially useful in a spreadsheet format, where there are a number of items (either in a payment application or a claim to be considered).

Essentially, a Scott schedule is a tool for comparing each item in a claim, alongside the recipient's response. The key element of a Scott schedule is that not only are all the issues listed but columns are also provided in which the parties state their summarised cases.

This enables the adjudicator, arbitrator or judge to quickly establish each party's position for each item in dispute from just one line in the schedule. The final columns might be left for the adjudicator, arbitrator, judge or adjudicator to fill in with the amounts as decided with reasons if appropriate.

Items that are in dispute and not in dispute can be clearly identified and perhaps 'filtered', so only the relevant items are shown. Categories of each disputed item can easily be applied, so they can quickly be located in Scott schedule. It is essential that then entries for each item only take up a single row in a spreadsheet, so if the spreadsheet schedule is sorted (e.g. in decreasing value of difference order), no stray items or partial descriptions for an item fall outside the 'sort' or the 'filter' applied to the spreadsheet.

# Part II

# 11

# Extensions of Time Under JCT 2016 Contracts

## 11.1 Introduction

Whilst this chapter is headed 'Extensions of time under the JCT 2016 contracts', a more accurate title could be 'Adjustment of the Completion Date (s) under JCT 2016 contracts'. Indeed, the clauses governing extensions of time in JCT SBC/Q 2016, JCT DB 2016 and JCT IC 2016 are set out under the heading 'Adjustment of Completion Date'. This is because the contractor's obligations to complete the works 'on time' may be adjusted as follows:

- An extension of time may be given so the contractor has more time in which to complete the works (including any variations);
- Work may be omitted from the overall scope so the contractor has to complete less work within the same amount of time; or
- The contract may permit the contractor to agree to accelerate his rate of progress to finish the work in less time.

The extension of time clauses in JCT SBC/Q 2016 and JCT DB 2016 contain a similar level of procedural detail – given that they are intended for use with relatively more complex projects (compared to JCT IC 2016 and JCT MW 2016). As can be expected for a contract intended for use on simpler contracts, the provisions for extensions of time under JCT IC 2016 are similar, but simpler than those under its larger counterparts. This principle is carried down to JCT MW 2016, where the extension of time procedure is reduced to a single clause (clause 2.7 – simply headed 'Extension of time'). Owing to its relative simplicity, not all of the tables in this chapter will include references to JCT MW 2016's clauses. The 'simplicity' in JCT MW 2016 is that the only ground for extending time is in the event of '*reasons beyond the control of the Contractor*' (clause 2.7) and not the numerous relevant events that are found in the other JCT 2016 contracts.

### 11.1.1 Definitions and Interpretation

Table 11.1 summarises the governing definitions that apply to the extension of time procedures:

*Building Contract Claims*, Sixth Edition. Guy Higginbottom and David Chappell.

**Table 11.1** JCT 2016 contracts: Definitions and interpretations for extensions of time.

| Clause description: | JCT SBC/Q 2016 | JCT DB 2016 | JCT IC 2016 |
|---|---|---|---|
| 'Delay' or 'extension of time' also means 'further delay' or 'further extension of time' | 2.26.1 | 2.23.1 | 2.19.5 |
| 'Pre-agreed Adjustment[a)]' means a revised Completion Date under: | | | |
| • A variation quotation[a)]; or | 2.26.2 | 2.23.2 | — |
| • An acceleration quotation[b)] | 2.26.2 | 2.23.2 | — |
| Meaning of 'relevant omission' | 2.26.3 | 2.23.3 | — |

a) Refer to Section 5.4.2.
b) See Footnote 2.

The clauses summarised in Table 11.1 provide further clarification to the relevant definitions that are already provided in clause 1.1 of those contracts. Firstly, any reference to a 'delay or extension of time' also refers to any further delays or extensions of time. Secondly, JCT SBC/Q 2016 and JCT DB 2016 both include the facility to for the employer's representative and the contractor to reach 'pre-agreed adjustments' to the completion date (or to a section[1]) via either a variation quotation[2] or an acceleration quotation.[3] JCT IC 2016 does not contain any provisions for 'quotations' in these respects, so it contains no similar clauses. Thirdly, both JCT SBC/Q 2016 and JCT DB 2016 permit the reduction of an extension of time that has been previously given[4] if work or obligations[5] are to be omitted. This is called a 'Relevant Omission'. It is surprising that JCT IC 2016 does not contain a similar provision, so the employer has the option of either granting an extension of time or (in this case) omitting work so the contractor may be able to complete the works earlier than he possibly would have done if he had to complete the full scope or quantity of work.

1 It should be noted that references to extending time and fixing a new date for completion is taken to mean the date for completion of the Works or, if the Works are divided in the contract into sections, of any section. Unless there is a particular reason to do so, reference will not constantly be made to sections in this commentary.
2 'Variation Quotation' – JCT SBC/Q 2016 – see clause 5.3 and Paragraph 1 of Schedule 2. For JCT DB 2016 – Supplemental Provision 2 must apply in the Contract Particulars (see Part 1 of Schedule 2).
3 'Acceleration Quotation' – JCT SBC/Q 2016 – see Paragraph 2 of Schedule 2. For JCT DB 2016 – Supplemental Provision 4 must apply in the Contract Particulars (see Part 2 of Schedule 2).
4 JCT SBC/Q 2016 – clause 2.28.3.2, and JCT DB 2016 – clause 2.25.3.2.
5 For example, if the contractor was no longer obliged to use a particular site entrance that was further away from its workfaces.

### 11.1.2 The Contractor's Obligation to Notify a Delay to Progress

**Table 11.2** JCT 2016 contracts: Contractor's notification of delay.

| **Table 11.2 summarises the contractor's obligations to notify the employer's representative (or the employer) of delays to progress: Contractor's obligations to notify delays** | **JCT SBC/Q 2016** | **JCT DB 2016** | **JCT IC 2016** |
|---|---|---|---|
| Notices must be in writing | 1.7.1 | 1.7.1 | 1.7.1 |
| Contractor's to notify '*if and whenever it becomes apparent*' | 2.27.1 | 2.24.1 | 2.19.1 |
| Progress of the Works '*is being or is likely to be delayed;*' | 2.27.1 | 2.24.1 | 2.19.1 |
| Progress of a Section '*is being or is likely to be delayed;*' | 2.27.1 | 2.24.1 | 2.19.1 |
| Notice to: | | | |
| • Include the 'material circumstances' | 2.27.1 | 2.24.1 | — |
| • Include the cause or causes of delay | 2.27.1 | 2.24.1 | 2.19.1 |
| • Identify which cause or causes is a 'Relevant Event[a)]' | 2.27.1 | 2.24.1 | — |
| Further particulars: | | | |
| • Can be provided with the 'notice' or '*as soon as possible thereafter;*' | 2.27.2 | 2.24.2 | — |
| • Of the delay(s) expected effects; and | 2.27.2 | 2.24.2 | — |
| • An estimate of the expected delay beyond the relevant Completion Date | 2.27.2 | | — |
| Contractor to provide further notifications: | | | |
| • Forthwith, and if there is any '*material change*' in the estimated delay or other particulars | 2.27.3 | 2.24.3 | — |
| • And provide further information '*at any time*' if the employer/his representative '*reasonably requires*'. | 2.27.3 | 2.24.3 | 2.19.4.2 |

a) 'Relevant Event' – see JCT SBC/Q 2016 – clause 2.29, JCT DB 2016 – clause 2.26 and JCT IC 2016 – clause 2.20.

## 11.2 The Contractor's Notifications

### 11.2.1 The Notice

The contractor's obligation to notify delays is perhaps misunderstood, or put more accurately, is often not fully appreciated or is sometimes overlooked. The contractor is required to give notice to the employer or his representative of every delay, irrespective of whether it is caused by a 'Relevant Event'.[6] This means that if the contractor is responsible for the delay, he must also notify. This may sound an ominous prospect for a contractor – having to admit delaying the progress of the works, but the purpose of this clause is not to 'trip up' the contractor, it is intended to give the employer (or his representative) time to monitor progress,[7] consider and implement any measures that could mitigate the delay.

6 Or if the … *delay occurs for reasons beyond the control of the Contractor…*' – JCT MW 2016 – clause 2.7.
7 *London Borough of Merton v Stanley Hugh Leach Ltd* (1985) 32 BLR 51.

This principle continues as the contractor must not only give notice when the progress of the Works is being delayed but also when it becomes reasonably apparent that progress is likely to be delayed in the future. It has to be obvious to a contractor, acting in a reasonable manner, that the progress of the Works is being or is likely to be delayed.

The reference to progress seems to refer to actual progress. However, measuring whether progress has been delayed may not be a straightforward task – given that the JCT 2016 contracts contain either little or no reference to a 'programme' (see Section 3.5.1). If a programme is provided, it is probably relevant to compare the contractor's actual progress to the progress indicated in the contractor's programme although a failure to comply with the programme would be unlikely, without more detail (such as regular updates), to conclusively demonstrate delay to progress. Strange as it may seem, the contractor has no obligation to comply with its own programme (see Section 3.1.3) and it may, for example, progress the Works faster than shown on the programme.[8] 'Apparent[9]' in this sense, means that something is clearly seen or understood, and that is the criterion which will determine whether or not the contractor has complied with its obligation to notify. Once it becomes reasonably apparent that the progress is actually being delayed or is likely to be delayed, the contractor must notify. The contractor is to give the notice 'forthwith,' which the courts have variously defined as: 'as soon as it is reasonable to do so[10]' or 'without delay or loss of time'.[11] It appears that its meaning may be adjusted depending on the context. In most building contracts, it conveys the fact that the action required must not be delayed, as it may not necessarily mean 'immediately'.[12] Clause 1.7.1 of each contract provides that all notices must be in writing.[13]

Under each contract, the contractor's notice must specify the cause of delay. Although often overlooked, it is advisable for the contractor to identify the precise activity (or activities) which is delayed together with its relation to the project's critical path, and it is certainly in the contractor's interest to do so. This notice was held not to be a condition precedent to giving an extension of time by the architect under JCT 63[8] and this decision applies to the JCT 2016 contracts as follows:

- For JCT SBC/Q 2016 and JCT DB 2016:
  - The employer (or his representative) may only consider an extension of time after practical completion if the contractor has failed to issue a notification of delay[14] (e.g. on the review of the completion date not later than the expiry of 12 weeks from the date of practical completion whether or not the contractor has notified the relevant);
  - It seems clear that the employer (or his representative) cannot give an extension of time before practical completion in the absence of written notice because his obligation to

8 *Glenlion Construction Ltd v The Guiness Trust* (1987) 39 BLR 89.

9 'Apparent' can be defined as 'clearly visible', 'understood' or 'obvious' – Oxford Dictionary of English, 3rd Edition (2010).

10 *London Borough of Hillingdon* v *Cutler* (1967) 2 All ER 361.

11 *Roberts v Brett* (1865) 11 HLC 337.

12 'Immediately' normally means that an action must be performed with all reasonable speed: *Alexiadi v Robinson* (1861) 2 F & F 679.

13 JCT MW 2016 – see clause 1.6.1.

14 In any event the employer (or his representative) has power to give an extension in the absence of the contractor's notice once the date for completion has passed (see JCT SBC/Q 2016 – clause 2.28.5, JCT DB 2016 – clause 2.25.5 and JCT IC 2016 – clause 2.19.3).

do so is triggered by the receipt of the 'notice and particulars'[15] (see Section 11.2.2); however

- For JCT IC 2016 and JCT MW 2016, the employer's representative's duty to give an extension of time does not seem to be triggered by receiving the contractor's notification of delay. Instead, it is triggered by
  - JCT IC 2016 – the employer's representative's opinion that completion will be delayed beyond the completion date by any of the relevant events[16]; and
  - JCT MW 2016 – when the employer's representative considers it 'reasonable' to do so.[17]

The contractor's notice must state the cause or causes of the delay, and for JCT SBC/Q 2016 and JCT DB 2016, it must also state the material circumstances of those delays. It is important that the notice should go into some detail regarding why the delay is occurring (or is likely to occur) and describe the form of such delay. The cause of the delay should be interpreted as meaning all the factors giving rise to the delay. The 'material circumstances' will include such things as the progress and the proposed order of Works and anything else which might affect progress at the time of the delay. Any knock-on effects of the delay (see Section 2.5), with consequent likely further disruption, would also be a material circumstance.

The information must be provided even if it is uncertain whether the current completion date will be affected. Further, if the contractor fails to give notice of a delay which he clearly should have been able to anticipate, the employer or his representative may argue that the contractor has not used his best endeavours to prevent delay in progress, which he is bound to do.[18] Anecdotal evidence suggests that it is rare for a contractor to give notice of his own delays unless he believes that a relevant event giving rise to an extension of time is involved.

A simple example will make clear how the principle works. Take the case of an employer's representative (an architect) to whom the contractor has given a specific written request for information at the appropriate time - i.e. neither too early nor too late. The architect has overlooked the request. A week before the information is required, it is reasonably apparent to the contractor that the information is unlikely to arrive, but he does nothing. On the day the information is required, it has still not arrived, and it is not until a week later (when progress is now seriously delayed) that the contractor sends the architect a notice of delay. By the time the architect has diverted staff onto the task and the information is produced, a further two weeks have passed making a total of three-week delay. The contractor has clearly breached his notification obligations because he failed to notify the architect that the work was likely to be delayed a week before the delay occurred. In fixing a new completion date, the architect is entitled to take the contractor's breach into account by asking the question: what would have been the length of delay had notification been received at the correct time? It would still have taken two weeks to produce the information, but the preparation would have been able to start a week before it was required. The contractor would have been delayed one week and that is all the architect need consider in fixing a new completion date. This is shown in Figure 11.1:

15 Refer to JCT SBC/Q 2016 – clause 2.28.1, and JCT DB 2016 – clause 2.25.1.
16 JCT IC 2016 – clause 2.19.1.
17 JCT MW 2016 – clause 2.7.
18 JCT SBC/Q 2016 – clause 2.28.6.1, JCT DB 2016 – clause 2.25.6.1, and JCT IC 2016 – clause 2.19.4.1.

| Line | Name | Start | Finish | Duration |
|---|---|---|---|---|
| 1 | Data when the Contractor needs the information from the architect | 20 Jun 22 | 20 Jun 22 | |
| | | | | |
| | Notification in compliance with JCT SBC/Q 2016 | | | |
| 2 | Contractor's Notice: when it should have been 'reasonably' apparent' that the progress of work was likely to be delayed (JCT SBC/Q 2016-clause 2.27.1) | 13 Jun 22 | 13 Jun 22 | |
| 3 | 2 week period for Architect to provide the requested information | 13 Jun 22 | 24 Jun 22 | 2w |
| 4 | 1 week delay (had the Contractor notified in accordance with clause 2.27.1) | 20 Jun 22 | 24 Jun 22 | 1w |
| | | | | |
| | Contractor's actual notification | | | |
| 5 | Contractor's notice of delay | 27 Jun 22 | 27Jun 22 | |
| 6 | Architect produces information | 27 Jun 22 | 08 Jun 22 | 2w |

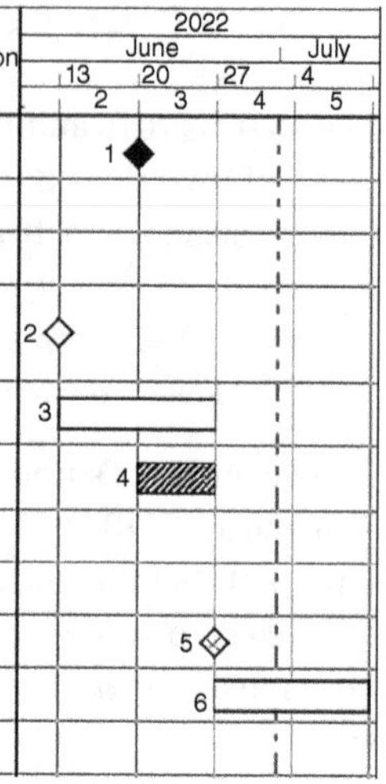


**Figure 11.1** JCT SBC/Q 2016 delay notification example.

Under JCT SBC/Q 2016 and JCT DB 2016, the contractor must identify in its notice of delay any event which in it believes is a 'Relevant Event'.[19] Whilst the notification clause (2.19.1) in JCT IC 2016 mentions 'Relevant Events', it does not require the contractor to identify them, only for the employer's representative to form an opinion for which Relevant Event has caused the delay.

### 11.2.2 Particulars of the Delay(s) - Expected Effects

JCT SBC/Q 2016 and JCT DB 2016 require the contractor to give '*particulars of [the] expected effects*' on the progress of the Works for each event notified (but seemingly, not of other delaying factors). In this context, 'event' is taken to mean 'relevant event'. The contract can provide these 'particulars' either, in its original notice or, where that is not practicable, as soon as possible after the notice. The contractor must also provide the employer's representative with an estimate of the expected delay in completion of the Works beyond the completion date. On a strict interpretation of the contract, this is a particularly onerous task. The contractor must address each expected delay separately, and its effects even if two or more delays are acting together. Contractors often promote their views of the delay, but rarely split them as to deal with each delay separately. The contractor must give enough information to enable the employer's representative to form an opinion. Therefore, strictly speaking, the contractor cannot merely estimate the delays' overall effect on the completion date; he must also show the effect of every delay between the event and the completion date. Given that JCT 2016 contracts hardly contain any requirements for the contractor to provide a programme, it is difficult to imagine how a contractor can comply with this requirement without using planning software.

19 'Relevant Events' – see JCT SBC/Q 2016 – clause 2.29, JCT DB 2016 – 2.26 and JCT IC 2016 – 2.20. JCT MW 2016 does not refer to any 'Relevant Events', only 'reasons beyond the control of the Contractor' (clause 2.7).

### 11.2.3 Updates: Material Change in Estimated Delay

The contractor is required to notify the employer or his representative of any material change in both of the submitted particulars and the estimate of delay, and give whatever further information the employer or his representative may reasonably require, keeping them updated with developments on site which are relevant to the notified delay. For example, there may be an ongoing delay for which the contractor must provide updates on a regular basis. The contractor's duty is not dependent upon the request of the employer or his representative, and failure to do so will amount to a breach of contract on the part of the contractor (which the employer or his representative may take into account in estimating any extension of time). Under JCT SBC/Q 2016 and JCT DB 2016, the obligation for the employer or his representative to deal with extensions of time only begin when the delay notice and required particulars have been received from the contractor. These provisions are intended to provide sufficient information to allow the employer or his representative to arrive at an informed opinion. Although the employer or his representative may not have been on the site at the time of the delay, there is little doubt that they must use whatever records are available from any source.[8] Therefore, the employer or his representative cannot avoid dealing with the contractor's delay notices because the contractor has failed to provide information which they already have in their possession, or of which they were well aware.

Importantly, it should be noted that it is expressly stated that the employer or his representative may require the further information at any time. The effect of that provision appears to be that the employer or his representative may require further information at any time up to the date on which he must give a definitive decision on extensions of time. The employer or his representative is required to review the extension of time position up to 12 weeks after the date of practical completion (refer to Section 9.1.9). Therefore, he may reasonably require further information up to that time.

## 11.3 Fixing a New Completion Date

### 11.3.1 Introduction

The employer or his representative can give an extension of time to the contractor by 'fixing a later date as the Completion Date' under JCT SBC/Q 2016 and JCT DB 2016 or give such extension of time as may be reasonable under JCT IC 2016 and JCT MW 2016. It is suggested that 'fixing a later date' is the more precise way of expressing an extension of time as the contractor will clearly know the later date by which he is to complete the works. If an extension of time is simply expressed in weeks or days, then it may not be immediately clear when such a period precisely expires.

Table 11.3 summarises the requirements and conditions applicable to the employer or his representative considering and giving an extension of time to the contractor under the JCT 2016 contracts[20]:

---

20 JCT MW 2016 clause 2.7 simply states that the employer's representative shall '*give such extension of time for completion as may be fair and reasonable;*

**Table 11.3** JCT 2016 contracts: Extension of time decisions.

| Employer's (or his representative's) obligations to give an extension of time | JCT SBC/Q 2016 | JCT DB 2016 | JCT IC 2016 |
|---|---|---|---|
| *First fixing of a later completion date:* | 2.28.1 | 2.25.1 | — |
| Obligation to give extension of time ('EoT') is triggered by receipt of the contractor's delay notice and 'particulars' | 2.28.1 | 2.25.1 | — |
| Cause(s) of delay(s) are relevant events | 2.28.1.1 | 2.25.1.1 | 2.19.1 |
| Completion of Works (or section) is likely to be delayed beyond the completion date | 2.28.1.2 | 2.25.1.2 | 2.19.1 |
| *Extension of time:* | | | |
| • Given by fixing a later date as the Completion Date | 2.28 | 2.25 | — |
| • Simply notified to the contractor | — | — | 2.19.1 |
| • To be 'estimated as fair and reasonable' | 2.28.1 | 2.25.1 | 2.19.1 |
| *Employer's, or his representative's decision:* | | | |
| • Must be given in any event (even if no extension of time is due) | 2.28.2 | 2.25.2 | —[a] |
| • Given 'as soon as is reasonably practicable' | 2.28.2 | 2.25.2 | — |
| • In any event, within 12 weeks of receipt of the contractor's 'particulars' | 2.28.2 | 2.25.2 | — |
| • If 'particulars' received less than 12 weeks before the completion date, the EoT decision should be issued before the completion date | 2.28.2 | 2.25.2 | — |
| • Must state the EoT attributable to each relevant event | 2.28.3.1 | 2.25.3.1 | —[b] |
| • Must state the reduction in that time for a relevant omission | 2.28.3.2 | 2.25.3.2 | — |
| *After the first fixing of a later completion date:* | | | |
| • May fix a completion date that is earlier than one previously fixed, accounting for relevant omissions | 2.28.4 | 2.25.4 | — |
| *After the completion date (if occurring before practical completion):* | | | |
| • Not later than 12 weeks after practical completion; | 2.28.5 | 2.25.5 | 2.19.3 |
| • Fix a later completion date that is fair and reasonable having regard to relevant events, of his own volition; or | 2.28.5.1 | 2.25.5.1 | 2.19.2 |
| • Fix an earlier completion date than previously fixed that is fair and reasonable having regard to relevant omissions, of his own volition; or | 2.28.5.2 | 2.25.5.2 | — |
| • Confirm the completion date previously fixed; | 2.28.5.3 | 2.25.5.3 | — |
| *Provided always that:* | | | |
| • Contractor constantly uses his best endeavours to prevent delay; | 2.28.6.1 | 2.25.6.1 | 2.19.4.1 |
| • Contractor to satisfy employer's representative that it is proceedings with the Works or a section; | 2.28.6.2 | 2.25.6.2 | 2.19.4.1 |

**Table 11.3** (Continued)

| Employer's (or his representative's) obligations to give an extension of time | JCT SBC/Q 2016 | JCT DB 2016 | JCT IC 2016 |
|---|---|---|---|
| • Neither the employer nor his representative can fix an earlier date for completion than the 'relevant date for completion'; and | 2.28.6.3 | 2.25.6.3 | — |
| • No EoT decision can alter the length of an EoT from an accepted variation quotation or acceleration quotation, unless by another variation quotation for a relevant omission | 2.28.6.4 | 2.25.6.4 | — |

a) JCT IC 2016 – clause 2.19.1 does not appear to require the employer's representative to notify a decision if, in his opinion, no extension of time is due.
b) JCT IC 2016 – clause 2.19.1 provides that an extension of time can only be given for a Relevant Event, but the employer's representative is not obliged to say which ones when giving an extension of time.

### 11.3.2 Fixing a New Completion Date During the Course of the Work

Under JCT SBC/Q 2016 and JCT DB 2016, the employer and his representative's duties to consider and give extensions of time differ slightly before the Completion Date has passed and after the Completion Date.

The employer's representative's duty to form an opinion about an extension of time does not arise until the contractor has provided both the notice of delay and the particulars (but not the estimate of expected delays in completion). If the particulars are received before the date for completion, the employer's representative must still consider them. He has to decide two important things:

- Whether any of the causes of delay specified by the contractor in the notice is in fact a Relevant Event[19], and
- Whether completion of the Works is in fact likely to be delayed by the specified relevant event beyond the completion date.

Obviously, the employer or his representative may disagree that the specified cause of delay is a Relevant Event. If that is the case, he need go no further. Contractors often specify relevant events by clause number without specifying any occurrence on site in relation to it. Where a contractor fails to identify the correct, or indeed any, Relevant Event, then strictly speaking, it is doubtful that the employer or his representative has any duty to identify it. There is nothing in the contract which suggests that he has such a duty, as it simply seems to check whether the contractor has named the correct Relevant Event and whether the notified cause falls under that Relevant Event.

An employer or employer's representative who concludes that the delay, however long it may be, has no effect on the completion date, must decide that no extension of time is applicable – as it simply cannot be a critical delay (see Section 3.3). That situation can occur when a delay occurs to an activity which is not on the critical path or which has a considerable amount of float to absorb the delay's effects, without the non-critical activity becoming critical. It is very easy for an employer's representative to fall into the trap of thinking that a delay of two weeks equals two weeks extension of time, but that is simply not what the contract calls for – it must be a delay to the completion date (either of the Works or a Section).

Delay to a non-critical activity which may or may not have an effect on the completion date must be differentiated from a critical delay that will affect the completion date. This issue was considered in a case[21] where the architects and project managers were alleged to have been negligent in relation to some of the extensions of time given to the contractor. The case against the project managers failed because the court held that it was no part of the duty of the project managers to second guess the architect's decisions. Moreover, there was no evidence to show that, if the project managers had expressed a contrary view to that of the architects, the architects would have taken any notice. The architects were not generally held to be negligent in the giving of extensions of time except in one concerning a delay which did not affect the date for completion. The court summed up the situation as follows:

> *'… had [the architect] directed its mind, when considering the question whether to grant a second extension of time on the Hydrotite Ground, to the issue whether the progress of the Works, as opposed to the activity "Wall and Floor Finishes", had been further delayed since the grant of the first extension of time on the Hydrotite Ground, it could only have concluded that it had not. It is thus, in my judgment, clear that in relation to the second grant of an extension of time on the Hydrotite Ground [the architect] negligently failed to direct its mind to the correct issue, and, if it had directed its mind to the correct issue, it could only have concluded that no further extension of time was appropriate.'*[24]

It is entirely a matter for the employer's or his representative's opinion, whether a delay in the contract completion date is likely to occur, or has occurred, and also whether the cause of delay falls into the category of a relevant event, and is therefore something for which an extension should be granted. However, the employer or his representative does not have a completely free rein in the matter, and must exercise his opinion in the proper way (see Section 9.1.8). As soon as he is notified by the contractor of a delay, either occurring or predicted to occur, it is for the employer or his representative to carefully monitor the position. The situation has been summarised in this way:

> *'Clause 23 imposes on the architect the duty of considering whether completion of the works is likely to be or has been delayed beyond the date for completion by way of the causes there set out and if it has whether any and if so what extension should be granted. That duty is owed both to the contractor and the building owner. The architect is entitled to rely on the contractor to play his part by giving notice when it has become apparent to him that the progress of the works is delayed. If the contractor fails to give notice forthwith upon it becoming so apparent he is in breach of contract and that breach can be taken into account by the architect in deciding whether he should be given an extension of time. But the architect is not relieved of his duty by the failure of the contractor to give notice or to give notice promptly. He must consider independently in the light of his knowledge of the contractor's programme and the progress of the works and of*

21 *The Royal Brompton Hospital National Health Service Trust v Frederick Alexander Hammond and Others (No. 7)* (2001) 76 Con LR 148 at 214 per Judge Seymour.

*his knowledge of other matters affecting or likely to affect the progress of the works … whether completion is likely to be delayed by any of the stated causes. If necessary he must make his own inquiries, whether from the contractor or others.'*[8]

The extension of time clause under the form of contract there being considered was clause 23 of JCT 63. If the contractor feels that the employer or his representative has been unreasonable in reaching an opinion, his recourse is to adjudication, litigation or arbitration. On receipt of the contractor's written notice the employer's representative must decide if the cause of delay is a relevant event, and if it is not, then, subject to the contractor's right to challenge that opinion by adjudication, litigation or arbitration, that is the end of the matter.

Of course, the employer or his representative must not arrive at the decision on a whim. The position should be carefully analysed and the effect of individual delays must be considered.[22] However, having reached a decision, that decision has a considerable status under the contract, as considered by the court:

> *'Lambeth was in my view entitled to criticise BB's case without putting forward an alternative. Since BB had not justified its case Lambeth was not obliged to justify the architect's extensions of time or certificates of non-completion. It was entitled to rely on them as they were apparently valid decisions by the architect and the parties by adopting the JCT conditions have agreed to be bound by them (subject to review by an Adjudicator or arbitrator). BB had to persuade the Adjudicator that the architect's decisions were wrong. Lambeth were not obliged to prove that they were right (although it is often prudent to do so).'*[23]

The employer or his representative must then decide whether or not the delay is going to mean a likely failure to complete by the date for completion (either for the Works or of a Section). He is also entitled to consider if the contractor has constantly used its best endeavours to prevent delay. The contractor's duty is to prevent delay, so far as he reasonably can, so for example, a delay to the progress at an early stage may be reduced or even eliminated by the contractor using its best endeavours during the works' remaining duration.

An interesting situation arises if the contractor is actually ahead of its own programme when a delay occurs. The correct position appears to be as follows:

> *'Provided the contractor has given written notice of the cause of delay, the obligation to make an extension appears to rest on the architect without the necessity of any formal request for it by the contractor. Yet he is required to do this only if the completion of the works "is likely to be or has been delayed beyond the Date for Completion", or any extended time for completion previously fixed. If a contractor is well ahead with his works and is then delayed by a strike, the architect may nevertheless reach the conclusion that completion of the works is not likely to be delayed beyond the date of completion. Under condition 21 (1), the contractor is under a double obligation: on being given possession of the site, he must "thereupon begin the works and regularly and diligently*

22 *John Barker Construction Ltd v London Portman Hotels Ltd* (1996) 83 BLR 31.
23 Balfour Beatty Construction Ltd v The Mayor and Burgesses of the London Borough of Lambeth [2002] EWHC 597 (TCC)

> *proceed with the same", and he must also complete the works "on or before the Date for Completion", subject to any extension of time. If a strike occurs when two-thirds of the work has been completed in half the contract time, I do not think that on resuming work a few weeks later the contractor is then entitled to slow down the work so as to last out the time until the date for completion (or beyond, if an extension of time is granted) if thereby he is failing to proceed with the work "regularly and diligently".'*[24]

Although these observations were *obiter*, it is thought that this sensible analysis accurately represents the law. It appears that the employer or his representative may take account of where the contractor actually is in terms of progress when compared with its programme and if so, the employer or his representative may account for that that in deciding upon the appropriate extension of time. The passage also reinforces the point that the contractor does not own the float in the programme[25] (see Section 3.6). In a situation where causes of delay overlap or where they are concurrent (see Section 3.7), the employer's representative must consider each cause separately. The cumulative effect on progress must be taken into account: it is delay to progress and the completion date which are the important factors.

An employer's representative who decides that one or more causes are relevant events and that the completion date is likely to be delayed as a result should give an extension of time to the contractor, but not necessarily loss and expense.

The employer's representative is required to grant the extension of time by fixing as a new completion date for the Works a later date which the architect estimates to be fair and reasonable. It should be noted that he is only expected to estimate the length of extension (and not to ascertain) which is a difficult concept to pin down. It has been examined by the court when considering the extension of time clause under JCT 80 as follows:

> *'The architect is not expected to use a coldly logical approach in assessing the relative significance of contractor's risk events and non-contractor's risk events; instead, as the wording of both clause 25.3.1 and clause 25.3.3.1 makes clear, the architect is to fix such new completion date as he considers to be "fair and reasonable". That wording indicates that the architect must look at the various events that have contributed to the delay and determine the relative significance of the contractor's and non-contractor's risk events, using a fairly broad approach. Judgment is involved. It is probably fair to state that the architect exercises discretion, provided that it is recognized that the architect's decision must be based on the evidence that is available and must be reasonable in all the circumstances of the case.'*[26]

---

24 *London Borough of Hounslow v Twickenham Garden Developments Ltd* (1970) 7 BLR 81 at 113 per Megarry J.

25 *How Engineering Services Ltd v Lindner Ceilings Partitions PLC* 17 May 1995 unreported; *Ascon Contracting Ltd v Alfred McAlpine Construction of the Isle of Man Ltd* (2000) 16 Const LJ 316 at 338. Unless the contract contains a 'float sharing' clause, it is understood that the project 'owns' the float, and it is used up by whatever delays come along on a 'first come, first served' basis – irrespective of whether they are caused by the employer, the contractor or by a neutral event.

26 *City Inn Ltd v Shepherd Construction Ltd* (2007) CSOH 190 and Paragraph 13 per Lord Drummond Young.

Once he has made his decision, the employer's representative must notify the contractor of whether or not he's decided that an extension is to be given. That is important, as he cannot simply sit back and say nothing if there is no extension. In respect of each notification of delay and provision of particulars, the employer's representative must notify the contractor in writing where the decision is not to fix a later completion date as a new completion date. It is also important because the employer's representative's decisions are required before the provisions restricting the level of fluctuations or formula adjustment can be operated if the contractor is in default over completion.[27] The same period of time applies in which the employer's representative must make his decision whether it is positive or negative.

The employer or his representative must inform the contractor in writing of the decision as soon as 'reasonably practicable' under JCT SBC/Q 2016 and JCT DB 2016. This term has a narrower meaning that whether something is actually physically possible and may be narrower even than 'practicable' alone. In essence, the obligation is to inform the contractor within a time period which is not delayed, but which takes all the circumstances into account and applies the test of reasonableness. In any event, the employer or his representative has a maximum of 12 weeks (see Section 9.1.9) in which to notify the decision[28] (running from receipt of the required particulars).

However, under JCT SBC/Q 2016 and JCT DB 2016, if there are less than 12 weeks left between receipt of the contractor's particulars and the completion date, the employer or his representative must endeavour reach a decision and fix a new date for completion no later than the current completion date.[32] The intention clearly is that the contractor should always have a date for completion towards which it can work. A more unscrupulous contractor may attempt to intimidate employer or his representative by waiting until the last moment to provide the required particulars and then maintaining that he is obliged to notify the contractor of his decision before the completion date, no matter how little time is left. Such a course of action is inadvisable. Firstly, the JCT SBC/Q 2016 and JCT DB 2016 contracts are clear – the employer or his representative must *endeavour* to notify the contractor of the new date for completion before the current completion date. 'Endeavour' means '*to try hard to achieve* something',[29] so the employer or his representative must strive or attempt to do it. Therefore, even if there is a very short period left before the completion date, it may not be reasonably practicable for him to come to a decision in time no matter how hard he tries. However, after the completion date, the employer or his representative has power to make a decision either as part of the review or of their own volition after the completion date has passed (this latter point, does not rely upon receiving the contractor's prior notice - see Section 11.3.3 below).[30] Secondly, and in the event that the contractor attempted such a 'last minute' tactic to provide a notification and particulars for the first fixing of a later completion, then not only will he have to wait another 12 weeks, but he will also have to rely upon the employer or his representative deciding that it's 'fair and reasonable' to fix a later completion date.

---

27 JCT SBC/Q 2016 and JCT DB 2016 – Schedule 7 – Fluctuations Option A, Paragraph A.9.2.2, Fluctuations Option B – Paragraph B.10.2.2, Fluctuations Option C – Paragraph C.6.2.2 and JCT IC 2016 – Schedule 4 – Fluctuations etc., Paragraph 9.

28 Under JCT IC 2016, clause 2.19.1 - the employer's representative's notification must be given '*as soon as he is able*'.

29 Oxford Dictionary of English – (3rd Ed) 2010.

30 JCT SBC/Q 2016 – clause 2.28.5.1, JCT DB 2016 – clause 2.25.5.1 and JCT IC 2016 – clause 2.19.3.

If it seems that the employer or his representative would be able to make a decision if a further week or so was available, he may decide to make the best decision practicable (which may well be precautionary) before the completion date and then use the additional period thus created to come to a more considered decision. Thus, an employer or his representative faced with making a decision just one week before the completion date may be able to give two weeks extension of time and the extra two weeks may enable him, on further reflection, to give a further one week. However, he is not obliged to act in this way.

Some employers have adopted the practice of amending the contract so as to do away with the 12-week time limits. This may be unwise, particularly if the fluctuations provisions apply. Fluctuations are only to be frozen at completion date if the printed text of the relevant clauses[31] is unamended. In the absence of a fixed period, an adjudicator, court or arbitrator, if called upon, may decide that the architect ought to come to a decision in less than 12 weeks.

If the employer or his representative decides that an extension of time (in fixing a later completion date) is to be given, he must state two things:

- He must state the amount of extension of time given in respect of each relevant event and
- Any reduction in time attributed to each Relevant Omission (see Section 11.1.1).

Previous JCT contracts (such as the 1998 editions) did not require the employer or his representative to state the period of time attributable to each relevant event. That was sensible because to allocate time periods against relevant events is generally not in the employer's interests. The contractor always, of course, demands these details because without them it is difficult to challenge an extension of time decision unless it is obviously incorrect (see Section 9.1.8).

When giving an extension of time, the employer or his representative often sets out the reasoning behind it in considerable detail. Of course, this then allows the contractor to respond, pointing out where he considers the employer or his representative was wrong, and often in a similar level of detail! The exchange can be a useful way of narrowing the differences in dispute, as it reflects the steps advocated by the Pre-Action Protocol for Construction and Engineering Disputes (refer to Section 10.1.2). Unfortunately, relatively few employers or their representatives provide the contractor with meaningful reasons for an giving an extension of time or for rejecting it. This statement should not be taken as a slight on the competence of architects, contract administrators or employer's agents to assess extensions of time. It is intended to (1) emphasise the disparity in supporting information that is often held by the employer's representative, compared to that held by the contractor, and (2) reflect that the contractor is likely to have a dedicated planning resource, whereas an architect, for example, may not. It is sadly commonplace for employer's representatives not to keep their own records of progress, and instead, solely rely upon those provided by the contractor (see Section 10.1.3). Furthermore, if an employer's representative is likely to be the cause of a delay, he may deliberately avoid ascribing an extension of time to the correct relevant event (e.g. saying the delay was caused by a neutral event such as weather) or deny that the contractor is entitled to an extension of time at all. Given that the employer's

31 JCT SBC/Q 2016 and JCT DB 2016 – Schedule 7 – Fluctuations Option A, Paragraph A.9.2.1, Fluctuations Option B – Paragraph B.10.2.1, Fluctuations Option C – Paragraph C.6.2.1 and JCT IC 2016 – Schedule 4 – Fluctuations etc., Paragraph 9.2.1.

representative would be obliged to reveal his reasoning and calculations during an adjudication, litigation or arbitration if the decision had to be defended, there is no good practical reason why both parties' time an expense cannot be save by the employer or his representative explaining the decision. However, it must be borne in mind that the JCT 2016 contract do not require such a detailed explanation to be provided.

It seems that the employer or his representative must list all the relevant events notified by the contractor even if he has discounted some of them. The employer or his representative then has to allocate the extension of time (or its constituent parts) to the notified relevant events, allocating 'nil' to relevant events which have been notified but for which the architect has not given any extension of time. Any reductions to the allocated time must can only be for a relevant omission. Note the allocation of reductions is not to relevant events, but to relevant omissions.

Under JCT SBC/Q 2016 and JCT DB 2016, if instructions to omit works or obligations (amounting to 'relevant omissions' – see Section 11.1.1) are issued after the first extension of time (by fixing a later completion date), the employer or his representative may reduce the first, or a previous extension of time to reflect the reduced scope of work that the contractor must now carry out. After the first (or a previous) extension of time (or after a revision to the completion date stated in a confirmed acceptance of a schedule 2 variation or acceleration quotation – see Section 5.4.2), the employer or his representative may notify the contractor that he's fixing an earlier Completion Date, owing to a 'Relevant Omission'. His notification must provide the same level of detail as he provides in his decision to grant or not to grant and extension of time.[32]

It is important to understand that each extension is deemed to take account of all omissions instructed from the date of the previous extension up to the date of the new extension. The employer or his representative can reduce extensions previously granted, and even extinguish them completely so as to return to the original date for completion, but no matter how much work or how many obligations are omitted, he cannot fix any earlier date than the original completion date.[33] Therefore, if he has issued instructions which result in a relevant omission (i.e. of work or obligations), then this may be taken into account and a new completion date (earlier than that previously fixed) may then be fixed, provided that it is fair and reasonable to do so having regard to the relevant omission. It should be noted that the employer's or his representative's ability to reduce a previous extension of time for a relevant omission is not dependent upon the contractor giving a notice of delay notice of delay. He can simply issue an instruction requiring a relevant omission and then fix a new date for completion earlier than previously fixed, taking the relevant omission into account.

It is suggested that employers or their representatives desiring to exercise the power to reduce extensions previously granted by taking relevant omissions into account should fix the new, earlier date for completion and notify the contractor as soon as possible. This will allow the contractor more time to organise his resources (e.g. cancel deliveries or trades that would otherwise be carrying out the omitted work) to meet the new completion date (earlier than that previously fixed). However, experience suggests that employers and their representatives may not appreciate the urgency for giving extensions of time because they have up to 12 weeks after practical completion in which to review any extensions of time.

32 JCT SBC/Q 2016 – clause 2.28.3 and JCT DB 2016 – clause 2.25.3.
33 JCT SBC/Q 2016 – clause 2.28.6.3, and JCT DB 2016 – clause 2.25.6.3.

Although that approach is understandable, especially if the employer may be litigious, it is not a good practice. The ideal time is when instructing the relevant omission instruction and not to leave it until the next extension of time.

### 11.3.3 Fixing a New Completion Date After the Completion Date Has Passed

This section describes the procedure for the employer or his representative to assess extensions of time after the completion date which is quite separate from what has gone before. For clarity, 'completion date' in this context means the latest completion date by which the contractor is obliged to complete the Works or a Section. It may be the completion date as set out in the contract particulars (if no later completion date has been fixed, or extension of time given), or it may be the last later date fixed as the date for completion by the employer or his representative. It gives employer, or his representative the opportunity to consider extensions of time of his own volition (without the contractor having to provide a notice or particulars) and make a final decision in this regard. Some commentators believe that the requirement to do so within 12 weeks after practical completion is not mandatory. A more precise view is that the 12-week period is '*directory only*' *as to time and is not something which would invalidate the calculation and payment of liquidated damages*'[34] (refer to Section 9.1.9). This may pose the question 'what if the contractor provides no information at all to the architect until after the 12 weeks has expired?'

The employer or his representative is encouraged to review the need to adjust the completion date. This review may be carried out after the completion date has passed, and ideally, before practical completion has occurred, or within 12 weeks thereafter[30]. In doing this, the employer or his representative must take account of any known relevant events, whether or not specifically notified by the contractor (either on time or late). It is clear that the employer or his representative must take account of any relevant events which have occurred since the commencement of the contract and is his final opportunity to consider extensions of time and possibly (although unlikely) prevent time becoming at large. It has been suggested that the employer or his representative may exercise this power to review only once. However, a more common-sense view would be that he may exercise it a number of times before either (1) the expiry of the 12 -week period after the practical completion, or (2) prior to the final certificate[35] or relevant (final) statement[36] becoming conclusive evidence in respect of extensions of time. In practice, an employer or his representative will often wait until after practical completion has occurred to carry out his review. When he notifies the contractor, the JCT SBC/Q 2016 and JCT DB 2016 contracts are clear that the employer, or his representative must include the same details required when notifying the contractor for fixing a later date as the completion date[37] as would be contained in the equivalent notification before practical completion (i.e. the allocation of extensions of time to each relevant event and any reduction in time to each relevant omission). In carrying out the review, the employer or his representative must do one of three things, either:

34 *Temloc Ltd v Errill Properties Ltd* (1987) 39 BLR 30 at page 39.
35 JCT SBC/Q 2016 and JCT IC 2016 – clause 1.9.1.3.
36 JCT DB 2016 – clause 1.8.1.2.
37 JCT SBC/Q 2016 – clause 2.28.3 and JCT DB 2016 – clause 2.25.3.

- Fix a completion date later than the one previously fixed. It is clear that the employer or his representative must fix the later date whether that is as a result of reviewing a previous decision or 'otherwise' (i.e. where there was no previous decision to fix a different date as the completion date). Thus, the employer or his representative must decide if it is fair and reasonable to fix a completion date later than previously fixed. One wonders if this interpretation is really what was intended, as the employer or his representative has a very wide scope to fix a new completion date.
- Again, if it is fair and reasonable to so, fix a completion date earlier than that previously fixed to account for any relevant omissions instructed after the previous completion date was fixed; or
- Confirm to the contractor the completion date previously fixed.

In practice, the employer or his representative should write to the contractor soon after practical completion, reminding it of the 12-week period, and that the employer may consider that his representative may have no power to make any extension of time after the expiry of the period. Ideally, the contractor should be invited to make any further submission within the 12 week period.

### 11.3.4 Other Conditions for Extensions of Time

The JCT 2016 contracts contain a number of important conditions regarding extensions of time, that apply to both the contractor and the employer (or his representative). The relevant clause's introductory wording[38] ('*Provided always that…*') suggests that compliance with the first two[39] provisos (the contractor using his best endeavours and doing everything reasonably required to satisfy the employer or his representative) is a condition precedent to the valid award of extensions of time. However, in practice, it is often relatively difficult to prove that those conditions have not been satisfied.

The first obligation[40] placed on the contractor is a substantial one. He must use his 'best endeavours' to prevent delay to progress and to the completion date. The Court of Appeal held, (in a case concerning the obtaining of planning permission), that 'best endeavours' obliged a person to take *'all those reasonable steps which a prudent and determined man, acting in his own best interests and desiring to achieve that result would take'*.[41] Seemingly, the 'best endeavours' obligation applies to preventing 'likely' delays,[42] as well as preventing an existing delay from becoming worse. Employers or their representatives should not assume that 'best endeavours' means that the contractor should take every imaginable step, irrespective of its cost.[43] However, it seems that the contractor should neither remain passive, nor make a minimal or token effort. He should take prudent steps such as advising the employer of delays, along with rescheduling the works and the programme if appropriate. However, it seems that the extent of using 'best endeavours' is measured by reference

38 JCT SBC/Q 2016 – clause 2.28.6, JCT DB 2016 – clause 2.25.6, and JCT IC 2016 – clause 2.19.4.
39 The first two obligations (for the contractor) apply to JCT SBC/Q 2016, JCT DB 2016 and JCT IC 2016, whereas the third and fourth obligations (for the employer or his representative) do not apply to JCT IC 2016.
40 JCT SBC/Q 2016 – clause 2.28.6.1, JCT DB – clause 2.25.6.1 and JCT IC 2016–2.19.4.1.
41 *IBM (UK) Ltd v Rockware Glass Ltd* (1980) FSR 335.
42 When read alongside JCT SBC/Q 2016 – clause 2.27.1, JCT DB 2016 clause 2.24.1 and JCT IC 2.19.1.
43 Keating on Construction Contracts, 10th Edition at 20-152.

to someone acting in their own interests[44] and not fulfilling a contractual obligation. In other words, the contractor's best endeavours are focussed on helping himself avoid delays. The scope of the contract's variation clause provides a useful guide to whether the contractor has used constant 'best endeavours' as he will be entitled to additional payment if the 'endeavours' fall within the scope of the variation clause.[45]

It therefore appears to be a lesser obligation than to 'ensure' or to 'secure', which impart an absolute liability to perform the duty set out[46], but a higher obligation than using 'reasonable endeavours'. Where there may be many courses of action that it would be reasonable for the contractor to take, an obligation to use reasonable endeavours requires it to take any one of the reasonable courses. In a practical sense, it seems likely that 'best endeavours' simply means that the contractor must continue to work regularly and diligently and has not contributed to the delay through its fault. The addition of the word 'constantly' clearly provides that the contractor's use of its best endeavours should not be employed selectively, or from time to time. If the employer (or his representative) can clearly demonstrate that a contractor's failure to constantly use its best endeavours, it is likely to disqualify him from any extension of time for the particular relevant event. However, proving that a contractor has failed or is failing to use its best endeavours to prevent delay may not be straightforward. The circumstances in which the progress has been delayed may be contentious and unclear. A contractor may easily be able to employ the defence that his 'best endeavours' may not be immediately apparent, as its effects have been obscured by delays and interruption to progress, caused by the employer (or his representative).

The second obligation[47] imposed on the contractor also applies at all times, if there is a delay. In this case, the contractor must do everything reasonably required to satisfy the employer or his representative so he can proceed with the Works (or a section). This probably forms part of the contractor's obligation to proceed regularly and diligently, but adds the subjective element of satisfying the employer or his representative. The key word in this second obligation is 'reasonably'. It should be read alongside the contractor's first obligation (to use 'best endeavours'[44]). It seems clear that it would be unreasonable for the employer or his representative to require the contractor to take disproportionate steps, or expend large additional sums in order to obtain his satisfaction. However, it does seem to cover such measures as the contractor rescheduling its programme of work, moving operatives from one part of the Works to another and notifying any delays to progress.

The third obligations[48] is placed upon the employer or his representative, and make clear that he cannot fix a completion date earlier than that stated in the contract particulars (see Section 5.4.2). This obligation is only found in JCT SBC/Q 2016 and JCT DB 2016 and is not included in JCT IC 2016. This is probably because JCT IC 2016 does not permit the contractor's entitlement to extensions of time to be reduced by omitting work from the overall scope.[49] What that means for JCT SBC/Q 2016 and JCT DB 2016 is that no matter how much work is omitted, the contract period as set out by reference to the date of possession

44 Noble A, "Contractors' and Sub-Contractors' Constant Best Endeavours to Prevent Delay," (1997) 13 Const LJ 293.

45 JCT SBC/Q 2016, JCT DB 2016 and JCT IC 2016 - clause 5.1.2.

46 *John Mowlem & Co v Eagle Star Insurance Co Ltd* (1995) 62 BLR 126.

47 JCT SBC/Q 2016 – 2.28.6.2, JCT DB 2016 – clause 2.25.6.2, and JCT IC 2016 – clause 2.19.4.1.

48 JCT SBC/Q 2016 – clause 2.28.6.3, and JCT DB 2016 – clause 2.25.6.3.

49 Refer to 'Relevant Omissions' in JCT SBC/Q 2016 – clauses 2.26.3, 2.28.3.2, 2.28.4 and 2.28.5.2, and JCT DB 2016 – clauses 2.23.3, 2.25.3.2, and 2.25.5.2.

and the date for completion in the contract cannot be shortened. It is suggested that as JCT IC 2016 only considers positive 'extensions of time' (i.e. increases to the period for completion), and whilst the employer's representative can instruct omissions,[50] he has no express contractual authority to account for omissions when assessing an extension of time.

The fourth obligation[51] also applies to the employer or his representative, and is closely related to the third. Again, it refers to the reduction of extensions of time being adjusted by a 'Relevant Omission' so it only applies to JCT SBC/Q 2016 and JCT DB 2016. In this case, it prevents the employer or his representative from altering (the implication is by reducing) the length of an extension of time that has been agreed by either:

- A 'pre-agreed adjustment'[52]; or
- An acceleration quotation[53];

unless the variation, change or acceleration quotation was itself, the subject of a relevant omission.[53]

## 11.4 Relevant Events – Grounds for Extensions of Time

The grounds for extensions of time in all of the JCT 2016 contracts (save for JCT MW 2016) are called 'Relevant Events'. Owing to its simpler nature, JCT MW 2016 only provides a broad ground of '*reasons beyond the control of the Contractor*'[54] (see Section 11.1). Table 11.4 summarises the Relevant Events in the JCT 2016 contracts:

Table 11.4 summarises the grounds on which the employer or his representative is entitled to revise the Completion Date (the 'Relevant Events'). They may be divided into the two categories shown in Table 11.5:

The grounds for extending time are as follows:

### 11.4.1 Variations or Changes

This ground includes all matters, (including instructions from the employer's representative, or employer under JCT DB 2016) which are to be treated as requiring a Variation (or 'Change' under JCT DB 2016), whether or not so intended. For a detailed exploration of these grounds, refer to Section 5.2 and Table 5.3.

### 11.4.2 Instructions

Instructions given rise to an extension of time must be in writing and all of the JCT 2016 contracts provide for their issue – refer to Sections 5.2.3, 5.2.3.1 and 5.2.3.2 and Tables 5.4, 5.5, and 5.7.

50 JCT IC 2016 – omissions are a Variation under clause 5.1.1.1.
51 JCT SBC/Q 2016 – clause 2.28.6.4 and JCT DB 2016 – clause 2.25.6.4.
52 'Pre-Agreed Adjustment': for JCT SBC/Q 2016 – via the 'Confirmed Acceptance of a Variation Quotation' – refer to clause 5.3 and Paragraphs 1.2.2 (Variation Quotation) and 4 (Acceptance of the quotation) of Schedule 2 (Variation Quotations), and for JCT DB 2016 – 'Value of Changes – Contractor's Estimates' – Paragraph 2.3.4 of Part 1 of Schedule 2 (Supplemental Provisions).
53 'Acceleration Quotation' – for JCT SBC/Q 2016 – refer to Paragraph 2 (Acceleration Quotation) of Schedule 2, and for JCT DB 2016 – refer to Paragraph 4 (Acceleration Quotation) of Part 2 of Schedule 2.
54 JCT MW 2016 – clause 2.7 (Extension of time).

**Table 11.4** JCT 2016 contracts 'Relevant Events'.

| Relevant Event | JCT SBC/Q 2016 | JCT DB 2016 | JCT IC 2016 |
|---|---|---|---|
| 'Variations' or 'Changes' | 2.29.1 | 2.26.1 | 2.20.1 |
| Instructions for: | | | |
| • Resolving a discrepancy[a)] | 2.29.2.1 | 2.26.2.1 | 2.20.2.1 |
| • Postponement of work[b)] | 2.29.2.1 | 2.26.2.2 | 2.20.2.2 |
| • Provisional Sums[c)] | 2.29.2.1 | 2.26.2.2 | 2.20.2.2 |
| • Opening up/inspection/testing[d)] | 2.29.2.2 | 2.26.2.3 | 2.20.2.3 |
| • Objecting to a 'Named specialist'[e)] | 2.29.2.2 | — | — |
| • Dealing with a 'Named Sub-Contractor' | — | — | 2.20.1, 2.20.2.2[f)] |
| Deferment of possession | 2.29.3 | 2.26.3 | 2.20.3 |
| Dealing with 'Antiquities'[g)] | 2.29.4 | 2.26.4 | — |
| Inaccurate Approximate Quantity | 2.29.5 | — | 2.20.4 |
| Contractor suspension for non-payment[h)] | 2.29.6 | 2.26.5 | 2.20.5 |
| Employer's impediment, prevention or default | 2.29.7 | 2.26.6 | 2.20.6 |
| Work by or failure of statutory undertaker | 2.29.8 | 2.26.7 | 2.20.7 |
| Exceptionally adverse weather | 2.29.9 | 2.26.8 | 2.20.8 |
| Loss or damage by Specified Perils[i)] | 2.29.10 | 2.26.9 | 2.20.9 |
| Civil commotion or terrorism | 2.29.11 | 2.26.10 | 2.20.10 |
| Strike, lock out etc | 2.29.12 | 2.26.11 | 2.20.11 |
| New legislation after the Base Date | 2.29.13 | 2.26.12 | 2.20.12 |
| Delay in receiving necessary permission | — | 2.26.13 | — |
| Insolvency of 'Named Specialist' | 2.19.14 | — | — |
| Force majeure | 2.29.15 | 2.26.14 | 2.20.13 |

a) JCT SBC/Q 2016 – clause 2.15 (Notice of discrepancies etc), JCT DB 2016 – clause 2.13 and JCT IC 2016 – clause 2.13 (instructions on errors, omissions and inconsistencies).
b) JCT SBC/Q 2016 – clause 3.15 (Postponement of work), JCT DB 2016 – clause 3.10, and JCT IC 2016 – clause 3.12.
c) JCT SBC/Q 2016 – clause 3.16 (Instructions on Provisional Sums), JCT DB 2016 – clause 3.11 and JCT IC 2016 – clause 3.13.
d) JCT SBC/Q 2016 – under clause 3.17 (Inspections, tests) or 3.18.4 (unless work not in accordance with the contract), JCT DB 2016 – clause 3.12 and 3.13.3, and JCT IC 2016 – clauses 3.14 and 3.15.1.
e) JCT SBC/Q 2016 – Paragraph 9 (Named Specialists) of Schedule 8 (Supplemental Provisions).
f) JCT IC 2016 – includes provisions for 'Named Sub-Contractors' (see clause 3.7 and Schedule 2). The provisions for extensions of time associated with Named Sub-Contractors are not included as a specific 'Relevant Event', but nonetheless apply via the Relevant Event for the instructions of 'Variations' (clause 2.20.1) and for the expenditure of 'Provisional Sums' (clause 2.20.2.2) – see Section 5.2.3.2.
g) JCT SBC/Q 2016 – clause 3.22.1 or instructions under clauses 3.22.2, JCT DB 2016 – clauses 3.15.1 and 3.15.2,
h) JCT SBC/Q 2016 – clause 4.13, JCT DB 2016 – clause 4.11, and JCT IC 2016 – clause 4.14.
i) 'Specified Perils' – JCT SBC/Q 2016, JCT DB 2016 and JCT IC 2016 – clause 6.8.

**Table 11.5** Relevant Events for which (1) the employer is responsible and (2) nobody is responsible.

| Category | JCT SBC/Q 2016 | JCT DB 2016 | JCT IC 2016 |
|---|---|---|---|
| Relevant Events for which the employer or his representative are responsible | 2.29.1, 2.29.2.1, 2.29.2.2, 2.29.3, 2.29.5, 2.29.6, 2.29.7, 2.19.14[a)] | 2.26.1, 2.26.2.1, 2.26.2.3, 2.26.3, 2.26.5, 2.26.6, | 2.20.1, 2.20.2.1, 2.20.2.2, 2.20.2.3, 2.20.3, 2.20.4, 2.20.5, 2,20.6 |
| Relevant Events for which neither the employer, his representative or the contractor are responsible (often called 'neutral events') | 2.29.4, 2.29.8, 2.28.9, 2.29.10, 2.29.11, 2.29.12, 2.29.13, 2.29.15 | 2.26.4, 2.26.7, 2.26.8, 2.26.9, 2.26.10, 2.26.11, 2.26.12, 2.26.13, 2.26.14 | 2.20.7, 2.20.8, 2.20.9, 2.20.10, 2.20.11, 2.20.12, 2.20.13 |

a) JCT SBC/Q 2016 – clause 2.19.14. Whilst the employer may argue that insolvency of a 'Named Specialist' is not his responsibility, the contractor will argue that he has no control over the Named Specialist and the employer should have carried out a 'due diligence' exercise regarding the Named Specialist's financial position.

### 11.4.3 Deferment of Possession

Failure by the employer to give possession of the site by the date of possession remains fairly common. There may be a number of reasons for this, such as he may be waiting for permissions to allow the contractor to commence work, or he may be waiting for funding. The relevant clauses[55] dealing with possession of the site were amended accordingly, to enable the employer to defer giving the contractor possession of the site for a period of up to six weeks (default), unless a shorter period was stipulated in the contract particulars (refer to Section 3.1.2).

Possession refers to the whole of the site and, in the absence of sectional possession, the employer is not entitled to give possession in 'parcels'.[56] It is sometimes thought that the employer can get away with giving the contractor what is referred to as 'sufficient possession', in other words, enough possession of the site to enable the contractor to start work and to proceed for a period until possession of more of the site becomes necessary. However, this more relaxed view must be treated with caution. The contractor's right to possession is an express term of the contract[57] and in any event there is an implied term in any construction contract that the employer will give possession of the site to the contractor in time to enable it to carry out and complete the work by the contractual date for completion.[58]

In a 1970 case,[59] Mr. Justice Megarry said '*The contract necessarily requires the building owner to give the contractor such possession, occupation or use as is necessary to enable him to perform the contract*'. Even if the 'sufficient possession' argument has any validity, sufficient possession will usually mean possession of the whole of the site (or section, if applicable).

55 JCT SBC/Q 2016 and JCT IC 2016 – clause 2.5, DB 2016 – clause 2.4, the employer may defer for a maximum period of 6 weeks.

56 *Whittal Builders v Chester-Le-Street District Council* (1987) 40 BLR 82.

57 JCT SBC/Q 2016 and JCT IC 2016 – clause 2.4, JCT DB 2016 – clause 2.3, JCT MW 2016 – clause 2.2.

58 *Freeman & Son v Hensler* (1900) 64 JP 260.

59 *London Borough of Hounslow v Twickenham Garden Developments Ltd* (1970) 7 BLR 81.

That is because the contractor may wish to do very many things on the site other than simply constructing the building. It may be necessary for him to investigate or check the condition of other parts of the site, or decide the best place for his site accommodation, along with appropriate areas for storing plant and materials. Unless full possession is given, the contractor is deprived of the ability to consider the site as a whole.

Accordingly, subject to the employer's right to defer possession, if activated, any failure by the employer to give possession on the due date is a breach of contract, entitling the contractor to an extension of time, loss and expense (refer to Section 12.4.1) or alternatively, a claim for damages at common law in respect of any loss that it suffers as a consequence.[63] In an Australian case,[60] it was held that, where there was failure to give possession of the building site to a contractor, this constituted a breach of contract and on the facts the contractor was entitled to treat the contract as repudiated. Instructions to defer possession have also been held to arise as a matter of fact.[61]

It is important to emphasise that the employer only has this power if it is expressly so stated in the contract particulars. The default deferment period is stated to be 6 weeks or whatever shorter period is agreed by the parties in the contract particulars. In view of the often unpredictable nature of such matters as demolition, site clearance works and occupying tenants, to say nothing of unlawful squatters, it seems risky to reduce this period. Indeed, in some instances, employers would be wise to increase the six-week period.

It is considered that deferment is a positive activity which the employer should signal by giving written notice although none of the JCT 2016 contracts required it. It should be noted that, on a strict reading of the relevant clauses, an extension of time can only be given where the employer has actually exercised the right to defer. This relevant event probably does not apply if the employer, without formally deferring possession, has simply failed to give possession on the due date – as the contractor's entitlement to an extension of time in such circumstances would be under the ground of the employer's impediment, prevention, etc. (refer to Section 11.4.7).

### 11.4.4 Antiquities

Only JCT SBC/Q 2016 and JCT DB 2016 contain an entitlement for the contractor to receive an extension of time in the event that '*fossils, antiquities and other objects of interest or value are found on the site*'.[62] Upon discovering any of the aforementioned, the contractor must immediately carry out the following three 'preservation steps':

- Use his best endeavours not to disturb the discovery, and even stop work if it's likely to endanger it or hinder its excavation[63];
- Preserve the discovery's exact position and condition[64]; and

60 *Carr v Berriman Pty Ltd* (1953) 27 ALJR 273.
61 See *M Harrison & Co (Leeds) Ltd v Leeds City Council* (1980) 14 BLR 118, where an instruction expressed as a variation order was held to be in fact an order for postponement and *Holland Hannen & Cubitts (Northern) v Welsh Health Technical Services Organisation* (1981) 18 BLR 80, where a notice which was apparently intended to notify the contractor of defective work was held to instruct postponement.
62 JCT SBC/Q 2016 – clause 3.22.1 and JCT DB 2016 – clause 3.15.1 'Antiquities' as a Relevant Event was introduced with the JCT 2011 SBC/Q and DB contracts.
63 JCT SBC/Q 2016 – clause 3.22.1.1, and JCT DB 2016 – clause 3.15.1.1.
64 JCT SBC/Q 2016 – clause 3.22.1.2 and JCT DB 2016 – clause 3.15.1.2.

- Inform the employer or his representative[65] of its discovery and exact location.[66]

The employer or his representative shall instruct what action the contractor should take, including allowing a third party (such as an archaeologist) to examine or remove the discovery.[67] Seemingly, this instruction is not dependent upon the contractor's prior notification under the relevant clause; however, it is important for the contractor to notify because his entitlement to an extension of time can include the time spent complying with:

- The three 'preservation steps' set out above; and
- The instruction issued by the employer or his representative.

### 11.4.5 Inaccurate Approximate Quantities

This ground proceeds on the perfectly reasonable bases that a contractor will plan to carry out the works using, among other things:

- The quantities in the contract bills; and
- That those quantities will be accurate, and in the case of 'approximate quantities', they will be a reasonably accurate forecast of the quantity of work so described.

Where such quantities are described as 'approximate', it is because the employer's representative and/or the quantity surveyor either does not know, or is yet to decide, the amount required. All the contractor can do is to use the approximate quantities as if they gave a reasonably accurate forecast of the quantities required. However, if the Approximate Quantity (or 'forecast') is significantly lower than the actual quantity required, the contractor is likely to need additional time to carry out the work.

The question that inevitably arises in these circumstances is what is a reasonable forecast? The answer is likely to be one of common sense.

It will be an unusual situation if the contractor is entitled to an extension of time because the actual quantities are less than the forecast. However, the wording of the clause leaves that possibility open.

### 11.4.6 Contractor's Suspension for Nonpayment

This ground is included to comply with the Housing Grants, Construction and Regeneration Act 1996 ('HGCRA 1996')[68] which entitles a contractor to suspend performance of its obligations if the employer does not pay a sum due in full by the final date for payment. The contractor must give the employer 7 days' prior notice[69] before suspending. The relevant section of the HGCRA 1996 is s112(4), which states:

> *'Any period during which performance is suspended in pursuance, or in consequence of the exercise of the right conferred by this section shall be disregarded in computing*

65 Including an appointed clerk of works (see JCT SBC/Q 2016 – clause 3.4).
66 JCT SBC/Q 2016 – clause 3.22.1.3 and JCT DB 2016 – clause 3.15.1.3.
67 JCT SBC/Q 2016 – clause 3.22.2 and JCT DB 2015 – clause 3.15.2.
68 The Housing Grants Construction and Regeneration Act 1996 (as amended).
69 S112(2) HGCRA 1996.

*for the purposes of any contractual time limit the time taken, by the party exercising the right or by a third party, to complete any work directly or indirectly affected by the exercise of the right. Where the contractual time limit is set by reference to a date rather than a period, the date shall be adjusted accordingly.'*

JCT 2016 contracts all provide an the right[70] for the contractor to suspend performance, which echoes that right provided by s112(4) of the HGCRA 1996.

However, JCT contracts seem more generous than the contractor's ordinary right to suspend performance of his obligations under the HGCRA 1996, as s112(4) provides that if a party suspends performance for a period, that period will be disregarded when calculating any contractual time limit – effectively giving an extension of time for the period of suspension. However, this does not account for any time that the contractor may need to get ready to recommence – which may be the case after a long period of suspension. The wording of the Relevant Event clauses[71] (particularly when read along the relevant clauses entitling the contractor to compensation[72]) imply that the employer or his representative should consider all the delay (including costs and expenses reasonably incurred[76]) including remobilisation and not just for the period of suspension. In any event, any further delay occasioned by the contractor exercising his right of suspension is likely to fall under the Relevant Event for an act of impediment, prevention and/or default by the employer (see Section 11.4.7).

### 11.4.7 Impediment, Prevention or Default by the Employer

This relevant event was added to the 1998 editions of the JCT contracts in 2002 and was intended as a 'catch all clause' to reduce the possibility of time becoming at large (see Section 3.1.5) due to an act of prevention or the like on the part of the employer. However, it excludes any parts that were caused or contributed to by the default of the contractor, its servants, agents or sub-contractors. The key words in this Relevant Event seem strange bedfellows:

- 'Impede' means to 'delay' or 'prevent by obstruction'[73];
- 'Prevent' means to 'keep something from happening'[77] or 'to stop someone from doing something'[77]; and
- 'Default' means 'failure to fulfil and obligation'.[77] It certainly covers a breach of contract, but it may go further than that in some circumstances.[74]

In its 2005, 2011 and 2016 editions, the JCT has reduced the number of Relevant Events because many are now covered by this 'catch all' clause.[75] However, the specific

70 JCT SBC/Q 2016 – clause 4.13, JCT DB 2016 – clause 4.11, JCT IC 2016 – clause 4.14 and JCT MW 2016 – clause 4.7.

71 JCT SBC/Q 2016 – clause 2.29.6, JCT DB 2016–2.26.5 and JCT IC 2016 – clause 2.20.5. An extension of time for suspension under JCT MW 2016 (under clause 4.7) would be equivalent to that provided under s112(4) of the HGCRA 1996, so the contractor would have to claim any time re-mobilise under clause 2.7 (extensions of time).

72 JCT SBC/Q 2016 – clause 4.13.2, JCT DB 2016 – clause 4.11.2 and JCT IC 2016 – clause 4.14.2.

73 Oxford Dictionary of English, (3rd Ed) 2010.

74 See Chappell, Cowlin and Dunn, *Building Law Encyclopaedia*, (2009) Wiley-Blackwell p. 146 for a fuller consideration of the term.

75 JCT SBC/Q 2016 – clause 2.29.7, JCT DB 2016 – clause 2.26.7 and JCT IC 2016 – clause 2.20.6.

circumstances giving rise to the pre-2005 relevant events are still common. It will be useful to list the former relevant events so covered, since they are likely to be the most common reasons why a contractor will cite this relevant event. They are as follows:

#### 11.4.7.1 Late Provision of Information

There are essentially two parts to this former relevant event – depending upon which JCT 2016 contract is used. Firstly, under JCT SBC/Q 2016 and JCT IC 2016, if an Information Release Schedule[76] is used, the employer's representative's failure to comply with it would be a 'default'. This means that if he does not provide the information (which also includes 'consents'[77]) as set out in the schedule, the contractor has a ground for an extension of time provided other criteria are met. That is very straightforward, easy to understand and to operate.

The second part of the former relevant event (which may also apply under JCT DB 2016) refers to the employer's representative's failure to provide or further information (including 'consents')[81] that is necessary to '*explain and amplify*' the design information[78] provided under the contract. The employer's representative must provide this information to allow the contractor to carry out the works in accordance with the contract – which obviously includes him completing by the Completion Date. Assessing delays under this second part is more difficult than considering the extent to which the employer's representative has failed to comply with an Information Release Schedule. This is because there are no dates to act as benchmarks against which information should have been provided. It is rather a matter of judgment by the employer, or his representative, who must consider the appropriate time to provide the necessary information to either:

- Not delay the contractor's progress of the Works or a Section beyond its Completion Date; or
- Not to create any further delays (e.g. if the Completion Date has passed but practical completion is yet to occur).

If the contractor's rate of progress is such that it will not finish by the due date, the employer or his representative may have 'regard' to this fact. Seemingly, this means that the employer or his representative is entitled to postpone, or slow the rate of providing information to match the contractor's progress. That is not a practice to be advocated, not least, because months later it may be unclear whether the contractor's slow progress provoked the slower delivery of information, or if it was a result of it.

If the contractor looks likely to finish before the completion date, the architect is not obliged to furnish information to allow this to happen.[79]

76 'Information Release Schedule' – see JCT SBC/Q 2016 see the Fifth Recital and clauses 2.11 and 2.12, and JCT IC 2016 see the Sixth Recital and clauses 2.10 and 2.11.

77 For 'consents' see JCT SBC/Q 2016 & JCT IC 2016 – clause 1.11.1, and JCT DB 2016 – clause 1.10. The employer's failure to provide consents was considered an 'impediment, prevention or default' by the Court in *Clin v Walter Lilly & Co Ltd* (2018) EWCA Civ 490.

78 Such as the 'Contract Drawings' under JCT SBC/Q 2016, and JCT IC 2016, and although not expressly mentioned, or any further information required to explain or amplify a 'Change' in the Employer's Requirements under JCT DB 2016.

79 This echoes the judgment in *Glenlion Construction Ltd v The Guinness Trust* (1987) 39 BLR 89.

The question of the time within which to provide any necessary further information has been considered by the courts[80] (although not in relation to a JCT contract). It has been stated that such information must be given within a reasonable time, but it has been made clear that this is a limited duty. Although the case was concerned with the duty of an engineer, it is thought that the principle applies equally to other employer's representatives (such as an architect, contract administrator or project manager):

> *'What is a reasonable time does not depend solely upon the convenience and financial interests of the [contractors]. No doubt it is in their interest to have every detail cut and dried on the day the contract is signed, but the contract does not contemplate that. It contemplates further drawings and details being provided, and the engineer is to have a time to provide them which is reasonable having regard to the point of view of him and his staff and the point of view of the [employer] as well as the point of view of the contractors.'*[80]

This is a sensible approach to the matter. On the one hand, the employer's representative must take into account the time necessary to enable the contractor to organise his resources (e.g. including long lead-in items) to have everything ready on site when needed to complete the Works in accordance with the contract. On the other hand, the contractor cannot expect to ask for information to be immediately issued upon request. He must allow the employer's designer team sufficient time to produce the information.

The current date for completion must always be borne in mind when analysing the particular circumstances. The court's view[81] of the provision of information clause in JCT 63 was:

> *'What the parties contemplated by these provisions was first that the architect was not to be required to furnish instructions, drawings, etc., unreasonably far in advance from the date when the contractor would require them in order to carry out the work efficiently nor to be asked for them at a time which did not give him a reasonable opportunity to meet the request. It is true that the words "on a date" grammatically govern the date on which the application is made. But they are … capable of being read as referring to the date on which the application is to be met. That construction seems to me to give effect to the purpose of the provision – merely to ensure that the architect is not troubled with applications too far in advance of the time when they will be actually needed by the contractor … and to ensure that he is not left with insufficient time to prepare them. If that is right then there seems … to be no reason why an application should not be made at the commencement of the work for all the instructions etc which the contractor can foresee will be required in the course of the works provided the date specified for delivery of each set of instructions meets these two requirements. Of course if he does so and the works do not progress strictly in accordance with this plan some modification may be required to the prescribed timetable and the subsequent furnishing of instructions and the like … It does not follow that the programme was a sufficiently specified application made at an appropriate time in relation to every item*

80 *Neodox v. Borough of Swinton & Pendlebury* (1958) 5 BLR 34 at 42 per Diplock J.
81 *London Borough of Merton v Stanley Hugh Leach Ltd* (1985) 32 BLR 51, which included a more substantial clause calling for a 'specific written application' to be made.

*of information required, more particularly in light of the delays and the rearrangement of the programme for the work.'*

This is another sensible analysis and, although it refers to JCT 63 and the wording of the JCT 2016 contracts are now somewhat different, it gives a useful indication to the way the courts are likely to deal with this kind of question. If there were not express terms of the contract requiring timely provision of information, there would almost certainly be an implied term to the same general effect (refer to Section 6.2.3).

However, JCT SBC/Q 2016 and JCT IC 2016 provide that if the contractor has reasonable grounds for believing that the employer's representative is unaware when the contractor should receive information, the contract must provide an 'advance notice', giving the employer's representative sufficient time to provide the required information.[82] The contractor need only notify the employer's representative so far as is 'reasonably practicable', but it is difficult to envisage many circumstances where it would not be practicable. It is suggested that if the contractor requires further information from the employer under JCT DB 2016, he should adopt a similar course of action. To that extent, the fulfilling of this requirement by the contractor may prove a hurdle to some claims for delays under this head.

#### 11.4.7.2 Work Not Forming Part of the Contract

The JCT 2016 contracts[83] permit the employer (or others acting on his behalf) to carry out work that is not part of the scope of works that the contractor is carrying out.[84] Such work may include fitting out an area of the works. The JCT 2016 contracts consider two separate situations:

- The first is where the employer engages others to carry out part of the Works and the contract documents contain information to enable the contractor to understand the extent and nature of the work concerned.[85] The work could be anything, but in practice, it is usually more convenient if such work is easy to identify and preferably separate from other work (e.g. in a different location). The contractor must permit the employer to carry out the work.
- The second is similar, but the contract documents do not contain any information to assist the contractor. In that instance, the employer may only arrange for the work to be done after obtaining the contractor's consent.

However, the contracts are clear that the contractor's consent must not be delayed or withheld unreasonably.[81] A delay being caused in the first situation is arguably less likely where the contract documents inform the main contractor from the time of tender what

82 JCT SBC/Q 2016 – clause 2.12.3 and JCT IC 2016 – clause 2.11.3.

83 'Work not forming part of the Contract' – JCT SBC/Q 2016 and JCT IC 2016 – clause 2.7 and JCT DB 2016 – clause 2.6.

84 The meaning of 'work not forming part of the contract' has been defined as '*For some purposes the work does form a part, literally, of the contract; but for other purposes it does not. It is not work which the Employer can require the Contractor to do. All that he can require is that the Contractor affords attendance etc. on those who do the work … [and that] I take the pragmatic view that the relevant work is work not forming part of the contract*'. *Henry Boot Construction Ltd v Central Lancashire New Town Development Corporation* (1980) 15 BLR 8 at 19 per Judge Fay.

85 JCT SBC/Q 2016 and JCT IC 2016 – clause 2.7.1, and JCT DB 2016 – clause 2.6.1.

another contractor will be undertaking. However, in the second situation, where the main contractor has not been pre-warned about the other work to be carried out by directly employed labour, he has less chance of mitigating or accommodating any delay that the other contractor has caused.

#### 11.4.7.3 Provision of Materials or Goods by the Employer

Unlike the execution of work by others, there is no contractual term which entitles the employer to provide materials or goods. If the employer is to supply materials or goods, it will usually be in the belief that by directly engaging a supplier, he will benefit from a financial saving. However, it is submitted that the contract documents must clearly state that it is the employer who will be providing the relevant materials and/or goods, in order for the contractor to be entitled to an extension of time by the 'impediment, prevention or default' clauses.[79] If, in the contract documents, the employer (or his representative) has simply specified particular materials or goods (e.g. by product type or by manufacturer), then it the contractor will remain responsible for their quality in the usual way.[86] However, in that situation, and if it was clear that the employer would the supplier directly for those materials or goods (to be subsequently delivered to the contractor) and has failed to do so, then the contractor may be protected by the same clauses.[79]

An interesting scenario would be created if the materials subsequently were found to be defective. It can reasonably be argued that the supply of materials and goods can only refer to materials and goods which are in accordance with the contract. Therefore, the supply by the employer of materials and goods which are not in accordance with the contract amounts to a failure to supply because they should be rejected by the contractor.

However, if the employer has elected to supply, that will almost certainly have been before tendering and certainly prior to executing the contract. The contract documents may refer to the employer-supplied materials or goods as a 'Prime Cost Sum' ('PC Sum'). If so, it is unlikely that they will be specified in detail, as the precise quality may be unknown. A PC Sum is a 'supply only' rate, and excludes all other elements of cost such as fixing, installing, design, ancillary items, contractor's preliminaries along with his overheads and profit.[87] In the absence of a specification and if they were to be supplied by the contractor, there are certain terms which would be implied, such as that such materials and goods will be appropriate for their purpose. Where the employer would otherwise be responsible for the specification, and if no specification was given, it is considered that the contractor would not be liable if an inferior product was supplied because the employer would either be responsible himself or relying on his representative and not on the contractor, in such matters.[88] In the event that the employer-supplied materials or goods were obviously so inferior that they would seriously jeopardise the project or even become a danger, then it is likely that the contractor would have a duty to warn the employer.

But a difficult problem would arise if either the materials and goods initially appeared to be in accordance with the contract or, if there was no specification, they appeared to be

86 'Materials, goods and workmanship': JCT SBC/Q 2016 – clause 2.3.1, JCT DB 2016 and JCT IC 2016 – clause 2.2.1, and JCT MW 2016 – clause 2.1.2.

87 See the definition of a 'Prime cost sum (PC Sum) at Section 1.6.3 of the New Rules of Measurement – NRM2: Detailed measurement for building works – (1st Ed) 2013 RICS.

88 *Rotherham Metropolitan Borough Council v Frank Haslam Milan and Co Ltd* (1996) 59 Con LR 33.

satisfactory and were incorporated into the works, but were subsequently found to be defective and required replacement causing delay and additional expense. The employer would have to instruct the contractor to supply replacement materials or goods, which would have the effect of adding these items to the contract at a very late stage. The cost of the materials or goods could be dealt with by a Variation or Change, and all the other costs would amount to direct loss and/or expense.[89] An appropriate extension of time would be indicated under 'Instructions' requiring a variation or possibly late instructions. If the contract documents contained a specification and the employer's materials or goods were shown not to comply, an extension under this ground would also be appropriate.

#### 11.4.7.4 Failure to Give Access, Ingress or Egress

Under the JCT 2016 contracts, the employer's failure in these regards would certainly fall within the 'impediment, prevention, default' ground for an extension of time.[79] However, it may not cover a situation where protestors prevented access to a site.[90] They should include where the employer has undertaken to obtain access across land which is not in the employer's possession and control. In addition, such a failure by the employer would give rise to a claim at common law.[91]

A delay notified under the current relevant events would simply specify the delay and that it was caused by the employer's failure to provide ingress to or egress from the site. Whereas the former relevant event might be seen as limiting the circumstances in which an extension of time could be given, it is likely that claims for lack of ingress or egress are now possible on a much broader basis.

#### 11.4.7.5 Compliance or Non-compliance with CDM Regulations 2015

The JCT 2016 contracts all refer to the parties' (in this context, the employer's) obligation to comply with the CDM Regulations 2015.[92] The employer's obligation to ensure compliance is onerous. The problem for the employer was (and is) that the 'Principal Designer' has duties under the regulations which may have to be carried out after the issue of any instructions, although they could attract an extension of time in any event under the 'Instructions' clause.[93] Under the JCT 2016 contracts, the question for the contractor will be simply whether the employer has complied with the CDM clauses[96] and if not, has any delay to progress of the Works been caused?

The employer's compliance with the CDM Regulations 2015 also extends to the provision of the 'Pre-Construction Information' to the contractor.[94] The HSE's guidance states:

> *'The client has the main duty for providing preconstruction information. This must be provided as soon as practicable to each designer (including the principal designer) and*

89 JCT SBC/Q 2016 – clause 4.20.5, JCT DB 2016 – clause 4.21.5 and JCT IC 2016 – clause 4.17.4 (refer to Section 12.4.4).
90 *LRE Engineering Services Ltd v Otto Simon Carves Ltd* (1981) 24 BLR 127.
91 Refer to Sections 6.2.3 (co-operation) and 6.6 (Possession of the site).
92 JCT SBC/Q 2016 – clause 3.23, JCT DB 2016 – clause 3.16, JCT IC 2016 – clause 3.18 and JCT MW 2016 – clause 3.9.
93 Refer to Table 11.4.
94 Regulation 4(4) *'A client must pre-construction information as soon as is practicable to every designer and contractor appointed or being considered for appointment to the project'*.

*contractor (including the principal contractor) who is bidding for work on the project or has already been appointed. For projects involving more than one contractor, the client should expect the principal designer to help bring the preconstruction information together and provide it to the designers and contractors involved.'*[95]

The preconstruction information must provide information about the health and safety hazards of the site, include design and construction hazards, be relevant to the particular project, have an appropriate level of detail and be proportionate to the to the risks involved. It is information that the employer either has in his possession or can reasonably obtain.[96] Typically, this information could include surveys and assessments (e.g. asbestos) and information from earlier construction work (such as as-built structural, architectural and building services drawings).

If the pre-construction information produced by the employer is shown to be inaccurate, particularly to the extent that but for the inaccuracy, the contractor would have avoided a delay, then it is arguable that this falls within the 'impediment, prevention or default' ground for an extension of time. Employers should be wary of introducing 'no reliance' clauses in their contracts, which place the onus of responsibility on the contractor for verifying the accuracy of any reports or statements issued by, or on behalf of the employer. It is doubtful whether such 'no reliance' clauses would absolve the employer of his duty to provide accurate preconstruction information under the CDM Regulations.

### 11.4.8 Statutory Undertaker's Work

This ground deals with delay caused by the carrying out by a statutory undertaker of work in in relation to the Works, or its failure to do so. Traditionally, statutory undertakers were the nationalised organisations such as the water, gas and electricity boards – suppliers that were authorised by statute to construct, maintain and operate public utilities. However, in more recent times, a statutory undertaker is a private company, albeit authorised and regulated to carry out work of the same nature.

Whilst the scope of a statutory undertaker's work may be precisely defined and known from the outset, it is not uncommon for them to take (what is often perceived to be) and inordinate period of time to execute that work. Take the example of the 'statutory undertaker' authorised to provide electrical connections to the national electricity grid supply. The contractor may engage this 'statutory undertaker' as soon as he enters into the building contract with the employer (i.e. at the earliest possible time) and make a sensible forecast for the timing of the statutory undertaker's work in his programme. However, the statutory undertaker may be unable to commit to a precise date or period for carrying out the relevant work. This is neither the fault of the contractor nor the employer, so it's considered a 'neutral event' – i.e. one that provides a ground for an extension of time (thereby relieving the contractor of any liability for liquidated damages), but not an entitlement to loss and expense.

Although it is not beyond doubt, it is unlikely that a local authority could be classed as a statutory undertaker except, perhaps, when carrying out certain specific activities in relation to roadworks.

95 Health & Safety Executive: Managing Health & Safety in Construction (2015) para 43.
96 Health & Safety Executive: Managing Health & Safety in Construction (2015) at Appendix 2.

In a 1980 case,[97] the court was concerned with a forerunner to the current clauses, and the question of whether or not statutory undertakers were artists, tradesmen or others engaged by the employer under a JCT 63 contract.[98] The court upheld the decision of an arbitrator that, in that instance, the statutory undertakers carrying out particular work were carrying it out under a direct contract with the employer. In other words, they were not executing the work because relevant legislation obliged them to do so, but because they had contracted with the employer to carry it out. Therefore, the court held that the undertakers were engaged by the employer to carry out work which did not form part of the main contract.[102] Having reached that decision, it followed that extensions of time should be granted to the contractor on the basis of delays by artists and tradesmen engaged by the employer (who happened to be the statutory undertakers). That enabled the contractor to claim direct loss and/or expense from the employer under the same grounds.

The case is mentioned because statutory undertakers often carry out work other than under a statutory obligation. Where that occurs, if they have been directly engaged by the employer, the contractor would be entitled to an extension of time under the 'impediment, prevention or default' grounds (also entitling him to direct loss and/or expense – which could not be claimed if the contractor relied upon the 'statutory undertaker' Relevant Event). Obviously, no extension of time would be applicable if the statutory undertakers had been engaged directly by the contractor to carry out other work than they would undertake in pursuance of their statutory obligations – as they would be acting as a sub-contractor. Therefore, whether or not or under which particular clause an extension of time should be given for delays caused by statutory undertakers may often depend on the nature of the work being done and the surrounding circumstances. For example, if a statutory undertaker lays electricity supply cables (not for the purposes of the contract Works but for another site nearby) in the road which provides the only access to the site, there would be no grounds for extension of time. It makes no difference that the statutory undertaker concerned might be under a statutory obligation to lay the service because it would not be carrying out his work in relation to the Works.

### 11.4.9 Exceptionally Adverse Weather Conditions

It is not unknown for employers, their representatives and contractors to disagree over this ground. Despite the change in wording in the 1980 Form from 'inclement' to 'adverse,' many contractors and consultants still refer to 'inclement weather' when seeking an extension of time under the JCT 2016 contract. The change of wording to 'exceptionally adverse' was intended to make it clear that the ground was intended to cover any kind of adverse conditions including unusual heat or drought. Adverse weather is any kind of weather that impedes the progress of the Works. For example, where tower cranes are being used on site, wind may be adverse, and above a certain wind speed, it will be positively dangerous to operate the cranes at all.

A crucial factor is the kind of weather that ought to be expected at the site at the time of year when the delay occurs. Employers and their representatives will often request the

97 *Henry Boot Construction Ltd v Central Lancashire Development Corporation* (1980) 15 BLR 8.
98 JCT 2016 contracts would consider the equivalent situation as 'work not forming part of the contract' – refer to Footnote 100.

contractor to provide meteorological reports for previous years to make a comparison to show that the current weather being complained of was 'exceptionally adverse' for the site's location at that time of year. In that instance exceptional refers to exceeding what may be reasonably expected based on the evidence of past years. Even if by comparing appropriate historical records, the contractor can demonstrate that the weather is exceptionally adverse for the time of year it must also be such that it interferes with the particular stage of the Works. Obviously, even subzero temperature blizzard conditions in the United Kingdom in July are unlikely to affect the progress of internal fit-out works! Clearly, it matters not that the Works have been affected only because they have been delayed through the contractor's own fault.[99]

The contractor is expected to programme the Works making proper allowance for the adverse weather that could reasonably expected for the time of year and the Works location. The contractor's programme for particular work operations should reflect that weather interruptions are likely to occur and should allow for them (the 'tower crane' example above). If the contractor is aware at the time of tender that it is tendering for a project which is to be constructed during the winter and if it is constructed in the winter, it is no use requesting an extension of time on the ground of exceptionally adverse weather when the Works are delay on account of snow. The contractor will be expected to have allowed for snow and ice in winter and higher, hotter temperatures in the summer and perhaps heavier rainfall in spring. However, if the contractor can show that the winter weather was more severe than it could have reasonably anticipated, an extension of time should be given if the Completion Date was delayed as a result.

The contractor should provide details of 'how' the exceptionally adverse weather has affected his operations. Whilst extremes in temperature, rainfall and wind often cause delays to progress, other factors such as humidity may affect the contractor's work. This can be the case for items of specialised work, such as particular types of finishes. If work has been affected, the contractor should provide information regarding the recommended conditions for installing or applying the work. In the event that the exceptionally adverse weather is outside those recommended conditions, then it is easier for the contractor to demonstrate the effects of delay.

On a strict reading of the relevant event, it is only the 'exceptional' aspect of the adverse weather which will attract an extension of time. Thus, if ten days of snow in January is just on the borderline between usual and exceptional and a project suffers two weeks' adverse weather, the contractor may face more difficulty in demonstrating an entitlement to an extension of time for the consequences on the completion date for the whole two weeks.

Weather data is widely available with many companies (including JCT) providing both historical and current weather data, that's both detailed and relevant to many locations.

### 11.4.10 Loss or Damage Occasioned by One or More of the Specified Perils

The purpose of this ground is to give the contractor sufficient additional time to fulfil its obligations to repair damage caused by one of the Specified Perils,[100] which are insurable events under the JCT 2016 contracts:

99 *Walter Lawrence v Commercial Union Properties* (1984) 4 Con LR 37.
100 JCT SBC 2016, JCT DB 2016 and JCT IC 2016 – defined in clause 6.8. For JCT MW 2016, see clauses 1.1 and 5.4B.

- Fire,
- Lightning,
- Explosion,
- Storm, flood, tempest,
- Escape of water from a water tank, apparatus or pipe,
- Earthquake,
- Aircraft or other aerial devices, or articles dropped therefrom,
- Riot and civil commotion,

but exclude what are called the Excepted Risks.[101]

It appears that the contractor is entitled to an extension of time if the events are caused by the default or negligence of the contractor's own employees.

### 11.4.11 Civil Commotion or the Use or Threat of Terrorism

'Civil commotion' means, for insurance purposes, 'a stage between a riot and a civil war.'[102] There must be an element of turbulence, and it is thought that the activities of protesters in public places may amount to civil commotion.

There are three possibilities in relation to terrorism. The first is that the Works are delayed because terrorism has been used. For example, part of the Works may be damaged by an explosion.

The second is that terrorist action may be threatened, and, as a result, the area of the site may have to be cleared. It is thought that the threat of terrorism would have to be more substantial than just the fact that other terrorist incidents have occurred in the area. A specific terrorist threat directed at the project or a threat to an area which, if it was to be carried out, would affect the project would qualify. It seems that an extension of time would be applicable where the contractor or its operatives received direct threats of injury if work did not cease and the contractor stopped work accordingly.

The third is that terrorist action may have been carried out or threatened and the relevant authorities, who may be government, police or army, may cause delay to the works as a direct result of the way in which the act or threat is dealt with. The activities of the relevant authorities which would qualify under this ground would include such measures as evacuation of premises and the restriction of access. For example, there may be a threat to a particular building and the police may evacuate the surroundings for a period. If the building site is part of the evacuated area, there will be a delay to the Works and, to the extent that the date for completion is affected, an extension of time must be given. This ground is not restricted to the site of the Works and, therefore, it is likely that any such threat or action which affected the execution of the Works in any way (such as the forced evacuation or destruction of the contractor's offices) would give entitlement to extension of time.

---

101 The 'Excepted Risks' are contamination by radioactivity, supersonic pressure waves from aircraft or other aerial devices, and acts of terrorism (not covered by the contract's 'Terrorism Cover' provisions). See Footnote 117 for clause references.

102 *Levy v Assicurazioni Generali* (1940) 3 All ER 427 at 431 per Luxmore LJ, approving an extract from Welford and Otterbarrys *Fire Insurance*, 3rd edition, at p. 64.

### 11.4.12 Strikes, Lock Out etc.

The list of events under this clause includes:

- Strike
- Lock-out; or
- Local combination of workmen affecting any trade engaged:
  - Either upon the Works; or
  - In the preparation, manufacture or transport of goods or materials required for the works; or
  - Affecting any persons carrying out design for the Works.

So far as strikes are concerned, extension of time may be given for any delaying circumstance which affects the contractor and its work on the site or any persons preparing design including a strike affecting any trade involved in preparing or transporting any goods and materials which are required for the Works. The reference to design is relevant because of the contractor's designed portion options in JCT SBC/Q 2016 and JCT ICD 2016, and obviously for JCT DB 2016. It is not immediately obvious how a strike might affect designers employed in the contractor's office or, indeed, an independent firm of designers engaged by the contractor to provide such services. The clause is drafted to cover all strikes whether official or unofficial, but is unlikely to cover 'working to rule' or any other obstructive practice which is not actually a strike. An unofficial strike has been described as any strike or other industrial action which is not authorised or endorsed by a trade union.[103] A strike or other event referred to in the sub-clause must be one in which the trades mentioned in it are directly involved. It was held that a strike by workers employed by statutory undertakers which are directly engaged by the employer to execute work which did not form part of the Works was not covered by the forerunner of this clause in JCT 63.[104] The reference to local combination of workmen is an antiquated phrase which may possibly be held to cover activities which fall short of a strike and occur in a specific area. However, it is thought more likely to refer to a small localised strike.

It is probable that a situation where deliveries to site are delayed, not due to a strike, but due to some form of secondary picketing does not fall under this relevant event.

### 11.4.13 New Legislation After the Base Date

The relevant event refers to the exercise of a statutory power by the UK Government or a local authority after the Base Date.[105]

This provision might, for example, be relied upon whenever the Government, or a local or public authority, exercises any statutory power as set out in this ground. This may range from introducing a new act or Parliament, a regulation (or Statutory Instrument), or in a more local sense, the closure of essential means of access to the site of the Works.

A particular significance of this ground is that this relevant event prevents the contract being brought to an end by frustration. The matter is simply dealt with by an extension of time.

103 S.237(2) of The Trade Union and Labour Relations (Consolidation) Act 1992.

104 *Boskalis Westminster Construction Ltd v Liverpool City Council* (1983) 24 BLR 83.

105 The 'Base Date' is that date written into the contract particulars. The "Base Date" is effectively the cut-off date for changes in legislation etc., which may entitle the contractor to adjust the Contract Sum.

### 11.4.14 Delay in Receiving the Necessary Permissions (JCT DB 2016 only)

This Relevant Event is exclusive to JCT DB 2016[106] and deals with delay in the receipt of any necessary permission of any statutory body. There is a requirement that the contractor must have taken all practicable steps to reduce the delay, and it must be taken seriously. Realistically, this will probably amount to little more than that the contractor must have made any necessary applications in good time, replied promptly to queries and used its best endeavours to obtain the permissions or approvals. An employer (or his representative) in the position of making applications to statutory bodies cannot guarantee the result and neither can the contractor. The contractor will be entitled to an extension of time under this relevant event if it can show that the delay was not due to its fault. This relevant event refers to any kind of statutory permission or approval. Virtually, all buildings require planning permission, and they must satisfy the Building Regulations. There are, however, many other possible controls over such things as fire, water and entertainment (licenses for the intended premises).

### 11.4.15 Insolvency of a 'Named Specialist' (JCT SBC/Q 2016 only)

This Relevant Event is exclusive to JCT SBC/Q 2016 and only when the provisions relating to a 'Named Specialist' apply.[107] Schedule 8 to the contract contains the relevant provisions for the employer to select a 'Named Specialist' by naming him either in the Contract Documents[108] or in an instruction (including for the expenditure of a provisional sum).[109] As a 'Pre-Named Specialist' is named in the Contract Documents, the contractor has no right of objection to that specialist (as he's not only accepted that the supplemental provision applies, but has had an opportunity to carry out some 'due diligence' checks on the pre-named specialist during the tender process). However, where a 'post named specialist' is instructed, the contractor is able to make a reasonable objection,[110] to which the employer's representative can instruct:

- Select another named specialist[111];
- Direct the contractor to carry out the relevant work himself[112]; or
- Omit the named specialist works as a variation.[113]

The employer's representative may also give these instructions,[114] in the event that (for whatever reason) the contractor notifies him that he is unable to enter into a sub-contract with a prenamed specialist.

106 JCT DB 2016 – clause 2.26.13.
107 'Named Specialists' provisions apply in JCT SBC/Q 2016 if 'Supplemental Provision 9 (Named Specialists)' is selected in the 'Contract Particulars' (entry against the Eighth Recital).
108 Supplemental Provision 9, Paragraph 9.1.1 – as a 'Pre-Named Specialist'.
109 Supplemental Provision 9, Paragraph 9.1.2 – as a 'Post-Named Specialist' and carrying out 'Post-Named Specialist Work'.
110 Supplemental Provision 9, Paragraph 9.4.
111 Supplemental Provision 9, Paragraph 9.3.2.
112 Supplemental Provision 9, Paragraph 9.3.3.
113 Supplemental Provision 9, Paragraph 9.3.4.
114 The employer's representative may also instruct to remove the ground that prevents the Pre-Named Specialist entering into a sub-contract with the contractor (Supplemental Provision 9, Paragraph 9.3.1).

The Relevant Event clearly recognises that the contractor's control over a Named Specialist is relatively limited by Supplemental Provision 9, as he may have little option to proceed with him.[115] Therefore, in the event that the Named Specialist (who has effectively been imposed by the employer) becomes or has become insolvent, then the contractor is entitled to an extension of time if this affects the Completion Date.

### 11.4.16 'Named Sub-Contractors' (JCT IC 2016 Only)

The provisions for 'Named Sub-Contractors' are exclusive to JCT IC 2016 and in many ways are procedurally similar to the use of 'Named Specialists' in JCT SBC/Q 2016 (refer to Section 4.11.5). However, there are no dedicated 'relevant events' that cover 'named sub-contractors' separately because the relevant events for 'variations' and for instructions for the expenditure of provisional sums will apply.

Clause 3.7 (Named Sub-Contractors) provides that if so stated in the 'contract bills, specification or work schedules', the contractor must engage, as sub-contractors, any persons or firms who are named either in the contract documents or in an instruction for the expenditure of a provisional sum, to carry out particular elements of the Works. He must do this within 21 days of entering into the JCT IC 2016 contract.[116] The provisions of schedule 2 (named sub-contractors) will then apply, and the 'naming' process involves standard forms of invitation to tender, tender, articles of agreement and standard conditions of sub-contract.[117]

Named Sub-Contractors effectively become domestic sub-contractors to the contractor. Once the 'naming' procedure has been completed, it is unusual for the employer's representative to have any further involvement with the Named Sub-Contract's administration. He neither directs the contractor as to the payments to be made to the named sub-contractors nor is he required to certify practical completion of the Named Sub-Contractor's work. Matters such as extensions of time and settlement of its account are solely matters between the named sub-contractor and the contractor.

However, the employer's representative may become involved particular problems occur as specified in Schedule 2. Where a sub-contractor has been named in the contract documents but the contractor is unable to enter into a 'Named sub-contract' with him, the contractor must notify the employer's representative immediately, specifying the reason for such.[118] If reasonably satisfied with the contractor's reasons, he may instruct the contractor as follows:

115 For example, if the contractor has no grounds for objecting to a Post-Named Specialist, or a Pre-Named Specialist readily enters into a sub-contract with the contractor. See also Supplemental Provision 9, Paragraph 9.7.

116 JCT IC 2016 – clause 3.7.

117 JCT IC 2016 – 'Named Sub-Contractor' documents include (1) Intermediate Named Sub-Contract Tender and Agreement 2016 that comprises (a) Invitation to Tender (ICSuB/NAM/IT), (b) Tender (ICSub/NAM/T), and (c) Agreement (ICSub/NAM/A), (2) Intermediate Named Sub-Contract Conditions 2016 (ICSub/NAM/C – incorporated by the 'Agreement' ICSub/NAM/A), and (3) the Intermediate Named Sub-Contractor/Employer Agreement 2016 (ICSub/NAM/E 2016).

118 JCT IC 2016 – Schedule 2, para 2.

- Change the particulars in order to remove the problem[119]; or
- Omit the work[120]; or
- Omit the work and substitute a provisional sum[121] (allow the option of naming another sub-contractor).

Instructions issued by the architect under either of the first two options are treated as a 'Variation' and entitle the contractor[122] to both (1) an extension of time and (2) the loss and expense.[123] Such effects are likely to include the time (and any ensuing cost) of the employer engaging others (including a replacement 'Named Sub-Contractor) to do the relevant work. If the third option is followed, the expenditure of a provisional sum requires a replacement 'Named' Sub-Contractor:

- Any delay caused to the regular progress may become grounds for an extension of time under clause 2.19, as a Relevant Event under clause 2.20.2.1 (instruction for expending a Provisional Sum); and
- Any loss and expense caused to the contractor by its effect on regular progress of the Works is recoverable under clause 4.17.2.1 which expressly refers to clause 3.13 (instructions for Provisional Sums).

For example, if the named sub-contractor's availability does not correlate with the contractor's programme, it's likely that the regular progress will be affected, resulting in:

- If the Completion Date is likely to be affected, an extension of time and loss and expense (prolongation costs); or
- If the Completion Date is not affected, the contractor may incur loss and expense in the form of disruption costs.

The main contractor may make a 'reasonable objection' to entering into the 'Named Sub-Contract[124] within 14 days of the instruction, and if he does so, the employer's representative must instruct another sub-contractor but not under the 'Named' procedure. Therefore, provided that the contractor's objection is 'reasonable', there is no provision which allows the architect to nevertheless instruct the contractor to proceed.

If no objection is raised and the contractor enters into the 'Named Sub-Contract', he may still be entitled to an extension of time and loss and expense if the compliance with the employer's representative affects the regular progress of his work.[125]

If the contractor terminates the 'Named Sub-Contractor's' employment because of the sub-contractor's default or insolvency,[126] the contractor notifies the employer's representative who may issue the following instructions:

---

119 JCT IC 2016 – Schedule 2, para 2.1.
120 JCT IC 2016 – Schedule 2, para 2.2.
121 JCT IC 2016 – Schedule 2, para 2.3.
122 JCT IC 2016 – Schedule 2, para 3 - '... *and the provisions of clauses 2.19 (Adjustment of Completion Date) and 4.15 and 4.16 (Loss and Expense) shall as relevant apply*'.
123 Refer to Section 12.4.4.
124 JCT IC 2016 – Schedule 2, para 5.3.
125 JCT IC 2016 – Schedule 2, para 3: owing to the time taken, or cost incurred for either (i) the para 2.1 instruction to remove the impediment, (ii) to omit the work, or (iii) instruct the expenditure of a Provisional Sum.
126 JCT IC 2016 – Schedule 2, para 6.

- 'Name' another sub-contractor to complete the work,[127] or
- Require the contractor to make its own arrangements for completion of the work, in which case the contractor may sub-let the remaining work to a sub-contractor of its own choice,[128] or
- Omit the remainder of the work from the contract.[129]

If the original sub-contractor was 'Named' in the contract documents, the exercise of the first option by the employer's representative will entitle the contractor to an extension of time for any resulting delay, but not to loss and expense.[130] Instructions issued under the two options, do entitle the contractor to an extension of time and loss and expense.[131]

If the first option is adopted, the contract sum is to be adjusted to take account of the increase or decrease in price between what would have been payable to the original sub-contractor and what is now payable (save for the cost of rectifying defects in the work of the original named sub-contractor) to the replacement.[134]

If the second option is pursued, the contractor will be entitled to be paid in accordance with the valuation of variations. The contractor must pursue the original sub-contractor,[132] and provided that the employer is prepared to indemnify the contractor against reasonable legal costs,[133] the contractor can invoke the sub-contract's dispute resolution procedures[134] in order to recover the additional expense, for which he must account to the employer.[135]

### 11.4.17 Force Majeure

*Force majeure* is a term originating in French law. It is wider in meaning than the often-referred 'Act of God,' which has been described as '*an overwhelming superhuman event*.'[136] It seems that the event relied upon as *force majeure* must make the performance of the contract wholly impossible. In this sense, there are marked similarities to the doctrine of frustration of contract. In practice, however, all the surrounding circumstances must be taken into account. The dislocation of business caused by the general coal strike of 1912 has been held to be covered by the term and also covered the breakdown of machinery, but not delay caused by bad weather, football matches or a funeral.

> '*These are the usual incidents interrupting work and the defendants, in making their contract, no doubt took them into account.*'[137]

127 JCT IC 2016 – Schedule 2, para 7.1.
128 JCT IC 2016 – Schedule 2, para 7.2.
129 JCT IC 2016 – Schedule 2, para 7.3.
130 JCT IC 2016 – Schedule 2, para 8.1. Entitlement to both extension of time and loss and/or expense would apply to the first option also if the sub-contractor was named in an instruction regarding the expenditure of a provisional sum (see also Schedule 2, para 9).
131 JCT IC 2016 – Schedule 2, para 8.2.
132 JCT IC 2016 – Schedule 2, para 10.2.1.
133 JCT IC 2016 – Schedule 2, para 10.2.3.
134 JCT IC 2016 – Schedule 2, para 10.2.3. In the absence of the employer's indemnity, the contractor is still obliged under Paragraph 10.2.1 to take whatever other reason action is available to recover such expense.
135 JCT IC 2016 – Schedule 2, para 10.2.2.
136 *Oakley v Portsmouth & Ryde Steam Packet Co* (1856) T1 Exchequer Reports 6 1F.
137 *Matsoukis v Priestman & Co Ltd* (1915) 1 KB 681.

There seem to be no reported cases which deal expressly with *force majeure* in the context of JCT contracts although there are cases that simply consider the term itself. A useful English authority is *Lebeaupin v Crispin & Company*, which throws light on the way in which this term should be interpreted. Mr. Justice McCardie said:

> *'This term is used with reference to all circumstances independent of the will of man, and which it is not in his power to control …. Thus war, inundations and epidemics are cases of force majeure; it has even been decided that a strike of workmen constitutes a case of force majeure … [But] a force majeure clause should be construed in each case with a close attention to the words which precede or follow it and with due regard to the nature and general terms of the contract. The effect of the clause may vary with each instrument.'*[138]

On that authority, care has been taken when interpreting *force majeure* in the JCT 2016 contracts, particularly having regard to the other relevant events. Where the term *force majeure* is used in these contracts, its meaning will effectively be restricted because many things which would normally fall under the category of *force majeure* are included under more specific headings. Such matters as strikes, fire and exceptional weather are examples.

The effect on construction (and commerce in general) caused by the global COVID-19 pandemic, unsurprisingly saw a significant increase in interest in force majeure clauses and the Court's interpretation of 'epidemic' in the *Lebeaupin*[138] case. In a 2022 case,[139] the Court clearly agreed that 'pandemic' is covered by the 'epidemic' in the context of force majeure.

---

138 *Lebeaupin v Crispin & Company* (1920) 2 KB 714.

139 *European Professional Club Rugby v RDA Television LLP* [2022] EWHC 50 (Comm) '*It is common ground and in any event I find that the onset of the Pandemic was a Force Majeure event … This is so since 'epidemic' includes a pandemic in the context of use of that word in the definition of a Force Majeure Event and in any event because the Pandemic was a circumstance '… beyond the reasonable control of a party affecting the performance by that party of its obligations under this Agreement…'* para 17 per HH Judge Pelling QC (as he then was).

# 12

# Loss and/or Expense under JCT 2016 Contracts

## 12.1 Introduction

### 12.1.1 General Points

There are a number of provisions in the JCT 2016 contracts that govern the contractor's entitlement to loss and/or expense. These provisions are separate from those dealing with 'Variations' (see Chapter 5) and with extensions of time (see Chapter 11). For JCT SBC/Q 2016, JCT DB 2016 and JCT IC 2016, the direct loss and/or expense clauses (for brevity, referred to as 'loss and expense') are contained in the payment clauses and deal with loss and/or expense caused by matters materially affecting regular progress of the Works (refer to Section 4.1.2). Owing to its relatively more simplistic operation, JCT MW 2016 provides that any 'loss and expense' should be included within the valuation of a 'variation,' and does not contain the separate, and more detailed provisions of its larger counterparts. Therefore, references to the loss and expense procedures are predominantly to those found in JCT SBC/Q 2016, JCT DB 2016 and JCT IC 2016.

Once the contractor decides to trigger the loss and expense clause, it imposes specific obligations not only on the contractor but also on the employer or his representative, which is not something that they always fully appreciate. These clauses confer on the contractor a clear and enforceable right to financial reimbursement for 'direct loss and/or expense' suffered or incurred as a direct result of certain relevant matters, provided that the contractor complies with the procedures laid down by the provisions. The contractor's entitlement to recovery under the clause depends on the following overarching and 'governing matters':

- Deferment of possession of the site; or
- That regular progress of the Works has been materially affected by one or more of the 'relevant matters' (see Table 12.1 and Section 12.3); or

It is therefore vitally important that all those concerned (employers, contractors, architects, contract administrators and/or quantity surveyors) should fully understand the way in which these clauses are intended to work. It must always be borne in mind that the detailed provisions in the loss and expense clauses in JCT SBC/Q 2016, JCT DB 2016 and JCT IC 2016 are not simply a matter of meaningless procedure, as there are very clear purposes behind them. The clauses are both procedural, in the sense of instructing the parties

*Building Contract Claims*, Sixth Edition. Guy Higginbottom and David Chappell.

**Table 12.1** JCT 2016 contracts – 'relevant matters'.

| Relevant Matter | JCT SBC/Q 2016 | JCT DB 2016 | JCT IC 2016 |
|---|---|---|---|
| Deferment of Possession | 4.20.1 | 4.19.1 | 4.15.1 |
| Variations (or Changes) | 4.22.1 | 4.21.1 | 4.17.1 |
| Instructions: | — | — | — |
| • Postponement of work | 4.22.2.1 | 4.21.2.1 | 4.17.2.1 |
| • Provisional Sums | 4.22.2.1 | 4.21.2.1 | 4.17.2.2 |
| • 'Named Sub-Contractors' | — | — | 4.17.2.2 |
| • Opening up/testing of compliant work | 4.22.2.2 | 4.21.2.2 | 4.17.2.3 |
| • Resolving a discrepancy or divergence | 4.22.2.3 | — | — |
| • Errors, omissions, inconsistencies in Contract Documents | — | — | 4.17.2.4 |
| • Objection to 'Named Specialist' | 4.22.2.4 | — | — |
| • Insolvency of 'Named Specialist' | 4.22.2.5 | — | — |
| Antiquities | 4.22.3 | 4.21.3 | — |
| Contractor suspending for non-payment[a)] | 4.13.2 | 4.11.2 | 4.14.2 |
| Inaccurate Approximate Quantities | 4.22.4 | — | 4.17.3 |
| Delay in receiving permissions | — | 4.21.4 | — |
| Employer's Impediment, prevention, or default | 4.22.5 | 4.21.5 | 4.17.4 |

a) Also see JCT MW 2016 – clause 4.7.2

what they should do at each stage of the process, and contractual, in the sense that they set out the respective rights and obligations which the parties have assumed in respect of one another.

It is important to note that like many of the 'Relevant Events' (entitling the contractor to an extension of time – refer to Section 11.4 and Table 11.5), many of the 'Relevant Matters' to which these clause refers are not breaches of contract by the employer, the Employer's Persons or by his representative.

Table 12.1 sets out the 'Relevant Matters' and Table 12.2 shows which are the employer's responsibility (i.e. a breach), and which are 'no fault' events:

From Table 12.2, it can be seen that the first line shows the 'category (1) – employer responsible' 'relevant matters', that would be breaches of the contract terms (either express or implied[1]) and would entitle the contractor to claim common law damages.

However, none of the relevant matters in 'category (2) – no fault' of Table 12.2 are breaches of contract which would entitle the contractor to recover damages at common law – they are all matters which the contract expressly permits the employer (or his representative) to put into effect. Importantly, these clauses provide the only right to

1 Refer to Section 6.2

**Table 12.2** Employer's responsibility for 'relevant matters'.

| Category | JCT SBC/Q 2016 | JCT DB 2016 | JCT IC 2016 |
|---|---|---|---|
| Category (1) – Employer responsible<br>Relevant matters for which the employer, employer's person or his representative are responsible | 4.22.2.2,<br>4.22.2.3,<br>4.22.2.5,<br>4.22.4,<br>4.22.5 | 4.21.2.2,<br>4.21.5 | 4.17.2.3,<br>4.17.2.4,<br>4.17.3<br>4.17.4 |
| Category (2) – 'No fault'<br>Relevant Matters for which neither the employer, the employer's persons or his representative or the contractor are responsible | 4.20.1,<br>4.22.1,<br>4.22.2.1,<br>4.22.2.4<br>4.22.3 | 4.19.1,<br>4.21.1,<br>4.21.2.1,<br>4.21.3,<br>4.21.4 | 4.15.1,<br>4.17.1,<br>4.17.2.1,<br>4.17.2.2 |

reimbursement for such matters. Therefore, if the contractor should lose its right to reimbursement under these clauses, it will be unable to recover any money at common law despite the contract expressly preserving all the contractor's other rights and remedies.[2]

The loss and/or expense provisions consist of the following important stages:

- The matters affecting the contractor's regular progress.[3]

  These clauses provide the contractor's entitlement to claim loss and expense. They summarise the two 'governing matters' (i.e. deferment of possession and the Relevant Matters – see Section 12.1.1)[4] and clarify that the contractor may have no entitlement to loss and expense under these clauses if he is entitled to financial compensation under any other provisions of the contract.[5]
- Claiming and assessing loss and expense[6]

  This is the 'engine room' of the loss and expense provisions. The machinery which must be operated if the contractor wishes to recover loss and expense is set out together with responses required from the employer and/or his representative. Failure to operate these clauses correctly and timeously may preclude the contractor from recovery of loss and or expense under the contract.
- The 'Relevant Matters'

  This clause lists the relevant matters which are the grounds that entitle a contractor to loss and/or expense (refer to Table 12.1) – and includes that arising from deferment of possession (although that entitlement is curiously listed separately).

2 JCT SBC/Q 2016 – clause 4.24, JCT DB 2016 – clause 4.23 and JCT IC 2016 – clause 4.19.

3 JCT SBC/Q 2016 – clause 4.20, JCT DB 2016 – clause 4.19, JCT IC 2016 – clause 4.15, and JCT MW 2016 – clause 3.6.3.

4 JCT SBC/Q 2016 – clause 4.20.1, JCT DB 2016 – clause 4.19.1 and JCT IC 2016 – clause 4.15.1.

5 JCT SBC/Q 2016 – clause 4.20.2, JCT DB 2016 – clause 4.19.2 and JCT IC 2016 – clause 4.15.2.

6 'Notification and ascertainment' – JCT SBC/Q 2016 – clause 4.21, JCT DB 2016 – clause 4.20, and JCT IC 2016 – clause 4.16.

- Inclusion of loss and expense in payments to the contractor

  These clauses are highly significant and require amounts to be added to the Contract Sum[7] and included in both interim[8] and final payments[9] to the contractor. The employer or his representative are not entitled to wait until the whole of the contractor's claim has been ascertained. This has the important effect of making payment of amounts found due to the contractor as soon as the amount is ascertained.
- Preservation of the contractor's common law rights.[2]

  This clause preserves all the contractor's rights and remedies so that it is not confined to the rights and remedies expressly stated in the contract.[10] However, as noted above, the contractor may not be able to claim at common law in respect of some of the relevant matters (see the 'Category (1)' matters in Table 12.2).

### 12.1.2 No Relationship Between Relevant Matters and Relevant Events

There is no connection between extensions of time (relevant events) and loss and expense (relevant matters) other than that some of the grounds for extending time are echoed in the grounds for claiming loss and expense. It by no means follows that an extension of time is necessary before an application for loss and expense can be made. However, there is the common but mistaken belief that there is some automatic connection between the giving of an extension of time and the contractor's entitlement to reimbursement. For the avoidance of doubt, there is not, and there never was, any such connection.[11] This difference can clearly be explained as follows:

- The contractor's entitlement to an extension of time depends upon whether any delay to progress will (or is likely to) delay completion beyond the Completion Date (of the Works or a Section), whereas
- His entitlement to loss and expense depends upon whether the regular progress of the Work (or a Section) has been 'materially affected'.

Sections 4.4.2 and 4.4.3 distinguished the contractor's financial claims arising from extensions of time (prolongation costs) and uneconomic working (disruptions costs), respectively.

7 JCT SBC/Q 2016 – clause 4.3.4, JCT DB 2016 – clause 4.2.4 (for both SBC/Q and DB – excluding any loss and expense included under the Confirmed Acceptance of a Variation Quotation or Acceleration Quotation – see Section XX) and JCT C 2016 – clause 4.3.3.

8 JCT SBC/Q 2016 – clause 4.14.2.3, JCT DB 2016 – clause 4.12.2.3 (Alternative A) or 4.13.2.3 (Alternative B) and JCT IC 2016 – clause 4.9.2 (100% of loss and expense)

9 JCT SBC/Q 2016 – clause 4.25.2.1, JCT DB 2016 – clause 4.24.2 (referring to clause 4.2) and JCT IC 2016 – clause 4.20.1.

10 This is important in view of the Court of Appeal decision in *Lockland Builders Ltd v John Kim Rickwood* (1995) 77 BLR 38, which seems to suggest that in the absence of express provision, contract machinery and common law rights can co-exist only in circumstances where the contractor displays a clear intention not to be bound by the contract. The more recent Court of Appeal decision in *Strachan & Henshaw Ltd v Stein Industrie (UK) Ltd* (1998) 87 BLR 52 takes a different view and, in any event, the relevant JCT 2016 clauses put the matter beyond doubt.

11 See *H Fairweather & Co Ltd v London Borough of Wandsworth* (1987) 39 BLR 106; *City Inn Ltd v Shepherd Construction Ltd* (2007) CSOH 190 upheld on appeal [2010] Scot CS CSIH 68.

An extension of time has only one primary effect – it extends the period allowed to the contractor for carrying out and completing the Works. Obviously, in doing so, its secondary effect postpones the date from which the contractor becomes liable to pay liquidated damages to the employer. Contrary to popular belief, an extension of contract time does not in itself entitle the contractor to any extra money.[12]

Table 12.3 is essentially a repeat of Table 11.4 (relevant events), but cross-referenced to the relevant matters in Table 12.1.

The essential difference between the relevant events and relevant matters was neatly summed up in a Scottish case:

> *'Although it took various forms the essential argument presented by counsel for the pursuers, as I understood it, was that a certificate of extension of time issued under [extension of time] clause 25 had no direct bearing on a claim based on disruption under [loss and expense] clause 26. While there might, indeed, be many situations on the ground which would result in both clauses being invoked, the purpose of an extension of time certificate was to avoid a claim for liquidate damages rather than found a claim for disruption.*
> *In my opinion one has only to look at the terms of the two clauses to see that this argument is self evidently correct. The operation of clause 25 depends on the occurrence of a "Relevant Event", as there defined whereas the operation of clause 26 depends on whether "the regular progress of the Works or any part thereof has been or is likely to be materially affected by any one or more of the matters referred to in clause 26.2..."'*[13]

A 1997 case[14] highlighted the difference between loss and expense claimed for disruption and loss and expense claimed for prolongation (i.e. an extension of time). The judge clarified that the contractor's claim was for loss and expense due to disruption. He upheld the essential argument of the claimants that a certificate for an extension of time had no bearing on a claim for disruption:

> *'In the end, [Counsel] for the pursuers, was, I think constrained to accept that there was no essential link between [the extension of time] clause 25 and [the loss and expense] clause 26, but he nonetheless sought to persuade me that if reference was made to the Notification Certificates themselves and to the claim document (all of which were lodged in processes and incorporated in the pleadings) it could be seen that the extensions of time granted and the claim proceeded on exactly the same "architect's instructions". Having looked at these documents with [counsel], however, I regret that I am quite unable to take that view. And even if they did, I am not sure that I fully understand the significance bearing in mind the distinct purposes of clause 25 and clause 26, respectively.'*

12 *'[The contractor] is not entitled to claim even items set out in "Preliminaries" for the extended period.'* Parris, J *The Standard Form of Building Contract*, 2nd edition, 1985, Blackwell Science, in Section 10.03: The relationship of JCT 80 clause 25 to clause 26.

13 *Methodist Homes Housing Association Ltd v Messrs Scott & McIntosh* 2 May 1997 Unreported.

14 *Methodist Homes Housing Association Ltd v Messrs Scott & McIntosh* 2 May 1997, unreported. This case concerned a contract under the JCT Standard Building Contract 1980 Edition: clause 25 referring to extensions of time, and clause 26 referring to loss and expense.

**Table 12.3** 'Relevant Events' with corresponding 'Relevant Matters'.

| Relevant Event | Corresponding Relevant Matter under: | | |
|---|---|---|---|
| | JCT SBC/Q 2016 | JCT DB 2016 | JCT IC 2016 |
| 'Variations' or 'Changes' | 4.22.1 | 4.21.1 | 4.17.1 |
| Instructions for: | | | |
| • Resolving a discrepancy | 4.22.2.3 | — | 4.17.2.4 |
| • Postponement of work | 4.22.2.1 | 4.21.2.1 | 4.17.2.1 |
| • Provisional Sums | 4.22.2.1 | 4.21.2.1 | 4.17.2.1 |
| • Opening up/inspection/testing | 4.22.2.2 | 4.21.2.2 | 4.17.2.3 |
| • Objecting to a 'Named specialist' | 4.22.2.4 | — | — |
| Deferment of possession | 4.20.1 | 4.19.1 | 4.15.1 |
| Dealing with 'Antiquities' | 4.22.3 | 4.21.3 | — |
| Inaccurate Approximate Quantity | 4.22.4 | — | 4.17.3 |
| Contractor suspension for nonpayment[a)] | 4.13.2 | 4.11.2 | 4.14.2 |
| Employer's impediment, prevention or default | 4.22.5 | 4.21.5 | 4.17.4 |
| Work by or failure of statutory undertaker | — | — | — |
| Exceptionally adverse weather | — | — | — |
| Loss or damage by Specified Perils | — | — | — |
| Civil commotion or terrorism | — | — | — |
| Strike, lock out, etc | — | — | — |
| New legislation after the Base Date | — | — | — |
| Delay in receiving necessary permission | — | 4.21.4 | — |
| Insolvency of 'Named Specialist' | 4.22.2.5 | — | — |
| Force majeure | — | — | — |

a) The JCT SBC/Q, DB and IC 2016 contracts include the contractor's entitlement for loss and expense ('...*costs and expenses reasonably incurred*...') under the relevant payment provisions, not as Relevant Matters. See also clause 4.7.2 of JCT MW 2016.

## 12.2 The Contractor's Application and Assessment

### 12.2.1 The Key Loss and Expense Clause

The key clauses governing loss and expense in each of the JCT 2016 contracts are as follows:

- JCT SBC/Q 2016 – clause 4.20;
- JCT DB 2016 – clause 4.19;

**Table 12.4** Summary of loss and expense entitlement (JCT 2016 contracts).

| **Entitlement to loss and expense** | **JCT SBC/Q 2016 (cl 4.20.1)** | **JCT DB 2016 (cl 4.19.1)** | **JCT IC 2016 (cl 4.15.1)** |
|---|---|---|---|
| If the contractor: | | | |
| • Incurs loss and expense | ✓ | ✓ | ✓ |
| • Is 'likely to incur' loss and expense | ✓ | ✓ | ✓ |
| Loss and expense results from: | | | |
| • Deferment of possession (of the Site or a Section) | ✓ | ✓ | ✓ |
| • A 'Relevant Matter' ('governing matters') | ✓ | ✓ | ✓ |
| Compliance with the 'Notification and Ascertainment' clauses | ✓ | ✓ | ✓ |
| Subject to the 'exceptions' (below) | ✓ | ✓ | ✓ |
| **Exceptions for claiming loss and expense** | **JCT SBC/Q 2016 (cl 4.20.2)** | **JCT DB 2016 (cl 4.19.2)** | **JCT IC 2016 (cl 4.15.2)** |
| If the Conditions say no entitlement | ✓ | ✓ | ✓ |
| If loss and expense is reimbursed under any other Condition of the contract | ✓ | ✓ | ✓ |

- JCT IC 2016 – clause 4.15; and
- JCT MW 2016 – clause 3.6.3.

The clauses for the JCT SBC/Q, DB and IC 2016 contracts are identical and come in two parts. The first part sets out what Section 12.1.1 of this book described as the 'governing matters' (that entitle the contractor to loss and expense), whereas the second part summarises the 'exclusions' where no such entitlement arises. Table 12.4 summarises these clauses:

### 12.2.2 Commentary on the Key Loss and Expense Clause

Whilst it is in the employer's interests that he, or his representative, gives the contractor an extension of time where the contractor has been delayed by a relevant events (see Section 3.2.1), there is no similar benefit so far as loss and expense is concerned. Neither an employer nor his representative has any duty to advise the contractor to claim for loss and expense and, indeed, an employer's representative if giving such advice, may be in breach of their duty to the employer. Whether or not the contractor decides to make such a claim is entirely up to him.

First, the following factors must be considered before the contractor makes his application for loss and expense:

- Is the contractor likely to incur loss and expense (caused by deferment or one or more of the relevant matters), or the loss and expense has actually been incurred; and have the 'relevant matters' (but not deferment) 'materially affected' the regular progress of the Works (or a Section)?

  By specifically mentioning that the 'relevant matters' (but not deferment) must 'materially affect' the contractor's progress, suggests that incurring loss and expense is relatively, a more straightforward matter if possession has been deferred. It doesn't appear that the contract requires the contractor to show that deferment affected his regular progress of work.

Second, none of the exceptions (see Table 12.4) should apply. This means that

- If the contract conditions provide that there shall be no addition to the Contract Sum, then the contractor has no entitlement to loss and expense; and
- The contractor must comply with the contract's 'notification and ascertainment' clauses.

The first 'exception' to the contractor claiming loss and expense is where the relevant contract says that there will be no addition to the contract sum. Unfortunately, the JCT 2016 contracts' drafting could be clearer in this regard, as not only do some clauses refer to the contractor not being entitled to any 'addition to the Contract sum', but other clauses refer to certain matters as being 'at no cost to the Employer' (or in other words, at the contractor's 'own cost'). Practically, these terms have the same effect – removing the contractor's entitlement to additional payment, which may otherwise be added to the Contract Sum. For the purposes of this Section 12.2.2, both terms are considered to have the same meaning and are summarised in Table 12.5:

The second 'exception' (except under JCT MW 2016) will apply if the contractor fails to comply with the clauses governing his notification and assessment. The contractor must notify the employer (or his representative) and submit details of his assessments as follows:

- The notification[15] must be submitted within what is seemingly intended to be a relatively short time – '...as soon as the likely effect...becomes (or should have become) reasonably apparent to him' [16] after either:
  - The effect that the Relevant Matter is having on the regular progress of the works; or
  - The 'likely nature and extent' of loss and expense resulting from deferred possession of the Works (or a Section); and may
  - Be accompanied by the contractor's initial loss and expense assessment.

15 JCT SBC/Q 2016 – clause 4.21.1, JCT DB 2016 – clause 4.20.1 and JCT IC 2016 – clause 4.16.1.

16 Considering similar wording for the contractor's 'application' for loss and expense under the former JCT 63 contract: *'to require applications to be made "as soon as it has become, or should reasonably have become apparent to him [the contractor] that the regular progress of the works or any part thereof has been or is likely to be affected" by specified events... and to state "that he has incurred or is likely to incur direct loss and/or expense"...' (Italics in judgment)' F G Minter Ltd v Welsh Health Technical Services Organisation* (1980) 13 BLR 7 at 20 per Stephenson LJ

**Table 12.5** Circumstances where 'no addition to the Contract Sum' is permitted – JCT 2016 contracts.

| Circumstances where 'no addition to the Contract Sum' permitted | JCT SBC/Q 2016 | JCT DB 2016 | JCT IC 2016 |
|---|---|---|---|
| Levels and setting out | 2.10 | — | 2.9 |
| Contract documents: | — | — | — |
| • Errors | 2.14.4 | — | — |
| • Discrepancies | 2.14.4, 2.16.1 | 2.14.1 | — |
| • Divergencies | 2.14.4 | — | — |
| (inc from statutory requirements) | 2.17.2.1 | 2.15.1 | — |
| Schedules of defects and instructions | 2.38 | 2.35.2 | 2.30 |
| Inspection/tests in the contract documents | 3.17 | 3.12 | 3.14 |
| Work not in accordance with the Contract | 3.18.3, 3.18.4 | 3.13.2, 3.13.3 | 3.15.1 |
| Workmanship not in accordance with the Contract | 3.19 | 3.14 | — |
| Instructions for the removal of work | — | — | 3.16.2 |
| Compliance with CDM Regulations | 3.23.3 | 3.16.4 | 3.18.3 |
| Variation Quotation – non-compliance | | Schedule 2, para 2.5, para 2.6 | — |
| Code of Practice | Schedule 4, para 1 | Schedule 4, para 1 | — |
| Fluctuations Option A | Schedule 7, para A.8 | Schedule 7, para A.8 | — |

- If the contractor's notification does not include his initial loss and expense assessment, then it must be provided 'as soon as is reasonably practicable' afterwards. The initial assessment[17] must include:
  - The loss and expense either incurred and/or
  - Further amounts likely to be incurred along with
  - Information[18] that is reasonably necessary for the employer or his representative to 'ascertain' the loss and expense.

17 JCT SBC/Q 2016 – clause 4.21.2, JCT DB 2016 – clause 4.20.2 and JCT IC 2016 – clause 4.16.2.
18 See also Section 10.1.3 (Requests for further information)

- The contractor must provide 'updates' to his initial assessment of loss and expense 'at monthly intervals' until the employer or his representative has all the information that they 'reasonably require' to complete their ascertainment.[19]

The apparent urgency of the contractor's notification is not to impose a 'time bar' clause to disentitle the contractor,[20] but to allow the employer or his representative (which may include the quantity surveyor) a better opportunity to mitigate the relevant matter and to begin their ascertainment at the earliest possible stage. If the employer is unable to mitigate, then he may need to secure additional finance or reduce the cost of the project (e.g. by omitting work).

### 12.2.3 Content of the Contractor's Application

The 'contractor's application' in this section is that is meant to include:

- His initial notification[15];
- The initial assessment of loss and expense[17]; and
- Any 'updates.'[19]

The JCT 20106 contracts do not provide any set format for the contractor's application, so it is left to the contractor to devise a format to suit itself. However, the contractor's application must be in writing and must fulfil the requirements set out in Section 12.2.2 above. It may be that the application is sufficient if it refers to the applicable relevant matter, identifies the occurrence, and states that loss and expense is being, or is likely to be incurred. By specifying that the contractor's initial assessment of loss and expense may be provided separately from (although soon after) its notification, and that the further 'updates' (including information requested by the employer or his representative) are subsequently provided, this suggests that the contractor only needs to provide a single notification for each cause of loss and expense. Indeed, this is particularly the case where possession of the works is deferred, so a single notification should suffice.

An employer or his representative may first become aware that there is likely to be a claim for loss and expense when it accompanies a notification of delay and further particulars. The contractor may often simply include additional words to the effect that it is also seeking loss and expense on the same grounds. Although it may be argued that what appears to be a casual approach is insufficient to comply with the relevant loss and expense clause, a prudent employer (or employer's representative) will examine whether the notice, brief though it may be, sufficiently complies with the contract. Whether it does so or not, will be a question of fact. Notices from the contractor giving no other information save that a claim for loss and expense may be provided are unlikely to satisfy the loss and expense clauses. Such a plainly deficient notification can be countered by the employer or his representative in the following ways:

- He can simply respond and reject the purported notification by pointing out its inadequacies (i.e. what's missing);

19 JCT SBC/Q 2016 – clause 4.21.3, JCT DB 2016 – clause 4.20.3 and JCT IC 2016 – clause 4.16.2.
20 As is the case with NEC4, clause 61.3 (see Section 2.9.2)

- He can respond, pointing out that the purported notification does not include the information that is 'reasonably necessary' for them to begin their ascertainment[17]; or
- In a subsequent payment notice (or pay-less notice), he can include a note that the contractor's loss and expense claim cannot be considered (or 'ascertained' – see Section 12.3) because the contractor has failed to meet the relevant notification requirements. In doing so, the employer, or his representative should not include 'loss and expense' as a new item, and value it at '£Nil,' as this could imply that some kind of 'ascertainment' has been carried out. Such a 'note' should be clear that any ascertainment has not been possible because of the deficiencies in the contractor's notice.

The employer (or his representative) should carefully consider the wording of such a response. Given that 'any difference', including one of opinion, constitutes a 'dispute' that the contractor could immediately refer to adjudication,[21] it is suggested that any response should be couched in terms of 'owing to the contractor's failure to comply with the contract's notification requirements , the employer is unable to consider the claim for loss and expense', rather than including words that reject the claim or value it at £Nil. Until the contractor complies with the contract's notification requirements (and provides his initial assessment), the employer's obligation to consider (or ascertain) any loss and expense claim will not commence.

The contractor's notification must clearly specify whether (1) deferment of possession or (2) which relevant matter has caused the incurrence of, or likelihood of incurring loss and expense. The contractor should be encouraged to provide as much information as he can about the surrounding circumstances, including details of the actual events which are the grounds for the claim. This 'encouragement' is good practice in any event, and if the employer (or his representative) find the information useful, then they're likely to include a value for loss and expense in the contractor's interim payments sooner rather than later. Therefore, it is clearly in the contractor's commercial interests to do so. In any event, the contractor is seemingly obliged to provide an explanation if the employer (or his representative) request one.

A contractor may issue standard correspondence (i.e. generated from a specific email, or letter template – covering purported notices of delay, extension of time, and loss and expense), every time something occurs which might fall under one of the relevant events and/or relevant matters. If this standard correspondence (in the form of a loss and expense application) contains the information required by the contract, then it must be considered, irrespective of how many such standard communications are received. However, if the standard communication does not satisfy the contract's requirements, it will not trigger the employer's or his representative's obligation to 'ascertain' the loss and expense. Where such standard correspondence plainly does not satisfy the contract, submitting it is not only a fruitless exercise, but may bring the contractor's credibility into question. The requirements of the contractor's notification and assessment of loss and expense are summarised in Table 12.6:

### 12.2.4 Timing of the Contractor's Notification ('Relevant Matters').

The JCT 2016 contracts all state that the contractor shall comply with the contract's relevant notification and assessment clauses (see Table 12.4). These kinds of provisions are often

21 See s108(1) HGCRA 1996 (as amended)

**Table 12.6** Summary of requirements for contractor's notification/assessment of loss and expense.

| Requirements of Contractor's notification and assessment of loss and expense | JCT 2016 contract: | | |
|---|---|---|---|
| | SBC/Q | DB | IC |
| Initial notification of loss and expense made when the following becomes apparent (or should become apparent): | — | — | — |
| • Regular progress affected by Relevant Matter; or | 4.21.1 | 4.20.1 | 4.16.1 |
| • Likely nature and extent of loss and expense | 4.21.1 | 4.20.1 | 4.16.1 |
| Notification must specify loss and expense incurred by: | — | — | — |
| • Deferment of possession | 4.20.1, 4.21.1 | 4.19.1, 4.20.1 | 4.15.1, 4.16.1 |
| • A 'relevant matter' | 4.21.1 | 4.20.1 | 4.16.1 |
| Contractor provides initial assessment of loss and expense: | — | — | — |
| • With the notification; or | 4.21.2 | 4.20.2 | 4.16.2 |
| • As soon as practicable, thereafter | 4.21.2 | 4.20.2 | 4.16.2 |
| Contractor's assessment of loss and expense includes: | — | — | — |
| • Loss and expense already incurred and | 4.21.2 | 4.20.2 | 4.16.2 |
| • Loss and expense likely to be incurred | 4.21.2 | 4.20.2 | 4.16.2 |
| • Include information reasonably necessary to facilitate ascertainment | 4.21.2 | 4.20.2 | 4.16.2 |
| Contractor to provide updates: | — | — | — |
| • In the form requested and | 4.21.3 | 4.20.3 | 4.16.3 |
| • At monthly intervals (until fully ascertained) | 4.21.3 | 4.20.3 | 4.16.3 |

referred to as 'conditions precedent' because one party has to fulfil a function in order for the contractual machinery to proceed to the next step. However, the effect of the 'condition precedent' in these JCT clauses[22] is not to impose a 'time bar' to remove the contractor's entitlement to loss and expense (if the fails to notify within a particular period[23]), but to only trigger the employer's obligation to ascertain the loss and expense, once the contractor's notification is provided.

Therefore, employers and their representatives must remember that their duty to ascertain loss claims only begins once the contractor has provided a compliant notification of loss and expense. It is not that they may ignore such claims until then, but that they simply have no power to consider them. The relevant 'notification and ascertainment' clauses[24] require the contractor's notification is made as soon as it becomes *'reasonably apparent'* that:

22 JCT SBC/Q 2016 – clause 4.20.1, JCT DB 2016 – clause 4.19.1 and JCT IC 2016 – clause 4.15.1
23 E.g. NEC4 – clause 61.3 (see Sections 2.9.2 and 12.2.2)
24 JCT SBC/Q 2016 – clause 4.21.1, JCT DB 2016 – clause 4.20.1 and JCT IC 2016 – clause 4.16.1

- *'...the likely nature and extent of any loss and expense'* resulting from deferment of possession; or
- *'...the likely effect of a Relevant Matter'* on regular progress;

is known.

A 2010 case[25] dealt with a standard trade contract under the JCT trade contract terms (TC/C 2002), in which the loss and/or expense clause (4.21) was broadly similar to those in the JCT 2016 contracts, save that a 'long stop' of 2 months had been inserted – before which, the trade contractor should submit his loss and expense application. The court concluded that the obligation to make application timeously was not so strict:

> *'It does seem however that the wording is such that the two-month long-stop period and indeed general periods run from one of two stages, namely either when it has become apparent or when it should reasonably have become apparent that the progress of the works was or was likely to be affected. There is no reason to construe this part of the sub-sub-clause in any way other than in effect giving the contractor the option of making its application under cl 4.21 at the later of the two alternative stages, if as a matter of fact they turn out to be different. Thus, the date when the regular progress of the works was actually affected may well be later than the date when it became reasonably apparent that the regular progress of the works was likely to be affected. Depending on the facts, it may be that the time for making any given application under cl 4.21 can await a time when actual delay to the relevant part or the whole of the works has materialised.'*

In a 2014 case,[26] the court considered a similarly worded notice provision:

> *'311. ...It is clear and indeed was unequivocally and properly accepted...that clause 20.1 imposes a condition precedent:*
> *"20.1 If the contractor considers himself to be entitled to any extension of the time for completion...under any clause of these conditions or otherwise in connection with the contract, the contractor shall give notice to the engineer, describing the event or circumstance giving rise to the claim. The notice shall be given as soon as practicable[27]...after the Contractor became aware, or should have become aware, of the event or circumstance..."*
> *312. Properly construed and in practice, the "event or circumstance giving rise to the claim" for extension must first occur and there must have been either awareness by the Contractor or the means of knowledge or awareness of that event or circumstance before the condition precedent bites...*

25 *WW Gear Construction Ltd v McGee Group Ltd* (2010) EWHC 1460 (TCC) at 75 per Akenhead J.
26 *Obrascon Huarte Lain SA v HM AG for of Gibraltar* (2014) EWHC 1028 (TCC). The case concerned a contract let under the FIDIC Conditions of Contract for Plant and Design Build 1st Edition 1999 ('FIDIC Yellow Book 1999'), which (in clause 20.1 'Claims') contained a similar provision to that in the JCT 2016 contracts for claiming loss and expense.
27 *Obrascon Huarte Lain SA v HM AG for of Gibraltar* (2014) EWHC 1028 (TCC). Clause 20.1 of FIDIC Yellow Book 1999 contains a time-bar of 28 days for the contractor's notification, but that is not relevant here.

> *313. Additionally, there is no particular form called for in Clause 20.1 and one should construe it as permitting any claim provided that it is made by notice in writing to the Engineer, that the notice describes the event or circumstance relied on and that the notice is intended to notify a claim for extension (or for additional payment or both) under the Contract or in connection with it. It must be recognisable as a "claim". The notice must be given as soon as practicable … after the Contractor has become or should have become aware. The onus of proof is on the Employer …here to establish that the notice was given too late.'*

The decision in *Obrascon* is helpful as it not only provides further assistance in clarifying that complying with the 'Notification and ascertainment' clauses[26] are a condition precedent to the contractor's entitlement to loss and expense, but that the contractor's notification should be sufficiently clear and be recognisable as such[28].

It is suggested that the inclusion of the words 'as soon as' in these clauses provide no excuse if the contractor's written application is made after the regular progress has been affected – if it was reasonable for the contractor to make the application earlier. Even if it was not reasonable for the contractor to apply before the occurrence began to affect progress, it is clear that the notification must be made as soon as the effects of a relevant matter on progress, or the extent and nature of loss and expense caused by deferment are reasonably apparent, and not within a reasonable time thereafter. The words 'within a reasonable time' are not included in these clauses.

Notification under these clauses should be the result of a considered decision made by the contractor and it should not be simply an automatic response to every instruction or occurrence on site (see the paragraph discussing 'standard correspondence' above). Obviously, in some instances, it will be difficult for the contractor to determine whether progress is likely to be affected, particularly if progress has already been disrupted, and it is not immediately apparent that a further Relevant Matter has not added to the effect. There may also be other factors:

> *'Notice of intention to claim, however, could not well be given until the intention had been formed . . . [and] it seems to me that the contractors must at least be allowed a reasonable time in which to make up their minds. Here the contractors are a limited company, and that involves that, in a matter of such importance as that raised by the present case, the relevant intention must be that of the board of management [i.e. directors] ... in determining whether a notice has been given as soon as practicable, all the relevant circumstances must be taken into consideration .... One of the circumstances to be considered in the present case is the fact that it was not easy to determine whether the engineer's orders . . . did or did not involve additional work ...'.*[29]

---

28 *Obrascon Huarte Lain SA v HM AG for of Gibraltar* [2014] EWHC 1028 (TCC) para 313. See also *Caledonian Modular Limited v MarCity Developments Ltd* (2015) EWHC 1855 (TCC) concerning an interim payment claim, but the principle is relevant: para 37, as per Coulson J '*…if contractors want the benefit of these provisions, they are obliged, in return, to set out their interim payment claims with proper clarity.'*

29 *Tersons Ltd v Stevenage Development Corporation* (1963) 5 BLR 54 at 68 per Wilmer LJ. It should be noted that the court was there concerned with the ICE Conditions (2nd Edition) and the question of whether certain notices were given 'as soon as practicable'.

Similar circumstances can be envisaged in relation to instructions, although it is doubtful that a contractor will need to consult its board of directors before every application for loss and/or expense. It is far more likely that matters that are likely to cause the contractor to incur loss and expense can be dealt with at project level, by the contractor's staff engaged on the project. If the employer or his representative have any doubt whether the contractor's application has been made in due time, a useful test is to consider whether the alleged lateness of the application prejudices the employer's interests in any way.

### 12.2.5 Meaning of Loss and Expense

The clause refers to the contractor incurring 'direct loss and/or expense,' so consequential losses are not covered. What the contractor is claiming under this clause may be equated with the common law right to damages for losses which are 'direct' but not 'indirect' or 'consequential'[30] (see Sections 2.3, 4.2 and 4.3). This relatively simple concept is often not fully understood by many contractors, employer or their representatives.

The loss and expense provisions in the JCT 2016 contracts preserve the contractor's right to claim common law damages (see Section 12.2.6 below).

### 12.2.6 Reimbursement Under Other Conditions of the Contract

To the extent to which he is reimbursed under any other clause, the contractor shall not be reimbursed under the loss and expense clauses.[31] This is to avoid duplication (also called 'double counting'). An obvious example is that if the contractor is entitled to additional payment under the variation (or change) clauses,[32] he cannot firstly claim additional costs as part of the variation (or change), and again as loss and expense. The loss and expense clauses do not stipulate that the contractor's entitlement is mutually exclusive to either (1) the loss and expense clauses or (2) the clauses covering the valuation of variations (or other entitlement for additional payment); they are intended to distinguish between the *'extent that the Contractor is reimbursed'*[31] under either of these categories. Table 12.7 summarises the circumstances (other than for loss and expense) under each JCT 2016 contract from which the contractor can be reimbursed. Clause references without brackets refer to the payment mechanism clause, whereas clauses in brackets refer to the clauses entitling the contractor to payment.

The loss and expense provisions in the JCT 2016 contracts also enable the contractor to be reimbursed for the 'Relevant Matters', which include both breaches of contract (by the employer), and for other matters, which are not breaches (such as deferment of possession, or instructing variations or changes – see Section 12.1 and Tables 12.1 and 12.2). However, the 'loss and expense' recoverable against the relevant matters is only those costs which the contractor has incurred directly (see Section 12.2.5). If the contractor has incurred 'indirect' or 'consequential' losses, and wants to recover these, then he must attempt to claim them as common law damages. The JCT 2016 contracts permit recovery of common law damages under the clauses entitled 'Reservation of the Contractor's

30 *F G Minter Ltd v Welsh Health Technical Services Organisation* (1980) 13 BLR 1

31 JCT SBC/Q 2016 – clause 4.20.2, JCT DB 2016 – clause 4.19.2 and JCT IC 2016 – clause 4.15.2.

32 Clause 5 under JCT SBC/Q 2016, (as a 'Change') under JCT DB 2016 and JCT IC 2016.

**Table 12.7** JCT 2016 contracts – reimbursement other than for loss and expense.

| Contractor reimbursed under other clauses (other than loss and expense): | JCT SBC/Q 2016 | JCT DB 2016[a)] | JCT IC 2016 |
|---|---|---|---|
| Interest for late payments | 4.14.1 | 4.9.6 | 4.12.6 |
| Value of completed work in 'stages' | — | 4.12.1.1 | — |
| Work properly executed (including materials and 'Listed Items'[b)] | 4.14.1 | 4.13.1.1<br>4.12.1.3 | 4.9.1 |
| Value of 'Changes' | — | 4.12.1.2 | — |
| Fluctuations | 4.14.1 | 4.12.1.4 | 4.9.1 |
| Acceleration Quotation | 4.14.1 | 4.12.2.3<br>4.13.2.3 | — |
| Value of 'activity' in Activity Schedule | 4.14.1 | — | 4.9.1 |
| Adjustments to the Contract Sum: | 4.14.2.1 | 4.12.2.1<br>4.13.2.1 | 4.9.2 |
| • Insurance for early use by employer | (2.6.2) | (2.5.2) | (2.6) |
| • Fees, charges inc for patent rights etc | (2.21) | — | (2.3) |
| • Instructions for patent rights | (2.23) | (2.20) | — |
| • Inspection/test for compliant work | (3.17) | (3.12) | (3.14) |
| • Inspection/text – unreasonable instruction | — | — | (3.15.3) |
| • Contractor insuring for employer | (6.5) | — | (6.5) |
| • Contractor insuring – Terrorism cover | (6.10.2)<br>(6.10.3) | (6.10.2)<br>(6.10.3) | (6.10) |
| • Contractor insuring – alternative Terrorism cover | (6.11.3) | (6.11.3) | (6.11) |
| • Employer fails to evidence insurance | (6.12.2) | (6.12.2) | (6.12.2) |
| • Joint fire code amendments | (6.20) | (6.20) | (6.18) |
| Contractor's suspension | 4.14.2.2<br>(4.13.2) | 4.12.2.2<br>4.13.2.2<br>(4.11.2) | 4.9.2<br>(4.14) |
| Insurance – Options B and C | 4.14.2.4<br>(6.13.5.3)<br>(6.13.4) | 4.12.2.4<br>4.13.2.4<br>(6.13.5.3)<br>(6.13.4) | 4.9.2<br>(6.13.4) |
| Other Fluctuations | 4.14.2.5 | 4.12.2.5<br>4.13.2.5 | 4.9.2<br>(4.4) |

a) JCT DB 2016 – Clause references beginning with '4.12' are for Payment Alternative A (Stage Payments), whereas those beginning with '4.13' are for Alternative B (Periodic Payments)

b) For JCT DB 2016, this excludes any value for 'Changes' which is included in clauses 4.12.1.2, Payment Option A (Stage Payments)

rights and remedies.'[33] However, in the event that the contractor seeks reimbursement of 'indirect' or 'consequential' losses, he must show that such losses were within the reasonable contemplation of the employer when he entered into the contract.[34]

### 12.2.7 Quotations and Estimates for Loss and Expense and Acceleration

JCT DB 2016, in the 'Supplemental Provisions' (at paragraph 3 of Part 1) contains a procedure for the contractor to submit estimates for loss and expense. The relevant entry in the contract particulars must be made for the 'loss and expense' estimate procedure to apply.[35] The procedure and content for the contractor submitting an estimate of his loss and expense under JCT DB 2016 is set out in Figure 12.1:

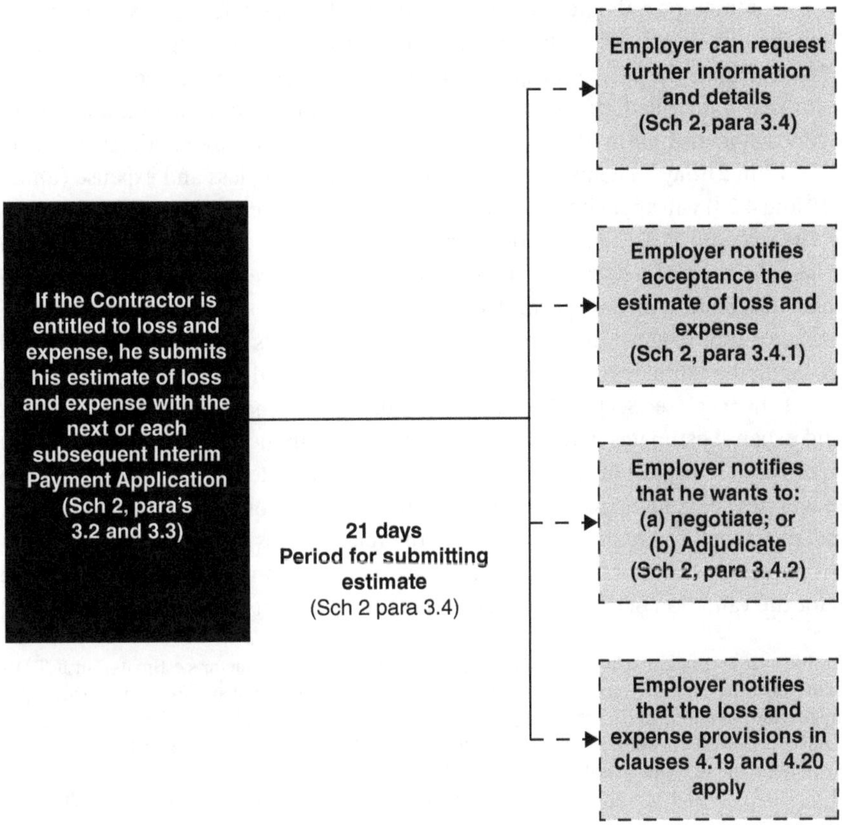


**Figure 12.1** JCT DB 2016 – Loss and expense – Contractor's estimates.

33 JCT SBC/Q 2016 and JCT DB 2016 – clause 4.23, JCT IC 2016 – clause 4.19. For JCT MW 2016, the contractor's right to claim for loss and expense is limited to those matters arising from a variation under clause 3.6.3, so claiming for damages is outside the contract's provisions, so the employer's representative will not have the authority to consider them.
34 See section 4.2 and the 'second limb' under *Hadley v Baxendale* (1854) 9 Ex 341
35 JCT DB 2016 – Contract Particulars, entry against 'Seventh Recital and Part 1 of Schedule 2' (Supplemental Provisions) for 'Supplemental Provision 3'.

JCT DB 2016's 'Loss and expense – Contractor's estimates' procedure is similar to the provisions[36] for the contractor to submit quotations or estimates for the following:

- JCT SBC/Q 2016 – quotations for:
  - Variations[37] and
  - Acceleration[38]
- JCT DB 2016 – estimates for:
  - Valuation of Changes[39]
  - Acceleration.[40]

The contents of, and procedures for submitting, accepting or not accepting the quotations under JCT SBC/Q 2016 and the estimates for the 'valuation of changes' and 'acceleration' under JCT DB 2016 are set out in detail in Section 5.4.2 in Chapter 5.

Each of the contractor's estimates of loss and expense shall include the loss and expense incurred for the period preceding the subsequent Interim Payment Application[41] and is repeated for as long as the loss and expense is incurred.[42] Once the contractor's loss and expense estimate is provided, the employer may request further information and details from the contractor, and notify the contractor of one of the first three options shown in Figure 12.1 within 21 days. The usual provisions for ascertaining loss and expense (under clauses 4.19 and 4.20) will apply if the employer either notified the contractor[43] or if negotiations are unsuccessful.[44] Any amount of loss and expense that is either accepted or agreed following negotiation is added to the contract sum, but is effectively in full and final settlement, as no further additional are permitted.[45]

The contractor should ensure he complies with the time limits set out in 'supplemental provisions' for giving estimates of 'loss and expense' – as he should also do when submitting a 'valuation of changes' (see Section 5.4.2). If the contractor does not comply with providing the loss and expense estimates with each interim payment application, then any addition to the contract sum in respect of an estimate that was not submitted in accordance with the procedure will not be included in any interim payments.[46] The contractor will have to wait for the final payment[47] to become due to receive the relevant loss and expense. This provision is less onerous for the contractor, when compared to the contractor failing to provide estimates for the valuation of changes (refer to Section 5.4.2) as it will only be the loss and

36 For JCT SBC/Q 2016 – refer to clause 5.3. For the applicability of the 'Contractor's estimates' in JCT DB 2016 – refer to the entries in the Contract Particulars against the Seventh Recital and Part 1 of Schedule 2 for the estimates for 'Valuation of Changes' (Supplemental Provision 2 will apply), and 'Loss and Expense' (Supplemental Provision 3 will apply), and the Seventh Recital and Part 2 of Schedule 2 for the 'Acceleration Quotation' (Supplemental Provision 4 will apply).
37 JCT SBC/Q 2016 – see clause 5.3, and Schedule 2 (Variation and Acceleration Quotation Procedures) paragraph 1.2.3
38 JCT SBC/Q 2016 – see clause 5.3, and Schedule 2, paragraph 2.1.1
39 JCT DB 2016 – see Schedule 2 (Supplemental Provisions) – paragraph 2.3.1
40 JCT DB 2016 – see Schedule 2, paragraph 4.1.2
41 JCT DB 2016 – Schedule 2, paragraphs 3.2 and 3.3.
42 JCT DB 2016 – Schedule 2, paragraph 3.3.
43 JCT DB 2016 – Schedule 2, paragraph 3.4.2. See the fourth option in Figure 12.1.
44 JCT DB 2016 – Schedule 2, paragraph 3.4.
45 JCT DB 2016 – Schedule 2, paragraph 3.5
46 JCT DB 2016 – Schedule 2, para 3.6.
47 JCT DB 2016 – clause 4.24.5

expense within the estimate that is submitted late that won't be included in the relevant interim payment and not the entirety of the loss and expense. The contractor will be entitled to have loss and expense included in interim payments for all other estimates of loss and expense that are submitted correctly.

### 12.2.8 Effect on Regular Progress

The bases of the loss and expense clauses[48] are that deferment of possession has given rise to loss and expense or that the regular progress of the Works, or any part has been or is likely to be materially affected by any one or more of the relevant matters. In other words, it is the effect that the deferment or relevant matter had upon the regular progress of the Works, for example, any significant delay to or disruption to the regular progress of the contract.

Some commentators misunderstand the clause, saying that it is only when the completion date is delayed that a claim is possible, but that is not what the clause says – 'Relevant Events' do not always correspond with 'Relevant Matters' (see Section 12.1.2). An obvious example is that regular progress can be affected other than by matters than just delays (refer to Section 4.4.3). To exclude other adverse effects on progress, pays little attention to the words used and goes against common sense and the straightforward commercial intention of the contract. There can be a disturbance to regular progress, resulting in disruption and loss of productivity in working, without causing a critical delay to completion of the Works.

It would have been simple to use clear words to confine the entitlement to delays to the regular progress affecting the completion date, but the JCT chose not to do so; preferring a broader expression. The clauses[47] cannot be interpreted so as to confine the contractor's right to reimbursement solely to circumstances that delay the completion date. They cover circumstances that may give rise, for instance, to reduced efficiency of working without progress as a whole being delayed.

It should be clear that this is not saying that just because the work has cost more or has taken longer to complete than was anticipated, that the contractor is entitled to additional payment. The contractor must demonstrate that the effect upon regular progress of the Works is directly attributable to deferment and/or to one (or more) of the Relevant Matters set out in the relevant clauses.[49]

The words 'regular progress' have caused difficulty, as their common interpretation relates to the contractors obligation to proceed with the Works regularly and diligently,[50] which the Court of Appeal defined as:

> *'What particularly is supplied by the word "regularly" is not least a requirement to attend for work on a regular daily basis with sufficient in the way of men, materials and plant to have the physical capacity to progress the works substantially in accordance with the contractual obligations.*

48 JCT SBC/Q 2016 – clause 4.20.1, JCT DB 2016 – clause 4.19.1, and JCT IC 2016 – 4.15.1.
49 JCT SBC/Q 2016 – clause 4.22, JCT DB 2016 – clause 4.21and JCT IC 2016 – clause 4.17.
50 JCT SBC/Q 2016 and JCT IC 2016 – clause 2.4, and JCT DB 2016 – clause 2.3. Whilst JCT MW 2016 does not expressly refer to proceeding 'regularly and diligently' in clause 2 (Contractor's Obligations), failing to do so is a default that entitles the employer to begin to terminate the contractor's employment under the contract (clause 6.4.1.2). Failing to proceed 'regularly and diligently' is also a 'specified default' entitling the employer to begin the termination procedures under clause 8.4.1.2 of JCT SBC/Q 2016, JCT DB 2016 and JCT IC 2016.

> *What in particular the word "diligently" contributes to the concept is the need to apply that physical capacity industriously and efficiently towards the same end.*
> *Taken together the obligation upon the contractor is essentially to proceed continuously, industriously and efficiently with appropriate physical resources so as to progress the works steadily towards completion substantially in accordance with the contractual requirements as to time, sequence and quality of work.'*[51]

The questions of whether or not the contractor has maintained regular progress and whether or not such progress has been (or is likely to be) materially affected must be answered by the employer or his representative. Their judgment must be exercised objectively and according to principles laid down by law. In carrying out this duty, the employer (or his representative) is likely to be assisted by the contractor's programme, along with any subsequent programmes (and he should be encouraged to record progress himself – see Section 9.18) that either show actual progress, or from which a comparison between planned and actual progress can be established (see Section 3.5).

The contractor's progress may already be irregular, partly due to factors within its control, or which do not entitle it to loss and expense. That may not be fatal to a loss and expense claim, although it will present the contractor with some evidential problems. Among other things, he will have to demonstrate what regular progress should have been (and further prove that, irrespective of its own failures in this respect, regular progress would have been) affected by the matter specified.

The reference to any part of the Works clearly emphasises the distinction between the extension of time and loss and expense clauses. Extensions of time must relate to critical delays – in completion of the contract as a whole or to a section. Loss and expense for the effect on regular progress may relate to circumstances affecting any part of the Works, even down to individual work operations – irrespective of whether those operations are critical activities.

Regular progress must have been or be likely to be *'materially'* affected. 'Materially' has been defined as 'in a significant way' or 'considerably',[52] and it is suggested that this definition is applicable here. Trivial disruptions, which are bound to occur on even the best-run projects, are clearly excluded. The circumstances must be such as to affect regular progress of the work to a significant or considerable[52] degree. The particular point at which disruption becomes significant or considerable, is impossible to define in general terms as it must depend upon the circumstances of each case, and is likely to be relative to other periods of progress that were unaffected.

### 12.2.9 Provision of Further Information

The loss and expense clauses[53] require the contractor to supply such information as is reasonably necessary to enable the employer or his representative to ascertain the

---

51 *West Faulkner Associates v London Borough of Newham* (1994) 71 BLR 1.
52 The Oxford Dictionary of English (3rd Ed) 2010. 'Considerably' is defined as 'a notably large amount, or to a notably large extent; greatly'
53 JCT SBC/Q 2016 – clause 4.21.2, JCT DB 2016 – clause 4.20.2, and JCT IC 2016 – clause 4.16.2.

loss and expense. In addition, and after the contractor's initial notification[54] and initial assessment[55] of the loss and expense, the employer or his representative may request further information '*in such a form and* manner' for the contractor to provide at monthly intervals.[56] It is clear that these clauses do not come into effect until the employer or his representative request the 'further information' from the contractor. It is in the contractor's interest to provide as much relevant information as possible at the time of its loss and expense application, and not to wait until the employer or his representative to request it. The information which the contractor is obliged to provide is that which should reasonably enable the employer or his representative to ascertain the loss and expense.

In requesting further information, it is thought that the employer or his representative should be precise in terms of the information required. For example: requesting a class of information, requesting specific dates for the site diary, or for labourer's time sheets, rather than simply requesting 'information that requires the contractor to 'prove' its claim'. Vague requests of this kind are often used by employers or their representatives as a delaying tactic, but they should exercise caution, as the contractor may simply lose patience, and get to a point when it sincerely believes that it has provided everything the employer or his representative reasonably need and simply adjudicate and refuse to provide anything further. However, there is a danger with the latter approach, and the contractor should be sure of its grounds and not simply be tired of digging out old records. It has been said:

> *'If [the contractor] makes a claim but fails to do so with sufficient particularity to enable the architect to perform his duty or if he fails to answer a reasonable request for further information he may lose any right to recover loss or expense under those sub-clauses and may not be in a position to complain that the architect was in breach of his duty.'*[57]

These are sensible words which highlight not only the contractor's responsibility, but also the consequences if it refuses to help itself. A question that often arises is whether there is any time limit on the contractor providing information, or whether a contractor, dissatisfied with the results of the ascertainment, can continue to submit further information right up to the issue of the final certificate[58] or the relevant final statement,[59] and expect it to be considered by the employer or his representative. It appears that there is no express restriction on the provision of information until such time as the final certificate (or the relevant final statement) becomes conclusive evidence of the amount of loss and expense[60] due to the contractor. In a 2003 case, the court, when considering a similar JCT contract provision, said:

54 JCT SBC/Q 2016 – clause 4.21.1, JCT DB 2016 – clause 4.20.1 and JCT IC 2016 – clause 4.16.1.
55 JCT SBC/Q 2016 – clause 4.21.2, JCT DB 2016 – clause 4.20.2 and JCT IC 2016 – clause 4.16.2.
56 JCT SBC/Q 2016 – clause 4.21.3, JCT DB 2016 – clause 4.20.3 and JCT IC 2016 – clause 4.16.3.
57 *London Borough of Merton v Stanley Hugh Leach Ltd* (1985) 32 BLR 51 at 104
58 JCT SBC/Q 2016 – clause 4.25 and JCT IC 2016 – clause 4.21
59 JCT DB 2016 – clause 4.24 (either the Final Statement or the Employer's Final Statement)
60 JCT SBC/Q 2016 – clause 1.9.1.4, JCT DB 2016 – clause 1.8.1.3, and JCT IC 2016 – clause 1.9.1.4, in each case the contractor must contest the Final Certificate or relevant Final Statement by commencing proceedings not later than 28 days after either (1) the Final Certificate is issued (JCT SBC/Q 2016 and JCT IC 2016 – clause 1.9.2), or (2) the relevant Final Statement is issued (JCT DB 2016 – clause 1.8.2)

> *'I cannot accept that the contract terms, properly construed, prohibit the provision and receipt of further information, documentation or details about direct loss and expense after the six month period following practical completion. Such a stringent time-bar would in my view require to be expressed in clear and unambiguous language, which I have been unable to find in the contract terms. On the contrary, the wording of [the clause] suggests that the [contractor] are correct in their contention that the contractual provisions simply provide a time table to which the parties are expected to adhere.'*[61]

This suggests that, although the provisions relating to the "final adjustment and final payment'[62] or to the 'final statement and final payment'[63] require information to be respectively submitted no later than 6 months or 3 months after practical completion, are not to be strictly enforced when new information is submitted. That would include considering whether the contractor has already had adequate opportunity to submit more information, whether the information is truly 'new' or simply a repackaging of old information, and whether a new argument based on existing information is being submitted.

The employer or his representative have the following 'self-help' remedies in these situations:

- If the contractor has failed to provide all the information necessary to assess the loss and expense:
  - For interim payments, then the employer or his representative will be justified in only including those sums which can be reasonably substantiated by the information that the contractor has provided; and
  - Within the time limits for the 'final adjustment'[64] of the contract sum, then the employer or his representative can adjust the contract sum based upon the documents in their possession, if the contractor fails to provide the required information after being notified to do so.[65]
- If the contractor continues to provide information beyond 6 months (JCT SBC/Q 2016 and JCT IC 2016) or 3 months (JCT DB 2016) after practical completion, the employer or his representative can simply proceed to issue:
  - the 'statement and any ascertainment[66]' and/or the final certificate[59] (JCT SBC/Q 2016 and JCT IC 2016) or
  - the relevant final statement[60] (JCT DB 2016).

This would leave the contractor having to dispute the loss and expense included in the final adjustment of the contract sum.

---

61 *Skanska Construction UK Ltd v The ERDC Group Ltd* (2003) SCLR 296 at paragraph 29 per Lady Paton.
62 JCT SBC/Q 2016 – clause 4.25 and JCT ID 2016 – clause 4.20.
63 JCT DB 2016 – clause 4.24.
64 Refer to Footnote no's 61 (JCT SBC/Q 2016 and JCT IC 2016) and 66 (JCT DB 2016). JCT SBC/Q 2016 and JCT IC 2016 require the contractor to provide the relevant information within 6 months after practical completion, whereas under JCT DB 2016, the contractor must provide the information within 3 months after practical completion.
65 JCT SBC/Q 2016 – clause 4.25.3, and JCT IC 2016 – clause 4.20.3: the contractor has 1 month from the employer's representative's notification to provide the required documents. JCT DB 2016 – clause 4.24.3: the contractor has 2 months from the employer's notification to provide the required documents.
66 JCT SBC/Q 2016 – clauses 4.25.2 and 4.25.3, and JCT IC 2016 – clauses 4.20.2 and 4.20.3.

### 12.2.10 'Form and Manner' of Requested Information

The "form and manner" of the information to be provided[55] to the employer or his representative may be very important (see Section 10.2.4). For example, if the request does not specify the form of the information to be provided, then the contractor may provide this in a form that may be more difficult to assess. For example, the employer or his representative may request the contractor provide accounts ledger details of all the labour costs expended during a particular period. The contractor is likely to have that information very quickly to hand in electronic format (such as an Excel spreadsheet), but unless the form is specified, he may provide this information as a hard copy printout or as a scanned .pdf document. In such cases, the employer or his representative would then have to manually process that information and possibly transfer it into a spreadsheet (to aid its ascertainment). If they requested the electronic (e.g. Excel spreadsheet) version, then it would be far easier for them to investigate the contractor's labour expenditure.

The contractor may not necessarily submit an elaborately formulated and detailed claim, although if the facts and losses are complex, it may be appropriate to do so. Whilst it is the duty of the employer or his representative to ascertain the loss and expense and they can look to the contractor to provide them with relevant factual information, there are not many contractors who would entrust the employer or his representative to compile their loss and expense claim for them. Given that the contracts first require the contractor's initial,[67] then updated assessments each month[68] to be provided to the employer or his representative, it is clear that they must ascertain the loss and expense from the contractor's assessments.

Contractors should think carefully before either rejecting requests for further information from the employer or his representative or providing that information in a form or manner that would likely to make it difficult or time-consuming to analyse.

## 12.3 Ascertainment by the Employer or His Representative

### 12.3.1 Criteria for Ascertaining Loss and Expense

As we have seen in Section 12.2, the contractor initiates the loss and expense process by first submitting a written notification, and second providing his 'initial assessment' of the loss and expense (if not provided with the notification), and then any 'updates' to that assessment (refer to Sections 12.2.2 and 12.2.3). If the contractor correctly submitted those documents in accordance with the terms of the contract, then the employer or his representative must act by first ascertaining an amount for loss and expense and then notifying that amount to the contractor. In previous editions, namely JCT SBC/Q 2011 and JCT IC 2011, the employer or his representative had to first, form an opinion over whether or not the regular progress of the contractor's Works had been or is likely to be materially affected, before ascertaining the loss and expense.[69] However, there is no express requirement for either the employer's representative or the employer to form such an opinion in any of the JCT 2016

67 JCT SBC/Q 2016 – clause 4.21.2, JCT DB 2016 – clause 4.20.2 and JCT IC 2016 – clause 4.16.2.
68 JCT SBC/Q 2016 – clause 4.21.3, JCT DB 2016 – clause 4.20.3 and JCT IC 2016 – clause 4.16.3.
69 JCT SBC/Q 2011 – clause 4.23, and JCT IC 2011 – clause 4.17. There was no requirement for the employer to form the same 'opinion' in JCT DB 2011.

contracts. The requirement for the employer or is representative to form an 'opinion' is now implied, and is confined to the following matters:

- If the contractor has complied with the contract's requirements for providing the relevant notifications, initial assessment and updates; and
- The amount (if any) of loss and expense that's ascertained.

It has been suggested that the contractor is not obliged to make a claim under the loss and expense clause, but merely to provide information to the employer or his representative, from who they will assess the loss and expense. If what is being suggested is that the contractor is entitled simply to submit large bundles of documents and expect the employer or his representative to effectively produce the claim for him, then that suggestion is misconceived. It has already been seen that the contractor's notification must identify the relevant matters, which will be the nub of the claim. Without that, and until they receive the contractor's '*initial assessment and information,*[70]' the employer or his representative can do nothing. Taking some useful guidance from litigation:

- The Pre-Action Protocol for Construction and Engineering Disputes (2nd Ed) provides clear guidance for making a claim (refer to Section 10.1.2) and
- A 1994 case considered the position of a party receiving pleadings in litigation:

> '*The basic purpose of pleadings is to enable the opposing party to know what case is being made in sufficient detail to enable that party properly to prepare an answer to it.*'[71]

The employer or his representative is in a similar position here. The basic purpose of the contractor's application may accurately be characterised as to enable the employer or his representative to understand the contractor's claim in sufficient detail to enable either of them to form an opinion. The High Court and the Court of Appeal analysed the contractual machinery of the claims provisions of JCT 63 by arranging the steps in a chronological order.[72] We can relate these stages to the 2016 JCT contracts as shown in Table 12.4 (in Section 12.2.1) along with:

- Ascertaining the loss and expense;
- Certification (or notification) of the amount properly due; and
- Ultimately, payment by the employer.

As the contractor's application must comply with the 'notification and ascertainment' provisions,[73] if it does not, then the employer or his representative has no authority under the contract to consider it. Moreover, an employer's representative who proceeds to consider a non-compliant application may be liable to the employer.

Under each of JCT 2016 contracts, the employer or his representative must carry out the following steps (shown in their logical order) set out Table 12.8:

70 JCT SBC/Q 2016 – clause 4.21.4, JCT DB 2016 – clause 4.20.4, and JCT IC 2016 – clause 4.16.4. See Sections 10.1.3 (Requests for further information), and Section 12.2.2).

71 *British Airways Pension Trustees Ltd v Sir Robert McAlpine & Sons Ltd and Others* (1994) 72 BLR 26 at 33 per Saville LJ.

72 *F G Minter Ltd v Welsh Health Technical Services Organisation* (1980) 13 BLR 7.

73 Refer to Sections 12.2.1 and 12.2.2

**Table 12.8** JCT 2016 contracts – requirements for 'ascertaining' loss and expense.

| **Steps the employer or his representative must take to 'ascertain' loss and expense** | **JCT SBC/Q 2016 4.21.4** | **JCT DB 2016 4.20.4** | **JCT IC 2016 4.16.4** |
|---|---|---|---|
| Ascertainment made by reference to the information supplied by the contractor | Yes | Yes | Yes |
| Notify the contractor;<br>• Within 28 days of the contractor's initial assessment or<br>• Within 14 days from each 'update' | —<br>4.21.4<br>By Architect, CA or QS | —<br>4.20.4<br>By employer | —<br>4.16.4<br>By Architect, CA or QS |
| Notification must contain the 'ascertained amount' of loss and expense | Yes | Yes | Yes |
| Ascertainment to be in sufficient detail to allow contractor to identify differences | Yes | Yes | Yes |

### 12.3.2 Ascertainment

The first logical step is that the 'ascertainment' of loss and expense must be carried out by the employer or his representative. Provided that the contractor has fulfilled his obligations to notify,[74] submit the initial assessment[75] and any updates,[76] along with providing any further information,[75] the employer or his representative must consider the information he's received and proceed to make his ascertainment.

The word 'ascertainment' is defined as meaning *'find out (for certain), make sure of.'* [77] Ascertainment is not simply something which can be left to the unfettered judgment of the employer or his representative (such as an architect or quantity surveyor). They have a duty to find out the amount of the loss and expense for certain, and must not make a 'general assessment'[78] (e.g. to an estimate or a 'best guess'). The loss and expense that has to be ascertained should be that which is being or has been actually incurred unless the parties have agreed some other (often more convenient) method of calculation (e.g. such as the use of a formula for head office overheads).[79] Often, claims for loss and expense are settled or assessed by the quantity surveyor on the basis of figures included in the contract's pricing document.[80] Such figures may have little, or no relationship to the actual costs that the contractor incurs. In many instances, claims assessed or settled on this basis may give

74 JCT SBC/Q 2016 – clause 4.21.1, JCT DB 2016 – clause 4.20.1, and JCT IC 2016 – clause 4.16.1
75 JCT SBC/Q 2016 – clause 4.21.2, JCT DB 2016 – clause 4.20.2, and JCT IC 2016 – clause 4.16.2
76 JCT SBC/Q 2016 – clause 4.21.3, JCT DB 2016 – clause 4.20.3, and JCT IC 2016 – clause 4.16.3
77 Oxford Dictionary of English (3rd Ed) 2010.
78 *Alfred McAlpine Homes North Ltd v Property and Land Contractors Ltd* (1995) 76 BLR 59, at page 88 per Judge Lloyd QC.
79 *Alfred McAlpine Homes North Ltd v Property and Land Contractors Ltd* (1995) 76 BLR 59 at page 63. See Section 4.4.2.5.
80 JCT SBC/Q 2016: the 'Contract Bills – Second Recital, JCT DB 2016: the 'Contract Sum Analysis' – Second Recital, and JCT IC 2016: Bill of Quantities, Specification or Work Schedules and the 'Activity Schedule' – Third and Fourth Recitals.

the contractor somewhat less recompense than he would otherwise be entitled because the figures in the pricing document may have been discounted in order to offer a more competitive price at tender stage.

Obviously, there may be instances where the contractor has poor records and an assessment is the best that can be done – which the JCT 2016 contracts appear to anticipate by stating that *'ascertainment being made by reference to the information'* that the contractor has provided.[81] It is thought that the employer or his representative cannot, in principle, refuse to certify anything on the ground that only general (and not precise) information has been provided if it is clear that the contractor has incurred loss and expense. In such circumstances, the employer or his representative should be careful (and possibly conservative) in their ascertainment.

The JCT SBC/Q 2016 and JCT IC 2016 contracts permit either the employer's representative or the quantity surveyor to ascertain the loss and expense. In the corresponding 2011 editions of these contracts, the employer's representative had to instruct the quantity surveyor to do so[82] but that is no longer the case in the 2016 editions. Whether the ascertainment of loss and expense is carried out by the employer's representative or by the quantity surveyor is now a matter for the employer, which can be achieved in the following ways:

- The 'services' to be provided under the conditions of engagement for the employer's representative, and the quantity surveyor expressly state who is to ascertain any loss and expense; or
- If the employer's representative is engaged in a 'project management' or 'project lead' role and provided he has the required authority, he may instruct the quantity surveyor to carry out the ascertainment, provided that the conditions of engagement for the quantity surveyor oblige him to comply with the employer's representative's instructions.

These examples are summarised in Table 12.9:

The assessment or ascertainment of loss and expense is usually an 'additional' or 'supplementary' service for the employer's representative and quantity surveyor, respectively. This is for the simple reason that neither employer nor contractor embarks on a project expecting to have loss and expense claims. Therefore, whilst the employer may select the 'ascertainment' services as options within the relevant conditions of engagement, it's highly unlikely that the employer or quantity surveyor will have included amounts within their fees for ascertaining the contractor's loss and expense claims. In the event that ascertainment services are required, then the employer will usually pay for them on a time-related (i.e. hourly rate) basis.

In circumstances where the 'ascertainment' is carried out by the quantity surveyor, it will be difficult for the employer's representative to certify (or include for payment) anything other than the amount ascertained by the quantity surveyor. However, responsibility for certification of the amount lies with the employer's representative who may be held to be negligent if he certifies an amount for loss and expense, without taking reasonable steps to be satisfied as to its correctness.[83] It is therefore advisable for the employer to review the ascertainment with the quantity surveyor to be satisfied that the correct principles

81 JCT SBC/Q 2016 – clause 4.21.4, JCT DB 2016 – clause 4.20.4, and JCT IC 2016 – clause 4.16.4.
82 JCT SBC/Q 2011 – clause 4.23 and JCT IC 2011 – clause 4.17.
83 *Sutcliffe v Thackrah* (1974) 1 All ER 859.

**Table 12.9** Relevant professional services for ascertaining loss and expense.

| Service | RIBA PSC 2020 | RICS Services (May 2022) | |
|---|---|---|---|
| | Contract Administrator ('CA')/Project Lead ('PL') | Contract Administrator | Quantity Surveyor |
| Issue instructions to other consultants | PL – stages 5, 6 | — | — |
| Certify/notify payments in accordance with the building contract | CA – stage 5 | 5.1, 6.2 | 1.7.6, 1.7.7, 1.7.8 |
| Assist/discuss with the employer to agree the final account | CA – stage 6 | 7.3 | 1.7.9[a)] |
| Issue the final certificate | CA – stage 6 | 5.1 | — |
| Assist with contractor's claims | Additional service | Additional, 8.1 | Supplementary, 2.2.4, 2.2.5 |

a) RICS Quantity Surveyor Services (May 2022) – Construction Phase Services: 1.7.9 (final account) excludes the assessment of loss and expense.

have been applied. Older authorities[84] suggest there may be nothing in the contract which obliges the employer's representative to accept the quantity surveyor's ascertainment. It is respectfully suggested that such views should be disregarded when considering the ascertainment under the JCT SBC/Q 2016 and JCT IC 2016 forms of contract. The ascertainment provisions strongly suggest that only a single assessment of loss and expense is carried out for the purposes of inclusion within the interim[85] or final payments[86] and that ascertainment is carried out by either the employer's representative or by the quantity surveyor.

### 12.3.3 Timing, Content and Detail of the Ascertainment

The employer or his representative are given far less time to ascertain the contractor's loss and expense, than the 12 weeks[87] for notifying the decision for an extension of time (i.e. fixing a later date as the Completion Date – see Section 11.3.1). As briefly described in Section 9.1.10, there are two relevant time limits that relate to the ascertainment of loss and expense:

- The time limits for the employer or his representative notifying the contractor of the amount of loss and expense ascertained:
  - Within 28 days[88] after receiving the contractor's initial assessment[89]; or
  - Within 14 days[88] of receiving each subsequent update[90]; and

84 *R B Burden Ltd v Swansea Corporation* (1957) 3 All ER 243.
85 JCT SBC/Q 2016 – clause 4.14.2.3, JCT DB 2016 – clause 4.12.2.3 (Alternative A) or 4.13.2.3 (Alternative B) and JCT IC 2016 – clause 4.9.2 (100% of loss and expense)
86 JCT SBC/Q 2016 – clause 4.25.2.1, JCT DB 2016 – clause 4.24.2 (referring to clause 4.2) and JCT IC 2016 – clause 4.20.1.
87 JCT SBC/Q 2016 – clause 2.28.2, and JCT DB 2016 – clause 2.25.2. There is no corresponding 12 week period under JCT IC 2016 or JCT MW 2016.
88 JCT SBC/Q 2016 – clause 4.21.4, JCT DB 2016 – clause 4.20.4 and JCT IC 2016 – clause 4.16.4
89 JCT SBC/Q 2016 – clause 4.21.2, JCT DB 2016 – clause 4.20.2 and JCT IC 2016 – clause 4.16.2
90 JCT SBC/Q 2016 – clause 4.21.3, JCT DB 2016 – clause 4.20.3 and JCT IC 2016 – clause 4.16.3

- Within the periods for notifying the contractor of the amounts he is to be paid in an interim payment or the final payment;
  - In a 'payment notice' (or relevant certificate)[91] and/or a
  - 'Pay-less notice.[92]'

When considered together, the 14 or 28 period for notifying the contractor is likely to correlate with the monthly interim (or to a lesser degree, the stage[93]) payment cycles. Taking the example of monthly interim payments,[94] it is likely that the employer or his representative will have to ascertain and notify the contractor of the amount of loss and expense during a monthly interim payment cycle. Understandably, the contractor will expect the notified amount of loss and expense to be included within the sum certified (or notified) as the interim payment. It may seem like a significant task for the employer or his representative to ascertain the loss and expense on a monthly basis, but often in reality, it may be no more involved that carrying out a complicated interim valuation (e.g. valuing a significant amount of work, including numerous variations carried out or claimed during the last month). The relative urgency for ascertaining the loss and expense is reflected in the contract's provisions, implying that ideally, the contractor should regularly receive all of his financial compensation (as loss and expense is not subject to retention[95]).

In the event that an employer (or his representative) notified the contractor of the ascertained amount of loss and expense, but failed to include it within the amount of an interim payment, then he runs the risk of the contractor adjudicating over the unpaid loss and expense. Conversely, and in the event that employer or his representative failed to notify the contractor of the ascertained loss and expense within the 28- or 14-day time limit, but nevertheless included the loss and expense (and in sufficient detail – see below) in the subsequent payment notice, then provided that the employer paid the amount in the payment notice, the contractor would not have much cause for complaint.

The content of and detail to be provided with the loss and expense notification is described in the clauses[90] as follows:

- Being in sufficient detail so the contractor can identify differences between (1) the amount ascertained by the employer or his representative and (2) the amount of loss and expense claimed in the contractor's assessment or update and
- The identifiable differences between (1) the amount of ascertained loss and expense and (2) the amount claimed in the contractor's assessment or update can be seen by reference to the information that the contractor has provided.

It is fairly common for contractors to submit financial documents such as applications for interim payments or loss and expense assessments and updates in spreadsheet format.

91 JCT SBC/Q 2016 – clauses 4.9.1 (Interim Certificate) and 4.26.1 (Final Certificate), JCT DB 2016 – clauses 4.7.5 (Payment Notice) and 4.8 (Final Payment Notice), and JCT IC 2016 – clauses 4.8.2 (Interim Certificate) and 4.21.1 (Final Certificate)

92 JCT SBC/Q 2016 – clause 4.11.5, JCT DB 2016 – clause 4.9.5, JCT IC 2016 – clause 4.12.5

93 JCT DB 2016 – Contract Particulars, entry for clause 4.7.1 – Alternative A: Stage Payments

94 Contract Particulars' following entries: JCT SBC/Q 2016 – clause 4.8, JCT DB 2016 – clause 4.7.2 (periodic payments), and JCT IC 2016 – clause 4.8.1

95 JCT SBC.Q 2016 – clause 4.14.2.3, JCT DB 2016 – clauses 4.12.2.3 (Alternative A) or 4.13.2.3 (Alternative B), and JCT IC 2016 – clause 4.9.2.

This enables the employer or his representative to simply add a number of additional columns and insert the figures for their valuation or ascertainment against the individual figures that comprise the contractor's claim. It is also common for the employer or his representative to provide a 'comments' column in which he adds information that usually explains his valuation or ascertainment, and/or requests further information from the contractor (such as a Scott schedule – see Section 10.2.8). In such instances, and particularly if the contractor's loss and expense assessment or update consists of a number of individual items (each on a separate row in a spreadsheet), it is straightforward for the employer (or his representative) to set out his ascertainment, and for the contractor to know not just any overall difference between the totals claimed and ascertained but also for each individual item making up the loss and expense claim.

The contractor is clearly encouraged to provide as much information as possible, and as regularly as he can in support of his loss and expense claims. If his loss and expense assessment or update consists of a single line describing 'loss and expense' and a round figure, unsupported by any information, then it is unlikely that the employer or his representative are able to ascertain any significant amount in respect of the contractor's claim, irrespective of how obviously the regular progress of work was affected by a relevant matter or by deferment of possession.

There is a cautionary tale for submitting financial documents in spreadsheet format. Often, spreadsheets are exchanged between the contractor and the employer or his representative (especially, if a quantity surveyor). Such spreadsheets may clearly be 'working documents' that may fulfil a number of functions, such as draft applications for payment, quotations, cashflow forecasts or forecasts of the contractor's likely final account. If either party submits a spreadsheet in order to fulfil an important contractual function, such as that of an application for interim payment, payment or pay-less notice, or an assessment, update or notification of ascertainment of loss and expense, then the party submitting that spreadsheet must clearly set out its intended function. In a 2015 case[96] considering the exchange of spreadsheets for the purposes of interim payments and their notifications, the Court stated that a contractor should '*...set out [the purpose of its] claims with proper clarity.*' It is therefore advisable when submitting contractually significant financial documents in spreadsheet form to:

- Clearly state the purpose of the spreadsheet:
  - In the covering email;
  - In the spreadsheet's filename; and
  - In the header or footer of each page (as it may not be evident from looking at the 'worksheet' itself) if the spreadsheet is to be printed; and
- Send an accompanying hard copy or .pdf copy of the spreadsheet's printed form which clearly identifies its purpose.

### 12.3.4 Matters within the Employer's or His Representative's Knowledge

Contractors often argue that the employer (or his representative) already has sufficient knowledge of the circumstances giving rise to a loss and expense claim, and that complying with the contracts' notification clauses is merely providing additional

96 *Caledonian Modular Limited v Mar City Developments Ltd* (2015) EWHC 1855 (TCC)

confirmation of what is already known. If the employer's representative is regularly on site, or is kept abreast of matters regarding the regular progress of the works by either a clerk of works or by the contractor's regular site reports (particularly if they include programmes that measure progress[97]), then there may be some practical justification for the contractor's argument. However, irrespective of the employer's or his representative's actual knowledge, by signing the contract, the contractor has agreed to comply with the loss and expense clauses.

The employer or his representative may not have more than a general knowledge of what is happening on site, particularly if there is no permanent, or visiting clerk of works. This view has some judicial as well as practical support:

- *'the architect is not permanently on the site but appears at intervals, it may be of a week or a fortnight ...'*[98]
- The RIBA's guidelines[99] for the frequency of inspections:
  - *'The timing of periodic visits is not so crucial; they need simply to take place regularly…'* but
  - Inspections should be made before important parts of the work are covered up and
  - *'…shortly before issuing a certificate for payment.'*

The practical effect of the contractor's obligation to notify as soon as the regular progress is likely to be materially affected is quite significant.[100] The employer or his representative is entitled to assume, unless notified to the contrary, that work is progressing smoothly and efficiently and that there are no current or anticipated problems. For instance, if the contractor is instructed to carry out extra work and he does so without comment, the employer or his representative is probably entitled to assume that the effects of that instruction can be absorbed into the contractor's programme of work without any consequential delay or disruption.[101] However, that may not always be the case:

> *'Although I accept that the architect's contact with the site is not on a day to day basis, there are many occasions when an event occurs which is sufficiently within the knowledge of the architect for him to form an opinion that the contractor has been involved in loss or expense.'*[102]

That is undoubtedly true, and it is particularly so where the employer or his representative have first-hand knowledge of the problem. A clear example of that is when an architect is late providing important information to the contractor, knowing that it will create a delay to a critical activity. Other examples are if the architect instructs substantial variations to the Works, or issues postponement instructions.

---

97 Such as showing actual progress, either compared against the 'baseline' activities, or shown by using a 'dropline' – refer to Section 3.5.3 (Time Impact Analysis)

98 *East Ham Corporation v Bernard Sunley & Sons Ltd* (1965) 3 All ER 619 at 636 per Lord Upjohn when speaking of the architect's duty to 'supervise' work.

99 Good Practice Guide: Inspecting Works. RIBA Publishing. 2009

100 *Jennings Construction Ltd v Birt* [1987] 8 NSWLR 18.

101 *Doyle Construction Ltd v Carling O'Keefe Breweries of Canada* (1988) Hudson's Building and civil Engineering Contracts, 1995, 4.133.

102 *London Borough of Merton v Stanley Hugh Leach Ltd* (1985) 32 BLR 51 at 96 per Vinelott J, quoting the interim award of the arbitrator.

However, contractors should only use this line of argument as a last resort. Their primary focus should be on complying with the contract's notification provisions (for both loss and expense,[103] and informing the employer or his representative of any delay to the progress of the Works[104]). Many employers amend their JCT 2016 contracts so that any such 'notification' can only be in the form derived from their relevant clauses.[103, 104] The effect of such amendments allows the employer or his representative to disregard any other document (such as a contractor's progress report, or programme, or even an admission of delay by or on the employer's behalf) so his obligations to fix a later Completion Date (i.e. grant and extension of time) or ascertain any loss are not triggered.

## 12.4 The 'Relevant Matters'

The JCT 2016 contract contain up to 7 categories which set out grounds for entitlement to loss and expense (see Section 12.1.1 and Table 12.1), which are as follows:

- Deferment of possession and
- The following 6 categories of 'Relevant Matters':
  - Variations (or Changes);
  - The effects of various instructions;
  - Discovery of 'Antiquities'
  - Contractor's suspension for non-payment of interim payments by the employer;
  - Delay in receiving the relevant permissions (JCT DB 2016 only); and
  - Any impediment, prevention or default by the employer

Whilst the first category ('deferment of possession') is not referred to as a Relevant Matter,[105] it is clearly a ground for the contractor to recover loss and expense.[106]

### 12.4.1 Deferment of Possession

Each of the JCT 2016 contracts includes the option for the parties to agree that the employer may defer giving the contractor possession of the site for a maximum period of up to six weeks unless a shorter period was stipulated in the contract particulars.[107] A period of deferment applies from the 'Date of Possession', or from the 'relevant Date of Possession' if the Works are to be completed in Sections. In the latter case, a different period of deferment may apply to each Section. In the event that the parties do not agree that deferment 'applies', but the employer nevertheless delays giving possession of the site (or of a Section), then the contractor is likely to be entitled to loss and expense under the 'impediment, prevention or

103 JCT SBC/Q 2016 – clause 4.21.1, JCT DB 2016 – clause 4.20.1 and JCT IC 2016 – clause 4.16.1
104 JCT SBC/Q 2016 – clause 2.27.1, JCT DB 2016 – clause 2.24.1 and JCT IC 2016 – clause 2.19.1. See section 11.2.1.
105 'Deferment of possession' sits outside the 'Relevant Matter' clauses in JCT SBC/Q 2016 – clause 4.22, JCT DB 2016 – clause 4.21, and JCT IC 2016 – clause 4.17.
106 JCT SBC/Q 2016 – clause 4.20.1, and JCT DB 2016 – clause 4.19.1, and JCT IC 2016 – clause 4.15.1
107 Refer to the following clauses (and their corresponding entries in the Contract Particulars – 'applies/does not apply'): JCT SBC/Q 2016 and JCT IC 2016 – clause 2.5, and JCT DB 2016 – clause 2.4. There is no provision in JCT MW 2016 for deferment of possession.

default' Relevant Matter.[108] Alternatively in such circumstances, the employer or his representative could instruct the 'postponement of work[109]'. This appears to be a riskier option for the employer as not only does an instruction for postponement entitle the contractor to loss and expense[110] but also, if the period of postponement is continuous and exceeds either the default 2 months[111] or any other period agreed by the parties, then the contractor is entitled to terminate his employment under the contract.[112]

It is considered that deferment requires a positive action, which the employer should take by giving written notice (although that is not expressly stated). On a strict reading of the deferment clauses and the corresponding Relevant Event,[113] this ground can only apply where the employer has actually exercised his right to defer. It should be noted that the relevant clauses[105] refer to the contractor incurring loss and expense due to deferment of possession, so it appears to follow that in order for the relevant matters to apply, the deferment must have had a material effect on regular progress. Obviously, deferment is likely to have an immediate effect on regular progress, but establishing the incurrence of loss and expense may be less straightforward. Employers and their representatives should appreciate that the contractor's resources may have been scheduled to commence on the relevant 'Date of Possession', and by postponing their deployment, the contractor may incur charges for hire, or storage of materials, plant and equipment, and possibly even retaining (or booking) fees for specialist items of plant and equipment (see Section 4.4.2 for likely 'time-related costs'). It is unlikely that a contractor can efficiently redeploy resources elsewhere within such a short period of time (maximum of 6 weeks), particularly if the employer has given either no (or short) notice of its intention to defer possession. If the contractor is given early notice of deferment, it is likely to incur fewer costs than if the deferment is only notified a few days before the planned start on site.

### 12.4.2 Variations or Changes

This 'relevant matter' includes all Variations (or 'Changes' under JCT DB 2016). For brevity, the sole term 'Variations' will be used in this section.

Section 5.2 and Table 5.3 describe and summarise the various matters in the JCT 2016 contracts that constitute a 'Variation', which are as follows:

- Alteration of the design, quality or quantity of the Works (see Section 5.2.1)
- Changes or restrictions imposed by the employer (see Section 5.2.2)
- Other circumstances giving rise to a Variation or Change (see Section 5.2.3.1)
- Instructions (see Section 5.2.3.2)
- Complying with comments from the employer's representative for Design Submission Procedures (see Section 5.2.3.3)

108 JCT SBC/Q 2016 – clause 4.22.5, JCT DB 2016 – clause 4.21.5 and JCT IC 2016 – clause 4.17.5.

109 JCT SBC/Q 2016 – clause 3.15, JCT DB 2016 – clause 3.10, and JCT IC 2016 – clause 3.12.

110 A 'Relevant Matter' under the following: JCT SBC/Q 2016 – clause 4.22.2.1, JCT DB 2016 – 4.21.2.1, and JCT IC 2016 – clause 4.17.2.1

111 Refer to the 'Contract Particulars' entries against clause 8.9.2 of JCT SBC/Q 2016, JCT DB 2016 and JCT IC 2016

112 JCT SBC/Q 2016 and JCT IC 2016 – clause 8.9.2.1, and JCT DB 2016 – clause 8.9.2

113 JCT SBC/Q 2016 – clause 2.29.3, JCT DB 2016 – clause 2.26.3 and JCT IC 2016 – clause 2.20.3

- Insurance claims
- Remedial measures for Joint Fire Code compliance; and
- Named Sub-contractors (see section 11.4.16).

Valuations of Variations are covered under clause 5 of the JCT 2016 contracts,[114] which is discussed in Section 5.4.

The most obvious Variations (or Changes) are those that alter the design, quality or quantity of the physical works being carried out. The imposition of changes or restrictions that affect the carrying out of the physical works is also likely to affect the efficiency with which the contractor can complete the works.

### 12.4.3 Other Circumstances giving rise to a Variation (or Change)

Sections 5.2.3 and 5.2.3.1, along with Tables 5.4, 5.5 and 5.6, contain a detailed commentary on various matters that result in a Variation (or a Change) under JCT SBC/Q 2016, JCT DB 2016 and JCT IC 2016.

### 12.4.4 Instructions

Section 5.2.3.2, along with Tables 5.7, 5.8 and 5.9, provide a detailed commentary that covers the effects of instructions under the various JCT 2016 contracts:

- Postponement of work

  The clauses[115] which enable the employer or his representative to instruct a postponement are sometimes misinterpreted. They empower the employer or his representative to issue instructions for the postponement of any work to be executed under the provisions of the contract. Instructions for postponement should therefore be distinguished from 'deferment of possession' (i.e. which is deferring the 'start' date in the contract particulars for the contractor to take possession of the site, or of a section – refer to Section 11.4.3).

  Whether a postponement instruction gives rise to any loss and expense, then its extent (if any at all) must be the subject of careful investigation by the employer or his representative (including the quantity surveyor – see Section 12.3.2). For example, if the postponement instruction is issued relatively early, it may allow the contractor a greater opportunity to use its best endeavours to prevent any delay. Additionally, if this postponement is short, the effect upon regular progress may be relatively minor.
- Provisional Sums

  Essentially, the contract treats this as an instruction for additional work (i.e. a variation or change), and it is dealt with accordingly (see Section 5.2.3.2). Compliance with an instruction for the expenditure of a provisional sum for defined work is expressly excluded. That is because the contractor has been given sufficient information to enable it to make appropriate allowance in programming its work at tender stage.

---

114 Save for JCT MW 2016 – 'variations' are valued under clause 3.6, and include loss and expense.

115 JCT SBC/Q 2016 – clause 3.15, JCT DB 2016 – clause 3.10, and JCT IC 2016 – clause 3.12. There is no provision in JCT MW 2016 for the employer or his representative to instruct a postponement.

- Named Sub-contractors[116] (JCT IC 2016 only)

  In the event that either a 'named sub-contractor' has to be replaced, or the employer's representative instructs the expenditure of a 'Provisional Sum', specifying that the instructed work is to be carried out by a 'Named Sub-Contractor', then any resulting delays to the regular progress of the contractor's work will entitle the contractor to loss and expense (refer to Section 11.4.16).
- Opening Up/Testing of Compliant Work

  The JCT 2016 contracts permit the employer or his representative to instruct the contractor to open up competed work for inspections and tests,[117] to ascertain if that work, materials or goods comply with the contract. It should be noted that the default position is that the cost of such opening up and testing is to be added to the contract sum, unless the inspection or tests show that the materials or work are non-compliant. The wording appears to lay the burden of proving that materials or work are not in accordance with the contract on the employer or his representative. Normally, that should not be a problem because demonstrating compliance ought to be a matter of fact. However, it emphasises the need to have representatives of both the contractor and the employer on site when the opening up takes place.[118]

  An unusual situation may arise if the contract documents provide that work must not be covered up until after inspection by the employer or his representative. In such a case, the quality of the contractor's work would clearly be of greater importance to the employer, than the need for the contractor to complete the works as early as possible. Failure to observe that provision would place the contractor in breach of contract and, as a consequence, the employer or his representative may instruct that the work is to be opened up. It is doubtful that the contractor would be entitled to payment under these circumstances for the costs of any opening up, even if the work is found to be in accordance with the contract. This is simply because a party should not be permitted to take advantage of its own wrong.[119]
- Resolving discrepancies and divergences, correcting errors or inconsistencies in the contract documents.

  A major ground for reimbursement of loss and expense is often arises when correcting:
  - Discrepancies and inadequacies in or between the contract documents[120] (other than the Agreement[121] and Conditions[122]); and/or
  - Divergences from and emergency compliance with the Statutory Requirements.

---

116 JCT IC 2016 – clause 3.7 and Schedule 2.

117 JCT SBC/Q 2016 – clause 3.17, JCT DB 2016 – clause 3.12 and JCT IC 2016 – clause 3.14. There is no such provision in JCT MW 2016

118 Refer to section 12.3.4 and Footnote 99

119 *Mackay v Dick* (1881) 6 App Cas 251, reaffirmed by the Court of Appeal in *King Crude Carriers SA and others v Ridgebury November LLC and others* [2024] EWCA Civ 719, at para 79 per Popplewell LJ

120 Refer to Tables 5.1 and 5.4 – including correction of the 'Contract Bills' – JCT SBC/Q 2016 and JCT IC 2016.

121 Comprising the Recitals, Articles, and the Contract Particulars.

122 Clauses 1 to 9 in JCT SBC/Q 2016, JCT DB 2016 and JCT IC 2016, and clauses 1 to 7 in JCT MW 2016, along with the applicable Schedules.

Instructions, along with further drawings and documents, are usually issued during the progress of the Works, so if the following differences arise, the deficiency is likely to be discovered after they're issued:

- An instruction (or other document such as a drawing or schedule) may create a discrepancy or an inadequacy between itself and the contract documents or between itself and another instruction;
- There may be a discrepancy between two or more of the contract documents.

The discrepancy or inadequacy should then be promptly corrected by an instruction (refer to Tables 5.3 and 5.5); ideally without causing any material effect on the progress of the Works (see Section 12.2.2). The contractor's obligation is not to find discrepancies,[123] but merely to notify the employer or his representative if he does find them.[124] Therefore, the contractor may not discover a discrepancy until construction of a portion of the work is underway. That does not prevent the contractor from claiming under this ground, provided it complies with the relevant clauses (see Section 12.2.3).

- Objections to and insolvency of Named Specialists (JCT SBC/Q 2016 only)

In the event that the contractor objects to a 'named specialist' or a 'named specialist' becomes insolvent, any delays to the regular progress of the contractor's work will entitle the contractor to loss and expense (refer to Section 11.4.15).

### 12.4.5 Antiquities

This ground is only a relevant matter under JCT SBC/Q 2016 and JCT DB 2016 and concerns the action to be taken in regard to the discovery of *'all fossils, antiquities and other objects of interest or value'* on the site. The action expected of the contractor under the relevant clauses[125] (to inform the employer or his representative prior to receiving an instruction) creates the obligation for the employer or his representative to instruct in this regard. This instruction[126] for dealing with antiquities is included separately,[127] rather than under the other relevant matters that are expressly devoted to instructions.[128]

Claims under this ground are not uncommon. Modern technology, and especially ground penetrating radar, or drone surveys may indicate the presence of archaeological sites. However, the definition includes *'other objects of interest or value'*, so this may be wide enough to include examples of urban art or graffiti (maybe even a piece by Banksy!).

The provisions of the Ancient Monuments and Archaeological Areas Act 1979, which came into force in 1982, affect contractors working in 'areas of archaeological importance', which is a concept introduced by Part II of the Act. Moreover, if an 'ancient monument'

---

123 For example, it is the employer who checks the Contractor's design and confirms it meets his requirements: JCT DB 2016 – Third Recital, JCT SBC/Q 2016 – Twelfth Recital, and JCT ICD 2016 – Seventh Recital.

124 *London Borough of Merton v Stanley Hugh Leach Ltd* (1985) 32 BLR 51.

125 JCT SBC/Q 2016 – clause 3.22.1.3, and JCT DB 2016 – clause 3.15.1.3.

126 JCT SBC/Q 2016 – clause 3.22.2, and JCT DB 2016 – clause 3.15.2

127 JCT SBC/Q 2016 – clause 4.22.3 and JCT DB 2016 – clause 4.21.3

128 JCT SBC/Q 2016 – clause 4.22.2 and JCT DB 2016 – clause 4.21.2

was suddenly discovered in the course of contract Works, where one was unexpected, the possibility of its being scheduled under the Act (and thus protected) could not be completely discounted. In fact, this has happened only rarely. Instead, the Department of the Environment might make financial help available for the relocation of piling or the rafting-over of the remains, and in such a case, the contractor would obviously have a claim.

### 12.4.6 Contractor Suspending Performance of His Obligations

This ground is almost identical to its equivalent relevant event clause[129] (refer to Section 11.4.6). However, there are following notable differences. It is not a relevant matter within the loss and expense clause; it is included separately.[130]

### 12.4.7 Inaccurate Approximate Quantities

This matter only applies under JCT SBC/Q 2016 and JCT IC 2016 and is intended to cover the situation where the bills of quantities contain an approximate quantity, but the quantity of work actually executed under that item is significantly different, i.e. it is either greater or less. As long as the approximate quantity is reasonably accurate, the contractor has no claim. What is 'reasonably accurate' will depend upon all the circumstances. The contractor's entitlement will usually be based on the extra time it requires over and above the time it has allowed for doing the additional quantity of work, that exceeds the 'approximate quantity' set out in the bills.

The approximate quantity may also be an unreasonable estimate because the actual quantity is substantially less than in the bills. It may be the case that the reduction in quantity means that the work becomes more intricate or must be carried out by hand (instead of a larger quantity being undertaken by a machine). It should be noted that this clause expressly refers to *'the execution of work'* – i.e. the costs for any additional time taken when executing the relevant work, and not the costs of executing the work itself. Whilst this may sound a little unclear, clarity is found in the 'valuation of variations and provisional sum work' clauses.[131] Therefore, increases in materials, costs and labour of carrying out the works will be recoverable when valuing the work for which an 'approximate quantity' is included in the bills (i.e. under the 'valuation' clauses – see section 5.4). There may be circumstances where increases in the quantity of materials may result in additional expenses, such as off-site storage costs, design costs for temporary work, and resources such as materials or plant required for un-quantified work. However, it appears that the contractor will have to recover such costs within the valuation this work (as it would with a variation). It appears that only the costs for an increase in time taken to carry out the work will be recoverable as loss and expense (in the same vein as for a prolongation or disruption claim).

129 JCT SBC/Q 2016 – clause 4.13, JCT DB 2016 – clause 4.11, JCT IC 2016 – clause 4.14 and JCT MW 2016 – clause 4.7

130 JCT SBC/Q 2016 – clause 4.13.2, JCT DB 2016 – clause 4.11.2, JCT IC 2016 – clause 4.14.2 and JCT MW 2016 – clause 4.7.2

131 JCT SBC/Q 2016 – clause 5.2.1.4 and JCT IC 2016 – clause 5.2.4.

### 12.4.8 Delay in Receiving Permissions (JCT DB 2016 only)

This relevant matter is only found in JCT DB 2016[132] (see also Section 11.4.14) and entitles the contractor to loss and expense if there is any delay in receipt of permission or approval for the purposes of development control requirements.[133] There are two important qualifications to this entitlement:

- The permission must be necessary for the Works to start or to be carried out; and
- The contractor must have taken all practicable steps to avoid or reduce the delay.

The practical steps available to a contractor are rather limited. All he can do is to make the appropriate submission for approval or permission as soon as possible. Development control requirements have a broader meaning than simply planning requirements,[132] as it appears to encompass all ('any') kinds of statutory requirements. The contractor will usually formulate its application under this head once it has made its submission and approval has been delayed. For example, planning approval may be delayed for many reasons, and it is always safer for the employer to obtain it before letting the contract (e.g. during the pre-contract stages). Attempts to avoid the problem by deleting this ground can be fraught with difficulty because several clauses refer to this issue, and there is a danger that such a deletion may prevent the contract from dealing with this matter consistently.[134]

### 12.4.9 Impediment, Prevention or Default by the Employer

This ground[135] is identical to the equivalent relevant event (refer to Section 11.4.7), and the remarks there are generally applicable to loss and expense. This ground is very broadly drafted to include breaches of contract on the part of the employer. It brings many circumstances that would otherwise have been a common law claim, but are now a 'relevant matter', and therefore the loss and expense falls within the ascertainment powers of the employer or his representative However, such claims are still subject to any provision as to notice, etc, which are imposed by the contract (refer to Sections 12.2.4, 12.2.9 and 12.2.10). Typical employer default falling under this relevant matter are:

- Late provision of information (such as instructions drawings, details or levels.
- Work not forming part of the contract. The contractor may be able to claim loss and expense on work carried out by statutory undertakers, but only if such work is a contractual obligation (e.g. if an appropriate provision is made in the contract

132 JCT DB 2016 – clause 4.21.4

133 'Development Control Requirements' are defined in JCT DB 2016 clause 1.1 as *'any statutory provisions and any decision of a relevant authority thereunder which control the right to develop the site'* and form part of the 'Statutory Requirements' definition in clause 1.1.

134 Other clauses referring to the Development Control Requirements are clause 2.15.2.2 (Divergences from Statutory Requirements) and 8.11.1.6 (Termination by either Party etc). See also *Update Construction Pty Ltd v Rozelle Child Care Ltd* (1990) 20 NSWLR 251.

135 JCT SBC/Q 2016 – clause 4.22.5, JCT DB 2016 – clause 4.21.5 and JCT IC 2016 – clause 4.17.5

documents[136]), and not a statutory duty.[137] When statutory undertakers are acting in accordance with their statutory duties, the contractor is precluded from making any claim for loss and expense (even if their work does adversely affect the contractor's progress), but such work usually entitles the contractor to an extension of time.[138] However, in one case[139] a dispute arose and the arbitrator found as a fact that in the circumstances of that contract the statutory undertakers were engaged under contract and that most, if not all, of the work being carried out by the statutory undertakers was not being performed as part of their statutory duties, but was being executed by them expressly at the employer's request and expense and under contract.
- Provision of materials or goods by the employer
- Failure to give access, ingress to or egress from the site.
- Compliance or non-compliance with the CDM Regulations.

### 12.4.10 Amounts Ascertained to Be Added to the Contract Sum

So far as the contractor is concerned, these provisions[140] are very important, as they ensure that any loss and expense ascertained under the relevant clauses[141] is added to the contract sum. By referring to the 'ascertainment' clauses,[140] it is clear that the employer or his representative is not entitled to wait until the entirety of the contractor's loss and expense claim has been provided before including any ascertained amount in an interim payment. Clearly, the employer (or his representative) must notify the contractor of the ascertained loss and expense on a periodic and progressive basis (see Section 12.3.3). As parts of the contractor's eventual total claim are progressively ascertained, the employer or his representative must add them to the contract sum and include such a sum in the next interim payment[142] (without any deduction for retention). Not only is this an excellent provision from the point of view of the contractor's cash flow, it also acts as a means of reducing the employer's possible liability for interest or financing charges.[143]

---

136 JCT SBC/Q 2016 and JCT IC 2016 – clause 2.7.1, and JCT DB 2016 – clause 2.6.1

137 This follows from the decision in *Henry Boot Construction Ltd v Central Lancashire New Town Development Corporation* (1980) 15 BLR 8, which is of relevance to SBC regarding the meaning of words in this subclause 'Work not forming part of the contract'. The usual legal position of statutory undertakers is that they are under a statutory duty to carry out certain work and, when so doing, their obligation, if any, 'depends on statute and not upon contract' – *Clegg Parkinson & Co v Earby Gas Co* (1896) 1 QB 56

138 JCT SBC/Q 2016 – clause 2.29.7, JCT DB 2016 – clause 2.26.7 and JCT IC 2016 – clause 2.20.6 (refe to section 11.4.7).

139 [139]*Henry Boot Construction Ltd v Central Lancashire New Town Development Corporation* (1980) 15 BLR 8 at 20 per Judge Edgar Fay: *'These statutory undertakers carried out their work in pursuance of a contract with the employers; that is a fact found by the arbitrator and binding on me .... In carrying out [their] statutory obligations they no doubt have statutory rights of entry and the like. But here they were not doing the work because statute obliged them to; they were doing it because they had contracted with the [employers] to do it.'*

140 [140]JCT SBC/Q 2016 – clause 4.23, JCT DB 2016 – clause 4.22, and JCT IC 2016 – cluse 4.18. There is no singular, similar clause in JCT MW 2016, but the operation of clauses 3.6 and 4.3.1 have the same effects.

141 JCT SBC/Q 2016 – under clause 4.21, JCT DB 2016 – under clause 4.20, and JCT IC 2016 – under clause 4.16.

142 JCT SBC.Q 2016 – clause 4.14.2.3, JCT DB 2016 – clauses 4.12.2.3 (Alternative A) or 4.13.2.3 (Alternative B), and JCT IC 2016 – clause 4.9.2

143 See *F G Minter Ltd v Welsh Health Technical Services Organisation* (1980) 13 BLR 7, and the discussion in Chapter 7, Section 7.3.10.

### 12.4.11 Contractor's Other Rights and Remedies

This clause acts to preserve the contractor's common law and other rights.[144] This is in addition to the rights expressly given under this clause which provide a particular remedy or set of remedies for deferment, and the relevant matters, but subject to the contractor complying strictly with the contract's provisions. In particular, the contractor's right to claim damages for breach of contract on the same grounds are preserved.[145] The right to claim reimbursement for loss and expense under the terms of the contract is not connected to any rights or remedies which the contractor may possesses in law. If it was not for this clause,[144] there would be many instances when the contractor would certainly suffer loss and expense but be quite unable to recover it. Occasionally, it may be that because of the limitations imposed by the contract machinery the contractor will be advised to pursue remedies outside the contract.[146] However, it has already been remarked that many of the relevant matters do not fall into the category of breaches of contract.[147]

144 JCT SBC/Q 2016 – clause 4.24, JCT DB 2016 – clause 4.23, and JCT IC 2016 – clause 4.19. The Court of Appeal decision in *Lockland Builders Ltd v John Kim Rickwood* (1995) 77 BLR 38, suggested that where a party's common law rights were not expressly reserved, they could co-exist with the contractual machinery only where the other party displayed a clear intention not to be bound by the contract. This view seems to ignore earlier contrary authority: *Modern Engineering (Bristol) Ltd v Gilbert Ash (Northern) Ltd* (1974) 1 BLR 73; *Architectural Installation Services Ltd v James Gibbons Windows Ltd* (1989) 46 BLR 91. The more recent Court of Appeal decision in *Strachan & Henshaw Ltd v Stein Industrie (UK) Ltd* (1998) 87 BLR 52 appears to set the matter straight.

145 *London Borough of Merton v Stanley Hugh Leach Ltd* (1985) 32 BLR 51.

146 See Chapter 6.

147 Chapter 1, Section 1.3.2.

# 13

# Claims Under NEC 4

## 13.1 Introduction

The first three editions of this book were published between 1983 and 1998, before the NEC contracts were written. By the time the fifth edition of this book was published in 2011, NEC3 (first being published in 2005) had become increasingly popular. Whereas the fifth edition of this book included a single chapter (Chapter 18) dedicated to claims under NEC3, this sixth edition addresses the various elements of 'claims' (now under NEC4) in a number of chapters. This is because of NEC's approach of including claims for what would be the separate matters under a JCT contract (i.e. variations, extensions of time, and loss and expense), altogether under the umbrella of 'compensation events'. Therefore, many of the relevant principles for notifying, making and assessing 'compensation events' have been covered in detail in the previous chapters listed below:

- An introduction to NEC contracts in Section 1.2;
- Notices in Section 2.9;
- The importance and content of the 'programme' in Section 3.5.1 and Table 3.4;
- Methods of revising the 'programme' in Sections 3.5.2 to 3.5.6 (including 'float' in Section 3.6.2);
- 'Key Dates' in Section 3.10 and Table 3.17;
- NEC4 contracts – Definition of 'compensation events' (including variations) – Section 5.3:
  - Compensation events – Section 5.3.1;
  - Compensation events procedure – Section 5.3.2, Figure 5.1 and Figure 5.2;
  - Valuation under NEC4 contracts – Section 5.5;
  - Assessing changes to the Prices – Section 5.5.1; and
  - Acceleration under NEC4 – Section 5.6.2.

This Chapter 13 provides the following:

- A summary of:
  - Compensation events (Section 13.1.1);
  - The 'compensation events' procedure (Section 13.1.2); and
  - Assessing changes to the Prices (Section 13.2)

*Building Contract Claims*, Sixth Edition. Guy Higginbottom and David Chappell.

that are covered in previous chapters of this book that are set out below;

- Assessments of a delay to the Completion Date or a Key Date (Section 13.3);
- The Project Manager's assessments (Section 13.4); and
- A commentary for each compensation event (Section 13.5).

The contractor's 'claim' (to use a simplified term) for a compensation event is then assessed as one or both of the following:

- A 'change to the Prices' and/or
- A delay to the Completion Date and/or a Key Date.

NEC4 appears to be quite specific insofar as neither the Client nor the contractor are entitled to any common-law claims because the compensation event procedure provides an exclusive remedy.[1]

### 13.1.1 Compensation Events

The majority of compensation events in the NEC4 contract are set out in 'Core Clause' 60.1, and additional compensation events are added under the applicable 'Main Option Clauses' (i.e. for Options A, B, C, D or E). A full summary of all compensation events applicable to all of the Main Option Clauses is set out in Section 5.3.1 (NEC4 – compensation events) and summarised in Table 5.10 (NEC4 Compensation Events cross-reference table). A further comparison between NEC4's compensation events and their similar or equivalent 'Relevant Events' and 'Relevant Matters' (under JCT SBC/Q 2016) is set out in in Section 2.1.1 (Contractual entitlement) and Table 2.1 (Comparing 'claim' events under JCT SBC/Q 2016 with NEC4).

A commentary for each of the compensation events is provided in Section 13.5.

### 13.1.2 Compensation Events Procedure

The procedures for the contractor notifying and quoting for compensation events are described in Section 5.3.2 (NEC4 – Compensation events procedure) and in Figure 5.1.

## 13.2 Assessing Changes to the Prices

Assessing compensation events in respect of the changes to the Prices (i.e. an addition or reduction thereto) is set out in Sections 5.5 (Valuation under NEC4 contracts – including 'contractor's share') and 5.5.1 (Assessing changes to the Prices) and illustrated in Figure 5.2.

## 13.3 Assessing a Delay to the Completion Date and/or a Key Date

The general principles for the contractor claiming an extension of time in respect of a delay to the Completion Date, or to a Key Date are covered in more detail in Section 5.3.2 and

1 NEC4 – clause 63.6. However, other 'breaches' of the contract that are not mentioned elsewhere in clause 60.1 (compensation events), are compensation events under clause 60.1(18).

Figure 5.1 (compensation event procedure), and in Sections 3.5.1 to 3.5.6. This Section 13.3 focuses on the procedure that the Project Manager must use to assess those delays, as part of his responsibility to assess compensation events (see Figure 5.2).

The importance of the contractor's Accepted Programme cannot be understated, as it is central to the process for assessing changes to the Completion Date (and changes to any Key Dates), and how NEC4 requires those changes to be expressed. NEC4's requirements for the contractor's programme (and the Accepted Programme) are set out in clause 31, and these are summarised in Section 3.5.1 and Table 3.4. In terms of the matters (Completion Date and Key Dates) for which changes due to compensation events are assessed, the contractor's programme must show the following:

- Completion Date and 'planned Completion'; and
- Key Dates and the 'planned date' when the Condition for satisfying the Key Date is likely to be met.

The use of 'planned dates' is central to NEC4's ethos of good management practice, and these dates are considered independently of the contractual dates (namely, the Completion Date and any Key Dates) because, in practice, the 'actual' dates when the contractor completes the works, or hits a Key Date (when compared to the relevant dates stipulated by the contract) may be significantly different. As seen in Section 3.1.3, the contractor is obliged to complete the works, and the work for any Key Dates 'on or before' the completion dates or key dates set out in Contract Data Part One.[2]

NEC4's close monitoring and management of the programme (see Section 3.5.1) enables the Client to be kept abreast of when the contractor is likely to complete the works and achieve the Key Dates from the following:

- The contractor's revised and Accepted Programmes; and
- The benefit of the Project Manager's advice regarding the same,[3] i.e. accepting or rejecting the contractor's revised programme.

This, not only allows the Client to form an objective view of progress (so as not to solely rely on what the contractor is telling him), but also to manage himself, and any responsibilities he has for the project. If the contractor is likely to complete the project (or stages of it) early, then by virtue of the NEC4's management procedures, such as reviewing and accepting the revised programmes, early warnings and notifying compensation events, the Client should be able to arrange for his own early occupation or use of the project. Similarly, if the contractor is likely to satisfy the Condition for meeting a Key Date early, and the Client is responsible for necessary works that follow on from that, then he can make arrangements for his own work (or that of others) to be brought forward. Conversely, if the contractor is likely to complete the works (and satisfy the Key Date conditions) later than the dates in Contract Data Part One, then the Client has a greater opportunity to mitigate any delay and its effects.

The contractor's programme is NEC4's key document for monitoring progress, so it is unsurprising that the method of assessing delays to the Completion Date, and to any

2 NEC4 – clause 30.1 (completion date) and clause 30.3 (Key Date). If Option X5 (Sectional Completion) is selected, the "completion date" will be for each section of the works.

3 See 8.6.3 and the reasons for the Project Manager not accepting the contractor's programme.

Key Dates involves the contractor's programme. Clause 63.5 provides that delays to either the Completion Date or a Key Date are established by the following steps:

- First, the 'dividing date'[4] is established (see Section 5.5.1 above);
- Second, the following periods of time (as at the 'dividing date') are established:
  - That the compensation event has caused either 'planned Completion' to be later than the planned completion on the Accepted Programme; or
  - That the compensation event has caused the 'planned date for meeting a Condition' for a Key Date to be delayed beyond those respective 'planned' dates shown on the latest Accepted Programme;
- Third, and in order to ensure the assessment of delay is accurate as at the 'dividing date', it must allow for:
  - any events (that have hindered or assisted the works' progress) that are already accounted for the in latest Accepted Programme[5] (see clause 62.2); and
  - Likewise, any such events that occurred between the date of the last Accepted Programme and the 'dividing date'; and
- Finally, the same relevant period of delay to either 'planned Completion' or to the 'planned date for meeting a Condition for a Key' dates is then applied to the Completion Date or the Key Date. NEC4's 'User Guide' neatly captures this principle as follows:

  > *'By taking this approach, if planned Completion is delayed the Completion Date is delayed by the same period. If planned Completion is not delayed the Completion Date is unchanged. The same principle applies to the planned achievement of a Condition required by a Key Date…'*[6]

Clause 63.5 sets out a general method for assessing the length of time for delays by comparing the differences between one or both of the following:

- 'planned Completion' and the 'Completion Date'; and/or
- When the 'Condition' for achieving a Key Date is later than the date shown on the Accepted Programme.

The delays are assessed as the length of time that the compensation event has caused 'planned completion' to be later than 'planned completion' shown on the Accepted Programme. Likewise, it is the difference between the planned key date on the programme and the planned date when it will be met. The exercise will necessitate careful examination of the programmes in the normal way, but with regard to the planned completion, it is only when a decision has been reached about the delay to the planned completion that the length of that delay is transferred to the completion date. Similarly, and regarding delays to a key date, the contractor should be able to demonstrate its reasoning by means of the revised programme that it is obliged to provide to the Project Manager under clause 62.2.

In technical terms, the actual method of assessing the delays is not proscribed, but it is likely to be one of the accepted CPA methods, which are covered in more detail

4 The "dividing date" is either (i) the date of an instruction from the Project Manager or Supervisor; or (ii) the date when the contractor issued the Project Manager with his notification of a compensation event under clause 61.3.

5 For example, any events that were the subject of an Early Warning (clause 15).

6 NEC4 'User Guide – Managing an engineering and construction contract' (2017), p64.

in 3.5 (Methods of delay analysis), unless the parties agree to the method beforehand. The Project Manager's duties in making his assessment of delays are summarised in Section 9.1.8. Whilst it may not be essential that the Project Manager uses planning software to either assess delays arising from compensation events or considering the contractor's programme before notifying[7] the contractor of accepting (or reasons for not accepting) a revised programme, it is advisable for him to either do so, or have access to the means for doing so.

A potential difficulty in assessment may be caused by clause 63.8 which states the assessment of the effect of a compensation event must include time risk allowances. The matters which are at the contractor's risk are those which do not entitle it to a compensation event and which it would make an allowance when tendering.[8] For example, including additional time in the programme for encountering physical conditions which an experienced engineering contractor should have anticipated – i.e. which fall outside compensation event 60.1(12). Additionally, it is clear that certain risks[9] under clause 81.1 (Contractor's liabilities) must be borne by the contractor from the starting date of the contract until the defects certificate has been issued.

As with many aspects of NEC4, it assumes that everyone is carrying out their functions in accordance with the contract, particularly at the required times. Therefore, it is essential to have an up-to-date Accepted Programme.

## 13.4 The Project Manager's Assessments

### 13.4.1 Clause 64 – the Project Manager's Assessments

This clause sets out stipulations for the Project Manager to assess a compensation event. The Project Manager's assessment should be the exception, because the contractor ought to have provided an acceptable quotation (under clause 62.3) using the principles set out in clause 63. The Project Manager may only assess the compensation event himself if the contractor fails to satisfy all the criteria in clause 64.1. Of course, it is initially a matter for the project manager to decide whether such failure has occurred so as to justify his intervention. There are seven circumstances where the Project Manager may assess the compensation event set out in clauses 64.1 and 64.2:

- The contractor has not submitted his quotation and/or details within the time allowed i.e. if the contractor does not comply with the three-week time period set out in clause 62.3 or such extension of that time as agreed under clause 62.5;
- The Project Manager decides that the contractor has failed to correctly[10] assess the compensation event.

7 NEC4 – clause 31.3.
8 NEC4 'User Guide – Managing an engineering and construction contract' (2017), p65.
9 NEC4, clause 81.1 (Contractor's liabilities). Other relevant examples of 'risk allowances' for time to include under clause 63.8 could be (i) loss or damage to the works, plant, materials or equipment, or (ii) loss or damage to other property occupied by the Client (e.g. if the works are confined to one part of a building or to one out of a number of buildings for which the Client is responsible).
10 'Correctly' in this case means in accordance with the contract. It is unlikely to have a subjective meaning from the Project Manager's perspective.

- The contractor fails to submit a programme to accompany its quotation (as required by the contract). Obviously, if the programme for the remaining work is unaffected by the compensation event, the requirement to provide a revised programme under clause 62.2 falls away, as the contractor is not claiming for an extension of time to the Completion Date or to a Key Date;
- The contractor has submitted a quotation, but the Project Manager has not accepted the accompanying programme for one of the reasons stated in clause 31.3 (see Section 3.5.1 and Table 3.6);
- There is no Accepted Programme;
- The contractor has failed to submit a revised programme for acceptance; or
- The Project Manager has not accepted the latest programme that the contractor submitted for his acceptance for one of the reasons stated in the contract.[11]

It is not quite clear why the last two reasons[12] are separated from the others in clause 64. The only difference is that the Project Manager seemingly has a free hand in assessing the compensation event for the reasons in clause 64.1, but in clause 64.2, the Project Manager's own assessment of the programme is used. Since clause 64.2 deals with those circumstances where there is no Accepted Programme, it seems that the Project Manager must assess the programme for remaining work in any event.

Clause 64 provides a default mechanism if the Project Manager does not provide his assessment (and details) within the stipulated period.[13] In the same way as he may notify the Project Manager for failing to respond to the contractor's compensation event notification,[14] or to the contractor's quotation,[15] the contractor may notify the Project Manager of his failure to make his assessment (clause 64.4). If the contractor has submitted more than one quotation, he must state which quotation it proposes should be accepted, but if only one quotation was submitted, the notification need say nothing more.

If the Project Manager does not reply within two weeks (presumably receipt) of the contractor's clause 64.4 notice, the contractor's quotation or nominated quotation (refer to Section 5.3.2 and Figure 5.2) is treated as accepted (by default).

## 13.5 The Compensation Events

A commentary on each compensation event follows in Sections 13.5.1 to 13.5.30. The compensation events are found in:

- Core clauses 60.1(1) to 60.1(21)
- Main Option clauses:
  - A and D – clause 60.4;
  - B and D – clauses 60.5, 60.6 and 60.7
- Secondary Option clauses X2.1, X12.3(6), X15.2, X15.2 and Y2.5.

---

11 NEC4 – clause 31.3.

12 In NEC4 clause 64.2, the first and second bullet points.

13 NEC4 clause 64.3. The 'default' period for the contractor to submit his quotation is 3 weeks (clause 62.3) unless either (i) the Project Manager extends the time, or (ii) the Project Manager and the Contractor agree (clause 62.5).

14 NEC4 – the contractor may notify a compensation event under clause 61.3.

15 NEC4 – clause 62.6.

### 13.5.1 Project Manager's Instructions Changing the Scope – Clause 60.1(1)

The Scope is defined in clause 11.2(16) as information specifying or describing the Works, or stating any constraints imposed upon the contractor when carrying them out. Essentially, it appears that it will comprise drawings, specifications and other relevant documents describing the work to be done. Where the contractor is to carry out some of the design, that will also form part of the Scope.[16] The project manager may issue instructions to change the Scope under clause 14.3. The second part of clause 11.2(16) states that the Scope is to be found in the documents where the Contract Data says it is, or in an instruction which is given in accordance with the contract (clause 21.2). It is therefore essential, from the Client's point of view, that the Contract Data lists all the relevant design documents, unless the design is to be provided pursuant to a clause 21.2 instruction.

This compensation event covers what (under a JCT 2016 contract) would otherwise be called a 'variation' to the physical works (see Sections 5.1.1 and 5.1.2). It will also include any instruction changing the Scope (i.e. as contained in the documents specified by the contract data), albeit an instruction that may have little or no effect on the Date for Completion (or a Key Date) or on the Prices.

Table 13.1 sets out the various instructions that change the Scope.

There are the following exceptions, where an instruction will not create a compensation event:

- If it is made under clause 44.2, which is merely accepting a defect to avoid correcting it; or
- If it is made to the Scope; either
  - provided by the contractor at its request; or
  - in order to make it comply with the Scope provided by the Client (i.e. the Scope referred to in Contract Data Part One).

The reasons for the exceptions are so that the contractor cannot benefit from its own defects or from failing to ensure his design meets the Scope provided by the Client (nor from its decision to change its design later). A consequence of the latter is that the contractor may be entitled to keep any savings which it can make to its design. The clause does not expressly address the position if the Project Manager fails to issue an instruction. This is partially addressed in clause 63.1, which refers to the date when the Project Manager should have instructed the contractor to submit a quotation.

### 13.5.2 The Client's Failure to Allow Access to and Use of Part of the Site by Whichever is the Later of the Access Date and the Date Shown on the Accepted Programme – Clause 60.1(2)

This clause concerns the Client providing access to the contractor[17] to all, or parts of the site by the dates set out by the Client in Contract Data Part One.[18] The contractor then

16 If the contractor is providing the design, the Scope described in Contract Data Part One (Data provided by the Client) will include the Client's requirements for the contractor's design (akin to the 'employer's Requirements' in JCT DB 2016). The design actually provided by the contractor (akin to the 'Contractor's Proposals' in JCT DB 2016) is either contained in the Scope referred to in Contract Data Part Two (Data provided by the Contractor), and/or in the design submitted to the Project Manager for approval under clause 21.2 (The Contractor's design).

17 NEC4 – clause 33.1 (Access to and use of the Site).

18 NEC4 – Contract Data Part One – Section 3 (Time).

**Table 13.1** NEC4 – instructions that change the Scope.

| Description/category of instruction | NEC4 Clause |
|---|---|
| An instruction forms part of the Scope | 11.2(16) |
| Project Manager instructs a change to the Scope or a Key Date | 14.3 |
| Instructs change a Scope change arising from an early warning meeting | 15.4 |
| Instructs a Scope change arising from a contractor's proposal | 16.2 |
| Instructs a Scope change to remove illegal or impossible requirements | 17.2 |
| Instructs the contractor on how to deal with an act of prevention[a)] | 19.1 |
| Contractor obeys an instruction from the Project Manager or Supervisor | 27.3 |
| Instructs the removal of work from the Scope | 34.1 |
| Instructs a change in the Scope to reflect an accepted defect[b)] | 45.2 |
| Options B and D: Instruction to correct the Bill of Quantities[c)] | 60.6 |
| Instruct the contractor to provide a 'proposed instruction' | 65.1 |
| Instruction dealing with objects of value, historical or other interests | 73.1 |
| Client is liable for the Project Manager changing the Scope | 80.1 |
| Option X22: Instructs a change in the Client's Requirements (as stated in the Scope) | X22.6(1) |

a) NEC4 – clause 19.1. Whilst this clause does not expressly refer to a change in the Scope, it may be that an illegality or impossibility may have arisen regarding some components for the works (e.g. if they are subject to import laws, or may no longer be available), so the Scope could be changed to substitute or omit those components.
b) NEC4 – clause 45. The acceptance of a defect and the implied reduction in the Scope (e.g. changed to a lesser standard) will cause a reduction of the Prices.
c) NEC4 – Options B and D clause 60.6. An instruction correcting the Bill of Quantities may result in additional work for the contractor, requiring a change in the Scope.

inserts those access dates into the programme that is either identified in Contract Data Part Two, or submitted thereafter to become the Accepted Programme (see Section 3.5.1). However, if the contractor prepares a subsequent programme, which indicates later dates for access, and this programme becomes the Accepted Programme, then these later 'access dates' will apply. The same will apply every time the contractor revises its programme with revised 'access dates', which then becomes the Accepted Programme. It appears that clause 31.2 precludes the contractor from showing earlier 'access dates' than the dates shown in Contract Data Part One.[19]

A party which enters onto land is normally taken to have possession of the land, although, in the case of building contracts, a contractor is said to have an express or implied licence to be on the land. The extent of the contractor's occupation of the site under NEC4 may not give him exclusive possession (refer to Sections 3.1.2 and 3.9). That is also confirmed

19 NEC4 – clause 31.2, seventh bullet point refers to each programme showing the dates when the contractor will need access to part of the Site *'if later than its access date'*.

by clause 25.1, which states that it 'shares' the Working Areas[20] with Others[21] as provided for in the Scope. A continual failure to give access is a serious breach of contract, and may amount to an event that entitles the contractor to terminate.[22]

### 13.5.3 The Client Fails to Provide Something by the Date in the Accepted Programme – Clause 60.1(3)

Clause 31.2 contains a mandatory list of the contents for any programme that the contractor submits for acceptance (see Sections 3.5.1 and Table 3.4). The purpose of this compensation event is to ensure clarity over when the Client is to provide something to the contractor. By stating his requirements for what the Client is to provide and when it is to be provided on the programme, revised programme and ultimately the Accepted Programme, the contractor is ensuring that the project's key management tool consistently records important information such as this.[23]

It should be clear that this compensation event refers to the situation where the parties have agreed what the Client will provide to the contractor (such as materials, plant, equipment, etc.). Having come to an agreement, the contractor has quite reasonably included it (and the date by which it must be provided if the Works are not to be delayed) in its programme for acceptance by the Project Manager.

It is noteworthy that, under clause 31.3, the Project Manager has very limited scope to refuse to accept a programme (refer to Section 3.5.1 and Table 3.6). If the contractor has inserted either the wrong date for the Client to provide something, or the wrong thing to be provided, then the only grounds for rejection appear to be that either the contractor's plans are not practicable, or it does not represent the contractor's plans realistically. The contractor is entitled to revise the programme and it may be in the contractor's interests to bring forward some of the dates when something is required from the Client. There appear to be no other grounds on which the Project Manager can refuse to accept such a programme and the Client may be placed in a difficult situation. It is seemingly inequitable that the contractor can revise the programme so as to place the Client in breach of its obligations. However, there appears to be no easy solution, because the Project Manager's refusal to accept a programme for reasons other than those listed in clause 31.3 is itself ground for a compensation event (clause 60.1(9)).

---

20 NEC4 – clause 11.2(20) defines the 'Working Areas' as those areas necessary for providing the Works (so they can be on- or off-site), that are only used for the works, unless subsequently changed.

21 NEC4 – clause 11.2(12) defines 'Others' as not being the Client, or any of his representatives, the Contractor, his sub-contractors or suppliers, or any official dispute resolution personnel (including the adjudicator).

22 NEC4 – clause 91.6 – if the Project Manager firstly instructs the contractor to stop or not any substantial work because of a Client default (e.g. failure to give access under clause 60.1(2)), and an instruction to re-start has not been given for 13 weeks.

23 NEC4 'Managing Reality' Book 4 'Managing Change' (2018) *'The wording is to make clear that where a Contractor requires the Client to provide something that this is shown on the Accepted Programme and not lost in supporting information'*. p7.

### 13.5.4 The Project Manager's Instruction to Stop Work or Not to Start Work or Changing a Key Date – Clause 60.1(4)

Clause 34.1 authorises the Project Manager to instruct the contractor to stop, or not to start work. This appears to be similar to a postponement instruction under a JCT contract.[24] Clause 14.3 authorises changes to key dates.

Firstly, in relation to clause 34.1, this clause has been slightly amended from its equivalent under NEC3. Whilst clause 34.1 retains the Project Manager's ability to firstly instruct the contractor to stop, or not to start any work, and subsequently instruct him to start or re-start that work, NEC4 now gives the Project Manager the ability to remove the relevant work from the Scope. The reasons for the Project Manager's instructions in this regard are likely to be caused by the Client not being ready for the relevant work to either start, or proceed any further for the time being. Some examples are that the Client may not have been granted permission or access for those works, or may be waiting for funding to pay for them.

Secondly, the provisions of clause 14.3 (instruction to change a Key Date) are unchanged in NEC4 from the wording in NEC3 (save that the 'Works Information' is now referred to as the 'Scope').

Some commentators consider that compensation event 60.1(4) only permits a change in the Prices, the Completion Date or a Key date in the event of the Project Manager's instruction to stop or not to start any work, but not his subsequent instruction to re-start or start the relevant work. However, when considering the practical terms of assessing a compensation event under clause 60.1(4), it is difficult to imagine how the subsequent instruction's effect would not be included. Taking a simple example of the contractor incurring time-related costs relating to the affected work (e.g. tower crane hire), the incurrence of cost would begin with the Project Manager's clause 34.1 instruction (not to start or to stop) along with the entitlement to a compensation event under clause 60.1(4), and end when, or soon after the subsequent (re-start) instruction is received, and the work resumes.

Obviously, any instructions under clauses 14.3 or 34.1 would not entitle the contractor to a compensation event under clause 60.1(4) if they necessarily arose from a fault of the contractor.[25]

### 13.5.5 The Failure of the Client or Others to Work (a) Within the Times on the Accepted Programme or (b) the Conditions Stated in the Scope or (c) Carrying Out Work on the Site that's Not in the Scope – Clause 60.1(5)

This compensation event is split into the following grounds that broadly cover the breaches by the Client, or 'Others' where they;

- Do not carry out work:
  - At the times specified on the Accepted Programme; or
  - Within conditions specified in the Scope; or
- Carry out work on site that is not provided for in the Scope.

24 Refer to Section 5.2.3.2 and Table 5.7 (Relevant Event) and Section 12.4.4 (Relevant Matters).
25 NEC4 – clause 61.4: in which case (see the third bullet point) the Project Manager notifies the contractor that no changes will be made to the Prices, the Completion Date or to a Key Date.

The key documents are the Accepted Programme and the Scope. The Accepted Programme must state when the Client or Others (E.g. An organisation carrying out work necessary to the project, such as services connections) are to carry out the relevant work – as stated in the Scope, and also sets out what work, and under what conditions this work is to be carried out. From clause 31.2 (fourth and seventh bullet points), the Accepted Programme should show the timing of this work. By reference to the Scope, it should be fairly straightforward to ascertain if the Client or Others are carrying out work that is not provided for in the Scope, and by operation of clause 31.2, is also not provided for in the Accepted Programme.

This is a sensible compensation event, and the Project Manager, having agreed on appropriate dates with the contractor by accepting the revised programme, would be wise to keep the Client or Others updated, to ensure that the dates for their works are not changed to something that cannot easily be achieved. The difficulties of achieving this have been noted earlier when dealing with the compensation event under clause 60.1(3) – see Section 13.5.3. The Project Manager also has a duty to discuss any changes to the programmed dates with the Client.

NEC4 also includes the third ground, which enlarges the scope of this compensation event to include any work carried out by the Client and/or Others that is neither referred to in the Scope, and by implication, nor in the Accepted Programme. Although the contractor does not have exclusive possession of the site, clause 25.1 expressly states that the contractor shares the 'Working Areas' with Others as stated in the Scope. In other words, the Contract states precisely the parties with whom the contractor must share. Therefore, if the Client sees fit to introduce another party, who is not included in the Scope, the contractor may suffer inconvenience and disruption for which it now has a contractual remedy under this compensation event. The carrying out of work by the Client or Others that isn't contained in the Scope is limited to the site (bullet point 3), whereas it appears that the relevant work that is described in the Scope (bullet points 1 and 2) can be anywhere in the Working Areas.[26]

### 13.5.6 The Project Manager or the Supervisor Fails to Respond to a Communication Within the Agreed Period – Clause 60.1(6)

This is the sanction if the Project Manager or the Supervisor fail to comply with clauses 13.3 and 13.4, which require them (and the contractor) to reply to communications[27] within the 'period for reply' stated in Contract Data Part One (unless a particular clause provides for a different period[28]). However, despite 'Managing Reality'[29] stating there is a '*default period*

26 NEC4 – clause 11.2(20) defines the 'Working Areas' as including the Site, and other parts of the working areas (such as off-site areas).

27 NEC4 – clause 13.7 provides that a 'notification' or a 'certificate' is communicated separately from other communications.

28 NEC4 – Some notable examples: (1) clause 31.3 states that the Project Manager must notify the contractor that he's accepted the programme within '2 weeks' and not within the 'period for reply' set out in Contract Data Part One, (2) clause 36.1 – the Project Manager accepts an acceleration quotation within 3 weeks, or (3) clause 65.2 – accepting a quotation for a proposed instruction within the period set out in Project Manager's instruction under clause 65.1.

29 NEC4 'Managing Reality' Book 4: Managing Change (2018) p8.

*for reply*' in Contract Data Part One, this is not the case. In addition, many of the examples where the contractor is to receive a 'reply' will come from the Project Manager and not from the Supervisor.

The Client must ensure he has completed the relevant entries for each 'period for reply'. In the event that he does not, and the entries are left blank, it is doubtful whether this compensation event would be available to the contractor. Instead, he may be left with trying to argue that the Project Manager or the Supervisor failed to respond to a communication within a reasonable time, and having to rely upon the compensation event in clause 60.1(18). Even this seems fraught with difficulty, as establishing whether or not the contractor has complied with the notification time bar in clause 61.3 would be far from straightforward.

Table 13.2 shows the clauses in NEC4 which require the Project Manager or the Supervisor to reply to 'communications' within the period(s) for reply in Contract Data Part One.

Claims for this compensation event are likely to include the time-related costs of delays or disruption, particularly if the contractor cannot proceed for want of the Project Manager's reply under clauses 21.2, 23.1, 24.1 or 26.1.

**Table 13.2** NEC4 – Project Manager or Supervisor's reply to communications.

| Project Manager or Supervisor 'reply' to: | NEC4 Clause |
|---|---|
| Contractor's proposal to enlarge the Working Areas | 16.3 |
| Contractor's notification of an ambiguity or inconsistency | 17.2 |
| Accept the contractor's design | 21.2 |
| Accept contractor's equipment design | 23.1 |
| Accept contractor's personnel | 24.1 |
| Accept sub-contractors | 26.1 |
| Accept sub-contract documents | 26.3 |
| Accept the quality policy statement and quality plan | 40.2 |
| Accept the contractor's reduced price quotation for defects | 45.2 |
| Option A: Accept the revised Activity Schedule | 55.3 |
| Accept the contractor's insurance certificate | 84.1 |
| X4: Accept an alternative guarantor | X4.2 |
| X10: Accept the Information Execution Plan | X10.4 |
| X13: Accept bank or insurer (performance bond) | X13.1 |
| X14: Accept the proposed bank or insurer (advanced payment) | X14.2 |
| X16: Accept the proposed bank or insurer (retention bond) | X16.3 |
| X21: Accept quotation (whole life cost) | X21.3 |
| X22: Accept contractor's 'stage two' design proposals | X22.3 |

### 13.5.7 The Project Manager gives Instructions about an Object of Value, or Historical or Other Interest Found Within the Site – Clause 60.1(7)

This is similar to the JCT 2016 provisions regarding what are termed 'antiquities' in those contracts (see Sections 11.4.4 and 12.4.5). NEC4's definition in clause 73.1 is very broad, and similar to JCT 2016's definition as it includes objects of 'historical or other interest' and/or value. The only constraint seems to be that it must be an 'object' – which can be defined as '*a material thing that can be seen or touched: he was dragging a large object | small objects such as shells*'.[30] That suggests that curious or even significant rock strata may be covered by this clause, along with more obvious objects such as pieces of jewellery, old weapons and the foundation of a Roman villa. The example of 'shells' within the definition of 'object' is interesting. It seems fairly clear that 'shells' means spent, and certainly unexploded ordinance would fall within the definition. However, this implies that natural, organic or living things may also fall within the definition – for example, discovery of rare flora or fauna, or the presence of nesting protected animals (such as bats or birds).

### 13.5.8 The Project Manager or Supervisor Changes a Decision Previously Communicated to the Contractor – Clause 60.1(8)

NEC4 does not provide a definition of 'decision', so it is not clear if it has any special, broad or restricted meaning; for example, a formal decision given under particular circumstances. Therefore, it is perhaps appropriate to consider the ordinary meaning[31] of 'decision' alongside the following duties of the Project Manager and the Supervisor, such as:

- For the Project Manager:
  - Issuing instructions;[32]
  - Withdrawing or changing instructions;
  - Notifying acceptances or non-acceptances;[33] or
  - Deciding if the contractor has complied with his obligations.
- For the Supervisor:
  - Deciding if tests and inspections have been passed;[34]
  - Deciding if the contractor has corrected defects;[35]
  - Deciding if equipment, plant or materials outside the Working Area can be correctly marked.[36]

Understandably, a change of 'decision' is likely to cause the contractor to carry out additional work.

30 Oxford Dictionary of English – 3rd Ed (2010).

31 Decision – '*a conclusion or resolution reached after consideration*', Oxford Dictionary of English – (3rd Ed) 2010.

32 Refer to Section 13.5.1 and Table 13.1.

33 Refer to Section 9.1.6 and Table 9.4.

34 NEC4 – clause 41.3 and 41.5.

35 NEC4 – clause 44.3.

36 NEC4 – clause 71.1.

### 13.5.9 The Project Manager Withholds an 'acceptance' for a Reason Not Stated in the Contract – Clause 60.1(9)

This compensation event is concerned with the Project Manager rejecting (or notifying the contractor of his 'non-acceptance') of a contractor's submission for invalid reasons. It does not cover the instance where the Project Manager fails to respond, or fails to respond within the time allowed (this would be a compensation event under clause 60.1(6) – see Section 13.5.6). For many administrative functions where the contractor makes a submission, the Project Manager must notify his 'acceptance' or 'non-acceptance'. NEC4 provides reasons why he may refuse (or notify 'non-acceptance) – see Section 9.1.6 and Table 9.4.

The exceptions where the Project Manager may 'withhold' an acceptance of a 'submission' (made by the contractor) are governed by clause 13.8. He may withhold acceptance of a quotation for acceleration (clause 36.1) or a quotation for not correcting a defect (clause 45.2). There are two elements to note about this power:

- The first is that it must be implied that the contractor's submission concerns something which the contractor is entitled to submit. Examples are to be found throughout the contract and its options, together with contractually valid reasons for non-acceptance by the Project Manager;[37] and
- The second is that although the Project Manager often has a very broad power,[37] and the exercise of it may result in a compensation event if the reason for non-acceptance is not one of the reasons stated in the contract (see Table 9.4). Therefore, whenever the contractor makes a submission under the contract, it is imperative that the Project Manager carefully considers the valid reasons for non-acceptance before coming to his decision.

There may be instances when the Project Manager does not wish to accept a programme, particulars of a design or a quotation, etc., for reasons not included in the contract. The potential effect of the resultant compensation event requires a careful weighing of the balance between the benefits to the employer compared to the likely cost in terms of time and money.

### 13.5.10 No Defect is Found After the Supervisor Instructs the Contractor to Search – Clause 60.1(10)

Clause 43.1 allows the Supervisor to instruct the contractor to search for a Defect[38] – but only until the 'defects date'.[39] Of course, the contractor is obliged to correct a Defect in any event,[40] irrespective of whether the Supervisor has notified it or not. The Scope may require

37 NEC4 – clauses 16.2, and X21.3 provide that the Project Manager may give 'any reason' for his non-acceptance.
38 NEC4 – clause 11.2(6) defines a 'Defect'.
39 NEC4 – the 'defects date' is set out in Section 4 (Quality management) of Contract Data Part One, and is described as *'The period between Completion of the whole of the works and the defects date is …'* and the period is then set out. Typically, the period from Completion to the 'defects date' is likely to be 6 or 12 months from completion of the 'whole' of the Works – i.e. if Option X5 (Sectional Completion) is selected, the period will run from completion of the final section.
40 NEC4 – clause 44.1.

some tests and inspections to be carried out, but this ground deals with the situation under clause 42.1, where the Supervisor instructs (giving reasons for doing so) the contractor to search for Defects. The meaning of 'searching' in this context is very broad, and includes the following:

- Dismantling and re-erecting work;
- Uncovering and re-covering work;
- Providing facilities for the Supervisor to use when he carries out tests and inspections (including samples and materials); and
- Carrying out additional tests or inspections that are not set out in the Scope.

The Supervisor's instruction probably goes further than the similar 'opening up for inspection' under other building contracts.[41] The compensation event occurs if the contractor searches for a defect, but no defects are found. In the same way as under other building contracts[41], this compensation event protects the contractor from being liable for additional costs and the time taken for searching, if there is no defect.

There is a proviso that excludes the situation where the contractor gave insufficient notice before doing work which obstructed a required test or inspection. Such notice is required under the provisions of clause 40.3.

### 13.5.11 A Test or Inspection Carried Out by the Supervisor Causes Unnecessary Delay – Clause 60.1(11)

This ground is a reference to clause 41.5 which requires the Supervisor to carry out tests and inspections without causing unnecessary delay. 'Unnecessary delay' is a subjective term and presupposes that some delay is necessary. However, given that the 'tests and inspections' have to be stated in the Scope,[42] and the contractor's programme ought to provide for 'time risk allowances' and 'procedures set out in the contract',[43] then his programme should include allowances for such tests and inspections. Further, and given that the Project Manager must reject the contractor's programme if it shows impracticable or unrealistic plans, or fails to comply with the Scope,[44] then if the Accepted Programme identifies the 'tests or inspections', and includes 'time risk allowances' for them, then by accepting the contractor's programme (or revised programme), the Project Manager has effectively stated what a 'necessary delay' would be for such tests and inspections. Therefore, any additional time (to that shown in the Accepted Programme) expended for tests and inspections may, on the face of it, be considered an unnecessary delay.

By implication,[45] it appears that the only reason that would remove his entitlement to a compensation event under clause 60.1(11) is if the contractor caused the delay to the Supervisor's tests and inspections.

41 For example, see JCT SBC/Q 2016 – clause 3.17.
42 NEC4 – clause 41.2.
43 NEC4 – clause 31.2.
44 NEC4 – clause 31.3.
45 NEC4 – clauses 41.5 and 41.6 imply that any delays (possibly including the discovery of a defect) which are the contractor's fault remove the entitlement to a payment (if it's conditional upon passing a Supervisor's test).

### 13.5.12 Physical Conditions Encountered by the Contractor Within the Site – Clause 60.1(12)

This compensation event represents a departure from how other building contracts deal with the contractor encountering 'physical conditions'[46] within the site. For example, none of the JCT 2016 contracts covered in this book include any entitlement (whether a Relevant Event or a Relevant Matter) for the contractor to receive an extension of time or any loss and expense for encountering adverse physical conditions (such as unfavourable ground conditions). However, this compensation event does not arise every time that the contractor encounters conditions that he considers unfavourable or adverse, the following conditions apply:

- The 'physical conditions' must be within the Site;[47]
- 'Weather conditions' are excluded;[48]
- No experienced contractor would have thought it reasonable to make allowances (to that extent) for such physical conditions at the Contract Date;[49] and
- The compensation event is assessed only by considering the difference between (i) the actual physical conditions the contractor encounters and (ii) those physical conditions for which it would have been reasonable for the contractor to make an allowance.

Part of assessing this compensation event must consider that the contractor should have made some allowance[50] for encountering physical conditions when providing the Works, but not to the full extent that such allowances would absorb the effects[51] of encountering the actual physical conditions (would be unlikely[52]).

That is perfectly sensible and may well have been implied in any event.

It is notable that there is no reference to 'site', 'ground' or 'soil' conditions, which would narrow the effect of this compensation event. However, by virtue of the definition of the 'Site',[47] it the compensation even is likely to have a much wider effect, as it includes the 'volumes' above and below the area marked by the '*boundaries of the site*'[53] as defined in Contract Data Part One. So far as the soil conditions are concerned, pre-existing permanent

46 This ground is similar, but not the same as clause 12(1) of the ICE 7th edition 1999 form of contract. Both use the expression 'physical conditions'. Also see *Atlantic Civil Pty Ltd v Water Administration Ministerial Corp (NSW)* (1997) 83 BLR 113 – in which the contractor was invited to use material from an on-site draw pit, but large quantities of the material excavated therefrom were unsuitable. It was held that the presence of unsuitable material was a 'physical condition'.

47 NEC4 – clause 11.2(17) defines the Site as being within the defined boundaries, along with any 'volumes' above or below it which are affected by any of the Works. The 'Site' is within the 'Working Areas' – see clause 11.2(20).

48 NEC4 – Weather conditions (referred to as 'weather measurements') may be a compensation event under clause 60.1(13) – see Section 13.5.13.

49 See *Atkins Ltd v Secretary of State for Transport* (2013) EWHC 139 (TCC). NEC4 – clause 11.2(4) defines the 'Contract Date' as the date when the contract came into existence. Given that NEC4 does not include an attestation, execution (either under hand or as a deed) or signature page, establishing the Contract Date may not be straightforward.

50 Including 'time risk allowances' in the programme under clause 31.2.

51 By implication from clause 61.2, that such allowances would ensure the 'Defined Cost', 'Completion Date', or meeting a 'Key Date' would remain unaffected.

52 'Unlikely' in the sense that that the encountered physical conditions would *'have such a small chance of occurring'* when making allowances at the Contract Date.

53 Refer to section '1 General' of Contract Data Part One (Data Provided by the Client) – *'The boundaries of the site are…'*.

conditions are included,[54] but so are conditions which may change for one reason or another. It is often overlooked that the expression also refers to physical conditions above ground.

Weather conditions are excluded because the clause means exactly what it says: physical conditions which '*are not weather conditions*'.

The conditions regarding the contractor's reasonable allowances could cause difficulties as they must be read in conjunction with clauses 60.2 and 60.3. Clause 60.2 stipulates that, in judging the physical conditions, the contractor is deemed to have taken the following into account:

- The Site Information;[55]
- Any publicly available information noted therein, which the contractor can obtain from a visual inspection; and
- Any other information that could reasonably be expected to have or to acquire.

For projects in the United Kingdom, where the Construction (Design and Management) Regulations 2015 ('the CDM Regulations') apply, the Client is required[56] to provide the 'pre-construction information' to the contractor. It is suggested that the relevant parts of the Site Information, along with the other categories of information that the contractor ought to obtain, should already be contained in the 'pre-construction information' that the Client must provide. The Health & Safety Executive (HSE) provide the following guidelines:[57]

- 'Pre-construction information' is defined as '*information about the project that is already in the client's possession or which is reasonably obtainable by or on behalf of the client.*'
- It must be:
  - Relevant to the particular project
  - Include an appropriate level of detail; and
  - Be proportionate to the risks involved.
- It must include proportionate information[58] regarding:
  - The project, such as the client's brief, and key dates for the construction phase (e.g. RIBA stage 5);
  - The planning and management of the project, including the time and resources being allocated to each stage, and arrangements to ensure that the work is coordinated, and co-operation between 'Dutyholders;'[59] and
  - The site's health and safety hazards, including design and construction hazards, and how they will be addressed.

These categories of 'proportionate information' (the last one, in particular) should go some way to assist the contractor when assessing the risks associated with the site and

54 *Humber Oil Terminals Trustee v Harbour and Genera Works (Stevin) Ltd* CA (1992) 59 BLR 1.

55 NEC4 – clause 11.2(18) defines the Site Information, which describes the 'Site' and its surroundings. Refer to section '1 General' of Contract Data Part One (Data Provided by the Client).

56 CDM Regulations – Section 4(4).

57 'Managing health and safety in construction' Guide on Regulations. Health and Safety Executive (HSE) (2015). Appendix 2.

58 Including any information in an existing health and safety file.

59 'Dutyholders' under the CDM Regulations are the Client, Designers, Principal Designers, Principal Contractors, Contractors and Workers. 'Managing health and safety in construction' Guide on Regulations. Health and Safety Executive (HSE) (2015) p6 (Table 1).

make the appropriate allowances – considering the chances that certain risks of 'physical conditions' have of occurring. For example, it is suggested[60] that the presence of natural and man-made conditions (such as radiation, pollution or other contaminants) falls within the definition of 'physical conditions'. Clearly, such conditions should be identified in the pre-construction information under the CDM Regulations and in the Site Information. In the event that the Client includes a list of risks in the Site Information,[61] then the contractor is obliged to take those into account when judging the physical conditions.

In doing so, there are three major considerations for the contractor when making this judgement:

1) Physical conditions which will not occur;
2) Physical conditions which have such a small chance of occurring that it would be unreasonable for the contractor to allow for them; and
3) Physical conditions which are very likely to occur.

Effectively, the contractor is assumed to have made its decision based upon both the Site Information (this might well include ground investigation reports) and any other available information, which in his experience, is likely to be relevant. Therefore the job of deciding whether situation (1) or (3) apply should be fairly easy.

The task of deciding whether physical conditions have so small a chance of occurring that it would be unreasonable to allow for them, requires the contractor to calculate the risk. Whether it would be unreasonable is clearly to be decided based on what the contractor would view as unreasonable, taking all factors into account. In practice, it is likely to be difficult to argue that a contractor was being unreasonable unless the matter was something which another contractor used to carrying out similar work (and possibly on a site in a similar area) would not have done. Obviously, and for commercial purposes, when preparing its tender, a contractor is unlikely to make allowances for unlikely risks. It is suggested that a contractor faced with this kind of risk assessment should carry it out in a formal way, so as to have evidence of the method of assessment if needed in the future. However, a contractor is unlikely to receive risk information that's as comprehensive as in an Early Warning Register[62] at the tender stage.

Clause 60.3 provides that if there is an inconsistency within the Site Information, the contractor is assumed[63] to have taken account of the conditions more favourable to doing the work. This appears to be a statement of what otherwise would be implied as a result of the operation of the *contra proferentem* rule.

### 13.5.13 A 'weather measurement' is Recorded – Clause 60.1(13)

At first glance, clause 60.1(13) simply provides the contractor with a compensation event in the event of adverse weather. This is a similar entitlement to that provided by certain

60 Keating on NEC Contracts, (2nd Ed) 2022 at 7–047.
61 NEC4 – clause 60.2 provides that the contractor must take information contained in the Site Information into account when judging physical conditions.
62 NEC4 – clause 15.2 – The Project Manager provides the contractor with the Early Warning Register within 1 week of the starting date.
63 It should be noted that 'assumed' used in this context appears to be similar to 'deemed' as used in many contracts. Meaning that circumstances are to be treated as existing even if manifestly they are not: *Re Cosslett (Contractors) Ltd, Clark, Administrator of Cosslett (Contractors) Ltd (in Administration) v Mid Glamorgan County Council* (1997) 4 All ER 115.

Relevant Events in the JCT 2016 contracts (see Section 11.4.9), but with two significant differences:

- As clause 60.1(13) is a compensation event, unlike the JCT 2016 contracts, when encountering 'exceptionally adverse weather', NEC4 also provides an entitlement to loss and expense, which the JCT 2016 contracts do not; and
- Clause 60.1(13) provides some objectivity to assessing the extent to which the 'weather event' will affect the Accepted Programme, considering all the allowances for 'float' or 'time risk allowances' that NEC4 encourages the contractor to include in his first, and every subsequent programme submitted for acceptance (see Section 3.5.1).

NEC4 stipulates that only the difference between the 'weather measurement' (i.e. the actual weather encountered) and the weather recorded at the location identified in Contract Data Part One[64] (that occurs less frequently than once in 10 years[65]) is taken into account. Weather conditions only qualify to rank as a compensation event if it exceeds what is in the contract data. The following weather is to be recorded each calendar month:

- Rainfall;
  - Cumulative rainfall (mm); and
  - The number of days in that month when rainfall exceeds 5 mm;
- The minimum air temperature is less than 0 °C; or
- When snow is lying at the time specified in Contract Data Part One (e.g. sometime early in the working day).

Contract Data Part One also includes provisions for the following:

- Additional weather criteria, for example:
  - A higher temperature range;
  - Humidity; or
  - Windspeed (e.g. if tower cranes are being used or the project requires exposed working at height)
- If no recorded weather data is available, the 'assumed values'[66] for each calendar month's 'weather measurement' can be inserted.

Weather reports that are location-specific, and cover stages such as (i) planning the project and (ii) construction downtime are widely available.[67]

It appears that the intention is to provide a simpler and objective system of deciding when the contractor is entitled to further time and or money without having to make a more subjective judgment if the weather conditions are particularly adverse. However, a more subjective assessment may feel more appropriate in some cases, as the following examples illustrate:

---

64 NEC4 – Contract Data Part One (Data provided by the Client) – Section 6 'Compensation Events'.
65 Clause 60.1(13) states that the 'weather measurement' is *'shown to occur on average less frequently than once in ten years'*.
66 NEC4 – Contract Data Part One – Section 6 (Compensation events) *'Assumed values for the ten-year weather return weather data for each weather measurement for each calendar month are…'*.
67 See https://www.jctltd.co.uk/jct-met-office-weather-reports or https://www.metoffice.gov.uk/services/business-industry/construction/downtime-report-for-extension-of-time

- Reliance on weather data that is more than 10 years old. Two examples:
  1) If the contractor substantiates his claim for a compensation event under clause 60.1(13) by providing the last 10 years' weather records for the relevant location, and they only record a single similar 'weather measurement' that occurred 6 years ago, then it appears that the contractor has satisfied clause 60.1(13). However, if weather data going back 11 years is examined, and a similar weather measurement was recorded 11 years ago, then the average occurrence of that weather measurement will be twice in 11 years, which is more frequent than once every 10 years; or
  2) If a '1 in a hundred year' flood'[68] occurred within the last 10 years, but its previous occurrence was more than 10 years ago, this highly exceptional flood event would not satisfy the requirements of clause 60.1(13).
- By setting precise parameters for 'weather measurements', then, on the face of it, minor deviations from such parameters may satisfy the requirements of clause 60.1(13). For example, if the cumulative rainfall each February for the last 10 years has not exceeded 20 mm, but the current February's cumulative rainfall is 21 mm, then this would appear to satisfy the threshold for a compensation event.

### 13.5.14 An event which is 'Client's liability' Occurs – Clause 60(14)

In the previous NEC3 contract, Client's risks (then described as 'Employer's risks') were set out in clause 80.1. However, 'risks' were used in a number of contexts throughout NEC3, and this could cause some confusion. NEC4 now uses the term 'Client liabilities' which precisely refers to the matters in clause 80.1. If such a liability arises, it becomes a compensation event. The Client's liabilities are as follows:

- Claims, proceedings, compensation and costs to and from third parties,[69] arising from:
  - The use or occupation of the Site by, or for the purposes of the works, which is an unavoidable result therefrom; and
  - Negligence, breach of statutory duty or interference of a legal right by the Client (or any person employed on the Client's behalf, save for the contractor).
- An event which is the Client's fault (or someone employed on the Client's behalf, save for the contractor);
- A design fault found in either:
  - The design set out in the Scope described in Contract Data Part One; or
  - A Project Manager's instruction changing the Scope.[70]
- Loss or damage to Plant and Materials[71] supplied by the Client until the point at which the contractor has accepted them;
- Loss or damage to Plant and Materials[71] caused by:
  - Military conflicts such as war, civil war, rebellion, revolution, insurrection or usurped power;

68 The phrase really means a flood that has a 1% chance of occurring in any year.
69 NEC4 – clause 80.1 refers to the third parties as 'Others' – as defined in clause 11.2(12).
70 NEC4 – clause 80.1. An instruction from the Project Manager changing the design in both the Scope referred to in Contract Data Part One (Section 1 – General), and in Contract Data Part Two (Section 2 – The Contractor's main responsibilities) if it causes a design fault.
71 NEC4 – clause 11.2(14) defines 'Plant and Materials'.

- Civil disturbances that involve others than the contractor's employees, such as strikes, civil commotion and riots; and
- Radioactive contamination.

- Loss or damage to parts of the Works taken over by the Client, except the loss or damage that occurs before the Defects Certificate, because of:
  - A Defect which was present before the Client's taking over;
  - An event occurs (which is not a Client's liability under clause 80.1) before the Client's taking over; or
  - An event on the site caused by the contractor after the Client's taking over.
- Loss or damage to the Works, including any Equipment,[72] Plant and Materials,[71] etc., kept on site by the Client after termination unless damaged by the contractor on site after termination; and
- Any additional Client's liabilities stated in Contract Data Part One.

It appears that the contractor has two remedies if a Client liability (under clause 80.1) occurs. First, he can seek additional time and money via the compensation event in clause 60.1(14), or under clause 82.2, he can recover any cost that he's either paid, or will pay to Others from the Client.

### 13.5.15 Taking Over of Part of the Works is Certified by the Project Manager before Both Completion and the Completion Date – Clause 60.1(15)

This compensation event relates to 'Take over' of the works (which is dealt with in clause 35). Under clause 35.3, the Project Manager must certify the date when, and extent to which the Client takes over any part of the Works within 1 week of the Client doing so. It should be noted, however, that the Client may use any part of the Works before completion has been certified.[73] In that case, the Client has taken over that part of the Works when it begins to use it; unless it is for a reason stated in the Scope or the use is to suit the contractor's method of working (clause 35.2). Take over is not quite the same as practical completion under the JCT 2016 contracts, as clause 35 clearly permits the Client's use of the works, whereas 'practical completion' is focused on the extent that the works are complete and the significance of any remaining defects.

The relevant definitions regarding 'Completion' are in clause 11.2:

- 'Completion' is defined in clause 11.2(2) as being when the contractor has carried out the work in the Scope stated to be carried out by the completion date, and it has corrected the defects which would have stopped the Client from using the Works. The contractor may complete the Works before the Completion Date;[74] and
- The 'Completion Date' is defined in clause 11.2(3) as the completion date (as stated in Contract Data Part One) unless it has been changed later in accordance with the contract (e.g. by a compensation event).

Therefore, this compensation event only arises if the following occur:

- 'Completion' has not taken place;

72 NEC3 – clause 11.2(9) defines 'Equipment'.
73 NEC4 – clause 35.2.
74 NEC4 – clause 30.1.

- the 'Completion Date' remains in the future so that the contractor is not in culpable delay; and
- The Client has taken over the Works (not to help the contractor or for a reason set out in the Scope), or a part of them; but
- The Project Manager has certified such take over.

The contractor is unlikely to suffer any delays, or disruption or incur additional costs if the Client has taken over for a reason in the Scope (as he is deemed to have included such allowances in his programmes), or to assist him.

But what if the project manager fails to certify? On the face of it, there can be no compensation event, unless the contractor adjudicates over the Project Manager's failure to certify. Once the completion date has passed and the contractor is in culpable delay, it appears that the Project Manager may certify take over and the Client may use some parts or all of the Works with impunity.

### 13.5.16 The Client Fails to Provide Materials, Facilities and Samples for Tests in Accordance with the Scope – Clause 60.1(16)

This is a straightforward breach of the Client's obligation in clause 41.2 which requires him and the contractor to provide these things. Of course, it is only the Client's breach which is of consequence here. The breach specified is a failure to provide and not a failure to provide by any particular date. Therefore, if the employer provides materials but three weeks after the date stipulated on the Accepted Programme, that will be a compensation event under clause 60.1(3) – see Section 13.5.3.

### 13.5.17 The Project Manager Notifies a Correction to an Assumption Regarding a Compensation vent – Clause 60.1(17)

When the effects of a compensation event cannot be reasonably forecast, the Project Manager states his assumptions under clause 61.6 (refer to Section 5.3.2). If these assumptions are subsequently found to be wrong, the Project Manager must notify the correction. The importance of this event is related to clause 66.3, which provides that the assessment of a compensation event is not revised except as provided by the NEC4 conditions. Therefore, the revised assessment can only be carried out under clause 60.1(17) or the contractor may be deprived of adequate compensation.

### 13.5.18 The Client Commits a Breach of the Contract that's Not a Compensation Event – Clause 60.1(18)

This is NEC4's catch-all clause.[75] Its idea is to allow all breaches of contract to be dealt with under the contractual mechanism rather than having to be dealt with at common law. It also

75 Equivalent to the JCT 2016 Relevant Events for 'impediment, prevention or default' under JCT SBC/Q 2016 – clause 2.29.7, JCT DB 2016 – clause 2.26.6 and JCT IC 2016 – clause 2.20.6 (see Section 11.4.7), and Relevant Matters for the same under JCT SBC/Q 2016 – clause 4.22.5, JCT DB 2016 – clause 4.21.5 and JCT IC 2016 – clause 4.17.4 (see Section 12.4.9).

prevents time from becoming at large if an employer's action or default is responsible for delay, but for which, the contract does not expressly provide a compensation event. Clause 63.6 expressly restricts the contractor's rights in respect of changing the Prices, the Completion Date or a Key Date to those provided by compensation events and their associated procedures. This is a clear example of an 'exclusive remedy' clause, which has been found in a number of other building contracts.[76]

Reading the NEC4 contract as a whole, the following two issues are raised:

- Would the Client's breach of any terms that are normally implied into a building contract constitute a compensation event; and
- What is the likely position regarding consequential losses?

The first issue can be answered by reference to clause 12.4, which states that the contract is the entire agreement between the parties. The purpose of an entire agreement clause is to exclude liability[77] for any statements and representations (sometimes called warranties) made prior to executing the contract that may have induced the other party to enter into that contract. They may also operate to exclude implied terms (see Section 6.2) and extrinsic evidence proving additional terms.[78] However, the Law Commission has had this to say about an entire agreement clause:

> *'It may have a strong persuasive effect but if it were proved that, notwithstanding the clause, the parties actually intended some additional term to be of contractual effect, the court would give effect to that term because such was the intention of the parties.'*[79]

For the second point, it may be arguable (and probably from the Client's position), that the 'exclusive remedy' clause 63.6 operates as an exclusion clause to exclude damages for consequential losses. However, some commentators[80] point out that firstly, exclusion clauses are construed strictly against the party relying on them and secondly, clause 63.6's reference to rights could be construed as meaning rights under the contract, so a common law claim for consequential losses may be unaffected.

### 13.5.19 An Event that Prevents the Contractor Completing the Works in Accordance with the Contract Occurs – Clause 60.1(19)

There are two grounds under which the contractor is entitled to a compensation event under this clause:

- An event occurs which prevents the contractor from completing the whole of the works; or

---

76 A notable example of an 'exclusive remedy' clause was found in the MF/1 Contract (now in its 7th edition) published by the Institution of Engineering and Technology. Amendments having a similar effect are often proffered by employers procuring work under JCT contracts – e.g. so that the 'Relevant Matters' contain the contractor's exclusive remedies for recovering loss and expense.

77 *Inntrepreneur Pub Co Ltd v East Crown Ltd* (2000) 2 Lloyd's Rep 611.

78 *McGrath v Shah* (1989) 57 P & CR 452.

79 Law Commission, Law Com 154, Cmnd. 9700 (1986) para 2.15.

80 Keating on NEC Contracts (2022) 2nd Ed at 7–182.

- If that event prevents the contractor from completing the whole of the works by 'planned completion' date (as shown on the Accepted Programme).

The following three criteria apply to whichever of the consequences the event has caused:

- Neither the contractor nor the Client could prevent the event;
- An experienced contractor would have judged it unreasonable to make any allowance (in his programme or the Prices) for such an event at the Contract Date; and
- The event is not one of the other compensation events stated in the contract.

NEC4's Managing Reality refers to clause 60.1(19) as a 'force majeure' clause,[81] and cites the example of the 2001 'foot and mouth epidemic' in the United Kingdom, as one out of a class of events that neither party could prevent. Since then, 'force majeure' has been the principal ground for seeking extensions of time for delays arising out of the Covid-19 pandemic in 2020. However, despite the seemingly wide variety of events that could fall within a legal definition of 'force majeure' (as the term is not set out in the NEC4 contract), clause 60.1(19) contains the following limitations:

- The compensation event only arises if the event prevents the 'whole' of the works from being completed (or completed by the 'planned completion date'). Therefore, if Option X5 (sectional completion)[82] applies and sections were completed prior to the event occurring, then clause 60.1(19) would not apply to the remaining, uncompleted sections;
- An experienced contractor would find it unreasonable to make any allowance for the event at the Contract Date (see Section 3.5.1);[83] and
- The event does not fall into any of the other compensation events in the contract. Compensation events do not only include those set out in clause 60.1.[84] Contract Data Part One enables the Client to add additional compensation events[85] – which may include events of a 'force majeure' nature. Certainly, many Clients are now including provisions that deal with the occurrence of pandemics, or other events of a 'force majeure' nature.

### 13.5.20 The Contractor is Notified by the Project Manager that a Quotation for a Proposed Instruction is Not Accepted – Clause 60.1(20)

Clause 65 permits the Project Manager to instruct the contractor to provide a quotation for a 'proposed instruction'. From the Client's and Project Manager's perspective, a 'proposed instruction' is essentially an exercise to determine the cost or programming feasibility of the works described in the 'proposed instruction'. Clause 65.1 is very clear that the contractor

81 NEC4 – Managing Reality, Book 4 – Managing Change (2018) p14. 'Force majeure' is a Relevant Event which entitles the contractor to an extension of time (and relief from liquidated damages) under the JCT 2016 contracts (see Section 4.11.17), but it is not a Relevant Matter that entitles the contractor to loss and expense.

82 NEC4 – clause X5.1 states that references to the 'whole of the works' do not apply.

83 See also Sections 13.5.12 and 13.5.13.

84 Or for Options A and D, clause 60.4, B and D – clauses 60.5, 60.6 and 60.7, Option X2 – clause X2.1, Option X12 – clause X12.3(6), Option X14 – clause X14.2, Option X15 – clause X15.2, and Y(UK)2 – clause Y2.5.

85 NEC4 - Contract Data Part One – entry in Section 6 (Compensation events).

does not carry out the works set out in the 'proposed instruction' unless its quotation is formally accepted.[86] In addition to accepting the quotation for the 'proposed instruction', the Project Manager's reply (clause 65.2) may also instruct a revised quotation, or confirm that the quotation is not accepted. The contractor's quotation is assessed as a compensation event if it is accepted.

The compensation event under clause 60.1(20) permits the contractor to recover any time and costs incurred in compiling the quotation for the 'proposed instruction' if it is not accepted – as this clearly follows the NEC ethos that the contractor should not 'lose out' (see Section 1.2.2). It follows that if the quotation is accepted, such costs and time for its compilation would be included when it's assessed. However, a prudent contractor will show the changes in the Prices, Completion Date and a Key Date associated with compiling the quotation separately from changes to the Prices, Completion Date and a Key Date arising from carrying out the physical works described in the accepted 'proposed instruction'. Therefore, if the contractor's quotation for the 'proposed instruction' is not accepted, the changes to the Prices, Completion Date or a Key Date arising under clause 60.1(20) can be easily identified.

### 13.5.21 Additional Compensation Events in Contract Data Part One – Clause 60.1(21)

Clause 60.1(21) expressly recognises the 'additional compensation events' that the Client may have set out in Contract Data Part One.[85] NEC4's Managing Reality[87] suggests to Clients that better 'value for money' may be obtained from the contractor by including 'additional compensation events', by defining a '*grey area…or simply taking on board a risk*'. It goes on to suggest the following as 'additional compensation events':

- Discovery of objects and materials such as:
  - Asbestos;
  - Archaeological objects;
  - The presence of wildlife;
  - Unexploded ordinance; and
- External matters such as:
  - Planning approval; and
  - Transport delays (e.g. late flights, trains)

With regard to the first category, discovery of all of these matters ought to be covered by a Project Manager's instruction under clause 73.1, and therefore should become a compensation event under clause 60.1(7) – refer to Section 13.5.7. Similarly, matters in the second category may also already be compensation events in the following circumstances:

- Planning permission:
  - If this is to be obtained by the Client, then it is highly likely that the Accepted Programme will show the date when permission should be obtained. If the Client

86 NEC4 – clause 65.2 provides that the contractor's quotation for the 'proposed instruction' is accepted by the Project Manager's 'reply' which is (i) a reissue of the instruction, and (ii) notification that the quotation has been accepted.

87 NEC4 – Managing Reality, Book 4 – Managing Change (2018) p14.

fails to meet this date, then the contractor should be entitled to a compensation event under clause 60.1(5); or
  - If planning permission is to be obtained by the contractor, and (i) he has made reasonable and appropriate allowances in its price and on the Accepted Programme, and (ii) planning permission is unduly delayed by the local planning authority, it is likely to be a compensation event under clause 60.1(19).
- Transport delays. The 'transport' in this example (late trains or flights) is meant to be provided by a third party and not by the contractor, the Client or Others. The implication is that there may be haulage delays (rail) or flight delays, the latter caused by weather conditions (such as the 2010 Icelandic volcanic ash cloud) or by industrial action. These circumstances could amount to a 'force majeure' event under clause 60.1(19), but only if the requirements of that clause are met (see Section 13.5.19).

### 13.5.22 Main Options B and D – A Difference Between the Final Quantity of Work Done and the Quantity for an Item in the Bill of Quantities – Clause 60.4

This compensation event is only found in Main Options B and D. It arises from a specific difference between the actual (and final) quantity of work for an item in the Bill of Quantities and the quantity of work stated for that item in the Bill of Quantities. When reading the NEC4 contract for Options B and D as a whole (and in tandem with clause 60.5 – see Section 13.5.23), it appears that clause 60.4 only affects the Defined Cost per unit rate of the relevant work, that arises from any difference in 'final' quantity, and the quantity stated in the Bill of Quantities. Therefore, clause 60.4 appears to provide a procedure more akin to the JCT Valuation Rules (see Section 5.4.1) where 'rates' for work can be adjusted appropriately for significant changes in quantities. It does not compensate the contractor (either in time or money) for any resulting changes to the Completion Date or to a Key Date (e.g. for the extra time required to execute a significantly greater quantity of work), as this is covered in clause 60.5 (see Section 13.5.23). This is clear from the wording of clause 60.4 that '*the difference does not result from a change to the Scope*'; instead, it is effectively a correction of the Bill of Quantities and valuation of the change in quantities.

It provides objective criteria for dealing with this common situation,[88] in which, work may have been 'under-measured' or 'over-measured' when compiling the Bill of Quantities. Three criteria relating to these differences must be satisfied before a compensation event can be said to have occurred:

- The difference in quantity cannot result from a change to the Scope – in other words, a compensation event under clause 60.1(1);
- The difference must be the cause of the defined cost[89] per unit changing; and
- A compensation event will not arise unless the item's rate in the Bill of Quantities, when multiplied by the final quantity of work for that item, exceeds a positive or negative sum equivalent to 0.5% of the total of the 'Prices at the Contract Date'.

---

88 JCT SBC/Q 2016 and JCT IC 2016 both include similar provisions for valuing differences in the 'actual v Bill of Quantities/Approximate Quantities' set out in the pricing document (refer to Sections 5.1.3 and 5.2.3.1 and Tables 5.3 and 5.12).
89 Refer to Section 5.5.1.

NEC4's Managing Reality[90] gives two examples ('Scenario A' and 'Scenario B') of the sum equivalent to 0.5% of the total of the Prices being a positive or negative difference. In both cases, the 'total of the Prices' is £1,000,000 and the two scenarios are shown in Figures 13.1 and 13.2 as follows:

| 'Scenario A' description | Final Total | BoQ Total |
|---|---|---|
| Excavation of unacceptable material: | | |
| Final quantity: 10,000m$^3$ x £10/m$^3$ = | £100,000 | |
| BoQ quantity: 5,000m$^3$ x £10/m$^3$ = | | £50,000 |
| **Difference (%): (£100,000–£50,000)/£1,000,000 = +0.05%** | | |

**Figure 13.1** Clause 60.4 compensation event – 'Scenario A'.

| 'Scenario B' description | Final Total | BoQ Total |
|---|---|---|
| Excavation of unacceptable material: | | |
| Final quantity: 5,000m$^3$ x £10/m$^3$ = | £50,000 | |
| BoQ quantity: 10,000m$^3$ x £10/m$^3$ = | | £100,000 |
| **Difference (%): (£50,000–£100,000)/£1,000,000 = –0.05%** | | |

**Figure 13.2** Clause 60.4 compensation event – 'Scenario B'.

Clause 60.4 states that if the defined cost per unit is reduced, the relevant Bill of Quantities rate is also reduced. This appears to be to clarify that the compensation event is not only assessed on the basis that the relevant rate either remains as per the Bill of Quantities or increases. The examples given in Managing Reality[90] consider both increases and reductions[91] arising from the change in the defined cost for the excavation work:

- The increase in the final quantity of excavation in 'Scenario A' means that the work can be carried out more efficiently (e.g. by larger and more efficient excavating plant), therefore the defined cost per unit (m$^3$) reduces, and the rate reduces from £10/m$^3$ to £9/m$^3$; whereas
- The reduction in the final quantity in 'Scenario B' has the opposite effect of the contractor working less efficiently, so the defined cost per unit (m$^3$) increases from £10/m$^3$ to £12/m$^3$

When reading the contract when either Main Option B or D is selected, a compensation event under clause 60.4 will only concern a change in the Prices, as changes in the Completion Date and a Key date arising from a difference between the final quantity for an item, and the quantity for that item in the Bill of Quantities are covered in clause 60.5 (refer to Section 13.5.23).

90 NEC4 – Managing Reality, Book 4 – Managing Change (2018) p17.
91 NEC4 – Managing Reality, Book 4 – Managing Change (2018) p17. No calculation or further detail is given in either 'Scenario A', or 'Scenario B' for the adjustments to the unit rate.

### 13.5.23 Main Options B and D. A Difference Between the Final Total Quantity of Work Done and the Quantity for an Item in the Bill of Quantities at the Contract Date which Delays Completion or a Key Date – Clause 60.5

A compensation event under clause 60.5 of Main Options B or D is triggered in the same way as under clause 60.4 (refer to Section 13.5.22) – by a difference between the final quantity of work for an item in the Bill of Quantities, and the quantity stated therein. However, whereas clause 60.4 seems solely to be a vehicle for changing the Prices (the Defined Cost per unit), clause 60.5 is intended to compensate the contractor[92] in both terms of time and money, for any effect the difference in quantity has on the Completion Date and/or a Key Date.

Then contractor's entitlement is similar to that in clause 60.4, save that there is no criteria applied to the items of work to be considered (e.g. the 0.05% difference does not apply). That is clearly sensible, because even items which are relatively small in quantity may have a significant effect if they are delayed. The key criterion is that the difference, of whatever amount, is likely to delay completion or affect a Key Date. This is a straightforward clause which, in a similar fashion to JCT SBC/Q 2016 and JCT IC 2016, amounts to a warranty on the part of the Client that the Bill of Quantities is accurate. Essentially, if the bills are not accurate, the contractor cannot include the appropriate allowances in respect of his price and/or programme, and the difference causes a delay, the contractor has a claim.

### 13.5.24 Main Options B and D. Project Manager Corrects Mistakes in the Bill of Quantities which are Corrected by the Project Manager – Clause 60.6

As with compensation events under clauses 60.4 and 60.5, clause 60.6 only applies to Main Options B and D, and concerns the correction of mistakes in the Bill of Quantities. The mistakes must be either:

- Departures from the 'method of measurement' stipulated in Contract Data Part One[93]; or
- Ambiguities or inconsistencies.

The effects of departing from the stated method of measurement can be significant. If the Bill of Quantities' descriptions contain less information than the 'method of measurement' prescribes, then the contractor may have to carry out additional (undescribed) works in order to complete a particular item of work.

Ambiguity[94] and inconsistency[95] are ordinary English words, and the incidence of both may cause the Bill of Quantities to lack certainty or clarity. In order to correct such a mistake, the Project Manager must issue an instruction, and this is likely to provide the following directions to the contractor:

---

92 JCT SBC/Q 2016 and JCT IC 2016 both compensate the contractor in terms of an extension of time (refer to Sections 11.4.2 and 11.4.5) and loss and expense for the same (refer to Sections 12.4.2, 12.4.4 and 12.4.7) for correcting the quantities in the Bill of Quantities for an over-measure or under-measure, or for an inaccurate Approximate Quantity.

93 See Contract Data Part One – Section 6 (compensation events).

94 'Ambiguity' *'the quality of being open to more than one interpretation, inexactness'* Oxford Dictionary of English (2010) Third Ed.

95 'Inconsistency' *'not staying the same throughout …not compatible or in keeping with'* Oxford Dictionary of English (2010) Third Ed.

- In the event of a departure from the stated 'method of measurement', the Project Manager's instruction should make good any discrepant descriptions, or provide the correct description in accordance with the stated method of measurement;
- To correct an ambiguity, the Project Manager's instruction must clearly identify the ambiguity, and then clarify which meaning or interpretation in the Bill of Quantities is to be relied upon; and similarly
- In order to correct an inconsistency, the Project Manager must clearly elect which one of the inconsistent elements prevails over the others.

There appears to be no restriction on whether the Project Manager's instruction, correcting the mistake, results in a greater or lesser obligation being imposed on the contractor. In any event, the correction is dealt with as a compensation event, and it may result in reduced prices (e.g. if the Project Manager's instruction enables the contractor to work in a significantly more efficient manner). However, it seems clear that clause 60.5 does not allow the Project Manager to correct other mistakes in Bills of Quantities, such as contractor's pricing errors.

### 13.5.25 Main Options B and D. Correcting an Inconsistency Between the Bill of Quantities and Another Document When Assessing a Compensation Event – Clause 60.7

Clause 60.7 is also only found in Main Options B and D. It provides that where a compensation event is to be assessed as a result of the correction of an inconsistency between (i) the Bill of Quantities and (ii) any other document, the contractor has assumed that the Bill of Quantities are correct. This clause refers to 'another document', which may include the following:

- The Scope (including any specification or drawings, and amendments to any contractor's design that the Project Manager did not initially accept for want of compliance with the Scope);
- The Site Information;
- An instruction from the Project Manager or Supervisor.

### 13.5.26 Secondary Option X2.1. The Applicable Law Changes After the Contract Date

Where Secondary Option X2 (Changes in the law) applies, the 'law' is the 'law of the country in which the Site is located' and not the 'law of the contract' set out in Contract Data Part One.[96] Therefore, the clause is very broad and it appears that any change in any aspect[97] of the applicable law where the site is located is a compensation event. As highlighted,

96 NEC4 – Contract Data Part 1 – Section 1 (General) and clause 12.2. NEC4's 'Managing Reality' Book Four (2018) p17 appears to suggest that the Option X2 refers to changes to the 'law of the contract' which is not the same as the law of the country containing the site.

97 For example, in the United Kingdom, the 'change in the law' could be a change arising from an Act of Parliament, a Statutory Instrument (or Regulation) such as the 'Approved Documents' under the Building Regulations 2010, applicable standards (such a British Standards).

the law of the contract[96] may be different from the law in the country where the site is located. For example, it is not unknown for a project where the site is located in Europe to be carried out under English law. For example, a project in Germany will have to comply with German building regulations, and it is unlikely that any change in UK building regulations (unless the change is following a change in EU regulations) would be relevant to a project in Germany.

The principle of entitling the contractor to a compensation event if there are changes in the law of the country containing the site may benefit by being extended in certain circumstances. Given NEC4's use as an engineering contract, it may be used where a significant component (e.g. a turbine for a power station) may be manufactured in and transported from a country that is different from the country containing the site, and even from the country where the 'law of the contract' applies.

The effect of the change in the law may be to reduce the total defined cost, in which case the prices are also to be reduced. Perhaps, for that reason, the clause empowers the project manager to give notice to the contractor and to instruct it to submit quotations.

Care should be taken when referring to NEC4's guidance notes.[98] Whilst they may be helpful, particularly with procedural matters, they do not have the force of law nor do they bind the parties. It is likely that an adjudicator, arbitrator or judge (or even the parties' legal advisors), when trying to decide the meaning of the clauses, may only find the NEC4 guidance to be informative.

### 13.5.27 Secondary Option X12.3(6) An Instruction Given by the Core Group to the Partners to Change the Partnering Information

This compensation event X12.3(6) only applies when Option X12 (Multiparty collaboration) is selected. The 'collaboration' is between a 'Core Group' of 'Partners'[99] (which can include the contractor) and the 'Promotor' (usually the Client) who are listed in Contract Data Part One (under option X12). This also includes the 'Partnering Information'.[100] Not all of the Partners and the Promotor may be in contract with each other, so Option X12 can set out a collaborative approach between them. An instruction from the 'Core Group' to the Partners which changes the Partnering Information then constitutes a compensation event under clause X12.3(6). The clause states that it may result in reduced prices.

### 13.5.28 Secondary Option X14.2 – Delay in Making the Advanced Payment

A compensation event under clause X14.2 arises only where Option X14 (Advance payment to the contractor) applies. The Client must make an advanced payment to the contractor of the amount, and at the time stated in Contract Data Part One. If an advanced payment bond

98 See Footnote 99. NEC4's 'Managing Reality' (2018) volumes 1 to 5 and the 'User Guide' (2020).
99 NEC4 – Option X12. The 'Schedule of Partners' is listed in Contract Data Part One. Clause X12.1(3) states that the 'Schedule of Core Group Members' is a list of the Partners forming the Core Group, who (clause X12.1(4)) comprise the list of Partners appointed to make decisions on their behalf.
100 NEC4 – Option X12 – clause X12.1(6) defines the 'Partnering Information' as a document specifying how the Partners collaborate with each other. It is either identified in Contract Data Part One or issued as an instruction under the NEC4 contract.

is required, the payment may be delayed until the next assessment date[101] after the employer receives the bond. If the employer is late in making the payment, it is a compensation event. Clearly, a delay in receiving the advanced payment may result in serious financial and time consequences for a contractor.

### 13.5.29 Secondary Option X15.2 – The Correction of a Defect for Which it was Not Liable by the Contractor

Option X15 refers to the contractor's design and limits the standard for design to one of reasonable skill and care.[102] Clause X15.1 states that the contractor is not liable for a Defect[103] arising from its design if it used reasonable skill and care when producing that design. Therefore, it appears that this compensation event not only applies to the correction of Defects arising from design (carried out with reasonable skill and care) but also to Defects for which the contractor was not liable.

### 13.5.30 Secondary Option Y(UK)2 Y2.5 Suspension of Performance by the Contractor under the Housing Grants, Construction and Regeneration Act 1996 ('the Act')

Section 112 of the Act entitles the contractor to suspend performance of all its obligations (including design) if the Client has not paid the 'notified sum' by the final date for payment. Under NEC4's provisions, the 'notified sum' (as defined in s111(2) of the Act) is:

- The amount stated in the Project Manager's 'certificate' under clauses 51.1 and Y2.2;[104] or
- The amount stated as due at the due date in the contractor's 'application for payment' under clause 50.2 in the event that the Project Manager fails to certify payment under clause 51.1.[105]

Once the 'notified sum' has been established, the Client may only pay less than the notified sum if the Client[106] (not the Project Manager) issues a pay-less notice to the contractor in accordance with clause Y2.3.

---

101 NEC4 – clause 50.1. The Project Manager decides the first 'assessment date' within the 'assessment interval' after the 'starting date' set out in Contract Data Part One – Data provided by the Client, in Section 5. Subsequent 'assessment dates' occur at the end of each subsequent 'assessment interval'.
102 Compared to the higher standard of ensuring the design is 'fit for purpose'. Given that NEC4 is often used for engineering projects (including process engineering) where a project's output may be its primary function, it is perhaps a little surprising that Contract Data Part One does not give the Client the option of selecting a 'fitness for purpose' or 'reasonable skill and care' standard for the contractor's design.
103 Clause 11.2(6) defines a 'Defect' as (a) work that does not comply with the Scope, and (b) work designed by the contractor that does not comply with (i) the design accepted by the Project Manager (under clause 21.2) or (ii) the applicable law (see Section 13.5.26).
104 NEC4 – clause Y2.2 provides that the 'Project Manager's certificate' is the 'payment notice' that complies with s110A(2)(a)(i) of the Act.
105 S110A(3) of the Act. s110(A)(4) provides that the contractor's 'application for payment' under clause 50.2 will constitute a notice under s110A(3) of the Act.
106 NEC4 – clause &2.3 states that the 'pay less notice' must be given by the 'Parties', i.e. the 'Client' and (in the case where a pay-less notice is given by him) 'the contractor' (see clause 11.2(13). It seems that if the Client delegates issuing a 'pay less notice' to the Project Manager (or another), then the issuer will have to clearly set out that the 'pay less notice' is issued on the Client's behalf and with his authority.

This clause turns such suspension into a compensation event. By doing so, it entitles the contractor to a change in the Completion Date,[107] as well as '*a reasonable amount in respect of costs and expenses*'.[108] It appears that the compensation event may provide an equivalent remedy in respect of a change to Key Date, because s112(4) of the Act states, '*Where the contractual time limit is set by reference to a date rather than a period, the date shall also be adjusted accordingly.*'

107 s112(4) of the Act.
108 s112(3A) of the Act.

# Part III

# 14

# Sub-Contract Claims

## 14.1 Introduction

Claims made by sub-contractors against main contractors are often made on similar bases to claims made by main contractors against employers. Indeed, the sub-contract may contain equivalent entitlements (relevant events, relevant matters or compensation events) to those under its corresponding main contract. Important differences to note are

- Under the usual 'main contractor/sub-contractor' contractual arrangement, the sub-contractor will not be able to claim directly against the employer (in the event that one of the employer's actions – such as a variation, have caused the sub-contractor to suffer loss or delays), so he will have to claim against the main contractor;
- The employer's representative will usually have no involvement in notifying, assessing and/or certifying claims made by the sub-contractor; and
- The sub-contractor may be entitled to claim against the main contractor even where the main contractor has no equivalent claim against the employer (e.g. in the event that the main contractor, or another sub-contractor for whom the main contractor is responsible, has adversely affected the progress of the sub-contractor's work).

The legal principles which apply to claims under main contracts often remain good under sub-contracts. This is something which is often misunderstood, and claims under sub-contracts are sometimes treated as though they are subject to a completely different set of rules to those governing claims under main contracts. For example, there is nothing to prevent a sub-contractor bringing its claims as claims for damages under the common law, and reference should be made to Chapter 6 for general principles (see also Section 14.4.3). However, only claims made under the express provisions of the sub-contracts are considered in Sections 14.1 to 14.4.2. Indeed, many of the sub-contract provisions are now very similar to the equivalent provisions of the relevant main contracts.[1]

Therefore, many of the matters that entitle a main contractor to additional time or money (Relevant Events and Relevant Matters under the JCT 2016 main contracts, and compensation events under the NEC4 main contract) will also be found in the

1 Refer to Tables 11.4 and 14.8 to compare the delay events under the JCT 2016 main and sub-contracts.

*Building Contract Claims*, Sixth Edition. Guy Higginbottom and David Chappell.

corresponding sub-contracts. Where there is no material difference between the source of a sub-contractor's entitlement to the equivalent matter or compensation event under its corresponding main contract, this chapter will refer to those matters in the chapters (summarised below) that deal with particular claims under the main contracts:

- Chapter 5 – *Claims for Variations under the Contract, Acceleration, Bonus clauses, and Contractor's Share*;
- Chapter 11 – *Extensions of time under JCT 2016 contracts*;
- Chapter 12 – *Loss and expense under JCT 2016 contracts*; and
- Chapter 13 – *Claims under NEC4*.

The corresponding sub-contracts that are considered in this chapter are as set out in Section 14.1.1 (JCT 2016 Sub-Contracts) and 14.1.2 (NEC4 Sub-Contract) below.

### 14.1.1 JCT 2016 Sub-Contracts

The JCT 2016 Sub-Contracts and their corresponding JCT 2016 main contracts are shown in Table 14.1 below.

With the exception of the JCT MWSub/D 2016 and the JCT ShortSub 2016 sub-contracts, the project specific information (i.e. the equivalent to the information set out in the Agreement[2] under the other JCT 2016 main contracts) is set out in the following 'Agreement' documents (that are separate from their corresponding sub-contracts' 'Conditions') (Table 14.2).

Each of the JCT 2016 sub-contract 'Agreement' consists of the following sections:

- Recitals. There are typically between 6 and 10 recitals, depending upon the extent of design provided by the sub-contractor, and if any framework or supplemental provisions apply;
- Articles. The Articles typically set out the following:
  - The 'Sub-Contract Documents' comprising the sub-contract, including the Agreement, the Conditions, the Main Contract Information Schedule[3] and the Numbered Documents[4] (refer to Table 14.3);
  - The sub-contract sum, which will either be:
    1. The Sub-Contract Sum and Final Sub-Contract Sum if the sub-contract works are not to be re-measured; or

2 Under the JCT 2016 main contracts, the 'Agreement' comprises the Recitals, the Articles and the Contract Particulars to which the corresponding main contract 'Conditions' are annexed.

3 The 'Main Contract Information Schedule' is defined in clause 1.1 of the JCT 2016 Sub-Contracts as providing information relating to the Main Contract (such as selected information from the main contract's 'Contract Particulars') and is annexed to the sub-contract. It includes details of (i) the employer, (ii) the employer's representative (e.g. Architect, Contract Administrator, Employer's Agent), (iii) the Principal Designer, (iv) the Principal Contractor, (v) the Main Contract Conditions, (vi) Main Contract Particulars and Construction Phase plan, (vii) alterations to the Main Contract Particulars, (viii) any division of the Main Contract Works into Sections, (ix) other documents relating to the Main Contract and (x) 'Programme Information' – earliest and latest start dates for the sub-contract works, holiday periods, and site opening times.

4 The 'Numbered Documents' are defined in clause 1.1 of the JCT 2016 sub-contracts as the documents listed in the 'Sub-Contract Particulars' which are set out in the relevant sub-contract Agreement.

**Table 14.1** JCT 2016 Sub-Contract Conditions.

| Corresponding Sub-Contract Conditions | JCT SBC/Q 2016 | JCT DB 2016 | JCT IC 2016 | JCT MW 2016 |
|---|---|---|---|---|
| JCT SBCSub/C 2016<br>Standard Building Sub-Contract | ✓ | | | |
| JCT SBCSub/D/C 2016[a)]<br>Standard Building Sub-Contract with sub-contractor's design | ✓ | | | |
| JCT DBSub/C 2016<br>Design and Build Sub-Contract | | ✓ | | |
| JCT ICSub/C 2016<br>Intermediate Sub-Contract | | | ✓ | |
| JCT ICSub/D/C 2016[a)]<br>Intermediate Sub-Contract with sub-contractor's design | | | ✓ | |
| JCT ICSub/NAM/C 2016[b)]<br>Intermediate Named Sub-Contract | | | ✓ | |
| JCT MWSub/D 2016[c)]<br>Minor Works Sub-Contract with sub-contractor's design | | | | ✓ |
| JCT ShortSub 2016[d)]<br>Short Form of Sub-Contract | ✓ | ✓ | ✓ | ✓ |

a) For use with JCT 2016 main contract with 'Contractor's Designed Portion' (CDP).
b) Although the sub-contractors engaged under the JCT ICSub/NAM/C 2016 are often referred to as 'named', they must not be confused with 'nominated sub-contractors' that were often appointed under JCT contracts until their 1998 editions. There are considerable differences. For example, the employer's representative cannot certify amounts for payment to the sub-contractor, there is no provision for direct payment by the employer if the main contractor defaults, and the employer's representative is not required separately to certify practical completion of the 'nominated' sub-contractor's work. The employer's representative may only be involved if the sub-contractor or main contract are entitled to terminate the sub-contractor's employment. Sub-contractors engaged under JCT ICSub/NAM 2016 form have more in common with domestic sub-contractors under the other forms of JCT 2016 sub-contract.
c) For use with JCT MW 2016 and JCT MWD 2016.
d) For use where the main contract is a JCT 2016 contract and where the sub-contract works are small in quantity, of straightforward content and with little (or low) risk involved.

2. The Sub-Contract Tender Sum and Final Sub-Contract Sum (if the sub-contract works are to be re-measured);
   - Dispute resolution provisions (i.e. Adjudication, arbitration or legal proceedings); and
- The 'Sub-Contract Particulars'.

One of the most important categories of document in the sub-contract will be the Numbered Documents.[4] These are listed in the last section of the sub-contract 'Agreement' and will comprise:

- Information for the sub-contract works (such as drawings, specifications and work schedules);

**Table 14.2** JCT 2016 Sub-Contract Conditions and Agreements.

| JCT 2016 Sub-Contract Conditions | JCT 2016 Sub-Contract Agreement |
|---|---|
| JCT SBCSub/C 2016 | JCT SBCSub/A 2016 |
| JCT SBCSub/D/C 2016 | JCT SBCSub/D/A 2016 |
| JCT DBSub/C 2016 | JCT DBSub/A 2016 |
| JCT ICSub/C 2016 | JCT ICSub/A 2016 |
| JCT ICSub/D/C 2016 | JCT ICSub/D/A 2016 |
| JCT IC/NAM/C 2016 | JCT IC/NAM/A 2016[a)] |

a) The JCT IC/NAM/A 2016 'Agreement' forms part of the document entitled 'JCT ICSub/NAM 2016', along with JCT ICSub/NAM/T 2016 (invitation to tender), and JCT ICSub/NAM/T (Tender).

- Information for any design carried out by the sub-contractor, such as:
  - The 'Contractor's Requirements'; and
  - The 'Sub-Contractor's Proposals'
- Pricing information, such as:
  - Bills of Quantities;
  - Priced activity schedule;
  - 'Sub-Contractor's Designed Portion Analysis' (SCDP Analysis);

**Table 14.3** JCT 2016 Sub-Contracts: Sub-Contract Documents.

| JCT 2016 Sub-Contract | Main Contractor Document | Sub-Contractor Document | Pricing Document |
|---|---|---|---|
| SBCSub/C | Numbered Documents | — | Bills of Quantities<br>Priced Activity Schedule |
| SBCSub/C/D | Numbered Documents<br>Contractor's Requirements | Sub-Contractor's Proposals | Bills of Quantities<br>SCDP Analysis |
| DBSub/C | Numbered Documents<br>Contractor's Requirements | Sub-Contractor's Proposals | Bills of Quantities<br>Sub-Contractor's Designed Works Analysis |
| ICSub/C | Numbered Documents | — | Bills of Quantities<br>Priced Activity Schedule. |
| ICSub/C/D | Numbered Documents<br>Contractor's Requirements | Sub-Contractor's Proposals | Bills of Quantities<br>SCDP Analysis |
| ICSub/NAM | ICSub/NAM<br>Numbered Documents<br>NAM Design Requirements | Design for the NAM Designed Works | Bills of Quantities<br>NAM Design Requirements |
| MWSub/D | Sub-Contract Documents<br>Contractor's Requirements | Sub-Contractor's Designed Portion | Pricing Documents |
| ShortSub | Sub-Contract Documents<br>Contractor's Requirements | — | Pricing Documents |

  - Schedules of rates/dayworks; and
  - Any fluctuations
- Programme information[5] (if not practicable to include in the entries in the Sub-Contract Particulars); including
  - Period for procurement, and
  - Start or finish dates for the sub-contract work, or sections;
- Collateral warranties and bonds;
- BIM protocol and communications; and
- Relevant information from the main contract (referred to in the Main Contract Information Schedule).

### 14.1.2 NEC4 Sub-Contract

The NEC4 Sub-contracts are:

- The NEC4 engineering and construction sub-contract ('NEC4 Sub-contract'); and
- The NEC4 engineering and construction short sub-contract ('NEC4 Short Sub-contract').

The NEC4 Sub-contract is for use where a sub-contractor is engaged by the main contractor, whom the Client has appointed under one of the NEC4 main contracts. The NEC4 Sub-contract includes provisions for Main Option Clauses A to E,[6] Options W1 and W2 (resolving and avoiding disputes), and Secondary Option Clauses X1 to X18, X20 and X21,[7] along with Options Y(UK)1 to Y(UK)3.[8]

The NEC4 Short Sub-contract is an abridged version of the NEC4 Sub-contract. It is intended for use where the sub-contract work does not require sophisticated management procedures, and the nature of the sub-contract work is of low risk, and straightforward execution. It neither contains any of the Main Option or Secondary Option Clauses; however, it does include provision for the HGCRA 1996, which is the equivalent of Option Y(UK)2 under the NEC4 Sub-contract.

---

5 Also refer to the Main Contract Information Schedule – see Footnote 3.

6 NEC4 Subcontract Main Option Clauses: A – Priced subcontract with activity schedule, B – Priced subcontract with bill of quantities, C – Target subcontract with activity schedule, D – Target subcontract with bill of quantities, and E – Cost reimbursable subcontract.

7 NEC4 Subcontract Secondary Option Clauses: X1 – Price adjustment for inflation (Options A to D), X2 – changes in the law, X3 – Multiple currencies (Options A and B), X4 – Ultimate holding company guarantee, X5 – Sectional Completion, X6 – Bonus for early completion, X7 – Delay damages, X8 – Undertakings to the Client or Others, X9 – Transfer or rights, X10 – Information modelling, X11 – Termination by the Contractor, X12 – Multiparty collaboration, X13 – Performance bond, X14 – Advanced payment to Subcontractor, X15 – The Subcontractor's design, X16 – Retention, X17 – Low performance damages, X18 – Limitation of liability, X20 – Key performance indicators, and X21 – Whole life cost.

8 NEC4 Subcontract Options Y(UK)1 – Project Bank Account, Y(UK)2 – The Housing Grants, Construction and Regeneration Act 1996 ('HGCRA 1996') and Y(UK)3 – The Contracts (Rights of Third Parties) Act 1999.

## 14.2 Variations Under the Sub-Contracts

### 14.2.1 Variations Under the JCT 2016 Sub-Contracts

Variations under the corresponding JCT 2016 main contracts are covered in Chapter 5. Variations under the JCT 2016 Sub-Contracts are summarised in Table 14.4. The categories of sub-contract 'Variation' (summarised in Table 14.4) are as follows:

- Alteration, modification of design, quality or quantity of the Sub-Contract Works. The circumstances of a Variation arising under this head are set out in Section 5.2.1.
- Changes or restrictions imposed by the main contractor. The equivalent changes or restrictions under the JCT 2016 main contracts are set out in Section 5.2.2 (save that they are imposed by the employer).
- Main contractor's directions. The equivalent to 'directions' to the sub-contractor under the JCT 2016 main contracts are 'Instructions' from the employer (the effects of which are set out in Section 5.2.3, and 5.2.3.2). Under the JCT 2016 Sub-Contracts, the contractor's power to issue directions is described in a more general way,[9] which oblige[10] the sub-contractor to act on any 'reasonable direction' given by the main contractor. The general nature of main contactor's directions cover more specific matters for which particular instructions may be given under the JCT 2016 main contracts,[11] such as for:
  - Postponing work;
  - Expenditure of Provisional Sums;
  - Non-compliant work, including materials or goods;
  - Non-compliant workmanship; and
  - Antiquities.[12]

  All of the above may be covered under the main contractor's 'reasonable direction'.

- Other circumstances giving rise to a variation. The equivalent to some of these circumstances arising under the JCT 2016 main contracts are set out under Section 5.2.3:
  - Corrections to the contract documents (under the main contracts) are covered in Section 5.2.3.1, save that under sub-contracts requiring sub-contractor's design,[13] the 'Contractor's Requirements' will be the stepped-down equivalent of the 'Employer's Requirements' under the corresponding main contract. In Section 5.2.3.1, Table 5.4 summarises the 'Contract Document's under each of the JCT 2016 main contracts and Table 14.3 summarises the equivalent documents under the JCT 2016 Sub-Contracts.

- Resolving discrepancies in the Sub-Contract Documents will include discrepancies in and between the Contractor's Requirements (and 'NAM Design Documents'[14]), any

9 For example, see clause 3.4 of JCT SBCSub/C 2016 and its equivalent under other JCT 2016 Sub-Contracts.

10 For example, in JCT SBCSub/C 2016, see the reference to clause 3.4 in clause 2.1 (Sub-Contractor's Obligations, General obligations).

11 Refer to Section 5.2 and Table 5.3.

12 See also JCT SBCSub/C 2016 – clause 2.5 (Compliance with main contract and indemnity) and its equivalent.

13 See JCT SBCSub/D/C 2016, JCT DBSubC 2016, JCT ICSubD/C 2016 and JCT MWSub/D 2016.

14 JCT ICSub/NAM 2016 sub-contract (defined in clause 1.1) if the Named Sub-Contractor is required to carry out any design.

**Table 14.4** Variations under JCT 2016 Sub-Contracts.

| | JCT SBC | | JCT DB | JCT IC | | | JCT MW | All |
|---|---|---|---|---|---|---|---|---|
| **Grounds for Variation:** | **Sub/C** | **Sub/ D/C** | **Sub/C** | **ICSub/C** | **ICSub/ D/C** | **ICSub/ NAM/C** | **MW Sub/D** | **Short Sub** |
| Alteration, modification of design, quality or quantity* of the Sub-Contract Works: | | | | | | | | |
| Addition, omission or substitution of work | 5.1.1.1 | 5.1.1.1 | 5.1.1.1 | 5.1.1.1 | 5.1.1.1 | 5.1.1.1 | 12.1 | 11.1 |
| Altering kinds or standards of materials or goods | 5.1.1.2 | 5.1.1.2 | 5.1.1.2 | 5.1.1.2 | 5.1.1.1 | 5.1.1.2 | | |
| Removal of compliant work, materials or goods | 5.1.1.3 | 5.1.1.3 | 5.1.1.3 | 5.1.1.3 | 5.1.1.3 | 5.1.1.3 | | |
| Changes in 'other works' conditions | 5.11 | 5.11 | 5.11 | 5.7 | 5.7 | 5.7 | | |
| Changes or restrictions imposed by the Contractor: | | | | | | | | |
| Access to the site, or parts of it | 5.1.2.1 | 5.1.2.1 | 5.1.2.1 | 5.1.2.1 | 5.1.2.1 | 5.1.2.1 | 12.1 | 11.1 |
| Working space limitations | 5.1.2.2 | 5.1.2.2 | 5.1.2.2 | 5.1.2.2 | 5.1.2.2 | 5.1.2.2 | | |
| Working hours limitations | 5.1.2.3 | 5.1.2.3 | 5.1.2.3 | 5.1.2.3 | 5.1.2.3 | 5.1.2.3 | | |
| Undertaking or completing work in a particular order | 5.1.2.4 | 5.1.2.4 | 5.1.2.4 | 5.1.2.4 | 5.1.2.4 | 5.1.2.4 | | |
| Contractor's directions (generally) | 3.4 | 3.4 | 3.4 | 3.4 | 3.4 | 3.4 | 11.1 | 10.1 |
| Other circumstances giving rise to a Variation: | | | | | | | | |
| Correction, alteration or modification of Bills of Quantities | 2.9.2 | 2.9.3 | 2.9.3 | 2.9.1 (2.7.2) | 2.9.1 (2.7.2) | 2.9.1 (2.7.2) | — | — |
| Discrepancy in Contractor's Requirements | — | 2.11.2 | 2.11.2 | — | 2.9.1 (2.8.3.1) | 2.9.1 (2.8.3.1) | — | — |
| Divergence between Statutory Requirements and the sub-contract documents after the Base Date | 2.12.2 | 2.12.2.1, 2.12.2.2 | 2.12.2.1, 2.12.2.2 | 2.10.2 | 2.10.2.1, 2.10.2.2 | 2.10.2.2 | 8.1 | — |
| Restoration, replacement etc. costs (damage to sub-contract works) | 4.9.1.6 (6.8.3.2) | 4.9.1.6 (6.8.3.2) | 4.9.1.6 (6.8.3.2) | 4.7.1.1 (6.8.3.2) | 4.7.1.1 (6.8.3.2) | 4.7.1.1 (6.8.3.2) | — | — |

Main Contract documents (found in either the sub-contract's Numbered Documents or in the Main Contract Information Schedule). The general rule is that if the discrepancy is caused by a document produced by the sub-contractor, then no Variation arises and the sub-contractor must resolve the discrepancy at its own time and cost (see Section 5.2.3.1).
- Divergences between the sub-contract documents and Statutory Requirements operate under the same principles under the JCT 2016 main contracts (see Section 5.2.3.1).
- Restoration of damage to sub-contract works. A Variation occurs if the sub-contract works have be restored, replaced or repaired if damaged by terrorism, on the proviso that the employer does not terminate the main contractor's employment under the Main Contract.
- Design Submission Procedure. This is a reproduction of the same procedure found in the corresponding main contracts and is described in Section 5.2.3.3. The design submission procedure is found in the following JCT 2016 Sub-Contracts:
  - JCT SBCSub/C/D 2016 (in Schedule 7).
  - JCT DBSub/C 2016 (in Schedule 7); and
  - JCT ICSub/C/D 2016 (in Schedule 4).

### 14.2.2 NEC4 Sub-Contracts

In keeping with the NEC ethos, the NEC4 Sub-contract and Short Sub-contract each operate a 'compensation event' procedure. The number of compensation events (refer to Sections 5.3 and 5.3.1) and the procedures for notifying and assessing them are generally similar to those in the NEC4 Main Contract (refer to Section 5.3.2), but there are some important differences. In Section 5.3.1, Table 5.10 set out the compensation events under the NEC4 Main Contract, and this has been reproduced to include the compensation events under the NEC4 Sub-contract and NEC4 Short Sub-contract in Table 14.5.

The relevant compensation events found in the NEC4 Sub-contract and NEC4 Short Sub-contract are covered in detail in Chapter 13, in Sections 13.5.1 to 13.5.30.

The procedures for notifying and assessing compensation events under the NEC4 Sub-contract and the NEC4 Short Sub-contract are similar in nature to the procedure under the NEC4 Main Contract (refer to Figures 5.1 and 5.2) but contain a number of differences to firstly reflect that the main contractor may be relying upon the sub-contractor to notify a compensation event (in circumstances where the sub-contractor's information is required for the main contractor's notification and/or quotations for the equivalent compensation event under the NEC4 main contract), and secondly, the procedure (in particular for the Short Sub-contract) is simplified. The compensation event procedure under the NEC4 Sub-contract is shown in Figure 14.1:

When compared to Figure 5.1 (in Section 5.3.2), clause 61.3 of the NEC4 Sub-contract only allows the sub-contractor 7 weeks in which to notify the main contractor. This provides a 'window' of 1 week for the main contractor to utilise the sub-contractor's notification, as he has 8 weeks[15] in which to notify the compensation event to the Project Manager under the main contract. In the event that the sub-contractor fails to notify the main contractor, then clause 61.3 of the NEC4 Sub-contract provides that neither the Prices, the Sub-contract

15 NEC4 main contract – clause 61.3.

**Table 14.5** NEC4 compensation events across the Main Contract, Sub-contract and Short Sub-contract.

| NEC4 Main Contract compensation event – clause 60.1 (for all Main Option Clauses B and D and Secondary Options) | | NEC4 Sub-contract | NEC4 Short Sub-contract |
|---|---|---|---|
| (1) | Project Manager's[a] Instruction ("PMI") to the Contractor[b] to change the Scope[c] | 60.1(1) | 60.1(1) |
| (2) | The Client[d] prevents access to parts of the site by the "access date" or relevant dates on the Accepted Programme | 60.1(2) | 60.1(2) |
| (3) | The Client[d] fails to provide something by the Accepted Programme's date | 60.1(3) | 60.1(3) |
| (4) | PMI[a] not to start, or stop work | 60.1(4) | 60.1(4) |
| (4) | PMI[a] to change a Key Date | 60.1(4) | — |
| (5) | The Client (or other contractors)[d] do not do their work in accordance with the Accepted Programme | 60.1(5) | — |
| (5) | The Client (or other contractors)[d] do not work in accordance with the Scope[e] | 60.1(5) | 60.1(5) |
| (5) | The Client (or other contractors)[d] undertake work that is not set out in the Scope.[e] | 60.1(5) | — |
| (6) | The PM[f] fails to respond to a communication within the period for reply | 60.1(6) | 60.1(6) |
| (7) | PMI[a] dealing with an object of archaeological value or that is of interest found on the Site | 60.1(7) | — |
| (8) | The PM[a] changes a previous decision | 60.1(8) | 60.1(7) |
| (9) | The PM's[a] reason(s) for not accepting something is/are not set out in the contract | 60.1(9) | — |
| (10) | The contractor[b] is instructed to search for a defect but there isn't one | 60.1(10) | — |
| (11) | Testing causes an otherwise avoidable delay | 60.1(11) | — |
| (12) | The contractor[b] encounters unforeseen physical conditions within the site, which are not weather conditions | 60.1(12) | 60.1(8) |
| (13) | The contractor[b] encounters adverse weather conditions within the meaning of the contract | 60.1(13) | 60.1(9) |
| (14) | The Client[d] causes an event for which he is liable | 60.1(14) | — |
| (15) | PM[a] certifies "taking over" before "Completion" and the "Completion Date" | 60.1(15) | — |
| (16) | Client[d] fails to provide necessary facilities, equipment and materials for tests | 60.1(16) | — |
| (17) | The PM[a] corrects an assumption he made about a previous 'compensation event' | 60.1(17) | 60.1(10)[g] |
| (18) | Client[d] commits a breach of the contract which is not a Compensation Event | 60.1(18) | — |
| (19) | An event occurs (which is not another Compensation Event) which prevents the contractor[h] from completing the works[i] that is outside of each party's control, or that the contractor could not have reasonably foreseen | 60.1(19) | 60.1(11) |
| (20) | The PM[a] does not accept a contractor's[b] quotation for a proposed instruction | 60.1(20) | — |

*(Continued)*

**Table 14.5** (Continued)

| NEC4 Main Contract compensation event – clause 60.1 (for all Main Option Clauses B and D and Secondary Options) | NEC4 Sub-contract | NEC4 Short Sub-contract |
|---|---|---|
| (21) Any additional Compensation Events set out in the Contract Data Part One occur | 60.1(21) | — |
| Options B and D – differences between the final quantities of work and the quantities in the Bill of Quantities: | | |
| • If it does not arise from a change in the Scope; | 60.4 | — |
| • If it causes the Defined Cost to change (see Tables 5.12 and 5.13 in Section 5.5.1); or | 60.4 | — |
| • If the Bill of Quantities for that rate exceeds 0.5% of the total of the Prices. (All clause 60.4) | 60.4 | — |
| If the difference in quantity delays Completion or a Key Date (clause 60.5) | 60.5 | — |
| • If the Project Manager[a)] corrects a mistake in the Bill of Quantities arising from a departure from the method of measurement, ambiguity or inconsistency (clause 60.6) including between the Bills of Quantities and another document (clause 60.7) | 60.6 60.7 | 60.1(12)[j)] |
| Secondary Option X2 (Changes in the law) clause X2.1 | X2.1 | — |
| Secondary Option X12 (Multiparty Collaboration): | | |
| • Instruction to change the Partnering Information: reduced prices – clause X12.3(6) | X12.3(6) | — |
| • Change of programme to match the revised timetable: reduced prices – clause X12.3<br>• (7) | X12.3(7) | — |
| Secondary Option X14 (Advanced Payment) clause X14.2 | X14.2 | — |
| Secondary Option X15 (Contractor's[b)]Design) clause X15.2 | X15.2 | — |
| Y2.5 Contractor[b)] suspension | Y2.5 | — |

a) NEC4 Subcontract and NEC4 Short Subcontract – the 'Contractor' acts (e.g. issues an instruction) instead of the Project Manager.
b) NEC4 Subcontract and NEC4 Short Subcontract – the 'Subcontractor' acts instead of the 'Contractor'.
c) NEC4 Subcontract – including (a) an instruction to change the Subcontract Scope to resolve an ambiguity or inconsistency under clause 63.10 (or clause 63.8 of the NEC4 Short Subcontract) or (b) correcting a change in the Conditions for a Key Date under clause 63.11.
d) NEC4 Subcontract and NEC4 Short Subcontract – the 'Contractor' is in default instead of the 'Client'.
e) NEC4 Subcontract and NEC4 Short Subcontract – the 'Subcontract Scope' applies in place of the 'Scope'.
f) NEC4 Subcontract and NEC4 Short Subcontract – the 'Contractor' fails to respond instead of the Project Manager, or those working on the his behalf – such as the Supervisor.
g) NEC4 Short Subcontract – clause 60.1(10) either party (the contractor or the subcontractor) may notify the other of an change of assumption.
h) NEC4 Subcontract and NEC4 Short Subcontract – the 'Subcontractor' acts instead of the 'Contractor'.
i) NEC4 clause 60.1(19) – the event can also prevent the contractor from completing by the Accepted Programme's planned completion date.
j) NEC4 Short Subcontract – clause 60.1(12): the contractor issues an instruction to correct a mistake in the 'Price List' which is set out in the 'Subcontractor's Subcontract Data' (comprising columns for 'item number', 'description', 'unit', 'quantity', 'rate' and 'price'. Clause 11.2(13) refers to the 'Price List' as containing the 'Prices' as the amounts listed in the 'Price' column. Where an item in the Price List has a stated 'quantity', the 'Price' for that item is the product of the quantity multiplied by the rate.

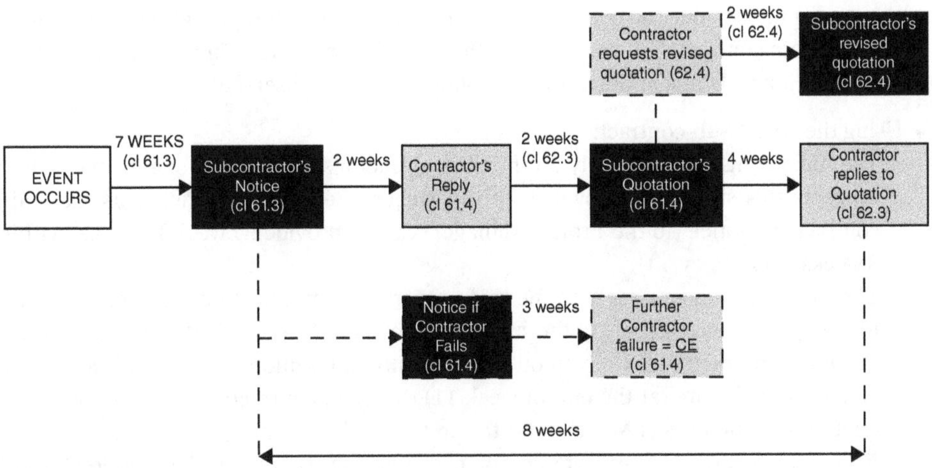


**Figure 14.1** NEC4 Sub-contract compensation event procedure.

Completion Date nor any Key Dates are to be changed (unless the event arises from the main contractor issuing an instruction, notification or certificate, or changing an earlier decision[16]).

The timetable for the NEC4 Short Sub-contract is far shorter (e.g. the sub-contractor only has 3 weeks in which to notify a compensation event to the main contractor), and in the event that the 'period for reply' is set to 1 week,[17] may only last for an overall period of 4 weeks, as shown in Figure 14.2:

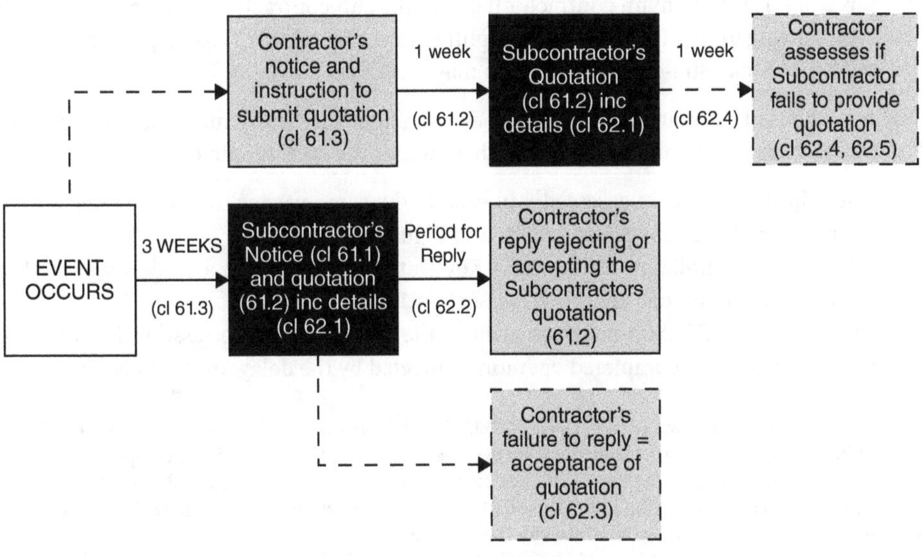


**Figure 14.2** NEC4 Short Sub-contract compensation event procedure.

16 NEC4 Subcontract – a compensation event under clause 60.1(8).

17 NEC4 Short Subcontract – a 'period for reply' applying to the Contractor and the Subcontractor is set out in 'The Contractor's Subcontract Data' (also refer to clause 13.2).

It is useful to consider the overall periods for dealing with compensation events under the NEC4 main contract, Sub-contract and the Short Sub-contract. Figure 14.3 shows these procedures alongside one another and the following can be observed:

- Using the NEC4 Sub-contract:
  - The main contractor may rely on the sub-contractor's notification of a compensation event as this should be provided up to 7 weeks after the event, whereas the main contractor's notice (to the Project Manager) can be provided 1 week later (i.e. within 8 weeks after the event);
  - The Project Manager's reply to the main contractor's notification may coincide with the main contractor's reply to the sub-contractor's notification (at the end of week 9);
  - The main contractor may rely upon the sub-contractor's quotation, as this may be provided 1 week before (at the end of week 11) the main contractor needs to submit his quotation to the Project Manager (at the end of week 12); and
  - Following the Project Manager's acceptance of the main contractor's quotation (end of week 14), the main contractor may have up to 1 week to reply to the sub-contractor's quotation (end of week 15).
- If using the NEC4 Short Sub-contract:
  - The main contractor may have another 5 weeks (from receipt of the sub-contractor's notification - at the end of week 3) in which to rely upon or incorporate it into his own notification to the Project Manager (end of week 8); however
  - By virtue of the relatively short timescale, the main contractor runs the risk of having to accept the sub-contractor quotation at the end of week 4, when the Project Manager may not have to accept the main contractor's quotation until the end of week 13. Whilst this could leave the main contractor liable to the sub-contractor for the compensation event, owing to the 'low risk and straightforward execution' of works procured under the NEC4 Short Sub-contract, this risk may not arise.

The methods of assessing compensation events under the NEC4 Sub-contract and Short Sub-contract are very similar to assessing them under the NEC4 main contract:

- Changes in the Prices are assessed[18] in exactly the same way as under the NEC4 main contract (refer to Sections 5.5 and 5.5.1, and Figure 5.2);
- A delay to the Completion Date or to a Key Date is assessed[19] in a similar way to that under the NEC4 main contract (refer to Section 13.3) save that:
  - Under the NEC4 Sub-contract, when the delay is being assessed, only the Sub-contractor's uncompleted operations affected by the delay are changed[20]; but

18 NEC4 Subcontract – clauses 63.1 to 63.4 and NEC4 Short Subcontract – clauses 63.1 to 63.5. The NEC4 Short Subcontract's assessment of compensation event is relatively simpler as (a) if the compensation event only changes the quantities in the Price List, the change to the Prices is assessed by multiplying the new quantity by the Price List's existing rate (clause 63.1), and (b) any new 'People Rates' may either be agreed, or failing agreement, assessed by the contractor (clause 63.5).

19 NEC4 Subcontract – clause 63.4, and NEC4 Short Subcontract 63.6.

20 NEC4 Subcontract – clause 63.5. The equivalent clause (63.5) in the NEC4 main contract does not contain this proviso. Instead, the NEC4 main contract states that assessment of changes to the Completion Date and/or to a Key Date consider (a) delays caused by the compensation event that are already shown on the Accepted Programme and (b) events occurring between (i) the date of the Accepted Programme and (ii) the dividing date (see Section 13.3).

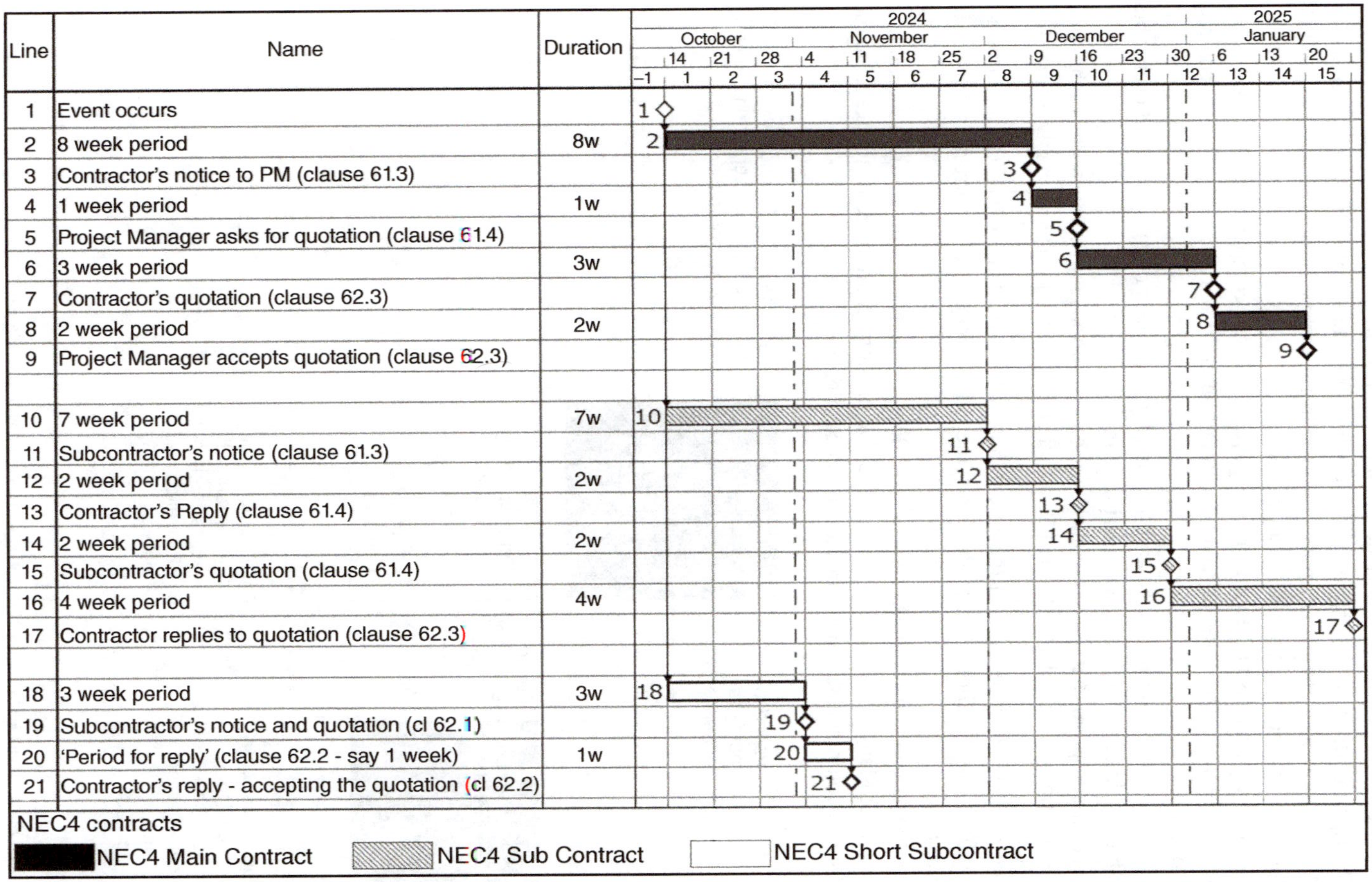


**Figure 14.3** NEC4 compensation event timetables (main contract, Sub-contract and Short Sub-contract).

- Under the NEC4 Short Sub-contract, the assessment only considers the length of time that by owing to the compensation, Completion is forecast to be delayed.[21]

- The rights of the contractor and the sub-contractor to change the Prices, the Completion Date and any Key Dates are limited to those in the compensation event clause (NEC4 Sub-contract only – clause 63.6, also see Section 5.3.2);
- The following are also taken into account when assessing compensation events:
  - Any failure by the sub-contractor to submit an early warning[22] (NEC4 Sub-contract only);
  - Any risk allowances for cost and time matters that are not compensation events but have a significant chance of occurring[23]; and
  - The assumption that the Sub-contractor reacts promptly and competently to the compensation event and any Defined Cost that is time-related was reasonably incurred.[24]

The assessment procedures for the NEC4 Short Sub-contract and the NEC4 Sub-contract are different. The shorter procedure for the NEC4 Short Sub-contract is shown in Figure 14.4:

The longer compensation assessment procedure for the NEC4 Sub-contract is set out in Figure 14.5. Many of the aspects of the NEC4 main contract are directly applicable to the NEC4 Sub-contract assessment procedure (see Section 5.3.2).

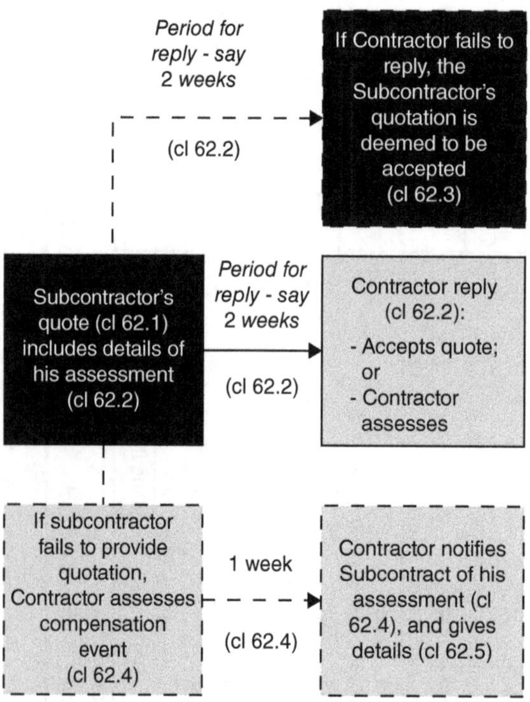


**Figure 14.4** NEC4 Short Sub-contract – compensation events assessment procedure.

21 NEC4 Short Subcontract – clause 63.6.
22 NEC4 Subcontract – clause 63.7 (Early warnings are covered in clause 15.1).
23 NEC4 Subcontract – clause 63.8 and NEC4 Short Subcontract 63.7 (first bullet point).
24 NEC4 Subcontract – clause 63.9, and NEC4 Short Subcontract – clause 63.7 (second bullet point).

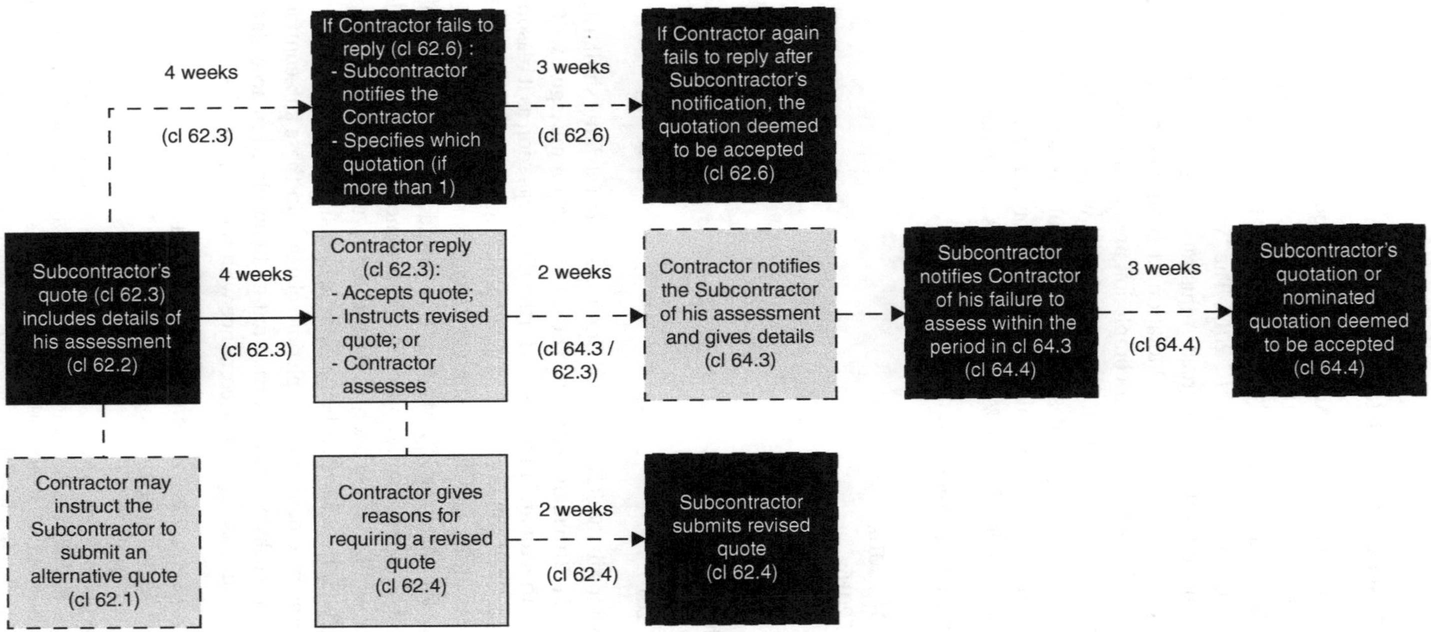


**Figure 14.5** NEC4 Sub-contract – compensation event assessment procedure.

## 14.3 Extensions of Time Under Sub-Contracts

### 14.3.1 JCT 2016 Sub-Contracts' Delay Notifications

Extensions of time (also referred to as 'Adjustment of Period for Completion') are dealt with by the dedicated clauses in the JCT 2016 Sub-Contracts. An extension of time can also be agreed under the 'Variation Quotation' procedure found in Schedule 2 of JCT SBCSub/C 2016, JCT SBCSub/D/C and JCT DBSub/C 2016. The 'Variation Quotation' procedure in these sub-contracts includes the 'Pre-Agreed Adjustment', used to revise the completion date, fixed by acceptance of a Variation Quotation. It should be noted that references to extending time and fixing a new date for completion are taken to mean the date for completion of the sub-contract works or, if the sub-contract works are divided into sections, completion of the relevant section. Unless there is a particular reason to do so, reference will not constantly be made to sections in this commentary. For clarity, the express terminology for 'extensions of time' under the JCT 2016 Sub-Contracts is as follows:

- 'Adjustment of Period for Completion':
  - JCT SBCSub/C 2016;
  - JCT SBCSub/D/C 2016;
  - JCT DBSub/C 2016;
  - JCT ICSub/C 2016
  - JCT ICSub/D/C 2016; and
  - JCT ICSub/NAM/C 2016
- 'Extension of time':
  - JCT MWSub/D 2016; and
  - JCT ShortSub 2016.

The notice provisions in the JCT 2016 Sub-Contracts bear a striking similarity to those in the corresponding JCT 2016 main contracts. The sub-contractor must give written notice of delay to the main contractor and must do this forthwith (i.e. as soon as it reasonably can[25]). The requirements for each sub-contractor's 'delay notice' are set out in Table 14.6.

The circumstances under which the Sub-Contractor is to notify delay under JCT 2016 Sub-Contracts (save for under the JCT MWSub/D 2016 and JCT ShortSub 2016) are similar, insofar as it must become 'reasonably apparent' that commencement and completion of the sub-contract works will be, or are likely to be affected by a delay. The same applies if the 'progress' of the sub-contract works is being delayed or is likely to be delayed.

The 'Time of commencement and completion' clause[26] in these sub-contracts provides that the sub-contract works are to be carried out and completed in accordance with the programme details, and reasonably in accordance with the following:

- The Sub-Contract Particulars' 'programme details'; and
- '*reasonably in accordance with the progress of the Main Contract Works*'.

25 *Hudson v Hill* (1874) 43 LJCP 273; *London Borough of Hillingdon v Cutler* (1967) 2 All ER 361.

26 JCT SBCSub/C 2016, JCT SBCSub/D/C 2016 and JCT DBSub/C 2016 – clause 2.3, and JCT ICSub/C 2016, JCT ICSub/D/C 2016 and JCT ICSub/NAM/C – clause 2.2. The Sub-Contractor's obligation under these clauses is subject to receiving the Contractor's notice to commence work on site and any 'Adjustment of Period for Completion'.

**Table 14.6** JCT 2016 Sub-Contracts' delay notice requirements.

| Requirements for the Sub-Contractor's 'Delay Notice' | JCT SBC | | JCT DB | JCT IC | | | JCT MW | All |
|---|---|---|---|---|---|---|---|---|
| | SBCSub/C | SBCSub/D/C | DBSub/C | ICSub/C | ICSub/D/C | ICSub/NAM/C | MWSub/D | ShortSub |
| Notice issued as soon as it becomes 'reasonably apparent' that 'commencement, progress or completion' of the Sub-Contract Works is being, or is likely to be affected | 2.17.1 | 2.17.1 | 2.17.1 | 2.12.1 | 2.12.1 | 2.12.1 | — | — |
| Notice if the Sub-Contractor is delayed in completing | — | — | — | — | — | — | 13.1 | 12.1 |
| Notice must be 'in writing' | 1.7.1 | 1.7.1 | 1.7.1 | 1.7.1 | 1.7.1 | 1.7.1 | 13.1 | 12.1 |
| Notice must be given 'forthwith' to the contractor | 2.17.1 | 2.17.1 | 2.17.1 | 2.12.1 | 2.12.1 | 2.12.1 | — | — |
| Notice must include: | | | | | | | | |
| • Material circumstances of delay/likely delay | 2.17.1 | 2.17.1 | 2.17.1 | — | — | — | — | — |
| • Cause or causes of delay; and | 2.17.1 | 2.17.1 | 2.17.1 | 2.12.1 | 2.12.1 | 2.12.1 | — | — |
| • Identify any Relevant Sub-Contract Event | 2.17.1 | 2.17.1 | 2.17.1 | 2.12.1 | 2.12.1 | 2.12.1 | — | — |
| For each 'event' cited in the notice, provide (if practicable): | | | | | | | | |
| • Particulars of its expected effects | 2.17.2 | 2.17.2 | 2.17.2 | — | — | — | — | — |
| • Estimate of expected delay to completion[a)] | 2.17.2 | 2.17.2 | 2.17.2 | 2.12.1 | 2.12.1 | 2.12.1 | — | — |
| Sub-Contractor provides the Contractor with updates: | | | | | | | | |
| • If any 'material change' in estimated delay, or in any 'particulars' | 2.17.3 | 2.17.3 | 2.17.3 | — | — | — | — | — |
| • Provide the update 'forthwith'; and | 2.17.3 | 2.17.3 | 2.17.3 | — | — | — | — | — |
| • Include further information reasonably required by the Contractor | 2.17.3 | 2.17.3 | 2.17.3 | 2.12.4.2 | 2.12.4.2 | 2.12.4.2 | — | — |

a) Expected delay beyond either (a) the period (or periods if the Sub-Contract Works are divided into Sections) as set out in the Sub-Contract Particulars, or (b) any period (or periods if completion by Sections) fixed previously.

The first point relates to the 'Programme' details at section 5 in the Sub-Contract Particulars – which is fairly self-explanatory insofar as the Sub-Contractor commences work on site within the stated period following the main contractor's notice, and completes the works within the period (or periods, if the works are divided into sections) described below (expressed in weeks):

- Any 'lead in' period for procurement of materials, fabrication and delivery to site before commencing the Sub-Contract Works on site; and
- The period of notice required to commence work on site (for the entire sub-contract works, or for each section thereof); and
- The period for carrying out the sub-contract works (or periods for each section if the sub-contract works are so divided); and
- Any '*further details or arrangements*' relevant to carrying out the Sub-Contract Works (e.g. a more detailed programme or phasing plan relevant to the Sub-Contract Works).

The words in the second point have been construed very broadly by main contractors over the years as obliging the sub-contractor to work in accordance with the main contractor's progress of the (main contract) Works, and that if the main contractor's progress slowed or increased, then the sub-contractor was obliged to follow suit. It is suggested that the sub-contractor may plan and perform the work as it pleases if there is no indication to the contrary, provided that it finishes within the period (or periods) of time fixed in the sub-contract. The sub-contractor's only obligation so far as programming requirements are concerned, are those requirements expressly contained in the sub-contract itself.[27]

Save for the JCT MWSub/D 2016 and the JCT ShortSub 2016 sub-contracts, the requirements for the sub-contractor's delay notices are qualified. The sub-contractor's delay notice must include the cause or causes of the delay (insofar as the sub-contractor is able to identify them) along with stating any Relevant Sub-Contract Events. In addition, JCT SBCSub/C 2016, JCT Sub/D/C 2016 and JCT SBSub/C 2016 also require the notice to include the 'material circumstances' of the delay, or likely delay. In practice, this qualification probably will make little difference, but it does offer the sub-contractor some limited protection against allegations that it has not included appropriate or sufficient information in its notice.

The requirements for the notice in JCT MWSub/D 2016 and JCT ShortSub 2016, when taken strictly, appear to impose a retrospective obligation to notify the main contractor, if '*completing the Sub-Contract Works within the Period for Completion*' triggers the obligation to notify. On one hand, it appears the obligation notify only arises if the work has not been completed by the time the 'Period for Completion' expires; however, this seems neither a pro-active nor commonsense approach and one that's hardly conducive to mitigating any delays by notifying the main contractor in good time.

Further guidance (under the JCT 2016 main contracts) for the notices is given in Section 11.2.1.

---

27 *Pigott Foundations v Shepherd Construction* (1996) 67 BLR 48.

### 14.3.2 JCT 2016 Sub-Contracts: Fixing Period for Completion

The main contractor gives an extension of time to the sub-contractor by fixing a revised period for completion (or periods, if the works are divided into sections). The main contractor's obligation to do so is principally triggered by receipt of the sub-contractor's notice of delay (see Section 14.3.1). Table 14.7 sets out the requirements for fixing a new period for completion in detail.

For JCT SBCSub/C 2016, JCT SBCSub/D/C 2016 and JCT DBSub/C 2016, the main contractor is given a time limit of 16 weeks from receiving the sub-contractor's notice and particulars in which to give an extension of time. It will be noted that this is four weeks longer than the corresponding 12 week period (for the employer, or his representative) under the JCT 2016 main contracts (refer to Section 11.3.3). The reason is probably to allow the main contractor to obtain the necessary extension of time from the employer (or his representative) before giving an extension of time to the sub-contractor. Although that may be sensible in theory, in practice, there are often arguments about whether the employer or his representative has received the full particulars and estimate of delay required. If less than 16 weeks remain between receipt of the notice, particulars and estimate and the currently fixed completion date, the main contractor must endeavour to give any extension no later than that date.

The extension of time provisions in the JCT ICSub/C 2016, JCT ICSub/D/C 2016 and JCT ICSub/NAM/C 2016 sub-contracts are shorter than in the larger JCT 2016 Sub-Contracts. Notable differences are as follows:

- The main contractor's obligation to give a 'fair and reasonable' extension of time does not appear to be triggered by the sub-contractor's notice. Instead, the main contractor's obligation in this regard, appears to arise only when:
  - The main contractor properly considers that completion is likely to be delayed (beyond the period in the Sub-Contract Particulars or any other period fixed);
  - The delays arise from Relevant Sub-Contract Events; and
  - As soon as the main contractor is able to estimate the period (or periods) of delay.
- It appears that the main contractor can give extensions of time of his own volition, up to 16 weeks after the date upon which the Sub-Contract Works reach practical completion.

The shortest JCT 2016 Sub-Contracts (JCT MWSub/D 2016 and the JCT ShortSub 2016) seemingly follow a similar process to their JCT ICSub 2016 counterparts, insofar as the sub-contractor's notice does not appear to trigger the main contractor's obligation to make any extension of time. The main contractor's obligation to do so appears to arise under the following circumstances:

- If the delay causes completion of the Sub-Contract Works to occur after the Period for Completion has expired; and
- The delay was caused by a variation or reasons outside of the sub-contractor's control.

In which cases, the main contractor shall give such extension of time as is reasonable.

It seems obvious that extensions of time cannot be reduced until time has been extended in the first place. A question which sometimes arises is whether the main contractor can 'set-off' (or account for) omitted work against the other grounds for extension in the first

**Table 14.7** JCT 2016 Sub-Contracts: Adjustment of Period for Completion.

| | JCT SBC | | JCT DB | JCT IC | | | JCT MW | All |
|---|---|---|---|---|---|---|---|---|
| **Requirements for the Contractor to fix a new Period for Completion** | **SBCSub/C** | **SBCSub/D/C** | **DBSub/C** | **ICSub/C** | **ICSub/D/C** | **ICSub/NAM/C** | **MWSub/D** | **ShortSub** |
| Contractor's obligation arises: | | | | | | | | |
| • On receipt of the sub-contractor's delay notice and particulars | 2.18.1 | 2.18.1 | 2.18.1 | — | — | — | — | — |
| • If any of the events are Relevant Sub-Contract Events; and | 2.18.1.1 | 2.18.1.1 | 2.18.1.1 | 2.12.1 | 2.12.1 | 2.12.1 | — | — |
| • It's likely that, or that Completion of the Sub-Contract Works (or sections) will occur after the expiry of the Period (or periods) for Completion | 2.18.1.2 | 2.18.1.2 | 2.18.1.2 | 2.12.1 | 2.12.1 | 2.12.1 | — | — |
| • As soon as he is able to estimate the length of delay | — | — | — | 2.12.1 | 2.12.1 | 2.12.1 | — | — |
| • If an extension of time is reasonable | — | — | — | — | — | — | 13.1 | 12.1 |
| If the Sub-Contractor is so entitled, the contractor fixes such revised Period (or periods) for Completion as is 'fair and reasonable' | | | | | | | | |
| Contractor notifies the Sub-Contractor of his decision: | 2.18.2 | 2.18.2 | 2.18.2 | 2.12.1 | 2.12.1 | 2.12.1 | — | — |
| • Irrespective if an extension of time is given | 2.18.2 | 2.18.2 | 2.18.2 | — | — | — | — | — |
| • Within 16 weeks of receiving the Sub-Contractor's notice and particulars, or less if reasonably practicable; or | 2.18.2 | 2.18.2 | 2.18.2 | — | — | — | — | — |
| • If the Period (or periods) for Completion expires in less than 16 weeks, before the expiry of that period or periods. | 2.18.2 | 2.18.2 | 2.18.2 | — | — | — | — | — |
| The Contractor's decision states: | | | | | | | | |
| • The extension of time for each Relevant Sub-Contractor Event; and | 2.18.3.1 | 2.18.3.1 | 2.18.3.1 | 2.12.2 | 2.12.2 | 2.12.2 | — | — |
| • Any reduction for a Relevant Sub-Contract Omission | 2.18.3.2 | 2.18.3.2 | 2.18.3.2 | — | — | — | — | — |

| | | | | | | | | |
|---|---|---|---|---|---|---|---|---|
| After the first fixing of revised Period for Completion: | | | | | | | | |
| • Contractor may reduce the extended Period, accounting for a Relevant Sub-Contract Omission | 2.18.4 | 2.18.4 | 2.18.4 | — | — | — | — | — |
| After the period for completion of the Sub-Contract Works has expired (if occurring before their practical completion) the Contractor may: | | | | | | | | |
| • Not later than 16 weeks after the Sub-Contract Works' practical completion; | 2.18.5 | 2.18.5 | 2.18.5 | 2.12.3 | 2.12.3 | 2.12.3 | — | — |
| • Extend the relevant period of periods previously fixed, considering any Relevant Sub-Contract, of his own volition; or | 2.18.5.1 | 2.18.5.1 | 2.18.5.1 | 2.12.3 | 2.12.3 | 2.12.3 | — | — |
| • Reduce the relevant period or periods previously fixed, considering any Relevant Sub-Contract Omission; or | 2.18.5.2 | 2.18.5.2 | 2.18.5.2 | — | — | — | — | — |
| • Confirm the period or periods that were previously fixed. | 2.18.5.3 | 2.18.5.3 | 2.18.5.3 | — | — | — | — | — |
| Provided always that: | | | | | | | | |
| • The Sub-Contractor constantly uses best endeavours to prevent delays, or further delays; | 2.18.6.1 | 2.18.6.1 | 2.18.6.1 | 2.12.4.1 | 2.12.4.1 | 2.12.4.1 | 13.2 | 12.2 |
| • The Contractor is provided with all information as is reasonably necessary | — | — | — | 2.12.4.2 | 2.12.4.2 | 2.12.4.2 | | |
| • Satisfy the Contractor that the Sub-Contractor is proceeding with his works; | 2.18.6.2 | 2.18.6.2 | 2.18.6.2 | — | — | — | — | — |
| • The Contractor cannot fix a period (or periods) for completion of the Sub-Contract Works that is shorter than is set out in the Sub-Contract Particulars; and | 2.18.6.3 | 2.18.6.3 | 2.18.6.3 | — | — | — | — | — |
| • No decision of the Contractor can alter the length of any Pre-Agreed Adjustment, save where a Variation Quotation where the Variation is the subject of a Relevant Sub-Contract Omission. | 2.18.6.4 | 2.18.6.4 | 2.18.6.4 | — | — | — | — | — |

instance so as to give a reduced first extension? Common sense suggests that there is no good reason why the main contractor cannot do so, but a strict reading of the relevant clause[28] (in JCT SBCSub/C 2016, JCT SBCSub/D/C 2016 and JCT DBSub/C 2016) indicates, that by the word 'After', the first extension must be given without accounting for any omitted work (referred to as a 'Relevant Sub-Contract Omission[29]'). However, after a first extension of time, the main contractor may (by its own volition), direct (instruct) any Relevant Sub-Contract Omission, and subsequently notify a reduction of that first (or any previous) extension of time to the sub-contractor. In practical terms, the reduction of a 'first' extension of time can be done soon after, and each subsequent extension of time must take account of Relevant Sub-Contract Omissions of work directed up to the date of the extension.

JCT SBCSub/C 2016, JCT SBCSub/D/C 2016 and JCT DBSub/C 2016 all clearly state that no matter how much work is omitted, the sub-contractor is always entitled to its original sub-contract period,[30] and the main contractor is prevented from fixing a shorter period for completion than the period stated in the sub-contract particulars.

The principle of 'Relevant Sub-Contract Omissions' to reduce extensions of time does not apply to the JCT ICSub/C 2016, JCT ICSub/D/C 2016, JCT ICSub/NAM/C 2016 sub-contracts, nor the JCT MWSub/D 2016 or JCT ShortSub 2016 sub-contracts. Therefore, given that it is doubtful if the main contractor can subsequently omit work in order to reduce extensions of time, he may be cautious in granting extensions of time in the first instance.

Under each of the JCT 2016 Sub-Contracts, the sub-contractor is obliged to use his best endeavours to prevent delay in the progress of the sub-contract works, and to prevent any such delay resulting in the completion of the sub-contract works being delayed or further delayed (also see Section 11.3.4). Under JCT SBCSub/C 2016, JCT SBCSub/D/C 2016 and JCT DBSub/C 2016, the sub-contractor is also required to do all that is reasonably required to the satisfaction of the main contractor, to proceed with the sub-contract works. They are essentially provisions requiring the main contractor to mitigate any losses (in this case - lost time), but it is suggested that these requirements do not require the sub-contractor to spend substantial sums of money.

Both the JCT SBCSub/C 2016[31] and the JCT DBSub/C 2016 conditions include provisions[32] for the sub-contractor to provide 'Variation Quotations' and 'Acceleration Quotations'. The corresponding main contracts (JCT SBC/Q 2016 and JCT DB 2016) also include similar provisions for the main contractor to provide equivalent quotations to the

---

28 JCT SBCSub/C 2016, JCT SBCSub/D/C 2016 and JCT DBSub/C 2016 – clause 2.18.4.

29 'Relevant Sub-Contract Omission' is defined in clause 1.1 of in JCT SBCSub/C 2016, JCT SBCSub/D/C 2016 and JCT DBSub/C 2016 by reference to clause 2.16.3 (*'the omission of any work or obligation through a direction for a Variation…or in regard to a Provisional Sum for defined work in any Bills of Quantities'*).

30 JCT SBCSub/C 2016, JCT SBCSub/D/C 2016 and JCT DBSub/C 2016 – clause 2.18.6.3 (see also JCT SBC/Q 2016 – clause 2.28.6.3 and JCT DB 2016 – clause 2.25.6.3 for the similar principle under the corresponding main contracts).

31 The Pre-Agreed Adjustment (via the Variation Quotation and Acceleration Quotation procedures) is also found in Schedule 2 of the JCT SBCSub/D/C 2016 sub-contract, however, only references to the JCT SBCSub/C 2016 and JCT DBSub/C 2016 sub-contracts in the text and Figure 14.6 for brevity (clause references also apply to JCT SBCSub/D/C 2016).

32 JCT SBCSub/C 2016 and JCT DBSub/C 2016 – 'Variation Quotation' and 'Acceleration Quotation' procedures are found in Schedule 2 to the Sub-Contract Conditions.

employer (refer to Section 5.4.2 and Figures 5.3 and 5.4). Under both of these sub-contracts, a 'Pre-Agreed Adjustment' (meaning fixing a revised period or periods for completion[33]) can be reached between the main contractor and the sub-contractor via a Variation Quotation or an Acceleration Quotation. After a first extension of time, the period for completion can be amended by a 'Pre-Agreed Adjustment',[34] and the length of any Pre-Agreed Adjustment cannot be altered,[35] unless by a further Pre-Agreed Adjustment which is itself, a Relevant Sub-Contract Omission.[29]

The procedures for the sub-contractor's Variation Quotation and Acceleration Quotation are shown in Figure 14.6:

The requirements for the main contractor's response to either the sub-contractor's notification of delay (or to the delay itself) vary, and are summarised in Table 14.8. It should be noted that the main contractor's obligation to consider and give an extension of item is only triggered by receipt of the sub-contractor's delay notice and 'particulars' under the JCT SBCSub/C 2016, JCT SBCSub/D/C 2016 and JCT DBSub/C 2016 sub-contracts. Despite the other sub-contracts requiring the sub-contractor to notify the main contractor of a delay, it appears that the sub-contractor's notification (no 'particulars' are required) is not conditional upon the main contractor's obligations to consider and given an extension of time.

The form and content of the main contractor's notification to the sub-contractor of whether or not an extension of time is to be given is set out in varying detail in all of the JCT 2016 Sub-Contracts, with the exception of JCT MWSub/D 2016 and JCT ShortSub 2016. Under these latter two, the main contractor must only 'make' such an extension

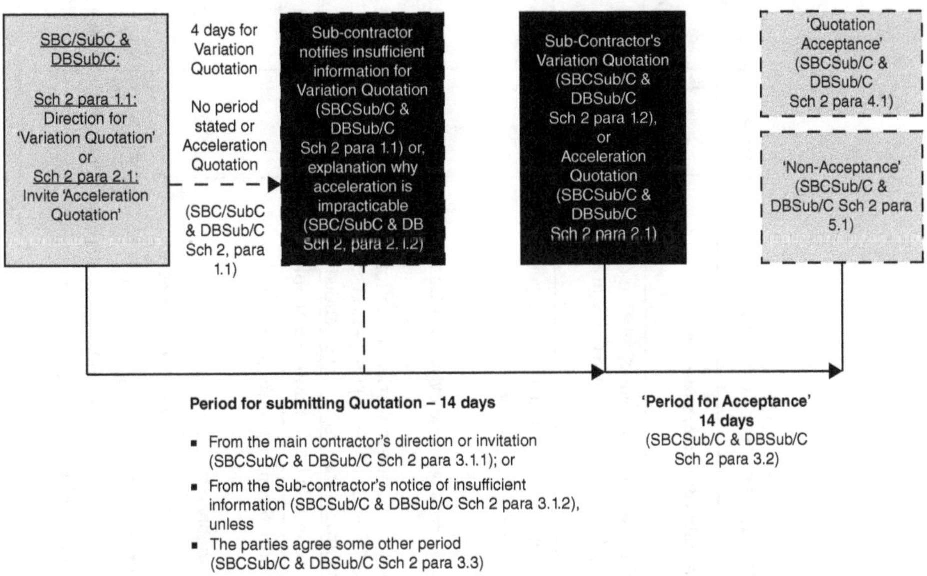


**Figure 14.6** JCT SBCSub/C 2016 and JCT DBSub/C 2016 – procedures for 'Variation Quotation' and 'Acceleration Quotation'.

33 JCT SBCSub/C 2016 and JCT DBSub/C 2016 – clause 2.16.2.
34 JCT SBCSub/C 2016 and JCT DBSub/C 2016 clause 2.18.4.
35 JCT SBCSub/C 2016 and JCT DBSub/C 2016 clause 2.18.6.4.

**Table 14.8** JCT 2016 Sub-Contracts: Contractor's extension of time obligations.

| Requirements for the main contractor's response for extensions of time | JCT SBC | | JCT DB | JCT IC | | | JCT MW | All |
|---|---|---|---|---|---|---|---|---|
| | SBCSub/C | SBCSub/D/C | DBSub/C | ICSub/C | ICSub/D/C | ICSub/NAM/C | MWSub/D | ShortSub |
| Main contractor's obligation to consider extension of time is triggered by receiving the sub-contractor's delay notice and 'particulars' | 2.18.1 | 2.18.1 | 2.18.1 | — | — | — | — | — |
| First fixing of an extension of time (such revised period [periods] for completion): Main contractor must consider: | | | | | | | | |
| • If any of the notified events are Relevant Sub-Contract Events; and | 2.18.1.1 | 2.18.1.1 | 2.18.1.1 | 2.12.1 | 2.12.1 | 2.12.1 | — | — |
| • Completion of the Sub-Contract Works (or section) is likely to be delayed beyond either (i) the period set out in the Sub-Contract Particulars, or (ii) any extended period. | 2.18.1.2 | 2.18.1.2 | 2.18.1.2 | 2.12.1 | 2.12.1 | 2.12.1 | — | — |
| Any extension of time should be 'fair and reasonable' | 2.18.1 | 2.18.1 | 2.18.1 | 2.12.1 | 2.12.1 | 2.12.1 | 13.1 | 12.1 |
| In any event, the contractor must notify the sub-contractor of his decision: | | | | | | | | |
| • As soon as is reasonably practicable; and | 2.18.2 | 2.18.2 | 2.18.2 | — | — | — | — | — |
| • At any time up to 16 weeks after practical completion | — | — | — | 2.12.3 | 2.12.3 | 2.12.3 | — | — |
| • Within 16 weeks of receiving the 'particulars'; or | 2.18.2 | 2.18.2 | 2.18.2 | — | — | — | — | — |
| • If less than 16 weeks remain until the relevant period(s) for completion expires, prior to the expiry of the relevant period(s) for completion | 2.18.2 | 2.18.2 | 2.18.2 | — | — | — | — | — |

| | | | | | | | | |
|---|---|---|---|---|---|---|---|---|
| • The extension of time for each Relevant Sub-Contract Event | 2.18.3.1 | 2.18.3.1 | 2.18.3.1 | — | — | — | — | — |
| • For a subsequent fixing of the period(s) for completion, the reduction of that period owing to a Relevant Sub-Contract Omission | 2.18.3.2 | 2.18.3.2 | 2.18.3.2 | — | — | — | — | — |
| Fixing a subsequent extension of time: | | | | | | | | |
| Main contractor may subsequently fix a shorter extension of time considering a Relevant Sub-Contract Omission | 2.18.4 | 2.18.4 | 2.18.4 | — | — | — | — | — |
| After the period(s) for completion, if it expires before the date of practical completion of the sub-contract works, the main contractor may (of his own volition): | | | | | | | | |
| • If fair and reasonable to do so, extend the period(s) for completion (inc those previously fixed) regarding any Relevant Sub-Contract Event; or | 2.18.5.1 | 2.18.5.1 | 2.18.5.1 | 2.12.2 | 2.12.2 | 2.12.2 | — | — |
| • If fair and reasonable to do so, reduce the relevant period(s) after considering a Relevant Sub-Contract Omission (issued after the last fixing of period(s) for completion); or | 2.18.5.2 | 2.18.5.2 | 2.18.5.2 | — | — | — | — | — |
| • Confirm the relevant period(s) previously fixed. | 2.18.5.3 | 2.18.5.3 | 2.18.5.3 | — | — | — | — | — |

of time, and it is suggested that the main contractor must confirm any extension of time to the sub-contractor by effective means.

### 14.3.3 JCT 2016 Sub-Contracts: Relevant Sub-Contract Events

The matters (or events) which entitle the sub-contractor to an extension of time are referred to as 'Relevant Sub-Contract Events' in all of the JCT 2016 Sub-Contracts, save for JCT MWSub/D 2016 and JCT ShortSub 2016. These events are effectively the equivalent of the 'Relevant Events' in the JCT 2016 main contracts that have been 'stepped down' from applying between the employer and main contractor, to apply between the main contractor and a sub-contractor. A detailed commentary on the JCT 2016 main contracts' Relevant Events is provided in Chapter 11, in Sections 11.4.1 to 11.4.17. Table 14.9 reconciles the 'Relevant Sub-Contract Events'[36] across all of the JCT 2016 Sub-Contracts.

The various Relevant Sub-Contract Events are described below. Many of these events correspond with equivalent 'Relevant Events' under the corresponding JCT 2016 main contracts. The commentary on the Relevant Sub-Contract Events may refer to their equivalent (in Section 11.4) events under the corresponding main contracts.

Owing to the way in which the JCT 2016 Sub-Contracts deal with the sub-contractor's obligations to complete the work, there is no equivalent entitlement for 'Deferment of Possession' (as is found under the relevant JCT 2016 main contracts - see Section 11.4.3). The JCT 2016 Sub-Contracts recognise that sub-contractors progressively start work as the main contract works progress. Therefore, unlike where the JCT 2016 main contracts give the main contractor a specific start (or 'possession') date and a 'completion' date, the sub-contractor's obligation to start work depends upon the main contractor notifying them to commence within a certain period, and agreeing an overall period in which to complete the sub-contract works.[37] Therefore, any 'deferment of possession' of the main contract works can be accommodated by the main contractor simply notifying the relevant sub-contractors to commence work after the period of deferment has ended.

- Variations. All of the JCT 2016 Sub-Contracts entitle the sub-contractor to an extension of time in the event of variations to the sub-contract works – provided, of course, that the delay caused by the variation will (or is likely to) result in the sub-contract works being completed after the period for completion has expired. Variations under the JCT 2016 Sub-Contracts are summarised in Table 14.4 and the equivalent 'Relevant Event' under the main contract is summarised in Section 11.4.1.
- Main contractor's directions. Section 14.2.1 pointed out that the JCT 2016 Sub-Contracts provide a more 'general' mechanism for the contractor to instruct the sub-contractor, compared to the employer (or his representative) instructing the main contractor under the corresponding main contract. The sub-contracts often refer to instructions under the

36 JCT MWSub/D 2016 and JCT ShortSub 2016 do not refer to 'Relevant Sub-Contract Events'.

37 For example, the JCT SBCSub/A 2016 'Agreement' (see Section 14.1.2) contains (i) the period for procurement of materials, fabrication and delivery to site before starting work on site, (ii) the period of notice to commence work on site and (iii) the period in which the sub-contract works must be completed. The 'Agreement' also contains provisions for carrying out the sub-contract works in sections as separate periods for completion are provided for each Section).

**Table 14.9** JCT 2016 Sub-Contracts: Relevant Sub-Contract Events.

| Relevant Sub-Contract Event | JCT SBC | | JCT DB | JCT IC | | | JCT MW | All |
|---|---|---|---|---|---|---|---|---|
| | SBCSub/C | SBCSub/D/C | DBSub/C | ICSub/C | ICSub/D/C | ICSub/NAM/C | MWSub/D | ShortSub |
| Matters beyond the Sub-Contractor's control | — | — | — | — | — | — | 13.1 | 12.1 |
| 'Variations' or matters deemed to be 'Variations' | 2.19.1 | 2.19.1 | 2.19.1 | 2.13.1 | 2.13.1 | 2.13.1 | 13.1 | 12.1 |
| Main contractor's directions (passing on main contract instructions): | | | | | | | | |
| • Resolving a discrepancy[a)] | 2.19.2.1 | 2.19.2.1 | 2.19.2.1 | 2.13.2.1 | 2.13.2.1 | 2.13.2.1 | — | — |
| • Postponement of work[b)] | 2.19.2.1 | 2.19.2.1 | 2.19.2.1 | 2.13.2.1 | 2.13.2.1 | 2.13.2.1 | — | — |
| • Provisional Sums[c)] | 2.19.21 | 2.19.2.1 | 2.19.2.1 | 2.13.2.1 | 2.13.2.1 | 2.13.2.1 | — | — |
| • Contractor's disagrees with a Variation Quotation[d)] | 2.19.2.1 | 2.19.2.1 | — | — | — | — | — | — |
| • Inspections and tests for compliant works[e)] | 2.19.2.2 | 2.19.2.2 | 2.19.2.2 | 2.13.2.2 | 2.13.2.2 | 2.13.2.2 | — | — |
| • Opening up for inspections etc. (compliant works)[f)] | 2.19.2.3 | 2.19.2.3 | 2.19.2.3 | 2.13.2.3 | 2.13.2.3 | 2.13.2.3 | — | — |
| Deferment of possession (of the Works, or Section)[g)] | 2.19.3 | 2.19.3 | 2.19.3 | 2.13.3 | 2.13.3 | 2.13.3 | — | — |
| Inaccurate Approximate Quantity | | | | | | | | |
| • Inaccurate Approximate Quantity in the Contract Bills[h)] | 2.19.4.1 | 2.19.4.1 | — | 2.13.4.1 | 2.13.4.1 | 2.13.4.1 | — | — |
| • Inaccurate Approximate Quantity in a Bill of Quantities provided by the Contractor | 2.19.4.2 | 2.19.4.2 | 2.19.4 | 2.13.4.2 | 2.13.4.2 | 2.13.4.2 | — | — |
| Antiquities[i)] | 2.19.5 | 2.19.5 | 2.19.5 | — | — | — | — | — |
| Sub-contractor's suspension for non-payment (by contractor) | 2.19.6 | 2.19.6 | 2.19.6 | 2.13.5 | 2.13.5 | 2.13.5 | — | — |
| Main contractor suspension for non-payment (by employer) | 2.19.7 | 2.19.7 | 2.19.7 | 2.13.6 | 2.13.6 | 2.13.6 | — | — |
| Impediment, prevention or default by employer, employer's representative (inc quantity surveyor) or Employer's Person[j)] | 2.19.8 | 2.19.8 | 2.19.8 | 2.13.7 | 2.13.7 | 2.13.7 | — | — |

(Continued)

**Table 14.9** (Continued)

| Relevant Sub-Contract Event | JCT SBC | | JCT DB | JCT IC | | | JCT MW | All |
|---|---|---|---|---|---|---|---|---|
| | SBCSub/C | SBCSub/D/C | DBSub/C | ICSub/C | ICSub/D/C | ICSub/NAM/C | MWSub/D | ShortSub |
| Impediment, prevention or default by the main contractor, or main contractor's Person[k)] | 2.19.9 | 2.19.9 | 2.19.9 | 2.13.8 | 2.13.8 | 2.13.8 | — | — |
| Work or failure of statutory undertaker | 2.19.10 | 2.19.10 | 2.19.10 | 2.13.9 | 2.13.9 | 2.13.9 | — | — |
| Exceptionally adverse weather | 2.19.11 | 2.19.11 | 2.19.11 | 2.13.10 | 2.13.10 | 2.13.10 | — | — |
| Loss or damage by any Specified Perils[l)] | 2.19.12 | 2.19.12 | 2.19.12 | 2.13.11 | 2.13.11 | 2.13.11 | — | — |
| Civil commotion or terrorism | 2.19.13 | 2.19.13 | 2.19.13 | 2.13.12 | 2.13.12 | 2.13.12 | — | — |
| Strike, lockout, etc. | 2.19.14 | 2.19.14 | 2.19.14 | 2.13.13 | 2.13.13 | 2.13.13 | — | — |
| New legislation after the Main Contract Base Date[m)] | 2.19.15 | 2.19.15 | 2.19.15 | 2.13.14 | 2.13.14 | 2.13.14 | — | — |
| Delay receiving permissions or approvals (design) | — | — | 2.19.16 | — | — | — | — | — |
| Force majeure | 2.19.16 | 2.19.16 | 2.19.17 | 2.13.15 | 2.13.15 | 2.13.15 | — | — |

a) JCT SBC/Q 2016 – clause 2.15 (Notice of discrepancies etc), JCT DB 2016 – clause 2.13 and JCT IC 2016 – clause 2.13 (instructions on errors, omissions and inconsistencies).
b) JCT SBC/Q 2016 – clause 3.15 (Postponement of work), JCT DB 2016 – clause 3.10, and JCT IC 2016 – clause 3.12.
c) JCT SBC/Q 2016 – clause 3.16 (Instructions on Provisional Sums), JCT DB 2016 – clause 3.11 and JCT IC 2016 – clause 3.13.
d) JCT SBC/Q 2016 - clause 5.3.2 (Variation Quotations).
e) JCT SBC/Q 2016 – under clause 3.17 (Inspections, tests) or 3.18.4 (unless work not in accordance with the contractor), JCT DB 2016 – clause 3.12 and 3.13.3, and JCT IC 2016 – clauses 3.14 and 3.15.1.
f) JCT SBC/Q 2016, JCT DB 2016 – generally instructed under clause 3.10, JCT IC 2016 – clause 3.9.
g) JCT SBC/Q 2016 and JCT IC 2016 – clause 2.5, JCT DB 2016 – clause 2.4.
h) 'Contract Bills' means the 'priced bills of quantities' prepared according to the 'Measurement Rules' (e.g. NRM2) referred to in the main contract: JCT SBC/Q 2016 – Second Recital, or in JCT IC 2016, in the Third and Fourth Recitals.
i) JCT SBC/Q 2016 – clause 3.22.1 or 3.22.2, JCT DB 2016 clause 3.15.1 or 3.15.2.
j) JCT SBC/Q 2016 – clause 1.1 defines 'Employer's Person' as *'all persons employed, engaged, or authorized by the Employer'* but excluding the employer's representative, or the quantity surveyor, or any statutory undertaker.
k) JCT SBC/Q 2016 – clause 1.1 defines 'Contractor's Persons' as *'the Contractor's employees and agents, all other persons employed or engaged on or in connection with the Works or any part of them [i.e. other sub-contractors], and any person properly on the site in connection therewith…'* excluding the employer, his representative, any Employer's Person (see Footnote 57) or statutory undertaker.
l) 'Specified Perils' are defined in JCT SBCSub/C 2016 – clause 6.1.
m) The 'Base Date' is that date written into the contract particulars. The "Base Date" is effectively the cut-off date for changes in legislation etc., which may entitle the contractor to adjust the Contract Sum.

main contract, without specifically identifying the nature of those instructions. However, the Relevant Sub-Contract Matters (refer to Table 14.8) confirm the main contract clauses under which the instructions originate (i.e. from the employer to the contractor, and then to the sub-contractor - that constitute Relevant Sub-Contract Matters). Examples of these (which are not expressly referred to in the JCT 2016 Sub-Contracts) are instructions to defer possession, postpone work or expend provisional sums. For the equivalent 'Relevant Events' under the JCT 2016 main contracts, refer to Section 11.4.2.

- Deferment of possession. This is self-explanatory and the equivalent 'Relevant Event' is described in Section 11.4.3.
- Inaccurate Approximate Quantities. An inaccurate forecast of an approximate quantity of work may delay a sub-contractor if the actual quantity of work is to be undertaken by that sub-contractor and significantly exceeds the approximately quantity (hence, taking longer to carry out). There are two Relevant Sub-Contract Events to consider:
  - First, where the inaccurate forecast of the Approximate Quantity is contained in the Contract Bills – i.e. the Bills of Quantities prepared on employer's behalf, and in accordance with the Measurement Rules set out in the main contract; or
  - Where the inaccurate forecast of the Approximate Quantity is contained in any Bills of Quantities prepared by the main contractor.

Both Relevant Sub-Contract Events may apply under the JCT SBCSub/C 2016, JCT SBC-Sub/D/C 2016, JCT ICSub/C 2016 and JCT ICSub/D/C 2016 sub-contracts, but only the second Relevant Sub-Contract Event applies under the JCT DBSub/C 2016 sub-contract. The equivalent Relevant Event is described at Section 11.4.5.

- Antiquities. This Relevant Sub-Contract Matter is only found in the JCT SBCSub/C 2016, JCT SBCSub/D/C 2016 and JCT DBSub/C 2016 sub-contracts and relates to instructions given under the corresponding main contract for dealing with objects falling within the ambit of the 'antiquities' clauses (refer to Section 11.4.4 for the Relevant Event).
- Sub-contractor's suspension for non-payment by the main contractor. In the event that the main contractor fails to pay the amount due to the sub-contractor, then following 7 days' prior written notice[38] to the main contractor, the sub-contractor may suspend performance of his obligations under the sub-contract. Section 11.4.6 sets out the statutory right to suspend performance.
- Sub-contractor's suspension for non-payment (by the employer) of the main contractor. The Relevant Event for the employer's breach in this regard is described in Section 11.4.6. In the event that the main contractor is entitled to suspend performance of his obligations under the main contract, this will include work (or other obligations) performed by his sub-contractors. As the main contractor will be compensated for the period of suspension and relieved from liquidated (or other delay) damages, the same entitlement passes down to the sub-contractor.
- Impediment, prevention or default by the employer (or by someone for whom the employer is responsible). Some of the ways in which the employer (or someone acting

38 JCT SBCSub/C 2016, JCT SBCSub/D/D 2016 and JCT DBSub/C 2016 – clause 4.8 (Sub-contractor's right of suspension), and JCT ICSub/C 2016, JCT ICSub/D/C 2016 and JCT ICSub/NAM/C 2016 – clause 4.10, and JCT MWSub/D 2016 – clause 16 and JCT ShortSub/C 2016 – clause 15.

on his behalf) could impede or prevent the main contractor from carrying out his obligations under the JCT 2016 main contracts are set out in Section 11.4.7.

- Impediment, prevention or default by the main contractor (or by someone for whom the main contractor is responsible). This Relevant Sub-Contract Event is effectively the 'stepped down' version of the above point, only it is the main contractor (or someone acting on his behalf - such as another sub-contractor) that is impeding or preventing the sub-contractor from progressing his works. As the above Relevant Sub-Contract Event is likely to include late instructions, drawings, work not forming part of the main contract, and the employer's failure to provide materials or goods, and failure to give access or comply with the CDM Regulations 2015, this Relevant Sub-Contract Event is likely to arise when another sub-contractor is delaying the sub-contractor's work. The overlapping of sub-contract trades or congested workfaces often mean that two or more sub-contractors are trying to carry out their respective works in close proximity, which may delay one or all of the sub-contractors concerned. Similarly, where a sub-contractor's work (such as a suspended ceiling fixer) is dependent upon the progress of preceding works (such as mechanical and electrical installations), a delay by the sub-contractor undertaking the preceding works is likely to delay the sub-contract works. This is the Relevant Sub-Contract Event that entitles the sub-contractor to an extension of time;
- Failure by statutory undertaker. Refer to Section 11.4.8 for how failure by a statutory undertaker may delay a sub-contractor's works. It should be borne in mind that not every sub-contract trades' work may be affected by the work of a statutory undertaker, so it is a question of fact and degree if their failure will cause the sub-contractor's work to be delayed beyond the expiry of the period for completion.
- Exceptionally adverse weather. Refer to Section 11.4.9 for more details relating to this Relevant Sub-Contract Event. This Relevant Sub-Contract Event may be deleted by some main contractors in the event that the sub-contractor's work is unlikely to be affected by exceptionally adverse weather – for example, if the sub-contractor's work is internal.
- Loss or damage occasioned by one or more of the Specified Perils. The 'Specified Perils' under the JCT 2016 Sub-Contracts are identical to those in the JCT 2016 main contracts (refer to Section 11.4.10).
- Civil commotion or terrorism. Refer to Section 11.4.11 for more details relating to this Relevant Sub-Contract Event.
- Strike, lockout, etc. Refer to Section 11.4.12 for more details relating to this Relevant Sub-Contract Event. For the JCT SBCSub/D/C 2016, JCT DBSub/C 2016 and JCT ICSub/D/C 2016 that the event has been enlarged so an extension of time can be given if the strike affects persons engaged to prepare the main contractor's designed portion. This could affect the sub-contractor itself if it prepares the design in-house but would apply also if the design was prepared by a sub-sub-contractor such as a firm of architects or engineers.
- New legislation after the main contract Base Date. Refer to Section 11.4.13 for more details relating to this Relevant Sub-Contract Event.
- Delay receiving permissions. This Relevant Sub-Contract Event is only found in the JCT DBSub/C 2016 sub-contract. Refer to Section 11.4.14 for more details relating to this Relevant Sub-Contract Event. There is a requirement that the sub-contractor must have taken all practicable steps to reduce the delay, and it must be taken seriously. Realistically, this will probably amount to little more than making any necessary applications in good time, replying promptly to queries and using its best endeavours to obtain the permissions or

approvals. An architect (or other design consultant) or main contractor in the position of making applications to statutory bodies cannot guarantee the result and neither can a sub-contractor. The sub-contractor will be entitled to an extension of time under this Relevant Sub-Contract Event if it can show that the delay was not due to its fault, as it refers to any kind of statutory permission or approval. Virtually all buildings require planning permission and they must satisfy the Building Regulations; however, there are many other possible controls over such things as fire, water and entertainment.
- Force Majeure. Refer to Section 11.4.17 for more details relating to this Relevant Sub-Contract Event.

### 14.3.4 NEC4 Sub-Contracts

The same compensation event procedures, covering the notification of, quotation for, and assessment of compensation events apply to extension of time (changing the Completion Date and/or a Key Date) as are set out in Section 14.2.2.

## 14.4 Loss and Expense Under the Sub-Contracts

### 14.4.1 JCT 2016 Sub-Contractor's Application and Main Contractor's Assessment

The principles governing the sub-contractor's entitlement to 'direct loss and/or expense' (for brevity, referred to as 'loss and expense') and the procedures for obtaining it, are very similar to those found in the JCT 2016 main contracts that are covered in Chapter 12. For example:

- The equivalent 'Relevant Matters' summarised in Table 12.1 that apply between the employer and the main contractor will apply between the main contractor and the sub-contractor as 'Relevant Sub-Contract Matters';
- Similarly, the category of 'Employer responsible' in Table 12.2 will now be 'contractor responsible' in relation to the Relevant Sub-Contract Matters; and
- The Sub-Contract Relevant Events (extensions of time) and Relevant Sub-Contract Matters (loss and expense) are not interrelated (refer to Section 12.1.2) in terms of a contractual cross-reference, but of course, and as a fact, an entitlement to an extension of time, as well as to loss and expense may be caused by the same event.

The JCT 2016 Sub-Contracts[39] all refer to Relevant Sub-Contract Matters, which are summarised in Table 14.10.

The entitlement to loss and expense and the procedures for claiming it under the JCT 2016 Sub-Contracts is similar to that required by their JCT 2016 main contract counterparts (refer to Sections 12.2.1, 12.2.2 and 12.2.3). These principles and procedures

39 Save for the JCT MWSub/D 2016 and JCT Short Sub 2016 sub-contracts, as their simpler content only provides for basic 'loss and expense' as part of agreeing a price for Variations (see clauses 12.3 and 11.3 respectively).

**Table 14.10** JCT 2016 Sub-Contracts – Relevant Sub-Contract Matters.

| | JCT SBC | | JCT DB | JCT IC | | | JCT MW | All |
|---|---|---|---|---|---|---|---|---|
| **Relevant Sub-Contract Matter** | **SBCSub/C** | **SBCSub/D/C** | **DBSub/C** | **ICSub/C** | **ICSub/D/C** | **ICSub/NAM/C** | **MWSub/D** | **ShortSub** |
| Variations | 4.16.1 | 4.16.1 | 4.16.1 | 4.13.1 | 4.13.1 | 4.13.1 | 12.3 | 11.3 |
| Main contractor's directions: | | | | | | | | |
| • Expenditure of Provisional sums | 4.16.2.1 | 4.16.2.1 | 4.16.2.1 | 4.13.2.1 | 4.13.2.1 | 4.13.2.1 | — | — |
| • Opening up, testing of compliant work (instruction under the main contract) | 4.16.2.2 | 4.16.2.2 | 4.16.2.2 | 4.13.2.2 | 4.13.2.2 | 4.13.2.2 | — | — |
| • Opening up, testing of compliant work (contractor's direction under the Sub-Contract) | 4.16.2.3 | 4.16.2.3 | 4.16.2.3 | 4.13.2.3 | 4.13.2.3 | 4.13.2.3 | — | — |
| • Discrepancy or divergence (error, omission or inconsistency[a)]) in the Numbered Documents, inc Contract Documents under the main contract) | 4.16.2.4 | 4.16.2.4 | 4.16.2.4 | 4.13.2.4 | 4.13.2.4 | 4.13.2.4 | — | — |
| • Postponement of work | 4.16.2.5 | 4.16.2.5 | 4.16.2.5 | 4.13.2.5 | 4.13.2.5 | 4.13.2.5 | — | — |
| Dealing with Antiquities | 4.16.3 | 4.16.3 | 4.16.3 | — | — | — | — | — |
| Suspension of works (Main contractor suspension) | 4.16.4 | 4.16.4 | 4.16.4 | 4.13.3 | 4.13.3 | 4.13.3 | — | — |
| Suspension of work (Sub-Contractor suspension)[b)] | 4.8.2 | 4.8.2 | 4.8.2 | 4.10.2 | 4.10.2 | 4.10.2 | 16.2 | 15.2 |
| Inaccurate forecast of Approximate Quantities (Contract Bills in the main contract) | 4.16.5 | 4.16.5 | — | 4.13.4 | 4.13.4 | 4.13.4 | — | — |
| Inaccurate forecast of Approximate Quantities (any other Bills of Quantities) | 4.16.6 | 4.16.6 | 4.16.5 | 4.13.5 | 4.13.5 | 4.13.5 | — | — |
| Delay in approval/permission necessary for the Development Control Requirements | — | — | 4.16.6 | — | — | — | — | — |
| Impediment, prevention or default caused by the employer or those acting on his behalf | 4.16.7 | 4.16.7 | 4.16.7 | 4.13.6 | 4.13.6 | 4.13.6 | — | — |
| Impediment, prevention or default caused by the main contractor or those acting on his behalf | 4.16.8 | 4.16.8 | 4.16.8 | 4.13.7 | 4.13.7 | 4.13.7 | — | — |

a) JCT ICSub 2016 sub-contracts refer to 'error, omission or inconsistency' in the Numbered Documents (inc with the main contracts's Contract Documents).
b) The JCT 2016 sub-contracts do not referred to the sub-contractor's suspension (in the event of contractor's non-payment) as a 'Relevant Sub-Contract Matter' but, in the event of such suspension, the sub-contractor remains entitled to any costs and expenses that are reasonably incurred.

are reproduced as applying to the JCT 2016 sub-contracts in the following paragraphs. JCT SBCSub/C 2016 and JCT DBSub/C 2016 also provide that loss and expense can be paid to the sub-contractor,[40] when such amounts have been included with an accepted 'Variation Quotation' or 'Acceleration Quotation' (refer to Section 14.3.2 and Figure 14.6).

Table 14.10 sets out the 'Relevant Sub-Contract Events' that entitle the sub-contractor to loss and expense under the JCT 2016 Sub-Contracts.

The sub-contractor's entitlement to loss and expense is summarised in Table 14.11. This sets out the criteria that must be satisfied to enable the sub-contractor to notify a valid loss and expense claim to the contractor. Similar to the contractor's entitlement to loss and expense under the JCT 2016 main contracts, there are some exceptions (see Section 12.2.2), and similar exceptions found in the JCT 2016 Sub-Contracts are first highlighted in Table 14.11 and then set out in detail in Table 14.12.

The procedures for (i) the sub-contractor to claim loss and expense and (ii) for the main contractor's ascertainment are identical across the JCT 2016 Sub-Contracts, with the exception of the JCT MWSub/D 2016 and JCT ShortSub 2016 sub-contracts. The sub-contractor's 'claim' for loss and expense may be made on a progressive basis and should contain:

- An initial 'notification' that progress of the sub-contract works is being or is likely to be affected by a Relevant Sub-Contract Matter;
- Either with its notification or as soon as is reasonably practicable thereafter provide the contractor with:
  - the sub-contractor's 'initial assessment' of loss and expense; and
  - the information that's reasonably necessary for the main contractor to ascertain the loss and expense;
- At monthly intervals after the initial assessment, provide the main contractor with 'updates':
  - In the manner and form as the main contractor may reasonably require[41]; and
  - Containing sufficient information for the main contractor to (1) fully ascertain the loss and expense and (2) to show the difference between his ascertainment and the sub-contractor's claim.

The above procedures are summarised in Table 14.13.

The main contractor's ascertainment of loss and expense is to be included in interim payments to the sub-contractor and in the final adjustment of the sub-contract sum. Given that the main contractor is not obliged[42] to notify his ascertainment of the sub-contractor's 'initial assessment' of loss and expense for 42 days after receiving it, the sub-contractor may

40 For Interim Payments under JCT SBCSub/C 2016 and JCT DBSub/C 2016 – the amount of loss and expense in a Variation Quotation can be included under clause 4.9.1 or clause 4.9.5 (to the extent not included under clause 4.9.1), and the loss and expense under an Acceleration Quotation under clause 4.9.5. For the Final Payment, the amounts for a Variation Quotation included under JCT SBCSub/C 2016 and JCT DBSub/C 2016 – clause 4.19.1 and for Acceleration Quotations, under clause 4.19.2.

41 For example, in order to clearly show the differences between (1) its ascertainment of loss and expense and (1) the amount of loss and expense that the sub-contractor has claimed (in its 'initial assessment' and/or 'updates'), contractor may reasonably require the sub-contactor to provide his 'updates' in a particular format – such as an Excel spreadsheet (or Scott Schedule).

42 JCT SBCSub/C 2016, JCT SBCSub/D/C 2016 and JCT DBSub/C 2016 – clause 4.15.4, and JCT ICSub/C 2016, JCT ICSub/D/C 2016 and JCT ICSub/NAM 2016 – clause 4.12.4.

**Table 14.11** Summary of loss and expense entitlement (JCT 2016 Sub-Contracts).

| | JCT SBC | | JCT DB | JCT IC | | | JCT MW | All |
|---|---|---|---|---|---|---|---|---|
| **Sub-Contractor's entitlement to loss and expense** | **SBCSub/C (4.14.1)** | **SBCSub/D/C (4.14.1)** | **DBSub/C (4.14.1)** | **ICSub/C (4.11.1)** | **ICSub/D/C (4.11.1)** | **ICSub/NAM/C (4.11.1)** | **MWSub/D (12.3)** | **ShortSub (11.3)** |
| If the sub-contractor: | | | | | | | | |
| • Incurs 'direct loss and/or expense' ('loss and expense') | ✓ | ✓ | ✓ | ✓ | ✓ | ✓ | ✓ | ✓ |
| • Is 'likely to incur loss and expense | ✓ | ✓ | ✓ | ✓ | ✓ | ✓ | | |
| Regular progress[a] of the Sub-Contract Works | | | | | | | | |
| • Has been affected by a Relevant Sub-Contract Matter; or | ✓ | ✓ | ✓ | ✓ | ✓ | ✓ | — | — |
| • Is likely to be affected by a Relevant Sub-Contract Matter | ✓ | ✓ | ✓ | ✓ | ✓ | ✓ | — | — |
| Sub-Contractor must comply with the notification and ascertainment clauses in the Sub-Contract | ✓ | ✓ | ✓ | ✓ | ✓ | ✓ | — | — |
| Sub-contractor must assist the Main Contractor | — | — | 4.14.3 | — | — | — | — | — |
| Subject to the 'exceptions' (below) | ✓ | ✓ | ✓ | ✓ | ✓ | ✓ | — | — |
| **Exceptions for claiming loss and expense** | **SBC/SubC (4.14.2)** | **SBCSub/D/C (4.14.2)** | **DBSub/C (4.14.2)** | **ICSub/C (4.11.2)** | **ICSub/D/C (4.11.2)** | **ICSub/NAM/C (4.11.1)** | **MWSub/D (12.3)** | **ShortSub (11.3)** |
| If the sub-contract's 'Conditions' say no entitlement | ✓ | ✓ | ✓ | ✓ | ✓ | ✓ | — | — |
| If loss and expense is reimbursed under any other Condition of the sub-contract | ✓ | ✓ | ✓ | ✓ | ✓ | ✓ | — | — |

a) Refer to Section 12.2.8 for the equivalent effect on progress under the JCT 2016 main contracts.

**Table 14.12** Circumstances where no addition to the Sub-Contract Sum is permitted (JCT 2016 Sub-Contracts).

| Circumstances where 'no addition to the Sub-Contract Sum' is permitted | JCT SBC | | JCT DB | JCT IC | | | JCT MW | All |
|---|---|---|---|---|---|---|---|---|
| | SBCSub/C | SBCSub/D/C | DBSub/C | ICSub/C | ICSub/D/C | ICSub/NAM/C | MWSub/D | ShortSub |
| Correcting error in the Sub-Contract design documents (Proposals and SCDP Analysis) | — | 2.9.4 | 2.9.4 | — | 2.9 | 2.9 | 8.2 | — |
| Compliance with Statutory Requirements | — | 2.12.2.1 | 2.12.2.1 | — | 2.10.2.1 | 2.10.2.1 | — | — |
| Making good defects | 2.22.1 | 2.22.1 | 2.22.1 | 2.16.1 | 2.16.1 | 2.16.1 | — | — |
| Contractor's direction to correct work not in accordance with the Sub-Contract | 3.11.2<br>3.11.3<br>3.13.1 | 3.11.2<br>3.11.3<br>3.13.1 | 3.11.2<br>3.11.3<br>3.13.1 | 3.10.1<br>3.11.2 | 3.10.1<br>3.11.2 | 3.10.1<br>3.11.2 | 11.3 | 10.3 |
| Contractor's direction regarding non-compliant work remaining | 3.12.3 | 3.12.3 | 3.12.3 | — | — | — | — | — |
| Compliance with health and safety requirements | 3.19 | 3.19 | 3.19 | 3.17 | 3.17 | 3.17 | — | — |
| Compliance with CDM Regulations | 3.20.3 | 3.20.3 | 3.20.3 | 3.18.3 | 3.18.3 | 3.18.3 | — | — |
| Fluctuations Option A | Schedule 4, para 8 | Schedule 4, para A8 | Schedule 4, para A8 | Schedule 2, para 8 | Schedule 2, para 8 | Schedule 2, para A8 | — | — |

**Table 14.13** Loss and expense procedures under the JCT 2016 Sub-Contracts.

| Sub-Contractor's 'claim' for loss and expense | JCT SBC | | JCT DB | JCT IC | | | JCT MW | All |
|---|---|---|---|---|---|---|---|---|
| | SBCSub/C | SBCSub/D/C | DBSub/C | ICSub/C | ICSub/D/C | ICSub/NAM/C | MWSub/D | ShortSub |
| Sub-contractor 'notifies' the main contractor in writing[a)] | 4.15.1 | 4.15.1 | 4.15.1 | 4.12.1 | 4.12.1 | 4.12.1 | | |
| Sub-contractor includes 'loss and expense' in variation claims | — | — | — | — | — | — | 12.3 | 11.3 |
| Sub-contractor's notification: | | | | | | | | |
| • If a Relevant Sub-Contract Matter is likely to affect regular progress; and | 4.15.1 | 4.15.1 | 4.15.1 | 4.12.1 | 4.12.1 | 4.12.1 | — | — |
| • Issued as soon as the effect of the Relevant Sub-Contract Matter becomes (or should have become) reasonably apparent | 4.15.1 | 4.15.1 | 4.15.1 | 4.12.1 | 4.12.1 | 4.12.1 | — | — |
| • Is accompanied by (or is provided soon after) the Sub-Contractor's initial assessment of loss and expense | 4.15.2 | 4.15.2 | 4.15.2 | 4.12.2 | 4.12.2 | 4.12.2 | — | — |
| Sub-Contractor's initial loss and expense assessment includes information reasonably necessary for the main contractor to ascertain | 4.15.2 | 4.15.2 | 4.15.2 | 4.12.2 | 4.12.2 | 4.12.2 | — | — |

| | | | | | | | | |
|---|---|---|---|---|---|---|---|---|
| Subsequent 'updates' of loss and expense: | | | | | | | | |
| • Be in the 'form and manner' as the contractor reasonably require | 4.15.3 | 4.15.3 | 4.15.3 | 4.12.3 | 4.12.3 | 4.12.3 | — | — |
| • Provided at monthly intervals; | 4.15.3 | 4.15.3 | 4.15.3 | 4.12.3 | 4.12.3 | 4.12.3 | — | — |
| • Until all reasonably necessary information for ascertainment has been provided | 4.15.3 | 4.15.3 | 4.15.3 | 4.12.3 | 4.12.3 | 4.12.3 | — | — |
| Contractor's assessment(s) of loss and expense: | | | | | | | | |
| • Provided 42 days after the Sub-Contractor's initial assessment; or | 4.15.4 | 4.15.4 | 4.15.4 | 4.12.4 | 4.12.4 | 4.12.4 | — | — |
| • 28 days after each 'update' | 4.15.4 | 4.15.4 | 4.15.4 | 4.12.4 | 4.12.4 | 4.12.4 | — | — |
| • Notifies the Sub-Contractor of loss and expense ascertained | 4.15.4 | 4.15.4 | 4.15.4 | 4.12.4 | 4.12.4 | 4.12.4 | — | — |
| • Ascertainment makes reference to information provided by the Sub-Contractor | 4.15.4 | 4.15.4 | 4.15.4 | 4.12.4 | 4.12.4 | 4.12.4 | — | — |
| • Must identify the differences between the Sub-contractor's 'claim' and the main contractor's ascertainment. | 4.15.4 | 4.15.4 | 4.15.4 | 4.12.4 | 4.12.4 | 4.12.4 | — | — |

a) Save for the JCT MWSub/D 2016 and JCT ShortSub 2016 sub-contracts, clause 1.7.1 of all of the JCT 2016 Sub-Contracts requires notices to be in writing.

have to wait until the following interim valuation after submitting his 'initial assessment'. However, it is advisable for the sub-contractor to provide the main contractor with its monthly 'updates' of loss and expense[43] to coincide with the date when the main contractor is obliged to issue its payment notice for interim payments, as the main contractor is obliged to notify its ascertainment of loss and expense (following the sub-contractor's 'update') within 28 days of receipt.[40] This 28-day period for ascertaining the sub-contractor's loss and expense 'update' is a shorter frequency than the monthly cycle for when interim payments become due.[44]

### 14.4.2 JCT 2016 Relevant Sub-Contract Matters

The Relevant Sub-Contract Matters that entitle a sub-contractor to loss and expense are very similar in nature to the 'Relevant Events' that provide the same entitlement for the main contractor under the corresponding JCT 2016 main contracts. The main contract 'Relevant Events' are explored in Section 12.4.

There is no equivalent entitlement under the JCT 2016 Sub-Contracts for claiming loss and expense for 'deferment of possession' under the main contract (see Section 14.3.3), but the remaining 'Sub-Contract Relevant Matters' fall into the rest of the categories which entitle the main contractor to loss and expense (see Section 12.4), namely:

- Variations. All of the JCT 2016 Sub-Contracts entitle the sub-contractor to loss and expense in the event of Variations to the sub-contract works (refer to Section 14.2.1);
- The effects of various 'contractor's directions' (instructions);
- Discovery of 'Antiquities'
- Main contractor's suspension for non-payment of interim payments by the employer;
- Delay in receiving the relevant permissions (JCT DBSub/C 2016 only); and
- Any impediment, prevention or default by the employer.

However, in addition to the above categories, there are the following Relevant Sub-Contract Matters which are defaults by the main contractor:

- The Sub-Contractor suspending performance of his obligations, owing to non-payment by the main contractor; and
- The main contractor causing an 'impediment, prevention or default' that affects the progress of the Sub-Contractor's works.

The 'Relevant Sub-Contract Matters' are explored in more detail below (refer to Table 14.10):

- Variations. All the JCT 2016 Sub-Contracts[45] include an entitlement for loss and expense in the event that 'Variations[46]' to the sub-contract works are instructed[47];

43 JCT SBCSub/C 2016, JCT SBCSub/D/C 2016 and JCT DBSub/C 2016 – clause 4.15.3, and JCT ICSub/C 2016, JCT ICSub/D/C 2016 and JCT ICSub/NAM 2016 – clause 4.12.3.

44 JCT SBCSub/C 2016, JCT SBCSub/D/C 2016 and JCT DBSub/C 2016, JCT ICSub/C 2016, JCT ICSub/D/C 2016 and JCT ICSub/NAM 2016 – all clause 4.6.1.

45 Including the JCT MWSub/D 2016 and JCT ShortSub 2016 sub-contracts.

46 'Variations' are defined in clause 1.1 and in clause 5 of the JCT 2016 Sub-Contracts, save for JCT MWSub/D 2016 (clause 12.1) and JCT ShortSub 2016 (clause 11.1).

47 Also see Sections 14.3.3 and 11.4.1 for the sub-contractor's and main contractor's respective entitlements to an extension of time and Section 12.4.2 for the main contractor's similar entitlement to loss and expense 'Relevant Matter'.

- Main contractor's directions. The main contractor may direct the sub-contractor to carry out a number of functions – which are summarised in Table 14.10 and below[48]:
  - Expenditure of Provisional Sums[49];
  - Opening up and testing of work (either arising from an instruction under the main contract or the sub-contract);
  - Correcting/resolving discrepancies in the sub-contract (and any applicable main contract) documents; and
  - Postponement of work.
- Antiquities. Whilst the JCT 2016 Sub-Contracts do not expressly deal with antiquities, they do entitle the sub-contractor to loss and expense if the main contractor so directs[50];
- Suspension of the main contract works (affecting the sub-contract works). This Relevant Sub-Contract Matter is triggered by the employer's failure to pay the notified sum to the main contractor under the main contract, and the main contractor's subsequent suspension of this performance (including the suspension of all sub-contractors' work)[51];
- Suspension of the sub-contract works (non-payment by contractor). This Relevant Sub-Contract Matter is provided by the HGCRA 1996 (refer to Section 11.4.6).[52] There is no equivalent 'Relevant Event' under the corresponding main contract as non-payment by the main contractor (resulting in the sub-contractor's suspension) is not a breach (by the main contractor) under the main contract;
- Inaccurate forecasts of Approximate Quantities. As pointed out in Section 14.3.3 (considering the Relevant Sub-Contract Events), there are two elements to this Relevant Sub-Contract Matter – each one concerning an inaccurate forecast of Approximate Quantities. One arising from an error in the Contract Bills (provided under the main contract) and the other under any 'Bills of Quantities' provided to the sub-contractor by the main contractor.[53]
- Delay in approval of Development Control Requirements. As with the corresponding Relevant Sub-Contract Matter (described in Section 14.3.3), this Relevant Sub-Contract Matter only applies under the JCT DBSub/C 2016 sub-contract.[54]
- Impediment, prevention and default. As with the 'suspension' Relevant Sub-Contract Matter (described above), the sub-contractor's becomes entitled to loss and expense if either (1) the employer causes an 'impediment, prevention or default' that first affects

---

48 As Footnote *a* in Table 14.13, but see Sections 14.2.1, 14.3.3 and 11.4.2, and 11.4.3 (extension of time) and 12.4.4 ('Relevant Matter').

49 The Relevant Sub-Contract Matter under JCT SBCSub/D/C 2016, JCT DBSub/C 2016 and JCT ICSub/D/C 2016 are enlarged to include directions of the contractor for the expenditure of provisional sums included in the 'Contractor's Requirements'. These are defined in clause 1.1 as the documents dealing with the sub-contractor's designed portion and included in the numbered documents.

50 As Footnote *a* in Table 14.13, but see Sections 14.2.1, 14.3.3 and 11.4.4 (extension of time) and 12.4.5 ('Relevant Matter').

51 As Footnote *a* in Table 14.13, but see Sections 14.2.1, 14.3.3 and 11.4.6 (extension of time) and 12.4.6 ('Relevant Matter').

52 Refer to Section 14.3.3 for the similar Relevant Sub-Contract Event (for an extension of time).

53 Refer to Sections 11.4.5 and 12.4.7 (respectively, the equivalent Relevant Event and Relevant Matter for the inaccurate forecast of Approximate Quantities in the 'Contract Bills' under the main contract).

54 Refer to Sections 11.4.14 and 12.4.8 for the Relevant Event and Relevant Matter under the JCT DB 2016 main contract.

the main contract works and second affects the sub-contract works,[55] or (2) the main contractor so defaults and this affects the sub-contract works. Clearly, there is no equivalent Relevant Event or Relevant Matter under the main contract for the latter event, as this is likely to amount to a breach of the main contract by the main contractor.

### 14.4.3 Reservation of Rights and Remedies

The JCT 2016 Sub-Contracts expressly state[56] that the rights of both the sub-contractor and main contractor (the parties) are not limited by the provisions of their respective loss and expense clauses. These provisions confirm that the parties remain entitled to raise common law claims against each other for damages (refer to Chapter 6). The presence of clauses that expressly preserve the right to common law damages are not necessary for a party to be entitled to those rights. Therefore, the parties to the JCT MWSub/D 2016 and JCT ShortSub 2016 sub-contracts are also eligible to raise common law claims, sometimes including the equitable right of set-off.[57] In order to exclude the right to raise common law claims, very clear words have to be used to restrict the parties solely to their rights as set out in the contract (refer to Section 4.6).

## 14.5 Main Contractor Claims Against Sub-Contractors

### 14.5.1 Main Contractor's Claims

The JCT 2016 Sub-Contracts contain express provisions[58] for the main contractor to make claims against sub-contractors. He may make such claims in respect of disturbance of regular progress of the main contract works by the sub-contractor or any of its persons (for example sub-sub-contractors).

The main contractor's claim begins with providing a written[39] notice to the sub-contractor within a reasonable time of the effect (on the regular progress of the main contract works) caused by the sub-contractor's default, becoming apparent. The relevant procedures are brief and are summarised in Table 14.14.

The main contractor's notification to the sub-contractor does not appear to be a condition precedent to deducting (or claiming) any loss and expense from monies that are either due, or that become due to the sub-contractor. Unlike the procedures under the JCT 2016 main

55 Refer to Section 11.4.14 and 12.4.9 for the Relevant Event and Relevant Matter under the corresponding JCT 2016 main contracts.
56 JCT SBCSub/C 2016, JCT SBCSub/D/D 2016, JCT DBSub/C 2016 – clause 4.18 and JCT ICSub/C 2016, JCT ICSub/D/C 2016 and JCT ICSub/NAM 2016 – clause 4.15.
57 See *Hermcrest plc v G Percy Trentham Ltd* (1991) 53 BLR 104. The case concerned the contractor (Trentham) attempting to set-off damages for late completion against its sub-contractor (Hermcrest) where the sub-contract (DOM/1) excluded any additional rights of set-off, restricting the contractor's only ability to set-off sums from those due to a sub-contractor to the set-off provisions in the sub-contract.
58 JCT SBCSub/C 2016, JCT SBCSub/D/C 2016 and JCT DBSub/C 2016 – clause 4.17, and JCT ICSub/C 2016, JCT ICSub/D/C 2016 and JCT ICSub/NAM 2016 – clause 4.14. The main contractor under either JCT MWSub/D 2016 or JCT ShortSub 2016 would have to raise common-law claims against the sub-contractor in the usual way.

**Table 14.14** JCT 2016 Sub-Contracts – Main contractor's reimbursement.

| **Main Contractor's claim against the Sub-Contractor** | **JCT SBCSub/C**<br>**JCT SBCSub/D/C**<br>**JCT DBSub/C** | **JCT ICSub/C**<br>**JCT ICSub/D/C**<br>**JCT ICSub/NAM** |
|---|---|---|
| If the regular progress of the main contract works is materially affected by an act, omission or default: | 4.17.1 | 4.14.1 |
| • Caused by the Sub-Contractor; or | 4.17.1 | 4.14.1 |
| • The Sub-Contractor's Persons, | 4.17.1 | 4.14.1 |
| Main contractor notifies the Sub-contractor; | | |
| • Giving reasonable particulars of effects, or likely effect on progress of the main contract works; | 4.17.1 | 4.14.1 |
| • Within a reasonable time of the material effect on the progress (of the main contract works) is reasonably apparent | 4.17.1 | 4.14.1 |
| Main contractor provides sub-contractor details of resultant loss and expense (as reasonably requested) | 4.17.1 | 4.14.1 |
| Amount due as loss and expense (caused to the Main contractor): | | |
| • Deducted from sums due to the Sub-Contractor | 4.17.2 | 4.14.2 |
| • Deducted from sums to become due to the sub-contractor; or | 4.17.2 | 4.14.2 |
| • Recovered as a debt from the sub-contractor | 4.17.2 | 4.14.2 |

contracts that permit the employer to deduct liquidated damages[59] (refer to Section 8.3.2 and Figure 8.1), the provisions covering the main contractor's notification may operate in isolation from the actual deductions made by the main contractor from the sub-contractor. Therefore, on the face of it, none of the provisions governing the main contractor issuing a payment notice, or a pay-less notice to the sub-contractor, also refer to compliance with the main contractor's notification provisions (summarised in Table 14.14). For each of the JCT 2016 Sub-Contracts, the deduction of loss and expense (contractor's reimbursement) is included as an 'amount deductible' in the Gross Valuation for interim payments,[60] without including any clear wording that any deduction may only be made if the main contractor has issued a compliant notification to the sub-contractor. The absence of a conditions precedent in this regard may be explained as follows:

- The main contractor's notice only has to include the effects of the sub-contractor's default upon the progress of the main contract works, and provide this detail as soon as it becomes reasonably apparent that 'the material effect' of the sub-contractor's default has affected the main contract works' regular progress; and
- There is no mention of the main contractor providing any details of the loss and expense itself in the notification; however
- If the sub-contractor reasonably so requests, the main contractor must provide him with details of the 'resultant' loss and expense.

59 For example, to deduct liquidated damages, the employer must issue (i) a non-completion certificate; (ii) a notice indicating that he may deduct liquidated damages, (iii) a further notice giving details of the liquidated damages' calculation and that the employer will make the deduction, and (iv) account for the deduction in a payment or pay-less notice.

60 JCT SBCSub/C 2016, JCT SBCSub/D/C 2016 and JCT DBSub/C 2016 – clause 4.9.8 and JCT ICSub/C 2016, JCT ICSub/D/C 2016 and JCT ICSub/NAM 2016 – clause 4.7.3.

The word 'resultant' appears to be significant. It means[61] something 'occurring' or 'produced' so seemingly, the loss and expense in this case, must either be in the process of being incurred, or has actually been incurred. On this interpretation, the main contractor only has to provide the sub-contractor with details of loss and expense being incurred or that has been incurred. It does not appear as if the main contractor has to provide any estimate of prospective loss and expense.

It is notable that the main contractor's entitlement under this clause is simply to 'loss and/or expense', whereas in contrast, the sub-contractor's entitlement is to '*direct* loss and/or expense'.[62] On the basis that in a contract the use of different words is to be construed as denoting different things,[63] it can be argued that the absence of the word 'direct' when referring to main contractor's claims in two separate clauses is a deliberate indication that the main contractor has a broader entitlement to recover loss and expense from a sub-contractor. It is suggested that the real distinction to be drawn between (1) what the main contractor can recover, and (2) the direct loss and/or expense which the sub-contractor can recover is that the main contractor may also recover consequential losses[64] under this provision in the correct circumstances.[65]

The main contractor may deduct the amount agreed from monies due or to become due to the sub-contractor or, if necessary, recover the sums due as a debt. Whilst the relevant clauses appear to state that only an 'amount agreed' can be deducted, by virtue of the JCT 2016 Sub-Contracts preserving the parties' common law rights,[66] the main contractor may set-off or deduct any amounts from that which are due, or to become due to the sub-contractor, provided that he provides a valid payment notice or pay-less notice prior to making the relevant deduction.

### 14.5.2 Delayed Completion by the Sub-Contractor

The main contractor's right to claim against a sub-contractor for delay in completion of the sub-contract works is set out in the JCT 2016 Sub-Contracts under the clauses set out in Table 14.15.

The main contractor's right to claim against the sub-contractor for failure to complete the sub-contract works on time is dealt with separately - on a different basis from its right to claim for the effect of any act, omission or default of the sub-contractor upon regular

**Table 14.15** JCT 2016 Sub-Contracts: Sub-Contractor's failure to complete on time.

| Main Contractor's claim against the Sub-Contractor: | JCT SBCSub/C<br>JCT SBCSub/D/C<br>JCT DBSub/C | JCT ICSub/C<br>JCT ICSub/D/C<br>JCT ICSub/NAM |
|---|---|---|
| Sub-Contractor's failure to complete on time | 2.21 | 2.15 |

61 Oxford Dictionary of English (2010) Third Ed.
62 Refer to Section 14.4.1 and Table 14.11.
63 *John Jarvis v Rockdale Housing Association* (1986) 36 BLR 48.
64 See Section 12.2.5.
65 *Millar's Machinery Co. Ltd v David Way & Son* (1934) 40 Com Cas 204.
66 Refer to Section 14.4.3.

progress of the main contract works.[67] In effect, the clauses in Table 14.15 set out the main contractor's entitlement to unliquidated damages. It is exceptionally rare for a sub-contract to include liquidated damages in respect of sub-contract work, the main reason being that, whereas under a main contract, the main contractor is in control of constructing the whole of the main contract works, the multitude of sub-contractors each capable of delaying others and being delayed themselves, creates significant difficulties in properly calculating rates of liquidated damages that represent a genuine pre-estimate[68] of the losses that could be applied to each sub-contractor.

Simply stepping down the full amount of liquidated damages from the main contract may occasionally represent an accurate representation of future loss, but more often, the main contractor's overall delay will be a complex interaction of individual delays, caused by sub-contractors, suppliers and even by the main contractor himself. For example, the main contractor could apportion the rate of main contract liquidated damages to each sub-contractor, based on the value of their sub-contract works against the contract sum in the main contract. However, this is likely to cause the following difficulties:

- Merely apportioning or 'stepping down' a portion of the main contract liquidated damages will not include any time-related costs, directly incurred by the main contractor for any resulting period of critical delay in completing the main contract works. The main contractor's weekly preliminary costs may exceed the weekly rate of main contract liquidated damages;
- If only one sub-contractor caused a critical delay to the main contract completion date, then it is likely that the main contractor would have to bear the majority of the liquidated damages himself as he could only recover a lesser (apportioned) amount from the culpable sub-contractor; and
- A sub-contractor failing to complete his works on time may not necessarily cause a delay to completion of the main contract works[69] and result in the employer deducting liquidated damages from the main contractor.

Therefore, dispensing with the need to calculate at an accurate amount of liquidated damages in respect of any particular sub-contractor, the main contractor can simply concentrate on a culpable sub-contractor's delay without risking any sum he sets as sub-contractor liquidated damages becoming a penalty and unenforceable.[62] By claiming unliquidated damages, the main contractor can deduct the actual losses caused by the sub-contractor. Even if a sub-contractor's failure to complete its sub-contract works on time does not critically delay completion of the main contract works, the main contractor may still incur disruption costs (e.g. suffered by other sub-contractors or costs for attendance on the culpable sub-contractor – such as craneage, power or storage costs).

The relevant JCT 2016 sub-contract clauses are very simple. If the sub-contractor does not complete the sub-contract works within the period for completion in the sub-contract

67 See the text regarding the substance of the clauses in Table 14.14.

68 Refer to Section 8.2.3.

69 For example, in its programme, the main contractor may have either allowed a (i) longer period than the sub-contract period, or (ii) included some float (see Section 3.6) in addition to the relevant sub-contract period, in order to reduce the risk of the sub-contractor critically delaying completion of the main contract works.

particulars, the sub-contractor '*shall pay or allow the [main] contractor the direct loss and/or expense*' caused by the failure to complete. The JCT 2016 sub-contracts state that in order to recover loss and expense, the main contractor must serve notice on the sub-contractor within a reasonable time of the expiry of the period for completion of the sub-contract works. In such circumstances, there appears to be no reason why a reasonable period should not be a few days, perhaps up to a couple of weeks at the most.

Unlike some other provisions considered in this chapter, there is no express stipulation that the amount of damages is to be agreed. At first sight, this appears to enable the main contractor to simply set-off the amount it considers to be due from any other payments due to the sub-contractor. However, clause 2.21 of JCT SBC/Sub C 2016 and JCT DBSub/C 2016 state that the sub-contractor must 'pay or allow' the amount of loss and/or expense. This phrase was considered by the Court of Appeal in 1991,[48] but it should be noted that in that instance, the main contractor's rights of set-off were limited to a particular clause, with which he failed to comply. It is suggested that given (1) that the JCT 2016 sub-contracts preserve the parties rights and remedies, (2) there is a requirement for payment and pay-less notices, and (3) the right to quickly resolve disputes by adjudication, the definition of 'pay or allow' would be given its plain and ordinary meaning.

The preceding clauses[70] establish the date of practical completion (of the sub-contract works), in one of the following two ways:

- In the absence of the main contractor's dissention (described below), if the sub-contractor notifies the main contractor of the date when, in his opinion, practical completion has taken place (and the reasonably necessary information – e.g. for design or for the health and safety file has been provided), practical completion is deemed to take place on the date notified; or
- If the main contractor notifies his dissent within 14 days of receiving the sub-contractor's notice (giving reasons), then by a further notice to the sub-contractor, states the date when he is reasonably satisfied that the sub-contract works are practically complete, then practical completion of the sub-contract works will be deemed to have taken place on the date set out in the main contractor's (second) notification.[71]

If the main contractor dissents, it is effectively in his hands when practical completion is deemed to have taken place. The only constraint is that the date cannot be later than the date of practical completion of the whole of the main contract Works.

### 14.5.3 Common Heads of Loss and Expense Claimed by the Main Contractor Against the Sub-Contractor

It is essential that, when entering into the sub-contract, the main contractor notifies the sub-contractor of the rate of liquidated damages that the employer intends to deduct in

70 'Date of practical completion' (of the sub-contract works): JCT SBCSub/C 2016, JCT SBCSub/D/C 2016 and JCT DBSub/C 2016 – clause 2.20, and JCT ICSub/C 2016, JCT ICSub/D/C 2016 and JCT ICSub/NAM – clause 2.14.

71 Unless a different date for practical completion of the sub-contract works is either agreed by the parties, or determined by the dispute resolution method(s) (such as adjudication) in the sub-contract.

the event that the main contract works are not completed by the main contract Completion Date. If the sub-contractor remains unaware of the rate of liquidated damages that the main contractor may attempt to 'pass on' (in the event of the sub-contractor causing a critical delay to the main contract completion), then the main contractor will fail to meet the 'second limb' of *Hadley v Baxendale*[72] (refer to Section 4.2):

> '*Where two parties have made a contract which one of them has broken, the damages which the other party ought to receive in respect of such breach of contract should be such as may fairly and reasonably be considered either …*
> *[1] arising naturally, i.e. according to the usual course of things from such breach of contract itself, or*
> *[2] such as may reasonably be supposed to have been in the contemplation of both parties at the time that they made the contract, as the probable result of the breach of it*'

The 'second limb' (denoted by '[2]' above) in this context, is providing the sub-contractor (i.e. allowing his 'contemplation') with knowledge of the rate of liquidated damages under the main contract, at the time that he entered into the sub-contract.

The typical costs that a main contractor may seek to recover from a sub-contractor may be as follows (refer to Section 4.4 and Table 4.1):

- Prolongation Costs: in the event that the sub-contractor's failure to complete the sub-contract works on time causes a commensurate critical delay to the main contract completion date:
  - Liquidated damages (i.e. passed down from the employer);
  - The main contractor's time-related costs (e.g. on-site establishment costs, management personnel costs and plant hire);
  - Prolongation costs of other sub-contractors;
  - Loss of overhead recovery (head office overheads);
  - Winter/uneconomic working;
  - Loss of profit;
  - Increased costs (fluctuations); and
  - Financing charges.
- Disruption Costs: if the sub-contractor's failure to complete the sub-contract works does not delay completion of the main contract:
  - Additional plant costs (e.g. prolonged period of crane hire);
  - Additional labour costs;
  - Other sub-contractor prolongation costs – if their sub-contract period has been prolonged (but not causing a critical delay to the main contract Completion Date); and
  - Other sub-contractor disruption costs.

72 *Hadley v Baxendale* (1854) 9 Exch 341.

# 15

# Professional Fee Claims

## 15.1 Introduction

Chapter 9 (*Duties of the Employer's Representative*) explored the general scope of the employer's representative's duties. However, whilst design duties commonly included within the scope of services for an employer's representative were briefly considered, Chapter 9 focussed on his administrative duties (such as assessing extensions of time, valuing works and variations, along with assessing loss and expense).

This chapter will consider various disciplines of professional consultants, and not just those who undertake administrative duties on the client's behalf - i.e. under the main contract between the 'Employer' (if a JCT 2016 contract) or the 'Client' (if a NEC4 contract) and the contractor. For brevity, any references to the 'client' in this chapter are to the party that engages the relevant consultants, and who is the 'Employer' or 'Client' under the building contract with the contractor. Therefore, additional fee claims from consultants such as project managers, architects, contract administrators, the supervisor, employer's agents, civil engineers, structural engineers, building services engineers, quantity surveyors, interior and landscape designers and the like will be considered.

### 15.1.1 Means of Appointing Professional Consultants

Consultants of all disciplines are usually appointed under a contract with their client, under which they provide their various services. These are often referred to as 'appointments' (or 'conditions of engagement'). Many clients choose bespoke appointments, but there are a number of standard forms of consultant's appointment that are produced by recognised professional bodies such as the Royal Institute of British Architects ('RIBA') and the Royal Institution of Chartered Surveyors ('RICS'). In addition, professional bodies such as the Association for Consultancy and Engineering ('ACE') produce a number of appointments under which clients can engage design engineers. NEC produce their own appointment document (see Section 9.1.2), but unlike the other appointments considered in this chapter, the NEC4 appointment has more in common with the NEC4 main and sub-contracts in terms of its operation and administration – save that the 'consultant'is engaged in places the 'contractor' or 'sub-contractor', and the 'Service Manager' carries out the functions in place of the 'Project Manager' (in the NEC4 ECC main contract), or the 'contractor' (in the NEC4 sub-contract or short sub-contract).

*Building Contract Claims*, Sixth Edition. Guy Higginbottom and David Chappell.

**Table 15.1** Conditions of appointment.

| Conditions of appointment/consultant discipline | Contract administration | Design | Lead designer | Separate services |
|---|---|---|---|---|
| RIBA Professional Services Contract ('PSC') 2020 | — | — | — | — |
| • Architect | ✓ | ✓ | ✓ | No |
| ACE Professional Services Agreement 2017 ('PSA') | — | — | — | — |
| • Civil engineer | No | ✓ | ✓ | ✓ |
| • Structural engineer | No | ✓ | ✓ | ✓ |
| • Mechanical Eng | No | ✓ | ✓ | ✓ |
| • Electrical Eng | No | ✓ | ✓ | ✓ |
| • Public health Eng | No | ✓ | ✓ | ✓ |
| RICS Standard Form of Consultant's Appointment ('SFCA') 2022 | — | — | — | — |
| • Contract Administrator | ✓ | No | No | ✓ |
| • Employer's Agent | ✓ | No | No | ✓ |
| • Project Manager | ✓ | No | No | ✓ |
| • Quantity Surveyor | No | No | No | ✓ |
| NEC4 Professional Services Contract ('PSC') (Options A, C and E)[a)] | — | — | — | — |
| • Any discipline | ✓[b)] | ✓[b)] | ✓[b)] | ✓[b)] |

a) NEC4 Professional Services Contract. Only operates with Option A (Priced contract with activity schedule), Option C (Target contract), or Option E (Cost reimbursable contract).
b) In addition to his design services.

Table 9.2 which summarised some of the appointments under which (who, in previous chapters was referred to as the 'employer's representative') has been expanded to form Table 15.1. This now includes the conditions of engagement for other consultants who may be engaged to provide design services but may not required to carry out any services that contribute towards the administration of the building contract between the client and the contractor.

The conditions of engagement in Table 15.1 all vary in terms of content. For example, the RIBA appointment includes the typical 'Services' that the consultant is to perform in a schedule contained within the appointment itself. Typically, the 'Services' (under all types of appointment) are generally arranged into the order set by the RIBA Plan of Work (either the 2013 or 2020 versions). The ACE and RICS appointments consist of the appointment document itself (i.e. containing the project-specific information and conditions applicable to the appointment) but include their published 'Services' in separate documents which are then appended to the appointment document. The only exception is the NEC4 appointment, as this only consists of the appointment document itself, to which the client has compiled and attached (or included by reference) the relevant 'Services' that the consultant is to perform. The 'Services' are then set out in the Scope.

Table 15.1 indicates that in many cases, where a consultant is engaged to provide design, he will not be required to carry out any 'contract administration' duties. Administering the building contract usually falls to a 'Contract Administrator', 'Employer's Agent' or 'Project Manager' (e.g. under the RICS appointments) or by selecting the contract administration services for an 'Architect', engaged under the RIBA appointment. However, appointments by the ACE and RICS require the relevant consultant to assist the person undertaking contract administration duties (i.e. the employer'srepresentative) by advising on matters of design, quality of work, the standard of workmanship, and project progress, as well as providing financial advice, such as valuing variations, and valuing work for interim payment and final account purposes. The contract administration, design and valuation services are often referred to as 'standard' services. Many conditions of appointment may contain additional or 'optional' services that the client may require the consultant to perform in the following circumstances:

- At the client's choice:
  - Carrying out surveys and studies[1, 2]
  - Public consultation services[1]
  - Carrying out the 'lead designer' function
  - Obtaining planning[1]
  - Conservation services[1]
  - Specialist works (such as sewer, road and rail design[2])
  - Grant or funding applications[1]
  - Two-stage tender / negotiating with the contractor[1, 2]
- In the event that a particular situation arises:
  - Assessing alternative designs[3]
  - Investigating defective work[3]
  - Dealing / assisting with contractor's claims[2, 3, 4]
  - Dispute resolution assistance[2, 3, 4]
  - Establish project database[4]

### 15.1.2 Consultants' Core Obligations

The consultant's core obligation will be to perform the 'services' for the client, which should be set out in the appointment. Clearly allied to the services themselves are the following consultant's obligations:

- The standard to which the services are to be performed; and
- When the services are to be performed.

Table 15.2 summarises the core obligations under each of the featured appointments.

1 See 'Other Services' in the RIBA PSC 2020.
2 See 'Optional Deliverables', and 'Additional Services' in the ACE PSA 2017.
3 See 'Additional Services' in the RIBA PSC 2020.
4 See 'Additional Services' for a Contract Administrator in the RICS SFCA 2022.

**Table 15.2** Consultants' core obligations.

| Appointment | Core clause(s) for performing the services |
|---|---|
| RIBA Professional Services Contract 2020 ('RIBA PSC') | Clause 3.1. In the performance of the Services and discharging all the obligations under the Contract, the Architect/Consultant will exercise the reasonable skill, care and diligence to be expected of an Architect/Consultant experienced in the provision of such services for projects of a similar size, nature and complexity to the Project. Notwithstanding anything that may appear elsewhere to the contrary, whether under this Contract or otherwise, the Architect/Consultant's duties and obligations shall be deemed to be subject to the exercise of such reasonable skill, care and diligence and nothing contained in this Agreement or elsewhere shall be construed as imposing on the Architect/Consultant any greater duty than the exercise of such reasonable skill, care and diligence<br>Clause 3.2. The Architect/Consultant shall:<br>(3.2.1) Perform the Services with due regard to the Project Brief.... |
| ACE Professional Services Agreement 2017 ('ACE PSA') | Clause 2.1(i). The Consultant shall provide the Services in accordance with this appointment. In providing the Services the Consultant shall exercise the reasonable skill, care and diligence appropriate to a consultant qualified in the relevant discipline engaged in the performance of such services for projects of a similar nature, size and complexity to the Project.<br>Clause 2.1(ii). Exercising the reasonable skill, care and diligence as set out in clause 2.1(i) the Consultant shall not specify or approve for use any products or materials which are generally known by consultants of the Consultant's discipline to be deleterious, in the particular circumstances in which they are specified for use, to health and safety and/or the durability of buildings or structures.<br>Clause 2.2. The Consultant shall use reasonable endeavours to perform the Services in accordance with any Programme or any variation to that Programme agreed with the Client from time to time. If there is no such Programme then in accordance with any programme submitted to and accepted by the Client. |
| RICS Standard Form of Consultant's Appointment 2022 ('RICS SFCA') | Clause 3.1. The Consultant exercises in the performance of the Services the Requisite Standard.[a] Notwithstanding any term or condition to the contrary in this Appointment, any documents referred to in this Appointment or otherwise, the Consultant has no higher standard of responsibility other than to exercise the Requisite Standard in the performance of its obligations under this Appointment.<br>Clause 3.2. The Consultant performs the Services upon and subject to the terms of this Appointment.<br>Clause 3.3. The Consultant performs the Services with due regard to the Client's Brief.<br>Clause 3.7. The Consultant performs the Services and provides the Documents so as to comply with the Programme, unless the Consultant is prevented from doing so by circumstances outside its reasonable control. |
| NEC4 Professional Services Contract ('NEC4 PSC') | Clause 10.1. The Parties and the *Service Manager* shall act as stated in this contract.<br>Clause 20.1. The Consultant Provides the Services in accordance with the Scope.<br>Clause 20.2. The Consultant's obligation is to use the skill and care normally used by professionals providing services similar to the *service*.<br>Clause 30.1. The Consultant does not start work until the *starting date* and does the work so that Completion is on or before the Completion Date. |

a) RICS SFCA – clause 1.1 defines the 'Requisite Standards' as *'The standard of reasonable skill and care to be expected of an appropriately qualified professional consultant of the discipline specified in the Appendix holding itself out as having the competence and resources to perform the Services, and who is experienced in providing services in connection with works of a similar size, scope, nature and complexity and value as the Project'.*

In each appointment, the consultant must exercise reasonable skill and care when performing the services and ought to possess appropriate qualifications and sufficient experience of projects similar to the project that they're engaged in. Whilst the consultants' services will either be expressly set out in[5] or be appended to the appointment, when performing those services, the consultant must consider the project itself and the client'srequirements. These requirements for the project and the client are often set out in in a document called the 'Brief' or 'Client's brief'.[6] The NEC4 PSC does not refer to a 'brief' as such, but it seems obvious that the 'Scope' should include all of the information commonly provided by the Client in the form of its 'brief'.

The consultants' obligations regarding the time for performing their services vary from appointment to appointment. They all refer to a 'programme' as summarised in Table 15.3 below.

The 'programme' references in the consultant's appointments should not be mistaken for the contractor's programme for the main contract works (i.e. RIBA Stage 5). This is because the consultants are often appointed before the contractor's involvement (usually from RIBA Stage 5 onwards) as their services often involve working during the earlier RIBA Stages 0 (Strategic Definition) to 4 (Technical Design). The 'programme' for the consultants can then be revised to account for the duration of the main contract works (for example, if the consultants' programme included a 10-month period for constructing the works, but when a 12-month programme for the main contract works was agreed between the client and the contractor, the period for the consultants' works would be increased to 12 months).

The significance of the 'programme' in the consultant's appointments is summarised below:

- If obliged to work in accordance with, or in consideration of a 'Programme':
  - The ACE PSA obliges the consultant to use 'reasonable endeavours' to work the programme[7] annexed to the appointment; and
  - The NEC4 PSC operates similar regime for the programme and 'Accepted Programme' as the NEC4 main contract (refer to Sections 3.5.1 and 13.3, 14.2.2 and 14.3.4). Whilst the 'Accepted Programme' must show 'planned completion',[8] the consultant's express obligation is to complete the services on or before the 'Completion Date' (clause 30.1); and
  - The RICS SFCA obliges the consultant to provide the services (and documents) to *'comply with the programme'*.[9] The SFCA suggests that the 'programme' is developed with all of the client's consultants (the 'Professional Team') and ultimately approved by the client; however
- The RIBA PSC, whilst recognising the importance of a 'Project Programme', expressly states[10] that the client acknowledges that the consultant does not promise (warrant) that

5 RIBA PSC 2020.
6 RIBA PSC – 'Project Brief', ACE PSA – the 'Brief', RICS SFCA – 'Client's Brief'.
7 ACE PSA – the 'programme' referred to is attached to the appointment. It may differ from the programme for the main contract works.
8 NEC4 PSC – clause 31.2.
9 RICS SFCA defines the programme.
10 RIBA PSC – clause 2.3 (Client's Responsibilities).

**Table 15.3** Programme in the consultants' appointments.

| **Appointment and reference to programme:** | **Obligation/reference** |
|---|---|
| RIBA PSC | — |
| Definition: | 'Project Programme' -*'the Client's initial programme for the Project'* |
| Reference / included in appointment: | • In 'Contract Details' Section D (project brief)<br>• Key project dates / 'target date' listed |
| Obligation to work to the 'programme' | No – see clauses 2.1.1 and 2.3.2 |
| ACE PSA | — |
| Definition: | 'Programme' – *'attached or included in Part 3 of the Schedule to this Agreement'* |
| Reference / included in appointment | In 'Schedule' Part 3: The Programme for the Services |
| Obligation to work to the 'programme' | Yes – see clause 2.2. |
| RICS SFCA | — |
| Definition: | 'Programme' – *'The programme agreed by the Professional Team and approved by the Client identifying key completion and other dates and time periods for the completion of key activities and the issue of Documents and information in relation to the Project and which may, from time to time, be adjusted by agreement between the Client and the Professional Team'.*[a)] |
| Reference / included in appointment | In Schedule 3 (Client's Brief), part 4 (Programme) |
| Obligation to work to the 'programme' | Yes – see clause 3.7 |
| NEC4 PSC | — |
| Definition: | 'Accepted Programme' – *'…is the programme identified in the Contract Data or is the latest programme accepted by the Service Manager. The latest programme accepted by the Service Manager supersedes previous Accepted Programmes'* – clause 11.2(1) |
| Reference / included in appointment | Contract Data – Part One (Data provided by the client), Part 3 (Time). |
| Obligation to work to the 'programme' | No – see clause 30.1. |

a) RICS SFCA – defines 'Professional Team' as *'The Consultant, the professionals listed in the Appendix and any other professionals engaged by the Client in relation to the Project and which the Client notifies to the Consultant'.*

he will comply with the programme, as it may need to be 'reviewed' for:
- Client-instructed variations;
- Delays caused by the contractor and / or other consultants;
- Changes in market prices; and
- Other factors (including those not reasonably foreseeable).

The consultant only appears to be obliged to notify the client if his progress (in performing the services) may '*materially affect...the Project Programme...*'.

However, the contractor's programme (for the construction works) is also likely to be a key factor in determining the consultants' time for performing their services (refer to Section 9.1.3). Ideally, the consultant's 'programme' and the contractor's programme should be aligned, so the relevant dates for consultants to provide information to the contractor[11] are consistent with the sequence and timing of the contractor's obligations to carry out and complete the work. Whilst there is nothing in the consultants' appointments that obliges them to perform their services in accordance with the contractor's programme, it would be expected for any consultant's programmes to be updated to reflect progress of the main contract works.

### 15.1.3 Client's Core Obligations

In Section 6.2.3, the implied obligations of co-operation and one party not to prevent the other from carrying out its obligations were discussed. These obligations are particularly important for the client to uphold, particularly where a consultant is responsible for providing the project's design, or significant elements of it. Ideally, the client will provide the consultant with a clear 'brief', and all of the information required for him to product that design. The client's timely provision of information is also important when he instructs changes to the project's design. Accommodating project change can be a difficult process and can lead to claims from contractors (for prolongation and disruption) and from consultants – particularly if the client instructs changes after a significant amount of the works has been completed.

The RIBA PSC, ACE PSA and RICS SFCA all contain express obligations for the client to perform. These are summarised in Table 15.4.

The NEC4 PSC also includes the usual 'NEC' obligation (clause 10.2) that the parties act in a spirit of mutual trust and co-operation.

### 15.1.4 Basis for Consultant's Fees

Consultant's fees are usually expressed in the following ways:

- Fixed fees for the known services to be provided;
- Expenses ancillary to the fixed fees that the client is obliged to pay; and
- Fees for any additional services.

A consultant will usually base his fixed fees upon the cost per hour of its staff for performing the services that the client requires under the terms of his appointment. The consultant

11 NEC4 – clause 31.2, the contractor's programme has to show '*the order and timing of the work of the Client and Others as last agreed with them by the Contractor or, if no so agreed, as stated in the Scope*'.

**Table 15.4** Client's obligations.

| Client obligation | RIBA PSC | ACA PSA | RICS SFCA |
|---|---|---|---|
| Provide necessary information to the consultant (e.g. the brief, budget and programme) | 2.1.1 | 3.2 | 8.1 |
| Provide information (free of charge) that the consultant requests that is necessary for the services to be performed in a timely and proper manner, including decisions | 2.1.2, 2.1.3 | 3.2, 3.5 | 8.1, 8.2, 8.8 |
| Engage other consultants to perform work and collaborate with the consultant, and hold them responsible | 2.1.4, 2.1.5 | 3.3, 3.6 | 8.7 |
| Ensure the contractor is responsible for carrying out and completing the construction works | 2.1.6 | 3.4 | — |
| Not to deal directly with the contractor where the consultant is the employer's representative (under the building contract) | 2.1.7 | — | — |
| Absolve the consultant of responsibility for instructions issued directly by the client to other consultants or the contractor | 2.1.8 | — | — |
| Pay fees/charges etc. for planning, building control and other consents | 2.1.9 | — | — |
| Pay the consultant in accordance with appointment's terms | 5 | 3.1 | 9.6 |
| Client warrants the accuracy of the 'brief' and any reports/surveys he provides | — | — | 8.4 |

(or firm) will normally allocate a particular member of staff, or members of staff to the services that the client requires, and estimates how many man-hours will be required to perform them. Ordinarily, the services are set out in the order that they appear in the RIBA Plan of Work (see Section 15.1.1), so the consultant will estimate the man-hours required to fulfil each of these services. Once the total number of man-hours (for all staff disciplines) has been reached, the consultant will normally apply the cost per hour for the relevant staff to reach a sub-total, and then add allowances for any contingencies, expenses (such as travel, accommodation, printing, etc.) and profit.

The consultant is likely to be able to estimate the man-hours for design tasks for each service fairly accurately, and his prior experience will be invaluable for doing so. He is likely to take the following into account:

- The complexity of the design;
- The number of elements of the design:
  - Structural
  - Civil engineering
  - Landscaping
  - Building services (mechanical, electrical and drainage);

- Any specialist services to be integrated (fire alarm, security installations and building management systems);
- Integration of any contractor-designed elements;

- The number of other design consultants;
- Any co-ordination responsibilities inherent with carrying out the 'lead designer' service; and
- Liaison with third parties, such as statutory bodies, BREEAM assessors, etc.

Where the consultant is to perform contract administration services, he is likely to estimate the man-hours required for performing a particular task (such as a quantity surveyor valuing the contractor's work and the client's (or employer's) representative certifying or notifying the contractor of the payment) and multiply the total man-hours by the estimated number of times that the consultant expects that service to be performed. For example, if a project's likely construction duration is 10 months, then the client's (or employer's) representative can expect to carry out at least 10 interim valuations, plus between another 3 and 6 interim valuations after completion (depending upon how long the contractor has to assemble his final account), and then the final valuation for the final payment. Similarly, a design consultant may be required to attend monthly progress meetings to assess the quality and progress of work that he's designed. He'll also be expected to contribute to witnessing testing and inspections, along with compiling any 'snagging lists' and signing off the works as 'defect free' after the rectification (or defects) period.

Expenses are usually charged at cost or at an agreed rate (e.g. vehicle mileage) and are charged as they are incurred by the consultant.

The additional services are usually charged at an hourly rate for each discipline of the consultant's staff or by seniority. The hourly rate is usually based on the cost per hour of the relevant staff, plus overheads and profit.

### 15.1.5 The Basis for Charging for Consultant's Fees

Consultants' services involving design, surveying work[12] and providing advice on building,[13] are considered to be 'construction contracts' for the purposes of the HGCRA 1996 (as amended); therefore, they will be entitled to 'stage' or 'periodic' payments,[14] even if their appointment doesn't expressly deal with payments.

The consultant's fee is usually broken down into amounts for each of the particular services falling within a fixed cost (e.g. for design) or a time-related cost (such as recurring activities such as chairing or attending meetings, producing monthly reports or carrying out interim valuation duties). There may also be additional services that the consultant charges on a time-related basis, with any expenses to be reimbursed. For convenience, many of the consultants' services (in particular, for those involving design) are split into the various RIBA Plan of Work stages and are charged based on the amount of work completed for each stage.

12 s104(2)(a) HGCRA 1996.
13 s104(2)(b) HGCRA 1996.
14 S109(1) HGCRA 1996 (provided that the duration for the 'services' exceeds 45 days).

**Table 15.5** Consultants' fees and payment.

| **Basis of fee charging (available options)** | **RIBA PSC** | **ACE PSA** | **RICS SFCA** | **NEC4 PSC** |
|---|---|---|---|---|
| Basic Fees expressed as: | — | — | — | — |
| • Lump sum | — | — | ✓[a)] | — |
| • As % of building (or project) cost | — | ✓[b)] | ✓[c)] | — |
| • Stages / instalments | ✓[d)] | ✓[e)] | ✓[f)] | — |
| • Price for Services Provided to Date | — | — | — | ✓[g)] |
| Additional Fees | — | — | — | — |
| • Unit fee (stages) | ✓ | — | — | — |
| • Hourly rate (time charge) | ✓[h)] | ✓[i)] | ✓[j)] | — |
| • Expenses | ✓[k)] | ✓[l)] | ✓[m)] | ✓[n)] |
| • Defined Cost | — | — | — | ✓[o)] |
| Payment | — | — | — | — |
| • Periodic (monthly) | ✓[p)] | ✓[d)] | — | ✓[q)] |
| • At the end of each stage | ✓[o)] | — | ✓[e)] | — |
| • In accordance with a schedule | ✓[o)] | — | — | — |
| • Other basis | ✓[o)] | — | — | — |

a) RICS SFCA Schedule 2 (Fees and payment), Part A - Fee.
b) ACE PSA 'Schedule' Part 4: Fees, Section 3.1.
c) See Footnote 14.
d) RIBA PSC Contract Details, Section F: Basic Fee – '*...the fee for the Services excluding VAT and any additional charges, such as expenses, disbursements, etc*'.
e) ACE PSA 'Schedule' Part 4: Fees, Section 3.2 – expressed as a percentage % of the Project Cost.
f) RICS SFCA Schedule 2 (Fees and payment), Part B – Fee Instalments (see clause 9.6).
g) NEC4 PSC definitions: Option A – clause 11.2(20), Options C and E– clause 11.2(21) including Contract Data Part One, section 5 (Payment inc share ranges), Option X1 (Price Adjustment for inflation).
h) RIBA PSC Contract Details, Section G: Time charges.
i) ACE PSA 'Schedule' Part 4: Fees, Section 5.1 (including fees for disruption).
j) RICS SFCA Schedule 2 (Fees and payment), Part D – Rates for Additional Services.
k) RIBA PSC Contract Details, Section H: Expenses.
l) ACE PSA 'Schedule' Part 4: Fees, Section 8.1.
m) RICS SFCA Schedule 2 (Fees and payment), Part C – Reimbursable Expenses.
n) NEC4 PSC, Contract Data Part Two, section 5 (Payment), Date for the Schedule (and Short Schedule) of Cost Components.
o) NEC4 PSC, Data for the Short Schedule of Cost Components – people rates.
p) RIBA PSC Contract Details, Section I: Payment.
q) NEC4 PSC, Contract Data Part One, option Y(UK)2.

Table 15.5 summarises how the consultants' fees can be expressed.

The basis for ascertaining and paying the consultant's fees under the NEC4 PSC will depend on which of the Main Options is selected. The consultant is paid the 'Price for Services Done to Date' (see Footnote *f* in Table 15.5), but that varies for the relevant 'Main Options' as shown in Table 15.6 below.

**Table 15.6** NEC4 PSC – definition of 'Price for Services Provided to Date'.

| NEC4 PSC Main Option | Definition |
|---|---|
| Option A – Priced Contract with activity schedule | Clause 11.2(20). The Price for Service Provided to Date is the total of the Prices for each completed activity.[a)] A completed activity is one without notified Defects the correction of which will delay following work. |
| Option C – Target contract | Clause 11.2(21). The Price for Service Provided to Date is the total Defined Cost[b)] which the Service Manager forecasts will have been paid by the Consultant before the next assessment date plus the Fee.[c)] |
| Option E – Cost reimbursable contract | Clause 11.2(21). The Price for Service Provided to Date is the total Defined Cost[d)] which the Service Manager forecasts will have been paid by the Consultant before the next assessment date plus the Fee[b)]. |

a) NEC4 PSC Option A. An 'activity' is set out in the 'Activity Schedule' which *'is the activity schedule unless later changed in accordance with these conditions of contract'* – clause 11.2(15).

b) NEC4 PSC Option C. 'Defined Cost' is *'the cost of the components in the Schedule of Cost Components'* – clause 11.2(17).

c) NEC4 PSC Options A, C and E, the 'Fee' is defined in clause 11.2(8) as the 'fee percentage' in Contract Data Part Two (Date provided by the Contractor), section 1 – General.

d) NEC4 PSC Option E. 'Defined Cost' is *'the cost of the components in the Schedule of Cost Components less Disallowed Costs'* – clause 11.2(17). 'Disallowed Cost' is cost which is (a) not substantiated by the contractor's accounts and records, (b) should not have been paid to a sub-contractor or supplier, (c) could have been avoided if the Consultant followed the Scope's acceptances and procedures, gave an early warning or notified the Service Manager of a formal dispute (adjudication, arbitration or litigation) between it and a sub-contractor or supplier, plus (d) correcting Defects after Completion, (e) correcting Defects caused by the consultant's failure to comply with a constraint set out in the Scope, (f) unused resources for providing the Services, and (g) costs for preparing an adjudication, litigation or arbitration against the Client.

## 15.2 Claims Made by Consultants

The nature of claims made by consultants under their professional appointments (such as the RIBA PSC, ACE PSA, and RICS SFCA) are often of a similar basis to the claims made by main contractors and sub-contractors, albeit, the grounds for such claims are likely to be expressed in far simpler terms than in most of the JCT 2016 main or sub-contracts.[15] Unsurprisingly, the NEC PSC operates the more detailed 'compensation event' regime.[16] Consultant's claims under the RIBA, ACE and RICS appointments are covered under the following sections:

- Claims for 'variations' or additional services;
- Extensions of time; and
- Additional fees.

Claims under the NEC4 PSC are covered in Section 15.2.3 below.

---

15 The grounds for the consultant claiming additional fees in the RIBA PSC, ACE PSA and RICS SFCA are similar to those found in the JCT MW 2016 contract, namely, for 'variations' and 'matters beyond the consultant's control' – see Sections 5.2 (variations), 11.4 (extensions of time) and 12.1.1 (loss and expense).

16 NEC4 PSC – the provisions and timetable for notifying compensation events is identical to that under the NEC4 main contract (see section 5.3.2).

**Table 15.7** Variations (additional services) under RIBA PSC, ACE PSA and RICS SFCA.

| Client's entitlement to vary 'services'Consultant's entitlement to additional fees | RIBA PSC | ACE PSA | RICS SFCA |
|---|---|---|---|
| Client can instruct variations | 2.2 2.3.2(a) | 2.3, 3.2, 8.1 | 3.9, 8.2, 10.2 |
| Restrictions on scope of additional Services | 3.2.7 | — | 11.1 |
| Consultants entitlement to additional payment: | 5.5, 5.7 | 8.1 | 9.2 |
| • Additional design for work not included in the 'Construction Cost' | 5.5.1 | — | 10.2 |
| • Client instructs a variation of any service already commenced or completed | 5.5.2 | — | — |
| • Carrying out design that was denoted as 'in abeyance', 'provisional', or 'approximate' | 5.5.3 | — | — |
| • Performing the services is delayed, disrupted or prolonged | 5.5.4 | 8.2 | 9.11 |
| • Client delays fulfilling his obligations | — | 8.2(i) | — |
| • Consultant delayed by others | — | 8.2(ii) | — |
| • Project or works damaged or destroyed | — | 8.2(iii) | — |
| • Reasons beyond the consultant's control | 5.5 | 8.2(iv) | 10.5 |

### 15.2.1 Claims for 'Variations' (Additional Services)

As shown in Section 15.1.5 and Table 15.5, the consultant is entitled to claim additional fees for 'variations' or additional services – i.e. further services than were originally contracted for with the client and expressly set out in the appointment.

The client's entitlement to instruct the consultant to carry out 'variations' (or additional services) is set out in Table 15.7:

### 15.2.2 Adjusting the Time (Programme) for the Consultants' Services

As shown in Table 15.3, the RIBA PSC, ACE PSA, RICS SFCA and NEC4 all refer to a 'programme'. This is the consultant's 'programme', and is either developed between the consultant and the employer, or in the case of the RICS SFCA, with the Professional Team[6]. Only the ACE PSA[17] and the RICS SFCA[18] oblige the consultant to work in accordance with the programme.

The RIBA PSC contains the least onerous and most flexible obligations in terms of the consultant performing the services to a deadline. The 'Project Programme' is agreed between the client and the consultant, but there is no obligation for the consultant to comply with it.[19] The programme simply appears to be a tool for the consultant to monitor

17 ACE PSA – clause 2.2.

18 RICS SFCA – clause 3.7.

19 RIBA PSC – see Section D (Project Brief), where 'Key project dates' are each expressed as a 'Target date', and clause 2.3.2 – the Client accepts that the consultant does not warrant (promise) to comply with the Project Programme.

the progress of performing his services. His only obligation is notify the client of any issue that may 'materially affect' the programme, and advise him of any actions (including decisions or information) to mitigate that effect.[20] It appears that any revisions to the programme can only be agreed with the client,[21] and this may entitle the consultant to adjust the 'Basic Fee'.[22]

The ACE PSA expressly obliges the consultant to perform his services in accordance with the Programme[23] and that the programme can be varied from time to time, with the client's agreement. The programme may also be adjusted by reference to the 'Early Warning' clause 4.2 of the ACE PSA, in the event of *'any matter'* occurring, which may affect the consultant providing the services. Under this clause, both the client and consultant are obliged to notify the other as soon as they become aware of any such a 'matter' and, within 7 days thereafter, meet or discuss to agree responsive actions to mitigate its effects. Under clause 4.2(ii) of the ACE PSA, the client and the consultant may agree adjustments to the programme. An adjustment to the programme may entitle the consultant to additional fees.[24] It also appears that the client may unilaterally adjust the programme[25] in the event of making an additional payment to the consultant arising from instructing additional work, the employment of additional resources and / or any disruption suffered.

The programme under the RICS SFCA differs from that employed in the RIBA PSC and ACE PSA, as it not only includes the timing of activities to be carried out by the consultant, but if he's acting as 'lead consultant' as part of his services, the programme should also include the activities of all other members of the Professional Team.[26] If the consultant becomes aware that performing the services[27] is likely to be delayed, he notifies the client, stating the cause of delay (if known), and provides an estimate of the delay's effect on the programme. The consultant is only excused from performing the services in accordance with the programme if he is *'prevented from doing so by circumstances outside its reasonable control'*.[28]

The RICS SFCA contains no provision by which the client or the consultant can adjust the programme. However, the consultant may be entitled to an adjustment[29] of the Fee[30] and to the fee for any additional services[31] for the adverse effects on performing the services and on the 'programme'. It therefore appears that the 'programme' is used as a reference

20 RIBA PSC – clause 3.2.2.
21 RIBA PSC – see definition of 'Project Programme'.
22 RIBA PSC – clause 5.4.1. See also Footnote *d* in Table 15.5.
23 ACE PSA – clause 2.2. Whilst this refers to *'any Programme'*, the capitalized definition clearly means the 'Programme' included in Part 3 of the Schedule to the PSA.
24 ACE PSA - clause 8.1.
25 ACE PSA – clause 8.2.
26 RICS SFCA. Refer to Footnote 7. If the consultant's services include acting as 'lead consultant', clause 3.4 provides that *'the Consultant schedules the activities so they are in conformity with the Programme and informs the Client of any failure by any member of the Professional Team to comply with the Programme.*
27 RICS SFCA – if acting as 'lead consultant', the consultant must also notify the Client if the services performed by other members of the Professional Team are being delayed.
28 RICS SFCA - clause 3.7.
29 RICS SFCA – clause 9.12b.
30 RICS SFCA – The 'Fee' is defined as *'The amount payable in accordance with this Appointment by the Client to the Consultant for the Basic Services and includes any adjustment of the Fee under this Appointment'*.
31 RICS SFCA – clause 10.6b.

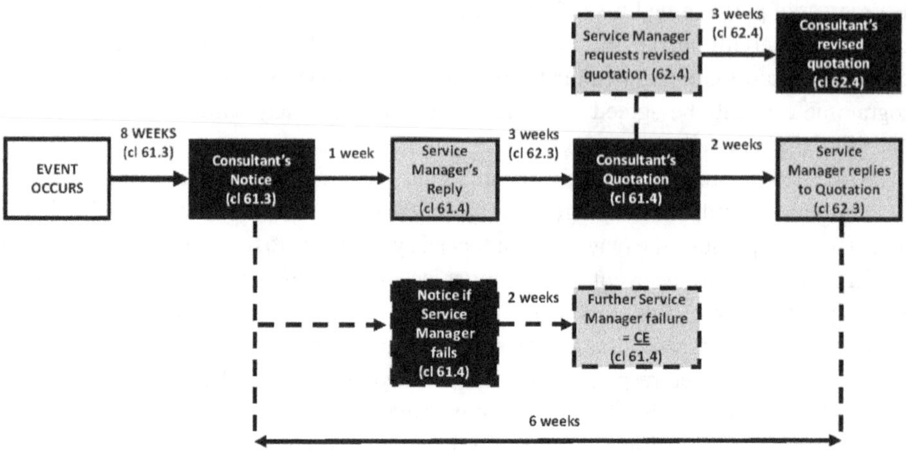


**Figure 15.1** NEC4 PSC compensation event procedure

point (or basis) to measure and assess any adjustment to the Fee in the event that delays prevent the consultant from carrying out the services in accordance with the programme.

Adjustments to the programme (or Accepted Programme) under NEC4 PSC are covered in Section 15.2.3 below.

### 15.2.3 NEC4 PSC Compensation Events

The NEC4 PSC operates a compensation event procedure, similar to that found under its counterpart contracts, the NEC4 ECC main contract (see Section 5.3, Chapters 13 and 14) and the NEC4 sub-contract (see Sections 14.2.2, 14.3.4 and 14.4.3). As stated previously, NEC4's compensation event procedure collects the consultant's entitlements to additional money and time under a single procedure, so there are no separate procedures that deal with those entitlements under variations, extensions of time or loss and expense. Table 15.8 shows all the compensation events under the NEC4 PSC:

The procedures for notifying and assessing compensation events under NEC4 PSC are similar to those found in the NEC4 Sub-contract, albeit some of the timescales for compliance differ. Figure 15.1 summarises the compensation event procedure under the NEC4 PSC:

Compensation events under the NEC4 PSC are assessed in a similar manner to that under the NEC4 Sub-contract, and the assessment process is summarised in Figure 15.2 (below):

Compensation events are assessed on the following bases:

- Changes in the Prices are assessed in exactly the same way as under the NEC4 ECC (main contract) and under the NEC4 Sub-contract and Short Sub-contract (refer to Sections 5.5, 5.5.1, and 14.2.2);
- A delay to the Completion Date or to a Key Date is assessed[32] in a similar way to that under the NEC4 main contract (refer to Section 13.3) as the length of time that the compensation

32 NEC4 – PSC, clause 63.5 refers to assessing delays to the Completion Date and to a Key Date.

**Table 15.8** NEC4 PSC compensation events.

| **NEC4 PSC compensation event – clause 60.1 (for all Main Option A, E and E and Secondary Options)** | **Similar under NEC4 main contract** |
|---|---|
| (1) Service Manager's Instruction ("SMI") to the Consultant to change the Scope[a)] | 60.1(1) |
| (2) The Client prevents access to parts of the site by the "access date" or relevant dates on the Accepted Programme | 60.1(2) |
| (3) The Client fails to provide something by the Accepted Programme's date | 60.1(3) |
| (4) SMI not to start, or stop work | 60.1(4) |
| (4) SMI to change a Key Date | 60.1(4) |
| (5) The Client (or Others[b)]) do not do their work in accordance with the Accepted Programme | 60.1(5) |
| (5) The Client (or Others[b)]) do not work in accordance with the Scope | 60.1(5) |
| (6) The Service Manager fails to respond to a communication within the period for reply | 60.1(6) |
| (7) The Service Manager changes a previous decision | 60.1(8) |
| (8) The Service Manager's reason(s) for not accepting something is/are not set out in the contract | 60.1(9) |
| (9) The Client causes an event for which he is liable | 60.1(14) |
| (10) The Service Manager corrects an assumption he made about a previous 'compensation event' | 60.1(17) |
| (11) Client commits a breach of the contract which is not a Compensation Event | 60.1(18) |
| (12) An event occurs (which is not another Compensation Event) which prevents the consultant from completing the services that is outside of each party's control, or that the consultant could not have reasonably foreseen | 60.1(19) |
| (13) The Consultant corrects a defect for which it is not liable | — |
| (14) SMI correcting information provided by the Client | — |
| (15) The Service Manager does not accept a quotation for a proposed instruction | 60.1(20) |
| (16) Any additional Compensation Events set out in the Contract Data Part One occur | 60.1(21) |
| Secondary Option X2 (Changes in the law) clause X2.1 | X2.1 |
| Secondary Option X12 (Multiparty Collaboration): | |
| • Instruction to change the Partnering Information: reduced prices – clause X12.3(6) | X12.3(6) |
| • Change of programme to match the revised timetable: reduced prices – clause X12.3(7) | X12.3(7) |
| Y2.5 Consultant's suspension | Y2.5 |

a) NEC4 PSC – including (a) an instruction to change the Scope to resolve an ambiguity or inconsistency under clause 63.10 or (b) correcting a change in the Conditions for a Key Date under clause 63.11.

b) NEC4 PSC defines 'Others' in clause 11.2(10) as '*…people or organisations that are not the Client, the Service Manager, the Adjudicator, the Consultant or any employee, Subcontractor or supplier of the Consultant*'.

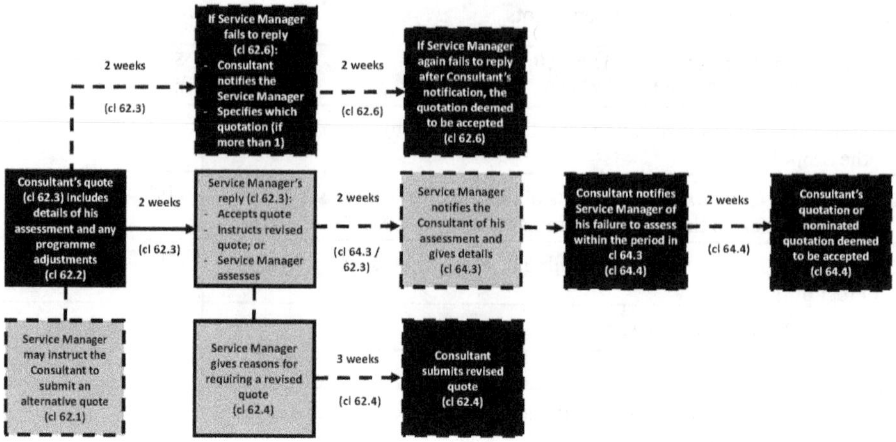


**Figure 15.2** NEC4 PSC compensation event assessment procedure

event has caused planned Completion (or the conditions for meeting a Key date) to be later than planned Completion (or the Key Date) on the Accepted Programme at the dividing date;

- The rights of the Client and the consultant to change the Prices, the Completion Date and any Key Dates are limited to those in the compensation event clause (clause 63.6, refer to Section 5.3.2);
- The following are also taken into account when assessing compensation events:
  - Any failure by the consultant to submit an early warning[33] notice;
  - Any risk allowances for cost and time matters that are not compensation events but have a significant chance of occurring[34]; and
- The assumption that the consultant reacts promptly and competently to the compensation event and any Defined Cost that is time-related was reasonably incurred.[35]

### 15.2.4 Consultant's Claims for Extensions of Time

Consultants' claims typically differ from those raised by main contractors and sub-contractors, as most only seek additional payment (fees), rather than an extension of time to avoid delay damages. With the exception of the NEC4 PSC, and then only if Option X7 (delay damages) is selected, it is fairly uncommon for a consultant to seek an extension of time to avoid a delay damages claim in the form of liquidated damages from the client. If a consultant has caused a project's construction to be critically delayed, then the client will be able to recover common law damages on the usual basis (refer to Chapter 6). In doing so, the client will not only seek recovery of the delay damages that, but for the consultant's breach, he would have otherwise recovered from the contractor, along with his liability for the contractor's time-related (and other) costs arising from the delay (refer

33 NEC4 PSC – clause 63.7 (Early warnings are covered in clause 15.1).
34 NEC4 PSC – clause 63.8.
35 NEC4 PSC – clause 63.9.

**Table 15.9** Consultants' further administrative tasks.

| Further administrative task examples | Employer's representative | Quantity surveyor | Other design consultant |
|---|---|---|---|
| Additional monthly site meetings: | — | — | — |
| • Chair meetings / compile minutes | ✓ | — | — |
| • Attend meetings | — | ✓ | ✓ |
| Additional reports to the Client | — | — | — |
| • Contractor's progress | ✓ | ✓ | ✓ |
| • Other consultant's progress | ✓ | — | — |
| • Budget / cash flow forecast | — | ✓ | — |
| Additional interim valuations | — | — | — |
| • Value the contractor's work | — | ✓ | — |
| • Issue payment notice / certificate | ✓ | — | — |
| • Advise Client for pay-less notices | ✓ | ✓ | — |
| Inspect quality of work (e.g. if work uncovered for a period of time) | ✓ | — | ✓ |

to Section 4.4) and any other time-related costs that the client has incurred during the relevant period of delay (see Table 15.9 for examples).

There are other obstacles to making a consultant's claim for an extension of time for performing their services:

- The consultant may not be obliged to perform the services in accordance with a programme (refer to Section 15.1.2 and Table 15.3):
  - He may have no obligation to perform the services, either to a completion date or in accordance with a programme, and the client may be obliged to ensure the consultant's work proceeds smoothly (RIBA PSC);
  - His obligation to work to a programme may be diluted to using 'reasonable endeavours' to work to the programme (ACE PSA);
- With the exception of the ACE PSA and NEC4 PSC, the consultant may not even have to notify the client of any delays to the programme (see Section 15.2.2),
- The relevant consultants' 'programme' may not have been updated (e.g. to include additional work for the consultant or to reflect delays in the project's construction) or may be in such a basic format that it is difficult to rely upon; and
- It may be difficult for the client to apportion delays to particular consultants if more than one consultant has contributed to a critical delay in constructing the project (see Section 15.2.3). In the cases of the RIBA PSC,[36] ACE PSA[37] and RICS SFCA,[38] these appointments contain net contribution clauses, so the consultant's damages recoverable from the consultant will only be proportional to their overall liability. The NEC4 PSC does

36 RIBA PSC – clause 7.3.
37 ACE PSA – clause 10.2.
38 RICS SFCA – clause 5.2.

not contain a net contribution clause, although the consultant's financial liability can be limited if Option X18 (Limitation of Liability) is selected.

In the event that a consultant is claiming for an extension of time as part of claiming additional fees, then unless the grounds are for one of the compensation events under the NEC4 PSC, the reasons for seeking an extension are likely to be the following:

- Carrying out additional work:
  - More work than could have been reasonably allowed for within the consultant's services or the 'Basic Fee' for those services (refer to Table 15.5); or
  - Further design for variations to the scope of the construction works (see Section 15.1.3 and Table 15.5); or
- The client is in breach of one or more of the appointment's terms; either
  - One of the express terms, setting out the client's obligations to the consultant (refer to Section 15.1.3 and Table 15.3); or
  - One of the implied terms of appointment (see Section 6.2).

The first sub-category of 'additional work' is where the consultant is not performing any additional or new services but is having to perform his existing services to a greater extent or more frequently than he would otherwise have done so. Some examples could be:

- If there has been a critical delay of 6 months, then certain consultants will have 6 months of further administrative tasks to complete, such as shown in Table 15.9:

For the examples shown in Table 15.9, which it may seem obvious to the client that he will have to pay the consultants to carry out more of their 'periodic' administrative tasks during a period of contractor-culpable delay, the consultant should observe all provisions for notifying the client of delays – including providing the client with any estimates of additional fees.

This is particularly important where the construction delay has been caused by the employer.[39] He may resent the employer's representative granting an extension of time to the contractor (which is likely to also entitle the contractor to loss and expense) and accuse him of doing so in order to increase his fees.

- Table 15.10 shows examples of services having to be carried out more frequently, in more depth than the consultant had reasonably foreseen:

The client is less likely to be aware of any delaying or disrupting effects that the examples in Table 15.10 may cause to the consultant performing the services. In some cases (e.g. such as reviewing the contractor's design[40]), the client may expect the consultant to perform this service as many times as it takes to obtain a compliant contractor's design, but for no extra fees. It is therefore advisable for all designing consultants (as well as the employer's representative) to qualify the number of times that they will review a contractor's design submission without the client incurring an additional fee.

---

39 For example, the employer may be a local authority and has engaged his in-house architect, but the architect has issued late or incorrect information.

40 For example, under the 'Design Submission Procedure' under Schedule 1 of the JCT SBC-Q 2016 and JCT DB 2016 contracts, or under clause 21.2 of NEC4.

**Table 15.10** More frequent/in-depth consultant tasks.

| More frequent / in-depth carrying out of services than consultant reasonably allowed for: | Employer's representative | Quantity surveyor | Other design consultant |
|---|---|---|---|
| Reviewing / reporting / commenting on contractor's / sub-contractor's design more times than ought reasonably be required | ✓ | — | ✓ |
| Client requesting progress/budget reports in significantly greater detail | ✓ | ✓ | — |
| Non-compliant contractor quotations - assessing compensation events under NEC4 | ✓ | ✓ | — |
| Inspecting / reporting on contractor's work that is repeatedly defective | ✓ | — | ✓ |
| Dealing with petty client claims that have no basis under the building contract (e.g. allegations that compliant work is defective) | ✓ | ✓ | ✓ |

Similarly, in the cases of financial or progress reports for the client, it is advisable for the consultant to show the client a typical example of the detail provided in such reports. In the event that the client has any special reporting requirements (or procedures for the consultant to follow), he should make the consultant fully aware of the same before the consultant provides a fee quote. If the client introduces such requirements after agreeing the consultant's appointment, then the consultant may be entitled to charge an additional fee.

Clients sometimes make unreasonable demands of their consultants and, in doing so, can place the consultant in a difficult position. This can occur if a client is attempting to impose an unreasonably high obligation (e.g. a level of quality or finish for the contractor's work that is far above the standard that he's obliged to achieve under the building contract). The consultant (in particular, if he's the employer's representative) must remain impartial and operate within the parameters of the building contract. He should clearly advise his client that the contactor is not obliged to achieve that high a standard of finish, and if the client persists with asking for more and more inspections, reports or snagging lists, then the consultant will seek additional fees for providing them, along with advising the client that he may be about to commit a breach of the building contract.

The second sub-category of 'additional work' is more straightforward, and a simple example is where the client instructs variations to the design. This category of 'additional work' may require the consultant to need more time to perform the services.

The consultants (e.g. architect, structural engineer, building services engineers, etc.) may have completed their services up to RIBA Stage 4 (Technical Design), only for the client to change his mind, and cause the design to be revised. The extent of any re-design and the additional time required could be significant. For example, if the building contract is NEC4 and the contractor encounters physical conditions[41] that require the foundations to be changed from strip foundations to piled foundations, then the Project Manager will have to instruct a change in the Scope. If the contractor is not responsible for the foundation's design, then the design consultants will have to re-design the building's foundations.

41 NEC4 – clause 60.1(12).

Where the client is in breach of the appointment's terms, breaching the express terms of the appointment is likely to be similar (if not identical) to breaching the appointment's implied terms. The 'express terms' set out in Section 15.1.3 and Table 15.4 are essentially mirrored by those implied by common law, as most focus on matters of co-operation and not preventing the consultant from performing his services (refer to Section 6.2.3).

When compiling his claim for additional time, the consultant should adopt one of the critical path analysis methods of delay analysis described in Section 3.5 when updating the relevant programme to demonstrate the effects of the delay. If the consultant is working under an appointment based on the ACE PSA conditions, then he should seek the client's agreement (under clause 4.2(ii)) of the adjustments to the programme.

If the consultant is working under the NEC4 PSC appointment, then he will have to comply with clauses 31 and 32 in terms of showing the required information on the programme, and then submitting it for the Service Manager to accept. The delay to the 'Completion Date', caused by a compensation event, will be assessed in the usual way for NEC4 under clause 63.5 (refer to Section 15.2.2).

### 15.2.5 Consultant's Claims for Additional Fees

Whether the consultant is claiming additional fees for carrying our additional work (either performing the services more frequently, or for longer, or performing different services – such as a variation), or any delay or interruption to the consultant's services, the consultant's entitlement to additional payment will fall under the following:

- NEC4 PSC – a compensation event under clause 60.1, entitling the consultant to a 'change to the Prices) under clause 63.1; or
- If the consultant is engaged under one of the other appointments:
  - RIBA PSC – clause 5.5 (extra work beyond the consultant's control)
  - ACE PSA – clause 8 (variation to and / or disruption to the services and valuation); and
  - RICS SFCA – clause 10 (additional services).

The consultant's entitlement and procedures to additional fees under the NEC4 PSC are set out in Section 15.2.3. Table 15.11 summarises the procedures and entitlements that apply to consultants seeking additional payment (additional fees) under the RIBA PSC, ACE PSA and RICS SFCA conditions:

The bases for the consultant's entitlement to additional fees for the RIBA PSC, ACE PSA and RICS SFCA are essentially 'additional work' (or variations) and delays, disruption and prolongation of those services (as summarised in Table 15.7).

All of the professional service appointments require the consultant to notify the client if incurring additional costs (or fees) becomes likely. The basis of calculating the 'additional fees' under all appointments is similar, insofar as the client and consultant can agree the amount, and in default, the additional fee is calculated in accordance with the rates or prices set out in the appointment. The RIBA PSC and ACE PSA provide that additional fees are principally calculated on the basis provided for by the appointment and, failing that, are ascertained by agreement with the client, as follows:

**Table 15.11** Consultant's entitlement to additional fees.

| Consultant's entitlement to additional fees and procedures | RIBA PSC clause | ACE PSA clause | RICS SFCA clause |
|---|---|---|---|
| Consultant must inform the client if additional fees are likely to apply | 5.6 | 8.2 | 10.1, 10.5(a) |
| Consultant to provide quotation (or estimate) to the client | — | 8.1, 8.2 | 10.3, 10.5(b) |
| Additional fees: | — | — | — |
| • Can be agreed with the client | 5.5 | 8.1 | 10.4, 10.5 |
| • Calculated by reference to rates in the appointment | 5.6 (Item G) | 8.1 (Sch Part 4) | 10.4, 10.5 (Part D, Sch2) |
| Consultant quotes or notifies additional fees | — | 8.1 | 9.11b, 10.3 |
| Consultant to retain records of time expended on additional services | 5.8 | — | 9.6 |

- The RIBA PSC assessment of additional fees is at the agreed hourly or daily rates for persons (either named or of a particular grade) – in accordance with 'Part G' of the appointment's 'Contract Details'); whereas
- The ACE PSA's assessment of additional fees is 'time-based' and carried out in accordance with Section 5 of Part 4 (the Fees) of the Schedule.[42] Section 5 provides grades of site staff and their hourly, daily, weekly or monthly rates, and states that the client shall pay 'time-based' fees on a weekly basis.

The RICS SFCA calculates the additional fee on a similar basis to the RIBA PSC and ACE PSA, but the primary focus is for the client and the consultant to agree. Failing agreement, the additional fee is calculated in accordance with appointment – Part D of Schedule 2. This sets out hourly or daily rates for either staff that are either named, or identified by role or position.

The RICS SFCA includes the following two categories of 'additional services fee', each calculated in the same way:

- Clause 10.3 is for the 'additional services fee' when the client has instructed additional services; and
- Clause 10.5 is for the 'additional services fee' when the services are *'materially delayed and / or disrupted…for [a] cause*[43] *outside the Consultant's reasonable control'.*

The RIBA PSC does not require the consultant to provide a quotation for any 'additional fees', whereas the ACE PSA and RICS SFCA do. The RIBA PSC and the RICS SFCA both require the consultant to keep records in support of the additional fee claims:

42 ACE PSA – Part 4: Fees, Section 5, clause 5.3 *'Such time-based fees shall be calculated by multiplying the hourly, daily or monthly rates applicable to the persons concerned by the number of hours, days or months (as the case may be) spent by such persons undertaking such additional work and / or suffering such disruption, including time spent in travelling in connection with such work'.*

43 RICA SFCA – clause 10.5. Such matters beyond the consultant's reasonable control include *'a change in the scope, size, complexity or duration of the Project'.*

- RIBA PSC – clause 5.8 provides that the consultant shall maintain such records of time (and of expenses) and these shall be made available to the client upon reasonable request. It seems to follow that the consultant's claim for additional fees does not need to be particularised when applying for payment,[44] but such information shall be provided only when the client reasonably requests it; whereas
- The RICS SFCA provides that each 'invoice' submitted by the consultant for payment, must contain *'such documents…as are reasonably necessary for checking each invoice.'* (clause 9.6).

The NEC4 PSC calculation and assessment for a 'change in the Prices' is set out in Section 15.2.3.[45]

### 15.2.6 Consultant's Claims for Common law Damages

Under most contracts, including consultants' conditions of appointment, a party may recover common law damages for the other's breach of contract. With the exception of the NEC4 PSC,[46] this is no different for these consultants' appointments. However, the principle of recovering common law damages is that the wronged party is compensated for his losses, but not rewarded. The rates for the relevant consultant's staff that form the basis of additional fee claims (described in Section 15.2.4) are likely to be 'all inclusive' rates which include for the following:

- The base cost of employing that staff member or grade of staff member for the relevant unit of time (hour, day, week or month);
- Other allowances such as holiday pay, bonuses, sick pay, or pension;
- The consultant's overheads; and
- Profit.

Therefore, the consultant may have to commute the relevant time-based rate, so it only reflects the cost to the consultant and excludes any profit or allowances that are unlikely to arise in the relevant circumstances.

## 15.3 Client Claims Against Consultants

Client's claims against professional consultants are usually based on an allegation that the consultant has breached his obligations relating to the standard of services being performed. In most cases (refer to Section 15.1.2) that standard is the *'reasonable skill and care'* expected of an appropriately qualified professional experienced in that particular discipline (such as architect, quantity surveyor, structural engineer, project manager, etc.).

44 RIBA PSC – clause 5.10, the consultant's application for payment is in the form of its 'Payment Notice' which must set out *'out any accrued instalments of the fee and other amounts due, less any amounts previously paid, and stating the basis of calculation of the amount specified'* (clause 5.12).
45 Changes in the Prices are assessed in exactly the same way as under the NEC4 ECC (main contract) and under the NEC4 Subcontract and Short Subcontract (refer to Sections 5.5, 5.5.1, and 14.2.2).
46 NEC4 PSC – clause 63.6 provides that the rights of both the client and consultant are those set out in the compensation event clause, so no claims for common-law damages can be raised by either party.

A client is likely to suffer losses or incur additional costs if consultants fail to perform their services to the required standard. In the case of design consultants, their breaches may include defective design, which may necessitate corrective design, and then remedial work by the contractor (which must be paid for by the client). If a consultant is responsible for administering the building contract, then failures to issue the proper notices regarding payments, completion, delay damages or termination may cause the client to incur significant losses. If a consultant (such as a quantity surveyor) is providing a financial service – such as valuing the works, he will have to perform this service carefully, otherwise the client may have to pay more to the contractor than he otherwise should have.

### 15.3.1 Establishing the Standard for the Consultant's Services

The consultant's appointment will set out the standard to which he must perform the services for his client (refer to Section 15.1.2), but this standard is of general application, and neither provides detailed guidance, nor a procedure which the consultant must follow in respect of specific services. Guidance is often provided by the professional bodies of which the consultant is a member, particularly if he is chartered, as is the case for architects, surveyors and engineers. These procedures are often in the form of checklists for particular services, or guidance notes (often, of mandatory application). If a consultant is employed by a large (often national or global) consultancy, then such firms are likely to have their own procedures, usually aligning with the standards of the professional bodies that regulate their staff. These procedures can be a valuable source of information for a client, when establishing the standard to which consultant's services are to be performed. These 'firms' procedures should, of course, reflect the prevailing standards applicable to the relevant consultants' discipline in their country of operation.

A consultant must also keep himself abreast with current developments and has a duty to warn a client, possibly even if his services and the project have been completed. This is why most professional bodies impose a mandatory requirement upon their members to engage in and report on the continuing professional development (CPD). A member failing to comply with his CPD requirements may find himself censured by his professional body or even expelled.

### 15.3.2 Common Claims by the Client Against his Representative

The employer's representative (such as the Architect, Contract Administrator or Project Manager) is responsible for administering the building contract. In order to do this correctly, he must ensure:

- Notifications to the contractor are issued on time, in the required format,[47] and contain the correct information:
  - Payment notices and pay-less notices;
  - Notices / statements of non-completion;
  - Notices / certificates of practical completion and making good of defects;
  - Notifications to permit the client to deduct delay damages;

47 For example, if the building contract requires notices to be sent by a particular means (such as by Recorded Delivery post or by actual delivery), the notice must be sent in that manner.

- He performs the relevant services in accordance with the building contract:
  - Responding to the contractor on time and in the correct manner;
  - Exercising discretion in the appropriate way and being fair and reasonable;
- Not exceeding his authority under the building contract.

If the employer's representative fails to issue a particular notice, it may have serious financial effects for the client. For example, if no payment nor pay-less notice is issued correctly, the client may be liable to pay the amount due (or 'notified sum') as claimed by the contractor. Whilst the employer can then take steps to recover any overpayment (either by ensuring that the subsequent round of payment or pay-less notices are correct or by embarking on a 'true value' adjudication), any overpayment may not recoverable if the contractor becomes insolvent. Similarly, if the employer's representative fails to issue the relevant notices or certificates in relation to non-completion, then the client will not be able to recover any delay damages from the contractor.

The employer's representative should also exercise his duties fairly and reasonably, and where he is permitted to use to exercise discretion, that he does so fairly (see Section 9.1.8).

### 15.3.3 Common Claims by the Client Against a Designer

The client will expect the design of his project to function and satisfy his requirements, and / or his brief. A project's design is often carried out by a number of consultants, such as those each providing the architectural, structural, building services, and other design (such as interior design or landscaping). In the majority of cases, a design consultant exercising reasonable skill and care will provide a functioning and workable design – one that may also be easy and safe for the contractor to construct, both on time and to budget. However, in the event that the relevant design consultant is in breach of his obligations, the client may raise claims for the following:

- Defective design resulting in extra costs for
  - The additional time taken to correct the design;
  - The contractor taking down and rebuilding part of the works;
  - The contractor having to adapt or alter the existing design;
  - Additional testing and inspections; and
  - An inoperable or unusable project (worst case).
- Late provision of design
  - Additional time-related costs from the contractor for the critical delays;
  - Loss of delay damages[48] (if the project is completed late); and
  - Additional consultant's costs for repeated services (see Table 15.9 and Section 15.2.3).

In projects where the majority of the design is produced by one discipline (most commonly, the architectural design), it is not uncommon for the leader of that discipline to also be appointed as the 'lead designer'. A lead designer's duties normally include the co-ordination and integration of all elements and disciplines into a coherent and buildable

48 Delay damages (whether liquidated or unliquidated) represent the Client's losses for the contractor not being able to complete the project on time (or in this example, the contractor is prevented from completing the project on time).

whole. Therefore, if further design (such as adaptations or fresh design work) has to be undertaken in order to overcome any failure to provide a co-ordinated design, then the client is likely to not only claim against the consultants providing the conflicting design but also against the 'lead designer' for their failure to ensure that the project's design was co-ordinated.

In order to avoid responsibility for any incorrect or late design or design that is not co-ordinated, the client will usually procure the project on a design and build basis – thereby transferring these responsibilities for design (either in whole[49] or in part[50]) to the contractor.

### 15.3.4 Common Claims by the Client Against a Quantity Surveyor

This section is intended to cover claims made by a client against a quantity surveyor, or any consultant that is carrying out services of a 'quantity surveying' nature. The quantity surveying services are directly related to the project's finances, and clearly, the higher value the project, the more money is at stake. The quantity surveyor will usually have to keep the client up to date on the project's cash flow, current values of work carried out and the value to complete, along with raising any concerns (over matters such as contractor claims) that may affect the project's budget. The client may also require this information to keep any funders abreast of the project's financial performance and any likely risk of increased expenditure. The client may also be obliged to regularly notify his funders under the terms of his funding agreement, and it is often the case that funders insist on having their own quantity surveyors monitor the project's funding.[51]

If the project's costs exceed the project's budget, then a client may decide to try and recover this perceived overspend from the quantity surveyor. However, the client may not find this to be a straightforward task because the quantity surveyor is likely to rely upon the following in his defence;

- The client has suffered no loss, as he would have incurred the (higher) project costs in any event, so he would always have had to pay a greater amount than the quantity surveyor may have been budgeted;
  - If the client had to take out an additional loan to make up the shortfall, the cost of the loan may be recoverable from the quantity surveyor if he had negligently prepared his budget.

49 By making the contractor responsible for the project's design – either by the JCT DB 2016 or NEC4 contract (where the Scope provides that the contractor is responsible for the entirety of the project's design).
50 By selecting JCT SBC-2016 (with a Contractor's Designed Portion), JCT ICD 2016 or JCT MW 2016 (where only part of the 'Works' is designed by the contractor), or NEC4 where the Scope only requires the contractor to design part of the overall works.
51 Quantity surveyors engaged to carry out 'fund monitoring' should satisfy themselves (even though the client's quantity surveyor may have valued the contractor's interim payment application), that the amount payable by the bank or funder has been properly substantiated and evidenced. *The Governor and Company of the Bank of Ireland v Faithful & Gould Ltd and CBRE Ltd* (2014) EWHC 2217 (TCC).

- The quantity surveyor's budget was exceeded for no fault of his, perhaps owing to:
  - Unexpected claims from the contractor that were submitted too late to be included in the latest or relevant quantity surveyor's budget (less likely under a NEC4 main contract[52]); and
  - Variations instructed by the client, the value of which reasonably exceeded the amount in the quantity surveyor's budget.

The quantity surveyor is often reliant upon the contractor providing financial information to allow him to compile his budget reports for the client. Clearly, the quantity surveyor will have an idea of project expenditure from the contractor's interim payment applications, and these applications often set out all anticipated adjustments to the contract sum – such as variations and any loss and expense. If he is keeping abreast of matters on site and accurately records any variations instructed by the client, then it's unlikely that the quantity surveyor will be negligent in preparing his budget.

Quantity surveyors are often careful when compiling budgets and financial reports for their clients. Once the main contract has been agreed, the 'contract sum' (or 'total of the prices') will form the 'foundation' of the quantity surveyor's budget reports. A prudent quantity surveyor should advise his client over the possible effects of 'provisional sums' (JCT 2016 contracts) or the effect of 'Defined Cost' in NEC4 projects – particularly those 'target cost' options (Main Options C and D). He should be alive to the timing of variations, and the likelihood of the contractor incurring prolongation or disruption costs. Many quantity surveyors use 'contingency' sums and other qualifications within their cost reporting, so even if a firm cost (or even an accurate estimate) cannot be included in his budget report, the client is made aware of the possibility that additional costs may be incurred.

52 Owing to NEC4's compensation event procedure and the 'time-bar' clause 61.3 (refer to Section 5.3.2).

# Appendix A

## Example of Contractor's Claim for Reimbursement of Direct Loss and/or Expense Under the JCT 2016 Contracts

Employer's representative's and quantity surveyor's assessment with commentary

## A.1 Introduction

In 1983, for the first edition of this book, the authors encountered difficulty in devising a suitable example of an ascertainment under the JCT Standard Form of Building Contract 1980 Edition ('JCT 80'). Even at the time of writing the second edition in 1988, relatively little practical experience of the JCT 80 contract had been gained, so there was limited data against which to examine the courts' attitudes towards JCT 80's provisions. Over the course of 18 amendments between 1984 and 1998, JCT 80 became JCT 98, which, in turn, developed into JCT 2005, JCT 2011 and JCT 2016 (refer to Section 1.2.1).

Importantly, the terms of JCT 2016 contracts provide that if the Employer (or his representative) and the contractor undertake their roles correctly, the contractor's entitlement to reimbursement under the clauses should be dealt periodically, either on a monthly basis, or on stage payments (as may apply) during the course of the contract. There should be no such thing as a 'claim' submission, consisting of several lever arch files, sometime after practical completion (see Section 1.1).

Though the clause numbers may have changed in the JCT contracts between JCT 80 and JCT 2016, the relevant clauses' content has not altered significantly. However, there is now a wealth of court decisions relating to the recovery of loss and expense under the JCT contracts that serve to clarify. There are, it seems, as many approaches to recovering loss and expense as there are contractors preparing them. It can often appear that the greatest factor constraining the constituents – and often the amount sought – can be the imagination of the claim's draughtsman.

To comprehensively discuss and provide examples of each such approach is certainly beyond the scope of this book. Moreover, an imaginative, and hypothetical, and sometimes seemingly emotive approach is likely to disguise one fundamental point - that the contractor, if entitled to payment at all, is entitled to reimbursement of *its* direct loss and/or expense[1] and not some hypothetical contractor's theoretical loss and/or expense[2].

1 See Chapter 5, Section 5.1.
2 Costain Ltd v Haswell & Partners Ltd (2009) 28 Con LR 154 at 212.

*Building Contract Claims*, Sixth Edition. Guy Higginbottom and David Chappell.

What follows is a simple claim for the contractor's prolongation costs. It is based upon the 'warehouse' example in Chapter 3, which was used to illustrate the various methods of delay analysis (see Sections 3.5.1 to 3.5.6). For brevity, the example loss and expense claim is based upon the 'as-built vs as-planned' method (see Section 3.5.4) and uses elements of the 'collapsed as-built' method (see Section 3.5.5).

## A.2 Example Particulars

The 'warehouse' example of a claim for loss and expense can be made under the following JCT 2016 contracts:

- JCT SBC/Q 2016 including certain Contractor's Designed Portion ('CDP') elements;
- JCT DB 2016 and
- JCT ICD 2016.

JCT MW 2016 has not been selected because it is unlikely that a project of the scope and value as used in this example would be procured under JCT MW 2016.

The warehouse example in Chapter 3 requires the reader to refer to the following:

- The contractor's intended (or 'as-planned') programme for the warehouse's construction that is shown in Figure 3.1;
- The actual (or 'as-built') durations for each activity (refer to Figure 3.9), which include
  - The various delays to progress (recorded in Figure 3.9), which are further analysed in Figure 3.12, show the delays to the project. In order to arrive at the delay programme in Figure 3.12, the following methods of delay analysis (described below) are applied to the Figure 3.1 programme; and
  - 'Alternative 2' (see Section 3.5.4) in which the Employer is responsible for (1) delayed possession of the site and (2) changing the car park layout, but the contractor was responsible for the delay in obtaining the cladding panels.

The 'collapsed as-built' method is then partially applied (see Section 3.5.5). This shows that but for the various delays (critical and non-critical), the contractor's 'as-planned' programme was realistic, so the unaffected activities proceeded in their planned sequence and durations.

Figure A.1 repeats Figure 3.13, but instead of 'collapsing' the programme to remove the critical delays (first, employer delays and second, contractor delays), it highlights and summarises the durations and dates of those critical delays.

In this example, the contractor has given the appropriate notifications of delay, and applications for loss and expense are summarised in Tables 11.2 and 12.6, respectively.

| Line | Name | Start | Finish | Duration |
|---|---|---|---|---|
| 1 | *Employer delay in giving possession* | 06 Jun 22 | 24 Jun 22 | 3w |
| 2 | *Date of Possession / starting date* | 27 Jun 22 | 27 Jun 22 | |
| 3 | *Site set up* | 27 Jun 22 | 01 Jul 22 | 1w |
| 4 | *Strip topsoil / remove trees* | 27 Jun 22 | 01 Jul 22 | 1w |
| 5 | *Excavate to reduce levels* | 01 Jul 22 | 14 Jul 22 | 2w |
| 6 | *Excavate foundations* | 14 Jul 22 | 22 Jul 22 | 1w 2d |
| 7 | *Concrete foundations* | 25 Jul 22 | 29 Jul 22 | 1w |
| 8 | *Bolt boxes* | 29 Jul 22 | 29 Jul 22 | 1d |
| 9 | *Steelwork procurement and manufacture* | 06 Jun 22 | 17 Jun 22 | 2w |
| 10 | *Steel frame erection* | 01 Aug 22 | 12 Aug 22 | 2w |
| 11 | *Roof safety netting* | 15 Aug 22 | 17 Aug 22 | 3d |
| 12 | *Profiled roof sheet and skylights* | 18 Aug 22 | 01 Sep 22 | 2w |
| 13 | *Wall cladding (secondary steelwork)* | 25 Aug 22 | 26 Aug 22 | 2d |
| 14 | *Contractor delay in obtaining cladding panels* | 30 Aug 22 | 23 Sep 22 | 3w 4d |
| 15 | *Wall cladding (installing cladding panels)* | 26 Sep 22 | 05 Oct 22 | 1w 3d |
| 16 | *Pre-cast concrete mezzanine floor* | 30 Aug 22 | 05 Sep 22 | 1w |
| 17 | *Pre-cast concrete stairs* | 06 Sep 22 | 08 Sep 22 | 3d |
| 18 | *Ground floor slab* | 09 Sep 22 | 19 Sep 22 | 1w 2d |
| 19 | *Internal blockwork walls* | 16 Sep 22 | 29 Sep 22 | 2w |
| 20 | *External windows and doors* | 06 Oct 22 | 12 Oct 22 | 1w |
| 21 | *1st fix M&E* | 13 Oct 22 | 19 Oct 22 | 1w |
| 22 | *Plastering to offices* | 20 Oct 22 | 28 Oct 22 | 1w 2d |
| 23 | *Ceilings* | 27 Oct 22 | 02 Nov 22 | 1w |
| 24 | *2nd fix M&E* | 03 Nov 22 | 07 Nov 22 | 3d |
| 25 | *Employer instructs to change to electrical layout* | 08 Nov 22 | 17 Nov 22 | 1w 3d |
| 26 | *2nd fix M&E* | 18 Nov 22 | 21 Nov 22 | 2d |
| 27 | *Decorations* | 08 Nov 22 | 23 Nov 22 | 1w |
| 28 | *Form external roads and car park* | 10 Oct 22 | 28 Oct 22 | 3w |
| 29 | *Surfacing and lining* | 31 Oct 22 | 31 Oct 22 | 1d |
| 30 | *Employer changes car park and road layout* | 01 Nov 22 | 14 Nov 22 | 2w |
| 31 | *Resume surfacing and lining* | 15 Nov 22 | 18 Nov 22 | 4d |
| 32 | *Approval under s278 Highways Act 1980* | 21 Nov 22 | 21 Nov 22 | |
| 33 | *Planned completion (on site)* | 24 Nov 22 | 24 Nov 22 | |
| 34 | *Testing, commissioning & snagging* | 24 Nov 22 | 30 Nov 22 | 1w |
| 35 | *Date for completion* | 01 Dec 22 | 01 Dec 22 | |
| 36 | Employer delays | 06 Jun 22 | 17 Nov 22 | 4w 3d |
| 37 | Contractor delays | 30 Aug 22 | 23 Sep 22 | 3w 4d |

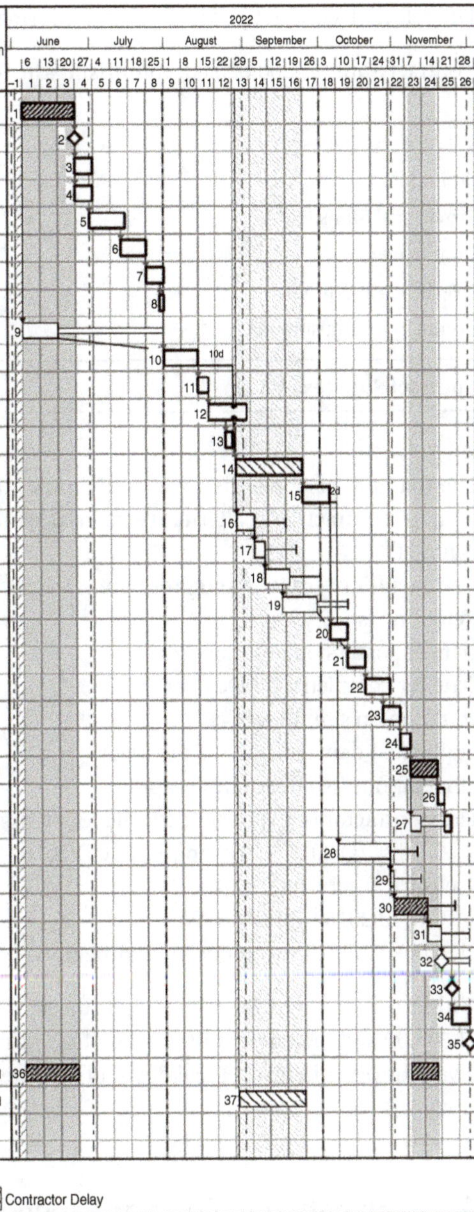


**Figure A.1** Warehouse programme showing critical delays.

The relevant 'contract particulars' and information are as follows:

- Contract Sum: £3,200,000.00, comprising the following:

| | |
|---|---:|
| Value of works/design | £2,610,848.26 |
| Preliminaries | £171,315.00 |
| Provisional Sums | £175,000.00 |
| Contingency | £40,000.00 |
| Overheads and profit | £226,941.24 |
| Commercial adjustment | (£24,104.50) |
| **Contract Sum (exc VAT)** | **£3,200,000.00** |

- The example in Appendix 1 can apply under the JCT SBC/Q 2016, JCT DB 2016 or JCT IC 2016 contract, and alternative references to the relevant clauses are given for each contract.
- Contractor's overheads and profit are 7.5%: 5% overheads and 2.5% profit.
- Retention is 5%
- Date of possession: 6 June 2022
- Date for completion: 7 October 2022[3]
- Duration on site: 18 weeks
- Liquidated damages: £1,800 per week
- Fluctuations: No fluctuations apply
- The Contractor is responsible for the following design:
  - JCT SBC/Q 2016 (CDP) and JCT ICD 2016 – structural steelwork and mechanical and electrical services
  - JCT DB 2016 – all of the Works.
- The total value of variations, including provisional sum expenditure, is £168,561.25.

Within the Contract Sum of £3,200,000.00, the contractor's 'preliminaries' amount to £171,315.00 and include the time-related resources, such as site staff, site accommodation and various other items. A breakdown of the 'preliminaries' is shown in Table A.1.

## A.3 Explanation of Costs or Expenditure

Table A.1 shows the 'value' of the contractor's preliminaries items and the total preliminaries' value (£171,315.00) within the Contract Sum (£3.2 m). The figures for the time-related items, i.e. in the 'Rate/wk' and the 'Sub-total' columns, are those that were agreed with

3 Refer to Figure 3.1 in Chapter 3.

**Table A.1** Contractor's 'preliminaries'.

| Resource | No. | % Allocation | No. | Unit | Rate/week | Sub-total |
|---|---|---|---|---|---|---|
| *Pre-start obligations* | | | | | | |
| Project Manager | 1 | 20% | 4 | weeks | £2,480.00 | £1,984.00 |
| Contracts Manager | 1 | 40% | 2 | weeks | £2,320.00 | £1,856.00 |
| Quantity Surveyor | 1 | 50% | 4 | weeks | £2,015.00 | £4,030.00 |
| Planner | 1 | 50% | 2 | weeks | £1,980.00 | £1,980.00 |
| Design Manager | 1 | 100% | 4 | weeks | £1,990.00 | £7,960.00 |
| Site Manager | 1 | 100% | 2 | weeks | £1,820.00 | £3,640.00 |
| *Construction stage* | | | | | | |
| Project Manager | 1 | 10% | 18 | weeks | £2,480.00 | £4,464.00 |
| Contracts Manager | 1 | 20% | 18 | weeks | £2,320.00 | £8,352.00 |
| Site Manager | 1 | 100% | 18 | weeks | £1,820.00 | £32,760.00 |
| Engineer | 1 | 80% | 7 | weeks | £1,795.00 | £10,052.00 |
| Quantity Surveyor | 1 | 20% | 18 | weeks | £2,015.00 | £7,254.00 |
| Planner | 1 | 10% | 18 | weeks | £1,980.00 | £3,564.00 |
| Design Manager (weeks 1 to 9) | 1 | 100% | 9 | weeks | £1,990.00 | £17,910.00 |
| Design Manager (weeks 10 to 18) | 1 | 50% | 9 | weeks | £1,990.00 | £8,955.00 |
| CDM/HSEQ officer | 1 | 10% | 18 | weeks | £1,930.00 | £3,474.00 |
| *Site accommodation* | | | | | | |
| Site offices | 2 | 100% | 18 | weeks | £80.00 | £2,880.00 |
| Site welfare cabins | 1 | 100% | 18 | weeks | £52.00 | £936.00 |
| Site WC unit | 1 | 100% | 18 | weeks | £52.00 | £936.00 |
| Site WC consumables | 1 | 100% | 18 | weeks | £15.00 | £270.00 |
| Empty WC unit | 1 | 100% | 18 | weeks | £18.00 | £324.00 |
| Secure 20' storage containers | 2 | 100% | 18 | weeks | £20.00 | £720.00 |
| Transport (to/from site) | 6 | 100% | 2 | Item | £350.00 | £4,200.00 |
| Smoking shelter (temporary) | 1 | 100% | 1 | Item | £500.00 | £500.00 |
| Electrical connection/disconnection | 1 | 100% | 2 | Item | £500.00 | £1,000.00 |
| Water connection/disconnection | 1 | 100% | 2 | Item | £500.00 | £1,000.00 |
| IT services connection/disconnection | 1 | 100% | 1 | Item | £2,500.00 | £2,500.00 |
| Electricity supply | 1 | 100% | 18 | weeks | £15.00 | £270.00 |
| Water supply | 1 | 100% | 18 | weeks | £8.00 | £144.00 |
| IT services supply | 1 | 100% | 18 | weeks | £15.00 | £270.00 |
| Heras fencing and gates | 1 | 100% | 18 | weeks | £55.00 | £990.00 |
| Hoarding and signage | 1 | 100% | 18 | weeks | £118.00 | £2,124.00 |
| Making good site area | 1 | 100% | 1 | Item | £2,500.00 | £2,500.00 |

*(Continued)*

**Table A.1** (Continued)

| Resource | No. | % Allocation | No. | Unit | Rate/week | Sub-total |
|---|---|---|---|---|---|---|
| *Office equipment/resources* | | | | | | |
| Printer | 1 | 100% | 1 | Item | £450.00 | £450.00 |
| Wi-Fi router | 1 | 100% | 1 | Item | £150.00 | £150.00 |
| Printing costs | 1 | 100% | 18 | weeks | £45.00 | £810.00 |
| Office consumables | 1 | 100% | 18 | weeks | £60.00 | £1,080.00 |
| *Other Staff/Labourer* | | | | | | |
| General labourer | 1 | 100% | 19 | weeks | £680.00 | £12,920.00 |
| Traffic management | 2 | 100% | 2 | weeks | £1,240.00 | £4,960.00 |
| Banksman/crane attendance | 1 | 100% | 6 | weeks | £1,100.00 | £6,600.00 |
| *Plant/small tools* | | | | | | |
| (Major plant inc sub-contracts) | | | | | | |
| Small tools/consumables | 1 | 100% | 18 | weeks | £45.00 | £810.00 |
| Task lighting | 1 | 100% | 9 | weeks | £104.00 | £936.00 |
| Road sweeper | 1 | 100% | 8 | visits | £350.00 | £2,800.00 |
| **Total Preliminaries** | | | | | | **£171,315.00** |

the Employer when executing the building contract. They are 'all in' rates and prices that include the following:

- Overheads and profit
- For staff,[4]
  - Lodging and subsistence
  - Travelling costs
  - Company car costs
  - National Insurance contributions
  - Income tax payable
  - Holiday and sickness pay
  - Pension contributions
  - Other levies (such as CITB)
  - Staff equipment (such as laptops, mobile phones and broadband costs)
- Plant, equipment and tools.

However, JCT 2016 contracts do not provide for the contractor to recover the amount of its 'direct loss' or 'direct expense' when claiming loss and expense, so it is not appropriate to use these 'values' from the preliminaries element of the Contract Sum. This is because these 'preliminaries rates' may include for profit, overheads and other allowances that the contractor has built into the 'all in' rates for time-related items. It is more appropriate for the contractor to refer to preliminaries' rates for claiming loss and expense for an

4 For more information, see the RICS' Definition of Prime Cost of Daywork Carried Out Under A Building Contract (3rd ED) 2007.

interim payment application, but he should expect a significant reduction in these rates when the Employer or his representative assess the amount of loss and expense that is due. The contractor's preliminaries are only really of use when giving a guide to the categories of time-related resources, for which the contractor is likely to claim loss and expense.

As set out in Section 10.2.4 (Cost Records), the contractor should be able to readily obtain the actual cost of his time-related resources from his accounting software. The Employer or his representative should be aware that these costs may not be the same from week to week. In the event that the contractor's accounting software does not show the amounts paid or incurred by each staff member, for example, then the costs per staff member are likely to vary from week to week. If the accounting software provides more detail, such as providing the weekly amounts for each staff member (see Table A.1 above[5]), then any significant variances in that staff member's weekly costs can be easily explained.

## A.4 Loss and Expense for Employer Delay 1 – Details

'Employer delay 1' is the Employer's failure to give possession of the site to the contractor on 6 June 2022 (i.e. the 'Date of Possession'). The facts regarding this delay are set out as follows:

- The Employer did not warn the contractor that possession would be delayed or deferred[5], so on 6 June 2022, the contractor arrived at the site with its site setup;
- The Employer failed to give access so the contractor could not deposit the temporary buildings and plant onto the site, so they had to be transported away to be put into a storage facility until the Employer granted access to the site. The contractor incurred additional transport costs and abortive visit costs for site services;
- The contractor's site staff were due to attend a site meeting on 6 June 2022 but this was aborted through lack of access. The contractor was able to re-allocate some of the project's intended staff to other projects, but not all of them:
- The contractor's groundworks sub-contractor was due to commence work on 6 June 2022 but incurred additional and abortive costs.
- The Employer finally notifies the contractor that he may take possession of the site on 27 June 2022 (a delay of 3 weeks).
- The contractor has promptly notified the Employer (or his representative) on 6 June 2022 (i.e. when it became apparent that progress of the Works is likely to be delayed) and provided the relevant accompanying information, along with notifying updates on a weekly basis for the particular contract as set out in Table 11.2.
- Finally, the contractor issued his notifications and assessments of loss and expense in relation to the relevant contract, as set out in Table 12.6.

5 Deferment of possession is a Relevant Event under clause 2.29.3 of JCT SBC/Q 2016, clause 2.26.3 of JCT DB 2016 and clause 2.20.3 of JCT IC 2016. The Employer is entitled to defer possession up to 6 weeks under clause 2.5 of JCT SBC/Q 2016 and JCT IC 2016 and clause 2.4 of JCT DB 2016, but in this example, they did not do so and simply failed to give the contractor access to the site.

## A.5 Loss and Expense for Employer Delay 1 – Entitlement and Quantum

First, the contractor is entitled to an extension of time of 3 weeks under the following clauses.

As the Employer did not defer possession, but simply failed to give access, this is a relevant event under the clauses set out in Table A.2. Furthermore, the Employer's failure in this regard is a Relevant Matter under the clauses in Table A.3:

**Table A.2** Entitlement to an extension of time.

| Relevant Event | JCT SBC/Q 2016 | JCT DB 2016 | JCT IC 2016 |
|---|---|---|---|
| Employer's impediment, prevention or default | 2.29.7 | 2.26.6 | 2.20.6 |

**Table A.3** Entitlement to loss and expense.

| Relevant Matter | JCT SBC/Q 2016 | JCT DB 2016 | JCT IC 2016 |
|---|---|---|---|
| Deferment of Possession | 4.20.1 | 4.19.1 | 4.15.1 |

The loss and expense claim for Employer Delay 1 is set out in Table A.4.

## A.6 Loss and Expense for Employer Delay 2 – Details

Employer delay 2 occurs because the Employer instructs a change in the electrical services layout. This change is to the Employer's Requirements for the CDP works under a JCT SBC/Q 2016 contract or a JCT ICD 2016 contract, and to the whole of the contractor's designed work under a JCT DB 2016 contract.

Under all of these JCT 2016 contracts, the contractor will have to re-design the electrical services layout. As the electrical works sit on the critical path[6], the delay of 1 week and 3 days is a critical delay between the 1 and 14 November 2022. The contractor has submitted the relevant notifications, particulars, estimates, and applications and assessments as were summarised in Section A.4.

## A.7 Loss and Expense for Employer Delay 2 – Entitlement and Quantum

First, the contractor is entitled to an extension of time of 1 week and 3 days under the following clauses:

Furthermore, the Employer's instruction is a Relevant Matter under the clauses in Table A.6.

The loss and expense claim for Employer Delay 2 is set out in Table A.6.

6 Refer activity 25 on Figure A.1.

**Table A.4** Loss and expense for Employer Delay 1.

| Date (W/E) | Category | Description | Cost description/reference | Unit cost | Adjusted cost |
|---|---|---|---|---|---|
| | | | | | |
| 05-Jun-22 | Staff | Project Manager | Mileage and accommodation for abortive site meeting (to be held on 6 June 2022) | £238.45 | £238.45 |
| 05-Jun-22 | Staff | Contracts Manager | Mileage and accommodation for abortive site meeting (to be held on 6 June 2022) | £248.32 | £248.32 |
| 05-Jun-22 | Staff | Quantity Surveyor | Mileage and accommodation for abortive site meeting (to be held on 6 June 2022) | £212.11 | £212.11 |
| 05-Jun-22 | Staff | Planner | Mileage and accommodation for abortive site meeting (to be held on 6 June 2022) | £248.32 | £248.32 |
| 05-Jun-22 | Staff | Design Manager | Mileage for abortive site meeting (to be held on 6 June 2022) | £48.20 | £48.20 |
| 05-Jun-22 | Staff | CDM/HSEQ officer | Mileage and accommodation for abortive site meeting (to be held on 6 June 2022) | £234.25 | £234.25 |
| 10-Jun-22 | Staff | Engineer | Freelance engineer is £1,525/week (for 4 days or 80% of the week). Appointed on a fixed-duration contract for the first 7 weeks of the project, so under the terms of that contract, the engineer is entitled to be paid the full value of £1,575/wk for the entirety of his time engaged on the project. Engineer's invoice No 1 | £1,525.00 | £1,525.00 |
| 12-Jun-22 | Staff | Project Manager | Basic salary for week ending 12 June 2022 100% allocation for the week commencing 6 June 2022, attempting to mitigate costs incurred, delay notice approval, speaking with sub-contractors/suppliers | £789.43 | £789.43 |
| 12-Jun-22 | Staff | Contracts Manager | Basic salary for week ending 12 June 2022 100% allocation for the week commencing 6 June 2022, attempting to mitigate costs incurred, delay notice approval, speaking with sub-contractors/suppliers | £738.50 | £738.50 |

*(Continued)*

**Table A.4** (Continued)

| Date (W/E) | Category | Description | Cost description/reference | Unit cost | Adjusted cost |
|---|---|---|---|---|---|
| 12-Jun-22 | Staff | Site Manager | Basic salary for week ending 12 June 2022 60% allocation for the week commencing 6 June 2022 (attempting to mitigate costs incurred, delay notice approval, speaking with sub-contractors/suppliers), as re-allocated to another project until 19 June 2022 | £579.34 | £347.60 |
| 12-Jun-22 | Staff | Quantity Surveyor | Basic salary for week ending 12 June 2022 100% allocation for the week commencing 6 June 2022, attempting to mitigate costs incurred, delay notice approval, speaking with sub-contractors/suppliers | £641.41 | £641.41 |
| 12-Jun-22 | Staff | Planner | Basic salary for week ending 12 June 2022 20% allocation for the week commencing 6 June 2022. Allocated to other activities after 1 day during w/c 6 June 2022 | £630.27 | £126.05 |
| 12-Jun-22 | Staff | Design Manager | Basic salary for week ending 12 June 2022 60% allocation for the week commencing 6 June 2022 (including early steelwork procurement). Allocated to other activities after 3 days during w/c 6 June 2022 | £633.45 | £380.07 |
| 12-Jun-22 | Staff | CDM/HSEQ officer | Basic salary for week ending 12 June 2022 20% allocation for the week commencing 6 June 2022. Allocated to other activities after 1 day during w/c 6 June 2022 | £614.35 | £122.87 |
| 17-Jun-22 | Staff | Engineer | See above - Engineer's invoice No. 2 | £1,525.00 | £1,525.00 |
| 19-Jun-22 | Staff | Contracts Manager | Basic salary for week ending 19 June 2022 50% allocation for the week commencing 6 June 2022, allocated to other projects during week | £738.50 | £369.25 |
| 24-Jun-22 | Staff | Engineer | See above - Engineer's invoice No. 3 | £1,525.00 | £1,525.00 |
| 26-Jun-22 | Staff | Contracts Manager | Basic salary for week ending 26 June 2022 50% allocation for the week commencing 6 June 2022, allocated to other projects during week | £738.50 | £369.25 |

| | | | | | |
|---|---|---|---|---|---|
| 26-Jun-22 | Staff | Site Manager | Basic salary for week ending 26 June 2022 100% allocation for the week commencing 20 June | £579.34 | £579.34 |
| 26-Jun-22 | Staff | Quantity Surveyor | Basic salary for week ending 26 June 2022 100% allocation for the week commencing 20 June 2022, attempting to mitigate costs incurred, delay notice approval, speaking with sub-contractors/suppliers | £641.41 | £641.41 |
| 30-Jun-22 | Overheads | Main Contractor's overhead claim | Main contractor's application for payment No. 1 (30 June 2022) 3 weeks claim for overheads × 5% Hudson formula calculation: Contract Sum (£3.2M) less prov sums (£175k) = £3,025.000 Contribution to overheads (5%) from Variations/Changes: £168,561.25 × 5% = £8,428.06 ('OH from Variations') [(5/100) × (£3,025,000/18 weeks) × 3 weeks] - [OH from Variations] | £16,780.27 | £16,780.27 |
| 30-Jun-22 | Accom | Site Offices (2 No) | Accommodation supplier invoice No. 1 (June 2022). Contractor liable for full payment for postponed hire £38/wk × 2 office | £228.00 | £228.00 |
| 30-Jun-22 | Accom | Site welfare cabins (1No) | Accommodation supplier invoice No. 1 (June 2022). Contractor liable for full payment for postponed hire £48.10/wk | £144.30 | £144.30 |
| 30-Jun-22 | Accom | Site WC unit | Accommodation supplier invoice No. 1 (June 2022). Contractor liable for full payment for postponed hire £48.10/wk | £144.30 | £144.30 |
| 30-Jun-22 | Accom | Secure 20' storage containers (2 No) | Accommodation supplier invoice No. 1 (June 2022). Contractor liable for full payment for postponed hire £18.50/wk | £111.00 | £111.00 |
| 30-Jun-22 | Accom | Transport (to/from site) | Accommodation supplier invoice No. 1 (June 2022) Transport to, and from site (as unable to access site) on 6 June 2022: - 2 × site office - Welfare cabin - WC Unit - 2 × 20' storage containers Each item × £323.75 each way | £3,885.00 | £3,885.00 |

*(Continued)*

**Table A.4** (Continued)

| Date (W/E) | Category | Description | Cost description/reference | Unit cost | Adjusted cost |
|---|---|---|---|---|---|
| 30-Jun-22 | Accom | Storage of site accommodation | Accommodation supplier invoice No. 1 (June 2022) Storage of all allocated site accommodation: £1,800 per week | £5,400.00 | £5,400.00 |
| 30-Jun-22 | Accom | Water connection/disconnection | Abortive visit cost £500 | £500.00 | £500.00 |
| 30-Jun-22 | Accom | IT services connection/disconnection | Abortive visit cost £150 | £150.00 | £150.00 |
| 30-Jun-22 | Sub-Con | Groundworks Sub-Contractor | Sub-Contractor Application for Payment No. 1 (30 June 2022): (1) Abortive visit for operatives for stripping topsoil planned for w/c 13 June 2022 (unable to re-allocate to other projects): - 2 × JCB with Bucket inc drivers - 2 × dumper trucks inc drivers - Abortive cost for cart-away lorry - 4 groundworkers idle for the week Total cost £6,850.00 | £6,850.00 | £6,850.00 |
| 30-Jun-22 | Sub-Con | Groundworks Sub-Contractor | Sub-Contractor Application for Payment No. 1 (30 June 2022): (2) Abortive visit for operatives for excavating to reduce levels for w/c 20 June 2022 (unable to re-allocate to other projects): - 2 × JCB with Bucket inc drivers - 2 × dumper trucks inc drivers - 4 groundworkers idle for the week Total cost £6,150.00 | £6,150.00 | £6,150.00 |
| | Staff | | | | £10,909.84 |
| | Accom | | | | £10,562.60 |
| | Sub-Con | | | | £13,000.00 |
| | Overheads | | | | £16,780.27 |
| | | | | | |
| | | | **Total Loss and Expense claimed for Employer Delay 1** | | **£51,252.71** |

**Table A.5** Entitlement to an extension of time (Employer Delay 2).

| Relevant Event | JCT SBC/Q 2016 | JCT DB 2016 | JCT IC 2016 |
|---|---|---|---|
| 'Variations' or 'Changes' | 2.29.1 | 2.26.1 | 2.20.1 |

**Table A.6** Entitlement to loss and expense (Employer Delay 2).

| Relevant Matter | JCT SBC/Q 2016 | JCT DB 2016 | JCT IC 2016 |
|---|---|---|---|
| 'Variations' or 'Changes' | 4.22.1 | 4.21.1 | 4.17.1 |

## A.8 Observations for Loss and Expense for Employer Delay Claims 1 and 2

It should be immediately clear that the unit rates and prices included in the loss and expense claims (Tables A.4 and A.6) are based upon the actual cost incurred by the contractor for staff, accommodation, etc., and are not claimed at the rates and prices in the 'Preliminaries' (Table A.1). The 'preliminaries rates' are values agreed as part of the Contract Sum and include for overheads, profit and other possible financial contingencies. Therefore, the 'preliminaries rates' are not the amounts that the contractor has 'lost' or 'expended'.

For both loss and expense claims, the contractor has met with some success at mitigating his losses (or expenditure):

- Employer Delay 1 loss and expense – the contractor was able to re-allocate the following staff to other projects:
  - Site Manager
  - Planner
  - Design Manager
  - CDM/HSEQ Officer
  - Contracts Manager.
- Employer Delay 2 loss and expense – re-allocating the following staff for part of the delay:
  - Quantity Surveyor
  - Planner
  - Contracts Manager
  - Site Manager.

The claims for loss of contribution from head office overheads are based upon the 'Hudson Formula'[7], and credit is given for the overhead contribution that the contractor has recovered via 'Variations' or 'Changes'. Whilst 'Employer Delay 2' clearly arises from a 'Variation' (or 'Change') that would entitle the contractor to additional payment (i.e. for the change in design and it is assumed additional work), for ease of reference, the credit

7 See Section 4.4.2.5.

**Table A.7** Loss and expense for Employer Delay 2.

| Date (W/E) | Category | Description | Cost description/reference | Unit cost | Adjusted cost |
|---|---|---|---|---|---|
| | | | | | |
| 12-Nov-22 | Staff | Project Manager | Basic salary for week ending 12 Nov 2022 100% allocation for the week commencing 8 Nov 2022, working with project team to ascertain scope of M&E changes speaking with sub-contractors/suppliers | £789.43 | £789.43 |
| 12-Nov-22 | Staff | Contracts Manager | Basic salary for week ending 12 Nov 2022 100% allocation for the week commencing 8 Nov 2022, working with project team to ascertain scope of M&E changes speaking with sub-contractors/suppliers | £738.50 | £738.50 |
| 12-Nov-22 | Staff | Site Manager | Basic salary for week ending 12 Nov 2022 100% allocation for the week commencing 8 Nov 2022, working with project team to ascertain scope of M&E changes speaking with sub-contractors/suppliers | £579.34 | £579.34 |
| 12-Nov-22 | Staff | Quantity Surveyor | Basic salary for week ending 12 Nov 2022 40% allocation for the week commencing 8 Nov 2022, speaking with sub-contractors/suppliers, working out costs for the M&E variation. | £641.41 | £256.56 |
| 12-Nov-22 | Staff | Planner | Basic salary for week ending 12 Nov 2022 20% allocation for the week commencing 8 Nov 2022, speaking with sub-contractors/suppliers, scheduling/programming the M&E variation. | £630.27 | £126.05 |
| 12-Nov-22 | Staff | Design Manager | Basic salary for week ending 12 Nov 2022 100% allocation for the week commencing 8 Nov 2022, M&E variation - co-ordination, obtaining sub-contractor's design | £633.45 | £633.45 |
| 19-Nov-22 | Staff | Contracts Manager | Basic salary for week ending 19 Nov 2022 60% allocation (3 days) for the week commencing 14 Nov | £633.45 | £380.07 |
| 19-Nov-22 | Staff | Contracts Manager | Basic salary for week ending 19 Nov 2022 60% allocation (3 days) for the week commencing 14 Nov | £738.50 | £443.10 |

| | | | | | |
|---|---|---|---|---|---|
| 19-Nov-22 | Staff | Site Manager | Basic salary for week ending 19 Nov 2022 60% allocation (3 days) for the week commencing 14 Nov | £579.34 | £347.60 |
| 19-Nov-22 | Staff | Quantity Surveyor | Basic salary for week ending 19 Nov 2022 60% allocation (3 days) for the week commencing 14 Nov | £641.41 | £384.85 |
| 30-Nov-22 | Overheads | Main Contractor's overhead claim | Main contractor's application for payment No. 6 (30 Nov 2022) 3 weeks claim for overheads × 5% Hudson formula calculation: (5/100) × (£3,025,000/18 weeks) × 1.6 weeks NB: Overhead recovery form Variations deducted in claim for Employer Delay 1 | £13,444.44 | £13,444.44 |
| 30-Nov-22 | Accom | Site Offices (2 No) | Accommodation supplier invoice No. 1 (June 2022). Contractor liable for full payment for postponed hire £38/wk × 2 office × 1.6 weeks (1w 3 days) | £121.60 | £121.60 |
| 30-Nov-22 | Accom | Site Welfare cabins (1 No) | Accommodation supplier invoice No. 1 (June 2022). Contractor liable for full payment for postponed hire £48.10/wk × 1.6 weeks (1w 3 days) | £76.96 | £76.96 |
| 30-Nov-22 | Accom | Site WC unit | Accommodation supplier invoice No. 1 (June 2022). Contractor liable for full payment for postponed hire £48.10/wk × 1.6 weeks (1w 3 days) | £76.96 | £76.96 |
| 30-Nov-22 | Accom | Empty WC unit | 1.6 weeks (1 week 3 days) × cost of £16.65/wk | £16.65 | £26.64 |
| 30-Nov-22 | Accom | Secure 20' storage containers (1 No) | Accommodation supplier invoice No. 1 (June 2022). Contractor liable for full payment for postponed hire Only 1 no storage container left on site £18.50/wk | £29.60 | £29.60 |

*(Continued)*

**Table A.7** (Continued)

| Date (W/E) | Category | Description | Cost description/reference | Unit cost | Adjusted cost |
|---|---|---|---|---|---|
| 30-Nov-22 | Accom | Electricity supply | 1.6 weeks (1 week 3 days) × cost of £13.88/wk | £13.88 | £22.20 |
| 30-Nov-22 | Accom | Water supply | 1.6 weeks (1 week 3 days) × cost of £7.40/wk | £7.40 | £11.84 |
| 30-Nov-22 | Accom | IT services supply | 1.6 weeks (1 week 3 days) × cost of £13.88/wk | £13.88 | £22.20 |
| 30-Nov-22 | Accom | Heras fencing and gates | 1.6 weeks (1 week 3 days) × cost of £50.88/wk | £50.88 | £81.40 |
| 30-Nov-22 | Accom | Hoarding and signage | 1.6 weeks (1 week 3 days) × cost of £109.15/wk | £109.15 | £174.64 |
| 30-Nov-22 | Office Equip | Printing | 1.6 weeks (1 week 3 days) × cost of £41.63/wk | £41.63 | £66.61 |
| 30-Nov-22 | Office Equip | Office consumables | 1.6 weeks | | £21.32 |
| 30-Jun-22 | Sub-Con | M&E Sub-contractor | Sub-Contractor Application for Payment No. 2 (30 Nov 2022): (1) All 2nd fix M&E work standing idle whilst scope of variation is designed and issued: - Unable to re-allocate 3/6 operatives × £1,140/week - Storage container × £20/week | £3,440.00 | £5,504.00 |
| | Staff | | | | £4,678.96 |
| | Accom | | | | £.00 |
| | Office Equip | | | | £87.93 |
| | Sub-Cont | | | | £.00 |
| | Overheads | | | | £13,444.44 |
| | | | **Total Loss and Expense claimed for Employer Delay 2** | | **£18,211.33** |

for overheads has been shown against the head office overhead loss and expense claim under Employer Delay 1.

Regarding sub-contractor claims, both loss and expense claim examples include claims from sub-contractors that seem entirely reasonable. For Employer Delay 1, the groundworks sub-contractor suffers abortive visits for the activities[8] that were planned during the period in which the Employer failed to give possession. For Employer Delay 2, 3 out of 6 of the electrical services sub-contractor's operatives were standing idle for the duration of the delay to re-design the electrical services – so the sub-contractor was able to mitigate his losses by re-allocating 3 operatives to another project for the delay's duration. Section 4.4.6 highlighted the common difficulty that some sub-contractors are unable to particularise their loss and expense claims to the same level of detail that can be provided by a main contractor. All a main contractor can do is ask the sub-contractor to provide evidence of his labour costs (such as wages or payment slips), and (in these examples) plant-hire costs.

8 Refer to Figure A.1. Planned activities were 'stripping topsoil' (activity 5) and 'excavate to reduce levels' (activity 6). The planned sequence and timing of these activities is shown in Figure 3.1 (see activity no's 2 and 3) as occurring between 13 and 30 June 2022.

# Appendix B

# Example of Compensation Event Claim Under NEC4

## B.1 Introduction

This Appendix B contains a 'claim' for a compensation event under NEC4. As explained in Sections 5.1.1, 5.3 and 13.1, NEC4 refers to any event which would entitle the contractor to change the amounts of time and money otherwise due to him, as a 'compensation event' (or 'CE'). Section 5.3.1 describes all of the 'compensation events' that arise under all the NEC4 main contract options (A to E), and a detailed commentary is provided in Section 13.5. The following sections deal with how 'compensation events' are evaluated:

- Section 5.3.2 sets out the procedure for dealing with the notification of compensation events, along with the subsequent procedures for submitting and assessing quotations for compensation events;
- Section 5.5 covers the 'valuation' of 'compensation events' – i.e. Assessing the Change to the Prices (see Section 5.5.1); and
- Section 13.3 covers the assessment of a delay to the Completion Date and/or a Key Date.

### B.1.1 Details of the Compensation Event

The compensation event is based upon the 'warehouse' programme in Figure 3.1, but with the introduction of a pedestrian footbridge, linking the 'warehouse' site, over a small stream, to the Client's land on the other side. The example is intended to be relevant to any of the NEC4 main contract options (Options A to E), so either the Schedule of Cost Components or the Shorter Schedule of Cost components may apply. The Project Manager instructed the compensation event before the 'starting date' (6 June 2022) so the date of the instruction will be the 'dividing date', and invited the contractor to submit a quotation and the 'revised programme' to show the change in the Prices and the change to the completion date. As no work has commenced on site, the contractor's quotation and the Project Manager's assessment will be based upon the forecast Defined Cost.

In order to submit the revised programme, the contractor has inserted the 'footbridge' activities into the Accepted Programme (i.e. the programme in Figure 3.1 in Chapter 3). The 'Footbridge Design' (activity 24/F100) and 'Footbridge Fabrication' (activity 25/F200) can be carried out at any time before the footbridge needs to be constructed. However, owing

*Building Contract Claims*, Sixth Edition. Guy Higginbottom and David Chappell.

to logistical reasons, establishing the 'Footbridge site setup' (activity F250) can realistically only begin at the same time as 'Form external roads and car park' (activity 23) – i.e. after 'Wall cladding' (activity 12) has been completed. The introduction of the footbridge compensation event is shown in Figure B.1 – which is a version of the Accepted Programme that has not been re-scheduled:

| Line | Activity No | Name | Start | Finish | Duration |
|---|---|---|---|---|---|
| 1 | | Date of Possession / starting date | 06 Jun 22 | 06 Jun 22 | |
| 2 | | Site set up | 06 Jun 22 | 10 Jun 22 | 1w |
| 3 | | Strip topsoil / remove trees | 13 Jun 22 | 17 Jun 22 | 1w |
| 4 | | Excavate to reduce levels | 17 Jun 22 | 30 Jun 22 | 2w |
| 5 | | Excavate foundations | 23 Jun 22 | 01 Jul 22 | 1w 2d |
| 6 | | Concrete foundations | 01 Jul 22 | 07 Jul 22 | 1w |
| 7 | | Bolt boxes | 08 Jul 22 | 08 Jul 22 | 1d |
| 8 | | Steelwork procurement and manufacture | 06 Jun 22 | 24 Jun 22 | 3w |
| 9 | | Steel frame erection | 11 Jul 22 | 22 Jul 22 | 2w |
| 10 | | Roof safety netting | 25 Jul 22 | 27 Jul 22 | 3d |
| 11 | | Profiled roof sheet and skylights | 28 Jul 22 | 10 Aug 22 | 2w |
| 12 | | Wall cladding | 04 Aug 22 | 17 Aug 22 | 2w |
| 13 | | Pre-cast concrete mezzanine floor | 25 Jul 22 | 29 Jul 22 | 1w |
| 14 | | Pre-cast concrete stairs | 01 Aug 22 | 03 Aug 22 | 3d |
| 15 | | Ground floor slab | 04 Aug 22 | 12 Aug 22 | 1w 2d |
| 16 | | Internal blockwork walls | 15 Aug 22 | 26 Aug 22 | 2w |
| 17 | | External windows and doors | 18 Aug 22 | 24 Aug 22 | 1w |
| 18 | | 1st fix M&E | 30 Aug 22 | 05 Sep 22 | 1w |
| 19 | | Plastering to offices | 06 Sep 22 | 12 Sep 22 | 1w |
| 20 | | Ceilings | 13 Sep 22 | 15 Sep 22 | 3d |
| 21 | | 2nd fix M&E | 16 Sep 22 | 22 Sep 22 | 1w |
| 22 | | Decorations | 23 Sep 22 | 29 Sep 22 | 1w |
| 23 | | Form external roads and car park | 18 Aug 22 | 08 Sep 22 | 3w |
| 24 | F100 | Footbridge design | 04 Aug 22 | 10 Aug 22 | 1w |
| 25 | F200 | Fabricate footbridge | 11 Aug 22 | 17 Aug 22 | 1w |
| 26 | F250 | Footbridge site set-up | 18 Aug 22 | 24 Aug 22 | 1w |
| 27 | F260 | Enabling work | 25 Aug 22 | 01 Sep 22 | 1w |
| 28 | F270 | Footbridge foundations | 02 Sep 22 | 08 Sep 22 | 1w |
| 29 | F300 | Assemble footbridge on site | 09 Sep 22 | 15 Sep 22 | 1w |
| 30 | F400 | Brick wingwalls | 16 Sep 22 | 19 Sep 22 | 2d |
| 31 | F500 | Paint footbridge | 20 Sep 22 | 26 Sep 22 | 1w |
| 32 | F600 | Clear footbridge area | 29 Sep 22 | 05 Oct 22 | 1w |
| 33 | | Electric lighting | 27 Sep 22 | 28 Sep 22 | 2d |
| 34 | | Surfacing and lining | 09 Sep 22 | 15 Sep 22 | 1w |
| 35 | | Approval under s278 Highways Act 1980 | 30 Sep 22 | 30 Sep 22 | |
| 36 | | Planned completion (on site) | 30 Sep 22 | 30 Sep 22 | |
| 37 | | Testing, commissioning & snagging | 30 Sep 22 | 06 Oct 22 | 1w |
| 38 | | Date for completion | 07 Oct 22 | 07 Oct 22 | |

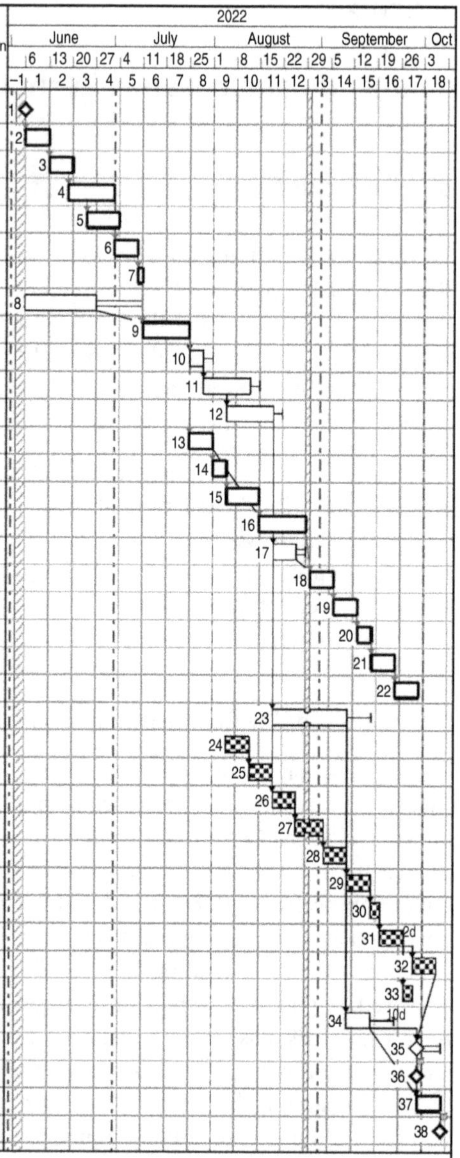


BCC 6 - Figure

Planned Activity | Compensation Event

**Figure B.1** Introduction of the footbridge CE activities into the accepted programme.

After adding the footbridge activities into the Accepted Programme, it must then be re-scheduled to show the effects (if any) upon the planned completion date and the Completion Date. Figure B.2 shows a re-scheduled version of the Accepted Programme:

As the s278 work approval (activity 35) depends upon completion of the footbridge, the date for planned completion has been postponed by one week and four days, from

| Line | Activity No | Name | Start | Finish | Duration |
|---|---|---|---|---|---|
| 1 | | Date of Possession / starting date | 06 Jun 22 | 06 Jun 22 | |
| 2 | | Site set up | 06 Jun 22 | 10 Jun 22 | 1w |
| 3 | | Strip topsoil / remove trees | 13 Jun 22 | 17 Jun 22 | 1w |
| 4 | | Excavate to reduce levels | 17 Jun 22 | 30 Jun 22 | 2w |
| 5 | | Excavate foundations | 23 Jun 22 | 01 Jul 22 | 1w 2d |
| 6 | | Concrete foundations | 01 Jul 22 | 07 Jul 22 | 1w |
| 7 | | Bolt boxes | 08 Jul 22 | 08 Jul 22 | 1d |
| 8 | | Steelwork procurement and manufacture | 06 Jun 22 | 24 Jun 22 | 3w |
| 9 | | Steel frame erection | 11 Jul 22 | 22 Jul 22 | 2w |
| 10 | | Roof safety netting | 25 Jul 22 | 27 Jul 22 | 3d |
| 11 | | Profiled roof sheet and skylights | 28 Jul 22 | 10 Aug 22 | 2w |
| 12 | | Wall cladding | 04 Aug 22 | 17 Aug 22 | 2w |
| 13 | | Pre-cast concrete mezzanine floor | 25 Jul 22 | 29 Jul 22 | 1w |
| 14 | | Pre-cast concrete stairs | 01 Aug 22 | 03 Aug 22 | 3d |
| 15 | | Ground floor slab | 04 Aug 22 | 12 Aug 22 | 1w 2d |
| 16 | | Internal blockwork walls | 15 Aug 22 | 26 Aug 22 | 2w |
| 17 | | External windows and doors | 18 Aug 22 | 24 Aug 22 | 1w |
| 18 | | 1st fix M&E | 30 Aug 22 | 05 Sep 22 | 1w |
| 19 | | Plastering to offices | 06 Sep 22 | 12 Sep 22 | 1w |
| 20 | | Ceilings | 13 Sep 22 | 15 Sep 22 | 3d |
| 21 | | 2nd fix M&E | 16 Sep 22 | 22 Sep 22 | 1w |
| 22 | | Decorations | 23 Sep 22 | 29 Sep 22 | 1w |
| 23 | | Form external roads and car park | 18 Aug 22 | 08 Sep 22 | 3w |
| 24 | F100 | Footbridge design | 06 Jun 22 | 10 Jun 22 | 1w |
| 25 | F200 | Fabricate footbridge | 13 Jun 22 | 17 Jun 22 | 1w |
| 26 | F250 | Footbridge site set-up | 18 Aug 22 | 24 Aug 22 | 1w |
| 27 | F260 | Enabling work | 25 Aug 22 | 01 Sep 22 | 1w |
| 28 | F270 | Footbridge foundations | 02 Sep 22 | 08 Sep 22 | 1w |
| 29 | F300 | Assemble footbridge on site | 09 Sep 22 | 15 Sep 22 | 1w |
| 30 | F400 | Brick wingwalls | 16 Sep 22 | 19 Sep 22 | 2d |
| 31 | F500 | Paint footbridge | 20 Sep 22 | 26 Sep 22 | 1w |
| 32 | F600 | Clear footbridge area | 29 Sep 22 | 05 Oct 22 | 1w |
| 33 | | Electric lighting | 27 Sep 22 | 28 Sep 22 | 2d |
| 34 | | Surfacing and lining | 09 Sep 22 | 15 Sep 22 | 1w |
| 35 | | Approval under s278 Highways Act 1980 | 06 Oct 22 | 06 Oct 22 | |
| 36 | | Planned completion (on site) | 30 Sep 22 | 30 Sep 22 | |
| 37 | | Testing, commissioning & snagging | 06 Oct 22 | 12 Oct 22 | 1w |
| 38 | | Date for completion | 07 Oct 22 | 07 Oct 22 | |
| 39 | | New planned completion date | 13 Oct 22 | 13 Oct 22 | |
| 40 | | Delay due to footbridge CE | 30 Sep 22 | 12 Oct 22 | 1w 4d |

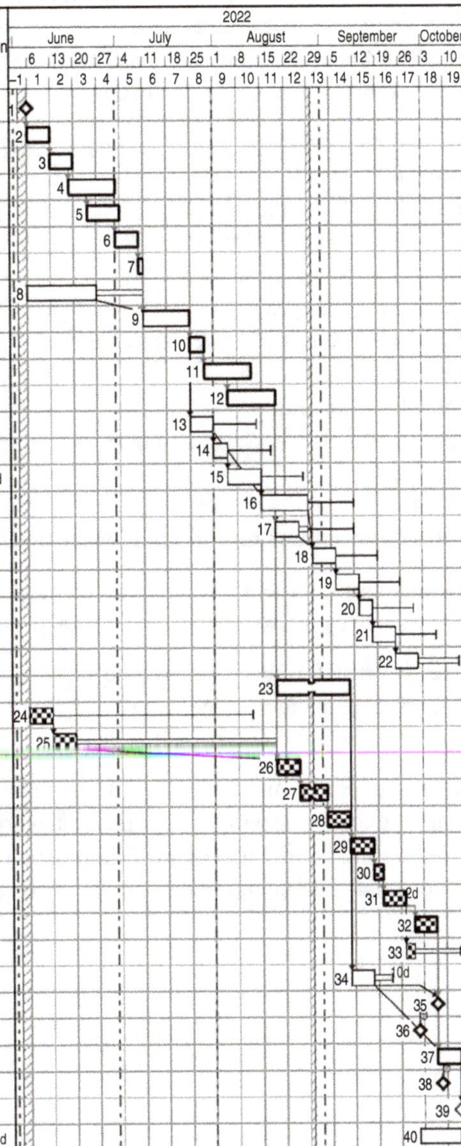


BCC 6 - Figure

Planned Activity | Compensation Event

**Figure B.2** Accepted programme inc footbridge CE – re-scheduled.

30 September 2022, until 13 October 2022. Therefore, and in accordance with clause 63.5 (refer to Section 13.3), the delay to the completion date is one week and four days.

The change to the Prices is assessed under clause 63.1. As the footbridge compensation event was instructed before the starting date, the change to the Prices will be quoted (and assessed) using the forecast Defined Cost for the work not done at the dividing date (the dividing date being the date of the Project Manager's instruction). As stated above, the contractor's quotation will use the rates and prices from either the Schedule of Cost Components or the Shorter Schedule of Cost components as this example is relevant to all of the NEC4 main contract options. The following tables set out the contractor's quotation for the footbridge compensation event:

- Table B.1 – Summary
- Table B.2 – People
- Table B.3 – Equipment[1]
- Table B.4 – Plant & Materials
- Table B.5 – Sub-contractors
- Table B.6 – Charges
- Table B.7 – Manufacture & Fabrication
- Table B.8 – Design
- Table B.9 – Insurance
- Table B.10 – clause 63.8 risk allowances

### B.1.2 Observations on the Compensation Event Assessment

The compensation event's financial assessment (i.e. the 'change in the Prices') includes the following two elements, which would be expressed separately under a JCT 2016 contract:

- The 'forecast Defined Cost' of the physical work (i.e. designing, manufacturing and constructing the footbridge); and
- The 'forecast Defined Cost' of the delay of one week and four days.

As can be seen from Tables B.2 to B.10, the majority of the 'change in the Prices' is taken up with the cost of carrying out the physical work. However, items 4 to 9 in the 'People' cost (Table B.2) and items 1 to 3 in the 'Equipment' cost (Table B.3) are clearly marked as being the time-related costs for the delay of one week and four days.

1 Some Equipment costs calculated using the CECA Schedules of Equipment Rates for use on Construction Contracts (2nd Ed) – April 2022.

**Table B.1** Footbridge compensation event – Summary.

| Item | Category | Project activity ref | SCC2 Section/ref | Description | No. of resource | Resource duration | Unit | Total resource | Rate | Sub-Total |
|---|---|---|---|---|---|---|---|---|---|---|
| | | B.2 | | People | | | | | | £26,316.90 |
| | | B.3 | | Equipment | | | | | | £20,726.45 |
| | | B.4 | | Plant & Materials | | | | | | £16,960.00 |
| | | B.5 | | Sub-contractors | | | | | | £7692.00 |
| | | B.6 | | Charges | | | | | | £650.00 |
| | | B.7 | | Manufacture/Fabrication | | | | | | £7200.00 |
| | | B.8 | | Design | | | | | | £6000.00 |
| | | B.9 | | Insurance | | | | | | £937.60 |
| | | B.10 | | cl 63.8 Risk Allowances | | | | | | £2850.00 |
| | | | | Total (excluding Fee) | | | | | | **£89,332.95** |
| | | | | Fee (10%) | | | | | | £8933.30 |
| | | | | **Change in the Prices: Forecast Defined Cost** | | | | | | **£98 266.25** |

2 'SCC' – Schedule of Cost Components (or can apply to the Short Schedule of Cost Components).

**Table B.2** Footbridge compensation event – 'People'.

| Item | Category | Project activity ref | Activity description | SCC Section/ref | Description | No. of resource | Resource duration | Unit | Total resource | Rate | Sub-Total |
|---|---|---|---|---|---|---|---|---|---|---|---|
| 1 | People | F250 | Site Setup | 1 | Foreman | 1 | 24 | hours | 24 | £15.00 | £360.00 |
| 2 | People | F250 | Site Setup | 1 | Labourer | 4 | 40 | hours | 160 | £10.51 | £1681.60 |
| 3 | People | F250 | Site Setup | 1 | Ganger | 1 | 40 | hours | 40 | £8.00 | £320.00 |
| 4 | People | F250 | Site Setup (Delay 1w 4d) | 1 | Site Manager | 1 | 72 | hours | 72 | £38.50 | £2772.00 |
| 5 | People | F250 | Site Setup (Delay 1w 4d) | 1 | Engineer | 1 | 30 | hours | 30 | £41.72 | £1251.60 |
| 6 | People | F250 | Site Setup (Delay 1w 4d) | 1 | Finishing Foreman | 1 | 72 | hours | 72 | £31.68 | £2280.96 |
| 7 | People | F250 | Site Setup (Delay 1w 4d) | 1 | Quantity Surveyor | 1 | 72 | hours | 72 | £43.62 | £3140.64 |
| 8 | People | F260 | Enabling Work | 1 | Foreman | 1 | 16 | hours | 16 | £15.00 | £240.00 |
| 9 | People | F260 | Enabling Work | 1 | Labourer | 3 | 40 | hours | 120 | £10.51 | £1261.20 |
| 10 | People | F270 | Footbridge foundations | 1 | Foreman | 1 | 40 | hours | 40 | £15.00 | £600.00 |
| 11 | People | F270 | Footbridge foundations | 1 | Labourer | 3 | 50 | hours | 150 | £10.51 | £1576.50 |
| 12 | People | F270 | Footbridge foundations | 1 | Ganger | 1 | 40 | hours | 40 | £8.00 | £320.00 |
| 13 | People | F300 | Assemble footbridge on site | 1 | Foreman | 1 | 50 | hours | 50 | £15.00 | £750.00 |

| | | | | | | | | | | | |
|---|---|---|---|---|---|---|---|---|---|---|---|
| 14 | People | F300 | Assemble footbridge on site | 1 | Ganger | 1 | 50 | hours | 50 | £8.00 | £400.00 |
| 15 | People | F300 | Assemble footbridge on site | 1 | Steel fixer | 4 | 50 | hours | 200 | £12.00 | £2400.00 |
| 16 | People | F300 | Assemble footbridge on site | 1 | Crane operator | 1 | 50 | hours | 50 | £12.00 | £600.00 |
| 17 | People | F300 | Assemble footbridge on site | 1 | Banksman | 1 | 50 | hours | 50 | £11.00 | £550.00 |
| 18 | People | F300 | Assemble footbridge on site | 1 | Labourer | 2 | 50 | hours | 100 | £10.51 | £1051.00 |
| 19 | People | F400 | Brick wingwalls | 1 | Foreman | 1 | 20 | hours | 20 | £15.00 | £300.00 |
| 20 | People | F400 | Brick wingwalls | 1 | Bricklayer | 4 | 20 | hours | 80 | £12.00 | £960.00 |
| 21 | People | F400 | Brick wingwalls | 1 | Labourer | 1 | 40 | hours | 40 | £10.51 | £420.40 |
| 22 | People | F500 | Paint bridge | 1 | Painters | 2 | 50 | hours | 100 | £12.00 | £1200.00 |
| 23 | People | F600 | Clear site | 1 | Foreman | 1 | 40 | hours | 40 | £15.00 | £600.00 |
| 24 | People | F600 | Clear site | 1 | Labourer | 2 | 50 | hours | 100 | £10.51 | £1051.00 |
| 25 | People | F260 | Enabling Work | 1 | Team leader (non-Working Area) | 1 | 1 | day | 1 | £80.00 | £80.00 |
| 26 | People | F260 | Enabling Work | 1 | Inspector (non-Working Area) | 1 | 1 | day | 1 | £150.00 | £150.00 |

*(Continued)*

**Table B.2** (Continued)

| Item | Category | Project activity ref | Activity description | SCC Section/ref | Description | No. of resource | Resource duration | Unit | Total resource | Rate | Sub-Total |
|---|---|---|---|---|---|---|---|---|---|---|---|
| | | | Total for each activity | | | | | | | | |
| | | F250 | Site Setup | | | | | | | | £11 806.80 |
| | | F260 | Enabling Work | | | | | | | | £1731.20 |
| | | F270 | Footbridge Foundations | | | | | | | | £2496.50 |
| | | F300 | Assemble footbridge on site | | | | | | | | £5751.00 |
| | | F400 | Brick wingwalls | | | | | | | | £1680.40 |
| | | F500 | Paint bridge | | | | | | | | £1200.00 |
| | | F600 | Clear site | | | | | | | | £1651.00 |
| | | | | | | | | | | | |
| | | | **Total for 1 - People** | | | | | | | | **£26 316.90** |

**Table B.3** Footbridge compensation event – 'Equipment'.

| Item | Category | Project activity Ref | Activity description | SCC Section/ref | CECA Ref 1 | Description | No | Resource duration | Unit | Total resource | Rate | Sub-Total |
|---|---|---|---|---|---|---|---|---|---|---|---|---|
| 1 | Equipment | F250 | Site Setup (Delay 1w 4d) | 2 | 16.1 | Contractor office | 1 | 1.8 | weeks | 1.8 | £78.53 | £141.35 |
| 2 | Equipment | F250 | Site Setup (Delay 1w 4d) | 2 | 16.10 | Store | 1 | 1.8 | weeks | 1.8 | £66.36 | £119.45 |
| 3 | Equipment | F250 | Site Setup (Delay 1w 4d) | 2 | 16.7 | Site toilet (double) | 1 | 1.8 | weeks | 1.8 | £47.49 | £85.48 |
| 4 | Equipment | F250 | Site Setup | 21 | 9.1 | Dumper (2.15 T) | 1 | 1 | weeks | 1 | £500.80 | £500.80 |
| 5 | Equipment | F250 | Site Setup | 21 | 14.2 | Lorry (7.5 T) | 1 | 5 | days | 5 | £177.28 | £886.40 |
| 6 | Equipment | F250 | Site Setup | 21 | | Hand saw | 1 | 1 | No | 1 | £20.00 | £20.00 |
| 7 | Equipment | F250 | Site Setup | 21 | | Shovel | 1 | 1 | No | 1 | £10.00 | £10.00 |
| 8 | Equipment | F260 | Enabling work | 21 | 9.1 | Dumper (2.15 T) | 1 | 1 | week | 1 | £500.80 | £500.80 |
| 9 | Equipment | F260 | Enabling work | 21 | 14.2 | Lorry (7.5 T) | 1 | 5 | days | 5 | £177.28 | £886.40 |
| 10 | Equipment | F260 | Enabling work | 21 | 10.5 | Excavator (11 T) | 1 | 5 | days | 5 | £385.76 | £1928.80 |
| 11 | Equipment | F270 | Footbridge foundations | 21 | 9.1 | Dumper (2.15 T) | 1 | 5 | days | 5 | £500.80 | £2504.00 |
| 12 | Equipment | F270 | Footbridge foundations | 21 | 5.3 | Concrete mixer (150 l) | 1 | 5 | days | 5 | £184.40 | £922.00 |
| 13 | Equipment | F270 | Footbridge foundations | 21 | | Wheel-barrow | 1 | 5 | days | 5 | £5.00 | £25.00 |
| 14 | Equipment | F300 | Assemble footbridge on site | 22 | 9.1 | Dumper (2.15 T) | 1 | 2 | days | 2 | £500.80 | £1001.60 |

*(Continued)*

**Table B.3** (Continued)

| Item | Category | Project activity Ref | Activity description | SCC Section/ref | CECA Ref 1 | Description | No | Resource duration | Unit | Total resource | Rate | Sub-Total |
|---|---|---|---|---|---|---|---|---|---|---|---|---|
| 15 | Equipment | F300 | Assemble footbridge on site | 22 | 14.2 | Lorry (7.5 T) | 1 | 1 | days | 1 | £177.28 | £177.28 |
| 16 | Equipment | F300 | Assemble footbridge on site | 22 | 28.3, 28.5 | Setting out equipment | 1 | 1 | days | 1 | £205.17 | £205.17 |
| 17 | Equipment | F300 | Assemble footbridge on site | 22 | 7.1 | Crane (13 T inc driver) | 1 | 2 | days | 2 | £559.52 | £1119.04 |
| 18 | Equipment | F400 | Brick wingwalls | 22 | 9.1 | Dumper (2.15 T) | 1 | 2 | days | 2 | £500.80 | £1001.60 |
| 19 | Equipment | F400 | Brick wingwalls | 22 | 14.2 | Lorry (7.5 T) | 1 | 2 | days | 2 | £177.28 | £354.56 |
| 20 | Equipment | F400 | Brick wingwalls | 22 | 5.1 | Concrete mixer 2 (90 l) | 1 | 1 | days | 1 | £158.00 | £158.00 |
| 21 | Equipment | F500 | Paint bridge | 22 | | Trestles | 1 | 5 | days | 5 | £10.00 | £50.00 |
| 22 | Equipment | F600 | Clear site | 22 | 9.1 | Dumper (2.15 T) | 1 | 7 | days | 7 | £500.80 | £3505.60 |
| 23 | Equipment | F600 | Clear site | 22 | 14.2 | Lorry (7.5 T) | 1 | 5 | days | 5 | £177.28 | £886.40 |
| 24 | Equipment | F250 | Site Setup | 23 | | Adjustable height restriction | 1 | 4 | weeks | 4 | £100.00 | £400.00 |
| 25 | Equipment | F300 | Assemble footbridge on site | 24 | | Bridge jacks | 4 | 4 | weeks | 16 | £121.17 | £1938.72 |
| 26 | Equipment | F300 | Assemble footbridge on site | 24 | | Bridge deck lifting eyes | 1 | 4 | weeks | 4 | £10.00 | £40.00 |
| 27 | Equipment | F250 | Site Setup | 25 | | Sacrificial framework | 10 | 1 | No. | 10 | £20.00 | £200.00 |
| 28 | Equipment | F250 | Site Setup | 25 | | Fuel for generator | 1 | 10 | Litres | 10 | £0.80 | £8.00 |

| | | | | | | | | | | | | |
|---|---|---|---|---|---|---|---|---|---|---|---|---|
| 29 | Equipment | F300 | Assemble footbridge on site | 26 | | Crane transport | 1 | 1 | No. | 1 | £100.00 | £100.00 |
| 30 | Equipment | F300 | Assemble footbridge on site | 26 | | Erect/ dismantle crane | 1 | 1 | No. | 1 | £100.00 | £100.00 |
| 31 | Equipment | F300 | Assemble footbridge on site | 26 | | Modify crane jib | 1 | 1 | No. | 1 | £500.00 | £500.00 |
| 32 | Equipment | F300 | Assemble footbridge on site | 27 | | Steel sections / RSAs | 1 | 1 | No. | 1 | £250.00 | £250.00 |
| 33 | Equipment | F300 | Assemble footbridge on site | 27 | | Lifting eye and chains | 1 | 1 | No. | 1 | £200.00 | £200.00 |
| | | | | | | | | | | | | |
| | | | Total for each activity | | | | | | | | | |
| | | F250 | Site Setup | | | | | | | | | £2371.48 |
| | | F260 | Enabling Work | | | | | | | | | £3316.00 |
| | | F270 | Footbridge Foundations | | | | | | | | | £3451.00 |
| | | F300 | Assemble footbridge on site | | | | | | | | | £5631.81 |
| | | F400 | Brick wingwalls | | | | | | | | | £1514.16 |
| | | F500 | Paint bridge | | | | | | | | | £50.00 |
| | | F600 | Clear Site | | | | | | | | | £4392.00 |
| | | | | | | | | | | | | |
| | | | **Total for Equipment** | | | | | | | | | **£20 726.45** |

**Table B.4** Footbridge compensation event – 'Plant & Materials'.

| Item | Category | Project activity Ref | Activity description | SCC Section/ref | Description | No. of resource | Resource duration | Unit | Total resource | Rate | Sub-Total |
|---|---|---|---|---|---|---|---|---|---|---|---|
| 1 | Plant & Materials | F200 | Fabricate Footbridge | 31 | Durasteel | 100 | 1 | m2 | 100 | £50.00 | £5000.00 |
| 2 | Plant & Materials | F200 | Fabricate Footbridge | 31 | RSA 45 × 45 mm | 100 | 1 | m | 100 | £5.00 | £500.00 |
| 3 | Plant & Materials | F200 | Fabricate Footbridge | 31 | Black bolts | 500 | 1 | no | 500 | £2.00 | £1000.00 |
| 4 | Plant & Materials | F200 | Fabricate Footbridge | 31 | Stainless steel handrails | 60 | 1 | m | 60 | £100.00 | £6000.00 |
| 5 | Plant & Materials | F270 | Footbridge Foundations | 31 | Concrete | 80 | 1 | m3 | 80 | £50.00 | £4000.00 |
| 6 | Plant & Materials | F270 | Footbridge Foundations | 31 | Formwork | 20 | 1 | sheets | 20 | £20.00 | £400.00 |
| 7 | Plant & Materials | F400 | Brick wing walls | 31 | Bricks | 2000 | 1 | no | 2000 | £0.05 | £100.00 |
| 8 | Plant & Materials | F400 | Brick wing walls | 31 | Storage boxes | 1 | 1 | no | 1 | £10.00 | £10.00 |
| 9 | Plant & Materials | F400 | Brick wing walls | 31 | Disposal of wrapping | 1 | 1 | no | 1 | £50.00 | £50.00 |
| 10 | Plant & Materials | F400 | Brick wing walls | 31 | Brick sample | 1 | 1 | no | 1 | £50.00 | £50.00 |

| | | | | | | | | | | | |
|---|---|---|---|---|---|---|---|---|---|---|---|
| 11 | Plant & Materials | F400 | Brick wing walls | 31 | Paint sample board | 1 | 1 | no | 1 | £50.00 | £50.00 |
| 12 | Plant & Materials | F400 | Brick wing walls | 32 | Restocking bricks | 1 | 1 | no | 1 | (£200.00) | (£200.00) |
| | | | | | | | | | | | |
| | | | Total for each activity | | | | | | | | |
| | | F200 | Fabricate Footbridge | | | | | | | | £12 500.00 |
| | | F250 | Site Setup | | | | | | | | £0.00 |
| | | F260 | Enabling Work | | | | | | | | £0.00 |
| | | F270 | Footbridge Foundations | | | | | | | | £4400.00 |
| | | F300 | Assemble footbridge on site | | | | | | | | £0.00 |
| | | F400 | Brick wingwalls | | | | | | | | £60.00 |
| | | F500 | Paint bridge | | | | | | | | £0.00 |
| | | F600 | Clear site | | | | | | | | £0.00 |
| | | | | | | | | | | | |
| | | | **Total for Plant & Materials** | | | | | | | | **£16 960.00** |

**Table B.5** Footbridge compensation event – 'Sub-contractor'.

| Item | Category | Project activity Ref | Activity description | SCC Section/ref | Description | No. of resource | Resource duration | Unit | Total resource | Rate | Sub-Total |
|---|---|---|---|---|---|---|---|---|---|---|---|
| 1 | Sub-contractors | F700 | Electricity to footbridge | 41 | Spring Electrics Ltd | 1 | 1 | No | 1 | £7692.00 | £7692.00 |
| | | | Total for each activity | | | | | | | | |
| | | F200 | Fabricate Footbridge | | | | | | | | £0.00 |
| | | F250 | Site Setup | | | | | | | £0.00 | |
| | | F260 | Enabling Work | | | | | | | £0.00 | |
| | | F270 | Footbridge Foundations | | | | | | | | £0.00 |
| | | F300 | Assemble footbridge on site | | | | | | | | £0.00 |
| | | F400 | Brick wingwalls | | | | | | | £0.00 | |
| | | F500 | Paint bridge | | | | | | | £0.00 | |
| | | F600 | Clear site | | | | | | | £0.00 | |
| | | F700 | Electricity to footbridge | | | | | | | | £7692.00 |
| | | | | | | | | | | | |
| | | | **Total for Sub-contractor** | | | | | | | | **£7692.00** |

**Table B.6** Footbridge compensation event – 'Charges'.

| Item | Category | Project activity Ref | Activity description | SCC Section/ref | Description | No. of resource | Resource duration | Unit | Total resource | Rate | Sub-Total |
|---|---|---|---|---|---|---|---|---|---|---|---|
| 1 | Charges | F250 | Site Setup | 51 | Temp water payment | 1 | 1 | no | 1 | £50.00 | £50.00 |
| 2 | Charges | F250 | Site Setup | 51 | Temp electricity | 1 | 1 | no | 1 | £50.00 | £50.00 |
| 3 | Charges | F250 | Site Setup | 52 | LA inspection charge | 1 | 1 | no | 1 | £50.00 | £50.00 |
| 4 | Charges | F250 | Site Setup | 53 | Payment for access | 1 | 1 | m | 1 | £500.00 | £500.00 |
| 5 | Charges | F250 | Site Setup | 54 | Payments rec'd for excavation | 1 | 1 | m3 | 1 | £0.00 | £0.00 |
| | | | | | | | | | | | |
| | | | Total for each activity | | | | | | | | |
| | | F200 | Fabricate Footbridge | | | | | | | | £0.00 |
| | | F250 | Site Setup | | | | | | | | £650.00 |
| | | F260 | Enabling Work | | | | | | | | £0.00 |
| | | F270 | Footbridge Foundations | | | | | | | | £0.00 |
| | | F300 | Assemble footbridge on site | | | | | | | | £0.00 |
| | | F400 | Brick wingwalls | | | | | | | | £0.00 |
| | | F500 | Paint bridge | | | | | | | | £0.00 |
| | | F600 | Clear site | | | | | | | | £0.00 |
| | | | | | | | | | | | |
| | | | **Total for 2 - Equipment** | | | | | | | | **£650.00** |

**Table B.7** Footbridge compensation event – 'Manufacture & Fabrication'.

| Item | Category | Project activity Ref | Activity description | SCC Section/ref | Description | No. of resource | Resource duration | Unit | Total resource | Rate | Sub-Total |
|---|---|---|---|---|---|---|---|---|---|---|---|
| 1 | Manufacture & Fabrication | F200 | Fabricate Footbridge | 61 | Foreman | 1 | 40 | hrs | 40 | £20.00 | £800.00 |
| 2 | Manufacture & Fabrication | F200 | Fabricate Footbridge | 61 | Fabricators | 8 | 40 | no | 320 | £20.00 | £6400.00 |
| | | | | | | | | | | | |
| | | | Total for each activity | | | | | | | | |
| | | F200 | Fabricate Footbridge | | | | | | | | £7200.00 |
| | | F250 | Site Setup | | | | | | | | £0.00 |
| | | F260 | Enabling Work | | | | | | | | £0.00 |
| | | F270 | Footbridge Foundations | | | | | | | | £0.00 |
| | | F300 | Assemble footbridge on site | | | | | | | | £0.00 |
| | | F400 | Brick wingwalls | | | | | | | | £0.00 |
| | | F500 | Paint bridge | | | | | | | | £0.00 |
| | | F600 | Clear site | | | | | | | | £0.00 |
| | | | | | | | | | | | |
| | | | **Total for Manufacture & Fabrication** | | | | | | | | **£7200.00** |

**Table B.8** Footbridge compensation event – 'Design'

| Item | Category | Project activity Ref | Activity description | SCC Section/ref | Description | No. of resource | Resource duration | Unit | Total resource | Rate | Sub-Total |
|---|---|---|---|---|---|---|---|---|---|---|---|
| 1 | Design | F100 | Design | 71 | Draughtsman | 4 | 50 | hrs | 200 | £30.00 | £6000.00 |
| 2 | Design | F100 | Design | 72 | Travel costs (inc in rate) | 4 | 0 | no | 0 | £0.00 | £0.00 |
| | | | Total for each activity | | | | | | | | |
| | | F100 | Design | | | | | | | | £6000.00 |
| | | F200 | Fabricate Footbridge | | | | | | | | £0.00 |
| | | F250 | Site Setup | | | | | | | | £0.00 |
| | | F260 | Enabling Work | | | | | | | | £0.00 |
| | | F270 | Footbridge Foundations | | | | | | | | £0.00 |
| | | F300 | Assemble footbridge on site | | | | | | | | £0.00 |
| | | F400 | Brick wingwalls | | | | | | | | £0.00 |
| | | F500 | Paint bridge | | | | | | | | £0.00 |
| | | F600 | Clear site | | | | | | | | £0.00 |
| | | | | | | | | | | | |
| | | | **Total Design** | | | | | | | | **£6000.00** |

**Table B.9** Footbridge compensation event – 'Insurance'.

| Item | Category | Project activity Ref | Activity description | SCC Section/ref | Description | No. of resource | Resource duration | Unit | Total resource | Rate | Sub-Total |
|---|---|---|---|---|---|---|---|---|---|---|---|
| 1 | Insurance | F200 | Insurance | 8 | Insurance by Contractor (Works' value × 1.5%) | 1 | 1 | No | 1 | £937.60 | £937.60 |
| | | | Total for each activity | | | | | | | | |
| | | F100 | Design | | | | | | | | £0.00 |
| | | F200 | Fabricate Footbridge | | | | | | | | £937.60 |
| | | F250 | Site Setup | | | | | | | | £0.00 |
| | | F260 | Enabling Work | | | | | | | | £0.00 |
| | | F270 | Footbridge Foundations | | | | | | | | £0.00 |
| | | F300 | Assemble footbridge on site | | | | | | | | £0.00 |
| | | F400 | Brick wingwalls | | | | | | | | £0.00 |
| | | F500 | Paint bridge | | | | | | | | £0.00 |
| | | F600 | Clear site | | | | | | | | £0.00 |
| | | | | | | | | | | | |
| | | | **Total for Insurance** | | | | | | | | **£937.60** |

**Table B.10** Footbridge compensation event – 'Clause 63.8 allowances'.

| Item | Category | Project activity Ref | Activity description | SCC Section/ref | Description | No. of resource | Resource duration | Unit | Total resource | Rate | Sub-Total |
|---|---|---|---|---|---|---|---|---|---|---|---|
| 1 | cl 63.8 Risk Allowances | F100 | Risk allowance | | Alternative design resource | 1 | 20 | Hrs | 20 | £30.00 | £600.00 |
| 2 | cl 63.8 Risk Allowances | F200 | Fabricate footbridge | | Increase in steel costs | 1 | 1 | It | 1 | £1500.00 | £1500.00 |
| 3 | cl 63.8 Risk Allowances | F260 | Enabling work | | Physical conditions <60.1(12) | 1 | 1 | It | 1 | £250.00 | £250.00 |
| 4 | cl 63.8 Risk Allowances | F270 | Footbridge foundations | | Physical conditions <60.1(12) inc in F260 | 1 | 1 | It | 1 | £0.00 | £0.00 |
| 5 | cl 63.8 Risk Allowances | F300 | Assemble footbridge | | Delay in assembly | 1 | 1 | It | 1 | £500.00 | £500.00 |

(Continued)

**Table B.10** (Continued)

| Item | Category | Project activity Ref | Activity description | SCC Section/ref | Description | No. of resource | Resource duration | Unit | Total resource | Rate | Sub-Total |
|---|---|---|---|---|---|---|---|---|---|---|---|
| | | | Total for each activity | | | | | | | | |
| | | F100 | Design | | | | | | | | £600.00 |
| | | F200 | Fabricate Footbridge | | | | | | | | £1500.00 |
| | | F250 | Site Setup | | | | | | | | £0.00 |
| | | F260 | Enabling Work | | | | | | | | £250.00 |
| | | F270 | Footbridge Foundations | | | | | | | | £0.00 |
| | | F300 | Assemble footbridge on site | | | | | | | | £500.00 |
| | | F400 | Brick wingwalls | | | | | | | | £0.00 |
| | | F500 | Paint bridge | | | | | | | | £0.00 |
| | | F600 | Clear site | | | | | | | | £0.00 |
| | | | | | | | | | | | |
| | | | **Total Clause 63.8 Allowances** | | | | | | | | **£2850.00** |

# Index



*Building Contract Claims*, Sixth Edition. Guy Higginbottom and David Chappell.

## d

## i



## k



## l

## m

## q



## r

***W***

www.ingramcontent.com/pod-product-compliance
Lightning Source LLC
LaVergne TN
LVHW062202200926
848484LV00006B/1301